PRINCIPLES OF

INFORMATION SYSTEMS

A Managerial Approach

Fifth Edition

PRINCIPLES OF

INFORMATION SYSTEMS

A Managerial Approach

Fifth Edition

Ralph M. Stair
Florida State University

George W. Reynolds
The University of Cincinnati

COURSE
TECHNOLOGY
™
THOMSON LEARNING

Australia • Canada • Mexico • Singapore • Spain • United Kingdom • United States

Managing Editor: Jennifer Locke
Senior Vice President, Publisher: Kristen Duerr
Project Management and Development: Elm Street Publishing Services, Inc.
Associate Product Manager: Matthew Van Kirk
Editorial Assistant: Janet Aras
Marketing Manager: Toby Shelton
Text Design: Elm Street Publishing Services, Inc.
Cover Design: Efrat Reis
Composition House: GEX, Inc.
Photo Researcher: Abby Reip

© **2001 by Course Technology-** I（T）P®

For more information contact:

Course Technology
25 Thomson Place
Boston, MA 02210
ITP Europe
Berkshire House 168-173
High Holborn
London WCIV 7AA
England
Nelson ITP, Australia
102 Dodds Street
South Melbourne, 3205
Victoria, Australia
ITP Nelson Canada
1120 Birchmount Road
Scarborough, Ontario
Canada M1K 5G4

International Thomson Editores
Colonia Polanco
11560 Mexico D.F. Mexico
ITP GmbH
Königswinterer Strasse 418
53227 Bonn
Germany
ITP Asia
60 Albert Street
Albert Complex
Singapore 189969
ITP Japan
Hirakawacho Kyowa Building, 3F
2-2-1 Hirakawacho
Chiyoda-ku, Tokyo 102
Japan

Library of Congress Cataloging-in-Publication Data
Stair, Ralph M.
    Principles of information systems: a managerial approach / Ralph M. Stair, George W. Reynolds.
       p. cm.
    Includes bibliographical references and index.
    ISBN 0-619-03357-6 (alk. paper)
     1. Management information systems. I. Reynolds, George Walter, 1944– II. Title.

    T58.6 .S72 2001
    658.4'038—dc21

                                            00-047561

Trademarks
Course Technology and the open book logo are registered trademarks of Course Technology.
The ITP logo is a registered trademark of International Thomson Publishing.
Microsoft, Windows 95, and Windows 98 are registered trademarks of Microsoft Corporation.
Some of the product names and company names used in this book have been used for identification purposes only and may be trademarks or registered trademarks of their manufacturers and sellers.
SAP, R/3, and other SAP product/services referenced herein are trademarks of SAP Aktiengesellschaft, Systems, Applications and Products in Data Processing, Neurottstasse 16, 69190 Walldorf, Germany. The publisher gratefully acknowledges SAP's kind permission to use these trademarks in this publication. SAP AG is not the publisher of this book and is not responsible for it under any aspect of press law.

Disclaimer
Course Technology reserves the right to revise this publication and make changes from time to time in its content without notice.

ISBN: 0-619-03357-6

Printed in the United States of America

For Lila and Leslie
—RMS

To Ginnie, Tammy, Kim, Kelly, and Kristy
—GWR

PREFACE

Education in information systems is critical for employment in almost any field. Today, information systems are used for business processes from communications to order processing to number crunching and in business functions ranging from marketing to human resources to accounting and finance. Chances are, regardless of your future occupation, you need to understand what information systems can and cannot do and be able to use them to help you accomplish your work. You will be expected to suggest new uses for information systems and participate in the design of solutions to business problems employing information systems. You will be challenged to identify and evaluate information systems options. To be successful, you must be able to view information systems from the perspective of business and organizational needs. For your solutions to be accepted, you must identify and address their impact on fellow workers. For these reasons, a course in information systems is essential for students in today's high-tech world.

Principles of Information Systems: A Managerial Approach, Fifth Edition, continues the tradition and approach of the previous editions. Our primary objective is to develop the best information systems text and accompanying materials for the first information technology course required of all business students. Through surveys, questionnaires, focus groups, and feedback that we have received from current and past adopters, as well as others who teach in the field, we have been able to develop the highest-quality set of teaching materials available.

Principles of Information Systems: A Managerial Approach, Fifth Edition, stands proudly at the beginning of the IS curriculum and remains unchallenged in its position as the only IS principles text offering the basic IS concepts that every business student must learn to be successful. In the past, instructors of the introductory course faced a dilemma. On one hand, experience in business organizations allows students to grasp the complexities underlying important IS concepts. For this reason, many schools delayed presenting these concepts until students completed a large portion of the core business requirements. On the other hand, delaying the presentation of IS concepts until students have matured within the business curriculum often forces the one or two required introductory IS courses to focus only on personal computing software tools and, at best, merely to introduce computer concepts.

This text has been written specifically for the principles course in the IS curriculum. *Principles of Information Systems: A Managerial Approach, Fifth Edition,* treats the appropriate computer and IS concepts together with a strong managerial emphasis.

APPROACH OF THE TEXT

Principles of Information Systems: A Managerial Approach, Fifth Edition, offers the traditional coverage of computer concepts, but it places the material within the context of business and information systems. The text stresses principles of IS, which are brought together and presented in a way that is both understandable and relevant. In addition, this book offers an overview of the entire IS discipline, as well as solid preparation for further study in advanced IS courses. It will serve both general business students and those who will become IS professionals. In particular, this book provides a solid groundwork from which to build advanced courses in such areas as Web site and systems development, programming, database management, Internet deployment, electronic commerce applications, and decision support.

The overall vision, framework, and pedagogy that made the previous editions so popular have been retained in the fifth edition, offering a number of benefits to students. We continue to present IS concepts with a managerial emphasis. While much of the fundamental vision of this market-leading text remains unchanged, the fifth edition more clearly highlights established principles and draws out new ones that have emerged as a result of corporate and technological change.

IS Principles First, Where They Belong

Exposing students to fundamental IS concepts provides a service to students who do not later return to the discipline for advanced courses. Since most functional areas in business rely on information systems, an understanding of IS principles helps students in other course work. In addition, introducing students to the principles of information systems helps future functional area managers avoid mishaps that often result in unfortunate consequences. Furthermore, presenting IS concepts at the introductory level creates interest among general business students who will later choose information systems as a field of concentration.

Author Team

Ralph Stair and George Reynolds have teamed up again for the fifth edition. Together, they have over fifty years of academic and industrial experience. Ralph Stair brings years of writing, teaching, and academic experience. He has written over twenty books and a large number of articles while at Florida State University. George Reynolds brings a wealth of computer and industrial experience to the project, with over thirty years of experience working in government, institutional, and commercial IS organizations. He has also authored nine texts and is an adjunct professor at the University of Cincinnati, teaching the introductory IS course. The Stair and Reynolds team brings a solid conceptual foundation along with practical IS experience to students.

GOALS OF THIS TEXT

Principles of Information Systems: A Managerial Approach, Fifth Edition, has three main goals:

1. To present a core of IS principles with which every business student should be familiar and to offer a survey of the IS discipline that will enable all business students to understand the relationship of advanced courses to the curriculum as a whole
2. To present the changing role of the IS professional
3. To show the value of the discipline as an attractive field of specialization

These goals help students, regardless of major, understand and use fundamental information systems principles so that they will efficiently and effectively function as future business employees and managers. Because *Principles of Information Systems: A Managerial Approach, Fifth Edition,* is written for all business majors, we believe it is important to present not only a realistic perspective on IS in business but also to provide students with the skills they can use to be effective leaders in their companies.

IS Principles

Principles of Information Systems: A Managerial Approach, Fifth Edition, although comprehensive, does not attempt to cover every aspect of the IS discipline. Instead, it offers an essential core of guiding IS principles for students to use as they face the career challenges ahead. From the opening vignettes to the end-of-chapter material, each chapter emphasizes these fundamental IS principles. These principles are highlighted in the chapter opener of each chapter. Think of principles as basic truths, rules, or assumptions that remain constant regardless of the situation. They provide strong guidance in the face of tough decisions. And the ultimate goal of *Principles of Information Systems* is to develop effective, thinking employees by instilling them with principles to help guide their decision making and actions.

This text not only offers the traditional coverage of computer concepts but also stresses the broad framework to provide students solid grounding in business uses of technology. In addition to serving general business students, this book offers an overview of the entire IS discipline and solidly prepares future IS professionals for advanced IS courses and their careers in the rapidly changing IS discipline.

Changing Role of the IS Professional

As business and the IS discipline have changed, so too has the role of the IS professional. Once considered a dedicated specialist, the IS professional now is often an internal consultant to all functional areas, knowledgeable about their needs and competent in bringing the power of information systems to bear throughout the business. The IS professional must exercise a broad perspective, encompassing the entire organization and the broader industry and business environment in which it operates.

The scope of responsibilities of an IS professional today ranges not only throughout the organization but also throughout the entire interconnected network of suppliers, customers, competitors, and other entities, no matter where they are located. This broad scope offers IS professionals a new challenge: how to help an organization survive in a highly interconnected, highly competitive global environment. In accepting that challenge, the IS professional plays a pivotal role in shaping the business itself and ensuring its success. To survive, businesses must now strive for ultimate customer satisfaction and loyalty through competitive prices and ever-improving product and service quality. The IS professional assumes the critical responsibility of determining the organization's approach to both overall cost and quality performance and therefore plays an important role in the ongoing survival of the organization. This new duality in the role of the IS employee—a professional who exercises a specialist's skills with a generalist's perspective—is reflected throughout the book.

IS as a Field for Further Study

The IS field is exciting, challenging, and rewarding, and it is important to show the value of the discipline as an attractive field of study for the average business student. The need to draw bright and interested students into the IS discipline is part of our ongoing responsibility—the IS graduate is no longer a technical recluse. Increasingly, we are seeing the brightest and most talented students enter the IS field. Upon graduation, IS graduates at many schools are among the highest paid of all business graduates. Throughout this text, the many challenges and opportunities available to IS professionals are highlighted and emphasized.

CHANGES IN THE FIFTH EDITION

Like the previous editions, the fifth edition retains the focus on IS principles and strives to be the most current text on the market. We are excited about a number of changes to the text, particularly those that were made in response to feedback on how the course is now being taught. Some of the highlights follow:

- *Theme for the Fifth Edition.* Placing IS concepts in a business context has always set us apart from general computer books. In this edition, we stress *electronic commerce and connectivity* as the major themes. As businesses of all sizes and types embrace electronic commerce, we are witnessing history in the making. This revolution in how business is conducted is profoundly changing businesses, markets, and society for decades to come. With years of service to the information systems discipline, this edition builds on the traditions and strengths of past successes while keeping an eye on the needs of future managers and decision makers.
- *Increased Emphasis on Principles.* Extensive effort was put into clarifying the fundamental information system principles and linking the chapter objectives to them. This revision helps drive home the importance of the principles and the ways they can be used in principle-based decision making.
- *New Chapter on Electronic Commerce.* Consistent with the theme, a new chapter has been added to the text devoted to electronic commerce. In addition, the importance of electronic commerce to other business processes and functions is stressed throughout the book. The latest technological developments are covered, including electronic markets, product identification, product selection, and electronic product distribution, as well as broader societal issues such as privacy, fraud, and on-line profiling. Current examples are included throughout the book, and the use of the Internet for electronic commerce is fully explored.
- *"E-Commerce" Boxes.* The "E-Commerce" supplemental interest boxes strengthen the text's theme by showing how IS professionals and organizations have used information systems to achieve their goals. Students are shown that the IS discipline is not only rewarding but fun!
- *Current Examples, Boxes, Cases, and References.* We take great pride in including the most recent examples, boxes, cases, and references throughout the text. Some were developed at the last possible moment, literally weeks before the book went into publication. Information on new hardware and software concepts, the latest operating systems, open source code, application service providers, virtual reality, the Internet, electronic commerce, and many other current developments can be found throughout the text. Our adopters have come to expect the best and most recent material. We have done everything we can to meet or exceed these expectations. Every effort was made to include the newest, freshest, and most relevant examples, boxes, cases, and references possible.
- *Revised End-of-Chapter Material.* The material at the end of each chapter has been thoroughly updated. Summaries keyed to the principles, key terms, review questions, discussion questions, problem-solving exercises, team activities, Web exercises, and cases have been replaced and revised to reflect the themes of the fifth edition and to give students the opportunity to explore the latest technology in a business setting. Chapter references are new and explore the latest developments in information systems. The number of Web exercises at the end of each chapter has been increased to further reinforce the themes of this text and help students explore this expanding technology.

TEACHING RESOURCE PACKAGE

The teaching tools that accompany this text offer many options for enhancing your course. In the fifth edition, we emphasize the importance of distance learning. And, as always, we are committed to providing one of the best teaching resource packages available in this market. Here are your options.

Instructor's Manual with Solutions

The *Instructor's Manual* is available in both electronic and printed formats. This all-new updated *Instructor's Manual* provides valuable chapter overviews; highlights key principles and critical concepts; offers sample syllabi, learning objectives, and discussion topics; and features possible essay topics, further readings or cases, and solutions to all of the end-of-chapter questions and problems as well as suggestions for conducting team activities. Additional end-of-chapter questions are also included, as well as the rationale, methodology, and solutions for each.

Course Test Manager and Test Bank

This cutting-edge, Windows-based testing software helps design and administer pretests, practice tests, and actual examinations. With *Course Test Manager*, students can randomly generate practice tests that provide immediate on-screen feedback and enable them to create detailed study guides for questions incorrectly answered. On-screen pretests help assess students' skills and plan instruction. *Course Test Manager* can also produce printed tests. In addition, students can take tests at the computer that can be automatically graded and can generate statistical information on students' individual and group performance.

Course Presenter

A CD-ROM–based presentation tool developed in Microsoft PowerPoint, *Course Presenter* offers a wealth of resources for use in the classroom. Instead of using traditional overhead transparencies, *Course Presenter* puts together impressive computer-generated screen shows including graphics and videos. All the graphics from the book (not including photos) have been included.

Distance Learning

Course Technology, the premiere innovator in Management Information Systems publishing, is proud to present online courses in WebCT and Blackboard, as well as at MyCourse.com, Course Technology's own course enhancement tool, to provide the most complete and dynamic learning experience possible. When you add online content to one of your courses, you're adding a lot: self tests, links, projects, glossaries, and, most of all, a gateway to the twenty-first century's most important information resource. We hope you will make the most of your course, both online and offline, and please, keep us posted! For more information on how to bring distance learning to your course, contact your Course Technology sales representative.

ACKNOWLEDGMENTS

A book of this size and undertaking is always a team effort. We would like to thank every one of our fellow teammates at Course Technology for their dedication and hard work. Many thanks to our Managing Editor, Jennifer Locke. A number of people behind the scenes made this book a reality; thanks to Matthew Van Kirk, associate product manager, and Christine Spillett, associate production manager. For their hard work on the manuscript, we would like to acknowledge and thank the team at Elm Street Publishing Services. Karen Hill helped with all stages of this project. Martha Beyerlein, Emily Friel, Barb Lange, Jonathan Lyzun, Jack Semens, and Abby Westapher helped with production and the final stages of the book.

Many thanks to the sales force at Course Technology. You make this all possible. You helped to get important feedback from current and future adopters. As Course Technology product users, we know how important you are.

Ralph Stair would like to thank the Department of Information and Management Sciences, College of Business Administration, at Florida State University for their support and encouragement. He would also like to thank his family, Lila and Leslie, for their support. George Reynolds thanks his family, Ginnie, Tammy, Kim, Kelly, and Kristy, for their patience and support in this major project.

TO OUR PREVIOUS ADOPTERS AND POTENTIAL NEW USERS

We sincerely appreciate our loyal adopters through the previous editions and welcome new users of *Principles of Information Systems: A Managerial Approach, Fifth Edition*. As in the past, we truly value your needs and feedback. We can only hope the fifth edition continues to meet your high expectations. In particular, we would like to thank the reviewers of the fifth edition, focus group members, and reviewers of previous editions.

Reviewers for the Fifth Edition

We are indebted to the following individuals for their perceptive feedback on early drafts of this text:

Roger Deveau, *University of Massachusetts—Dartmouth*
Al Maimon, *University of Washington*
Herb Snyder, *Fort Lewis College*
Amy Woszczynski, *Kennesaw State University*
Judy Wynekoop, *Florida Gulf Coast University*

Reviewers for the First, Second, Third, and Fourth Editions

The following people shaped the book you hold in your hands by contributing to previous editions:

Robert Aden, *Middle Tennessee State University*
A. K. Aggarwal, *University of Baltimore*
Sarah Alexander, *Western Illinois University*
Beverly Amer, *University of Florida*
Noushin Asharfi, *University of Massachusetts*
Yair Babad, *University of Illinois—Chicago*
Charles Bilbrey, *James Madison University*
Thomas Blaskovics, *West Virginia University*
John Bloom, *Miami University of Ohio*
Warren Boe, *University of Iowa*
Glen Boyer, *Brigham Young University*
Mary Brabston, *University of Tennessee*
Jerry Braun, *Xavier University*

Thomas A. Browdy, *Washington University*
Lisa Campbell, *Gulf Coast Community College*
Andy Chen, *Northeastern Illinois University*
David Cheslow, *University of Michigan—Flint*
Robert Chi, *California State University—Long Beach*
Carol Chrisman, *Illinois State University*
Miro Costa, *California State University—Chico*
Caroline Curtis, *Lorain County Community College*
Roy Dejoie, *USWeb Corporation*
Sasa Dekleva, *DePaul University*
Pi-Sheng Deng, *California State University—Stanislaus*
John Eatman, *University of North Carolina*
Juan Esteva, *Eastern Michigan University*
Badie Farah, *Eastern Michigan University*
Karen Forcht, *James Madison University*
Carroll Frenzel, *University of Colorado—Boulder*
John Gessford, *California State University—Long Beach*
Terry Beth Gordon, *University of Toledo*
Kevin Gorman, *University of North Carolina—Charlotte*
Costanza Hagmann, *Kansas State University*
Bill C. Hardgrave, *University of Arkansas*
Al Harris, *Appalachian State University*
William L. Harrison, *Oregon State University*
Dwight Haworth, *University of Nebraska—Omaha*
Jeff Hedrington, *University of Wisconsin—Eau Claire*
Donna Hilgenbrink, *Illinois State University*
Jack Hogue, *University of North Carolina*
Joan Hoopes, *Marist College*
Donald Huffman, *Lorain County Community College*
Patrick Jaska, *University of Texas at Arlington*
G. Vaughn Johnson, *University of Nebraska—Omaha*
Grover S. Kearns, *Morehead State University*
Robert Keim, *Arizona State University*
Karen Ketler, *Eastern Illinois University*
Mo Khan, *California State University—Long Beach*
Michael Lahey, *Kent State University*
Jan de Lassen, *Brigham Young University*
Robert E. Lee, *New Mexico State University—Carlstadt*
Joyce Little, *Towson State University*
Herbert Ludwig, *North Dakota State University*
Jane Mackay, *Texas Christian University*
James R. Marsden, *University of Connecticut*
Roger W. McHaney, *Kansas State University*
Lynn J. McKell, *Brigham Young University*
John Melrose, *University of Wisconsin—Eau Claire*
Michael Michaelson, *Palomar College*
Ellen Monk, *University of Delaware*
Bijayananda Naik, *University of South Dakota*
Leah R. Pietron, *University of Nebraska—Omaha*
John Powell, *University of South Dakota*
Maryann Pringle, *University of Houston*
John Quigley, *East Tennessee State University*
Mary Rasley, *Lehigh-Carbon Community College*
Earl Robinson, *St. Joseph's University*
Scott Rupple, *Marquette University*
Dave Scanlon, *California State University—Sacramento*

Werner Schenk, *University of Rochester*
Larry Scheuermann, *University of Southwest Louisiana*
James Scott, *Central Michigan University*
Vikram Sethi, *Southwest Missouri State University*
Laurette Simmons, *Loyola College*
Janice Sipior, *Villanova University*
Harold Smith, *Brigham Young University*
Alan Spira, *University of Arizona*
Tony Stylianou, *University of North Carolina*
Bruce Sun, *California State University—Long Beach*
Hung-Lian Tang, *Bowling Green State University*
William Tastle, *Ithaca College*
Gerald Tillman, *Appalachian State University*
Duane Truex, *Georgia State University*
Jean Upson, *Lorain County Community College*
Misty Vermaat, *Purdue University—Calumet*
David Wallace, *Illinois State University*
Michael E. Whitman, *University of Nevada—Las Vegas*
David C. Whitney, *San Francisco State University*
Goodwin Wong, *University of California—Berkeley*
Myung Yoon, *Northeastern Illinois University*

Focus Group Contributors for the Third Edition

Mary Brabston, *University of Tennessee*
Russell Ching, *California State University—Sacramento*
Virginia Gibson, *University of Maine*
Bill C. Hardgrave, *University of Arkansas*
Al Harris, *Appalachian State University*
Stephen Lunce, *Texas A & M International*
Merle Martin, *California State University—Sacramento*
Mark Serva, *Baylor University*
Paul van Vliet, *University of Nebraska—Omaha*

OUR COMMITMENT

We are sincerely committed to serving the needs of our adopters and readers. Like the field of IS itself, the writing and publishing process is an evolutionary and participatory one. We encourage participation in our endeavor to provide the freshest, most relevant information possible. We pride ourselves on listening to instructors and developing creative solutions to problems and needs. Numerous individuals in the IS discipline have given us their time and insight during the process of this revision. They have offered valuable feedback on outstanding features of the previous editions and potential improvements in the fifth edition. We have listened to their comments and thank them for their time.

As always, we welcome input and feedback. If you have any questions or comments regarding *Principles of Information Systems: A Managerial Approach, Fifth Edition*, please contact us through Course Technology or your local representative, via e-mail at **mis@course.com**, via the Internet at **www.course.com**, or address your comments, criticisms, suggestions, and ideas to:

Ralph Stair
George Reynolds
Course Technology
25 Thomson Place
Boston, MA 02210

BRIEF CONTENTS

CONTENTS

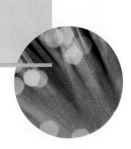

PART I

An Overview

An Introduction to Information Systems

Companies are more aggressive than ever in adopting and deploying leading-edge technology. The reason: fear of being left behind.

— Rick Whiting and Beth Davis summarizing results of a 1999 *Information Week* survey of 300 IT managers

Principles	Learning Objectives
The value of information is directly linked to how it helps decision makers achieve the organization's goals.	• *Distinguish data from information and describe the characteristics used to evaluate the quality of data.*
Models, computers, and information systems are constantly making it possible for organizations to improve the way they conduct business.	• *Identify four basic types of models and explain how they are used.* • *Name the components of an information system and describe several system characteristics.*
Knowing the potential impact of information systems and having the ability to put this knowledge to work can result in a successful personal career, organizations that reach their goals, and a society with a higher quality of life.	• *Identify the basic types of business information systems and discuss who uses them, how they are used, and what kinds of benefits they deliver.* • *Discuss why it is important to study and understand information systems.*
System users, business managers, and information systems professionals must work together to build a successful information system.	• *Identify the major steps of the systems development process and state the goal of each.*

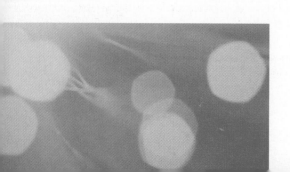

PHH Vehicle Management Services

*Using the Internet to Improve
Customer Service*

When was the last time your car was serviced? Can you remember what work was done? How much did it cost? Do you know how much you've paid for service since you first got the car? Now imagine being the fleet manager of a large company with responsibility for hundreds of trucks and cars—how would you obtain and track that information?

PHH Vehicle Management Services, one of the world's largest providers of fleet leasing solutions for corporate, government, and utility fleets, stores all that information and more about each of the 750,000 cars and trucks it leases to more than 5,000 clients. PHH enables customers to access data on-line through PHH Inter/Active, an Internet-based information system, currently used by PHH clients throughout North America. The result is a powerful tool for understanding, controlling, and reducing the cost of operating a fleet of vehicles.

To build this information source, PHH accumulates data during the lifetimes of its vehicles, including data on vehicle orders and resales, billing, fuel purchases, maintenance, accidents, and even driver safety. Much of the data on client cars is captured from charge card use. Constant updates during the day keep the data current. For emergency repairs or accidents, PHH provides a toll-free number to the driver to report the problem. Call center personnel record the information for future use.

PHH uses this information internally to analyze market trends and supplier quality issues and to use its purchasing volume to secure lower pricing for its clients. But the data is also valuable to the fleet managers of PHH's customers because leasing is one of the largest expenses for firms that depend on their vehicles to conduct business. PHH's goal is to provide improved customer service by enhancing its capability to advise its clients and adding value by helping them manage their vehicle leasing costs.

Clients can access the data via the Internet, call up detailed information on their bills, and analyze data from the past several years. By making information available to customers from their desktops, PHH enables them to analyze their fleets in ways they never could before. Customers are more satisfied, and PHH doesn't have to spend time doing the research and returning customer calls. The gain in productivity has enabled PHH to grow its business without increasing its call center or consulting staff.

Frequently, having the data helps trigger other actions. For example, if a vehicle is in the shop for one type of repair, a client might decide to do preventive maintenance at the same time. Or a fleet manager reviewing accident data might discover that one driver has had several accidents, triggering additional safety training for that driver.

The advantages provided by the Inter/Active system have set PHH apart from its competition. When PHH shows customers transactions that are only minutes old and can also produce a report on the cost of operation for each vehicle going back several years, it is quite impressive. Not only has this innovation dramatically boosted productivity and enhanced the role of the fleet manager, but it has also made it possible for PHH to expand its business geographically and enter new markets.

As you read this chapter, consider the following:

- If you were a fleet manager for a midsized company that leased 200 trucks and cars, what information would you want to be able to obtain for each vehicle? What information do you need to know about your entire fleet?

- Is it possible that errors could creep into the data? How might this happen? How serious of an impact might erroneous data have on the management of a large fleet of vehicles?

Information systems are everywhere. An advanced information system used by movie theaters provides patrons quick retrieval for advance ticket sales ordered through the telephone, Internet, or automated kiosks.
(Source: Courtesy of Radiant Systems, Inc.)

An **information system (IS)** is a set of interrelated components that collect, manipulate, and disseminate data and information and provide a feedback mechanism to meet an objective. We all interact daily with information systems, both personally and professionally. We use automated teller machines at banks, checkout clerks scan our purchases using bar codes and scanners, we access information over the Internet, and we get information from kiosks with touchscreens. We saw how an on-line information system is improving productivity and, as a result, improving the fleet-management services at PHH Vehicle Management Services. Major Fortune 500 companies are spending in excess of $1 billion per year on information technology. In the future, we will depend on information systems even more. Knowing the potential impact of information systems and having the ability to put this knowledge to work can result in a successful personal career, organizations that reach their goals, and a society with a higher quality of life.

Computers and information systems are constantly changing the way organizations conduct business. Today we live in an information economy. Information itself has value, and commerce often involves the exchange of information, rather than tangible goods. Systems based on computers are increasingly being used to create, store, and transfer information. Investors are using information systems to make multimillion-dollar decisions, financial institutions are employing them to transfer billions of dollars around the world electronically, and manufacturers are using them to order supplies and distribute goods faster than ever before. Computers and information systems will continue to change our society, our businesses, and our lives (see the "Ethical and Societal Issues" box). In this chapter, we present a framework for understanding computers and information systems and discuss why it is important to study information systems. This understanding will help you unlock the potential of properly applied information systems concepts.

INFORMATION CONCEPTS

information system (IS)

a set of interrelated components that collect, manipulate, and disseminate data and information and provide a feedback mechanism to meet an objective

data

consists of raw facts, such as an employee's name and number of hours worked in a week, inventory part numbers, or sales orders

information

a collection of facts organized in such a way that they have additional value beyond the value of the facts themselves

Information is a central concept throughout this book. The term is used in the title of the book, in this section, and in almost every chapter. To be an effective manager in any area of business, you need to understand that information is one of an organization's most valuable and important resources. This term, however, is often confused with the term *data*.

Data versus Information

Data consists of raw facts, such as an employee's name and number of hours worked in a week, inventory part numbers, and sales orders. As shown in Table 1.1, several types of data can be used to represent these facts. When these facts are organized or arranged in a meaningful manner, they become information. **Information** is a collection of facts organized in such a way that they have additional value beyond the value of the facts themselves. For example, a particular manager might find the knowledge of total monthly sales more suited to his purpose (i.e., more valuable) than the number of sales for individual sales representatives.

CHAPTER 1 ● An Introduction to Information Systems

5

ETHICAL AND SOCIETAL ISSUES
What Is Ethics?

Every society has its set of moral rules that establish the boundaries of acceptable behavior. Often the rules about such behavior are expressed in statements about what you should or should not do. These rules fit together, more or less consistently, to form the moral code by which a society lives.

Unfortunately, moral codes are seldom completely consistent. Our everyday lives raise moral questions that we cannot easily answer. Sometimes that is because there are contradictions among our different values and we are uncertain about which value should be given priority. For example, we witness a friend break a law and we are caught in a conflict between loyalty to our friend and the value of telling the truth. At other times, our traditional values do not cover new situations, and we have to work out how to extend them. For example, we believe in the value of personal privacy, but in a time in which information technology has had such a profound impact on society that some people call this the information age, what rules should we establish to govern access to information in computer databases about private individuals?

Furthermore, no two moral codes are identical. They often vary by cultural group, gender, ethnic background, and religion. In some countries, certain behaviors are frowned on, but in other cultures the opposite may be true. For example, attitudes toward the illegal copying of software (software piracy) vary from strong opposition to strong support.

When we step back and consciously reflect on our moral beliefs, we are engaging in ethical reflection. Ethics, then, is the conscious reflection on our moral beliefs with the aim of improving, extending, or refining those beliefs in some way.

Discussion Questions

1. Can you recall a situation in which you had to deal with a conflict in values? What process did you use to resolve this issue?
2. Identify three areas in which the changes that have been brought on by the expanded use of information systems may raise a conflict with our moral beliefs.

Critical Thinking Questions

3. Provide two other examples in which attitudes toward a particular behavior are viewed differently by people of different cultures, gender, ethnic backgrounds, or religions.
4. How are ethics learned?

Sources: Lawrence M. Hinman, *Ethics: A Pluralistic Approach to Moral Theory*, 2d ed. (Fort Worth, Tex.: Harcourt, Brace, 1997); and the "ISWorld Net Professional Ethics" Web site at http://www.cityu.edu.hk/is/ethics/ethics.htm, accessed February 16, 2000.

Data represents real-world things. As we have stated, data—simply, raw facts—has little value beyond its existence. For example, consider data as pieces of railroad track in a model railroad kit. In this state, each piece of track has little value beyond its inherent value as a single object. However, if some relationship is defined among the pieces of the track, they will gain value. By arranging the pieces of track in a certain way, a railroad layout begins to emerge (Figure 1.1a). Information is much the same. Rules and relationships can be set up to organize data into useful, valuable information.

The type of information created depends on the relationships defined among existing data. For example, the pieces of track could be arranged in a different way to form different layouts (Figure 1.1b). Adding new or different data means relationships can be redefined and new information can be created. For instance, adding new pieces to the track can greatly increase the value of the final product. We can now create a more elaborate railroad layout (Figure 1.1c)

TABLE 1.1

Types of Data

Data	Represented By
Alphanumeric data	Numbers, letters, and other characters
Image data	Graphic images and pictures
Audio data	Sound, noise, or tones
Video data	Moving images or pictures

process

a set of logically related tasks performed to achieve a defined outcome

knowledge

an awareness and understanding of a set of information and ways that information can be made useful to support a specific task or reach a decision

Likewise, our manager could add specific product data to his sales data to create monthly sales information broken down by product line.

Turning data into information is a **process**, or a set of logically related tasks performed to achieve a defined outcome. The process of defining relationships among data to create useful information requires knowledge. **Knowledge** is an awareness and understanding of a set of information and ways that information can be made useful to support a specific task or reach a decision. Part of the knowledge needed for building a railroad layout, for instance, is understanding how large an area is available for the layout and how many trains will run on the track. The act of selecting or rejecting facts based on their relevancy to particular tasks is also based on a type of knowledge used in the process of converting data into information. Therefore, information can be considered data made more useful through the application of knowledge.

In some cases, data is organized or processed mentally or manually. In other cases, a computer is used. In the earlier example, the manager could have manually calculated the sum of the sales of each representative, or a computer could calculate this sum. What is important is not so much where the data comes from or how it is processed but whether the results are useful and valuable. This transformation process is shown in Figure 1.2.

The Characteristics of Valuable Information

To be valuable to managers and decision makers, information should have the characteristics described in Table 1.2. These characteristics also make the information more valuable to an organization. If information is not accurate or complete, poor decisions can be made, costing the organization thousands, or even millions, of dollars. For example, if an inaccurate forecast of future demand indicates that sales will be very high when the opposite is true, an organization can invest millions of dollars in a new plant that is not needed. Furthermore, if information is not pertinent to the situation, not delivered to decision makers in a timely fashion, or too complex to understand, it may be of little value to the organization.

Useful information can vary widely in the value of each of these quality attributes. For example, with market-intelligence data, some inaccuracy and

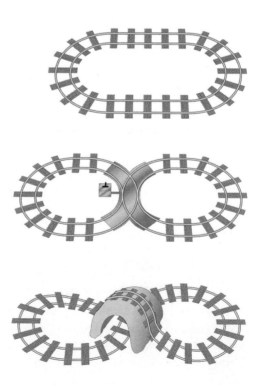

FIGURE 1.1

Defining and Organizing Relationships among Data Creates Information

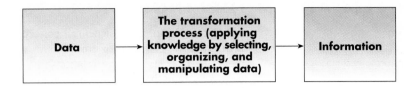

incompleteness is acceptable, but timeliness is essential. Market intelligence may alert us that our competitors are about to make a major price cut. The exact details and timing of the price cut may not be as important as being warned far enough in advance to plan how to react. On the other hand, accuracy, verifiability, and completeness are critical for data used in accounting for the use of company assets such as cash, inventory, and equipment.

The Value of Information

The value of information is directly linked to how it helps decision makers achieve their organizations' goals. For example, the value of information might be measured in the time required to make a decision or in increased profits to the company. Consider a market forecast that predicts a high demand for a new product. If market forecast information is used to develop the new product and the company can to make an additional profit of $10,000, the value of this information to the company is $10,000 minus the cost of the information. Valuable information can also help managers decide whether to invest in additional information systems and technology. A new computerized ordering system may cost $30,000, but it may generate an additional $50,000 in sales. The value added by the new system is the additional revenue from the increased sales of $20,000.

Characteristics	Definitions
Accurate	Accurate information is error free. In some cases, inaccurate information is generated because inaccurate data is fed into the transformation process (this is commonly called garbage in, garbage out [GIGO]).
Complete	Complete information contains all the important facts. For example, an investment report that does not include all important costs is not complete.
Economical	Information should also be relatively economical to produce. Decision makers must always balance the value of information with the cost of producing it.
Flexible	Flexible information can be used for a variety of purposes. For example, information on how much inventory is on hand for a particular part can be used by a sales representative in closing a sale, by a production manager to determine whether more inventory is needed, and by a financial executive to determine the total value the company has invested in inventory.
Reliable	Reliable information can be depended on. In many cases, the reliability of the information depends on the reliability of the data collection method. In other instances, reliability depends on the source of the information. A rumor from an unknown source that oil prices might go up may not be reliable.
Relevant	Relevant information is important to the decision maker. Information that lumber prices might drop may not be relevant to a computer chip manufacturer.
Simple	Information should also be simple, not overly complex. Sophisticated and detailed information may not be needed. In fact, too much information can cause information overload, whereby a decision maker has too much information and is unable to determine what is really important.
Timely	Timely information is delivered when it is needed. Knowing last week's weather conditions will not help when trying to decide what coat to wear today.
Verifiable	Information should be verifiable. This means that you can check it to make sure it is correct, perhaps by checking many sources for the same information.
Accessible	Information should be easily accessible by authorized users to be obtained in the right format and at the right time to meet their needs.
Secure	Information should be secure from access by unauthorized users.

SYSTEM AND MODELING CONCEPTS

system

a set of elements or components that interact to accomplish goals

Like information, another central concept of this book is that of a system. A **system** is a set of elements or components that interact to accomplish goals. The elements themselves and the relationships among them determine how the system works. Systems have inputs, processing mechanisms, outputs, and feedback. For example, consider an automatic car wash. Obviously, tangible *inputs* for the process are a dirty car, water, and the various cleaning ingredients used. Time, energy, skill, and knowledge are also needed as inputs to the system. Time and energy are needed to operate the system. Skill is the ability to successfully operate the liquid sprayer, foaming brush, and air dryer devices. Knowledge is used to define the steps in the car wash operation and the order in which those steps are executed.

The *processing mechanisms* consist of first selecting which cleaning options you want (wash only, wash with wax, wash with wax and hand dry, etc.) and then communicating that to the operator of the car wash. Liquid sprayers shoot clear water, liquid soap, or car wax, depending on where your car is in the process and which options you selected. The *output* is a clean car, which is a result of the interaction of independent elements or components (the liquid sprayer, foaming brush, and air dryer) of the system. The *feedback* mechanism involves your assessment of how clean the car is. Figure 1.3 shows a few systems with their elements and goals.

FIGURE 1.3

Examples of Systems and Their Elements and Goals
(Sources: Image copyright ©1998 PhotoDisc; courtesy of 3M Visual Systems Division; image copyright ©1998 PhotoDisc.)

System	Elements			Goal
	Inputs	Processing mechanisms	Outputs	
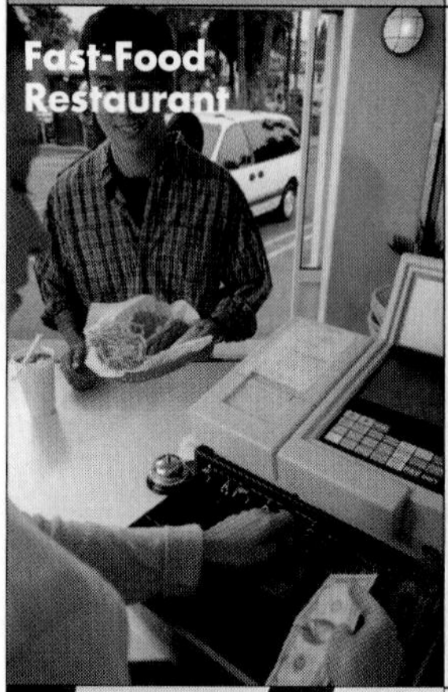 Fast-Food Restaurant	Meat, potatoes, tomatoes, lettuce, bread, drinks, labor, management	Frying, broiling, drink dispensing, heating	Hamburgers, french fries, drinks, desserts	Quickly prepared, inexpensive food
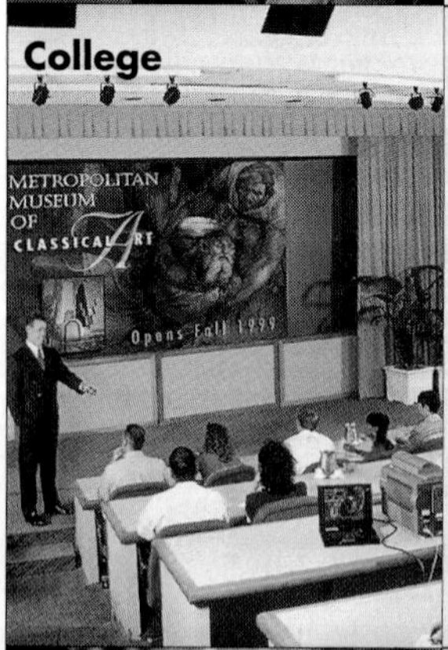 College	Students, professors, administrators, textbooks, equipment	Teaching, research, service	Educated students; meaningful research; service to community, state, and nation	Acquisition of knowledge
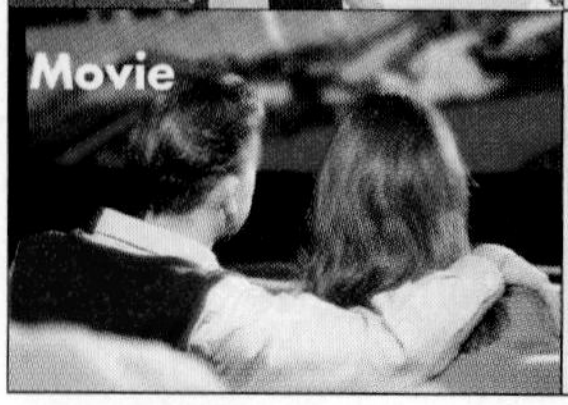 Movie	Actors, director, staff, sets, equipment	Filming, editing, special effects, film distribution	Finished film delivered to movie theaters	Entertaining movie, film awards, profits

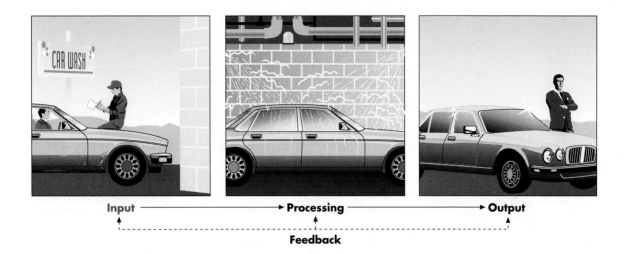

Input ——————————→ Processing ——————————→ Output

Feedback

FIGURE 1.4

Components of a System

A system's four components consist of input, processing, output, and feedback.

system boundary

defines the limits of a system and distinguishes it from everything else (the environment)

System Components and Concepts

Figure 1.4 shows a typical system diagram—that of a simple automatic car wash. The primary purpose of the car wash is to clean your automobile. The **system boundary** defines the system and distinguishes it from everything else (the environment).

The way system elements are organized or arranged is called the *configuration*. Much like data, the relationships among elements in a system are defined through knowledge. In most cases, knowing the purpose or desired outcome of a system is the first step in defining the way system elements are configured. For example, the desired outcome of our system is a clean car. Based on past experience, we know that it would be illogical to have the liquid sprayer element precede the foaming brush element. The car would be rinsed and then soap would be applied, leaving your car a mess. As you can see from this example, knowledge is needed both to define relationships among the inputs to a system (your dirty car and instructions to the operator) and to organize the system elements used to process the inputs (the foaming brush must precede the liquid sprayer).

System Types

Systems can be classified along numerous dimensions. They can be simple or complex, open or closed, stable or dynamic, adaptive or nonadaptive, permanent or temporary. Table 1.3 defines these characteristics.

TABLE 1.3

System Classifications and Their Primary Characteristics

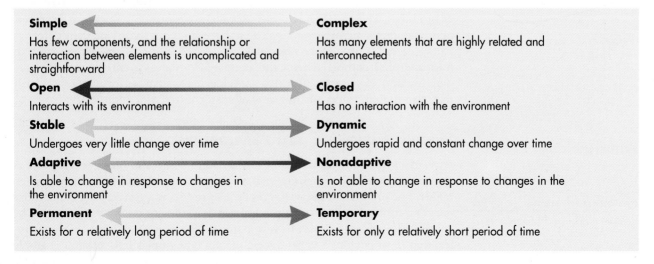

Simple	**Complex**
Has few components, and the relationship or interaction between elements is uncomplicated and straightforward	Has many elements that are highly related and interconnected
Open	**Closed**
Interacts with its environment	Has no interaction with the environment
Stable	**Dynamic**
Undergoes very little change over time	Undergoes rapid and constant change over time
Adaptive	**Nonadaptive**
Is able to change in response to changes in the environment	Is not able to change in response to changes in the environment
Permanent	**Temporary**
Exists for a relatively long period of time	Exists for only a relatively short period of time

Classifying Organizations by System Type

Most companies can be described using the classification scheme in Table 1.3. For example, a janitorial company that cleans offices after business hours most likely represents a simple, stable system because there is a constant and fairly steady need for its services. A successful computer manufacturing company, however, is typically complex and dynamic because it operates in a changing environment. If a company is nonadaptive, it may not survive very long. Many of the early computer companies, including Osborne Computer, which manufactured one of the first portable computers, and VisiCorp, which developed the first spreadsheet program, did not change rapidly enough with the changing market for computers and software. As a result, these companies did not survive. On the other hand, IBM was able to reinvent itself from a manufacturer of large, mainframe computers to a manufacturer of all classes of computers and a software and services provider.

System Performance and Standards

efficiency
a measure of what is produced divided by what is consumed

System performance can be measured in various ways. **Efficiency** is a measure of what is produced divided by what is consumed. It can range from 0 to 100 percent. For example, the efficiency of a motor is the energy produced (in terms of work done) divided by the energy consumed (in terms of electricity or fuel). Some motors have an efficiency of 50 percent or less because of the energy lost to friction and heat generation.

Efficiency is a relative term used to compare systems. For example, a gasoline engine is more efficient than a steam engine because, for the equivalent amount of energy input (gas or coal), the gasoline engine produces more energy output. The energy efficiency ratio (energy input divided by energy output) is high for gasoline engines when compared with that of steam engines.

effectiveness
a measure of the extent to which a system achieves its goals; it can be computed by dividing the goals actually achieved by the total of the stated goals

Effectiveness is a measure of the extent to which a system achieves its goals. It can be computed by dividing the goals actually achieved by the total of the stated goals. For example, a company may have a goal to reduce damaged parts by 100 units. A new control system may be installed to help achieve this goal. Actual reduction in damaged parts, however, is only 85 units. The effectiveness of the control system is 85 percent ($85 \div 100 = 85\%$). Effectiveness, like efficiency, is a relative term used to compare systems.

Efficiency and effectiveness are performance objectives set for an overall system. Meeting these objectives may involve trade-offs in terms of cost, control, and complexity.

system performance standard
a specific objective of the system

Evaluating system performance also calls for the use of performance standards. A **system performance standard** is a specific objective of the system. For example, a system performance standard for a particular marketing campaign might be to have each sales representative sell $100,000 of a certain type of product each year (Figure 1.5a). A system performance standard for a certain manufacturing process might be to have no more than 1 percent defective parts (Figure 1.5b). Once standards are established, system performance is measured and compared with the standard. Variances from the standard are determinants of system performance. Achieving system performance standards may also require trade-offs in terms of cost, control, and complexity.

System Variables and Parameters

system variable
a quantity or item that can be controlled by the decision maker

system parameter
a value or quantity that cannot be controlled, such as the cost of a raw material

Parts of a system are under direct management control, while others are not. A **system variable** is a quantity or item that can be controlled by the decision maker. The price a company charges for its product is a system variable because it can be controlled. A **system parameter** is a value or quantity that cannot be controlled, such as the cost of a raw material. The number of pounds of a chemical that must be added to produce a certain type of plastic is another example

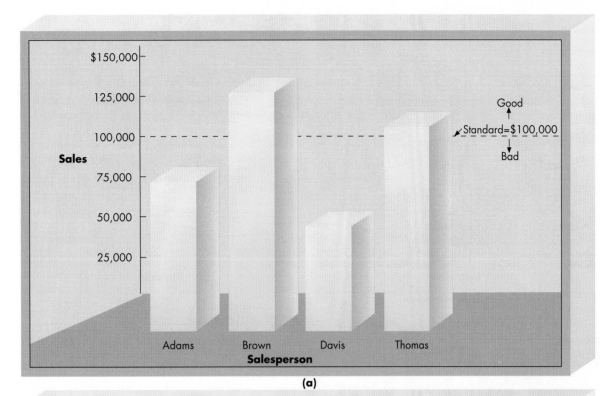

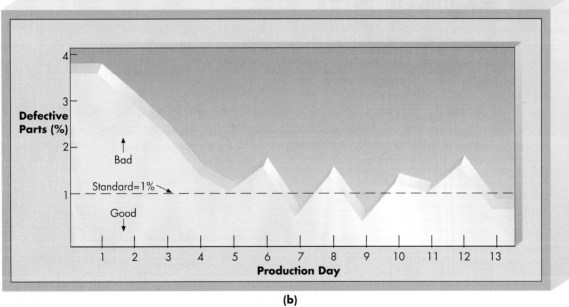

FIGURE 1.5

System Performance Standards

of a quantity or value that is not controlled by management; it is controlled by the laws of chemistry.

Modeling a System

The real world is complex and dynamic. So when we want to test different relationships and their effects, we use models of systems, which are simplified, instead of real systems. A **model** is an abstraction or an approximation that is used to represent reality. Models enable us to explore and gain an improved understanding of real-world situations.

Since the beginning of recorded history, people have used models. A written description of a battle, a physical mock-up of an ancient building, and the use of

model

an abstraction or an approximation that is used to represent reality

Narrative

Physical

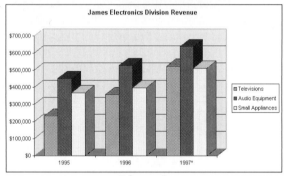

Schematic

Mathematical

symbols to represent money, numbers, and mathematical relationships are all examples of models. Today, managers and decision makers use models to help them understand what is happening in their organizations and make better decisions.

There are various types of models. The major ones are narrative, physical, schematic, and mathematical, as shown in Figure 1.6.

A narrative model, as the name implies, is based on words; thus, it is a logical and not a physical model. Both verbal and written descriptions of reality are considered narrative models. In an organization, reports, documents, and conversations concerning a system are all important narratives. Examples include the following: a salesperson verbally describing a product's competition to a sales manager, a written report describing the function of a new piece of manufacturing equipment, and a newspaper article explaining the economy or future sales of exports. Computers can be used to develop narrative models. Word processing programs, for example, can create written reports, and speech response software can store and play verbal messages like bank balances over the phone.

A physical model is a tangible representation of reality. Many physical models are computer designed or constructed. An engineer may develop a physical model of a chemical reactor to gain important information about how a large-scale reactor might perform. A builder may develop a scale model of a new shopping center to give a potential investor information about the overall appearance and approach of the development. In other examples, a marketing research department may develop a prototype of a new product, and a dentist may build a plastic tooth. These are all examples of physical models that can be used to provide information. Tupperware can produce a physical model (a plastic prototype) of a new product directly from a specialized computer system.

After a product, such as a new plastic container, has been designed, the computer system controls equipment that produces the physical model, saving days of development time and reducing costs.

A schematic model is a graphic representation of reality. Graphs, charts, figures, diagrams, illustrations, and pictures are all types of schematic models. Schematic models are used to a great extent in developing computer programs and systems. Such models are also called logical models. *Program flowcharts* show how computer programs are to be developed. *Data-flow diagrams* are used to reveal how data flows through the organization. A blueprint for a new building, a graph that shows budget and financial projections, electrical wiring diagrams, and graphs that show when certain tasks or activities must be completed to stay on schedule are examples of schematic models used in business. Graphics programs can be used to develop simple or complex schematic models.

A mathematical model is an arithmetic representation of reality. Computers excel at solving mathematical models. Again, these models are logical models and are used in all areas of business. For example, Sears has developed a mathematical model that uses standard templates to identify all the tasks, effort, and elapsed time associated with each task for planning, building, and opening a new store. This mathematical model enables Sears to forecast how long it will take to complete each new store. Sears's construction and remodeling process has become more efficient through the use of this model, which can be used to schedule delivery of construction items, fixtures, and store merchandise.[1]

In developing any model, it is important to make it as accurate as possible. An inaccurate model will usually lead to an inaccurate solution to a problem. In the mathematical model just presented, it is assumed that the length of time required to complete a given task is the same at all stores. Most models contain many assumptions, and it is important that they be as realistic as possible. It is also important that potential users of the model be aware of the assumptions under which the model was developed.

WHAT IS AN INFORMATION SYSTEM?

An information system (IS) is a specialized type of system and can be defined in a number of different ways. As mentioned previously, it is a set of interrelated elements or components that collect (input), manipulate (process), and disseminate (output) data and information and provide a feedback mechanism to meet an objective. (See Figure 1.7.)

Input, Processing, Output, and Feedback

Input

input

the activity of gathering and capturing raw data

In information systems, **input** is the activity of gathering and capturing raw data. In producing paychecks, for example, the number of hours worked for every employee must be collected before paychecks can be calculated or printed. In a university grading system, student grades must be obtained from instructors before a total summary of grades for the semester or quarter can be compiled and sent to the appropriate students.

FIGURE 1.7

The Components of an Information System

Feedback is critical to the successful operation of a system.

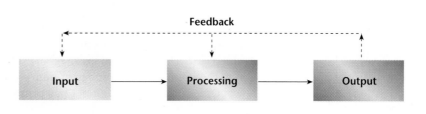

Input can take many forms. In an information system designed to produce paychecks, for example, employee time cards might be the initial input. In a 911 emergency telephone system, an incoming call would be considered an input. Input to a marketing system might include customer survey responses. Notice that regardless of the system involved, the type of input is determined by the desired output of the system.

Input can be a manual process, or it may be automated. A scanner at a grocery store that reads bar codes and enters the grocery item and price into a computerized cash register is a type of automated input process. Regardless of the input method, accurate input is critical to achieve the desired output.

Processing

processing

converting or transforming data into useful outputs

In information systems, **processing** involves converting or transforming data into useful outputs. Processing can involve making calculations, making comparisons and taking alternative actions, and storing data for future use.

Processing can be done manually or with the assistance of computers. In the payroll application, the number of hours worked for each employee must be converted into net pay. The required processing can first involve multiplying the number of hours worked by the employee's hourly pay rate to get gross pay. If more than 40 weekly hours are worked, overtime pay may also be determined. Then deductions are subtracted from gross pay to get net pay. For instance, federal and state taxes can be withheld or subtracted from gross pay; many employees have health and life insurance, savings plans, and other deductions that must also be subtracted from gross pay to arrive at net pay.

Output

output

production of useful information, usually in the form of documents and reports

In information systems, **output** involves producing useful information, usually in the form of documents and reports. Outputs can include paychecks for employees, reports for managers, and information supplied to stockholders, banks, government agencies, and other groups. In some cases, output from one system can become input for another. For example, output from a system that processes sales orders can be used as input to a customer billing system. Often output from one system can be used as input to control other systems or devices. For instance, manufacturing office furniture is complicated with many variables. Thus, the salesperson, customer, and furniture designer go through several iterations of designing furniture to meet the customer's needs. Special computer software and hardware is used to create the original design and rapidly revise it. Once the last design mock-up is approved, the design workstation software creates a bill of materials that goes to manufacturing to produce the order.

Output can be produced in a variety of ways. For a computer, printers and display screens are common output devices. Output can also be a manual process involving handwritten reports and documents.

Feedback

feedback

output that is used to make changes to input or processing activities

In information systems, **feedback** is output that is used to make changes to input or processing activities. For example, errors or problems might make it necessary to correct input data or change a process. Consider a payroll example. Perhaps the number of hours an employee worked was entered into a computer as 400 instead of 40 hours. Fortunately, most information systems check to make sure that data falls within certain predetermined ranges. For number of hours worked, the range might be from 0 to 100 hours. It is unlikely that an employee would work more than 100 hours for any given week. In this case, the information system would determine that 400 hours is out of range and provide feedback, such as an error report. The feedback is used to check and correct the input on the number of hours worked to 40. If undetected, this error would result in a very high net pay printed on the paycheck!

Feedback is also important for managers and decision makers. For example, output from an information system might indicate that inventory levels for a few items are getting low. A manager could use this feedback to decide to order more inventory. The new inventory orders then become input to the system. In this case, the feedback system reacts to an existing problem and alerts a manager that there are too few inventory items on hand. In addition to this reactive approach, a computer system can also be proactive by predicting future events to avoid problems. This concept, often called **forecasting**, can be used to estimate future sales and order more inventory before a shortage occurs.

forecasting

predicting future events to avoid problems

Manual and Computerized Information Systems

As discussed earlier, an information system can be manual or computerized. For example, some investment analysts manually draw charts and trend lines to help them make investment decisions. Tracking data on stock prices (input) over the last few months or years, these analysts develop patterns on graph paper (processing) that help them determine what stock prices are likely to do in the next few days or weeks (output). Some investors have made millions of dollars using manual stock analysis information systems. Of course, there are many excellent computerized information systems as well. For example, many computer systems have been developed to follow stock indexes and markets and to suggest when large blocks of stocks should be purchased or sold (program trading) to take advantage of market discrepancies.

Many information systems begin as manual systems and become computerized. For example, consider the way the U.S. Postal Service sorts mail. At one time most letters were visually scanned by postal employees to determine the ZIP code and were then manually placed in an appropriate bin. Today the bar-coded addresses on letters passing through the postal system are "read" electronically and automatically routed to the appropriate bin via conveyors. The computerized sorting system results in speedier processing time and provides management with information that helps control transportation planning. It is important to stress, however, that simply computerizing a manual information system does not guarantee improved system performance. If the underlying information system is flawed, the act of computerizing it might only magnify the impact of these flaws.

Computer-Based Information Systems

computer-based information system (CBIS)

a single set of hardware, software, databases, telecommunications, people, and procedures that are configured to collect, manipulate, store, and process data into information

technology infrastructure

all the hardware, software, databases, telecommunications, people, and procedures that are configured to collect, manipulate, store, and process data into information

hardware

computer equipment used to perform input, processing, and output activities

A **computer-based information system (CBIS)** is a single set of hardware, software, databases, telecommunications, people, and procedures that are configured to collect, manipulate, store, and process data into information. For example, a company's payroll system, order entry system, and inventory control system are examples of CBISs. The components of a CBIS are illustrated in Figure 1.8. A business's **technology infrastructure** includes all the hardware, software, databases, telecommunications, people, and procedures that are configured to collect, manipulate, store, and process data into information. The technology infrastructure is a set of shared IS resources that form the foundation of each individual computer-based information system.

Hardware

Hardware consists of computer equipment used to perform input, processing, and output activities. Input devices include keyboards, automatic scanning devices, equipment that can read magnetic ink characters, and many other devices. Processing devices include the central processing unit and main memory. There are many output devices, including secondary storage devices, printers, and computer screens.

FIGURE 1.8

The Components of a
Computer-Based Information
System

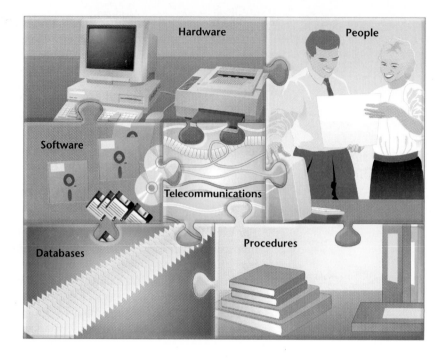

Software

software

the computer programs that govern
the operation of the computer

Software is the computer programs that govern the operation of the computer.
These programs allow the computer to, for example, process payroll, send bills
to customers, and to provide managers with information to increase profits,
reduce costs, and provide better customer service. There are two basic types of
software: system software (which controls basic computer operations such as
start-up and printing) and applications software (which allows specific tasks to
be accomplished, such as word processing and tabulating numbers). A program
(like Excel or Lotus) that allows users to create a spreadsheet is an example of
applications software.

Databases

database

an organized collection of facts
and information

A **database** is an organized collection of facts and information. An organization's database can contain facts and information on customers, employees,
inventory, competitors' sales information, and much more. Most managers and
executives believe a database is one of the most valuable and important parts of
a computer-based information system.

Telecommunications, Networks, and the Internet

telecommunications

the electronic transmission of
signals for communications; enables
organizations to carry out their
processes and tasks through
effective computer networks

networks

connected computers and computer
equipment in a building, around
the country, or around the world to
enable electronic communications

Internet

the world's largest computer network,
actually consisting of thousands of
interconnected networks, all freely
exchanging information

Telecommunications is the electronic transmission of signals for communications and enables organizations to carry out their processes and tasks through
effective computer networks. **Networks** are used to connect computers and
computer equipment in a building, around the country, or around the world to
enable electronic communications.

   Telecommunications and networks help people communicate using electronic mail (E-mail) and voice mail. These systems also help people work in
groups. The **Internet** is the world's largest computer network, actually consisting of thousands of interconnected networks, all freely exchanging information.
Research firms, colleges, universities, high schools, and businesses are just a few
examples of organizations using the Internet. But anyone who can gain access
to the Internet can communicate with anyone else on the Internet. It is estimated that about half of all U.S. households own personal computers, and perhaps a third of them are connected to the Internet.[2] The technology used to
create the Internet is now being applied within companies and organizations to

intranet

an internal network based on Web technologies that allows people within an organization to exchange information and work on projects

extranet

a network based on Web technologies that allows selected outsiders, such as business partners and customers, to access authorized resources of the intranet of a company

procedures

the strategies, policies, methods, and rules for using a CBIS

Engineers in this metro-Boston transportation control room rely on a sophisticated information system to make decisions to ease the flow of traffic. The system enables them to pinpoint bottlenecks and compensate by diverting traffic or increasing the number of open lanes.
(Source: Stone/Rich LaSalle)

create an **intranet**, which allows people within an organization to exchange information and work on projects. For example, KPMG, a management consulting firm, moved its entire body of knowledge to an intranet called KWorld. KWorld provides its consultants a central on-line repository of information and fosters more efficient collaboration among its consultants around the world.[3] The World Wide Web is a network of links to hypermedia (text, graphics, video, sound) documents. Information about the documents and access to them are controlled and provided by tens of thousands of Web servers. The Web is one of many services available over the Internet and provides access to literally millions of documents. An **extranet** is a network based on Web technologies that allows selected outsiders, such as business partners and customers, to access authorized resources of the intranet of a company. Many people use extranets everyday without realizing it—to track packaged goods, order products from their suppliers, or access customer assistance from other companies. Log on to the FedEx site to check the status of a package, for example, and you are using an extranet.

People

People are the most important element in most computer-based information systems. Information systems personnel include all the people who manage, run, program, and maintain the system. Users are any people who use information systems to get results. Users include financial executives, marketing representatives, manufacturing operators, and many others. Certain computer users are also IS personnel.

Procedures

Procedures include the strategies, policies, methods, and rules for using a CBIS. For example, some procedures describe when each program is to be run or executed. Others describe who can have access to facts in the database. Still other procedures describe what is to be done in case a disaster—such as a fire, an earthquake, or a hurricane—renders the CBIS unusable.

Now that we have looked at computer-based information systems in general, we briefly examine the most common types used in business today. These IS types are covered in more detail later in the book.

BUSINESS INFORMATION SYSTEMS

Workers at all levels, in all kinds of firms, and in all industries are using information systems to improve their own effectiveness. Today there are few workers who do not come into contact with a personal computer on at least a weekly (if not daily) basis to, for example, connect to a network, create presentations, prepare a memo, or develop a spreadsheet for analysis. At the corporate level, the most common types of information systems used in business organizations are electronic commerce systems, transaction processing systems, management information systems, and decision support systems. In addition, some organizations employ special-use systems such as artificial intelligence systems, expert systems, virtual reality systems, and geographic information systems. Together, these systems help employees in organizations accomplish

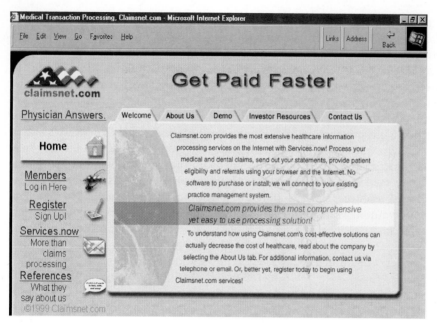

E-commerce is widely used for business-to-business (or B2B) transactions. Doctors can use a transaction processing system by Claimsnet.com to obtain payment from patients' insurance companies.

e-commerce

any business transaction executed electronically between parties such as companies (business-to-business), companies and consumers (business-to-consumer), business and the public sector, and consumers and the public sector

both routine and special tasks—from recording sales, to processing payrolls, to supporting decisions in various departments, to providing alternatives for large-scale projects and opportunities.

Electronic Commerce

E-commerce involves any business transaction executed electronically between parties such as companies (business-to-business), companies and consumers (business-to-consumer), business and the public sector, and consumers and the public sector. People perhaps assume that e-commerce is reserved mainly for consumers visiting Web sites for on-line shopping, but Web shopping is only a small part of the e-commerce picture. The major volume of e-commerce—and its fastest-growing segment—is business-to-business transactions that make purchasing easier for corporations (Figure 1.9).[4] This growth is being stimulated by increased Internet access, user confidence, better payment systems, and rapidly improving Internet and Web security. E-commerce offers opportunities for small businesses too, by enabling them to market and sell at a low cost worldwide, thus offering them an opportunity to enter the global market right from start-up.

Technically astute consumers who have tried on-line shopping appreciate the ease of e-commerce. They can avoid fighting the crowds in the malls, shop on-line at any time from the comfort of their home, and have goods delivered to them directly. In addition, under current laws governing on-line purchases, state sales taxes do not need to be paid. However, e-commerce is not without its downside. Consumers continue to have concerns about sending credit card information over the Internet to sites with varying security measures where high-tech criminals could obtain it. In addition, denial-of-service attacks that overwhelm the capacity of some of the Web's most established and popular sites have raised new concerns for the future growth of e-commerce.[5] Additional concerns involve the data gathered when a consumer visits a Web site and what companies do with the collected data; some have sold data to multiple sources, leading marketing companies to know more than we would like.

FIGURE 1.9

Growth of Business-to-Business E-Commerce
(Source: Gartner Group as published in Clinton Wilder, "Business Booms for Specialized Web Marketplaces," *Information Week*, February 7, 2000, p. 43. Reprinted by permission of CMP Media, Inc.)

Yet, in spite of the concerns, e-commerce offers many advantages for stream-lining work activities. Here is a brief example of how e-commerce can simplify the purchasing process for buying new office furniture from an office supply company (Figure 1.10). Under the manual system, a corporate office worker must get approval for a purchase that costs more than a certain amount. That request goes to the purchasing department, which generates a formal purchase order to procure the goods from the approved vendor. Business-to-business e-commerce automates that entire process. Employees go directly to the supplier's Web site, find the item in its catalog, and order what they need at a price prenegotiated by the employee's company. If approval is required, the approver is notified automatically.

As the use of e-commerce systems grows, companies will reduce their use of more traditional transaction processing systems. The resulting growth of e-commerce is creating many new business opportunities. One area of opportunity is discussed in the "E-commerce" special interest box.

Despite the dizzying growth rates of e-commerce, the percentage of total sales conducted on-line today, even in the leading e-commerce sectors such as books and music, is in the single digits. And wider questions remain: Are all business processes better off being on-line? What's the right mix of e-commerce and traditional business methods? Has channel conflict between brick-and-mortar outlets and Web sites been adequately addressed?

FIGURE 1.10

E-commerce greatly simplifies the purchasing process.

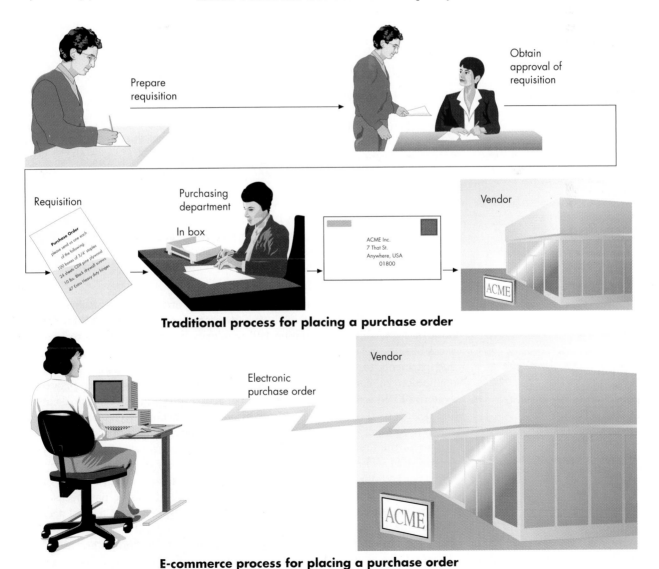

Traditional process for placing a purchase order

E-commerce process for placing a purchase order

The proliferation of e-commerce represents a profound and irrevocable shift in the way business is conducted. The leading shipping companies cite e-commerce as the single most important factor driving their growth. There's big money to be made supporting the back end of on-line shopping. By 2003, the number of people who buy over the Internet will have increased nearly sixfold—to 183 million from 32 million in 1998. Someone must handle the physical logistics—storing, packing, shipping, and then tracking hundreds of thousands of items for cybershoppers.

Manufacturers, accustomed to shipping goods on large pallets stacked to the ceiling of a truck and all going to a middleman for further distribution to customers, lack the logistical know-how and the physical infrastructure required to sell and ship directly to a fast-growing mass of Web-based consumers. So Federal Express, United Parcel Service, DHL Worldwide Express, and other major shippers are spending huge portions of their information systems budgets on systems-integration projects to support the burgeoning cyberlogistics business.

These projects include developing software tools and interfaces to directly link customers' ordering, manufacturing, and inventory systems with the shipper's network of highly automated warehouses, call centers, and worldwide shipping networks. The idea is to make the handoff of all information and inventory—from the manufacturer to the shipper to the consumer—totally seamless and lightning fast.

For example, when a customer orders a printer at the Hewlett-Packard Web site or over the telephone, that order actually goes to FedEx, which stocks all the products that H-P sells on-line at a dedicated "e-distribution facility" in Memphis. FedEx ships the order, which triggers an e-mail notification to the customer that the printer is on its way and an inventory notice to H-P that the FedEx warehouse now has one less printer in stock.

Discussion Questions

1. Trace the flow of a customer order for an H-P printer.
2. Briefly describe how product returns might be handled.

Critical Thinking Questions

3. Why are major manufacturers willing to turn over responsibility for order processing and logistics to the major shipping companies? Why don't they beef up their systems to be able to handle small customer orders?
4. What level of training and expertise must the shipping companies gain about the products they ship? Why is this level of understanding required to be successful?

Sources: Adapted from Clinton Wilder and Marianne Kolbasuk McGee, "Putting the 'E' Back in Business," *Information Week*, January 31, 2000, pp. 45–54; Jacqueline Emigh, "Vendor-Managed Inventory," *Computerworld*, August 23, 1999, p. 52; Jacqueline Emigh, "E-Commerce Strategies," *Computerworld*, August 16, 1999, p. 53; and Julia King, "Shipping Firms Exploit IT to Deliver E-Commerce Goods," *Computerworld*, August 2, 1999, p. 24.

Transaction Processing Systems, Workflow Systems, and ERP

Transaction Processing Systems

Since the 1950s, computers have been used to perform common business applications. The objective of many of these early systems was to reduce costs. This was done by automating many routine, labor-intensive business systems. A **transaction** is any business-related exchange such as payments to employees, sales to customers, and payments to suppliers. Thus, processing business transactions was the first application of computers for most organizations. A **transaction processing system (TPS)** is an organized collection of people, procedures, software, databases, and devices used to record completed business transactions. To understand a transaction processing system is to understand basic business operations and functions.

One of the first business systems to be computerized was the payroll system (Figure 1.11). The primary inputs for a payroll TPS are the numbers of employee hours worked during the week and pay rate. The primary output consists of paychecks. Early payroll systems were able to produce employee paychecks, along with important employee-related reports required by state and federal agencies, such as the Internal Revenue Service. Simultaneously, other routine processes, including customer billing and inventory control, were computerized as well. Because these early systems handled and processed daily business exchanges, or transactions, they were called transaction processing systems. In improved

transaction

any business-related exchange such as payments to employees, sales to customers, and payments to suppliers

transaction processing system (TPS)

an organized collection of people, procedures, software, databases, and devices used to record completed business transactions

A Payroll Transaction Processing System

The inputs (numbers of employee hours worked and pay rates) go through a transformation process to produce outputs (paychecks).

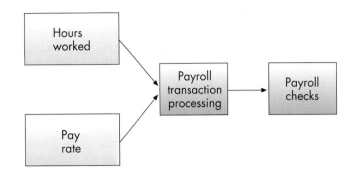

workflow system

rule-based management software that directs, coordinates, and monitors execution of an interrelated set of tasks arranged to form a business process

enterprise resource planning (ERP) system

a set of integrated programs capable of managing a company's vital business operations for an entire multisite, global organization

Enterprise resource planning software from vendors such as SAP helps Eastman Chemical Company build better customer relationships through its supply chain. SAP's software helps assess Eastman's customers' needs and plan production so that critical supplies are never depleted.

forms, these systems are still vital to most modern organizations. Consider what would happen if an organization had to function without its TPS for even one day. How many employees would be paid and paid the correct amount? How many sales would be recorded and processed? Transaction processing systems represent the application of information concepts and technology to routine, repetitive, and usually ordinary business transactions, but transactions that are critical to the daily functions of that business.

Workflow Systems

A **workflow system** is rule-based management software that directs, coordinates, and monitors execution of an interrelated set of tasks arranged to form a business process. The primary purpose of workflow systems is to provide employees with tracking, routing, document imaging, and other capabilities designed to improve business processes. Transactional workflow systems hold the promise of improving the productivity and dependability of business processes. Procter & Gamble, a major U.S. consumer goods manufacturer, implemented an expense reporting workflow application to enter, submit, process, and track expense reports. The system streamlines the reimbursement process by simplifying expense entries and automating the approval process. The system cuts the time employees spend filling out expense reports and reduces the amount of manual retyping and editing typically associated with expense report reconciliation.[6]

Enterprise Resource Planning

An **enterprise resource planning (ERP) system** is a set of integrated programs capable of managing a company's vital business operations for an entire multisite, global organization. Although the scope of an ERP system may vary from company to company, most ERP systems provide integrated software to support the manufacturing and finance business functions of an organization. In such an environment, a demand forecast is prepared that estimates customer demand for several weeks. The ERP system checks what is already available in finished product inventory to meet the projected demand. Any shortcomings then need to be manufactured. In developing the production schedule, the ERP system checks the raw material and packing material inventory and determines what needs to be ordered to meet the planned production schedule. Most ERP systems also have a purchasing subsystem that orders the items required. In addition to these core business processes, some ERP systems may be capable of supporting additional business functions such as human resources, sales, and distribution. The primary benefits of implementing an ERP system include easing adoption of improved work processes and improving access to timely data for operational decision making.

Management Information Systems and Decision Support Systems

The benefits provided by an effective transaction processing system are tangible and can be used to justify their cost in computing equipment, computer programs, and specialized personnel and supplies. They speed the processing of business activities and reduce clerical costs. Although early accounting and financial transaction processing systems were valuable, it has become clear that the data stored in these systems can be used to help managers make better decisions in their respective business areas, whether human resources, marketing, or finance. Satisfying the needs of managers and decision makers continues to be a major factor in developing information systems.

Management Information Systems

management information system (MIS)

an organized collection of people, procedures, software, databases, and devices used to provide routine information to managers and decision makers

A **management information system (MIS)** is an organized collection of people, procedures, software, databases, and devices used to provide routine information to managers and decision makers. The focus of an MIS is primarily on operational efficiency. Marketing, production, finance, and other functional areas are supported by management information systems and linked through a common database. Management information systems typically provide standard reports generated with data and information from the transaction processing system (Figure 1.12).

Management information systems began to be developed in the 1960s and are characterized by the use of information systems to produce managerial reports. In most cases, these early reports were produced periodically—daily, weekly, monthly, or yearly. Because they were printed on a regular basis, they were called *scheduled reports*. These scheduled reports helped managers perform their duties. For example, a summary report of total payroll costs might help an accounting manager control future payroll costs. As other managers learned the value of these reports, MISs began to proliferate throughout the management ranks. For instance, the total payroll summary report produced initially for the accounting manager might also be useful to a production manager to help monitor and

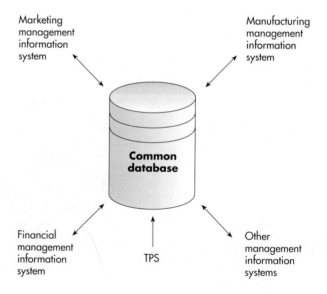

FIGURE 1.12

Functional management information systems draw data from the organization's transaction processing system.

control labor and job costs. Other scheduled reports could be used to help managers from a variety of departments control customer credit, payments to suppliers, the performance of sales representatives, inventory levels, and more.

Other types of reports were also developed during the early stages of management information systems. *Demand reports* were developed to give decision makers certain information upon request. For example, prior to closing a sale, a sales representative might seek a demand report on how much inventory existed for a particular item. This report would tell the representative if enough inventory of the item was on hand to fill the customer order. *Exception reports* describe unusual or critical situations, like low inventory levels. The exception report is produced only if a certain condition exists—in this case, inventory falling below a specified level. For example, in a bicycle manufacturing company, an exception report might be produced by the MIS if the number of bicycle seats is too low and more should be ordered.

Decision Support Systems

By the 1980s, dramatic improvements in technology resulted in information systems that were less expensive but more powerful than earlier systems. People at all levels of organizations began using personal computers to do a variety of tasks; they were no longer solely dependent on the information systems department for all their information needs. During this time, people recognized that computer systems could support additional decision-making activities. A **decision support system (DSS)** is an organized collection of people, procedures, software, databases, and devices used to support problem-specific decision making. The focus of a DSS is on decision-making effectiveness. Whereas an MIS helps an organization "do things right," a DSS helps a manager "do the right thing."

A DSS supports and assists all aspects of problem-specific decision making. It goes beyond a traditional management information system. A DSS can provide immediate assistance in solving complex problems not supported by a traditional MIS. Many of these problems are unique and not straightforward. For instance, an auto manufacturer might try to determine the best location to build a new manufacturing facility, or an oil company might want to discover the best place to drill for oil. Traditional MISs are seldom used to solve these types of problems; a DSS can help by suggesting alternatives and assisting final decision making.

Decision support systems are used when the problem is complex and the information needed to make the best decision is difficult to obtain and use. So, a DSS also involves managerial judgment. In addition, managers often play an active role in the development and implementation of the DSS. A DSS operates from a managerial perspective, and it recognizes that different managerial styles and decision types require different systems. For example, two production managers in the same position trying to solve the same problem might require different information and support. The overall emphasis is on supporting rather than replacing managerial decision making.

The essential elements of a DSS include a collection of models used to support a decision

decision support system (DSS)

an organized collection of people, procedures, software, databases, and devices used to support problem-specific decision making

EXPO, a DSS software application, helps professional traders, analysts, and portfolio managers in assessing investment risk.
(Source: Courtesy of Leading Market Technologies, Inc.)

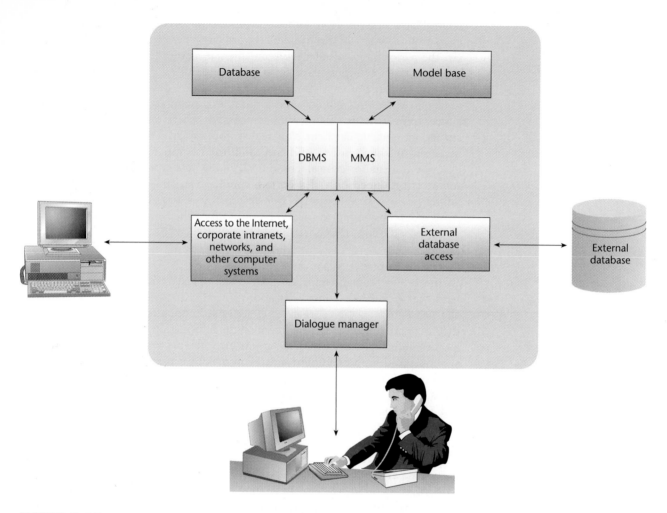

FIGURE 1.13

Essential DSS Elements

maker or user (model base), a collection of facts and information to assist in decision making (database), and systems and procedures (user interface) to help decision makers and other users interact with the DSS (Figure 1.13).

Special-Purpose Business Information Systems: Artificial Intelligence, Expert Systems, and Virtual Reality

In addition to TPSs, MISs, and DSSs, organizations often use special systems. One of these systems is based on the notion of **artificial intelligence (AI)**, whereby the computer system takes on the characteristics of human intelligence.

artificial intelligence (AI)

a field in which the computer system takes on the characteristics of human intelligence

Artificial Intelligence

The field of artificial intelligence includes several subfields (see Figure 1.14), including robotics, vision systems, natural language processing, learning systems, neural networks, and expert systems.

Robotics is an area of artificial intelligence in which machines take over complex, routine, or boring tasks, such as welding car frames or assembling computer systems and components. Vision systems allow robots and other devices to have "sight" and to store and process visual images. Natural language processing involves the ability of computers to understand and act on verbal or written commands in English, Spanish, or other natural languages. For example, United Parcel Service installed voice-recognition technology that lets customers make same-day pick-up requests by speaking their phone numbers when they make a telephone order, thus reducing the length of an average pick-up request call to

The Major Elements of Artificial Intelligence

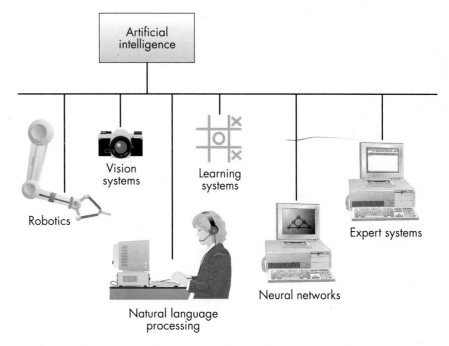

90 seconds from nearly four minutes.[7] Learning systems give computers the ability to learn from past mistakes and experiences, such as playing games or making business decisions, and neural networking is a branch of artificial intelligence that allows computers to recognize and act on patterns or trends. Some successful stock, options, and futures traders use neural networks to spot trends and make them more profitable with their investments.

Expert systems give the computer the ability to make suggestions and act like an expert in a particular field. The unique value of expert systems is that they allow organizations to capture and use the wisdom of experts and specialists. Therefore, years of experience and specific skills are not completely lost when a human expert dies, retires, or leaves for another job. Expert systems can be applied to almost any field or discipline. They have been used to monitor complex systems like nuclear reactors, perform medical diagnoses, locate possible repair problems, design and configure information system components, perform credit evaluations, and develop marketing plans for a new product or new investment strategies. The collection of data, rules, procedures, and relationships that must be followed to achieve value or the proper outcome is contained in the **knowledge base** of the expert system.

The 1980s and 1990s brought advances in artificial intelligence, including expert systems. More and more organizations are using these systems to solve complex problems and support difficult decisions. However, many issues remain to be resolved and more work is needed to refine their meaningful uses.

Virtual Reality

Originally, the term **virtual reality** referred to immersive virtual reality, which means the user becomes fully immersed in an artificial, three-dimensional world that is completely generated by a computer. The virtual world is presented in full scale and relates properly to the human size. It may represent any three-dimensional setting, real or abstract, such as a building, an archaeological excavation site, the human anatomy, a sculpture, or a crime scene reconstruction. Virtual worlds can be animated, interactive, and shared. Through immersion, the user can gain a deeper understanding of the virtual world's behavior and functionality.

A variety of input devices like head-mounted displays (Figure 1.15), data gloves (Figure 1.16), joysticks, and hand held wands allows the user to navigate through a virtual environment and interact with virtual objects. Directional sound, tactile

expert system

a system that gives a computer the ability to make suggestions and act like an expert in a particular field

knowledge base

the collection of data, rules, procedures, and relationships that must be followed to achieve value or the proper outcome

virtual reality

immersive virtual reality, which means the user becomes fully immersed in an artificial, three-dimensional world that is completely generated by a computer

FIGURE 1.15

A Head-Mounted Display

The head-mounted display (HMD) was the first device of its kind, providing the wearer with an immersive experience. A typical HMD houses two miniature display screens and an optical system that channels the images from the screens to the eyes, thereby presenting a stereo view of a virtual world. A motion tracker continuously measures the position and orientation of the user's head and allows the image-generating computer to adjust the scene representation to the current view. As a result, the viewer can look around and walk through the surrounding virtual environment.
(Source: Courtesy of Virtual Research Systems, Inc.)

systems development

the activity of creating or modifying existing business systems

systems investigation

a stage of systems development that has as its goal to gain a clear understanding of the problem to be solved or opportunity to be addressed

systems analysis

a stage of systems development during which the problems and opportunities of the existing system are defined

and force feedback devices, voice recognition, and other technologies are used to enrich the immersive experience. Several people can share and interact in the same environment. Because of this ability, virtual reality can be a powerful medium for communication, entertainment, and learning.

Virtual reality can also refer to applications that are not fully immersive, such as mouse-controlled navigation through a three-dimensional environment on a graphics monitor, stereo viewing from the monitor via stereo glasses, stereo projection systems, and others. Some virtual reality applications allow views of real environments with superimposed virtual objects. For instance, motion trackers monitor the movements of dancers or athletes for subsequent studies in immersive virtual reality. Telepresence systems (e.g., telemedicine, telerobotics) immerse a viewer in a real world that is captured by video cameras at a distant location and allow for the remote manipulation of real objects via robot arms and manipulators. Many believe that virtual reality will reshape the interface between people and information technology by offering new ways to communicate information, visualize processes, and express ideas creatively.

Useful applications of virtual reality include training (e.g., in the military, medical environments, and equipment operation), education, design evaluation (virtual prototyping), architectural walk-throughs, human factors and ergonomic studies, simulation of assembly sequences and maintenance tasks, assistance for the handicapped, study and treatment of phobias (e.g., fear of flying), entertainment, and, of course, games.

It is difficult to predict where information systems and technology will be in 10 to 20 years. It seems, however, that we are just beginning to discover the full range of their usefulness. Technology has been improving and expanding at an increasing rate; dramatic growth and change are expected for years to come. Without question, a knowledge of the effective use of information systems will be critical for managers both now and in the long term.

SYSTEMS DEVELOPMENT

Systems development is the activity of creating or modifying existing business systems. Developing information systems to meet business needs is so highly complex and difficult that it is common for information systems projects to overrun budgets and exceed scheduled completion dates. Business managers would like the development process to be more manageable, with predictable costs and timing. One strategy for improving the results of a systems development project is to divide it into several steps, each step with a well-defined goal and set of tasks to accomplish. These steps are summarized below.

Systems Investigation and Analysis

The first two steps of systems development are systems investigation and analysis. The goal of the **systems investigation** is to gain a clear understanding of the problem to be solved or opportunity to be addressed. Once this is understood, the next question to be answered is "Is the problem worth solving?" Given that organizations have limited resources—people and money—this question deserves careful consideration. If the decision is to continue with the solution, the next step, **systems analysis**, defines the problems and opportunities of the existing system.

Systems Design, Implementation, and Maintenance and Review

Systems design determines how the new system will work to meet the business needs defined during systems analysis. **Systems implementation** involves

FIGURE 1.16

A Data Glove

Realistic interactions with virtual objects via such devices as a data glove that senses hand position allow for manipulation, operation, and control of virtual worlds. (Source: Courtesy of Virtual Technologies, Inc.)

systems design

a stage of systems development that determines how the new system will work to meet the business needs defined during systems analysis

systems implementation

a stage of systems development during which the various system components (hardware, software, databases, etc.) defined in the design step are created or acquired and then assembled and the new system is put into operation

systems maintenance and review

a stage of systems development that has as its goal to check and modify the system so that it continues to meet changing business needs

computer literacy

knowledge of computer systems and equipment and the ways they function; it stresses equipment and devices (hardware), programs and instructions (software), databases, and telecommunications

information systems literacy

knowledge of how data and information are used by individuals, groups, and organizations

creating or acquiring the various system components (hardware, software, databases, etc.) defined in the design step, assembling them, and putting the new system into operation. The purpose of **systems maintenance and review** is to check and modify the system so that it continues to meet changing business needs.

WHY STUDY INFORMATION SYSTEMS?

Studies have shown that the involvement of managers and decision makers in all aspects of information systems is a major factor for organizational success, including higher profits and lower costs. A knowledge of information systems will help you make a significant contribution on the job. It will also help you advance in your chosen career or field. Managers are expected to identify opportunities to implement information systems to improve the business. They are also expected to be able to lead information system projects in their area of the business.

Information systems play a fundamental and ever-expanding role in all business organizations. If you are to have a solid understanding of how organizations operate, it is imperative that you understand the role of information systems within these organizations. Moreover, in this new century, we see continuing trends that business survival and prosperity are more difficult. For example, increased mergers among former competitors to create global conglomerates, continued downsizing of corporations to focus on their core businesses and to improve efficiencies, efforts to reduce trade barriers, and the globalization of capital all point to the increased internationalization of business organizations and markets. In addition, business issues and decisions are becoming more complex and must be made faster. An understanding of information systems will help you cope, adapt, and prosper in this challenging environment.

Regardless of your chosen field or the organization for which you may work, it is likely that you will use information systems. Why study information systems? A knowledge of information systems will help you advance in your career, solve problems, realize opportunities, and meet your own personal goals.

Computer and Information Systems Literacy

You must acquire both computer literacy and information systems literacy to be able to use information systems to meet personal and organizational goals. **Computer literacy** is knowledge of computer systems and equipment and the ways they function. It stresses equipment and devices (hardware), programs and instructions (software), databases, and telecommunications.

Information systems literacy goes beyond a knowledge of the fundamentals of computer systems and equipment. **Information systems literacy** is knowledge of how data and information are used by individuals, groups, and organizations. It includes not only a knowledge of computer technology but also aspects of the broader range of information technology. Most important, however, it encompasses *how* and *why* this technology is applied in business. Knowing about various types of hardware and software is an example of computer literacy. Knowing how to use hardware and software to increase profits, cut costs, improve productivity, and increase customer satisfaction is an example of information systems literacy. Information systems literacy can involve knowledge of how and why people (managers, employees, stockholders, and other individuals) use information technology; knowledge of organizations, decision-making approaches, management levels, and information needs; and knowledge of how organizations can use computers and information systems to achieve their goals. Knowing how to deploy transaction processing, management information, decision support, and expert systems to help an organization achieve its goals is a key aspect of information systems literacy.

Information Systems in the Functional Areas of Business

Information systems are used in all functional areas and operating divisions of business. In *finance* and *accounting*, information systems are used to forecast revenues and business activity, determine the best sources and uses of funds, manage cash and other financial resources, analyze investments, and perform audits to make sure the organization is financially sound and that all financial reports and documents are accurate. In *sales* and *marketing*, information systems are used to develop new goods and services (product analysis), determine the best location for production and distribution facilities (place or site analysis), determine the best advertising and sales approaches (promotion analysis), and set product prices to get the highest total revenues (price analysis).

In *manufacturing*, information systems are used to process customer orders, develop production schedules, control inventory levels, and monitor product quality. At Colgate-Palmolive, the maker of consumer goods deployed sophisticated business planning software that helped reduce its finished goods inventory by 50 percent and cut the time from order receipt to delivery to five days from twelve.[8] In addition, information systems are used to design products (*computer-assisted design*, or *CAD*), manufacture items (*computer-assisted manufacturing*, or *CAM*), and integrate multiple machines or pieces of equipment (*computer-integrated manufacturing*, or *CIM*). Office furniture makers provide their sales forces with three-dimensional software to allow a preview of what the customer's offices will look like, as well as an estimate of the total project's cost. Information systems are also used in *human resource management* to screen applicants, administer performance tests to employees, monitor employee productivity, and more. *Legal information systems* are used to analyze product liability and warranties and to develop important legal documents and reports.

Information Systems in Industry

Information systems are used in almost every industry and field. The *airline industry* employs Internet auction sites to offer discount fares and increase revenue.[9] *Investment firms* use information systems to analyze stocks, bonds, options, the futures market, and other financial instruments, as well as to provide improved services to their customers. Fidelity equips its best customers with specialized pagers that can be used to initiate trades and can let them know when stock prices rise or fall by a given amount.[10] *Banks and savings and loan companies* use information systems to help make sound loans and good investments. CUNA Mutual, which provides financial services and insurance, has implemented a database to track customer records better and provide salespeople with more information for cross-selling of services.[11] The *transportation industry* uses information systems to schedule trucks and trains to deliver goods and services at the least cost. Hertz automatically creates customized maps and uses wireless technology to process car returns.[12] *Publishing companies* use information systems to analyze markets and to develop and publish newspapers, magazines, and books. *Healthcare organizations* use information systems to diagnose illnesses, plan medical treatment, and bill patients. HMOs have begun to use Web technology to access patients' insurance eligibility and other information held in databases to cut patient costs.[13] *Retail companies* are using the Web to take customer orders and provide customer service support. *Power management* and *utility companies* use information systems to monitor and control power generation and usage. *Professional services* firms employ information systems to improve the speed and quality of services they provide to customers. Management consulting firms use intranets and extranets to provide information on products, services, skill levels, and past engagements to their consultants.[14] These industries will be discussed in more detail as we continue through the book.

● SUMMARY

PRINCIPLE • The value of information is directly linked to how it helps decision makers achieve the organization's goals.

Data consists of raw facts; information is data transformed into a meaningful form. The process of defining relationships between data requires knowledge. Knowledge is an awareness and understanding of a set of information and how that information can be made useful to support a specific task. To be valuable, information must have several characteristics: it should be accurate, complete, economical to produce, flexible, reliable, relevant, simple to understand, timely, verifiable, accessible, and secure. The value of information is directly linked to how it helps people achieve their organizations' goals.

PRINCIPLE • Models, computers, and information systems are constantly making it possible for organizations to improve the way they conduct business.

There are four basic types of models: narrative, physical, schematic, and mathematical. These models serve as an abstraction or an approximation that is used to represent reality. Models enable us to explore and gain an improved understanding of real-world situations. The narrative model provides a verbal description of reality. A physical model is a tangible representation of reality, often computer designed or constructed. A schematic model is a graphic representation of reality such as a graph, chart, figure, diagram, illustration, or picture. A mathematical model is an arithmetic representation of reality.

● ● ● ●

A system is a set of elements that interact to accomplish a goal or set of objectives. The components of a system include inputs, processing mechanisms, and outputs. Systems also contain boundaries that separate them from the environment and each other. Feedback is used by the system to monitor and control its operation to make sure it continues to meet its goals and objectives. Systems may be classified in many ways. They may be considered simple or complex. A stable, nonadaptive system does not change over time, while a dynamic, adaptive system does. Open systems interact with their environments; closed systems do not. Some systems exist temporarily; others are considered permanent.

System performance is measured by its efficiency and effectiveness. Efficiency is a measure of what is produced divided by what is consumed; effectiveness is a measure of the extent to which a system achieves its goals. A systems performance standard is a specific objective.

Information systems are sets of interrelated elements that collect (input), manipulate and store (process), and disseminate (output) data and information. Input is the activity of capturing and gathering new data; processing involves converting or transforming data into useful outputs; and output involves producing useful information. Feedback is the output that is used to make adjustments or changes to input or processing activities.

PRINCIPLE • Knowing the potential impact of information systems and having the ability to put this knowledge to work can result in a successful personal career, organizations that reach their goals, and a society with a higher quality of life.

Information systems play an important role in today's businesses and society. The key to understanding the existing variety of systems begins with learning their fundamentals. The types of systems used within organizations can be classified into four basic groups: e-commerce, TPSs, MISs and DSSs, and special-purpose business information systems.

E-commerce involves any business transaction executed electronically between parties such as companies (business-to-business), companies and consumers (business-to-consumer), business and the public sector, and consumers and the public sector. The major volume of e-commerce and its fastest-growing segment is business-to-business transactions that make purchasing easier for big corporations. E-commerce offers opportunities for small businesses by enabling them to market and sell at a low cost worldwide, thus enabling them to enter the global market right from start-up.

The most fundamental system is the transaction processing system (TPS). A transaction is any business-related exchange. The TPS handles the large volume of business transactions that occur daily within an organization. A workflow system is rule-based management software that directs, coordinates, and monitors execution of an interrelated set of tasks arranged to form a business process. The primary purpose of workflow systems is to provide end users with tracking, routing, document imaging, and other capabilities designed to improve business processes. An enterprise resource planning (ERP) system is a set of integrated programs capable of managing a company's vital business operations for an entire multisite, global organization.

The management information system (MIS) uses the information from a TPS to generate information useful for management decision making. Management information systems produce a variety of reports. Scheduled reports contain prespecified information and are generated regularly. Demand reports are generated only at the request of the user. Exception reports contain listings of items that do not meet a predetermined set of conditions.

A decision support system (DSS) is an organized collection of people, procedures, databases, and devices used to support problem-specific decision making. A DSS differs from an MIS in the support given to users, the decision emphasis, the development and approach, and system components, speed, and output.

The special-purpose business information systems include artificial intelligence systems, expert systems, and virtual reality systems. Artificial intelligence (AI) includes a wide range of systems, in which the computer system takes on the characteristics of human intelligence. Robotics is an area of artificial intelligence in which machines take over complex, routine, or boring tasks, such as welding car frames or assembling computer systems and components. Vision systems allow robots and other devices to have "sight" and to store and process visual images. Natural language processing involves the ability of computers to understand and act on verbal or written commands in English, Spanish, or other natural languages. Learning systems give computers the ability to learn from past mistakes or experiences, such as playing games or making business decisions, while neural networking is a branch of artificial intelligence that allows computers to recognize and act on patterns or trends. The expert system (ES) is designed to act as an expert consultant to a user who is seeking advice about a specific situation. Originally, the term *virtual reality* referred to immersive virtual reality, in which the user becomes fully immersed in an artificial, three-dimensional world that is completely generated by a computer. Virtual reality can also refer to applications that are not fully immersive, such as mouse-controlled navigation through a three-dimensional environment on a graphics monitor, stereo viewing from the monitor via stereo glasses, stereo projection systems, and others.

• • •

Our society is becoming dependent on information technology. Computer and information systems literacy are prerequisites for numerous job opportunities, not only in the IS field. Computer literacy is knowledge of computer systems and the way they function; information systems literacy is knowledge of how data, information, and information systems are used by individuals and organizations. Effective information systems can have a major impact on corporate strategy and organizational success. Businesses around the globe are enjoying better safety and service, greater efficiency and effectiveness, reduced expenses, and improved decision making and control because of information systems. Individuals who can help their businesses realize these benefits will be in demand well into the future.

PRINCIPLE • System users, business managers, and information systems professionals must work together to build a successful information system.

Systems development involves creating or modifying existing business systems. The major steps of this process and their goals include systems investigation (gain a clear understanding of what the problem is); systems analysis (define what the system must do to solve the problem); systems design (determine exactly how the system will work to meet the business needs); implementation (create or acquire the various systems components defined in the design step); and maintenance and review (maintain and then modify the system so that it continues to meet changing business needs).

● KEY TERMS

● REVIEW QUESTIONS

1. What is an information system? What are some ways information systems are changing our lives?
2. How would you distinguish data and information? Information and knowledge?
3. Identify at least six characteristics of valuable information.
4. Define the term *system*. What is the difference between an adaptive system and a nonadaptive system?
5. How have organizations changed as a result of the use of information systems?
6. How is system performance measured?
7. What is a model? What is the purpose of using a model?
8. What is a computer-based information system? What are its components?
9. Define efficiency and effectiveness as they relate to information systems.
10. What is a business's technology infrastructure?
11. What is the difference between an intranet and an extranet?
12. What is a workflow system? How is it different from a transaction processing system?
13. What are the most common types of computer-based information systems used in business organizations today? Give an example of each.
14. What are computer literacy and information systems literacy? Why are they important?
15. What are some of the benefits organizations seek to achieve through using information systems?
16. Identify the five steps in the systems development process and state the goal of each.

● DISCUSSION QUESTIONS

1. Why is the study of information systems important to you? What do you hope to learn from this course to make it worthwhile?
2. How could a workflow system simplify the approval of travel expense reports? What are the benefits of such a system?
3. Why are information and knowledge so critical to today's organizations?
4. Suppose you are a teacher assigned the task of describing the learning processes of preschool children. Why would you want to build a model of their learning processes? What kinds of models would you create? Why might you create more than one type of model?
5. Describe the "ideal" automated license plate renewal system for the drivers in your state. Describe the input, processing, output, and feedback associated with this system.
6. How is it that useful information can vary widely from the quality attributes of valuable information?
7. Discuss the potential use of virtual reality to enhance the learning experience for new automobile drivers. How might such a system operate? What are the benefits and potential disadvantages of such a system?
8. Discuss how information systems are linked to the business objectives of an organization.

● PROBLEM-SOLVING EXERCISES

1. Prepare a data disk and a backup disk for the problem-solving exercises and other computer-based assignments you will complete in this class. Create one directory for each chapter in the textbook (you should have 14 directories). As you work through the problem-solving exercises and complete other work using the computer, save your assignments for each chapter in the appropriate directory. On the label of each disk be sure to include your name, course, and section. On one disk write "Working Copy"; on the other write "Backup."

2. Search through several business magazines (*Business Week, Computerworld, PC Week*, etc.) for a recent article that discusses the use of information technology to deliver significant business benefits to an organization. Now use other resources to find additional information about the same organization (*Reader's Guide to Periodical Literature*, on-line search capabilities available at your school's library, the company's public relations department, Web pages on the Internet, etc.). Use word processing software to prepare a one-page summary of the different resources you tried and their ease of use and effectiveness.

3. Create a simple spreadsheet to help manage your "to do" list of tasks.

a. For each item on your "to do" list, define the date the task must be completed, briefly describe the task, and indicate whether the task is Urgent (important and it must be done by the due date), Pressing (important, but the due date can slide a day or two), or Trivial (not important and the due date is not critical). Your spreadsheet might look like the following:

Date Due	Task	Importance

b. Now enter the items for your "to do" list. Use the features of the spreadsheet software to sort all tasks by "importance" and within "importance," sort by "date due."

c. As you complete tasks, delete them from the spreadsheet. As you identify new tasks to be done, add them to the spreadsheet. Always sort the list by importance and due date.

4. Do some research to obtain estimates of the rate of growth of the Internet (e.g., number of computers connected to the Internet, number of Internet Web sites). Use the plotting capabilities of your spreadsheet or graphics software to produce a bar chart of that growth over a number of years. Share your findings with the class.

● TEAM ACTIVITIES

Before you can do a team activity, you need a team! The class members may self-select their teams, or the instructor may assign members to groups. Once your group has been formed, meet and introduce yourselves to each other. You will need to find out the first name, hometown, major, and e-mail address and phone number of each member. Find out one interesting fact about each member of your team, as well. Come up with a name for your team.

With the other members of your group, use word processing software to write a one-page summary of what your team hopes to gain from this course and what you are willing to do to accomplish these goals. Send the report to your instructor via e-mail.

● WEB EXERCISES

1. Throughout this book, you will see how the Internet provides a vast amount of information to individuals and organizations. We will stress the World Wide Web (the Web), which is an important part of the Internet. Most large universities and organizations have an address on the Internet, called a Web site or home page. The address of the Web site for this publisher is http://www.course.com. You can gain access to the Internet through a browser, such as Internet Explorer or Netscape. Using an Internet browser, go to the Web site for this publisher. What did you find? Try to obtain information on this book. You may be asked to develop a report or send an e-mail message to your instructor about what you found.

2. Access the ISWorld Net Professional Ethics Web site at http://www.cityu.edu.hk/is/ethics/ethics.htm. After visiting this site, write a one-page paper discussing the following issue: What is professional ethics and what are the advantages and disadvantages of being a member of an organization with a code of ethics? Access one of the Web sites identified below and read the Code of Ethics for a member of one of the many information systems societies or organizations. Find a code for another IT-related society or organization. What common issues do the two codes of ethics address?

• World IT Associations Database at http://www.esi.es/Information/ITAssociations
• Association for Computing Machinery at http://info.acm.org
• Association of Information Technology Professionals at http://www.aitp.org

3. Find a Web site whose subject is your favorite movie actor or actress, hobby, or your profession. After checking out this Web site, find at least two other sites containing information on the same subject.

● CASES

 ### Coors Ceramics Revamps Information Systems

Coors Ceramics was spun off from the Adolph Coors Company in December 1992. Today the company is one of the leading suppliers of ceramic materials and components to the semiconductor and laser industries and has developed a worldwide reputation for quality and precision.

Coors's old information systems took as long as two days to process new orders. Because of delays and inaccuracies in processing, there was no way a salesperson could track the exact status of a particular customer's order. With 1,500 orders coming in monthly, that was a huge problem. To compensate for the processing delays, Coors would produce more orders than it received so that it could build up inventory to meet customers' desired delivery dates. Although it did help meet customer demand, this approach raised inventory levels, production costs, and overhead costs. Customer delivery was also a problem. The old system could only track shipments on a weekly basis. If a customer wanted an order on Monday, and Coors shipped it by the following Saturday, the system logged that order as being on time. When customers called to complain, the salesperson would get no valid data from the system other than an incorrect "shipped on time" report.

It was clear that improvements were needed; however, before investing in the development of new information systems, Coors defined three key business goals that the new systems had to achieve. First, they had to increase customer satisfaction. Salespeople were under tremendous pressure to get information for customers—which in turn prevented them from developing new orders and selling product. Second, Coors wanted to reduce lead times. If work-in-progress, inventory, and delivery schedules could be reduced, then Coors could produce more customer orders. Third, Coors needed to reduce operating costs.

Coors's approach to meeting these goals required redoing and streamlining many fundamental business work processes. In other words, members of the project team focused first on how to meet the needs of the customers before they thought about how to update their outdated information systems.

This rethinking often required challenging fundamental assumptions about how the business should operate. Once the work processes were redesigned, the project team implemented an integrated set of information systems. These new systems automated the work processes associated with acquiring raw materials, transforming raw materials into top-quality products, and delivering them to customers within the shortest possible time.

The project proved to be highly successful. Since the systems were installed, Coors's product cycle has been cut from an average of twelve weeks to eight weeks and on-time shipments have improved to over 95 percent. Coors salespeople can now be confident that "shipped on time" means the order was delivered on time—not just that it shipped within a seven-day period.

The new information systems have also improved business decision making. Each morning, the general manager of sales and marketing meets with key people from manufacturing, engineering, and sales. They review the previous day's sales and requests for new products. They discuss how things are going and can check on the current status because everything that happened as of that morning is already in Coors's information systems, ready for decision making.

Discussion Questions

1. How has implementation of an integrated set of information systems enabled Coors to meet customer needs more effectively?
2. Did this system meet all three key business goals for new systems at Coors? Why or why not?

Critical Thinking Questions

3. Identify three key decisions that must be made at the business review meeting each morning. Identify six questions that are likely to be asked by the general manager at the morning business review meeting.
4. What additional features or benefits might you want this basic system to deliver?

Sources: Adapted from "Investor Relations" and "Products and Services" portions of the Coors Tek Web site, at http://www.coorstek.com, accessed February 9, 1999; "About QAD's Applications" portion of the QAD Web site, at http://www.qad.com/product, accessed February 9, 2000.

Sears Employs Modeling Software to Manage Growth

Through its network of 850 full-line stores and more than 2,100 specialty stores, Sears provides apparel, home, and automotive products and related services for nearly 60 million American households and generates revenues in excess of $40 billion. Sears's full-line stores, located in many of the nation's shopping malls, are at the heart of the company's retail strategy. Their dramatic transformation over the past five years as part of a $4 billion building and remodeling plan speaks to Sears's commitment to create a compelling shopping destination for the American consumer. Fifty-seven million square feet of space in these stores has been remodeled since 1993, with 65 new stores to be added between 1998 and 2001.

Sears has developed a planning and scheduling model that defines templates identifying all the tasks, effort, and elapsed time associated with planning, building, and opening a new store. This data enables Sears to forecast how long it will take to complete each new store. The construction and remodeling process has become more efficient by using this model to schedule the delivery of construction items, fixtures, and store merchandise. The software can combine several subprojects into one schedule, allowing the user to build a master schedule that maps out an overall store project. Users can also check for changes in upstream project tasks that will impact other project tasks and deadlines downstream.

The benefits of this model were apparent immediately. Under the old system, schedule delays were not communicated, causing urgent—and sometimes costly—requests for fixtures or merchandise to prepare the store for its opening. Merchandise was delivered too early or too late, resulting in last-minute changes to floor presentations. Likewise, crews were inappropriately scheduled. But the improved project tracking available through the new model keeps everyone informed of up-to-the-minute changes. If the construction deadline for a portion of a store or the installation of a display changes, the materials suppliers, construction teams, display builders, and merchandise buyers are all quickly informed. They then can adjust their tasks to avoid further complications or cost increases. Furthermore, as a result of this improved planning capability, Sears estimates that the company has reduced the time it takes to complete a new store by two weeks, bringing in additional revenue earlier.

Discussion Questions

1. What sort of reports do you think the model generates to simplify the effort of tracking the progress of opening each store? Would graphical output be of value? Why? What sort of graphs?
2. Identify the key benefits of using this model.

Critical Thinking Questions

3. How should the model handle variations in work rate, where crews at one store may work substantially faster than crews at another store?
4. Where would Sears derive the templates on the effort and elapsed time associated with each task for the planning, building, and opening of a new store? Is it possible that the planning, building, and opening of one store may require new tasks not in the template? Is it possible that certain tasks in the template could be eliminated for a given store? How would the model allow for these possibilities?

Sources: Adapted from Sears 1998 Annual Report; "Sears Announces Expansion of the Great Indoors," October 28, 1999, *PRNewswire*, http://www.prnewswire.com; "Sears Builds on Success of Softer Side Advertising Campaign: New Creative Campaign Asks Customers to 'Take Another Look,'" September 11, 1998, *PRNewswire*, http://www.prnewswire.com; "User Stories—Using Open Plan, Sears Is Making It Easier for Customers to Explore Its Many Sides," http://www.welcom.com/products/, accessed February 10, 2000; and "Products—Open Plan," http://www.welcom.com, accessed February 10, 2000.

NOTES

Sources for the opening vignette on p. 3: Adapted from "Avis Transfers Commercial MasterCard Program from PHH to Wright Express," January 25, 2000, PHH Media Center, http://www.phh.com/prsmithson.htm; Sam Dickey, "The Real Information Highway," *Beyond Computing*, November/December 1999, pp. 50–51; and "PHH Named One of *Information Week*'s 'E-Business 100,'" December 16, 1999, PHH Media Center, http://www.phh.com/prsmithson.htm.

1. User Stories—"Using Open Plan, Sears Is Making It Faster for Customers to Explore Its Many Sides," accessed at http://www.welcom.com/products/ February 10, 2000.
2. Stephen H. Wildstrom, "Age of the E-pliance," *Business Week*, January 17, 2000, p. 20.
3. John Madden, "KPMG Sharing Knowledge," *PC Week*, August 9, 1999, p. 51.
4. Clinton Wilder, "Business Booms for Specialized Web Marketplaces," *Information Week*, February 7, 2000, p. 43.
5. Jim Kerstetter and John Madden, "Web Security Breakdown," *PC Week*, February 14, 2000, pp. 1, 18.
6. "Procter & Gamble Selects Momentum's Boomerang: Plans to Roll," *Business Wire*, November 8, 1999.
7. Rick Whiting and Beth Davis, "More on the Edge," *Information Week*, August 23, 1999, pp. 35–48.
8. Rick Whiting and Beth Davis, " More on the Edge," *Information Week*, August 23, 1999, pp. 35–48.
9. Peter Coffee, "Math Sharpens IT's Competitive Edge," *PC Week*, August 30, 1999, p. 44.
10. Jagdish N. Sheth and Rajendra S. Sisodia, "Are Your IT Priorities Upside Down?" *CIO Enterprise*, November 15, 1999, pp. 84–88.
11. Rick Whiting and Beth Davis, "More on the Edge," *Information Week*, August 23, 1999, pp. 35–48.
12. Jagdish N. Sheth and Rajendra S. Sisodia, "Are Your IT Priorities Upside Down?" *CIO Enterprise*, November 15, 1999, pp. 84–88.
13. Matt Hicks, "The Post-Y2K Agenda," *PC Week*, December 20/27, 1999, pp. 81–83.
14. John Madden, "KPMG Sharing Knowledge," *PC Week*, August 9, 1999, p. 51.

CHAPTER 2

Information Systems in Organizations

*P*eople overestimate what will happen in two years and underestimate what will happen in 10. . . . You won't even think about where information really is. You'll just know that when you go to your home machine or your work machine, the files are there.

— Bill Gates, Microsoft

Principles	Learning Objectives
The use of information systems to add value to the organization is strongly influenced by organizational structure, culture, and change.	• *Identify the seven value-added processes in the supply chain and describe the role of information systems within them.* • *Provide a clear definition of the terms* organizational structure, culture, *and* change *and discuss how they affect the implementation of information systems.*
Because information systems are so important, businesses need to be sure that improvements or completely new systems help lower costs, increase profits, improve service, or achieve a competitive advantage.	• *Identify some of the strategies employed to lower costs or improve service.* • *Define the term* competitive advantage *and discuss how organizations are using information systems to gain such an advantage.* • *Discuss how organizations justify the need for information systems.*
Information systems personnel are the key to unlocking the potential of any new or modified system.	• *Define the types of roles, functions, and careers available in information systems.*

Farmers Insurance Group

Finding Higher Revenue and Lower Claims in Information

Large companies process millions of transactions every year and store huge amounts of data on these transactions. Often, the data is spread across a variety of different computer systems in different areas of the country or the world. This raw data, although needed for record keeping, has little value to managers and decision makers unless it can be filtered and processed into meaningful information. The results can be a staggering increase in revenues and profits. But doing this is the real challenge. Finding strategic information from a mountain of data can be like finding a needle in a haystack, but the effort is usually worth it. With today's fast computers and a knowledgeable IS staff, the possibility of turning raw data into useful and profitable information can become a reality. This was the case with Farmers Insurance Group.

Like other companies, Farmers Insurance Group was sitting on a huge amount of raw data. The data, however, was spread across different computer systems in different locations. As in all insurance companies, underwriting determines what insurance policies a company can offer and at what premiums. Farmers's underwriting business was responsible for assessing insurance risk, which can make the difference between profits and losses. The people who are responsible for determining insurance risk are called actuaries. According to Tom Boardman, an assistant actuary at Farmers, "As competition has gotten more intense in the insurance industry, the traditional ways of segmenting risk aren't good enough at providing you competitive advantage." Boardman was referring to how most insurance companies categorize risk. For example, high-powered sports cars are more likely to be involved in expensive accidents than ordinary sedans. Thus, insurance companies can put sports cars in a different risk category than sedans and charge customers who own them a higher premium. In assessing risk, an insurance actuary would traditionally have a hunch, such as that sports cars are more prone to accidents than sedans. Then the actuary would test his or her hunch using the computer. According to Boardman, this was like using the computer "to dig up data to prove or unprove those hunches." One disadvantage of this old approach is that small, but profitable, market niches may be ignored or not priced correctly. As a result, Farmers decided to look into a computer system to help it find profitable market niches.

The company found the help it needed through IBM, which developed a customized software product for Farmers called DecisionEdge. The computer system was an advanced decision support system that combined raw data from seven different databases on a staggering 35 million records. Consolidating the raw data into useful information took about twice as long as expected, but the additional wait was worth it. Farmers was able to locate market niches that it didn't see before it had the decision support system. For example, DecisionEdge helped Farmers determine that not all sports car owners are alike—those who were older and had at least one other car were less likely to be in an expensive accident. Once this market niche was identified, Farmers could offer that segment of the sports car market lower premiums. Using DecisionEdge to find the market niche resulted in millions of dollars of increased revenues for Farmers.

The approach used by Farmers is sometimes called "data scrubbing." It allows a company to consolidate important information and squeeze additional revenues and profits from it. After helping Farmers and seeing a market opportunity, IBM also decided to offer its DecisionEdge software to other insurance companies.

As you read this chapter, consider the following:

- How was Farmers able to transform its raw data into meaningful information and additional revenues?

- Describe how this approach could be used in other industries.

Technology's impact on business is growing steadily. Once used to automate manual processes, technology has now transformed the nature of work—and the shape of organizations themselves. During the late 1960s and early 1970s, many computerized information systems were developed to provide reports for business decision makers. The information in these reports helped managers monitor and control business processes and operations. For example, reports that listed the quantity of each inventory item in stock could be used to monitor inventory levels. Unfortunately, many of these early computer systems did not take the overall goals of the organization and managerial problem-solving styles into consideration. Some decision makers wanted detailed inventory reports of all the items, while others wanted a list of inventory items only when the number on hand was very low. Even more important, these early systems were not developed as part of the business process itself. As a result, many of the early systems failed or were not utilized to their potential. Today, businesses recognize that both important organizational concepts and processes must be considered and supported by effective information systems.

Business organizations use information systems for a number of purposes. Their use is strongly influenced by a business's organizational structure and the way particular businesses attempt to achieve their goals. For example, Farmers Insurance Group was able to use DecisionEdge to help it find new market niches. Finding one market niche resulted in millions of dollars in additional revenues for Farmers. To help provide an understanding of how information systems shape and are shaped by organizational structure, we now examine organizations and information systems.

ORGANIZATIONS AND INFORMATION SYSTEMS

organization

a formal collection of people and other resources established to accomplish a set of goals

An **organization** is a formal collection of people and other resources established to accomplish a set of goals. The primary goal of a for-profit organization is to maximize shareholder value, often measured by the price of the company stock. Nonprofit organizations include social groups, religious groups, universities, and other organizations that do not have profit as the primary goal.

An organization is a system. Money, people, materials, machines and equipment, data, information, and decisions are constantly in use in any organization. As shown in Figure 2.1, resources such as materials, people, and money are input to the organizational system from the environment, go through a transformation mechanism, and are output to the environment. The outputs from the transformation mechanism are usually goods or services. The goods or services produced by the organization are of higher relative value than the inputs alone. Through adding value or worth, organizations attempt to achieve their goals.

How does this increase in value occur? Within the transformation mechanism, various subsystems contain processes that help turn specific inputs into goods or services of increasing value. These value-added processes increase the relative worth of the combined inputs on their way to becoming final outputs of the organization. Let us reconsider our simple car wash example from Chapter 1 (Figure 1.4). The first value-added process might be identified as washing the car. The output of this system—a clean, but wet, car—is worth more than the mere collection of ingredients (soap and water), as evidenced by the popularity of automatic car washes. Consumers are willing to pay more for the skill, knowledge, time, and energy required to wash their cars. The second value-added process can be identified as drying—the transformation of the wet car into a dry one with no water spotting. Again, consumers are willing to pay more for the additional skill, knowledge, time, and energy required to accomplish this transformation.

In general, organizations establish these value-added processes to achieve their goals by exploiting opportunities and solving problems. In our chapter-opening vignette, we saw how Farmers Insurance Group was able to exploit a

FIGURE 2.1

A General Model of an Organization

Information systems support and work within all parts of an organizational process. Although not shown in this simple model, input to the process subsystem can come from internal and external sources. Just prior to entering the subsystem, data is external. Once it enters the subsystem, it becomes internal. Likewise, goods and services can be output to either internal or external systems.

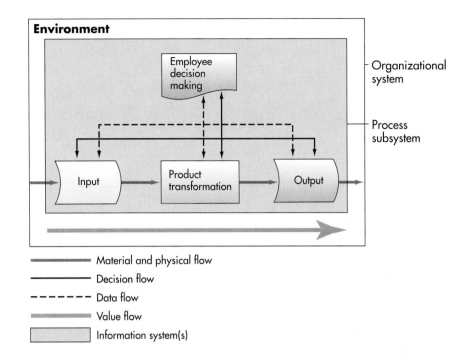

value chain

a series (chain) of activities that includes inbound logistics, warehouse and storage, production, finished product storage, outbound logistics, marketing and sales, and customer service

---

Information systems are an integral part of the value-added process of Garden.com. The Web site links customers with suppliers of gardening materials.

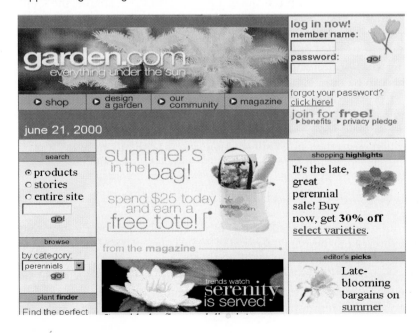

new opportunity by finding a new market niche. The result was millions of dollars of additional revenue.

All business organizations use a number of value-added processes. Providing value to a stakeholder—customer, supplier, manager, or employee—is the primary goal of any organization. The value chain, first described by Michael Porter in a 1985 *Harvard Business Review* article, is a concept that reveals how organizations can add value to their products and services. The **value chain** is a series (chain) of activities that includes inbound logistics, warehouse and storage, production, finished product storage, outbound logistics, marketing and sales, and customer service (Figure 2.2). Each of these activities is investigated to determine what can be done to increase the value perceived by a customer. Depending on the customer, value may mean lower price, better service, higher quality, or uniqueness of product. The value comes from the skill, knowledge, time, and energy invested by the company. By adding a significant amount of value to their products and services, companies will ensure further organizational success.

What role does an information system play in these value-added processes? A traditional view of information systems holds that they are used by organizations to control and monitor value-added processes to ensure effectiveness and efficiency. An information system can turn feedback from the value-added process subsystems into more meaningful information for employees' use within an organization. This information might summarize the performance of the systems and be used as the basis for changing the way the system operates. Such changes could involve using different raw materials (inputs), designing new assembly-line procedures (product transformation), or developing new

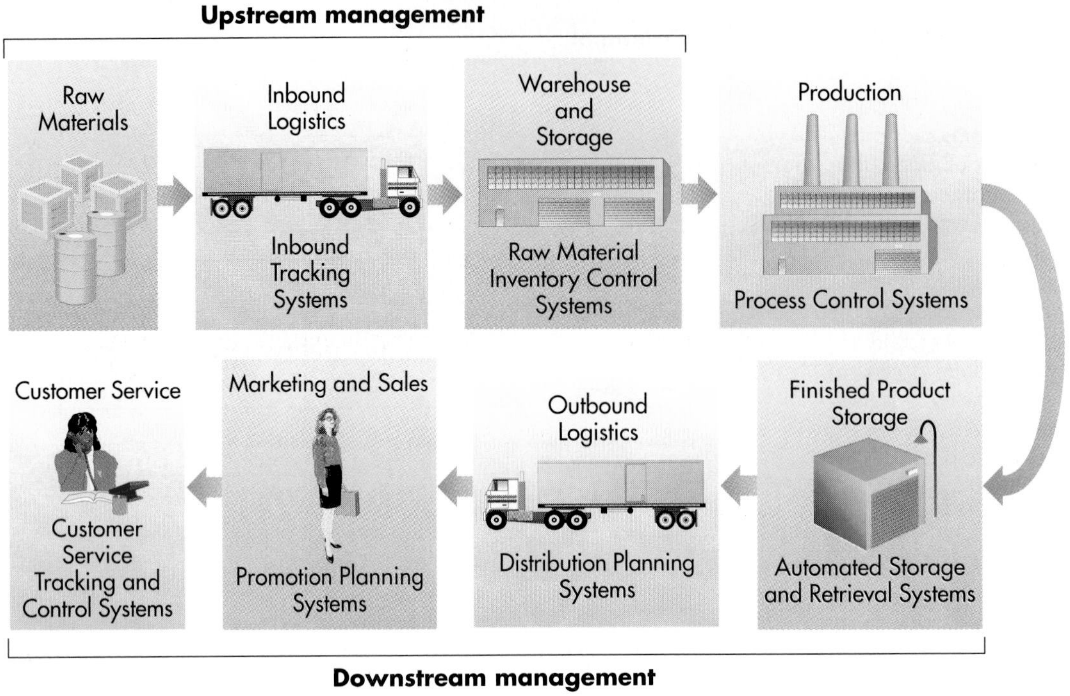

Upstream management

Downstream management

FIGURE 2.2

The Value Chain of a
Manufacturing Company

The management of raw
materials, inbound logistics,
and warehouse and storage
facilities is called upstream
management, and the
management of finished
product storage, outbound
logistics, marketing and sales,
and customer service is called
downstream management.

products and services (outputs). In this view, the information system is external to the process and serves to monitor or control it.

A more contemporary view, however, holds that information systems are often so intimately intertwined with the underlying value-added process that they are best considered part of the process itself.[1] From this perspective, the information system is internal to and plays an integral role in the process, whether by providing input, aiding product transformation, or producing output. Consider a phone directory business that creates phone books for international corporations. A corporate customer requests a phone directory listing all steel suppliers in Western Europe. Using its information system, the directory business can sort files to find the suppliers' names and phone numbers and organize them into an alphabetical list. The information system itself is an integral part of this process. It does not just monitor the process externally but works as part of the process to transform a product. In this example, the information system turns raw data input (names and phone numbers) into a salable output (a phone directory). The same system might also provide the input (data files) and output (printed pages for the directory).

The latter view brings with it a new perspective on how and why information systems can be used in business. Rather than searching to understand the value-added process independently of information systems, we consider the potential role of information systems within the process itself, often leading to the discovery of new and better ways to accomplish the process. Thus, the way an organization views the role of information systems will influence the ways it accomplishes its value-added processes.

Organizational Structure

organizational structure

organizational subunits and the
way they relate to the overall
organization

Organizational structure refers to organizational subunits and the way they relate to the overall organization. Depending on the goals of the organization and its approach to management, a number of structures can be used. An organization's structure can affect how information systems are viewed and what kind are used. Although there are many possibilities, organizational structure typically falls into one of these categories: traditional, project, team, or multidimensional.

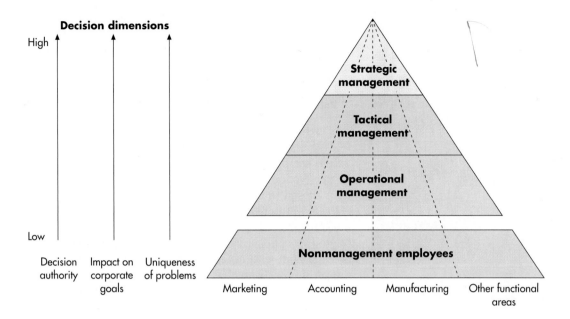

FIGURE 2.3

A simplified model of the organization, showing the managerial pyramid from top-level managers to nonmanagement employees.

traditional organizational structure

organizational structure in which major department heads report to a president or top-level manager

Traditional Organizational Structure

In the type of structure known as **traditional organizational structure**, a managerial pyramid shows the hierarchy of decision making and authority from the strategic management to operational management and nonmanagement employees. The strategic level, including the president of the company and vice presidents, has a higher degree of decision authority, more impact on corporate goals, and more unique and one-of-a-kind problems to solve (see Figure 2.3). In most cases, major department heads report to a president or top-level manager. The major departments are usually divided according to function and can include marketing, production, information systems, finance and accounting, research and development, and so on (Figure 2.4). The positions or departments that are directly associated with making, packing, or shipping goods are called *line positions*. A production supervisor who reports to a vice president of production is an example of a line position. Other positions may not be directly involved with the formal chain of command but may assist a department or area. These are *staff positions*, such as a legal counsel reporting to the president.

The traditional organizational structure is also referred to as a *hierarchical structure*, since it can be viewed as a series of levels, with those at higher levels having more power and authority within the organization. Today, the trend is to reduce the number of management levels, or layers, in the traditional organizational

FIGURE 2.4

A Traditional Organizational Structure

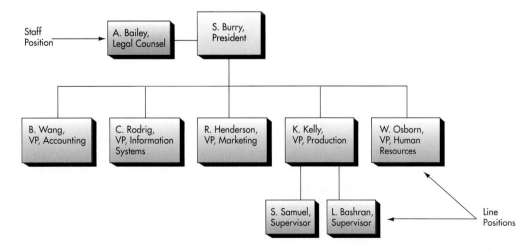

Employees in Samsung Electronics' manufacturing facilities are empowered to stop the production line if they detect a problem or defect. (Source: Courtesy of Samsung Electronics America, Inc.)

flat organizational structure

organizational structure with a reduced number of management layers

empowerment

giving employees and their managers more responsibility and authority to make decisions, take certain actions, and have more control over their jobs

project organizational structure

structure centered on major products or services

team organizational structure

structure centered on work teams or groups

multidimensional organizational structure

structure that may incorporate several structures at the same time

structure. A structure with a reduced number of management layers, often called a **flat organizational structure**, empowers employees at lower levels to make decisions and solve problems without needing permission from midlevel managers. **Empowerment** gives employees and their managers more responsibility and authority to make decisions, take certain actions, and in general have more control over their jobs. For example, an empowered salesclerk would be able to respond to certain customer requests or problems without needing permission from a supervisor. On the factory floor, empowerment can mean that an assembly-line worker has the ability to stop the production line to correct a problem or defect before the product is passed to the next station.

Empowerment usually results in faster action and quicker resolution of problems. It can also reduce costs and result in higher-quality products and services. It is usually less expensive, for example, to fix a problem with an automobile part on an assembly line than to wait until final inspection when the part could be very hard to reach. For instance, Saturn, a highly successful American car manufacturer, empowered its employees by turning assembly lines into dedicated workstations managed solely by the work team. Saturn employees make suggestions on how to improve processes; even the design process involves a high degree of employee participation. In an empowered organization such as Saturn, employees look beyond their own job, since they feel the responsibility to make the whole organization work better.

Information systems can be a key element in empowering employees. Often, information systems make empowerment possible by providing information directly to employees at lower levels of the hierarchy. The employees may also be empowered to develop or use their own personal information systems, such as a simple forecasting model or spreadsheet. For example, Atrium Empowerment is a series of self-service applications that allow employees and managers to handle many functions formerly managed through the human resource department. The applications can be accessed by telephone, touch-screen kiosk, desktop, and Web browser. All transactions are designed to allow for workflow and management approval. Atrium Empowerment is also integrated into human resource and payroll systems such as Ceridian Employer Services, PeopleSoft, MSA, Cyborg, and many other systems.[2]

Project Organizational Structure

A **project organizational structure** is centered on major products or services. For example, in a manufacturing firm that produces baby food and other baby products, each type is produced by a separate unit. Traditional functions like marketing, finance, and production are positioned within these major units (Figure 2.5). Many project teams are temporary—when the project is complete, the members go on to new teams formed for another project.

Team Organizational Structure

The **team organizational structure** is centered on work teams or groups. In some cases, these teams are small; in others, they are very large. Typically, each team has a team leader who reports to an upper-level manager in the organization. Depending on the tasks being performed, the team can be either temporary or permanent.

Multidimensional Organizational Structure

A **multidimensional organizational structure**, also called a *matrix organizational structure,* may incorporate several structures at the same time. For example, an organization might have both traditional functional areas and

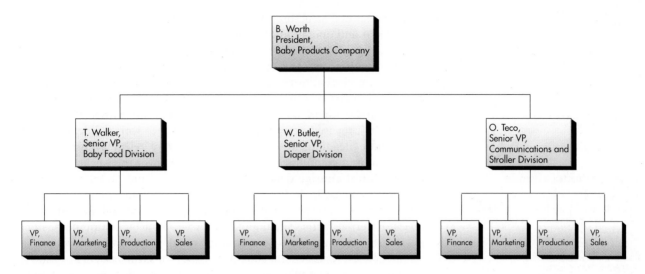

major project units. When diagrammed, this structure forms a matrix, or grid (Figure 2.6).

One advantage of the multidimensional organizational structure is the ability to simultaneously stress both traditional corporate areas and important product lines. A potential disadvantage is multiple lines of authority. Employees have two bosses or supervisors: one functional boss and one project boss. As a result, conflicts may occur when one boss wants one thing and the other boss wants something else. For example, the functional boss might want the employee to work on a new product in the next two days, while the project boss might want the employee to fly to a two-day meeting. Obviously, the employee cannot do both. One way to resolve this problem is to give one boss priority if there are problems or conflicts.

Organizational Culture and Change

culture

a set of major understandings and assumptions shared by a group

organizational culture

the major understandings and assumptions for a business, a corporation, or an organization

Culture is a set of major understandings and assumptions shared by a group—for example, within an ethnic group or a country. **Organizational culture** consists of the major understandings and assumptions for a business, a corporation, or an organization. The understandings, which can include common beliefs, values, and approaches to decision making, are often not stated or documented as goals or formal policies. Employees, for example, might be expected to be clean-cut, wear conservative outfits, and be courteous in dealing with all customers. Sometimes organizational culture is formed over years, or it can be formed rapidly by top-level managers—for example, implementation of a "casual Friday" dress policy.

	Vice President, Marketing	Vice President, Production	Vice President, Finance
Publisher, College Division	Marketing Group	Production Group	Finance Group
Publisher, Trade Division	Marketing Group	Production Group	Finance Group
Publisher, High School Division	Marketing Group	Production Group	Finance Group

Like organizational structure, organizational culture can significantly affect the development and operation of information systems within an organization. A procedure associated with a newly designed information system, for example, might conflict with an informal procedural rule that is part of organizational culture. Organizational culture might also influence a decision maker's perception of the factors and priorities that must be considered in setting objectives. For example, there might be an unwritten understanding that all inventory reports must be prepared before ten o'clock Friday morning. Because of this understood time deadline, the decision maker may reject a cost-reduction option that required compiling the inventory report over the weekend.

Organizational change deals with how for-profit and nonprofit organizations plan for, implement, and handle change. Change can be caused by internal or external factors. Internal factors include activities initiated by employees at all levels. External factors include activities wrought by competitors, stockholders, federal and state laws, community regulations, natural occurrences (such as hurricanes), and general economic conditions. Introducing or modifying an information system will also cause change. Improving an organizational process through information systems requires changing the activities and tasks related to the process. Often, this means changing the way individuals, groups, and the enterprise work.

Overcoming resistance to change can be the hardest part of bringing information systems into a business. Many potential improvements have failed because managers and employees were not prepared for change. Occasionally, employees even attempt to sabotage a new information system because they do not want to learn the new procedures and commands. In most of these instances, the employees were not involved in the decision to implement the change, nor were they fully informed about the reasons the change was occurring and the benefits that would accrue to the organization.

The dynamics of change can be viewed in terms of a change model. A **change model** is a representation of change theories that identifies the phases of change and the best way to implement them. Kurt Lewin and Edgar Schein proposed a three-stage approach for change (Figure 2.7). *Unfreezing* is the process of ceasing old habits and creating a climate receptive to change. *Moving* is the process of learning new work methods, behaviors, and systems. *Refreezing* involves reinforcing changes to make the new process second nature, accepted, and part of the job.[3] When a company introduces a new information system, a few members of the organization must become agents of change—champions of the new system and its benefits. Understanding the dynamics of change can help them confront and overcome resistance, so that the new system can be used to maximum efficiency and effectiveness.

Organizational learning is closely related to organizational change. According to the concept of **organizational learning**, organizations adapt to new conditions or alter their practices over time. This means that assembly-line workers, secretaries, clerks, managers, and executives learn better ways of doing business and incorporate them into their day-to-day activities. Collectively, these adjustments based on experience and ideas are called organizational learning. In some cases, the adjustments can be a radical redesign of business processes, often called reengineering. In other cases, these adjustments can be more incremental, a concept called continuous improvement.

Reengineering

To stay competitive, organizations must occasionally make fundamental changes in the way they do business. In other words, they must change the activities and tasks, or processes, that they use to achieve their goals. **Reengineering**, also called **process redesign**, involves the radical redesign of

organizational change

the responses that are necessary for for-profit and nonprofit organizations to plan for, implement, and handle change

change model

representation of change theories that identifies the phases of change and the best way to implement them

organizational learning

concept according to which organizations adapt to new conditions or alter their practices over time

reengineering (process redesign)

the radical redesign of business processes, organizational structures, information systems, and values of the organization to achieve a breakthrough in business results

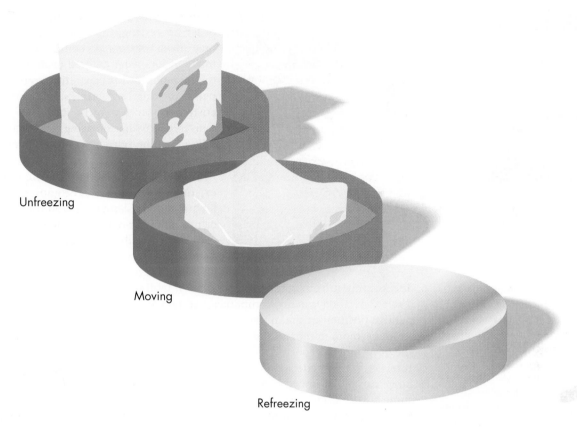

Unfreezing

Moving

Refreezing

FIGURE 2.7

A Change Model

business processes, organizational structures, information systems, and values of the organization to achieve a breakthrough in business results (Figure 2.8). Reengineering can reduce delivery time, increase product and service quality, enhance customer satisfaction, and increase revenues and profitability. Avon Products reengineered its worldwide operations to reduce its hundreds of incompatible computer systems to one platform.[4] Before the reengineering project, Avon had 700 different computer systems in South America, Asia, the United Kingdom, and Puerto Rico that didn't work together efficiently. So, Avon decided to standardize its hardware platform worldwide. In addition, the reengineering project also standardized the company's Internet usage to allow customers, suppliers, managers, and employees to share information.

A business process includes all activities, both internal (such as thinking) and external (such as taking action), that are performed to transform inputs into outputs. It defines the way work gets done. Despite the advent of the computer about 50 years ago, many organizations' sequential steps and assembly-line work processes have remained intact. In fact, many efforts to apply information systems have only further cemented the steps. For example, in some companies, a customer order is still processed by several different people. The order moves from one step to the next, allowing the potential for people to make numerous errors and create misunderstandings. Others now employ a customer service representative who oversees the entire process: taking the order, entering order processing data, and scheduling product delivery and setup. The customer service representative expedites and coordinates the process, and the customer has just one contact who always knows the status of the order.

This simple example illustrates the fundamental changes reengineering creates in the way things are done, often across multiple departments. But asking people to work differently often meets with stiff resistance, and change is difficult to maintain—the values of the organization and its employees must be

Before reengineering, Avon Products, a company that markets cosmetics to women in 137 countries, had 700 different computer systems. The reengineering project standardized the company's hardware platform and Internet usage, allowing the 3 million independent sales representatives to share information.
(Source: Adrian Bradshaw/SABA.)

changed also. In the previous example of order processing, the original work process may have evaluated employees on how many orders were entered each day. Under the reengineered process, they may be evaluated on different factors associated with customer service—percentage of orders delivered on time or accuracy of customer bills. Helping employees understand the benefits of the new system is a major hurdle.

In contrast to simply automating the existing work process, reengineering challenges the fundamental assumptions governing their design. It requires finding and vigorously challenging old rules blocking major business process changes. These rules are like anchors weighing a firm down and keeping it from competing effectively. Examples of such rules are given in Table 2.1. Today, many companies use reengineering to increase their competitive position in the market.

Many more companies are now doing business over the Internet. In 1999, the government reported its first formal stand-alone estimate of sales over the Internet.[5] It exceeded $5 billion for 1999. Customers, the company, and suppliers can all be linked through an effective Internet site. Although some consumers still aren't yet ready to buy goods over the Internet, people are becoming more and more receptive to e-commerce. As a result, companies are reengineering their sales processes to take advantage of this opportunity. Along the way they are learning that the cost to process an order can be cut in half, plus orders entered by customers themselves tend to have fewer errors than those dictated over the phone. This enables companies to offer price discounts to e-commerce customers. Amazon.com, which sells books, videotapes, and other items over the Internet, typically sells these items for 10 percent to 20 percent less than typical retail stores. In addition, some believe that Internet sales can reduce energy costs and pollution.[6] These changes to value-added business processes occurred because organizations considered information system components as integral parts of the business process.

FIGURE 2.8

Reengineering

Reengineering involves the radical redesign of business processes, organizational structure, information systems, and values of the organization to achieve a breakthrough in business results.

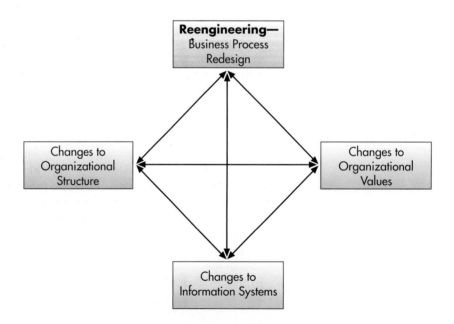

Rule	Original Rationale	Potential Problem
Small orders must be held until full-truckload shipments can be assembled.	Reduce delivery costs.	Customer delivery is slowed—lost sales.
No order can be accepted until customer credit is approved.	Reduce potential for bad debt.	Customer service is poor—lost sales.
All merchandising decisions are made at headquarters.	Reduce number of items carried in inventory.	Customers perceive organization has limited product selection—lost sales.

Continuous Improvement

continuous improvement

constantly seeking ways to improve
the business processes to add value
to products and services

The idea of **continuous improvement** is to constantly seek ways to improve the business processes to add value to products and services. This in turn will increase customer satisfaction and loyalty and ensure long-term profitability. Manufacturing companies make continual product changes and improvements. Service organizations regularly find ways to provide faster and more effective assistance to customers. By doing so, these companies increase customer loyalty, minimize the chance of customer dissatisfaction, and diminish the opportunity for competitive inroads.

To improve its operations, Bethlehem Steel adopted a continuous improvement plan.[7] One of its initiatives was to develop a closer relationship with its suppliers. "Value added through continuous improvement" was a key to the initiative. Bethlehem searched methodically and extensively to select suppliers that could assist the company in all aspects of its business. Information from 70 service providers was collected. This list was then narrowed to just 16, and a formal request for proposals (RFP) was sent out to the 16 firms. After the RFP process, the list was narrowed further to 6 firms. These firms underwent detailed analysis and investigation. From this analysis, Bethlehem selected 3 firms. One of the guiding principles in utilizing the 3 firms was to be flexible and to readjust when appropriate, which is what continuous improvement is all about.

Organizational commitment to goals such as continuous improvement can be supported by the strategic use of information systems. Continuous improvement involves constantly improving and modifying products and services to remain competitive and to keep a strong customer base. In doing so, companies can increase the quality of their products and services. Low-quality products can turn companies that once were the leaders in their industry into laggards that have lower profits and reduced market share. Without question, quality will continue to be an important factor for profitability and survival. Table 2.2 compares reengineering and continuous improvement.

Business Process Reengineering	Continuous Improvement
Strong action taken to solve serious problem	Routine action taken to make minor improvements
Top-down driven by senior executives	Worker driven
Broad in scope; cuts across departments	Narrow in scope; focus is on tasks in a given area
Goal is to achieve a major breakthrough	Goal is continuous, gradual improvements
Often led by outsiders	Usually led by workers close to the business
Information system integral to the solution	Information systems provide data to guide improvement team

Technology Diffusion, Infusion, and Acceptance

To be effective, reengineering and continuous improvement efforts must be accepted and used throughout an organization. The extent to which technology is used throughout an organization can be a function of technology diffusion, infusion, and acceptance. **Technology diffusion** is a measure of how widely technology is spread throughout an organization. An organization in which computers and information systems are located in most departments and areas has a high level of technology diffusion.[8] Some on-line merchants, such as Amazon.com, have a high level of diffusion and use computer systems to perform most of their business functions, including marketing, purchasing, and billing. **Technology infusion**, on the other hand, is the extent to which technology permeates an area or department. In other words, it is a measure of how deeply imbedded technology is in an area of the organization. Some architectural firms, for example, use computers in all aspects of designing a building or structure. This design area thus has a high level of infusion. Of course, it is possible for a firm to have a high level of infusion in one aspect of its operations and a low level of diffusion overall. The architectural firm, for example, may use computers in all aspects of design (high infusion in the design area) but may not use computers to perform other business functions, including billing, purchasing, and marketing (low diffusion).

Just because an organization has a high level of diffusion and infusion, with computers throughout the organization, it does not necessarily mean that information systems are being used to their full potential. Indeed, the assimilation and use of expensive computer technology throughout organizations varies greatly.[9] One reason is a low degree of acceptance and use of the technology among some managers and employees. Research has attempted to explain the important factors that enhance or hinder the acceptance and use of information systems.[10] A number of possible explanations of technology acceptance and usage have been studied. The **technology acceptance model (TAM)** specifies the factors that can lead to higher acceptance and usage of technology in an organization, including the perceived usefulness of the technology, the ease of its use, the quality of the information system, and the degree to which the organization supports the use of the information system.[11] Companies hope that a high level of diffusion, infusion, and acceptance will lead to greater performance and profitability.[12] Some experts believe that the booming stock market of the late 1990s was in part a result of the successful implementation and usage of technology to increase productivity and profitability.

Total Quality Management

The definition of the term *quality* has evolved over the years. In the early years of quality control, firms were concerned with meeting design specifications—that is, conformance to standards. If a product performed as designed, it was considered a high-quality product. A product can perform its intended function, however, and still not satisfy customer needs. Today, **quality** means the ability of a product (including services) to meet or exceed customer expectations.[13] For example, a computer that not only performs well but is easy to maintain and repair would be considered a high-quality product. This view of quality is completely customer oriented. A high-quality product will satisfy customers by functioning correctly and reliably, meeting needs and expectations, and being delivered on time with courtesy and respect. Read the "E-Commerce" box to see how Internet companies, like Drugstore.com, are implementing quality programs.

To help them deliver high-quality goods and services, some companies have adopted continuous improvement strategies that require each major business process to follow a set of total quality management guidelines. **Total quality management (TQM)** consists of a collection of approaches, tools, and techniques that

E-COMMERCE
Finding Quality through E-Commerce

Companies have long recognized that superior customer quality is a key to attracting and retaining customers. Slogans such as "Quality Is Job One" and "Our Customers Always Come First" are typical. Companies often spend thousands of dollars attracting new customers with special offers and one-time discounts but ignore the importance of keeping current customers satisfied. One dissatisfied customer can easily tell five to ten friends and family members of a bad experience. The result can be a massive loss, and the company may not even know what hit it. All the company sees is a drastic reduction in sales and profits. The lesson is that quality pays off in higher sales and profits, but have the new e-commerce companies learned this important lesson?

Internet-based companies have entered most markets that were once dominated by traditional companies. Travel, stock trading, investment advice, auctions, and shopping are just a few examples of traditional industries that are now thriving on the Internet. Yet, quality and customer satisfaction seem to be lacking in many of these industries. In a survey conducted by the NPD Group, it was revealed that fewer than one-third of the respondents "strongly agreed" that they got accurate information from Web shopping. Fewer than one-fourth thought that the customer service areas of on-line companies responded quickly or really understood their needs. Without question, quality is a major concern for on-line customers in general. Some Internet companies, however, are starting to get the message that quality does indeed count.

Many on-line companies, like Drugstore.com, are starting to realize the importance of high quality. Drugstore.com has instituted a quality program with two different types of quality specialists. One works with developing a better customer interface for the system and ensures that customers are satisfied with their on-line experience with Drugstore.com. The other quality specialists are the technical people that make sure that all the systems are running correctly with minimal downtime and interruptions. To achieve high quality at both the customer interface level and the technical level requires computer systems that work well at both the front end and the back end of the business.

The front-end systems for an Internet company are those the customer sees. They include the Internet Web site and systems that allow customers to get information and place orders. Once an order is placed, the back-end systems are used to fill the orders and ship products to customers. The back-end systems are more traditional uses of technology. So, for an existing company, the back-end computer systems may already be in place. The problem, though, is to make sure that the two systems are communicating with each other and working smoothly. Some Internet companies have great front-end systems that provide information and take orders, yet they have terrible back-end systems to fill orders and ship products. According to Steven Nevill, CIO of Gerald Stevens, which sells flowers and gifts on-line, "You want your customers—whether shopping on-line, calling an 800 number or walking into a store—to have the same experience. You want whoever is interacting with them to know everything about them, where they have been before, and be able to service them in the same way." This desire for a consistently high-quality experience is why some companies have waited to go on-line—they want to be sure that their sites meet expectations.

Many companies, like Wal-Mart, are taking their time to make their Web sites the best they can be. General Electric, which uses a rigorous quality program, has the same approach to quality for its e-business. Talking about the program, General Electric's e-business leader Camille Farhat said, "We cannot relax that standard." But this "go slow and get it right" approach may not bring in the highest revenues or profits. While companies are taking their time to get it right, other companies are launching on-line sites without careful quality planning. Companies like E*trade and eBay started quickly and had technical problems and crashes. But according to Thomas Eisenmann, who teaches a course on e-commerce at the Harvard Business School, "E*trade and eBay have both suffered many hiccups in terms of reliability, but it doesn't seem fundamentally to have hurt the trust in their brands."

Discussion Questions

1. What are the features of the quality program developed by Drugstore.com?
2. What is the difference between a front-end and a back-end system?

Critical Thinking Questions

3. Why is quality important for on-line companies? How would you increase quality for an on-line business?
4. Is it always best for a company to take its time to develop a high-quality Web site?

Sources: Adapted from Gary Anthes, "The Quest for E-Quality," *Computerworld,* December 13, 1999, p. 46; and Janet Rae-Dupree, "The Neupert Treatment," *Business Week,* February 14, 2000, p. 83.

offers a commitment to quality throughout the organization. TQM involves developing a keen awareness of customer needs, adopting a strategic vision for quality, empowering employees, and rewarding employees and managers for producing high-quality products. As a result, processes may be redefined and restructured. Companies like Boeing combine computer technology and total quality programs to both enhance quality and cut costs.[14] Boeing uses portable computing devices to monitor the quality of its airplane manufacturing process. After the quality data is entered into the portable computer, it is transferred to larger computers for analysis and report writing.

Information systems are fully integrated into business processes in organizations that adhere to continuous improvement or TQM strategies. Capturing and analyzing customer feedback and expectations and designing, manufacturing, and delivering quality products and services to customers around the world are only a few ways computers and information systems are helping companies pursue their goals of quality and continuous improvement.

Outsourcing and Downsizing

In an effort to control costs, organizations have looked at the number of people they have on the payroll. A significant portion of an organization's expenses go to hire, train, and compensate talented staff. So organizations today are trying to determine the number of employees they need to maintain high-quality goods and services. With fierce competition in the marketplace, it is critical for organizations to use their resources wisely. Two strategies to contain costs are outsourcing and downsizing (sometimes called rightsizing).

outsourcing

contracting with outside professional services to meet specific business needs

Outsourcing involves contracting with outside professional services to meet specific business needs. Often a specific business process—such as employee recruiting and hiring, development of advertising materials, product sales promotion, and global telecommunications network support—is outsourced. One reason organizations outsource a business process is to enable them to focus more closely on core business—and target limited resources to meet strategic goals. Other reasons for outsourcing are to obtain cost savings or to benefit from the expertise of the service provider.

Outsourcing not only permits a company to concentrate on its core business, but it allows the company to obtain excellent technical assistance it may not have with its own internal IS staff. To achieve both of these goals, FTD.com outsourced the technical operations of its on-line flower operations to Intira.[15] The outsourcing plan also allowed FTD to consolidate all IS operations to one vendor. Previously, FTD had used three outside companies to help with its IS operations. Intira owns all the equipment and network infrastructure, so it was able to offer a service agreement to ensure FTD system availability of up to 99.95 percent. FTD continues to have strong growth in sales revenues. The outsourcing agreement allows FTD to concentrate on growing its business, while allowing Intira to concentrate on delivering a quality Web site for customers.

Companies that are considering outsourcing to cut the cost of their IS operations need to review this decision carefully. A growing number of organizations are finding that outsourcing does not necessarily lead to reduced costs. One of the primary reasons for cost increases is poorly written contracts that result in additional charges from the outsourcing vendor for each additional task identified.

Downsizing

downsizing

reducing the number of employees to cut costs

Downsizing involves reducing the number of employees to cut costs. Rather than pick a specific business process to be downsized, companies usually look to downsize across the entire company. Downsizing clearly drives down wages. However, there are also often many bad side effects of downsizing. Employee morale hits rock bottom. Lines of communication within the company are weakened. Employee productivity drops. Often, high-priced consultants must be hired to help patch the business back together. The lost time, waning productivity, and devastated morale create hidden costs, which can far outweigh the usual cost savings predicted from a layoff.

Continued financial success or survival can depend on downsizing. Baan, a company that makes and markets enterprise resource planning (ERP) software, had disappointing financial results and turned to downsizing in an attempt to turn the company around.[16] The company closed 14 of its offices and cut about 4 percent of its workforce. In addition to the cuts, Baan also decided to realign its workforce. Its research and development efforts will now concentrate more on what it calls "e-enabled enterprise solutions." According to Jim Mooney, Baan's chief financial officer, "I feel good that we have paid close scrutiny to future development efforts, and that we have taken hard action to redirect efforts."

Employers need to be open to alternatives for reducing the number of employees, with layoffs viewed as the last resort. It's much simpler to encourage people to leave voluntarily through early retirement or other incentives. Following this approach, the downsizing effort is accompanied with a "buyout package" offered to certain classes of employees (e.g., those over 50 years old). The buyout package offers employees certain benefits and cash incentives if they voluntarily retire from the company. Other options are job-sharing and transfers.

The charge of age discrimination is frequently associated with downsizing, as older workers are most often affected. As a result, job discrimination lawsuits have been filed against companies such as Pacific Telesis, AT&T, and IBM. To avoid costly lawsuits, employers need to develop and apply neutral, nondiscriminatory criteria for their staff reduction policies. For example, employees cannot be selected by a lottery, nor can everyone over a certain age be downsized. Once an employer has established the criteria, they must be applied equally to all and with no exceptions. Otherwise, the employer runs the risk of a "disparate application" charge, which can lead to a costly judgment.

COMPETITIVE ADVANTAGE

competitive advantage

a significant and (ideally) long-term benefit to a company over its competition

A **competitive advantage** is a significant and (ideally) long-term benefit to a company over its competition. Establishing and maintaining a competitive advantage is complex, but a company's survival and prosperity depend on its success in doing so.

Factors That Lead Firms to Seek Competitive Advantage

A number of factors can lead to the attainment of competitive advantage. Michael Porter, a prominent management theorist, suggested a now widely accepted **five-force model**. The five forces include rivalry among existing competitors, the threat of new entrants, the threat of substitute products and services, the bargaining power of buyers, and the bargaining power of suppliers. The more these forces combine in any instance, the more likely firms will seek competitive advantage and the more dramatic the results of such an advantage will be.

five-force model

a widely accepted model that identifies five key factors that can lead to attainment of competitive advantage, including rivalry among existing competitors, the threat of new entrants, the threat of substitute products and services, the bargaining power of buyers, and the bargaining power of suppliers

Rivalry among Existing Competitors

The rivalry among existing competitors is an important factor leading firms to seek competitive advantage. Typically, highly competitive industries are characterized by high fixed costs of entering or leaving the industry, low degrees of product differentiation, and many competitors. Although all firms are rivals with their competitors, industries with stronger rivalries tend to have more firms seeking competitive advantage.

Threat of New Entrants

The threat of new entrants is another important force leading an organization to seek competitive advantage. A threat exists when entry and exit costs to the industry are low and the technology needed to start and maintain the business is commonly available. For example, consider a small restaurant. The owner does not require millions of dollars to start the business, food costs do not go down substantially for large volumes, and food processing and preparation equipment is commonly available. When the threat of new market entrants is high, the desire to seek and maintain competitive advantage to dissuade new market entrants is usually high.

Threat of Substitute Products and Services

The more consumers are able to obtain similar products and services that satisfy their needs, the more likely firms are to try to establish competitive advantage. Such an advantage often creates a "new playing field" in which "substitute" products are no longer considered as such by the consumer. Consider the personal computer industry and the introduction of low-cost computers. A number of consultants and computer manufacturers made much of the high cost of ownership associated with personal computers in the mid-1990s. They introduced low-cost network computers with minimal hard disk space, slower CPUs, and less main memory than some consumers desired, but at half the cost of a standard workstation. There was considerable interest in these new machines for a while, but traditional personal computer manufacturers fought back. They developed a class of powerful workstations and implemented new pricing strategies to make powerful workstations available at under $1,000. This eliminated the primary advantage of the stripped-down network computers and regained lost customers.

Bargaining Power of Customers and Suppliers

Large buyers tend to exert significant influence on a firm. This influence can be diminished if the buyers are unable to use the threat of going elsewhere to influence the firm. Suppliers can help an organization obtain a competitive advantage. In some cases, suppliers enter into strategic alliances with firms. When they do so, suppliers act like a part of the company. Suppliers and companies can use telecommunications to link their computers and personnel to obtain fast reaction times and the ability to get the parts or supplies when they are needed to satisfy customers.

Strategic Planning for Competitive Advantage

To be competitive, a company must be fast, nimble, flexible, innovative, productive, economical, and customer oriented. It must also align the information system strategy with general business strategies and objectives.[17] Given the five market forces just mentioned, Porter proposed three general strategies to attain

Our fast-moving society is highly competitive. To maintain a competitive advantage, companies continually innovate and create new products. Think Outside was the first to market the Stowaway keyboard, a full-size keyboard that folds to pocket-size for use with handheld computers.
(Source: Courtesy of Think Outside, Inc.)

competitive advantage: altering the industry structure, creating new products and services, and improving existing product lines and services. Subsequent research into the use of information systems to help an organization achieve a competitive advantage has confirmed and extended Porter's original work to include additional strategies—such as forming alliances with other companies, developing a niche market, maintaining competitive cost, and creating product differentiation.[18]

Altering the Industry Structure

Altering the industry structure is the process of changing the industry to become more favorable to the company or organization. This can be accomplished by gaining more power over suppliers and customers. Some automobile manufacturers, for example, insist that their suppliers be located close to major plants and manufacturing facilities and that all business transactions be accomplished using electronic data interchange (EDI, direct computer-to-computer communications with minimal human effort required). This helps the automobile company control the cost, quality, and supply of parts and materials.

A company can also attempt to create barriers to new companies entering the industry. An established organization that acquires expensive new technology to provide better products and services can discourage new companies from getting into the marketplace. Creating strategic alliances may also have this effect. A **strategic alliance**, also called a **strategic partnership**, is an agreement between two or more companies that involves the joint production and distribution of goods and services. An example of a successful strategic partnership is the merger of CMGI and Raging Bull.[19] David Wetherell, who runs CMGI as a successful mutual fund company, saw Raging Bull, one of the fastest-growing on-line investor chat rooms, while he was surfing the Web. He quickly purchased 50 percent of the company for $2 million. Raging Bull, which at the time was being run by three college students in a basement, quickly moved to CMGI's headquarters in Andover, Massachusetts, where the advantage of the strategic partnership started to be realized. The four-month-old Raging Bull site, which provided free investment chat rooms and had been financed by one of the founder's credit cards, started generating revenues immediately by using Internet ads. Whom did Raging Bull team up with to achieve these results? ADSmart, another CMGI company. The number of Internet hits for Raging Bull went from about 200,000 a day to more than 5 million a day in a few short months. This strategic partnership, like others, was a winning strategy for everyone involved.

strategic alliance (strategic partnership)

an agreement between two or more companies that involves the joint production and distribution of goods and services

The proposed strategic merger between AOL and Time Warner will create the world's first fully integrated media and communications company for the "Internet Century."
(Source: AP/World Wide Photos.)

Creating New Products and Services

Creating new products and services is always an approach that can help a firm gain a competitive advantage. This is especially true of the computer industry and other high-tech businesses. If an organization does not introduce new products and services every few months, the company can quickly stagnate, lose market share, and decline. Companies that stay on top are constantly developing new products and services. Vattenfall, Sweden's largest public utility company, for example, has developed a new wireless service that allows people to control machines on the factory floor and security and appliance devices at home and to gain access to the Internet through a cell phone.[20] Some believe that this technology could be worth billions of dollars when it is introduced in the United States in the future. New products are particularly important in the dynamic technology industry.

Improving Existing Product Lines and Services

Improving existing product lines and services is another approach to staying competitive. The improvements can be either real or perceived. Manufacturers of household products are always advertising new and improved products. In some cases, the improvements are more perceived than real refinements; usually, only minor changes are made to the existing product. Many food and beverage companies are introducing "healthy" and "light" product lines. A popular beverage company introduced "born on" dating for beer.

Using Information Systems for Strategic Purposes

The first IS applications were aimed at reducing costs and providing more efficient processing for accounting and financial applications, such as payroll and general ledger. These systems were seen almost as a necessary evil—something to be tolerated to reduce the time and effort required to complete previously manual tasks. As organizations matured in their use of information systems, enlightened managers began to see how they could be used to improve organizational effectiveness and support the fundamental business strategy of the enterprise. Combining the improved understanding of the potential of information systems with the growth of new technology and applications has led organizations to use IS to gain a competitive advantage. In simplest terms, competitive advantage is usually embodied in either a product or service that has the most added value to consumers and that is unavailable from the competition or is in an internal system that delivers benefits to a firm not enjoyed by its competition.

Although it can be difficult to develop information systems to provide a competitive advantage, some organizations have done so with success. A classic example is SABRE, a sophisticated computerized reservation system installed by American Airlines and one of the first CBISs recognized for providing competitive advantage. Travel agents used this system for rapid access to flight information, offering travelers reservations, seat assignments, and ticketing. The travel agents also achieved an efficiency benefit from the SABRE system. Because SABRE displayed American Airlines flights whenever possible, it also gave the airline a long-term, significant competitive advantage.

Quite often, the competitive advantage a firm gains with a new information system is only temporary—competitors are quick to copy a good idea. So although the SABRE system was the first on-line reservation system, other carriers soon developed similar systems. However, SABRE has maintained a leadership position in the past because it was the first system available, has been aggressively marketed, and has been continually upgraded and improved over time. Maintaining a competitive advantage takes effort and is not guaranteed. SABRE's competitive advantage, for example, is being challenged with the many Internet-based travel sites becoming popular with today's travelers. Many companies have even instituted a new position—chief knowledge officer—to help them maintain a competitive advantage.

The extent to which companies are using computers and information technology for competitive advantage continues to grow. Forward-thinking companies must constantly update or acquire new systems to remain competitive in today's dynamic marketplace. In addition to using information systems to help a company achieve a competitive advantage internally, companies are increasingly investing in information systems to support their suppliers and customers.[21] Investments in information systems that result in happy customers and efficient suppliers can do as much to achieve a competitive advantage as internal systems, such as payroll and billing. Table 2.3 lists several examples of how companies have attempted to gain a competitive advantage.

Factors That Lead to Attainment of a Competitive Advantage	Strategies		
	Alter Industry Structure	**Create New Products and Services**	**Improve Existing Product Lines and Services**
Rivalry among existing competitors	Blockbuster changes the industry structure with its chain of video and music stores.	Dell, Gateway, and other PC makers develop computers that excel at downloading Internet music and playing the music on high-quality speakers.	Food and beverage companies offer "healthy" and "light" product lines.
Threat of new entrants	AOL and Time Warner plan to merge to form a large Internet and media company.	Apple Computer introduces an easy-to-use iMac computer that can be used to create and edit home movies.	Starbucks offers new coffee flavors at premium prices.
Threat of substitute products and services	Ameritrade and other discount stockbrokers offer low costs and research on the Internet.	Wal-Mart uses technology to monitor inventory and product sales to determine the best mix of products and services to offer at various stores.	Cosmetic companies add sunscreen to their product lines.
Bargaining power of buyers	Ford, GM, and others require that suppliers locate near their manufacturing facilities.	Investors and traders of the Chicago Board of Trade (CBOT) put pressure on the institution to implement electronic trading.	Retail clothing stores require manufacturing companies to reduce order lead times and improve materials used in the clothing.
Bargaining power of suppliers	American Airlines develops SABRE, a comprehensive travel program used to book airline, car rental, and other reservations.	Intel develops SpeedStep, a chip for laptop computers, that operates at faster speeds when connected to an electrical outlet.	Hayworth, a supplier of office furniture, has a computerized-design tool that helps it design new office systems and products.

TABLE 2.3

Competitive Advantage Factors and Strategies

PERFORMANCE-BASED INFORMATION SYSTEMS

There have been at least three major stages in the business use of IS. The first stage started in the 1960s and was oriented toward cost reduction and productivity. This stage generally ignored the revenue side, not looking for opportunities to increase sales via the use of IS. The second stage started in the 1980s and was defined by Porter and others. It was oriented toward gaining a competitive advantage. In many cases, companies spent large amounts on IS and ignored the costs. Today, we are seeing a shift from strategic management to performance-based management in many IS organizations today. This third stage carefully considers both strategic advantage and costs. This stage uses return on investment (ROI), net present value, and other measures of performance.[22] Figure 2.9 illustrates these stages.

Productivity

Developing information systems that measure and control productivity is a key element for most organizations. **Productivity** is a measure of the output achieved divided by the input required. A higher level of output for a given level of input means greater productivity; a lower level of output for a given level of input means lower productivity. Consider a tax preparation firm, for example, where productivity can be measured by the tax returns prepared divided by the total hours the employee worked. For example, in a 40-hour week, an employee

productivity

a measure of the output achieved divided by the input required

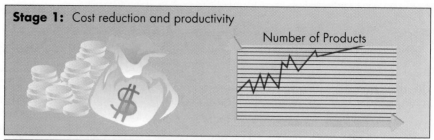

Stage 1: Cost reduction and productivity

Number of Products

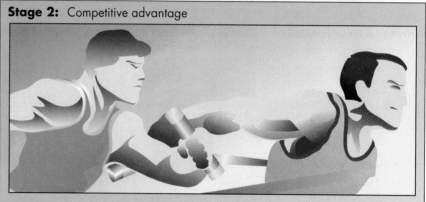

Stage 2: Competitive advantage

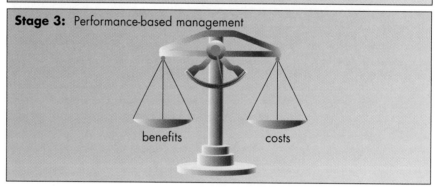

Stage 3: Performance-based management

benefits   costs

may have prepared 30 tax returns. The productivity is thus equal to 30/40, or
75 percent. With administrative and other duties, a productivity level of 75 may
be excellent. The numbers assigned to productivity levels are not always based
on labor hours—productivity may be based on factors like the amount of raw
materials used, resulting quality, or time to produce the goods or service. In any
case, what is important is not the value of the productivity number but how it
compares with other time periods, settings, and organizations.

$$\text{Productivity} = (\text{Output/Input}) \times 100\%$$

Once a basic level of productivity is measured, an information system can monitor and compare it over time to see whether productivity is increasing. Then, corrective action can be taken if productivity drops below certain levels. In addition to
measuring productivity, an information system can also be used within a process to
significantly increase productivity. Thus, improved productivity can result in faster
customer response, lower costs, and increased customer satisfaction.

Measuring productivity is important because improving productivity boosts
a nation's standard of living. In an era of intense international competition, the
need to improve productivity is critical to the well-being of any enterprise or
country. If a company does not take advantage of technological and management innovation to improve productivity, its competitors will. The ability to
apply information technology to improve productivity will separate successful
enterprises from failures.

It is important to understand that information technology is not productive by itself. It takes well-managed, superbly trained, and motivated people—with or without information technology—to deliver measurable gains in output. Many people think that real improvements in productivity come from a synergy of information technology and sweeping changes in management and organizational structure that redefines how work gets done. Largely produced in response to increasing global competition, these overhauls are known loosely as reengineering, previously discussed.

Once work has been redefined, information technology can be used to move information to the front lines—to give employees on the factory floor or in the customer service department the knowledge they need to act quickly. That is the formula for a productivity explosion. Such a productivity explosion leads to an increase in the world's standard of living.

Return on Investment and the Value of Information Systems

One measure of IS value is **return on investment (ROI)**. This measure investigates the additional profits or benefits that are generated as a percentage of the investment in information systems technology. A small business that generates an additional profit of $20,000 for the year as a result of an investment of $100,000 for additional computer equipment and software would have a return on investment of 20 percent ($20,000/$100,000).

Earnings Growth

Another measure of IS value is the increase in profit, or earnings growth, it brings. For instance, suppose a mail-order company, after installing an order processing system, had a total earnings growth of 15 percent compared with the previous year. Sales growth before the new ordering system was only about 8 percent annually. Assuming that nothing else affected sales, the earnings growth brought by the system, then, was 7 percent.

Market Share

Market share is the percentage of sales that one company's products or services have in relation to the total market. If installing a new on-line Internet catalog increases sales, it might help a company increase its market share by 20 percent.

Customer Awareness and Satisfaction

Although customer satisfaction can be difficult to quantify, about half of today's best global companies measure the performance of their information systems based on feedback from internal and external users. Some companies use surveys and questionnaires to determine whether the investment in information systems has increased customer awareness and satisfaction.

Total Cost of Ownership

In addition to such measures as return on investment, earnings growth, market share, and customer satisfaction, some companies are also tracking total costs. One measure, developed by the Gartner Group and explored in more detail in a case at the end of the chapter, is the **total cost of ownership (TCO)**.[23] This approach breaks total costs into such areas as the cost to acquire the technology, technical support, administrative costs, and end-user operations. Other costs in TCO include retooling and training costs. TCO can be used to get a more accurate estimate of the total costs for systems that range from small PCs to large mainframe systems.

The preceding are only a few measures that companies have used to plan for and maximize the value of their investments in information systems technology.

To compete successfully, airlines are turning to technology to improve their productivity and customer service.
(Source: Stone/Stewart Cohen.)

return on investment (ROI)

one measure of IS value that investigates the additional profits or benefits that are generated as a percentage of the investment in information systems technology

total cost of ownership (TCO)

measurement of the total cost of owning computer equipment, including desktop computers, networks, and large computers

In many cases, it is difficult to be accurate with ROI measures—for example, an increase in profits caused by an improved information system or other factors, such as a new marketing campaign or a competitor that was late in delivering a new product to the market. Regardless of the difficulties, organizations must attempt to evaluate the contributions information systems make to be able to assess their progress and plan for the future. Information technology and personnel are too important to leave to chance.

Justifying Information Systems

Because information systems are so important to the work in organizations, businesses need to be sure that improvements or completely new systems are worthwhile. The process for reviewing IS changes involves justification that the change is necessary and will yield gains.

To avoid waste, each potential information systems project should be reviewed to ensure that the project meets an important business need, is consistent with corporate strategy, and leads to attainment of specific goals and objectives. A second check should be made to assess the degree of risk or uncertainty associated with each project. It is not unusual for IS departments to formally analyze and manage risk.[24]

Risk can be assessed by answering questions such as:

1. How well are the requirements of the system understood?
2. To what degree does the project require pioneering effort in technology that is new to the firm?
3. Is there a risk of severe business repercussions if the project is poorly implemented?

The fundamental benefits for considering the project should be identified. Most IS projects fall into one of the following categories:

- **Tangible Savings.** Implementation of the project will result in hard-dollar savings to the company that can be quantified (e.g., reduced staff, lowered operating costs, increased sales).
- **Intangible Savings.** Implementation of the project will result in soft-dollar savings to the company, the magnitude of which will be difficult to measure (e.g., help managers make better decisions, improve control over the operations of the business).
- **Legal Requirement.** Implementation of the project is required to meet a state or federal regulation (e.g., reporting information on the employment of handicapped or minorities).
- **Modernization.** Implementation of the project is needed to keep current with changing business requirements (e.g., systems changes needed because of a conversion from English to metric units of measure or conversion from many European currencies to the use of the Euro) or technology requirements (e.g., computer upgrade to improve work with new software).
- **Pilot Project.** Implementation of the project is required to gain experience in a new technology to the existing business (e.g., the use of portable computers by salespeople to enhance customer presentations).

Most organizations today realize that they must look at both sides of the equation—benefits as well as costs—in evaluating potential information system investments. Furthermore, determining return on investment can help the IS organization prove its contribution to the organization and ensure that its efforts are aligned with the company's overall business objectives.

CAREERS IN INFORMATION SYSTEMS

Work with information systems continues to be an exciting career choice. Business demand for IS professionals has continued to grow, and numerous schools have degree programs with such titles as information systems, computer information systems, and management information systems. These programs are typically in business schools and within computer science departments. Degrees in information systems have provided high starting salaries for many students after graduation from college. For many schools and departments, information systems majors attain the highest starting salaries of all undergraduate business majors and show significant potential job growth. Information systems majors can start at salaries that are over $50,000.

In addition to having a relatively high starting salary, careers in information systems promise to expand even more than other business disciplines. Strong growth is projected for IS occupations as a result of the continuing spread of computer technology and information systems throughout government and business. The growth in Internet-based jobs remains explosive, as companies scramble to get effective and efficient Web sites up and running smoothly. There is also growth in the so-called packaged software segment of the business, which includes careers with companies such as Microsoft, IBM, Intuit, and all the other producers of software on sale at your local computer store. Systems analysts and computer scientists increasingly will be needed to meet the demand for the use of computers in offices, factories, and research agencies and to support the rapid growth of telecommunications technology. This rapid growth, however, has made it difficult for some companies to get the skilled people they need. Read the "Ethical and Societal Issues" box on page 61 to explore the difficulties organizations face in finding talented IS employees.

Roles, Functions, and Careers in the Information Systems Department

Information systems personnel typically work in an IS department that employs Web developers, computer programmers, systems analysts, computer operators, and a number of other information systems personnel. They may also work in other functional departments or areas in a support capacity. In addition to technical skills, information systems personnel also need skills in written and verbal communication, an understanding of organizations and the way they operate, and the ability to work with people (the system users). In general, information systems personnel are charged with maintaining the broadest perspective on organizational goals. For most medium- to large-sized organizations, information resources are typically managed through an IS department. In smaller businesses, one or more people may manage information resources, with support from outside services—outsourcing. As shown in Figure 2.10, the information systems organization has three primary responsibilities: operations, systems development, and support.

Operations

The operations component of a typical IS department focuses on the use of information systems in corporate or business unit computer facilities. It tends to focus more on the *efficiency* of information system functions rather than their effectiveness.

The primary function of a system operator is to run and maintain IS equipment. System operators are responsible for starting, stopping, and correctly operating mainframe systems, networks, tape drives, disk devices, printers, and so on. System operators are typically trained at technical schools or through on-the-job experience. Other operations include logging, scheduling, hardware maintenance, and preparation of input and output. Data-entry operators convert data into a form the computer system can use. They may use terminals or

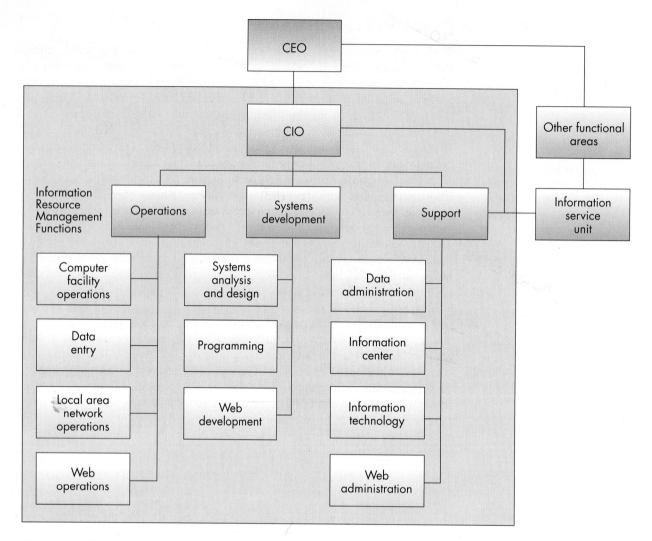

The Three Primary
Responsibilities of Information
Systems

Each of these elements—operations, systems development,
and support—contains subelements critical to the efficient
and effective operation of the
organization.

other devices to enter business transactions, such as sales orders and payroll
data. Increasingly, data entry is being automated—captured at the source of the
transaction rather than being entered later. In addition, companies may have
local area network and Web or Internet operators who are responsible for running the local network and any Internet sites the company may have.

Systems Development

The systems development component of a typical IS department focuses on
specific development projects and ongoing maintenance and review. Systems
analysts and programmers, for example, focus on these concerns.

The role of a systems analyst is multifaceted. Systems analysts help users
determine what outputs they need from the system and construct the plans
needed to develop the necessary programs that produce these outputs. Systems
analysts then work with one or more programmers to make sure that the appropriate programs are purchased, modified from existing programs, or developed.
The major responsibility of a computer programmer is to use the plans developed by the systems analyst to develop or adapt one or more computer programs that produce the desired outputs. The main focus of systems analysts and
programmers is to achieve and maintain information system effectiveness.

With the dramatic increase in the use of the Internet, intranets, and extranets,
many companies have Web or Internet developers who are responsible for developing effective and attractive Internet sites for customers, internal personnel, suppliers, stockholders, and others with a business relationship with the company.

ETHICAL AND SOCIETAL ISSUES
Should IS Managers Hire More Foreign Workers?

During the last several years, being a management information systems or computer science graduate typically resulted in a number of high-paying job offers. The entire information systems industry has been booming over the last several years. Stock prices of technology companies have grown beyond expectations, making many employees and investors instant millionaires. And there seems to be no end in sight. Traditional careers in such areas as programming and systems analysis are in high demand, as well as the new Internet careers. This boom in technology, however, has presented problems for some organizations. It can be hard to fill all the technology jobs on the market today. The Computing Technology Industry Association claims that at least 250,000 additional high-tech workers are needed and that a lack of qualified workers has cost American companies $4.5 billion in lost productivity.

Oliver Sharp, the chief technology officer of iTurf.com, a new Internet company aimed at teens, is constantly worried about where he will find skilled employees to fuel his company's growth. "When I came to iTurf, nothing mattered except assembling a team. It was an absolute occupation of mine, and it remains one today." Sharp, like many other IS managers, desperately needs more qualified people. To fill the gap, American high-tech companies are increasingly looking to foreign workers. One way of getting foreign workers to the United States is through the H-1B visa program.

The H-1B visa is a six-year temporary visa that can be granted to foreign workers with college degrees. Congress has increased the number of people that can be brought into the United States under the program, in part to meet the increasing demand of technology jobs. In 1998, the number of H-1B visas allowed was increased from 65,000 to 115,000. But many in the information systems industry believe this number is still too small. In addition, other nontechnical workers can qualify under the H-1B program. In fact, only about 50 percent of foreigners that participate in the program are skilled in high-tech areas. The others are lawyers, fashion models, health workers, and other professionals.

Some members of Congress and industry experts, however, believe that the push for more H-1B visas is a ploy by some high-tech companies to get high-tech employees at a lower cost. In addition, some believe that not every high-tech company is willing to sponsor H-1B visa holders for a green card. A green card would give foreign workers more freedoms, including the ability to seek other jobs.

Congress introduced the new T visa to address some of these concerns. The T visa, if adopted, would require technology companies to pay foreign workers at least $60,000 annually and would require a masters or doctorate degree. Even with the proposed T visa, some critics, including Rae Peppers, director of the Tribal Business Information Technology Center on the Cheyenne reservation in Montana, think that the United States should train its own citizens instead of looking to foreign countries to fill high-tech jobs.

Although the unemployment rate in the United States overall has been at a historic low of 4 percent, unemployment rates for some American Indian tribes have averaged about 50 percent. This has caused many to rethink the H-1B and the proposed T visas and to concentrate on better training at institutions such as Dull Knife, a two-year tribal college. Under a recent National Science Foundation grant, tribal colleges are getting funding for technology classrooms and distance-learning classes. According to Peppers, "A lot of our people don't want to leave home, and they're frightened of technology." There are also cultural barriers. "When the key to the culture has its roots in Mother Earth, the virtual world—using e-mails to talk to someone in Japan or selling a hand-woven blanket on the Internet—is a difficult concept to grasp," says Victor Chavez, director of the Sandia National Laboratories Small-Business Empowerment Program.

Discussion Questions

1. What is the H-1B visa program?
2. What is the difference between the H-1B and the proposed T visa?

Critical Thinking Questions

3. If you were a personnel manager for a high-tech company, how would you solve the problem of finding qualified people?
4. What would you recommend to Congress?

Sources: Adapted from Bradley Brooks, "Tech Industry Seeks More Immigration," *Denver Rocky Mountain News*, February 4, 2000, p. 8B; Bronwyn Fryer, "Neglected Workforce," *Computerworld*, December 20, 1999, p. 48; and Karen Cheney, "Trends," *Business Week,* October 11, 1999, p. F10.

Systems operators focus on the efficiency of information system functions rather than its effectiveness. Their primary function is to run and maintain IS equipment. (Source: Image provided by PhotoDisc © 2000.)

Support

The support component of a typical IS department focuses on providing user assistance in the areas of hardware and software acquisition and use, data administration, and user training and assistance. Because information systems hardware and software is costly, especially if mistakes are made, the acquisition of computer hardware and systems software is often managed by a specialized group within the support component. This group sets guidelines and standards for the rest of the organization to follow in making its purchases. Gaining and maintaining an understanding of available technology is an important part of the acquisition of information systems. Also, developing good relationships with vendors is important.

Firms may look to one outside source to supply part or all of their information systems needs—a single-vendor solution. There are advantages to this approach, such as potential cost savings and built-in compatibility. There are also risks to this approach, including lack of flexibility, vendor complacency due to lack of competitive bidding, and the possibility of missing out on new products from other vendors. Having an in-house specialist who focuses on the acquisition of information systems may also be wise when using the outsourcing approach.

The database administrator focuses on planning, policies, and procedures regarding the use of corporate data and information. For example, it is the role of the database administrator to develop and disseminate information about the corporate databases for developers of IS applications. In addition, the database administrator is charged with monitoring and controlling database use.

information center

a support function that provides users with assistance, training, application development, documentation, equipment selection and setup, standards, technical assistance, and troubleshooting

The support component typically operates the information center. An **information center** provides users with assistance, training, application development, documentation, equipment selection and setup, standards, technical assistance, and troubleshooting. Although many firms have attempted to phase out information centers, others have changed the focus of this function from technical training to helping users find ways to maximize the benefits of the information resource.

Information Service Units

information service unit

a miniature IS department

An **information service unit** is basically a miniature IS department attached and directly reporting to a functional area. Notice the information service unit shown in Figure 2.10. Even though this unit is usually staffed by IS professionals, the project assignments and the resources necessary to accomplish these projects are provided by the functional area to which it reports. Depending on the policies of the organization, the salaries of IS professionals staffing the information service unit may be budgeted by either the IS department or the functional area.

The growth of information service units may be directly attributed to the increased number of users doing their own computing. The increasing use of networks has put computers at nearly every desk. Communication between IS personnel and users is more effective the closer they work together. When such information service units are not part of the formal organizational structure, they tend to arise informally in organizations. That is, a particular functional manager might establish and maintain informal groups of employees who are more proficient with IS than other users. As more employees become computer users, such cooperation must be considered to manage resources properly. It is probably more productive to support and provide training to these informal groups than to attempt to interfere with or thwart their activities.

Typical IS Titles and Functions

The organizational chart shown in Figure 2.10 is a simplified model of an IS department in a typical medium- or large-sized organization. Many organizations have even larger departments, with increasingly specialized positions such as librarian, quality assurance manager, and the like. Smaller firms often combine the roles depicted in Figure 2.10 into fewer formal positions.

The Chief Information Officer

The overall role of the chief information officer (CIO) is to employ an IS department's equipment and personnel in a manner that will help the organization attain its goals. The CIO is usually a manager at the vice presidential level concerned with the overall needs of the organization. He or she is responsible for corporatewide policy, planning, management, and acquisition of information systems. Some of the CIO's top concerns include integrating information systems operations with corporate strategies, keeping up with the rapid pace of technology, and defining and assessing the value of systems development projects in terms of performance, cost, control, and complexity. The high level of the CIO position is consistent with the idea that information is one of the organization's most important resources. This individual works with other high-level officers of the organization, including the chief financial officer (CFO) and the chief executive officer (CEO), in managing and controlling total corporate resources.

Depending on the size of the information systems department, there may be several people at senior IS managerial levels. Some of the job titles associated with information systems management are the CIO, vice president of information systems, and manager of information systems. A central role of all these individuals is to communicate with other areas of the organization to determine changing needs. Often these individuals are part of an advisory or steering committee that helps the CIO and other IS managers with their decisions about the use of information systems. Together they can best decide what information systems will support corporate goals. CIOs must work closely with advisory committees, stressing effectiveness and teamwork and viewing information systems as an integral part of the organization's business processes—not an adjunct to the organization.

LAN Administrators

Local area network (LAN) administrators set up and manage the network hardware, software, and security processes. They manage the addition of new users, software, and devices to the network. They isolate and fix operations problems. LAN administrators are currently in high demand.

Internet Careers

Explosive growth in the use of the Internet to conduct business has caused a need for skilled personnel to develop and coordinate Internet usage. In 2000, the amount of business-to-business commerce exceeded $40 billion, and this number is expected to exceed $1 trillion in a few short years.[25] This surge in Internet use has resulted in career opportunities with traditional and on-line companies. As seen in Figure 2.10, these careers are in the areas of Web operations, Web development, and Web administration. As with other areas in IS, there are a number of top-level administrative jobs related to the Internet. These career opportunities are with traditional companies and companies that specialize in the Internet.

Internet jobs within a traditional company include Internet strategists and administrators, Internet systems developers, Internet programmers, and Internet or Web site operators. The Internet has become so important to some companies

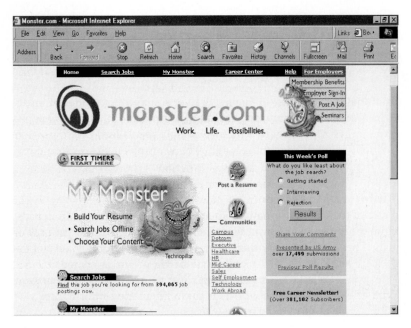

Internet job sites such as Monster.com allow job hunters to browse job opportunities and post their résumés.

certification

process for testing skills and knowledge that results in an endorsement by the certifying authority that an individual is capable of performing a particular job

that some have suggested a new position, chief Internet officer, with responsibilities and salary similar to the CIO's.

In addition to traditional companies, there are many exciting career opportunities in companies that offer products and services over the Internet.[26] These companies include Amazon.com, Yahoo!, eBay, and many others. Systest, for example, specializes in finding and eliminating digital bugs that could halt the operation of a computer system.[27] According to Christopher Hardesty, chief financial officer, "The Internet has come in and really revolutionized the kind of performance testing we do." Typically, these companies start as private operations and then are offered to the public through initial public offerings (IPOs). To achieve their initial funding, many start-up Internet companies seek funding from venture capital (VC) groups. These groups often provide seed money in return for stock and stock options. Unlike traditional companies, many startup Internet companies offer their employees low salaries but generous stock options that can be worth millions if the company becomes successful. Some top-level managers, politicians, and national newscasters have abandoned their traditional jobs to work for an Internet start-up company, hoping to become rich from stock options when the company goes public.

A number of Internet sites post job opportunities for Internet careers and more traditional careers, such as Monster.com. These sites allow prospective job hunters to browse job opportunities, job locations, salaries, benefits, and other factors. In addition, some of these sites allow job hunters to post their resumes.

Quite often, the people filling IS roles have completed some form of certification. **Certification** is a process for testing skills and knowledge resulting in an endorsement by the certifying authority that an individual is capable of performing a particular job. Certification frequently involves specific, vendor-provided or vendor-endorsed coursework. There are several popular certification programs, including Novell Certified Network Engineer, Microsoft Certified Professional Systems Engineer, Certified Project Manager, and others.

One of the greatest fears of every IS manager is spending several thousand dollars to help an employee get certified and then to lose that person to a higher-paying position with a new firm. As a consequence, some organizations request a written commitment from individuals to stay a certain length of time after obtaining their certification. Needless to say, this requirement can create some ill will with the employee. To provide newly certified employees with incentives to remain, other organizations provide salary increases based on additional credentials acquired.

Other IS Careers

In addition to working for an IS department in an organization, information systems personnel can work for one of the large consulting firms, such as Andersen Consulting, EDS, and others. These jobs often entail a large amount of travel, because consultants are assigned to work on various projects wherever the client is. Such roles require excellent people and project management skills in addition to IS technical skills.

Another IS career opportunity is to be employed by a hardware or software vendor developing or selling products. Such a role enables the individual to work on the cutting edge of technology and can be extremely challenging and exciting!

● SUMMARY

PRINCIPLE • The use of information systems to add value to the organization is strongly influenced by organizational structure, culture, and change.

Value-added processes increase the relative worth of the combined inputs on their way to becoming final outputs of the organization. The value chain is a series (chain) of activities that includes inbound logistics, warehouse and storage, production, finished product storage, outbound logistics, marketing and sales, and customer service.

Organizations use information systems to support organizational goals. Before deciding on an information system for an organization, managers should identify the firm's critical success factors, which must be supported by the system. Because information systems typically are designed to improve productivity, methods for measuring the system's impact on productivity should be devised.

. . . .

An organization is a formal collection of people and various other resources established to accomplish a set of goals. The primary goal of a for-profit organization is to maximize shareholder value. Nonprofit organizations include social groups, religious groups, universities, and other organizations that do not have profit as the primary goal. Organizations are systems with inputs, transformation mechanisms, and outputs.

Organizational structure refers to organizational subunits and how they are related and tied to the overall organization. Several basic organization structures exist: traditional, project, team, and multidimensional (also called matrix structure).

Organizational culture consists of the major understandings and assumptions for a business, corporation, or organization. Organizational change deals with how profit and nonprofit organizations plan for, implement, and handle change. Change can be caused by internal or external factors. The change model consists of these stages: unfreezing, moving, and refreezing. According to the concept of organizational learning, organizations adapt to new conditions or alter practices over time.

PRINCIPLE • Because information systems are so important, businesses need to be sure that improvements or completely new systems help lower costs, increase profits, improve service, or achieve a competitive advantage.

Business process reengineering involves the radical redesign of business processes, organizational structures, information systems, and values of the organization to achieve a breakthrough in results.

Continuous improvement involves constantly seeking ways to improve business processes to add value to products and services. Total quality management consists of a collection of approaches, tools, and techniques that offers a commitment to quality throughout the organization. The extent to which technology is used throughout an organization can be a function of technology diffusion, infusion, and acceptance. Technology diffusion is a measure of how widely technology is in place through an organization. Technology infusion is the extent to which technology is located in an area or department. The technology acceptance model (TAM) investigates factors—such as perceived usefulness of the technology, how easy the technology is to use, the quality of the information system, and the degree to which the organization supports the use of the information system—to predict information systems usage and performance. Outsourcing involves contracting with outside professional services to meet specific business needs. It allows a company to focus more closely on its core business and to target its limited resources to meet strategic goals. Downsizing involves reducing the number of employees to reduce payroll costs; however, it can lead to unwanted side effects.

. . . .

Competitive advantage is usually embodied in either a product or service that has the most added value to consumers and that is unavailable from the competition or is in an internal system that delivers benefits to a firm not enjoyed by its competition. A five-force model covers factors that lead firms to seek competitive advantage: rivalry among existing competitors, the threat of new market entrants, the threat of substitute products and services, the bargaining power of buyers, and the bargaining power of suppliers. Three strategies to address these factors and to attain competitive advantage include altering the industry structure, creating new products and services, and improving existing product lines and services.

The ability of the information system to provide or maintain competitive advantage should also be determined. Several strategies for achieving competitive advantage include enhancing existing products or services or developing new ones, as well as changing the existing industry or creating a new one.

. . . .

The objectives of each potential IS project are reviewed to ensure that the project meets an important business need, is consistent with corporate strategy, and leads to attainment of specific goals and objectives. A second check is made to assess the degree of risk or uncertainty associated with each project. The fundamental reason for considering the project should be identified. Developing information systems that measure and control productivity is a key element for most organizations. A useful measure of the value of an information system project is return on investment (ROI). This measure investigates the additional profits or benefits that are generated as a percentage of the investment in IS technology. Total cost of ownership (TCO) can also be a useful measure. Most IS projects fall into one of the following categories: tangible savings, intangible savings, legal requirement, modernization, or pilot project.

PRINCIPLE ● Information systems personnel are the key to unlocking the potential of any new or modified system.

Information systems personnel typically work in an information systems department that employs a chief information officer, systems analysts, computer programmers, computer operators, and a number of other information systems personnel. The overall role of the chief information officer (CIO) is to employ an IS department's equipment and personnel in a manner that will help the organization attain its goals. Systems analysts help users determine what outputs they need from the system and construct the plans needed to develop the necessary programs that produce these outputs. Systems analysts then work with one or more

programmers to make sure that the appropriate programs are purchased, modified from existing programs, or developed. The major responsibility of a computer programmer is to use the plans developed by the systems analyst to develop or adapt one or more computer programs that produce the desired outputs. Computer operators are responsible for starting, stopping, and correctly operating mainframe systems, networks, tape drives, disk devices, printers, and so on. LAN administrators set up and manage the network hardware, software, and security processes. There is also an increasing need for trained personnel to set up and manage a company's Internet site, including Internet strategists, Internet systems developers, Internet programmers, and Web site operators. Information systems personnel may also work in other functional departments or areas in a support capacity. In addition to technical skills, information systems personnel also need skills in written and verbal communication, an understanding of organizations and the way they operate, and the ability to work with people (users). In general, IS personnel are charged with maintaining the broadest enterprisewide perspective.

In addition to working for an information systems department in an organization, IS personnel can work for one of the large information systems consulting firms, such as Andersen Consulting, EDS, and others. Another IS career opportunity is to be employed by a hardware or software vendor developing or selling products.

● KEY TERMS

certification 64
change model 44
competitive advantage 51
continuous improvement 47
culture 43
downsizing 50
empowerment 42
five-force model 51
flat organizational structure 42
information center 62
information service unit 62
multidimensional organizational
 structure 42

organization 38
organizational change 44
organizational culture 43
organizational learning 44
organizational structure 40
outsourcing 50
productivity 55
project organizational structure 42
quality 48
reengineering (process redesign) 44
return on investment (ROI) 57
strategic alliance (strategic
 partnership) 53

team organizational structure 42
technology acceptance model
 (TAM) 48
technology diffusion 48
technology infusion 48
total cost of ownership (TCO) 57
total quality management (TQM) 48
traditional organizational
 structure 41
value chain 39

● REVIEW QUESTIONS

1. What is meant by organization structure?
2. What is a value-added process? Give several examples.
3. What role does an information system play in the value-added processes of an organization?
4. What is reengineering? What are the potential benefits of performing a process redesign? What is the difference between reengineering and continuous improvement?
5. What is quality? What is total quality management (TQM)?
6. What are organizational change and organizational learning?
7. List and define the four basic organizational structures.
8. Sketch and briefly describe the three-stage organizational change model.
9. What is downsizing? How is it different from outsourcing?
10. What are some general strategies employed by organizations to achieve competitive advantage?
11. What are the five common justifications for implementation of an information system?
12. Define the term *productivity*. Why is it difficult to measure the impact that investments in information systems have on productivity?
13. Briefly define *technology diffusion*, *infusion*, and the *technology acceptance model*.
14. What is the total cost of ownership?
15. What is an information systems unit?
16. How do you define quality?

● DISCUSSION QUESTIONS

1. You have been hired to work in the IS area of a manufacturing company that is starting to use the Internet to order parts from its suppliers and offer sales and support to its customers. What types of Internet positions would you expect to see at the company?
2. What sort of information systems career would be most appealing to you—working as a member of an IS organization, being a consultant, or working for an IT hardware or software vendor? Why?
3. As part of a TQM project initiated three months ago, you decided your company needed a new information system. The computer systems were brought in over the weekend. The first notice your employees received about the new information system was the computer located on each desk. How might the new system affect the culture of your organization? What types of behaviors might employees exhibit in response? As a manager, how should you have prepared the employees for the new system?
4. You have been asked to participate in the preparation of your company's strategic plan. Specifically, your task is to analyze the competitive marketplace using Porter's five-force model. Prepare your analysis, using your knowledge of a business you have worked for or have an interest in working for.
5. Based on the analysis you performed in Discussion Question 4, what possible strategies could your organization adopt to address these challenges? What role could information systems play in these strategies? Use Porter's strategies as a guide.
6. There are many ways to evaluate the effectiveness of an information system. Discuss each method and describe when one method would be preferred over another method.
7. Imagine that you are the CIO for a large, multinational company. Outline a few of your key responsibilities.
8. Discuss how the change model can be applied to breaking a bad habit—say, smoking or eating fatty foods. Some people have also related the stages in the change model to the changes one must go through to deal with a major life crisis—like divorce or the loss of a loved one. Explain.

● PROBLEM-SOLVING EXERCISES

1. Locate a firm that uses the traditional organization structure. Then, using a presentation graphics package, draw the organizational structure for this firm. The firm should have more than 15 employees. Do the same for a firm that has another type of structure discussed in the text. Use your word processor to create a document that describes the differences between these two firms.

2. A new IS project has been proposed that will produce not only cost savings but also an increase in revenue. The initial costs to establish the system are estimated to be $500,000. The rest of the cash flow data is presented in the following table.

	Year 1	Year 2	Year 3	Year 4	Year 5
Increased Revenue	$0	$100	$150	$200	$250
Cost Savings	0	50	50	50	50
Depreciation	0	75	75	75	75
Initial Expense	500				

Note: All amounts in 000's.

a. Using your spreadsheet program, calculate the return on investment for this project. Assume the cost of capital is 7 percent.
b. How would the rate of return change if the project were able to deliver $50,000 in additional revenue and generate cost savings of $25,000 in the first year?

● TEAM ACTIVITIES

1. With your team, interview one or more people who were either outsourced or downsized from a position in the last few years. Find out how the process was handled and what justification was given for taking this action. Also try to get information from an objective source (financial reports, investment brokers, industry consultants) on how this action has affected the organization.

2. Have your team interview a company, university, or government agency that is quality conscious. Write a brief report that describes the quality efforts taken and the impact of these quality initiatives on how successful the organization has been in achieving its goals.

● WEB EXERCISES

1. This book emphasizes the importance of information. You can get information from the Internet by going to a specific address, such as http://www.ibm.com, http://www.whitehouse.gov, or http://www.fsu.edu. These will give you access to the home page of IBM Corporation, the White House, and Florida State University. Note that "com" is used for businesses and commercial operations, "gov" is used for government offices, and "edu" is used for educational institutions. Another approach is to use a search engine. Yahoo!, developed by two Tulane University students, was one of the first search engines on the Internet. A search engine is a Web site that allows you to enter key words or phrases to find information. There are also lists or menus that can be used. The search engine will return other Web sites (hits) that correspond to a search request. Using Yahoo! at http://www.yahoo.com, search for information about a company or topic discussed in Chapter 1 or 2. You may be asked to develop a report or send an e-mail message to your instructor about what you found.

2. Use the Internet to search for information about a company that you think might make a good stock investment. You can use a search engine, like Yahoo!, or a database at your college or university. Write a brief report describing the company and why it might make a good stock investment over the next few years.

● CASES

 Total Cost of Ownership

For decades, companies have been investing millions of dollars in information systems in an attempt to increase revenues, reduce costs, or both. Finally, with the booming stock market and increased productivity, it seems that this investment in information systems has paid off. But it can be very difficult for a company to measure the impact of its information systems investment.

Using traditional measures, like net present value and internal rate of return, many companies have attempted and failed to measure the impact of information systems investments on profitability. Doctoral students doing research on the impact of information systems investments have also failed to show a significant relationship between IS investment and results. Having to deal with so many other factors, such as general economic conditions, the competition's activities, and corporatewide programs in such areas as marketing has made it almost impossible to determine whether an IS investment or some other factor has contributed to increased profits. Even with these difficulties, most firms agree that measuring the impact of any investment is important. As a result, a number of consulting organizations are proposing other measures, including Gartner Group's total cost of ownership (TCO) model.

The TCO model investigates the total cost of owning computer equipment or systems. The model attempts to include all costs, including direct and indirect costs, of owning a computer device. In addition, TCO can include what is called a "futz factor" to take into account other computing activities, such as nonbusiness uses of a computer system. The model was first applied to desktop computers. Since then, the approach has been applied to software, networks, telecommunications, larger mainframe computers, handheld organizers, and other computer devices. According to Bill Kirwin, vice president of the Gartner Group, "We

use TCO to look at the overall impact of the implementation. Cost is the numerator. The denominator might be service, customer satisfaction, quality levels or productivity, for example. . . . TCO can be very revealing. It might show that certain departments are too autonomous or that the IT department is being poorly managed."

A number of companies have successfully used the TCO approach. A pharmaceutical company, for example, was able to use the TCO approach to save about 30 percent of what it paid for basic computer services. In addition, some companies offer products and services that include the TCO approach. IBM, for example, offers TCO Baseline Analysis, TCO Advanced Analysis, and Universal Management Tools Validation Pilot. These services can cost from about $3,000 to more than $7,500.

Although TCO has been widely used and accepted, some people believe that the approach is not worth the effort. According to a spokesman for the National Computing Center (NCC), "We can't take TCO as a reliable guide to realistic expenditures. We've found from our members that it's a meaningless figure."

Discussion Questions

1. What is the TCO model?
2. What are the disadvantages of the TCO approach?

Critical Thinking Questions

3. Describe how TCO can be used to help a company determine its total computer-related expenditures.
4. If you had to measure the effectiveness of a complete computer system, what measures would you use?

Sources: Adapted from Jacqueline Emich, "Total Cost of Ownership," *Computerworld,* December 20, 1999, p. 52; and Simon Goodley, "Forget TCO," *Computing,* February 3, 2000, p. 1.

UPS Turns to Technology for a Strategic Advantage

People often claim that the saying "You can't teach an old dog new tricks" applies to old, traditional companies. It is often said that it takes a new, upstart company to take full advantage of changing times and the Internet age. Although this may be true for some old, traditional companies, it is not true for one of the oldest and most respected companies in America, United Parcel Service (UPS).

UPS began in the early 1900s by moving a limited number of packages in the Seattle area. The first vehicles used were Model-T Fords. With its coffee-brown uniforms and vehicles, UPS has not only survived for almost a century, it has thrived. Company income for 1999 exceeded $2 billion on revenues of $27 billion. In recent years, the company has seen annual growth rates that exceed 20 percent. Today, the company moves 13 million packages daily. UPS's ability to change and adapt is a key reason for its continued success for almost 100 years. According to Chief Executive James Kelly, "We have to be more adaptable. We have to know when to add and when to subtract." Clearly, Kelly, who started with UPS as a part-time driver, knows how to compute the way to success for his company. And that way is through technology. The success has reached all levels of the company. After a recent initial public offering, a number of long-term UPS truck drivers and other employees became instant millionaires as a result of their stock options.

This long-term success is a result of the staggering investment UPS has made in computer technology. Over the last 10 years, UPS has invested about $11 billion in computer systems and related equipment. In the past, UPS could have been categorized as a trucking company that used technology. Today, UPS thinks of itself as a technology company that uses trucks. All aspects of its business have been automated, with the Internet playing a central part in its long-term business strategy. Each driver, for example, uses an electronic tracking device, called a Delivery Information Acquisition Device (DIAD). Using this device, a company can track its shipment even before the UPS truck leaves its driveway. But UPS does much more than deliver packages. For example, UPS delivers Gateway computers to customers with a cash-on-delivery system, where UPS collects payments from customers receiving Gateway computers and deposits the payments directly into Gateway bank accounts.

UPS, however, hasn't always had an easy or successful time. A few years ago, the Teamsters walkout cost UPS about $200 million in lost sales. For many inside UPS, this was a wake-up call to be even more aggressive in using technology to propel the company into the next century. According to one observer of the impact of the Teamsters walkout on UPS, "You never want to wound a tiger. You want to kill it, because if you wound it, it only becomes more ferocious." From all accounts, UPS is becoming more ferocious in its use of technology to increase profits and give it a long-term competitive advantage.

Discussion Questions

1. Describe the history and success of UPS.
2. How was UPS able to use technology to its competitive advantage?

Critical Thinking Questions

3. How could the lessons of UPS be used in other industries?
4. If you were the CEO of another shipping company, such as FedEx, what would you do to keep your company competitive with UPS?

Sources: Adapted from Kelly Barron, "UPS Company of the Year," *Forbes*, January 10, 2000, p. 79; and Beth Bacheldor, "Ford–UPS Alliance," *Information Week*, February 7, 2000.

When Things Go Wrong

In this chapter and throughout the book, we will present a number of stellar examples of how technology can be used as a competitive weapon to increase revenues and profits while slashing costs. Today, most companies could not survive without computer systems and technology. When properly implemented, a computer system can make a tremendous difference to the bottom line. A poorly implemented system, however, can cause big problems.

A well-known candy company realized the negative side of poorly implemented technology one Halloween. A few months earlier, a distributor for the company had placed a large order of 20,000 pounds of candy to arrive just in time for the holiday. Unfortunately, the candy never arrived, store shelves were empty, and the candy company lost a huge opportunity.

The problem was traced to a new $112 million computer system that was installed months earlier to automate many of the functions of the candy company. The new computer system was to automate everything from order processing to loading candy onto trucks for delivery. Because of problems with the new computer system, the 20,000-pound candy order was never received. In addition, precious shelf space was turned over to competitors—what was a big loss for this candy company became a bonanza for other candy companies. As this candy company found out, a poorly implemented computer system can be as scary as Halloween itself.

Discussion Questions

1. What went wrong in this case?
2. In your opinion, what was the current and future loss of sales as a result of the computer problem?

Critical Thinking Questions

3. What steps would you follow to make sure a new computer system works properly before it is placed into operation?
4. Read newspapers and journals to find other examples of computer installations that went wrong.

Sources: Adapted from Emily Nelson, et al., "Trick or Treat," *The Wall Street Journal*, October 29, 1999, p. A1; and Peter Galuszka, "Managing," *Business Week*, November 8, 1999, p. 36.

● NOTES

Sources for the opening vignette on p. 37: Adapted from Thomas Hoffman, "Finding a Rich Niche," *Computerworld*, February 8, 1999, p. 44; and "Farmers Trim Paper Trail," *Future Banker*, September 6, 1999, p. 24.

1. Peter Keen, "IT's Value in the Chain," *Computerworld*, February 14, 2000, p. 48.

2. Atrium Web site, http://www.atrium-hr.com, accessed February 16, 2000.

3. E. H. Schein, *Process Consultation: Its Role in Organizational Development* (Reading, Mass.: Addison-Wesley, 1969). See also Peter G. W. Keen, "Information Systems and Organizational Change," *Communications of the ACM,* vol. 24, no. 1, January 1981, pp. 24–33.

4. Stacy Collet, "Avon Calls for Revamp of Its World Wide IT," *Computerworld*, July 12, 1999, p. 38.

5. Jeannine Aversa, "$5B Drop in the Bucket," *ABCNews Online*, March 2, 2000.

6. Christine Chen et al., "Internet Boot Camp," *Fortune*, March 20, 2000, p. 211.

7. Susan Avery, "Competing Suppliers Partner to Benefit Bethlehem Steel," *Purchasing Magazine*, August 12, 1999, p. 114.

8. Christoph Loch and Berndao Huberman, "A Punctuated-Equilibrium Model of Technology Diffusion," *Management Science*, February, 1999, p. 160.

9. Curtis Armstrong and V. Sambamurthy, "Information Technology Assimilation in Firms," *Information Systems Research*, December, 1999, p. 304.

10. Ritu Agarwal and Jayesh Prasad, "Are Individual Differences Germane to the Acceptance of New Information Technology," *Decision Sciences*, Spring 1999, p. 361.

11. Kwon et al., "A Test of the Technology Acceptance Model," *Proceedings of the Hawaii International Conference on System Sciences*, January 4–7, 2000.

12. Henry Lucas and V. K. Spitler, "Technology Use and Performance," *Decision Sciences*, Spring 1999, p. 291.

13. Van Dyke et al., "Cautions of the Uses of SERVQUAL Measures to Assess the Quality of Information Systems Services," *Decision Sciences*, Summer 1999, p. 877.

14. Matt Hamblen, "Handhelds Help Boeing Boost Quality Inspections," *Computerworld*, November 8, 1999, p. 38.

15. Larry Greenemeier, "FTD.com Consolidates IT Outsourcing," *Information Week*, February 7, 2000.

16. Jennifer Shah, "Baan Tackling Its Troubles," *Electronic Buyer's News*, January 10, 2000, p. 14.

17. Hellen Pukszta, "Don't Split IT Strategy from Business Strategy," *Computerworld*, January 11, 1999, p. 35.

18. M. Porter and V. Millar, "How Information Systems Give You Competitive Advantage," *Journal of Business Strategy,* Winter 1985. See Also M. Porter, *Competitive Advantage* (New York: Free Press, 1985).

19. Paul Judge, "One Big Happy Family—But for How Long," *Business Week*, October 25, 1999, p. 148.

20. Eva Sohlman, "The New Remote Control," *ABCNews Online*, April 18, 2000.

21. "The State of IT: 2010," *Computerworld*, January 17, 2000, p. 50.

22. Eileen Birge, "How to Measure IT Performance," *Beyond Computing*, January, 1999, p. 32.

23. Jacqueline Emigh, "Total Cost of Ownership," *Computerworld*, December 20, 1999, p. 52.

24. Mark Hall, "Risk Management," *Computerworld*, January 17, 2000, p. 58.

25. Peter Sappal, "E-Commerce Recruitment," *The Wall Street Journal*, January 25, 2000, p. B20.

26. Vicky Uhland, "Tech Jobs Abound This Year: Top Positions Involve E-Commerce," *Denver Rocky Mountain News*, February 27, 2000, p. J1.

27. Sandy Graham, "SysText Goes After Digital Bugs," *Denver Rocky Mountain News,* February 27, 2000, p. 3G.

PART II

Information Technology Concepts

CHAPTER 3

Hardware: Input, Processing, and Output Devices

*T*here is no such thing as a
personal computer in the future.
There are only available
appliances. You'll use your smart
card or your smart ring or some
sort of proximity device, so that
as you get near it, the device
knows who you are.

— Scott McNealy, CEO of
Sun Microsystems, in response to
a question asked by a reporter
from *Computerworld* about the
future of the PC

Principles

Assembling an effective, efficient
computer system requires an
understanding of its relationship to
the information system and the
organization. The computer system
objectives are subordinate to, but
supportive of, the information system and the organization.

When selecting computer
devices, you also must consider
the current and future needs of
the information system and the
organization. Your choice of a
particular computer system device
should always allow for later
improvements.

Learning Objectives

- *Describe how to select and organize computer system components to support information system objectives.*

- *Describe the power, speed, and capacity of central processing and memory devices.*
- *Describe the access methods, capacity, and portability of secondary storage devices.*
- *Discuss the speed, functionality, and importance of input and output devices.*
- *Identify popular classes of computer systems and discuss the role of each.*
- *Define the term* multimedia computer *and discuss common applications of such a computer.*

Ford Motor Company

A Technology Makeover

In an announcement that surprised many industry experts and Ford employees, Ford Motor Company has come up with an idea that will please its employees, while helping to bring the company into the computer age. The idea is to offer each of its employees a home PC and Internet access for the low fee of $5 per month. The offer will be made to all 300,000 employees. This new employee perk seems like a good move for Ford from many perspectives. The United Auto Workers has been asking Ford to enhance the technical skills of its employees. In addition, Ford CEO Jacques Nasser strongly believes that technology and the Internet will have a dramatic impact on Ford and the entire auto industry.

The PC offer, in addition to being very inexpensive, will offer employees other features and benefits. Each $5-per-month home PC and Internet connection will come with a host of technical support and assistance. Each PC includes a 500-MHz Celeron chip, 64 MB of memory, a 4.3 GB hard disk, and a color printer. Hewlett-Packard is not only supplying the computers and printers, but it will also assist in maintenance and repair issues that may arise. MCI WorldCom will supply the Internet access through its UUnet subsidiary. Employees will have unrestricted use of their home PC, printer, and Internet access without interference from Ford. Ford has stated that it will not be monitoring how employees use their home PCs. Of course, Ford is hoping that its employees will gain new and useful computer skills that can directly apply to their jobs.

In addition to the inexpensive PC offer, Nasser is also launching Ford into the new age of technology. He wants to convert Ford's image from a metal-bending manufacturing company into a Web-savvy consumer marketing company. To achieve this objective, Nasser is forming strategic partnerships with a number of technology companies. For example, Ford is developing a comprehensive Web site to allow it to take orders over the Internet. According to Nasser, "It isn't like we woke up one morning and said, 'Let's try this.' We thought that consumers were demanding it." Getting an effective Internet site up and running will be challenging, however. Ford has a large number of different hardware systems that must be made to communicate with each other. Internet orders coming into its Web site must flow smoothly from the front-end computer systems taking the orders to the back-end computers that do all the processing, manufacturing, scheduling, inventory control, and the other critical manufacturing operations required to get finished autos and trucks to customers.

On the supply side, Ford is reaching out to its suppliers to make parts delivery smoother. To help accomplish this objective, Ford has entered into a partnership with Oracle, a database company, to help coordinate communication between its suppliers' computers and its own computers.

With the new initiatives, customers will be pulling cars and trucks through the manufacturing process via their orders. The old model pushed ready-made cars and trucks to customers, hoping they would buy what Ford was producing. Of course, one of the most difficult problems is getting traditional Ford managers and employees on the technology bandwagon. Nasser knows the importance of these new technology initiatives. "Anyone who believes we are in a day of eternal stability just doesn't get it."

As you read this chapter, consider the following:

- How are companies using hardware and technology to help them compete?

- What technology, hardware, and systems would you recommend to Ford and other companies to help them increase sales and profits?

Appropriate use of technology can reap huge benefits in business, as seen with Ford Motor Company. Employing information technology and providing additional processing capabilities can increase employee productivity, expand business opportunities, and allow for more flexibility. As we already discussed, a computer-based information system (CBIS) is a combination of hardware, software, database(s), telecommunications, people, and procedures—all organized to input, process, and output data and information. In this chapter, we concentrate on the hardware component of a CBIS. **Hardware** consists of any machinery (most of which uses digital circuits) that assists in the input, processing, storage, and output activities of an information system. The overriding consideration in making hardware decisions in a business should be how hardware can be used to support the objectives of the information system and the goals of the organization.

hardware

any machinery (most of which uses digital circuits) that assists in the input, processing, storage, and output activities of an information system

COMPUTER SYSTEMS: INTEGRATING THE POWER OF TECHNOLOGY

A computer system is a special subsystem of an organization's overall information system. It is an integrated assembly of devices—centered on at least one processing mechanism utilizing digital electronics—that are used to input, process, store, and output data and information.

Putting together a complete computer system, however, is more involved than just connecting computer devices. In an effective and efficient system, components are selected and organized with an understanding of the inherent trade-offs between overall system performance and cost, control, and complexity. For instance, in building a car, manufacturers try to match the intended use of the vehicle to its components. Racing cars, for example, require special types of engines, transmissions, and tires. The selection of a transmission for a racing car, then, requires not only consideration of how much of the engine's power can be delivered to the wheels (efficiency and effectiveness) but also how expensive the transmission is (cost), how reliable it is (control), and how many gears it has (complexity). Similarly, organizations assemble computer systems so that they are effective, efficient, and well suited to the tasks that need to be performed.

Consider the hardware used by eBay, one of the most popular on-line auction Web sites.[1] In the summer of 1999, eBay was out of service for 22 hours because of technical problems. Users could not get access for almost a day. This massive failure was followed by several shorter outages. Meg Whitman was hired as CEO in part to help get a handle on the hardware and technological needs of the fast-growing company. In one incident, when the company again faced a technical problem, Whitman responded, "That's a hardware issue. This is what you'll see, and this is how long it will take." She was right, and the problem was solved.

Like other companies facing explosive growth, eBay is constantly working to keep its hardware working and in line with its business. The goal is often to upgrade hardware to keep pace with ever-increasing business. In eBay's case, the solution was to get a person who knew how to match hardware systems with the growing demands of the business.

To stay successful, eBay must ensure that its hardware system can support 7.7 million registered users. The site has more than 1.7 million visitors each day.

As seen with eBay, assembling a computer subsystem requires an understanding of its relationship to the information system and the organization. Although we generally refer to the computer subsystem as simply a computer system, we must remember that the computer system objectives are subordinate to, but supportive of, the information system and the organization.

The components of all information systems—such as hardware devices, people, procedures, and goals—are interdependent. Because the performance of one system affects the others, all of these systems should be measured according to the same standards of effectiveness and efficiency, given issues of cost, control, and complexity.

When selecting computer subsystem devices, you also must consider the current and future uses to which these systems will be put. Your choice of a particular computer system should always allow for later improvements in the overall information system. Reasoned forethought—a trait required for dealing with computer, information, and organizational systems of all sizes—is the hallmark of a true systems professional.

Hardware Components

Computer system hardware components include devices that perform the functions of input, processing, data storage, and output (Figure 3.1). To understand how these hardware devices work together, consider an analogy from a paper-based office environment. Imagine a one-room office occupied by a single individual. The human (the processor) is capable of organizing and manipulating data. The person's mind (register storage) and the desk occupied by the human (primary storage) are places to temporarily store data. Filing cabinets fill the need for a more permanent form of storage (secondary storage). In this analogy, the incoming and outgoing mail trays can be understood as sources of new data (input) or as places to put the processed paperwork (output).

The ability to process (organize and manipulate) data is a critical aspect of a computer system, in which processing is accomplished by an interplay between one or more of the central processing units and primary storage. Each **central processing unit (CPU)** consists of three associated elements: the arithmetic/logic unit, the control unit, and the register areas. The **arithmetic/logic unit (ALU)**

central processing unit (CPU)

the part of the computer that consists of three associated elements: the arithmetic/logic unit, the control unit, and the register areas

arithmetic/logic unit (ALU)

portion of the CPU that performs mathematical calculations and makes logical comparisons

FIGURE 3.1

Computer System Components

These components include input devices, output devices, communications devices, primary and secondary storage devices, and the central processing unit (CPU). The control unit, the arithmetic/logic unit (ALU), and the register storage areas constitute the CPU.

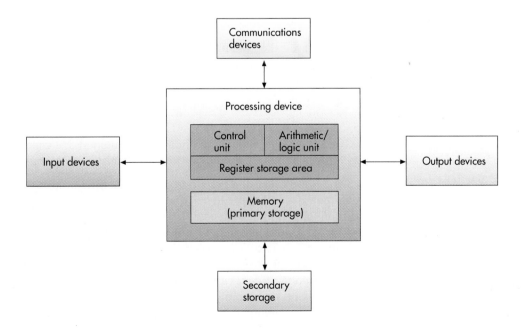

FIGURE 3.2

Execution of an Instruction

In the instruction phase, the computer's control unit fetches the instruction to be executed from memory (1). Then the instruction is decoded so the central processor can understand what is to be done (2). In the execution phase, the ALU does what it is instructed to do, making either an arithmetic computation or a logical comparison (3). Then the results are stored in the registers or in memory (4). The instruction and execution phases together make up one machine cycle.

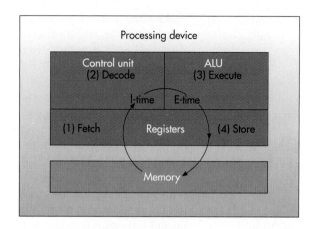

control unit

part of the CPU that sequentially accesses program instructions, decodes them, and coordinates the flow of data in and out of the ALU, the registers, primary storage, and even secondary storage and various output devices

register

high-speed storage area in the CPU used to temporarily hold small units of program instructions and data immediately before, during, and after execution by the CPU

primary storage (main memory; memory)

part of the computer that holds program instructions and data

instruction time (I-time)

the time it takes to perform the fetch-instruction and decode-instruction steps of the instruction phase

execution time (E-time)

the time it takes to execute an instruction and store the results

machine cycle

the instruction phase followed by the execution phase

pipelining

a form of CPU operation in which there are multiple execution phases in a single machine cycle

performs mathematical calculations and makes logical comparisons. The **control unit** sequentially accesses program instructions, decodes them, and coordinates the flow of data in and out of the ALU, the registers, primary storage, and even secondary storage and various output devices. **Registers** are high-speed storage areas used to temporarily hold small units of program instructions and data immediately before, during, and after execution by the CPU.

Primary storage, also called **main memory** or just **memory**, is closely associated with the CPU. Memory holds program instructions and data immediately before or immediately after the registers. To understand the function of processing and the interplay between the CPU and memory, let's examine the way a typical computer executes a program instruction.

Hardware Components in Action

The execution of any machine-level instruction involves two phases: the instruction phase and the execution phase. During the instruction phase, the following takes place:

- *Step 1: Fetch instruction.* The instruction to be executed is accessed from memory by the control unit.
- *Step 2: Decode instruction.* The instruction is decoded so the central processor can understand what is to be done, relevant data is moved from memory to the register storage area, and the location of the next instruction is identified.

Steps 1 and 2 are called the instruction phase, and the time it takes to perform this phase is called the **instruction time (I-time)**.

The second phase is the execution phase. During the execution phase, the following steps are performed:

- *Step 3: Execute the instruction.* The ALU does what it is instructed to do. This could involve making either an arithmetic computation or a logical comparison.
- *Step 4: Store results.* The results are stored in registers or memory.

Steps 3 and 4 are called the execution phase. The time it takes to complete the execution phase is called the **execution time (E-time)**.

After both phases have been completed for one instruction, they are again performed for the second instruction, and so on. The instruction phase followed by the execution phase is called a **machine cycle** (Figure 3.2). Some central processing units can speed up processing by using **pipelining**, whereby the CPU gets one instruction, decodes another, and executes a third at the same time. The Pentium processor, for example, uses two execution unit pipelines. This gives the processing unit the ability to execute two instructions in a single machine cycle.

PROCESSING AND MEMORY DEVICES: POWER, SPEED, AND CAPACITY

The components responsible for processing—the CPU and memory—are housed together in the same box or cabinet, called the system unit. All other computer system devices, such as the monitor and keyboard, are linked either directly or indirectly into the system unit housing. As discussed previously, achieving information system objectives and organizational goals should be the primary consideration in selecting processing and memory devices. In this section, we investigate the characteristics of these important devices.

Processing Characteristics and Functions

Because having efficient processing and timely output is important, organizations use a variety of measures to gauge processing speed. These measures include the time it takes to complete a machine cycle, and clock speed.

Machine Cycle Time

As we've seen, the execution of an instruction takes place during a machine cycle. The time in which a machine cycle occurs is measured in fractions of a second. Machine cycle times are measured in *microseconds* (one-millionth of one second) for slower computers to *nanoseconds* (one-billionth of one second) and *picoseconds* (one-trillionth of one second) for faster ones. Machine cycle time also can be measured in terms of how many instructions are executed in a second. This measure, called **MIPS**, stands for millions of instructions per second. MIPS is another measure of speed for computer systems of all sizes.

MIPS

millions of instructions per second

Clock Speed

Each CPU produces a series of electronic pulses at a predetermined rate, called the **clock speed**, which affects machine cycle time. The control unit portion of the CPU controls the various stages of the machine cycle by following predetermined internal instructions, known as **microcode**. You can think of microcode as predefined, elementary circuits and logical operations that the processor performs when it executes an instruction. The control unit executes the microcode in accordance with the electronic cycle, or pulses of the CPU "clock." Each microcode instruction takes at least the same amount of time as the interval between pulses. The shorter the interval between pulses, the faster each microcode instruction can be executed (Figure 3.3).

clock speed

a series of electronic pulses produced at a predetermined rate that affect machine cycle time

microcode

predefined, elementary circuits and logical operations that the processor performs when it executes an instruction

FIGURE 3.3

Clock Speed and the Execution of Microcode Instructions

A faster clock speed means that more microcode instructions can be executed in a given time period.

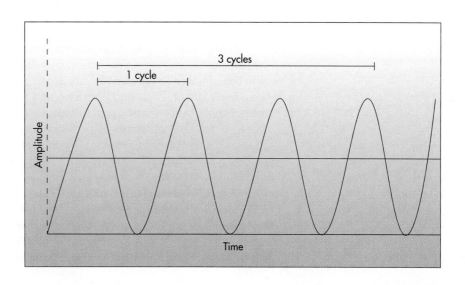

hertz

one cycle or pulse per second

megahertz (MHz)

millions of cycles per second

Clock speed is often measured in megahertz. As seen in Figure 3.3, a **hertz** is one cycle or pulse per second. **Megahertz (MHz)** is the measurement of cycles in millions of cycles per second. The clock speed for personal computers can range from 200 MHz to 700 MHz or more.[2] An AMD Athlon chip can reach speeds of 1,000 MHz.[3]

Because the number of microcode instructions needed to execute a single program instruction—such as performing a calculation or printing results—can vary, there is no direct relationship between clock speed measured in megahertz and processing speed measures such as MIPS and milliseconds.

Wordlength and Bus Line Width

bit

BInary digiT—0 or 1

wordlength

the number of bits the CPU can process at any one time

bus line

the physical wiring that connects the computer system components

Data is moved within a computer system not in a continuous stream but in groups of bits. A **bit** is a BInary digiT—0 or 1. Therefore, another factor affecting overall system performance—particularly speed—is the number of bits the CPU can process at any one time. This number of bits is called the **wordlength** of the CPU. A CPU with a wordlength of 32 (called a 32-bit CPU) will process 32 bits of data in one machine cycle.

Data is transferred from the CPU to other system components via **bus lines**, the physical wiring that connects the computer system components. The number of bits a bus line can transfer at any one time is known as bus line width. For example, a bus line with a width of 32 will transfer 32 bits of data at a time. Common wordlengths and bus line widths are 32 and 64. Bus line width should be matched with CPU wordlength for optimal system performance. It would be of little value, for example, to install a new 64-bit bus line if the system's CPU had a wordlength of only 16. Assuming compatible wordlengths and bus widths, the larger the wordlength, the more powerful the computer. Computers with larger wordlengths can transfer more data between devices in the same machine cycle. They can also use the larger number of bits to address more memory locations and hence are a requirement for systems with certain large memory requirements.

Because all these factors—machine cycle time, clock speed, wordlength, and bus line width—affect the processing speed of the CPU, comparing the speed of two different processors can be confusing. Although the megahertz rating has important consequences for the design of a PC system and is therefore important to the PC design engineer, it is not necessarily a good measure of processor performance, especially when comparing one family of processors with the next. As a result, chip maker Intel and others have developed a number of benchmarks for speed. Intel's benchmark is the **iCOMP (Intel COmparative Microprocessor Performance) index**. The most recent index, the iCOMP index 3, is a collection of benchmark measures that attempt to take into account typical processing needs, including the increasing use of 3D, multimedia, and Internet access. For example, the Intel Pentium III chip running at 500 MHz has an iCOMP index of 1,650, while the Pentium III running at 800 MHz has an iCOMP index of 2,690. A higher index value means faster performance for typical processing activities. Other chip benchmarks include the CPUmark, WinTune Integer Test, and SPECint. Some general computer system benchmarks include SYSmark, High End Winstone, and Business Winstone. In addition, there are benchmarks for specific applications, such as Internet access and multimedia. For less technical measures of performance, popular PC journals (such as *PC Magazine* and *PC World,*) often rate personal computers on price, performance, reliability, service, and other factors.

iCOMP (Intel COmparative Microprocessor Performance) index

Intel's speed benchmark

Physical Characteristics of the CPU

CPU speed is also limited by physical constraints. Most CPUs are collections of digital circuits imprinted on silicon wafers, or chips, each no bigger than the tip of a pencil eraser. To turn a digital circuit within the CPU on or off, electrical current must flow through a medium (usually silicon) from point A to point B.

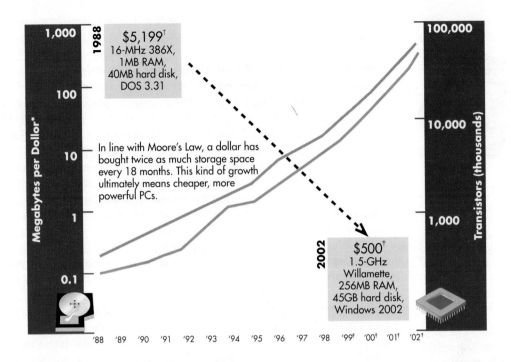

1988
$5,199[†]
16-MHz 386X,
1MB RAM,
40MB hard disk,
DOS 3.31

In line with Moore's Law, a dollar has bought twice as much storage space every 18 months. This kind of growth ultimately means cheaper, more powerful PCs.

2002
$500[†]
1.5-GHz
Willamette,
256MB RAM,
45GB hard disk,
Windows 2002

Megabytes per Dollar*

Transistors (thousands)

'88 '89 '90 '91 '92 '93 '94 '95 '96 '97 '98 '99[†] '00[†] '01[†] '02[†]

FIGURE 3.4

Moore's Law Affects More than Processor Speeds

(Source: Reprinted from "Moore's Law Will Continue to Drive Computing," *PC Magazine*, June 22, 1999, p. 146. Copyright © 1999, Ziff Davis Publishing Holdings, Inc. All rights reserved.

*Data from International Data Corp. for 2002; Disk/Trend Inc. for all other years.

[†]Data from *PC Magazine*, August 1988 (Compaq Deskpro 386S).

[‡]Projected data.

Moore's Law

a hypothesis that states that transistor densities on a single chip will double every 18 months

superconductivity

a property of certain metals that allows current to flow with minimal electrical resistance

optical processors

computer chips that use light waves instead of electrical current to represent bits

The speed at which it travels between points can be increased by either reducing the distance between the points or reducing the resistance of the medium to the electrical current.

Reducing the distance between points has resulted in ever-smaller chips, with the circuits packed closer together. In the 1960s, shortly after patenting the integrated circuit, Gordon Moore, former chairman of the board of Intel (the largest microprocessor chip maker), formulated what is now known as **Moore's Law**. This hypothesis states that transistor (the microscopic on/off switches, or the microprocessor's brain cells) densities on a single chip will double every 18 months. Moore's Law has held up amazingly well over the years. For Moore's Law to hold up over the long run, however, improved chip-fabrication techniques must be developed, and research into optical and laser chips must become successful.[4] For the short term, however, Moore's Law seems to remain in effect. In addition to increased speeds, Moore's Law has had an impact on costs and overall system performance. As seen in Figure 3.4, the number of megabytes per dollar and the number of transistors on a chip also continue to climb.

Another substitute material for silicon chips is superconductive metal. **Superconductivity** is a property of certain metals that allows current to flow with minimal electrical resistance. Traditional silicon chips create some electrical resistance that slows processing. Chips built from less resistant superconductive metals offer increases in processing speed.

To increase speed and performance, chip research continues. Intel and Hewlett-Packard have joined together to develop a superfast chip, called Merced. The companies have invested close to $1 billion, requiring hundreds of engineers. They hope to have the chip operational by the end of the year 2000.[5] Another chip by Intel, code-named McKinley, also shows promise. Other materials, including gallium arsenide (GaAs), are being investigated and show some promise. One company is even experimenting with a digital light processor.[6] This experimental chip uses microscopic mirrors instead of traditional electrical circuits. Other companies are experimenting with chips called **optical processors** that use light waves instead of electrical current to represent bits. The primary advantage of optical processors is their speed. Optical

The PowerPC750 microprocessor is a RISC processor that provides high levels of performance for desktop, workstation, and symmetric multiprocessing computer systems.
(Source: Permission. Motorola and the Motorola logo are registered trademarks of Motorola, Inc. PowerPC, the PowerPC logo, and the PowerPC750 are trademarks of IBM Corp. and are used by Motorola under license therefrom.)

complex instruction set computing (CISC)

a computer chip design that places as many microcode instructions into the central processor as possible

reduced instruction set computing (RISC)

a computer chip design based on reducing the number of microcode instructions built into a chip to an essential set of common microcode instructions

very long instruction word (VLIW)

a computer chip design based on further reductions in the number of instructions in a chip made by lengthening each instruction

byte (B)

eight bits together that represent a single character of data

processors have the potential of being 500 times faster than traditional electronic circuits.

Complex and Reduced Instruction Set Computing

Processors for many personal computers are designed based on **complex instruction set computing (CISC)**, which places as many microcode instructions into the central processor as possible. In the mid-1970s John Cocke of IBM recognized that most of the operations of a CPU involved only about 20 percent of the available microcode instructions. This led to an approach to chip design called **reduced instruction set computing (RISC)**, which involves reducing the number of microcode instructions built into a chip to this essential set of common microcode instructions. RISC chips are faster than CISC chips for processing activities that predominantly use this core set of instructions because each operation requires fewer microcode steps prior to execution. Most RISC chips use pipelining, which, as mentioned earlier, allows the processor to execute multiple instructions in a single machine cycle. With less sophisticated microcode instruction sets, RISC chips are also less expensive to produce and are quite reliable.

The PowerPC chip is a RISC processor created by Motorola. By almost any benchmark, RISC processors run faster than Intel's Pentium processor. And because RISC chips have a simpler design and require less silicon, they are cheaper to produce. The PowerPC chip is designed to provide portable and desktop personal computers the processing power normally associated with much more expensive computers. For example, the PowerPC has the ability to make functions such as voice recognition, dictation, pen input, and touch screens practical. Sun Microsystems' Sparc chip is another example of a RISC processor.

Further reductions in the number of instructions in a chip can be made by lengthening each instruction. This approach, called **very long instruction word (VLIW)**, is being developed by Intel, Hewlett-Packard, and other companies. Chips built using VLIW are potentially even faster than RISC-based chips. In the future, RISC and VLIW may gradually replace CISC chips.

When selecting a CPU, organizations must balance the benefits of speed with the issues of cost, control, and complexity. CPUs with faster clock speeds and machine cycle times are usually more expensive than slower ones. This expense, however, is a necessary part of the overall computer system cost, for the CPU is typically the single largest determinant of the price of many computer systems. CPU speed can also be related to complexity. Having a less complex code, as in the case of RISC chips, not only can increase speed and reliability but can also reduce chip manufacturing costs.

Memory Characteristics and Functions

Located physically close to the CPU (to decrease access time), memory provides the CPU with a working storage area for program instructions and data. The chief feature of memory is that it rapidly provides the data and instructions to the CPU.

Storage Capacity

Like the CPU, memory devices contain thousands of circuits imprinted on a silicon chip. Each circuit is either conducting electrical current (on) or not (off). By representing data as a combination of on or off circuit states, the data is stored in memory. Usually eight bits are used to represent a character, such as the letter *A*. Eight bits together form a **byte (B)**. Following is a list of storage capacity measurements. In most cases, storage capacity is measured in bytes, with one byte usually equal to one character.)

Name	Abbreviation	Number of Bytes
Byte	B	one
Kilobyte	KB	1,024
Megabyte	MB	1,024 × 1,024 bytes (about one million)
Gigabyte	GB	1,024 × 1,024 × 1,024 bytes (about one billion)
Terabyte	TB	1,024 × 1,024 × 1,024 × 1,024 bytes (about one trillion)

Types of Memory

random access memory (RAM)

a form of memory in which instructions or data can be temporarily stored

There are several forms of memory, as shown in Figure 3.5. Instructions or data can be temporarily stored in **random access memory (RAM)**. RAM is temporary and volatile—RAM chips lose their contents if the current is turned off or disrupted (as in a power surge, brownout, or electrical noise generated by lightning or nearby machines). RAM chips are mounted directly on the computer's main circuit board or in other chips mounted on peripheral cards that plug into the computer's main circuit board. These RAM chips consist of millions of switches that are sensitive to changes in electric current.

RAM comes in many different varieties. The mainstream type of RAM is extended data out, or EDO, RAM, which is faster than older types of RAM memory. Another kind of RAM memory is called dynamic RAM (DRAM). SDRAM, or synchronous DRAM, needs high or low voltages at regular intervals—every two milliseconds (two one-thousands of a second) if it is not to lose its information. SDRAM can exceed EDO RAM in performance. SDRAM also has the advantage of a faster transfer speed between the microprocessor and the memory.

Over the past decade, microprocessor speed has doubled every 18 months, but memory performance has not kept pace. In effect, memory has become the principal bottleneck to system performance. Thus, microprocessor manufacturers are working with memory vendors to keep up with the performance of faster processors and bus architectures. For example, Intel is working with Rambus to extend existing Rambus technology to improve total system performance. Developed in conjunction with Intel and in cooperation with other Rambus semiconductor partners, Direct Rambus (Direct RDRAM) technology is gaining industry support. The world's top 10 DRAM makers have announced their intention to develop Direct RDRAM products. Another two dozen companies representing the leaders in system-memory implementation products—including memory modules, connectors, clock chips, and test systems—announced their intention to support Direct Rambus technology. Planned applications include computer system memory multimedia and graphics memory, communications system memory, and consumer electronics memory. [7]

read-only memory (ROM)

a nonvolatile form of memory

Another type of memory, **ROM**, an acronym for **read-only memory**, is usually nonvolatile. In ROM, the combination of circuit states is fixed, and therefore its contents are not lost if the power is removed. ROM provides permanent storage for data and instructions that do not change, like programs and data from the computer manufacturer.

There are other types of nonvolatile memory as well. Programmable read-only memory (PROM) is a type in which the desired data and instructions—and hence the desired circuit state combination—must first be programmed into the memory chip. Thereafter, PROM behaves like ROM. PROM chips are used where the CPU's data and instructions do not change, but the application is so specialized or unique that custom manufacturing of a true ROM chip would be cost prohibitive. A common use of PROM chips is for storing the instructions to popular video games, such as those for Nintendo and Sega.

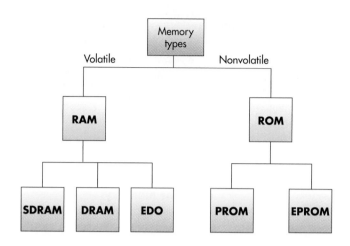

FIGURE 3.5

Basic Types of Memory Chips

Game instructions are programmed onto the PROM chips by the game manufacturer. Instructions and data can be programmed onto a PROM chip only once.

Erasable programmable read-only memory (EPROM) is similar to PROM except, as the name implies, the memory chip can be erased and reprogrammed. EPROMs are used where the CPU's data and instructions change, but only infrequently. An automobile manufacturer, for example, might use an industrial robot to perform repetitive operations on a certain car model. When the robot is performing its operations, the nonvolatility and rapid accessibility to program instructions offered by EPROM is an advantage. Once the model year is over, however, the EPROM controlling the robot's operation will need to be erased and reprogrammed to accommodate a different car model.

cache memory

a type of high-speed memory that a processor can access more rapidly than main memory

Cache memory is a type of high-speed memory that a processor can access more rapidly than main memory (Figure 3.6). Cache memory functions somewhat like a notebook used to record phone numbers. A person's private notebook may contain only 1 percent of all the numbers in the local phone directory, but the chance that the person's next call will be to a number in his or her notebook is high. Cache memory works on the same principle—frequently used data is stored in easily accessible cache memory instead of slower memory like RAM. Because there is less data in cache memory, the CPU can access the desired data and instructions more quickly than if it were selecting from the larger set in main memory. The CPU can thus execute instructions faster, and the overall performance of the computer system is raised. There are two types of cache memory present in the majority of systems shipped—the Level 1 (L1) cache is in the processor; the Level 2 (L2) cache memory is optional and found on the motherboard of most systems.

The main memory in your system that can move its information into your system's cache memory is called the cacheable memory. Memory in your system that is not cacheable performs as if your system is cacheless, moving information as needed directly to the processor without the ability to use the cache memory as a fast-retrieval storage bin. All systems have a main memory cacheable limit, typically 512 KB or greater.

Costs for memory capacity continue to decline. As we shall see, however, when considered on a megabyte-to-megabyte basis, memory is still considerably more expensive than most forms of secondary storage. Memory capacity can be important in the effective operation of a CBIS. The specific applications of a CBIS determine the amount of memory required for a computer system. For example, complex processing problems, such as computer-assisted product design, require more memory than simpler tasks like word processing. Also, because computer systems have different types of memory, other programs may be needed to control how memory is accessed and used. In other cases, the computer system can be configured to maximize memory usage. Before additional memory is purchased, all these considerations should be addressed.

FIGURE 3.6

Cache Memory

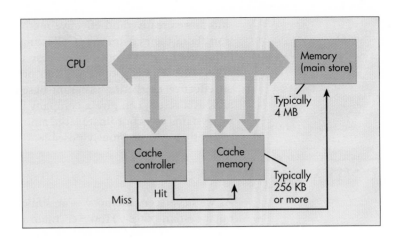

Processors can access this type of high-speed memory faster than main memory. Located near the CPU, cache memory works in conjunction with main memory. A cache controller determines how often the data is used and transfers frequently used data to cache memory, then deletes the data when it goes out of use.

Multiprocessing

multiprocessing

simultaneous execution of two or more instructions at the same time

coprocessor

part of the computer that speeds processing by executing specific types of instructions while the CPU works on another processing activity

A number of forms of **multiprocessing** involve the simultaneous execution of two or more instructions at the same time. One form of multiprocessing involves coprocessors. A **coprocessor** speeds processing by executing specific types of instructions while the CPU works on another processing activity. Coprocessors can be internal or external to the CPU and may have different clock speeds than the CPU. Each type of coprocessor best performs a specific function. For example, a math coprocessor chip can be used to speed mathematical calculations, and a graphics coprocessor chip decreases the time it takes to manipulate graphics.

Parallel Processing

parallel processing

a form of multiprocessing that speeds processing by linking several processors to operate at the same time, or in parallel

Another form of multiprocessing, called **parallel processing**, speeds processing by linking several processors to operate at the same time, or in parallel. The challenge is not connecting the processors but making them work effectively as a unified set. Accomplishing this difficult task requires software that can allocate, monitor, and control multiple processing jobs at the same time. With parallel processing, a business problem (such as designing a new product or piece of equipment) is divided into several parts. Each part is "solved" by a separate processor. The results from each processor are then assembled to get the final output (Figure 3.7).

The most frequent business uses for parallel processing are modeling, simulation, and analysis of large amounts of data. In today's marketplace, the requirements and array of services that consumers demand have forced companies to

FIGURE 3.7

Parallel Processing

Parallel processing involves breaking a problem into various subproblems or parts, then processing each of these parts independently. The most difficult aspect of parallel processing is not the simultaneous processing of the subproblems but the logical structuring of the problem into independent parts.

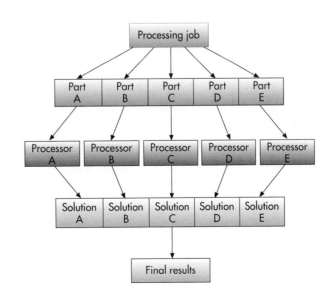

find more creative and effective methods for gathering and reporting information about their customers. Collecting and organizing the enormous amount of data about the consumer is no easy task. It is essential that the data be easily accessible. Parallel processing can provide the information necessary to build an effective marketing program based on existing consumer buying patterns and, as a result, can give a company a competitive advantage. Parallel processing systems can coordinate large amounts of data and access them with greater speed than was previously possible.

SECONDARY STORAGE

**secondary storage
(permanent storage)**

devices that store larger amounts of data, instructions, and information more permanently than allowed with main memory

FIGURE 3.8

Cost Comparison for Various Forms of Data Storage

All forms of secondary storage cost considerably less per megabyte of capacity than RAM, although they have slower access times. A diskette costs about 35 cents per megabyte, while RAM can cost around $4 per megabyte, 11 times more.
(Source: Data from CompUSA Direct Catalog, February 1998.)

As we have seen, memory is an important factor in determining overall computer system power. However, main memory provides only a small amount of storage area for the data and instructions required by the CPU for processing. Computer systems also need to store larger amounts of data, instructions, and information more permanently than allowed with main memory. **Secondary storage**, also called permanent storage, serves this purpose.

Compared with memory, secondary storage offers the advantages of nonvolatility, greater capacity, and greater economy. As previously noted, on a megabyte-per-megabyte basis, most forms of secondary storage are considerably less expensive than memory (see Figure 3.8). Because of the electromechanical processes involved in using secondary storage, however, it is considerably slower than memory. The selection of secondary storage media and devices requires an understanding of their primary characteristics—access method, capacity, and portability.

As with other computer system components, the access methods, storage capacities, and portability required of secondary storage media are determined by the information system's objectives. An objective of a credit card company's information system, for example, might be to rapidly retrieve stored customer data to approve customer purchases. In this case, a fast access method is critical. In other cases, such as salesforce automation via laptop computers, portability and storage capacity might be major considerations in selecting and using secondary storage media and devices.

Storage media that provide faster access methods are generally more expensive than slower media. The cost of additional storage capacity and portability varies widely, but it is also a factor to consider. In addition to cost, organizations must address security issues to allow only authorized people access to sensitive

Device	DAT tape	Hard drive	CD-RW	3.5" diskette	ZIP Plus Drive	RAM
Cost	$49.95	$230.00	$200.00	$.50	$100.00	$269.95
Storage	10,000MB	18GB	740MB	1.4MB	100MB–250MB	64MB
Cost per megabyte	$.005	$.08	$.27	$.35	$1.00	$4.21

data and critical programs. Because the data and programs kept in secondary storage devices are so critical to most organizations, all of these issues merit careful consideration.

Access Methods

sequential access

retrieval method in which data must be accessed in the order in which it is stored

direct access

retrieval method in which data can be retrieved without the need to read and discard other data

sequential access storage device (SASD)

device used to sequentially access secondary storage data

direct access storage device (DASD)

device used for direct access of secondary storage data

magnetic tape

common secondary storage medium, Mylar film coated with iron oxide with portions of the tape magnetized to represent bits

Data and information access can be either sequential or direct. **Sequential access** means that data must be accessed in the order in which it is stored. For example, inventory data stored sequentially may be stored by part number, such as 100, 101, 102, and so on. If you want to retrieve information on part number 125, it is necessary to read and discard all the data relating to parts 001 through 124.

Direct access means that data can be retrieved directly, without the need to pass by other data in sequence. With direct access, it is possible to go directly to and access the needed data—say, part number 125—without having to read through parts 001 through 124. For this reason, direct access is usually faster than sequential access. The devices used to sequentially access secondary storage data are simply called **sequential access storage devices (SASDs)**; those used for direct access are called **direct access storage devices (DASDs)**.

Devices

The most common forms of secondary storage include magnetic tapes, magnetic disks, and optical disks. Some of these media (magnetic tape) allow only sequential access, while others (magnetic and optical disks) provide direct and sequential access. Figure 3.9 shows some different secondary storage media.

Magnetic Tapes

One common secondary storage medium is **magnetic tape**. Similar to the kind of tape found in audio- and videocassettes, magnetic tape is a Mylar film coated with iron oxide. Portions of the tape are magnetized to represent bits. Magnetic tape is an example of a sequential access storage medium. If the computer is to read data from the middle of a reel of tape, all the tape before the desired piece

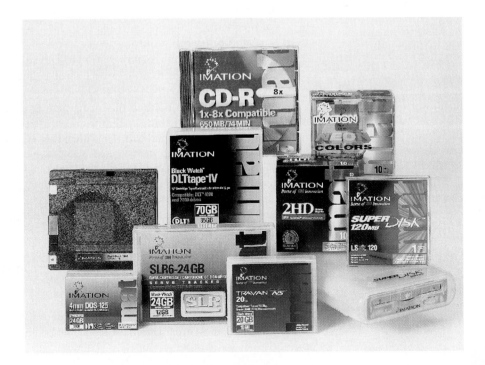

FIGURE 3.9

Types of Secondary Storage

Secondary storage devices such as magnetic tapes and disks, optical disks, and CD-ROMs are used to store data for easy retrieval at a later date.

(Source: Courtesy of Imation)

FIGURE 3.10

Hard Disk

Hard disks give direct access to stored data. The read/write head can move directly to the location of a desired piece of data, dramatically reducing access times, as compared with magnetic tape.

(Source: Courtesy of Seagate)

magnetic disk

common secondary storage medium, with bits represented by magnetized areas

redundant array of independent/inexpensive disks (RAID)

method of storing data that generates extra bits of data from existing data, allowing the system to create a "reconstruction map" so that if a hard drive fails, it can rebuild lost data

disk mirroring

a process of storing data that provides an exact copy that protects users fully in the event of data loss

of data must be passed over sequentially. This is one disadvantage of magnetic tape. When information is needed, it can take time for a tape operator to load the magnetic tape on a tape device and get the relevant data into the computer. Although access is slower, magnetic tape is usually less expensive than disk storage. In addition, magnetic tape is often used to back up disk drives and to store data off-site for recovery in case of disaster. One example, Procom Technology's Jetstream 1000 AIT Tape Array, provides a total capacity of 200GB, with a 20-MB-per-second data transfer rate and 72-GB-per-hour backup capability, making it an ideal backup device.

Magnetic Disks

Magnetic disks are also coated with iron oxide; they can be thin steel platters (hard disks; see Figure 3.10) or Mylar film (diskettes). As with magnetic tape, magnetic disks represent bits by small magnetized areas. When reading from or writing data onto a disk, the disk's read/write head can go directly to the desired piece of data. Thus, the disk is called a direct access storage medium. Although disk devices can be operated in a sequential mode, most disk devices use direct access. Because direct access allows fast data retrieval, this type of storage is ideal for companies that need to respond quickly to customer requests, such as airlines and credit card firms. For example, if a manager needs information on the credit history of a customer or the seat availability on a particular flight, the information can be obtained in a matter of seconds if the data is stored on a direct access storage device. If the data is stored on magnetic tape, it could take from a few minutes to over half an hour to load the tape and get the information.

Magnetic disk storage varies widely in capacity and portability. Standard diskettes are portable but have a slower access time and lower storage capacity (1.44 MB for some computers) than fixed hard disks. Hard disk storage, while more costly and less portable, has a greater storage capacity and quicker access time.

RAID

Companies' data storage needs are expanding rapidly. Today's storage configurations routinely entail many hundreds of gigabytes. However, putting the company's data on-line involves a serious business risk—the loss of critical business data can put a corporation out of business. The concern is that the most critical mechanical components inside a disk storage device—the disk drives, the fans, and other input/output devices—can break (like most things that move).

Organizations now require that their data storage devices be fault tolerant—the ability to continue with little or no loss of performance in the event of a failure of one or more key components. **Redundant array of independent/inexpensive disks (RAID)** is a method of storing data that generates extra bits of data from existing data, allowing the system to create a "reconstruction map" so that if a hard drive fails, it can rebuild lost data. With this approach, data is split and stored on different physical disk drives using a technique called stripping to evenly distribute the data. Since being developed at the University of Berkeley in 1987, RAID technology has been applied to storage systems to improve system performance and reliability.

RAID can be implemented in several ways. In the simplest form, RAID subsystems duplicate data on drives. This process, called **disk mirroring**, provides an exact copy that protects users fully in the event of data loss. However, if full copies are always to be kept current, organizations need to double the amount of storage capacity that is kept on-line. Thus, disk mirroring is expensive. Other RAID methods are less expensive because they only partly duplicate the data. This allows storage managers to minimize the amount of extra disk space (or overhead) they must purchase to protect data.

SAN

storage area network (SAN)

technology that uses computer servers, distributed storage devices, and networks to tie the storage system together

Storage area network (SAN) uses computer servers, distributed storage devices, and networks to tie everything together. To increase the speed of storing and retrieving data, fiber-optic channels are often used. Although SAN technology is relatively new, a number of companies are using SAN to successfully and efficiently store critical data. Burlington Coat Factory Warehouse Corporation uses three SANs that have a total capacity of 4 TB. Each SAN uses 12 processors and fiber optics to connect the storage devices. Morgan Stanley Dean Witter Trust uses SAN technology to eliminate paper mail.[8] When mail comes into Morgan Stanley's mailroom, it is opened and scanned into the computer and routed to the appropriate person for action. The scanned documents, however, require a huge amount of storage. Using SAN technology, Morgan Stanley can combine the resources of its local area network and mainframe computer to store 3 TB at its headquarters in New Jersey City and 1.5 TB of backup data in a center about 15 miles away.

Optical Disks

optical disk

a rigid disk of plastic onto which data is recorded by special lasers that physically burn pits in the disk

Another type of secondary storage medium is the **optical disk**. Similar in concept to a ROM chip, an optical disk is simply a rigid disk of plastic onto which data is recorded by special lasers that physically burn pits in the disk. Data is directly accessed from the disk by an optical disk device, which operates much like a stereo's compact disk player. This optical disk device uses a low-power laser that measures the difference in reflected light caused by a pit (or lack thereof) on the disk.

Each pit represents the binary digit 1; each unpitted area (called a land) represents the binary digit 0. Thus, the presence or lack of a pit determines the bit. Once a master optical disk has been created, duplicates can be manufactured using techniques similar to those used to produce music CDs.

compact disk read-only memory (CD-ROM)

a common form of optical disk on which data, once it has been recorded, cannot be modified

CD-writable (CD-W) disk

an optical disk that can be written upon but only once

CD-rewritable (CD-RW) disk

an optical disk that allows personal computer users to replace their diskettes with high-capacity CDs that can be written upon and edited over

A common form of optical disk is called **compact disk read-only memory (CD-ROM)**. Once data has been recorded on a CD-ROM, it cannot be modified—the disk is "read only." CD-ROM disks and hardware have moved from being unique add-ons in the mid-1980s to being a standard feature of today's personal computers. **CD-writable (CD-W)** disks allow data to be written once to a CD disk.[9] **CD-rewritable (CD-RW)** technology allows personal computer users to replace their diskettes with high-capacity CDs that can be written upon and edited over. The CD-RW disk can hold 740 MB of data—roughly 500 times the capacity of a 1.4-MB diskette.

Magneto-Optical Disk

magneto-optical disk

a hybrid between a magnetic disk and an optical disk

A **magneto-optical disk** is a type of disk drive that combines magnetic disk technologies with CD-ROM technologies. Like magnetic disks, MO disks can be read and written to. And like diskettes, they are removable. However, their storage capacity can be more than 200 MB, much greater than magnetic diskettes. In terms of data access speed, they are faster than diskettes but not as fast as hard disk drives. This type of disk uses a laser beam to change the molecular configuration of a magnetic substrate on the disk, which in turn creates visual spots. In conjunction with a photodetector, another laser beam reflects light off the disk and measures the size of the spots; the presence or absence of a spot indicates a bit. The disk can be erased by demagnetizing the substrate, which in turn removes the spots, allowing the process to begin again. Some magneto-optical drives can store more than a gigabyte on a single, removable disk. PowerMO by Olympus and the DynaMo by Fujitsu are two examples of magneto-optical devices.

The primary advantage of optical disks is their huge storage capacities, compared with other secondary storage media. Optical disks can store large applications and programs that contain graphics and audio data. They also allow for

FIGURE 3.11

Digital Video Disk and Player

DVD disks look like CDs but have a much greater storage capacity and can transfer data at a much faster rate.
(Source: Courtesy of Creative Labs)

digital video disk (DVD)

storage format used to store digital video or computer data

flash memory

a silicon computer chip that, unlike RAM, is nonvolatile and keeps its memory when the power is shut off

expandable storage devices

storage that uses removable disk cartridges to provide additional storage capacity

The PC memory card is like a portable hard disk that fits into any Type II PC Card slot and can store up to 1 gigabyte.
(Source: Courtesy of Kingston Technology)

storage of data on speculation; data not needed at a given moment can be easily stored for later possible use. Optical disks, however, suffer from some minor inconveniences, such as slow access time compared with diskettes and the lack of sufficient software to fully exploit the technology. In time, however, these inconveniences should diminish.

Digital Video Disk

The **digital video disk (DVD)** brings together the hitherto separate worlds of home computing and home video. A DVD disk is a five-inch CD-ROM look-alike (Figure 3.11) with the ability to store about 135 minutes of digital video. When used to store video, the picture quality far surpasses anything seen on tape, cable, or standard broadcast TV—sharp detail, true color, no flicker, no snow. The sound is recorded in digital Dolby, creating clear "surround" effects by completely separating all the audio channels in a home theater. DVDs cost less to duplicate and ship, take less shelf space, and deliver higher quality than videocassettes.

DVD can double as a computer storage disk and provide capacity of up to 17 GB. The physical disks resemble CD-ROMs, only they are thinner, so DVD players can also read current CD-ROMs, but current CD-ROM players cannot read the DVDs. Each DVD can hold at least 4.7 GB on a single side; some DVDs are double-layer disks, capable of holding 8.5 GB. Either type can be bonded back-to-back to create a two-sided disk with up to 17 GB of data. The access speed of a DVD drive is faster than the typical CD-ROM drive, with a data transfer rate clocked at 1.35 MB per second. DVD manufacturers include Sony, Philips, Toshiba, and others. These companies are also actively involved in making and improving standard CD-ROM drives. Newer DVD technology provides write-once disks and rewrite disks, often called DVD RAM.

Memory Cards

A group of computer manufacturers formed the Personal Computer Memory Card International Association (PCMCIA) to create standards for a peripheral device known as a PC memory card. These PC memory cards are credit-card-size devices that can be installed in an adapter or slot in many personal computers. To the rest of the system, the PC memory card functions as though it were a fixed hard disk drive. Although the cost per megabyte of storage is greater than for traditional hard disk storage, these cards are less failure prone than hard disks, are portable, and are relatively easy to use. Software manufacturers often store the instructions for their program on a memory card for use with laptop computers.

Flash Memory

Flash memory is a silicon computer chip that, unlike RAM, is nonvolatile and keeps its memory when the power is shut off. Flash memory chips are small and can be easily modified and reprogrammed, which makes them popular in computers, cellular phones, and other products. Flash memory is also used in some handheld computers to store data and programs, in digital cameras to store photos, and in airplanes to store flight information in the cockpit. Compared with other types of secondary storage, flash memory can be accessed more quickly, consumes less power, and is smaller in size. The primary disadvantage is cost. Flash memory chips can cost almost three times more per megabyte than a traditional hard disk. Nonetheless, the market for flash memory has exploded in recent years.

Expandable Storage

Expandable storage devices use removable disk cartridges (Figure 3.12). When your storage needs increase, you can use more removable disk cartridges.

Flash memory can store up to 160 MB of data. It's much faster than a hard drive, and unlike RAM, flash memory can retain data even when the power is turned off.
(Source: Courtesy of Kingston Technology)

The storage capacity can range from less than 100 MB to several gigabytes per cartridge. In recent years, the access speed of expandable storage devices has increased. Some devices are about as fast as an internal disk drive.

Expandable storage devices can be internal or external. A few personal computers are now including internal expandable storage devices as standard equipment. Zip by Iomega is an example. Of course, a CD-RW drive by Hewlett-Packard, Iomega, and others can also be used for expandable storage. With so much critical data stored on your hard drive, it is wise to make frequent backups. However, use of the standard diskette would require over 200 disks and several hours to back up even a small 300-MB hard drive. These expandable storage devices are ideal for backups. They can hold at least 80 times as much data and operate five times faster than the existing 1.44-MB diskette drives. Although more expensive than fixed hard disks, removable disk cartridges combine hard disk storage capacity and diskette portability. Some organizations prefer removable hard disk storage for the portability and control it provides. For example, a large amount of data can be taken to any location, or it can be secured so that access is controlled.

The overall trend in secondary storage is toward more direct-access methods, higher capacity, and increased portability. Organizations that select effective storage systems can greatly benefit from this trend. The specific type of storage should be chosen in consideration of the needs and resources of the organization. In general, the ability to store large amounts of data and information and access it quickly can increase organizational effectiveness and efficiency by allowing the information system to provide the desired information in a timely fashion. Table 3.1 lists the most common secondary storage devices and their capacities for easy reference.

INPUT AND OUTPUT DEVICES: THE GATEWAY TO COMPUTER SYSTEMS

A user's first experience with computers is usually through input and output devices. Through these devices—the gateways to the computer system—people provide data and instructions to the computer and receive results from it. Input and output devices are part of the overall user interface, which includes other hardware devices and software that allow humans to interact with a computer system.

As with other computer system components, the selection of input and output devices depends on organizational goals and information system objectives. For example, many restaurant chains use handheld input devices or computerized terminals that let waiters enter orders to ensure timely and accurate data input. These systems have cut costs by making inventory tracking more efficient and marketing to customers more effective.

FIGURE 3.12

Expandable Storage

Expandable storage drives allow users to add additional storage capacity by simply plugging in a removable disk or cartridge. The disks can be used to back up hard disk data or to transfer large files to colleagues.
(Source: Courtesy of Iomega)

Characteristics and Functionality

Rapidly getting data into a computer system and producing timely output can be very important for many organizations. The form of the output desired, the nature of the data required to generate this output, and the required speed and accuracy of the output and the input determine the appropriate output and input devices. Some organizations have very specific needs for output and input, requiring devices that perform specific functions. The more specialized the application, the more specialized the associated system input and output devices.

The speed and functions performed by the input and output devices selected and used by the organization should be balanced with their cost, control, and complexity. More

TABLE 3.1

Comparison of Secondary
Storage Devices

Storage Device	Year First Introduced	Maximum Capacity
3.5-inch diskette	1987	1.44 MB
CD-ROM	1990	650 MB
Zip	1995	100–250 MB
DVD	1996	17 GB

specialized devices might make it easier to enter data or output information, but they are generally more costly, less flexible, and more susceptible to malfunction.

The Nature of Data

Getting data into the computer—input—often requires transferring human-readable data, such as a sales order, into the computer system. Human-readable data is data that can be directly read and understood by humans. A sheet of paper containing adjustments to inventory is an example of human-readable data. By contrast, machine-readable data can be understood and read by computer devices (e.g., the universal bar code read by scanners at the grocery checkout) and is typically stored as bits or bytes. Data on inventory changes stored on a diskette is an example of machine-readable data.

Data can be both human readable and machine readable. For example, magnetic ink on bank checks can be read by humans and computer system input devices. Most input devices require some human interaction, because people most often begin the input process by organizing human-readable data and transforming it into machine-readable data. Every keystroke on a keyboard, for example, turns a letter symbol of a human language into a digital code that the machine can understand.

Data Entry and Input

data entry

process by which human-readable data is converted into a machine-readable form

data input

process that involves transferring machine-readable data into the system

Getting data into the computer system is a two-stage process. First, the human-readable data is converted into a machine-readable form through a process called **data entry**. The second stage involves transferring the machine-readable data into the system. This is **data input**.

Today, many companies are using on-line data entry and input—the immediate communication and transference of data to computer devices directly connected to the computer system. On-line data entry and input places data into the computer system in a matter of seconds. Organizations in many industries require the instantaneous update offered by this approach. For example, an airline clerk may need to enter a last-minute reservation. On-line data entry and input is used to record the reservation as soon as it is made. Reservation agents at other terminals can then access this data to make a seating check before they make another reservation.

Source Data Automation

source data automation

capturing and editing data whereby the data is originally created and in a form that can be directly input to a computer, thus ensuring accuracy and timeliness

Regardless of how data gets into the computer, it should be captured and edited at its source. **Source data automation** involves capturing and editing data where the data is originally created and in a form that can be directly input to a computer, thus ensuring accuracy and timeliness. For example, using source data automation, sales orders can be entered into the computer by the salesperson at the time and place the order is taken. Any errors can be detected and corrected immediately. If any item is temporarily out of stock, the salesperson can discuss options with the customer. Prior to source data automation, orders were written on a piece of paper to be entered later into the computer (often by someone other than the person who took the order). Often the handwritten information could not be read or, worse yet, got lost. If there were problems during data entry, it was necessary to contact the salesperson or the customer to "recapture" the data needed for order entry, thus leading to further delays and customer dissatisfaction.

Input Devices

Literally hundreds of devices can be used for data entry and input. These range from special-purpose devices used to capture specific types of data to more general-purpose input devices. Some of the special-purpose data entry and input devices will be discussed later in this chapter. First, however, we will focus on devices used to enter and input more general types of data, including text, audio, images, and video for personal computers.

Personal Computer Input Devices

A keyboard and a computer mouse are the most common devices used for entry and input of data such as characters, text, and basic commands. Some companies are developing newer keyboards that are more comfortable, adjustable, and faster to use. These keyboards, such as the split keyboard by Microsoft and others, are designed to avoid wrist and hand injuries caused by hours of keyboarding. Using the same keyboard, you can enter sketches on the touchpad and text using the keys.

A computer mouse is used to "point to" and "click on" symbols, icons, menus, and commands on the screen. This causes the computer to take a number of actions, such as placing data into the computer system.

Voice-Recognition Devices

voice-recognition device

an input device that recognizes human speech

Another type of input device can recognize human speech. Called **voice-recognition devices**, these tools use microphones and special software to record and convert the sound of the human voice into digital signals. Speech recognition can be used on the factory floor to allow equipment operators to give basic commands to machines while they are using their hands to perform other operations. Voice recognition can also be used by security systems to allow only authorized personnel into restricted areas. Voice recognition has been used in many industries, including medicine. Doctors at Quincy Medical Center in Massachusetts use voice recognition to create and update medical records for 32,000 patients per year.[10] The primary advantage is saving time. "I can handwrite a chart for a sprained ankle in four to five minutes," a doctor at Quincy Medical Center says. "I can dictate it in one minute and 10 seconds." In some cases, special dictation devices, such as the Dragon Naturally Speaking Mobile, can be used to capture memos and reports. Once captured, the speech can be easily uploaded through the serial port into the computer. Once stored in the computer, software is used to convert the speech to text.

Voice recognition has found its way into everyday products, such as autos.[11] The Clarion AutoPC, available on many Ford autos and trucks, is a voice-recognition system that allows a driver to activate radio programs and CDs. It can even tell you the time. Asking "What time is it?" will get a response such as, "Eleven thirty-four." The cost is about $1,000. The system includes a built-in computer.

Voice-recognition devices analyze and classify speech patterns and convert them into digital codes. Some systems require "training" the computer to recognize a limited vocabulary of standard words for each user. Operators train the system to recognize their voices by repeating each word to be added to the vocabulary several times. Other systems are speaker independent and allow a computer to understand a voice it has never heard. In this case, the computer must be able to recognize more than one pronunciation of the same word—for

An ergonomic keyboard is designed to be more comfortable to use.
(Source: Courtesy of Adesso)

FIGURE 3.13

A PC Equipped with a
Computer Camera

Digital video cameras make
it possible for people at
distant locations to conduct
videoconferences, thereby
eliminating the need for
expensive travel to attend
physical meetings.
(Source: Stone/Andreas Pollok
[Image #828420-015])

digital computer camera

input device used with a PC to
record and store images and video
in digital form

pixel

a dot of color on a photo image or
a point of light on a display screen

example, recognizing the use of the phrase "Please?" spoken by
someone from Cincinnati as meaning the same as "Huh?" spoken by someone from the Bronx, or "I beg your pardon?" spoken by an English person.

Digital Computer Cameras

Some personal computers work with **digital computer
cameras**, which record and store images and video in digital
form. These cameras look very similar to a regular camera.
When you take pictures, the images are electronically stored in
the camera. A cable is then used to connect the camera to the
parallel port on the computer and the images can be downloaded.
During the download process, the visual images are converted into digital codes by
a computer board. Once downloaded and converted into digital format, the images
can be modified as desired and included in other applications. For example, a
photo of the company office recorded by a digital computer camera can be captured and then pasted into a word processing document used in a company
brochure. You can even add sound and handwriting to the photo. Some digital
cameras, like the Sony Mavica, can store images on diskettes. The diskettes can
then be inserted into a computer to transfer the photos to a hard disk. Once
on the hard disk, the images can be edited, sent to another location, pasted
into another application, or printed. Some personal computers, as shown in
Figure 3.13, have a video camera that records full-motion video.

Although the first generation of digital cameras could create photos with a
resolution of 650×480 pixels, the current state-of-the-art cameras can deliver
more than 2 megapixel resolution, enough resolution to deliver snapshot-sized
photos that provide crisp, clear images. A **pixel** is a dot of color on a photo
image or a point of light on a display screen. The number one advantage of digital cameras is saving time and money by eliminating the need to process film.
Kodak is now allowing photographers to have it both ways. When Kodak print
film is developed, Kodak offers the option of placing pictures on a CD in addition to the traditional prints.[12] Once stored on the CD, the photos can be
edited, placed on an Internet site, or sent electronically to business associates
or friends around the world.

Terminals

Inexpensive and easy to use, terminals are input devices that perform data entry
and data input at the same time. A terminal is connected to a complete computer system, including a processor, memory, and secondary storage. General
commands, text, and other data are entered via a keyboard or mouse, converted
into machine-readable form, and transferred to the processing portion of the
computer system. Terminals, normally connected directly to the computer system by telephone lines or cables, can be placed in offices, in warehouses, and
on the factory floor.

Scanning Devices

Image and character data can be input using a scanning device. A page scanner
is like a copy machine. The page to be scanned is typically inserted into the
scanner or placed face down on the glass plate of the scanner, covered, and
scanned. With a handheld scanner, the scanning device is moved or rolled manually over the image to be scanned. Both page and handheld scanners can convert monochrome or color pictures, forms, text, and other images into
machine-readable digits. It has been estimated that U.S. enterprises generate
over one billion pieces of paper daily. To cut down on the high cost of using and
processing paper, many companies are looking to scanning devices to help them
manage their documents.

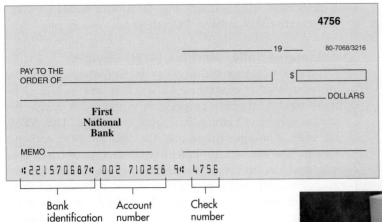

PAY TO THE
ORDER OF

First
National
Bank

MEMO

‖221570687‖ 002 710258 9‖ 4756

Bank
identification
number

Account
number

Check
number

FIGURE 3.14

MICR Device

Magnetic ink character recognition is a process by which data is coded on the bottom of a check or other form using special magnetic ink, which is readable by both computers and humans. For example, look at the bottom of a bank check or most utility bills.
(Source: Courtesy of NCR Corporation)

Optical Data Readers

A special scanning device called an optical data reader can also be used to scan documents. The two categories of optical data readers are for optical mark recognition (OMR) and optical character recognition (OCR). OMR readers are used for test scoring and other purposes when test takers use pencils to fill in boxes on OMR paper, which is also called a "mark sense form." OMR is used in standardized tests, including SAT and GMAT tests. In comparison, most OCR readers use reflected light to recognize various characters. With the use of special software, OCR readers can convert handwritten or typed documents into digital data. Once entered, this data can be shared, modified, and distributed over computer networks to hundreds or thousands of individuals.

Magnetic Ink Character Recognition (MICR) Devices

In the 1950s, the banking industry became swamped with paper in the form of checks, loan applications, bank statements, and so on. The result was the development of magnetic ink character recognition (MICR), a system for reading this data quickly. With MICR, data is placed on the bottom of a check or other form using a special magnetic ink. Data printed with this ink using a character set is readable by both people and computers (Figure 3.14).

Point-of-Sale (POS) Devices

point-of-sale (POS) device
terminal used in retail operations to enter sales information into the computer system

Point-of-sale (POS) devices are terminals used in retail operations to enter sales information into the computer system. The POS device then computes the total charges, including tax. Many POS devices also use other types of input and output devices, like keyboards, bar code readers, printers, and screens. A

Computer-readable bar codes on products provide retailers with data on the product, such as pricing, size, and color. The data is used to track sales and maintain adequate inventory levels.

(Source: Courtesy of PSC Inc.)

large portion of the money that businesses spend on computer technology involves POS devices.

Automated Teller Machine (ATM) Devices

Another type of special-purpose input/output device, the automated teller machine (ATM), is a terminal used by most bank customers to perform withdrawals and other transactions concerning their bank accounts. The ATM, however, is no longer used only for cash and bank receipts. Companies use various ATM devices to support their specific business processes. Some can dispense tickets for airlines, concerts, and soccer games. Some colleges use them to output transcripts. For this reason, the input and output capabilities of ATMs are quite varied. Like POS devices, ATMs may combine other types of input and output devices. Unisys, for example, has developed an ATM kiosk that allows bank customers to make cash withdrawals, pay bills, and also receive advice on investments and retirement planning.[13]

Pen Input Devices

By touching the screen with a pen input device, it is possible to activate a command or cause the computer to perform a task, enter handwritten notes, and draw objects and figures. Pen input requires special software and hardware. Handwriting recognition software can convert handwriting on the screen into text.

Light Pens

A light pen uses a light cell in the tip of a pen. The cell recognizes light from the screen and determines the location of the pen on the screen. Like pen input devices, light pens can be used to activate commands and place drawings on the screen.

Touch-Sensitive Screens

Advances in screen technology allow display screens to function as input as well as output devices. By touching certain parts of a touch-sensitive screen, you can execute a program or cause the computer to take an action. Touch-sensitive screens are popular input devices for some small computers because they preclude the necessity of keyboard input devices that consume space in storage or in use. They are frequently used at gas stations for customers to select grades of gas and request a receipt, at fast-food restaurants for order clerks to enter customer choices, at information centers in hotels to allow guests to request facts about local eating and drinking establishments, and at amusement parks to provide directions to patrons. They also are used in kiosks at airports and department stores.

A light pen is a type of input device that can be used to activate commands and place drawings on the screen.

(Source: Courtesy of MicroSpeed)

Bar Code Scanners

A bar code scanner employs a laser scanner to read a bar-coded label. This form of input is used widely in grocery store checkouts and in warehouse inventory control.

Output Devices

Computer systems provide output to decision makers at all levels of an organization to solve a business problem or capitalize on a competitive opportunity. In addition, output from one computer system can be used as input into another computer system within the same information system. The desired form of this output might be visual, audio, and even digital. Whatever the output's content or form, output devices function to provide the right information to the right person in the right format at the right time.

Display Monitors

The display monitor is a TV-screen-like device on which output from the computer is displayed. Because the monitor uses a cathode ray tube to display images, it is sometimes called a CRT. The monitor works in much the same way as a TV screen—one or more electron beams are generated from cathode ray tubes. As the beams strike a phosphorescent compound (phosphor) coated on the inside of the screen, a dot on the screen called a pixel lights up. The electron beam sweeps back and forth across the screen so that as the phosphor starts to fade, it is struck again and lights up again.

Many decision makers work directly off the monitor, or screen, of their computer systems. Even this fairly standard output device can provide organizational advantages. An upgrade from a 14-inch monitor to a 19-inch monitor, for example, can provide a productivity increase for managers working with certain applications. The larger display allows for easier, more precise reading of detailed reports. With today's wide selection of monitors, price and overall quality can vary tremendously.

Progress has been made in enabling one display device to display both TV and computer output. PC Theatre from Compaq is a consumer living room entertainment device that merges computing and traditional forms of media and entertainment content. This system combines the best features of a TV with a 36-inch monitor and multimedia capabilities. The consumer can watch TV, use the computer, or do both at the same time. The Gateway Destination Big Screen TV comes with a 31-inch monitor, which can be used with the personal computer or to watch TV.

The quality of a screen is often measured by the number of horizontal and vertical pixels used to create it. As mentioned previously, a pixel is a dot of color on a photo image or a point of light on a display screen. It can be in one of two modes: on or off. A larger number of pixels per square inch means a higher resolution, or clarity and sharpness of the image. For example, a screen with a $1,024 \times 768$ resolution (786,432 pixels) has a higher sharpness than one with a resolution of 640×350 (224,000 pixels). The distance between one pixel on the screen and the next nearest pixel is known as dot pitch. The common range of dot pitch is from .25 mm to .31 mm. The smaller the number, the better the picture. A dot pitch of .28 mm or smaller is considered good. Greater pixel densities and smaller dot pitches yield sharper images of higher resolution.

Display monitors can be either monochrome or color. Typically, characters in monochrome display screens appear in one of three colors: gray, green, or amber. Color monitors, also called RGB (red, green, and blue) monitors, have the ability to display the basic colors in a variety of shades. A monitor's ability to display color is a function of the quality of the monitor, the amount of RAM in the computer system, and the monitor's graphics adapter card. The color graphics adapter (CGA) was one of the first technologies to display color images on the screen. Today, super video graphics array (SVGA) displays are standard, providing vivid colors and superior resolution.

Liquid Crystal Displays (LCDs)

Because CRT monitors use an electron gun, there must be a distance of one foot between the gun and screen, causing them to be large and bulky. Thus, a different technology, flat-panel display, is used for portable personal computers and laptops. One common technology used for flat screen displays is the same liquid crystal display (LCD) technology used for pocket calculators and digital watches.

CRT monitors are large and bulky in comparison with LCD monitors (flat displays).
(Source: Courtesy of ViewSonic)

plotter

a type of hard-copy output device used for general design work

FIGURE 3.15

Laser Printers

Laser printers, available in a wide variety of speeds and price ranges, have many features, including color capabilities. They are the most common solution for outputting hard copies of information.
(Source: Courtesy of Epson America, Inc.)

LCD monitors are flat displays that use liquid crystals—organic, oil-like material placed between two polarizers—to form characters and graphic images on a backlit screen.

The primary choices in LCD screens are passive-matrix and active-matrix LCD displays. In a passive-matrix display, the CPU sends its signals to transistors around the borders of the screen, which control all the pixels in a given row or column. In an active-matrix display, each pixel is controlled by its own transistor attached in a thin film to the glass behind the pixel. Passive-matrix displays are typically dimmer, slower, but less expensive than active-matrix ones. Active-matrix displays are bright, clear, and have wider viewing angles than passive-matrix displays. Active-matrix displays, however, are more expensive and can increase the weight of the screen.

LCD technology is also being used to create thin and extremely high-resolution monitors for desktop computers. Although the screen may measure just 13 inches from corner to corner, the display's extremely high resolution—1,280 × 1,280 pixels—lets it show as much information as a conventional 20-inch monitor. And while cramming more into a smaller area causes text and images to shrink, you can comfortably sit much closer to an LCD screen than the conventional CRT monitor. Unfortunately, these monitors are expensive: a 14-inch flat-panel monitor costs around $650, while a top-of-the-line 21-inch CRT monitor costs around $700. However, prices are expected to continue dropping.

Printers and Plotters

One of the most useful and popular forms of output is called hard copy, which is simply paper output from a device called a printer. Printers with different speeds, features, and capabilities are available. Some can be set up to accommodate different paper forms such as blank check forms, invoice forms, and so forth. Newer printers allow businesses to create customized printed output for each customer from standard paper and data input using full color.

The speed of the printer is typically measured by the number of pages printed per minute (ppm). Like a display screen, the quality, or resolution, of a printer's output depends on the number of dots printed per inch. A 600-dpi (dots-per-inch) printer prints more clearly than a 300-dpi printer. A recurring cost of using a printer is the ink-jet or laser cartridge that must be replaced every few thousand pages of output. Figure 3.15 shows a laser printer and an example of its output.

Plotters are a type of hard-copy output device used for general design work. Businesses typically use these devices to generate paper or acetate blueprints, schematics, and drawings of buildings or new products onto paper or transparencies. Standard plot widths are 24 inches and 36 inches, and the length can be whatever meets the need—from a few inches to feet.

Computer Output Microfilm (COM) Devices

Companies that produce and store significant numbers of paper documents often use computer output microfilm (COM)

devices to place data from the computer directly onto microfilm for future use. The traditional photographic phase of conversion to microfilm is eliminated. Once this is done, a standard microfilm reader can access the data. Newspapers and journals typically place their past publications on microfilm using COM, giving readers the ability to view past articles and news items.

Music Devices

Music devices, such as Diamond Multimedia's Rio 500 portable MP3 player, are about the size of a cigarette pack and can be used to download music from the Internet and other sources.[14] These devices have no moving parts and can store hours of music. When you get tired of the music, you can always download new pieces. Music devices that use the MP3 standard can cost under $300. MP3 is an abbreviation for Motion Picture Experts Group Audio Layer 3, a popular music format for the Internet. This format requires about 1 MB of storage for every minute of music, which is about a tenth of the storage required for music on a standard CD. The MP3 standard allows for the music to be compressed, so it takes less time to download from the Internet. A piece of music in this format also requires less storage space on the MP3 player.

A number of computer manufacturers—including Dell, Hewlett-Packard, NEC, and others—offer computers that make downloading and playing music from the Internet in the MP3 format easier with higher sound quality. These specialized computers offer easy downloading and superior speakers for playing music. There are a number of Internet sites that allow people to share music using the MP3 format. Because musicians and production companies don't receive royalties from people sharing music in the MP3 format, there are also a number of legal and ethical issues that must be addressed.

Special-Purpose Input and Output Devices

Many additional input and output devices are used for specialized or unique applications. A **multifunction device** can combine a printer, fax machine, scanner, and copy machine into one device. Multifunction devices are less expensive than buying these devices separately, and they take less space on a desktop compared with separate devices. For example, Canon, Xerox, Hewlett-Packard and others make inkjet printers that can be converted to a color scanner for documents, photographs, and other images. Special-purpose hearing devices can be used to detect manufacturing or equipment problems. The Georgia Institute of Technology has developed a hardware device that can "listen to" equipment to detect worn or damaged parts. Voice-output devices, also called voice-response devices, allow the computer to send voice output in the form of synthesized speech over phone lines and other media. Some banks and financial institutions use voice recognition and response to give customers account information over the phone. The Smart Disk allows you to place a smart card, which is similar to a credit card in size and function, into the Smart Disk device. The Smart Disk is then inserted into a standard diskette drive on a PC, which allows you to complete a variety of financial transactions using your smart card.

In addition to multifunction machines, there are a number of remarkable special-purpose devices being developed. One person now uses special-purpose devices to help him see.[15] The person, who was blind for over 25 years, can now see spotty images. The special-purpose devices include special sunglasses that receive images, a small computer that processes the images from his glasses, and a plastic card the size of a postage stamp containing 64 electrodes that has been implanted into the visual cortex of his brain. Dr. William Heetderks, head of the National Institutes of Health, is optimistic that this type of special-purpose technology "will be able to provide significant function to blind people."

music device

a device that can be used to download music from the Internet and play the music

multifunction device

a device that can combine a printer, fax machine, scanner, and copy machine into one device

The Rocket eBook, a special-purpose display unit, allows people to read books and other documents anywhere. The device weighs about 22 ounces and has a paperback size of 5 inches wide by 7 inches high by 1 inch deep.[16] Books, reports, and other documents can be downloaded into the eBook and read at home, while traveling, or at any location. Once the book or document is read, another one can be downloaded into the device. Some bookstores, including Barnes and Noble, offer electronic books that can be downloaded into electronic book devices. Electronic book devices make turning the page easier. Many allow you to quickly move up or back in the document and to search for key terms or themes. In addition, new technology, such as ClearType from Microsoft, makes text displayed on e Book screens easier to read and closer to the quality of a printed page. Many authors, including Stephen King, have experimented with releasing their books over the Internet to be read on a computer or using an electronic book.

Xerox is experimenting with a form of digital paper.[17] The electronic paper, made of silicone rubber, is about as thin and flexible as poster board. Images can be sent from a computer to the electronic paper and displayed using tiny embedded plastic balls. The balls are like pixels in a normal display screen. Once sent to the electronic paper, the image can be maintained without power for months. The big advantage of the technology is the ability to reuse the electronic paper to display new information over time.

COMPUTER SYSTEM TYPES, STANDARDS, SELECTING, AND UPGRADING

special-purpose computers

computers used for limited applications by military and scientific research groups

In general, computers can be classified as either special purpose or general purpose. **Special-purpose computers** are used for limited applications by military and scientific research groups such as the CIA and NASA. Other applications include specialized processors found in appliances, cars, and other products. Special-purpose computers are increasingly being used by businesses. For example, automobile repair shops connect special-purpose computers to your car's engine to identify specific performance problems.

general-purpose computers

computers used for a wide variety of applications

General-purpose computers are used for a variety of applications and are the most common. The computers used to perform business applications discussed in this book are general-purpose computer systems. General-purpose computer systems combine processors, memory, secondary storage, input and output devices, a basic set of software, and other components. These systems can range from inexpensive personal computers to expensive supercomputers. These systems display a wide range of capabilities. Table 3.2 shows general ranges of capabilities for various types of computer systems.

Computer System Types

Computer systems can range from desktop (or smaller) portable computers to massive supercomputers that require housing in large rooms. Let's examine the types of computer systems in more detail.

Personal Computers

personal computer (PC)

relatively small, inexpensive computer system, sometimes called a microcomputer

As previously noted, **personal computers (PCs)** are relatively small, inexpensive computer systems, sometimes called microcomputers. Although personal computers are designed primarily for individual users, they are often tied into larger computer and information systems as well. Personal computers can be purchased from retail stores or on-line. Read the "E-Commerce" box to see how a very successful PC manufacturer sells PCs over the Internet.

Characteristic	Network Computer	Personal Computer	Workstation	Midrange Computer	Mainframe Computer	Super-computer
Processor Speed	1–5 MIPs	5–20 MIPs	50–100 MIPs	25–100 MIPs	40–4,550 MIPs	60 billion–3 trillion instructions per second
Amount of RAM	4–16 MB	16–128 MB	32–256 MB	32–512 MB	256–1,024 MB	8,192 MB+
Approximate Cost	$500–$1,500	$1,000–$5,000	$4,000 to over $20,000	$20,000 to over $100,000	$250,000 to over $2 million	$2.5 million to over $10 million
How Used	Supports "heads-down" data entry; connects to the Internet	Improves individual worker's productivity	Engineering; CAD; software development	Meets computing needs for a department or small company	Meets computing needs for a company	Scientific applications; marketing; customer support; product development
Example	Oracle Network computer	Dell Pentium computer	Sun Microsystems computer	Hewlett-Packard HP-9000	IBM ES/9000	Cray C90

TABLE 3.2

Types of Computer Systems
(Source: Photos courtesy of Wyse Technology, IBM Corporation, Los Alamos National Laboratory)

There are several types of personal computers. Named for their size (small enough to fit on an office desk), *desktop computers* are the most common personal computer system configuration. Increasingly, powerful desktop computers can provide sufficient memory and storage for most business computing tasks. Desktop PCs have become standard business tools; more than 30 million are in use in large corporations.

In addition to traditional PCs that use Intel processors and Microsoft software, there are other options. One of the most popular is the iMac by Apple Computer.[18] The iMac computer has excellent technology and is attractively priced and easy to use. The computer has excellent video and graphics capabilities, including the ability to create and edit home movies. The Power Mac G4, another Apple computer, has models that operate at speeds of 400, 450, and 500 MHz with a starting price of about $1,600.

Various smaller personal computers can be used for a variety of purposes. A *laptop computer* is a small, lightweight PC about the size of a briefcase. Apple, for example, has the iBook computer, which is a laptop system compatible with the iMac. The iBook has a price of about $1,500 with a 300-MHz chip and a 12-inch color monitor. The two-tone iBook comes in blueberry and white or tangerine and white. Many Apple users are excited that the original founder of Apple, Steve Jobs, is once again back at the helm of the company and directing its products and strategy.

E-COMMERCE

A Young Company Excels with E-Commerce

The company was started in a garage by a teenager. Today, the company has a market capitalization of $97 billion, which is almost 40 percent more than the total value of General Motors. The founder, who is still under 40 years old, is a billionaire. Companies with executives decades older are constantly seeking his advice on how best to get into e-commerce business. Andy Grove, the CEO of chip maker Intel, once said about this young entrepreneur, "He has done what a lot of people are aspiring to do."

The company is Dell, and the founder is Michael Dell. The company today has on-line sales that exceed $14 million daily. It is also one of the fastest-growing companies, averaging 40 percent growth in the last several years. Even more impressive, that increase has come largely from internal growth and not external acquisitions. This spectacular growth has made Dell one of the most successful companies in the hardware business and its founder one of the wealthiest. It is also why companies of all sizes and types are seeking Dell's advice. The CEO of Ford Motor Company, Jacques Nasser, sought Michael Dell's advice on how Ford could move into the Internet age with an effective e-commerce site. Other companies, including AlliedSignal, are also seeking time with the young founder of Dell. According to Louis Gerstner, CEO of IBM, "He's on the cutting edge of CEO-ship."

The demands on Dell's time are staggering. A few years ago, he had about 100 requests to speak per year.

The number is now over 1,500 requests every year. If he honored all these requests, he would be making more than four speeches every day. Yet, Michael Dell tries to keep his customers satisfied. "I get far more requests than I can possibly accommodate. If a very large customer of ours calls, I'm pretty likely to show up." When he does speak, Dell usually discusses his e-commerce business model and not a sales pitch for Dell computers. With his success, most senior executives pay close attention to what he says. Perhaps this is why Dell is also a finalist for the Chief Executive of the Year for Executive Focus International.

Discussion Questions

1. What was Dell's strategy for growth?
2. Visit the Dell Web site. What are your impressions?

Critical Thinking Questions

3. Could another young entrepreneur trying to start a PC company have success?
4. Describe what approach you would use if you were going to start a new PC manufacturing and sales company.

Sources: Adapted from Del Jones, "E-Commerce's Guru of Choice," *The Wall Street Journal*, April 15, 1999, p. 3B; and Bob Wallace, "Ford Turns to Dell," *Computerworld*, February 22, 1999, p. 6.

Newer PCs include the even smaller and lighter *notebook* and *subnotebook* computers that provide similar computing power. Some notebook and subnotebook computers fit into docking stations of desktop computers to provide additional storage and processing capabilities. Small PCs continue to rise in popularity because of their portability and performance. In the past, desktop computers typically ran at faster speeds than notebook computers. This was due primarily to the limitations of running a notebook computer from a battery. Today's chips, such as Intel's SpeedStep, can reduce the performance difference between notebook computers and desktop systems.[19] The SpeedStep allows a laptop to run at speeds that approach those of a desktop when using power from a wall outlet. When running on a battery, the SpeedStep runs at slower speeds that are typical of notebook computers.

Handheld (palmtop) computers are PCs that provide increased portability because of their smaller size—some are as small as a credit card. These systems often include a wide variety of software and communications capabilities. Boeing, for example, uses handheld computers to help improve quality inspections of aircraft and aircraft parts.[20] Boeing's plant in St. Louis manufactures F-14 and F-15 fighter planes. Using Palm IIIX computing devices, Boeing inspectors have been able to increase inspection quality while cutting inspection times in half. Boeing saves even more time when the data is uploaded into computer systems. The old approach used clipboards to manually record inspection data, which was then typed into the computer to produce reports.

Handheld (palm) computers include a wide variety of software and communication abilities, including the ability to connect to the Internet.
(Source: 3Com and the 3Com logo are registered trademarks. Palm IIIc™ and the Palm III™ logo are trademarks of Palm Computing, Inc., 3Com Corporation or its subsidiaries.)

network computer

a cheaper-to-buy and cheaper-to-run version of the personal computer that is used primarily for accessing networks and the Internet

workstation

computer that fits between high-end personal computers and low-end midrange computers in terms of cost and processing power

Many PC manufacturers, PC sellers, and Internet companies are aggressively marketing their systems, as prices continue to decline. Gateway and others, for example, are offering PCs, Internet connection, and other services for a fixed monthly fee. After a set number of years, the system can be traded in for a new one. Compuserve, an Internet company, offers a $400 rebate on a PC that is purchased with a three-year contract for Compuserve service, which costs about $20 per month.[21] Still other companies are offering free PCs to users who are willing to use an Internet connection with advertising for other products and services. These "free PCs" are usually offered by small start-up companies and are normally not built by one of the main PC makers. In most cases, these inexpensive or free PCs come with a number of restrictions and requirements.

Embedded computers are computers placed inside other products to add features and capabilities. In the case of automobiles, embedded computers can help with navigation, engine performance, braking, and other functions. Household appliances, stereos, and some phone systems also use embedded computers. Embedded computers and videos will be used in a "smart road" pilot project in Atlanta to send messages to oncoming vehicles to choose alternate routes if needed. In the Eisenhower Tunnel at the top of the Continental Divide in Colorado, embedded computers and video systems will monitor truck traffic and notify trucks to slow down if they are going too fast on the descent to Denver.

The **network computer** is a cheaper-to-buy and cheaper-to-run version of the personal computer that is used primarily for accessing networks and the Internet. These stripped-down versions of personal computers do not have the storage capacity or power of typical desktop computers, nor do they need it for the role they play. Unlike personal computers, network computers—or thin clients—download software from a network when needed. This can make it much easier and less expensive to manage the support, distribution, and updating of software applications. The initial target user is someone who performs what is called heads-down data entry—customer inquiry, phone order taking, and classic data entry. The network computer is designed to have no moving parts to avoid expensive equipment repairs. IBM, Oracle, and Sun Microsystems were the first companies to develop prototypes of such systems, with a purchase price in the $500 to $1,500 range.

Advocates of network computers argue that they not only cost less to purchase compared with a standard desktop PC but also cost less to operate. Thus, it is not so much the low cost of purchasing a network computer that makes it so attractive but its low maintenance cost. However, the network computer's flexibility is extremely limited when compared with the personal computer. In addition, PC companies have responded strongly to network computers with lower prices and more competitive products.

Workstations are computers that fit between high-end personal computers and low-end midrange computers in terms of cost and processing power. Workstation manufacturers use the RISC rather than the CISC computer chip to provide high computing power and reliability. They cost from $3,000 to $40,000. Workstations are small enough to fit on an individual's desktop. A workstation may be dedicated to support a single user or a small group of users. High-end personal computers are approaching the computing power of a workstation.

Workstations are used to support engineering and technical users who perform heavy mathematical computing, computer-aided design (CAD), and other applications requiring a high-end processor. Such users need very powerful CPUs, large amounts of main memory, and extremely high-resolution graphic displays to meet their needs. Engineers use CAD programs to create two- and three-dimensional engineering drawings and product designs. Although initially creating a design with CAD software may take as long as, if not longer than, creating a design in the traditional manner, CAD is much faster when it comes time to revise. Instead of redrawing an entire plan, CAD allows the engineers to modify it with a few clicks of a mouse. They can also easily pan the design or zoom in to magnify a particular part. They can also rotate the view to examine it from different perspectives.

Many companies, including Microsoft and Intel, are developing inexpensive **Web appliances**.[22] A Web appliance is a device that can connect to the Internet, typically through a phone line. It can be used to check stock prices, check e-mail messages, search the Internet for information, and more. Web appliances come in a variety of configurations. Some have a keyboard, a passive-matrix display, a 200-MHz processor, and Web and e-mail software. These devices can cost less than $200 to purchase and about $20 per month for an Internet connection.[23] Some Web appliances have the appearance of a cellular phone, with the capabilities of a standard phone and basic Internet connections. Other Web appliances are being attached to everyday products, such as TVs, stoves, and refrigerators. Once attached, the Web appliance will be able to get movie schedules, alert people when their stove may require maintenance, or advertise grocery specials. A number of companies are now making Web appliances, including Netpliance, Compaq, Dell, Microsoft, and others. In the future, Web appliances may be built into many of the products we use every day.

Midrange Computers

Midrange computers (formerly called *minicomputers*) are systems about the size of a small three-drawer file cabinet that can accommodate several users at one time. These systems often have secondary storage devices with more capacity than workstation computers and can support a variety of transaction processing activities, including payroll, inventory control, and invoicing. Midrange computers often have excellent processing and decision-support capabilities. Many small to medium-size organizations—like manufacturers, real estate companies, and retail operations—use midrange computers.

Mainframe Computers

Mainframe computers are large, powerful computers often shared by hundreds of concurrent users connected to the machine via terminals. The mainframe computer must reside in an environment-controlled computer room or data center with special heating, venting, and air-conditioning (HVAC) equipment to control the temperature, humidity, and dust levels around the computer. In addition, most mainframes are kept in a secured data center with limited access to the room through some kind of security system. The construction and maintenance of such a controlled access room with HVAC can add hundreds of thousands of dollars to the cost of owning and operating a mainframe computer. Mainframe computers also require specially trained individuals (called system engineers and system programmers) to care for them. Mainframe computers can crunch numbers at a rate of over 300 million instructions per second and start at $200,000 for a fully configured system.

The traditional role of the mainframe computer was as the large, centrally located computer of a firm. Mainframes have been the cornerstone of computing in large corporations for many years. From the early 1950s until the mid-1970s, virtually all commercial computer processing was mainframe based. Mainframe computers were acquired by many companies to automate accounting and finance processes, such as payroll, general ledger, accounts receivable, and accounts payable. Order processing, billing, and inventory control were other early computer applications.

Web appliance

a device that can connect to the Internet, typically through a phone line

midrange computer

formerly called minicomputer, a system about the size of a small three-drawer file cabinet that can accommodate several users at one time

mainframe computer

large, powerful computer often shared by hundreds of concurrent users connected to the machine via terminals

Today the role of the mainframe is undergoing some remarkable changes as lower-cost midrange computers, workstations, and personal computers become increasingly powerful. Many computer jobs that used to run on mainframe computers have migrated onto these smaller, less expensive computers. This information processing migration is called *computer downsizing*. The new role of the mainframe is a large information processing and data storage utility for a corporation—running jobs too large for other computers, storing files and databases too large to be stored elsewhere, and storing backups of files and databases created elsewhere (these large stores of data are sometimes called *data warehouses*). The mainframe is capable of handling the millions of daily transactions associated with airline, automobile, and hotel/motel reservation systems. It can process the tens of thousands of daily queries necessary to provide data to decision support systems. Its massive storage and input/output capabilities enable it to play the role of a video computer, providing full-motion video to users.

complementary metal oxide semiconductor (CMOS)

a semiconductor fabrication technology that uses special semiconductor material to achieve low power dissipation

Complementary metal oxide semiconductor (CMOS) is a semiconductor fabrication technology that uses special semiconductor material to achieve low power dissipation. IBM, Amdahl, and Data General have all made a substantial commitment to this new technology and have announced CMOS-based mainframe products. Over time, mainframes have been evolving into smaller, faster, less expensive systems as a result of CMOS processors and provide support for large packaged software products, Web technologies, and communications protocols, much like their smaller cousins, the midrange computers.

Table 3.3 lists the major mainframe computer manufacturers. IBM remains a dominant force in the mainframe market.

Supercomputers

supercomputers

the most powerful computer systems, with the fastest processing speeds

Supercomputers are the most powerful computer systems, with the fastest processing speeds. Military and research organizations trying to solve complex problems use these very expensive machines. They are also used by universities and large corporations involved with research or high-technology businesses. Some large oil companies, for example, use supercomputers to perform sophisticated analysis of detailed data to help them explore for oil.

Originally, supercomputers were used primarily by government agencies to perform the high-speed number crunching needed in weather forecasting and military applications. With recent improvements in the cost and performance (lower cost and faster speeds) of these machines, they are being used more broadly for commercial purposes. For example, supercomputers are used to perform the enormous number of calculations required to draw and animate Disney cartoons. To produce those special effects required handling a gigabyte (one billion characters) of data for every second of film time. Obviously, this required a very powerful computer. The S/390 G6 computer uses copper CPU technology to reach speeds of 1,600 MIPS, which is a 50 percent increase in speed over the G5 machine.[24] The system can run up to 16 processors. The copper processors are faster and require less electrical power compared with the traditional chip used in the G5 and earlier mainframe computers. The new

TABLE 3.3

Mainframe Computer Manufacturers

Company	Product	MIPS (Millions of Instructions per Second)
IBM	Systems/390 Parallel	178–5,000
Unisys	A Series	1–250
Amdahl	Millennium	122–350
Hitachi Data Systems	Pilot	40–350

Blue Mountain is a supercomputer located at the U.S. Department of energy's Los Alamos National Laboratory. At the heart of Blue Mountain are 48 commercially available Silicon Graphics® Cray® Origin2000 servers containing a total of 6,144 processors. Blue Mountain runs at 1.6 trillion operations per second (teraOps) and has one of the most advanced graphics systems in the world. With this visualization system, answers to complex scientific problems that would have taken weeks or more to display can now be displayed in minutes.
(Source: Courtesy of Los Alamos National Laboratory)

supercomputer is a successor to IBM's "Deep Blue" supercomputer that was used to beat chess champion Garry Kasparov in 1997.[25]

Scientists often use computer models to simulate the problems they are studying. These models typically omit certain details of the situation because they are not well understood or the effort required to include the details simply makes the model too difficult to create. Almost always, some degree of model completeness must be sacrificed to make the problem solvable in a reasonable amount of time. Lack of completeness leads to some loss of accuracy. For example, meteorologists studying how the oceans influence weather patterns know that ocean currents have a larger impact than the effect of sunlight reflected by the clouds. As a result, they may simplify their model to ignore the effect of the reflected sunlight to squeeze in more data on ocean currents. The degree of approximation is called the *granularity* of the computer model.

Although all of the above computer system types can be used for general processing tasks, they can also be used to serve a specific and unique purpose, such as supporting Internet and network applications. A *computer server* is a computer designed for a specific task, such as network or Internet applications. Servers typically have large memory and storage capacities, along with fast and efficient communications abilities. They can range in size from a PC to a mainframe system, depending on the needs of the organization. A Web server is used to handle Internet traffic and communications. An Internet caching server stores Web sites that are frequently used by a company. A file server, discussed in more detail in Chapter 6, stores and coordinates program and data files. A transaction server is used to store and process business transactions. As with general computers, there are benchmarks to help a company determine the performance of a server, such as ZD ServerBench, WebBench, and NetBench.

Multimedia Computers

Multimedia merges sound, animation, and digitized video. The technology to bring these media to the desktop has existed for a long time—the early Apple Macintosh had impressive capabilities. The current technology emphasizes delivery of multimedia applications that are rich in content and features while taking up less than 600 MB of storage space. This is possible because of video, image, and audio compression technology, which plays a key role in the development and success of the multimedia industry. A popular delivery medium for multimedia is the CD-ROM because of its low production cost and large storage capacity. Figure 3.16 shows the typical components of a multimedia computer.

Multimedia is being used in a wide variety of ways. For one, it is used to enhance presentations. Instead of a series of text and graphics slides, a multimedia business presentation includes video, sound, and animation. The possibilities for multimedia are restricted only by our imagination. In fact, many industry leaders believe that multimedia systems may become as widespread as the television and VCR are today.

Audio

Audio involves converting sound to a digital recording for storage on a magnetic disk or CD-ROM and then converting the digital recording to sound when the multimedia program is executed. On the input, or sensing, side of the system are audio input devices that record or play analog sound and then translate it for digital storage and processing. Audio devices include CD-audio (just

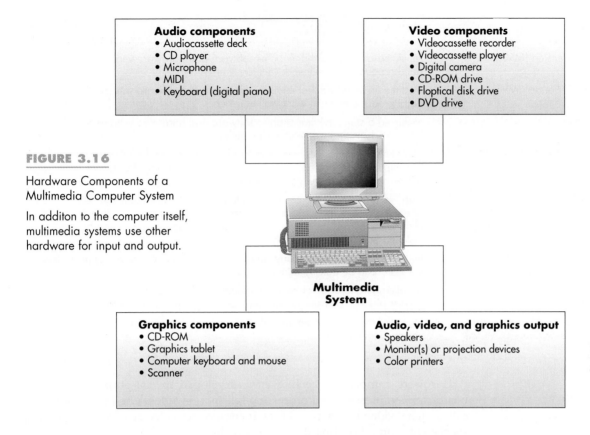

Audio components
- Audiocassette deck
- CD player
- Microphone
- MIDI
- Keyboard (digital piano)

Video components
- Videocassette recorder
- Videocassette player
- Digital camera
- CD-ROM drive
- Floptical disk drive
- DVD drive

Multimedia System

Graphics components
- CD-ROM
- Graphics tablet
- Computer keyboard and mouse
- Scanner

Audio, video, and graphics output
- Speakers
- Monitor(s) or projection devices
- Color printers

FIGURE 3.16

Hardware Components of a Multimedia Computer System

In additon to the computer itself, multimedia systems use other hardware for input and output.

like your home CD players) and cassette players. The signal is sent from the audiocassette to a special audio board in the computer that digitizes the sound and stores it for further processing. Alternatively, you can use an audio board to create music in digital form, store it, and play it back using a special board in your computer.

digital signal processor (DSP) chips

chips that improve the analog-to-digital-to-analog conversion process

Advanced sound systems use **digital signal processor (DSP) chips** to improve the analog-to-digital-to-analog conversion process. DSP chips take the signal conversion job from the personal computer's CPU chip, thereby improving the entire sound process. The resulting sound may be equivalent to that produced by audio CDs and amplifiers.

Video

Video brings multimedia alive and makes it a complete education, presentation, and entertainment system. Video is also the most difficult element to display because a single uncompressed frame requires almost 1 MB of storage. At this rate, each second of a 30-frame-per-second full-screen video requires 27 MB, yielding less than 24 seconds of video per CD-ROM disk. This limitation can be avoided by reducing the number of bits required to represent a single video frame by using mathematical formulas (called **video compression**), reducing the size of the video screen, reducing the rate at which the video is displayed, or a combination of all three methods. A common multimedia application displays compressed video in a quarter-screen window at 15 frames per second. Another approach is to add a video board to the personal computer, similar to a sound board for audio applications. This approach is more expensive but yields higher-quality video.

video compression

a process that reduces the number of bits required to represent a single video frame by using mathematical formulas

In the world of entertainment, movies can be converted to a multimedia digital format that allows viewers to control the order in which sections of the movie are watched. These multimedia products often include interviews, scripts, and a capability to search the movie for particular scenes, actors, and songs.

Standard	How Used
MultiMedia Extension (MMX)	Multimedia standard that enables software and hardware vendors to build products that will work well together
Multimedia PC Council (MPC)	Multimedia standard that enables software and hardware vendors to build products that will work well together
Ultimedia Solution	Multimedia standard that enables software and hardware vendors to build products that will work well together
Musical Instrument Digital Interface (MIDI)	Standard system for connecting musical instruments and synthesizers to computers; defines codes for musical events, including the start of a note and its pitch, length, volume, and other attributes
Plug 'n Play (PnP)	Hardware and software components that card, personal computer, and operating system manufacturers incorporate into their products to eliminate the need for manual configuration so that hardware can be installed and used immediately
Small Computer System Interface (SCSI)	Interface that ensures that any storage, input, or output device that meets this standard can be quickly added to a system
Fiber channel	An alternative to SCSI for connecting devices to a computer; allows a greater distance between devices and offers faster performance than SCSI
Personal Computer Memory Card International Association (PCMCIA)	Standard that ensures compatibility between PC memory and communications cards

TABLE 3.4

Industry Standards in Common Use

Hardware companies like Intel have developed special-purpose chips and boards for compressing and processing full-motion video. The Intel standard is called digital video interactive (DVI). It compresses video at a 150:1 ratio using special chips on a DVI board enabling one hour of video to be stored in about 720 MB. Without compression, over 110 GB would be required. To play back DVI video, the board decompresses the frames and plays them back through the color graphics card for display on your screen.

Standards

With the variety of multimedia products now available, the importance of hardware standards cannot be overemphasized. They diminish the cost of integration, help a developer determine which devices will be compatible with the rest of the system, provide increased options, and make upgrading a system less complex. Several common standards are summarized in Table 3.4. Note that in some cases there are competing standards.

In addition to industry standards, many large corporations also set their own internal standards by selecting specific computer configurations from a small set of manufacturers. The goal is to reduce hardware support costs and increase the organization's flexibility. Business units within an organization that adopt different hardware complicate future corporate information system projects. For example, the installation of software is made much easier if similar equipment from the same manufacturer is involved, rather than new and different systems at each installation site. Standards for computer system use can also be set. For example, some companies prohibit employees from using computers at work for nonbusiness activities. To enforce these usage standards, some companies monitor employee activities. Read the "Ethical and Societal Issues" box.

Selecting and Upgrading Computer Systems

computer system architecture

the structure, or configuration, of the hardware components of a computer system

The structure, or configuration, of the hardware components of a computer system is called the **computer system architecture**. This architecture can include a mixture of components, including processing, memory, storage,

ETHICAL AND SOCIETAL ISSUES
Employee Monitoring

Companies are supplying their managers and employees with PCs in record numbers. But how are these PCs being used? In the case of Ford Motor Company, the company is not monitoring how employees are using the PCs that the company has made available for employees' homes. In other cases, however, companies are increasingly monitoring how their employees are using company-purchased equipment. In some cases, employees are using corporate PCs at the office to research other companies and invest in stock. In other cases, employees are wasting time playing games, such as solitaire. Some employees have even visited pornographic Internet sites while at work. This was the case with a civil engineering company.

The information-technology manager for the engineering company noticed a high amount of Internet usage. The second-busiest Internet site company employees visited was not a common business site. The information-technology manager investigated further and found the site with the greeting "Surf in Style: The Sex Tracker." Realizing that one or more employees were spending a considerable amount of time at a pornographic site troubled the technology manager. And he disliked having to be a computer snoop, investigating employees' e-mails and Internet usage. But the technology manager often tells employees, "You live in a democracy, you don't work in one." The technology manager soon got so tired of seeing employees visiting pornographic, gambling, and other nonbusiness sites that he started blocking these sites from employee PCs. He also implemented monitoring software that could generate reports on which employees were the most frequent users of Internet sites. As a result, he found that one month a popular stock trading site had over 3 percent of the total Internet activity for the entire firm. He then referred to reports to see which employee was involved. This led to a detailed investigation of all the stock trading activities the employee performed.

With today's technology, it is becoming easier to snoop on employees. Specialized software is available that can monitor what employees are doing with their PCs. These programs can discover whether an employee is using his or her PC to hunt for a new job, buy stock, gamble on-line, or visit pornographic sites. Some believe that the surveillance power available today is staggering. Using these programs, it is possible to quickly pull up e-mails sent by a company employee, even if the e-mail is stored on the employee's own hard disk. Some employees believe that if they delete files on their PCs they are safe, but this is not the case. Using monitoring software, companies can review all the files that employees have deleted from their PCs. A company that has limited personal Internet usage can monitor virtually any activity performed on corporate PCs. Using monitoring software, an information technology manager can see that one employee is making retirement calculations on his PC, another employee is sending jokes via e-mail to friends and family, and another is using the Internet to search for other job opportunities.

While most information systems managers agree that monitoring employees' PCs is legal and needed for corporate security, there can be a fine line between corporate security and employee privacy.

Discussion Questions

1. Despite the legalities, do you think it is right for companies to monitor their employees' PC use?
2. What types of employee PC activities should a company monitor?

Critical Thinking Questions

3. If you were the information technology manager for this firm, what would you monitor and how would you respond to inappropriate use of corporate PCs?
4. Should there be different monitoring systems for employees and senior level managers?

Sources: Adapted from Michael McCarthy, "Web Surfers Beware," *The Wall Street Journal*, January 10, 2000, p. Q1; and John Morris, "Protect Your PC," *PC Magazine*, September 1, 1999, p. 107.

input, and output devices. As discussed in Chapter 2, organizations are adaptive systems that must respond to changes in their environment. A computer system that was once effective may need to be enhanced or upgraded to support new business activities and a changing environment. The ability to upgrade a system can be an important factor in selecting the best computer hardware.

Computer systems can be upgraded by installing additional memory, additional processors (such as a math coprocessor), more hard disk storage, a memory card, or other devices. When upgrading or expanding an existing computer system, it is usually necessary to reconfigure the system.

Considerations When Selecting or Upgrading the Hard Drive

The optimal hard drive is a function of several overlapping features. Since its main role is to serve as a long-term data store, capacity is a big plus. Most mobile PCs today come equipped with 2.5-inch removable hard drives. Look for something between 5 GB and 10 GB, depending on the type of data you'll be storing. Other considerations are access speed (look for a minimum of 10 to 12 milliseconds), RAM, and hard drive cache size. Access speeds should also be selected based on the type of data you'll be storing on your hard drive. Today's business software applications and large video, audio, and graphics files require several megabytes of storage.

Considerations When Selecting or Upgrading Main Memory

Main memory stores software code while the processor reads and executes the code. Having more RAM main memory means you can run more software programs at the same time. The minimum capacity you'll need to run most mainstream business software is 16MB to 64 MB. Systems with 64 MB are well suited to take advantage of today's advanced personal productivity software (word processing, spreadsheet, graphics, and database) and multimedia programs.

As discussed earlier, your system's processor, main memory, and cache memory are heavily dependent on each other to achieve optimal system functionality. The original manufacturer of your computer takes this into consideration when designing and choosing the parts for the system. If you plan to upgrade your system's main memory above 64 MB, you should consult your PC supplier to understand your system's main memory limits on size of cache and the implication of exceeding those limits.

As mentioned throughout this chapter, a computer system's components and architecture should be chosen to support fundamental objectives, current business processes, and future needs of the organization and information system. Each computer system component—processing, memory, storage, input, and output devices—has a critical role in the successful operation of the computer system, the information system, and the organization. A thorough understanding of the broader system goals and the characteristics of the hardware as they relate to these goals will be an important guide for the future IS professional.

● SUMMARY

PRINCIPLE ● Assembling an effective, efficient computer system requires an understanding of its relationship to the information system and the organization. The computer system objectives are subordinate to, but supportive of, the information system and the organization.

Hardware includes any machinery (often using digital circuitry) that assists with the input, processing, and output activities of a computer-based information system (CBIS). Hardware is a key component of a computer system, the heart of a CBIS. A computer system is an integrated assembly of physical devices with at least one central processing mechanism; it inputs, processes, stores, and outputs data and information. Computer system hardware performs many of these functions for a computer system.

Computer system hardware should be selected and organized to effectively and efficiently attain computer system objectives. These objectives should in turn support information system objectives and organizational goals. Balancing specific computer system objectives in terms of cost, control, and complexity will guide selection.

Hardware devices work together to perform input, processing, data storage, and output. Processing is performed by an interplay between the central processing unit (CPU) and memory. The CPU has three main components: the arithmetic/logic unit (ALU), the control unit, and register areas. The ALU performs calculations and logical comparisons. The control unit accesses and decodes instructions and coordinates data flow. Registers are temporary holding areas for instructions to be executed by the CPU.

Instructions are executed in a two-phase process. In the instruction phase, instructions are brought into the central processor and decoded. In the execution phase, the computer executes the instruction and stores the result. The completion of this two-phase process is a machine cycle. Processing speed is often measured by the time it takes to complete one machine cycle, which is measured in fractions of seconds.

Computer system processing speed is also affected by clock speed, which is measured in megahertz (MHz). Speed is further determined by a CPU's wordlength, the number of bits it can process at one time. (A bit is a binary digit, either 0 or 1.) A 32-bit CPU has a wordlength of 32 bits and will process 32 bits of data in one machine cycle. The iCOMP index, one measure of processing speed for Intel processors, averages the factors affecting processing speed into one number on a rating index. Other measures of processor speed include CPUmark, WinTune Integer, SYSmark, High End Winstone, and Business Winstone.

Moore's Law is a hypothesis that states that the number of transistors on a single chip will double every 18 months. This hypothesis has held up amazingly well.

Processing speed is also limited by physical constraints, such as distance between circuitry points and circuitry materials. Most CPUs are collections of digital circuits on silicon chips. Advances in gallium arsenide (GaAs) and superconductive metals will result in faster CPUs. Most processors are complex instruction set computing (CISC) chips, which have many microcode instructions placed in them. With reduced instruction set computing (RISC) chips, only essential instructions are included, so processing is faster.

Primary storage, or memory, provides working storage for program instructions and data to be processed and provides them to the CPU. Storage capacity is measured in bytes. A common form of memory is random access memory (RAM). RAM is volatile—loss of power to the computer will erase its contents—and comes in many different varieties. The mainstream type of RAM is extended data out, or EDO, RAM, which is faster than older types of RAM memory. Two other variations of RAM memory include dynamic RAM (DRAM) and synchronous DRAM. SDRAM also has the advantage of a faster transfer speed between the microprocessor and the memory. DRAM chips need high or low voltages applied at regular intervals—every two milliseconds (two one-thousandths of a second) or so—if they are not to lose their information.

Read-only memory (ROM) is nonvolatile and contains permanent program instructions for execution by the CPU. Other nonvolatile memory types include programmable read-only memory (PROM) and erasable programmable read-only memory (EPROM). Cache memory is a type of high-speed memory that CPUs can access more rapidly than RAM.

Together, a CPU and memory process data and execute instructions. Processing done using several processing units is called multiprocessing. One form of multiprocessing uses coprocessors; coprocessors execute one type of instruction while the CPU works on others. Parallel processing involves linking several processors to work together to solve complex problems.

• • •

Computer systems can store larger amounts of data and instructions in secondary storage, which is less volatile and has greater capacity than memory. The primary characteristics of secondary storage media and devices include access method, capacity, and portability. Storage media can implement either sequential access or direct access. Sequential access requires data to be read or written in sequence. Direct access means that data can be located and retrieved directly from any location on the media.

Common forms of secondary storage include magnetic tape, magnetic disk, and optical disk storage. Magnetic tape is an inexpensive sequential access storage medium. Magnetic disks are direct access media, including diskettes, fixed hard disks, and removable disk cartridges. Optical disks provide direct access storage and include compact disk read-only memory (CD-ROM), digital video disks (DVDs), and magneto-optical (MO) disks. Other storage alternatives are flash memory chips—silicon chips with nonvolatile memory—and PC memory cards, removable credit-card-size storage devices that function like fixed hard disk drives. Redundant array of independent/inexpensive disks (RAID) is a method of storing data that generates extra bits of data from existing data, allowing the system to more easily recover data in the event of a hardware failure. Storage area network (SAN) uses computer servers, distributed storage devices, and networks to provide fast and efficient storage.

• • •

Input and output devices allow users to provide data and instructions to the computer for processing and allow subsequent storage and output. These devices are part of a user interface through which humans interact with computer systems. Input and output devices vary widely, but they share common characteristics of speed and functionality.

Placing data into the computer system requires converting it from human-readable to machine-readable data. Data is thus placed in a computer system in a two-stage process: data entry converts human-readable data into machine-readable form; data input then transfers it to the computer. On-line data entry and input immediately converts and transfers data from devices to the computer system. Source data automation involves automating data entry and input so that data is captured close to its source and in a form that can be input directly to the computer.

Scanners are input devices that convert images and text into binary digits. Specialized scanners include magnetic ink character recognition (MICR) devices, optical mark recognition (OMR) devices, and optical character recognition (OCR) devices. Some input and output devices combine several into one. Point-of-sale (POS) devices are terminals with scanners that read and enter codes into computer systems. Automated teller machines (ATMs) are terminals with keyboards used for transactions.

Output devices provide information in different forms, from hard copy to sound to digital format. Display monitors are standard output devices; monitor quality is determined by size, color, and resolution. Other output devices include printers, plotters, and computer output microfilm. Printers are popular hard-copy output devices whose quality is measured by speed and resolution. Plotters output hard copy for general design work. They produce charts 24 inches or 36 inches wide and whatever length is needed. Computer output microfilm (COM) devices place data from the computer directly onto microfilm.

• • •

Computers may be classified as special purpose or general purpose. General-purpose computers are used for numerous applications and can be classified by processing speed, RAM capacity, and size. The six computer system types are network computer, personal computer, workstation, midrange computer, mainframe computer, and supercomputer. The network computer is a diskless, inexpensive computer used for accessing server-based applications and the Internet. Personal computers (PCs) are small, inexpensive computer systems. Two major types of PCs are desktop and laptop computers. Workstations are advanced PCs with greater memory, processing, and graphics abilities. Filing cabinet–size minicomputers have greater secondary storage and support transaction processing. Even larger mainframes have higher processing capabilities, while supercomputers are extremely fast computers used to solve the most intensive computing problems.

The configuration of computer system hardware components is the computer system architecture. Computer systems can be upgraded by changing or adding memory, processors, and other devices. Standards are being created to lower the cost and complexity of upgrades. A hardware standard called plug 'n play (PnP) consists of hardware and software components that card, personal computer, and operating system manufacturers incorporate into their products to eliminate the need for manual configuration so that hardware can be installed and used immediately. When attaching CD-ROM players, scanners, and other devices, a standard called Small Computer System Interface (SCSI) ensures compatibility. MMX is an emerging standard to provide high performance for communications and multimedia applications.

• • •

Computer systems that input (and output) a combination of text, audio, and video data are called multimedia systems. These systems use various input devices, including keyboards and mouses, voice-recognition devices that recognize human speech, digital computer cameras that record and store images and video in digital form, pen input devices that enable a user to input via handwriting, light pens, touch-sensitive screens, and bar code scanners. Terminals, by contrast, are used primarily for on-line data entry and input of text and character data.

Multimedia involves the marriage of sound, animation, and digitized video. A multimedia computer can deliver multimedia applications that are rich in content and features. Multimedia is being used, for example, to enhance presentations, help customers find the products they are looking for, and alter the ways lawyers manage evidence in court cases. The possibilities for multimedia are restricted only by our imagination.

● KEY TERMS

arithmetic/logic unit (ALU) 77
bit 80
bus line 80
byte (B) 82
cache memory 84
CD-rewritable (CD-RW) disk 89
CD-writable (CD-W) disk 89
central processing unit (CPU) 77
clock speed 79
compact disk read-only memory (CD-ROM) 89
complementary metal oxide semiconductor (CMOS) 105
complex instruction set computing (CISC) 82
computer system architecture 108
control unit 78
coprocessor 85
data entry 92
data input 92
digital computer cameras 94
digital signal processor (DSP) chips 107
digital video disk (DVD) 90
direct access 87
direct access storage device (DASD) 87
disk mirroring 88

execution time (E-time) 78
expandable storage devices 90
flash memory 90
general-purpose computers 100
hardware 76
hertz 80
iCOMP index 80
instruction time (I-time) 78
machine cycle 78
magnetic disk 88
magnetic tape 87
magneto-optical disk 89
mainframe computer 104
megahertz (MHz) 80
microcode 79
midrange computer 104
MIPS 79
Moore's Law 81
multifunction device 99
multiprocessing 85
music device 99
network computer 103
optical disk 89
optical processors 81
parallel processing 85
personal computer (PC) 100
pipelining 78
pixel 94

plotters 98
point-of-sale (POS) device 95
primary storage (main memory; memory) 78
random access memory (RAM) 83
read-only memory (ROM) 83
reduced instruction set computing (RISC) 82
redundant array of independent/inexpensive disks (RAID) 88
register 78
secondary storage (permanent storage) 86
sequential access 87
sequential access storage device (SASD) 87
source data automation 92
special-purpose computers 100
storage area network (SAN) 89
supercomputers 105
superconductivity 81
very long instruction word (VLIW) 82
video compression 107
voice-recognition device 93
Web appliance 104
wordlength 80
workstation 103

● REVIEW QUESTIONS

1. What is a computer system and what is the role of hardware in the system?
2. Describe a storage area network (SAN) system.
3. Why is it said that the components of all information systems are interdependent?
4. Explain the two-phase process for executing instructions.
5. Identify the three components of the CPU and explain the role of each.
6. What is the iCOMP index and how is it used?
7. What is Moore's Law?
8. What is the difference between CISC and RISC instruction sets?
9. What is the difference between wordlength and bus line width?
10. Describe the various types of memory.
11. Explain the difference between sequential and direct access.

12. Describe various types of secondary storage media in terms of access method, capacity, and portability.
13. Which secondary storage devices have the lowest cost per byte?
14. What is the difference between CD-R and CD-RW?
15. What is the difference between cache memory and main memory?
16. What is source data automation?
17. Discuss the speed and functionality of common input and output devices.
18. What are the computer system types? How do these types differ?
19. Discuss the methods of upgrading a computer system.
20. Discuss the role standards play in making it easier to use computer hardware.
21. Describe three special-purpose devices.
22. What is microcode?

● DISCUSSION QUESTIONS

1. What are the implications of Moore's Law—that is, continuing the trend of increased computing power at lower costs? Use Moore's Law to forecast the computing power that could be available in three years. What sort of applications could benefit from that level of computer power?

2. What functions and capabilities is it reasonable to expect to be performed by palmtop computers? Are there functions that they should be able to perform that a desktop computer cannot perform?

3. What are the trade-offs between main memory and cache memory?

4. How many different classes of computer would you expect to find in a small office? How about in a medium-size company with a couple hundred employees? Describe the types of functions the various classes of computers might perform in the medium-size company.

5. Imagine that you are the business manager for your university. What type of computer would you recommend for broad deployment in the university's computer labs—a standard desktop personal computer or a network computer? Why?

6. Imagine that you are responsible for determining the best secondary storage solutions for your medium-size company. Describe the options that are available and recommend a storage solution, which can include several secondary storage devices and systems.

7. If cost were not an issue, describe the characteristics of your ideal laptop computer.

8. What if you discovered that your favorite recording group composes, edits, and records all its music using multimedia computer technology? How do you feel? Does the use of computer technology to create original works of art or music diminish or enhance the accomplishment? Should such artists be considered as great as others who do not use computer technology?

● PROBLEM-SOLVING EXERCISES

1. Do research (read various trade journals and search the Internet) on companies that make CD-RW devices. Use your word processing program to write a short report summarizing your findings. Make sure to include the speed, features, and price of the various devices. Also summarize the difference between internal and external CD-RW devices. Develop a simple spreadsheet to compare the features and costs of the CD-RW devices you found.

2. Over the upcoming year, your department is expected to add eight people to its staff. You will need to acquire eight personal computer systems and two additional printers for the new employees to share. Standard office computers have a Pentium (266 MHz) processor with 32 MB of RAM, SVGA color monitor, and a minimum of 2.4 GB hard disk drive. At least four of the new people will use their computers more than three hours per day. You would like to provide larger monitors and special ergonomic keyboards for these people—if it fits within your budget. You are not sure if you want to upgrade the machines to 64 MB of RAM and 4.8 GB hard drive.

Your department budget will allow a maximum of $20,000 for computer hardware purchases this year, and you want to select only one vendor for all of the hardware. A price list from three vendors appears in the following table, with prices for a single unit of each component. Use a spreadsheet to find the department's best solution; write a short memo explaining your rationale. Specify which vendor to choose and which items to be ordered as well as the total cost.

Component	Expert Solutions Ltd.	Business Processing Enterprises	Super Systems Inc.
266 MHz Pentium with 32 MB RAM 2.4 GB hard drive	$1,245	$1,275	$1,200
Upgrade to 64 MB RAM	250	225	245
Upgrade to 4.8 GB hard drive	190	215	205
15-inch .28 dpi SVGA monitor	350	330	340
17-inch .28 dpi SVGA monitor	625	600	615
Ergonomic keyboard	55	50	50
8 ppm color ink-jet printer	325	325	320
Surge protector/power strip	35	32	35
Three-year warranty (parts and labor)	340	300	320

● TEAM ACTIVITIES

1. With two of your classmates, visit a major computer retail store (e.g., CompUSA, MicroCenter). Spend a couple hours during which each of you concentrates on identifying the latest developments in processing, input, and output devices. Write a brief report summarizing your findings.

2. With two or three of your classmates, visit the main computer facility of your college or university. Find out the manufacturer and model number as well as the specifications (speed of CPU, amount of main memory, disk drive capacity, etc.) of a mainframe or midrange computer. How long has it been in use? How much longer does the university plan to use it before replacing it with something different? What business processes have changed that spurred this alteration? Will your college or university upgrade the computer or buy a new one?

3. Identify a manager of a computer center and seek his or her permission to tour the facility. Make a list of various types of computers and input, output, and secondary storage devices that you see during your tour. Does the center have more than one class of computer? If so, why? What types of jobs are run on each class of computer?

● WEB EXERCISES

1. Visit the Web sites of Intel, Sun Microsystems, Hewlett-Packard, IBM, and other computer manufacturers. Identify as many as possible industry standards that these manufacturers are following in the development of their products. Also, identify other standardization efforts in which they are involved.

2. Search the Web for companies that make secondary storage devices and systems, including disk, tape, RAID, SAN, and others. Summarize your findings using your word processing device.

3. Visit the Web sites of three PC makers. Your search can include Dell, Gateway, IBM, Compaq, and others. From each of the three Web sites, summarize your ideal PC system. Develop a spreadsheet that compares features and prices. Which system and manufacturer do you prefer?

● CASES

 Electronic Ink

One early prediction of the computer revolution was the elimination of paper. Forecasters thought that offices and homes would not require it; everything would be done electronically, saving trees if not entire forests. This prediction, however, did not come true. Even with all the technological breakthroughs, it seems that paper is being used to a greater extent today in homes and offices than ever before. If the racks of paper at Office Depot, Staples, and Office Max are any indication, it seems that the use of paper is here to stay. Although the trend to using more paper is likely to remain, some advances are making paper less of a necessity. One of these is a result of a new printing technology, using what some call "electronic ink."

E-ink was first conceived in the 1970s by Xerox, the same company that pioneered copiers and the graphical user interface that is typical in Apple computers and PCs using Windows. All of these ideas came out of the famous Palo Alto Research Center (PARC). The initial work was done using microscopic balls that were half white and half black. When an electrical charge was applied, the balls would turn to show either white or black.

The idea of printing using electronic ink instead of traditional printing on paper was intriguing. Yet, Xerox decided to concentrate on copiers. Nick Sheridon's early ideas about electronic ink were put on hold. Although Xerox would restart its research into electronic ink in the 1990s, other companies would also investigate the technology.

In the mid-1990s, a young physicist at the MIT Media Lab hired some students to look into developing an electronic book. One of the students,

Barrett Comiskey, experimented with several approaches, including using balls painted half white and half black as was done at PARC. After many failures, he tried the idea of all white balls with a mixture of oil and dark dye. When the white balls were on the surface, you could see the white color. When the balls were submerged, you would see a dark black. These tests led to a workable solution, and Comiskey and a few others formed a new company, E-Ink Corporation.

Today, E-Ink is testing its electronic ink in stores such as J.C. Penney. A four-foot-by-four-foot foam board from E-Ink is being tested in the athletic-wear department. The new technology allows J.C. Penney to display messages such as, "Hello down there! You've got to try the Nike Air Quest. It's super comfortable and lightweight too." With a few keystrokes, the message can be changed. The success of the trials has resulted in financial backing from such companies as IBM and Motorola.

Discussion Questions

1. What other applications would be good for the printing technology developed by E-Ink?
2. What applications would still require paper?

Critical Thinking Questions

3. A similar idea is to develop electronic books. Do you think books that display text and figures electronically will ever replace today's textbooks and novels?
4. Compare electronic ink to a flat panel display. What are the advantages and disadvantages of each?

Sources: Adapted from Alec Klein, "Will the Future Be Written in E-Ink?" *The Wall Street Journal,* January 4, 2000, p. B1; and "Showing Off," *The Economist,* October 30, 1999.

2 Flash Chips

Sweden is known for its world-class Nordic skiers, people with fair skin and blond hair, and a telecommunications giant, Ericsson. The company has quarterly results of about $430 million in profits on about $6 billion in revenues. The president of Ericsson is a Harley-Davidson motorcycle fan and a hard-driving executive. He has been quoted as saying, "I will be here until they kick me out, I retire, or I die."

Not only is Ericsson a leader in networks, it is also a leader in cell phones and related equipment. Cell phone usage is expected to explode worldwide as technology improves and the cost of cellular service continues to decline. Today, there are about 275 million mobile phones in use worldwide, and the growth rate is expected to be about 30 percent per year for the next several years.

A key component of any cell phone is the memory device, which is used to store phone numbers and internal operating instructions for the phone. To get quality memory devices for its cell phones, Ericsson turned to Intel, the leading hardware maker of chips and memory devices. The move will guarantee that flash memory chips are available for Ericsson when the demand for these memory devices skyrockets.

In a recently announced deal, Ericsson agreed to purchase about $1.5 billion in flash memory chips from Intel over a three-year period. Unlike the standard volatile memory devices used in computers, flash chips are not volatile. They retain their contents even when the power is turned off. In addition to computers and mobile phones, flash chips are used in handheld personal organizers, digital cameras, and music players. According to Benny Ginman, Intel's director of European sales and marketing, "If you look at the growth in cell phones, you can see why we want to be a major supplier to this industry."

To provide high-quality, low-cost chips, Intel continues to innovate. Using new manufacturing and fabrication techniques, Intel has been able to streamline its operations and reduce costs. By mid- or late 2000, the company is expected to migrate from more traditional .25-micron to .18-micron flash chips. This should allow Intel to decrease its cost per megabit by about 30 percent annually. To meet increased demand, Intel is also acquiring additional manufacturing facilities. In a separate deal, Intel agreed to purchase two older chip factories from Rockwell International. Intel plans to completely renovate the chip factories for a total cost of about $1.5 billion.

In addition to flash memory, Intel continues to be a leader in Pentium processors used in personal computers. Annual sales for Intel are approximately $30 billion. About 80 percent of these revenues come from processors, like the Pentium. The remaining 20 percent comes from flash memory and other types of chips used in a variety of operations.

Discussion Questions

1. Describe the characteristics of flash memory compared with normal memory.
2. What types of devices are ideal for flash memory?

Critical Thinking Questions

3. What are the advantages to Ericsson of making a flash memory deal with Intel?
4. Will flash memory become more or less important in the future?

Sources: Adapted from Neal Boudette, "Intel to Supply Flash Chips for Ericsson's Wireless Phones," *The Wall Street Journal*, February 3, 2000, p. B13; and Stanley Reed, "Ericsson," *Business Week*, February 7, 2000, p. EB74.

CD-ROM Titles

Truck and auto repair shops can devote more space to manuals than parts. These large and sometimes confusing manuals can be as thick as large phone books and difficult to use. When needed, a mechanic can spend hours searching for the correct manual and the part information. If a page is ripped out or missing, it can be very frustrating. In addition, it can be very expensive to buy new manuals. With new car and truck models coming out every year, the cost can add up quickly. As a result, many auto and truck companies are starting to provide their repair manuals on CD-ROMs instead of using traditional paper manuals.

The CD-ROM manuals offer a number of advantages. They are much cheaper to produce and send through the mail. In addition, easy-to-use search routines can make it a snap for repairmen to find the information they need. Each year, the auto company can send a CD containing repair manuals for multiple car and truck models. Yet although the CD-ROM repair manuals offer numerous advantages over printed manuals, there are still problems and hassles. This was the case with car mechanic Charles Schultz.

Schultz specializes in foreign cars, such as BMWs. It was a blessing when BMW started to convert more of its shop repair manuals to CD-ROM. Using a laptop computer, Schultz could quickly find the information he needed by inserting the CD into the drive in his laptop and using the built-in search routines. Schultz also quickly realized that his growing library of CD repair manuals did have its own problems. Often, Schultz would have to scrub his hands and arms of work-related grease and grime before he inserted a new, clean CD into his laptop. Getting dirt and grime on a printed repair manual was no problem, but dirt and grime on a CD could destroy the CD and the CD drive in his laptop.

To solve this problem, Schultz found an inexpensive software solution. For $35, Schultz purchased Virtual Drive 2000, which simplified the task of downloading portions of a CD-ROM to a hard disk. Now Schultz can load only the information he needs from multiple CD-ROMs onto his hard disk. In addition to the advantage of not having to wash his hands every few minutes to load another CD-ROM into his laptop, Schultz also found that he could retrieve the needed information much faster from his hard disk than from the CD-ROM drive.

Discussion Questions

1. How was Charles Schultz able to take advantage of the convenience of repair manuals on CD-ROM while eliminating some of the problems?
2. With what other applications could this approach be used?

Critical Thinking Questions

3. What other features would Schultz likely find useful with the software solution described in this case?
4. Could Schultz's solution be used in other applications?

Sources: Adapted from Mitt Jones, "Hassle-Free Access to CD-ROM Titles," *PC World*, February 2000, p. 43, and Brian Quinton, "Ready to Play," *Telephony*, November 15, 1999.

● NOTES

Sources for the opening vignette on page 75: Adapted from Chris Murphy, "Ford's Better Idea: PCs for All," *Information Week,* February 7, 2000; and William Holstein, "The Dot Com within Ford," *US News & World Report,* February 7, 2000, p. 34.

1. Kathleen Melymuka, "CEO Meg Whitman Powers eBay," *Computerworld,* January 10, 2000.

2. "Desktops Hit 750 MHz," *PC Magazine,* January 18, 2000, p. 11.

3. Cade Metz, "A Cool 1,000 MHz," *PC Magazine,* February 22, 2000, p. 50.

4. Michael Miller, "Moore's Law Will Continue to Drive Computing," *PC Magazine,* June 22, 1999, p. 146.

5. Josh McHugh, "No Mercy for Merced," *Forbes,* September 20, 1999, p. 57.

6. Evan Ranstad, "TI's New Chip Does Its Tricks with Mirrors," *The Wall Street Journal,* September 20, 1999, p. B1.

7. "Rambus Completes Direct RDRAM Interface Design," press release on Rambus Web site, http://www.rambus.com, accessed February 17, 2000.

8. Mitch Wagner, "SAN Backs Up Financial Assets," *Internet Week,* April 26, 1999, p. 19.

9. Alfred Poor, "Optical-Drive Compatibility," *PC Magazine,* January 18, 2000, p. 124.

10. Gary Anthes, "Experiment with Voice Recognition," *Computerworld,* January 3, 2000, p. S20.

11. Stephen Manes, "Compute While You're Driving," *Forbes,* January 11, 1999, p. 112.

12. David Grotta, "New CD Writes Twice," *PC Magazine,* September 1, 1999, p. 30.

13. Unisys Web site, http://www.unisys.com, accessed February 17, 2000.

14. Michelle Campanale-Surkan, "Music Machines Serve Ear Candy for MP3 Devotees," *PC World,* January 2000, p. 90.

15. Daniel Rocks, "Now Electronic Eyes for the Blind," *Business Week,* January 31, 2000, p. 56.

16. Elizabeth Bukowski, "Rocket eBook: Toward a Paperless Future," *The Wall Street Journal,* June 15, 1999, p. A16.

17. David Orenstein, "How About Digital Paper?" *Computerworld,* January 25, 1999.

18. Andy Reinhardt, "Can Steve Jobs Keep His Mojo Working?" *Business Week,* August 2, 1999, p. 32.

19. Dominique Deckmyn, "Intel Technology Pushes Laptops to Desktop Speeds," *Computerworld,* January 17, 2000, p. 12.

20. Matt Hamblen, "Handhelds Help Boeing Boost Quality Inspections," *Computerworld,* November 8, 1999, p. 38.

21. Larry Armstrong, "The Free PC Game," *Business Week,* July 19, 1999, p. 80.

22. Dean Takahasi, "Intel Moves from Windows with Line of Web Devices," *The Wall Street Journal,* January 5, 2000, p. B6.

23. Stephen Wildstrom, "Age of the E-Pliance," *Business Week,* January 17, 2000, p. 20.

24. Mitch Wagner, "IBM Pumps Up Iron with Copper CPU,:" *Internet Week,* May 10, 1999, p. 9.

25. Jeff Bliss, "IBM's Supercomputer Uses Copper Wire for Speed," *The Wall Street Journal,* February 10, 2000, p. 7B.

CHAPTER 4

Software: Systems and Application Software

Software rates among the most poorly constructed, unreliable, and least maintainable technological artifacts invented by man. To be fair, software also shares credit for the most spellbinding advances of the 20th century. In today's world, banks, hospitals, and space missions would be inconceivable without it. The challenge of the next century will be to bring software quality to the same level as we expect from cars, televisions, and other relatively dependable hunks of hardware.

— Paul Strassman, former CIO for Xerox Corporation and former CIO for the U.S. Defense Department who now heads a private consulting firm

Principles

When selecting an operating system, you must consider the current and future needs for application software to meet the needs of the organization. In addition, your choice of a particular operating system must be consistent with your choice of hardware.

Do not develop proprietary application software unless doing so will meet a compelling business need that can provide a competitive advantage.

Choose a programming language whose functional characteristics are appropriate to the task at hand, taking into consideration the skills and experience of the programming staff.

Learning Objectives

- *Identify and briefly describe the functions of the two basic kinds of software.*
- *Outline the role of the operating system and identify the features of several popular operating systems.*

- *Discuss how application software can support personal, workgroup, and enterprise business objectives.*
- *Identify three basic approaches to developing application software and discuss the pros and cons of each.*

- *Outline the overall evolution of programming languages and clearly differentiate among the five generations of programming languages.*

Home Depot

Building Software to Meet Customers' Needs

Founded in 1978 in Atlanta, Georgia, Home Depot is the world's largest home improvement retailer, with over 1,900 stores planned for the United States, Canada, Puerto Rico, and Chile by 2004. Home Depot is an innovator in the home improvement retail industry, combining the economies of scale inherent in a warehouse format with a level of customer service unprecedented among warehouse-style retailers. The company's stores cater to do-it-yourselfers, as well as home improvement, construction, and building maintenance professionals. Each Home Depot store stocks over 40,000 different materials and products—building materials, cabinetry, flooring and countertop materials, plumbing and hardware supplies, and lawn and garden products.

Home Depot's Expo Design Centers are stand-alone stores that include interior design showrooms whose workers can plan and coordinate home improvement projects for customers. A one-stop full-service interior design center, each Expo store is filled with showroom after showroom of exciting ideas, hundreds of full-size displays, and lots of friendly, knowledgeable professionals who can help customers with any project, big or small. Expo Design Center customers have their very own project designer to work with them from the very beginning, assisting in the selection of products, styles, and colors that best suit their lifestyles. Once the details are settled, a project superintendent can take over and work with the contractor and installers to complete the job. The company plans to increase the number of Expo Design Centers from 12 now to 200 by 2005.

Until recently, the front-line work in the Expo Design Centers was done manually—a logistical nightmare for both customers and Home Depot staff. To order flooring, bath fixtures, appliances, and other items for customers, workers had to flip through paper catalogs and then enter special purchase orders into Home Depot's retail ordering system. That worked but not effectively because the existing retail system wasn't built to handle the number of nonstock orders that Expo creates. Meanwhile, interior designers and project coordinators who supervise installations had to use file folders and pieces of paper to track the many details.

The key to success of the Expo Design Centers is personal one-on-one attention and follow-up with customers. Home Depot recognized the need for a number of software applications to provide this high level of customer support. Software applications were built to enable viewing of products, creating purchase orders for customers, and tracking the progress of projects. Now products can be viewed easily and ordered directly using the new order entry application. Room measurements and installation schedules are also entered into the tracking system. All the details needed to track and follow up on home improvement projects are at the fingertips of the Home Depot project superintendents.

The new applications expand on Home Depot's heavy investment in the Java programming language, which it has used to develop systems for tool rentals and retail sales reports for district managers, among other applications. The company looked at installing standard off-the-shelf packaged software for the new applications, such as calendaring systems that could track projects; however, most of what the company's staff is doing is unique.

As you read this chapter, consider the following:

- What are the various kinds of software and how are they used?

- What are the sources for acquiring software, and what are the pros and cons of each approach?

FIGURE 4.1

The Importance of Software in Business

Since the 1950s, businesses have greatly increased their expenditures on software as compared with hardware.

computer programs

sequences of instructions for the computer

documentation

text that describes the program functions to help the user operate the computer system

systems software

the set of programs designed to coordinate the activities and functions of the hardware and various programs throughout the computer system

In the 1950s, when computer hardware was relatively rare and expensive, software costs were a comparatively small percentage of total information systems costs. Today, the situation has dramatically changed. Software can represent 75 percent or more of the total cost of a particular information system because of three major reasons: advances in hardware technology have dramatically reduced hardware costs, increasingly complex software requires more time to develop and so is more costly, and salaries for software developers have increased because the demand for these workers far exceeds the supply. In the future, as suggested in Figure 4.1, software is expected to make up an even greater portion of the cost of the overall information system. The critical functions software serves, however, make it a worthwhile investment.

AN OVERVIEW OF SOFTWARE

computer system platform

the combination of a particular hardware configuration and systems software package

application software

programs that help users solve particular computing problems

One of software's most critical functions is to direct the workings of the computer hardware. As we saw in Chapter 1, software consists of computer programs that control the workings of the computer hardware. **Computer programs** are sequences of instructions for the computer. **Documentation** describes the program functions to help the user operate the computer system. The program displays some documentation on screen, while other forms appear in external resources, such as printed manuals. There are two basic types of software: systems software and application software.

Systems Software

Systems software is the set of programs designed to coordinate the activities and functions of the hardware and various programs throughout the computer system. A particular systems software package is designed for a specific CPU design and class of hardware. The combination of a particular hardware configuration and systems software package is known as a **computer system platform**.

Application software has the greatest potential to affect processes that add value to a business because it is designed for specific organizational activities and functions.
(Source: Stone/Charlie Westerman)

Application Software

Application software consists of programs that help users solve particular computing problems. Both systems and application software can be used to meet the needs of an individual, a group, or an enterprise. Application software can support individuals, groups, and organizations to help them realize business objectives. Application software has the greatest potential to affect the processes that add value to a business because it is designed for specific organizational activities and functions, as we saw in the case of Home Depot. The effective implementation and use of application software can provide significant internal efficiencies and support corporate goals. Before an individual, a group, or an enterprise decides on the best approach for acquiring application software, goals and needs should be analyzed carefully.

Supporting Individual, Group, and Organizational Goals

Every organization relies on the contributions of individuals, groups, and the entire enterprise to achieve business objectives. The organization also supports individuals, groups, and the entire enterprise with specific application software and information systems. As the power and reach of information systems expand, they promise to reshape every aspect of our lives: how we work and play, how we are educated, how we interact with others, how our businesses and governments conduct their work, and how scientists perform research. Information systems will change virtually every method for capturing, storing, transmitting, and analyzing knowledge, including books, newspapers, magazines, movies, television, phone calls, musical recordings, and architectural drawings. One useful way of classifying the many potential uses of information systems is to identify the scope of the problems and opportunities addressed by a particular organization. This is called the **sphere of influence**. For most companies, the spheres of influence are personal, workgroup, and enterprise, as shown in Table 4.1.

Information systems that operate within the **personal sphere of influence** serve the needs of an individual user. These information systems enable their users to improve their personal effectiveness, increasing the amount of work they can do and its quality. Such software is often referred to as **personal productivity software**. There are many examples of such applications operating within the personal sphere of influence—a word processing application to enter, check spelling, edit, copy, print, distribute, and file text material; a spreadsheet application to manipulate numeric data in rows and columns for analysis and decision making; a graphics application to perform data analysis; and a database application to organize data for personal use.

A **workgroup** is two or more people who work together to achieve a common goal. A workgroup may be a large, formal, permanent organizational entity such as a section or department or a temporary group formed to complete a specific project. The human resource department of a large firm is an example of a formal workgroup. It consists of several people, is a formal and permanent organizational entity, and appears on a firm's organization chart. An information system that operates in the **workgroup sphere of influence** supports a workgroup in the attainment of a common goal. Users of such applications are operating in an environment where communication, interaction, and collaboration are critical to the success of the group. Applications include systems that support information sharing, group scheduling, group decision making, and conferencing. These applications enable members of the group to communicate, interact, and collaborate.

Information systems that operate within the **enterprise sphere of influence** support the firm in its interaction with its environment. The surrounding environment includes customers, suppliers, shareholders, competitors, special interest groups, the financial community, and government agencies. Every enterprise has many applications that operate within the enterprise sphere of influence. The input to these systems is data about or generated by basic business transactions with someone outside the business enterprise. These transactions include customer orders, inventory receipts and withdrawals, purchase orders, freight bills, invoices, and checks. One of the results of processing transaction data is that the records of the company are instantly updated. For example, the processing of employee time cards updates individuals' payroll records used to generate their checks. The order entry, finished product inventory, and billing information systems are examples of applications that operate in the enterprise sphere of influence.

sphere of influence

the scope of problems and opportunities addressed by a particular organization

personal sphere of influence

sphere of influence that serves the needs of an individual user

personal productivity software

software that enables users to improve their personal effectiveness, increasing the amount of work they can do and its quality

workgroup

two or more people who work together to achieve a common goal

workgroup sphere of influence

sphere of influence that serves the needs of a workgroup

enterprise sphere of influence

sphere of influence that serves the needs of the firm in its interaction with its environment

---

Lotus Notes is an application that enables a workgroup to schedule meetings and coordinate activities.
(Source: Courtesy of Lotus Development Corporation)

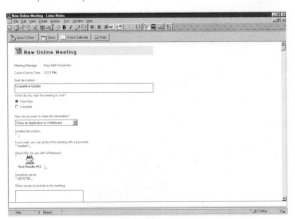

Software	Personal	Workgroup	Enterprise
Systems software	Personal computer and workstation operating systems	Network operating systems	Midrange computer and mainframe operating systems
Application software	Word processing, spreadsheet, database, graphics	Electronic mail, group scheduling, shared work	General ledger, order entry, payroll human resources

TABLE 4.1

Classifying Software by Type and Sphere of Influence

These applications support interactions with customers and suppliers.

SYSTEMS SOFTWARE

Controlling the operations of computer hardware is one of the most critical functions of systems software. Systems software also supports the application programs' problem-solving capabilities. Different types of systems software include operating systems and utility programs.

Operating Systems

operating system (OS)

a set of computer programs that controls the computer hardware and acts as an interface with application programs

An **operating system (OS)** is a set of computer programs that controls the computer hardware and acts as an interface with application programs (Figure 4.2). The operating system, which plays a central role in the functioning of the complete computer system, is usually stored on disk. After a computer system is started, or "booted up," portions of the operating system are transferred to memory as they are needed. The collection of programs, collectively called the operating system, executes a variety of activities, including:

- Performing common computer hardware functions
- Providing a user interface
- Providing a degree of hardware independence
- Managing system memory
- Managing processing tasks
- Providing networking capability
- Controlling access to system resources
- Managing files

kernel

the heart of the operating system, which controls the most critical processes

The **kernel**, as its name suggests, is the heart of the operating system and controls the most critical processes. The kernel ties all of the components of the operating system together and regulates other programs.

FIGURE 4.2

The role of the operating systems and other systems software is as an interface or buffer between application software and hardware.

Common Hardware Functions

All application programs must perform certain tasks. For example:

- Getting input from the keyboard or some other input device

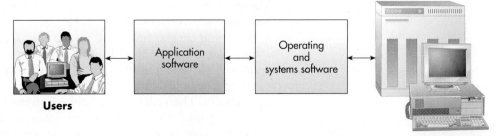

Users   Application software ↔ Operating and systems software   **Hardware**

- Retrieving data from disks
- Storing data on disks
- Displaying information on a monitor or printer

Each of these basic functions requires a more detailed set of instructions to complete. The operating system converts a simple, basic instruction into the set of detailed instructions required by the hardware. In effect, the operating system acts as intermediary between the application program and the hardware. The typical OS performs hundreds of such functions, each of which is translated into one or more instructions for the hardware. The OS will notify the user if input/output devices need attention, if an error has occurred, and if anything abnormal occurs in the system.

User Interface

One of the most important functions of any operating system is providing a **user interface**. A user interface allows individuals to access and command the computer system. The first user interfaces for mainframe and personal computer systems were command based. A **command-based user interface** requires that text commands be given to the computer to perform basic activities. For example, the command ERASE 00TAXRTN would cause the computer to erase or delete a file called 00TAXRTN. RENAME and COPY are other examples of commands used to rename files and copy files from one location to another. Many mainframe computers use a command-based user interface. In some cases, a specific job control language (JCL) is used to control how jobs or tasks are to be run on the computer system.

A **graphical user interface (GUI)** uses pictures (called **icons**) and menus displayed on screen to send commands to the computer system. Many people find that GUIs are easier to use because users intuitively grasp the functions. Today, the most widely used graphical user interface is Windows by Microsoft. Alan Kay and others at Xerox PARC (Palo Alto Research Center, located in California) were pioneers in investigating the use of overlapping windows and icons as an interface. As the name suggests, Windows is based on the use of a window, or a portion of the display screen dedicated to a specific application. The screen can display several windows at once. The use of GUIs has contributed greatly to the increased use of computers because users no longer need to know command-line syntax to accomplish tasks.

Hardware Independence

The applications make use of the operating system by making requests for services through a defined **application program interface (API)**, as shown in Figure 4.3. Programmers can use APIs to create application software without having to understand the inner workings of the operating system.

Suppose a computer manufacturer designs new hardware that can operate much faster than before. If the same operating system for which an application was developed can run on the new hardware, minimal (or no) changes are needed to the application to enable it to run on the new hardware. If APIs did not exist, the application software developers might have to completely rewrite the application program to take advantage of the new, faster hardware.

Memory Management

The purpose of memory management is to control how memory is accessed and to maximize available memory and storage. The memory management feature of many operating systems allows the computer to execute program instructions effectively and speed processing.

Controlling how memory is accessed allows the computer system to efficiently and effectively store and retrieve data and instructions and to supply them to the CPU. Memory management programs convert a user's request for

---

user interface

element of the operating system that allows individuals to access and command the computer system

command-based user interface

a user interface that requires that text commands be given to the computer to perform basic activities

graphical user interface (GUI)

an interface that uses icons and menus displayed on screen to send commands to the computer system

icon

picture

application program interface (API)

interface that allows applications to make use of the operating system

Application Program Interface Links Application Software to the Operating System

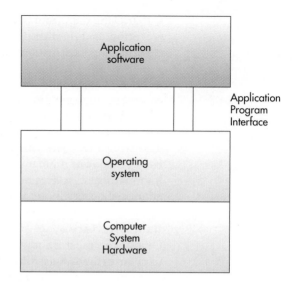

data or instructions (called a logical view of the data) to the physical location where the data or instructions are stored. A computer understands only the physical view of data—that is, the specific location of the data in storage or memory and the techniques needed to access it. This concept is described as logical versus physical access. For example, the current price of an item, say, a Texas Instruments BA-35 calculator with an item code of TIBA35, might always be found in the logical location "TIBA35$." If the CPU needed to fetch the price of TIBA35 as part of a program instruction, the memory management feature of the operating system would translate the logical location "TIBA35$" into an actual physical location in memory or secondary storage (Figure 4.4).

Memory management is important because memory can be divided into different segments or areas. With some computer chips, memory is divided into conventional, upper, high, extended, and expanded memory. In addition, some computer chips provide "rings" of protection. An operating system can use one or more of these rings to make sure that application programs do not penetrate an area of memory and disrupt the functioning of the operating system, which could cause the entire computer system to crash. Memory management features of today's operating systems are needed to make sure that application programs can get the most from available memory without interfering with other important functions of the operating system or with other application programs.

Most operating systems support **virtual memory**, which allocates space on the hard disk to supplement the immediate, functional memory capacity of RAM.

virtual memory

memory that allocates space on the hard disk to supplement the immediate, functional memory capacity of RAM

FIGURE 4.4

An Example of the Operating System Controlling Physical Access to Data

The user prompts the application software for specific data. The operating system translates this prompt into instructions for the hardware, which finds the data the user requested. Having successfully completed this task, the operating system then relays the data back to the user via the application software.

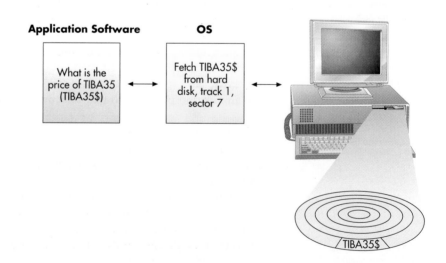

Virtual memory works by swapping programs or parts of programs between memory and one or more disk devices—a concept called **paging**. This reduces CPU idle time and increases the number of jobs that can run in a given time span.

Processing Tasks

Managing all processing activities is accomplished by the task management features of today's operating systems. Task management allocates computer resources to make the best use of each system's assets. Task management software can permit one user to run several programs or tasks at the same time (multitasking) and allow several users to use the same computer at the same time (time-sharing).

An operating system with **multitasking** capabilities allows a user to run more than one application at the same time. Without having to exit a program, you can work in one application, easily pop into another, and then jump back to the first program, picking up where you left off. Better still, while you're working in the *foreground* in one program, one or more other applications can be churning away, unseen, in the *background*, sorting a database, printing a document, or performing other lengthy operations that otherwise would monopolize your computer and leave you staring at the screen unable to get other work done. Multitasking can save users a considerable amount of time and effort.

Time-sharing allows more than one person to use a computer system at the same time. For example, 15 customer service representatives may be entering sales data into a computer system for a mail-order company at the same time. In another case, thousands of people may be simultaneously using an on-line computer service to get stock quotes and valuable business news.

Time-sharing works by dividing time into small CPU processing time slices, which can be a few milliseconds or less in duration. During a time slice, some tasks for the first user are done. The computer then goes from that user to the next. During the next time slice, some tasks for the next user are completed. This process continues through each user and cycles back to the first user. Because the CPU processing time slices are small, it appears that all jobs or tasks for all users are being completed at the same time. In reality, each user is sharing the time of the computer with other users.

The ability of the computer to handle an increasing number of concurrent users smoothly is called **scalability**. This is a critical feature for systems expected to handle a large number of users such as a mainframe computer or a Web server. Because personal computer operating systems usually are oriented toward single users, the management of multiple-user tasks often is not needed.

Networking Capability

The operating system can provide features and capabilities that aid users in connecting to a computer network. For example, Apple computer users have built-in network access through the AppleShare feature, and the Microsoft Windows operating systems come with the capability to link users to the Internet.

Access to System Resources

Computers often handle sensitive data that can be accessed over networks. The operating system needs to provide a high level of security against unauthorized access to the users' data and programs. Typically, the operating system establishes a log-on procedure that requires users to enter an identification code and a matching password. If the identification code is invalid or if the password does not go with the identification code, the user cannot gain access to the computer. The operating system also requires that user passwords be changed frequently—say, every 20 to 40 days. If the user is successful in logging onto the system, the operating system records who is using the system and for how long. In some organizations, such records are

paging
process of swapping programs or parts of programs between memory and one or more disk devices

multitasking
capability that allows a user to run more than one application at the same time

time-sharing
capability that allows more than one person to use a computer system at the same time

scalability
the ability of the computer to handle an increasing number of concurrent users smoothly

also used to bill users for time spent using the system. The operating system also reports any attempted breaches of security.

File Management

The operating system performs a file management function to ensure that files in secondary storage are available when needed and that they are protected from access by unauthorized users. Many computers support multiple users who store files on centrally located disks or tape drives. The operating system keeps track of where each file is stored and who may access it. The operating system must be able to resolve what to do if more than one user requests access to the same file at the same time. Even on stand-alone personal computers with only one user, file management is needed to keep track of where files are located, what size they are, when they were created, and who created them.

Personal Computer Operating Systems

Early operating systems for personal computers were very basic. In the last several years, however, more advanced operating systems have been developed, incorporating some features previously available only with mainframe operating systems. This section reviews selected popular personal computer operating systems. Table 4.2 classifies a number of operating systems by sphere of influence.

MS-DOS

MS-DOS (Microsoft Disk Operating System) was the Microsoft-marketed version of the first widely installed operating system in personal computers. It was essentially the same operating system that Bill Gates's young company developed for IBM as PC-DOS (Personal Computer DOS). Most users of either DOS system simply referred to their system as DOS. Like PC-DOS, MS-DOS was a nongraphical, line-oriented, command-driven operating system, with a relatively simple interface but not an overly "friendly" user interface. It was not capable of multitasking.

DOS with Windows

TABLE 4.2

Popular Operating Systems

The original Windows was not technically an operating system but an application that ran on top of the DOS operating system. On top of the DOS infrastructure,

Personal	Workgroup	Enterprise	Consumer
MS-DOS			
DOS with Windows			
Windows 95			
Windows 98			
			Pocket PC
Windows NT Workstation	Windows NT Server	Windows NT Server	
Windows 2000	Windows 2000	Windows 2000 Server	
Unix	Unix	Unix	
Linux	Linux		Linux
MAC OS 9			
MAC OS X	MAC OS X	MAC OS X	
Solaris	Solaris	Solaris	
	Netware		
	OS/390	OS/390	
	MPE/iX	MPE/iX	

Windows added a graphical operating environment and features like a simple multitasking scheme it used to run multiple Windows and DOS applications. Windows provided an intuitive GUI that was a major improvement over the MS-DOS interface.

Windows 95/98

With Windows 95, a Start button is used to launch applications or documents with their associated applications. As you launch programs, they appear on a task bar normally positioned at the bottom of the screen. Clicking on any task bar button switches to that program. This approach is a very intuitive style of task switching.

Windows 95 uses a desktop metaphor—your files are shown as icons within folders (directories). It also includes Windows Explorer, software that gives you a tree-based view of your computer and its connections. In addition, file names are no longer limited to eight characters plus a three-letter extension; you can use names up to 255 characters long. Windows 95 also comes bundled with Exchange, designed to be a universal mailbox. It works with Microsoft Mail for receiving electronic-mail messages and with Microsoft Fax for sending and receiving faxes. Windows 95 also comes with communications software to enable you to connect to the Internet. It is also a multitasking system. All these features can provide increased productivity and better control over the operation of the computer system.

Windows 98 is an improved version of Windows 95 with many end user productivity features, improved support for newer hardware devices, and additional enhancements. Several improvements were made to Windows 98 to improve end user productivity. A new image-preparation tool lets an organization's IS department install and configure the operating system and all applications on one machine, back them up, and then copy the image to a destination machine. Attempts were made to significantly shorten the time it takes for software applications to load from the hard disk drive. System shutdown was also speeded up.

Windows NT Workstation

The NT operating system is designed to take advantage of the newer 32-bit processors, and it features multitasking and networking capabilities. Using emulation software, NT can run programs written for other operating systems. When a computer tries to emulate another operating system, such as a Windows NT computer trying to emulate Unix, some translation of code must happen before any program can be run on the emulator. The Windows NT Workstation operating system and Unix are two very different operating systems, so they tend to execute computations in different ways. These ways of executing computations are known as instructions. Obviously, since the two operating systems have differing instruction sets, the emulation software must find a way to bridge the gap between the two. NT supports symmetric multiprocessing, the ability to make simultaneous use of multiple processors. It also has built-in networking to support several communications protocols. NT also has a centralized security system to monitor various system resources. The many features and capabilities of NT make it very attractive for use on many computers.

With Windows NT Workstation, Plug 'n Play support is finally present, although only systems based on the new Advanced Configuration and Power Interface (ACPI) standard can truly exploit it. Windows NT offers Windows 95 compatibility so that users can consider changing from a Windows 95 to Windows NT operating system rather than to Windows 98. Tighter security controls help NT administrators lock intruders out of the network. A new version of NT's native file format, called NTFS file system, significantly enhances performance and reliability. NTFS enhancements let workstation users encrypt data.

Windows NT Workstation also functions as a Web server and Web content authoring platform. Workstation users can tackle light publishing tasks with Personal Web Server (PWS), which integrates seamlessly with NT's security model. Additional features and capabilities included with the Personal Web Server make it powerful enough for use with a small workgroup over an intranet.

Windows 2000

Microsoft decided to rename the next release of the Windows NT line of operating systems—formerly known as Windows NT 5.0—as Windows 2000. This operating system, with 30 million lines of code, took four years to complete and cost Microsoft more than $1 billion to develop.[1] Windows 2000 Professional is a desktop operating system aimed at businesses of all sizes. Microsoft designed Windows 2000 Professional as the easiest to use Windows yet, with high-level security and significant enhancements for laptop users. The operating system is also designed to provide high reliability.

Unix

Unix is a powerful operating system originally developed by AT&T for minicomputers. Unix can be used on many computer system types and platforms, from personal computers to mainframe systems. Unix benefits companies using both small and large computer systems because it is compatible with different types of hardware and users have to learn only one operating system. This high degree of portability is one reason for Unix's popularity. A user can select a software application that runs under Unix and have a high degree of flexibility in choosing the hardware manufacturer. Unix also makes it much easier to move programs and data among computers or to connect mainframes and personal computers to share resources.

Unix is considered to have a complex user interface with strange and arcane commands. As a result, software developers have provided shells such as Motif from Open Systems Foundation and Open Look from Sun Microsystems. These shells do for Unix much of what older versions of Windows did for DOS—they provide a graphical user interface and shield the users from the complexity of the underlying operating system.

There are many variants of Unix—including HP/UX from Hewlett-Packard, AIX from IBM, UNIX SystemV from UNIX Systems Lab, Solaris from Sun Microsystems, and SCO from Santa Cruz Operations—which have addressed many of the original Unix shortcomings. These variants have created a lack of common standards, although most organizations do not find this to be an insurmountable problem.

Linux

Linux is a free Unix-type operating system (it looks like Unix but does not come from the same source code base) originally created by Linus Torvalds at the University of Helsinki with the assistance of developers around the world. He began his work in 1991 when he released version 0.02 and continued evolving the operating system until 1994, when version 1.0 of the Linux Kernel was released. The current full-featured version is 2.2 (released January 25, 1999), and development continues. (*Linux* is most often pronounced with a short "i" and with the first syllable stressed, as in LIH-nucks.)

Linux is only the kernel of the operating system, the part that controls hardware, manages files, separates processes, and so forth. There are several combinations of Linux with sets of utilities and applications to form a complete operating system. Each of these combinations is called a *distribution* of Linux. There are at least 15 English-language distributions of the operating system for the Intel platform—and none of them are exactly alike. Though they are similar, since they each start with some version of the same freely available source code base, there is no guarantee that a given commercial Linux application will run on all of them. Vendors porting commercial applications to Linux are finding that the subtle (and sometimes not so subtle) differences to Linux distributions can cause problems for customers and developers.[2]

Linux is developed under the GNU General Public License, and its source code is freely available to everyone. This doesn't mean, however, that Linux and its assorted distributions are free—companies and developers may charge money for it as long as the source code remains available. Linux includes true multitasking, virtual memory, shared libraries, demand loading, memory management, Internet networking, and other features consistent with other Unix systems. The Linux desktop and GUI are geared toward more technically oriented users, not the more casual business user. At first use it seems very different and a little cumbersome. Yet it has an innate intuitiveness that quickly becomes familiar.[3]

Linux proponents say that the Linux unconventional development model, in which thousands of far-flung programmers volunteer their efforts for the public good, is more effective at creating features and fixing bugs than proprietary methods of development. Users also like Linux's high reliability and ease of administration. Cendant, a $5.3 billion travel services, real estate, and direct marketing company, deployed Linux servers and workstations in about 60 percent of the 4,500 Super 8, Days Inn, and Howard Johnson hotels connected to its new Unix reservation management system. Stability and ease of administration have been remarkable.[4]

The biggest shortfall for Linux is the scarcity of application programs written for the operating system. Software developers are working to improve Linux's shortcomings as a platform for e-commerce, and more applications continue to become available.[5] The availability of service and support is also a concern. Now that Linux is gaining credibility among IS managers, they are demanding that Linux vendors offer more than repackaged freeware and telephone support. Vendors such as Caldera Systems, Red Hat, and VA Linux Systems offer Linux software packages with integrated tools, removing many of the do-it-yourself headaches of a downloaded operating system, while other vendors sell Linux on mainstream hardware platforms or are porting applications to it.[6]

Red Hat Linux 6.1 is a network operating system from Red Hat Software that has captured 16 percent of the 1998 total server market share.[7] Red Hat taps into tens of thousands of volunteer programmers who generate a steady

RedHat Linux 6.2 is a network operating system from RedHat Software.

(Source: Screen shot copyright © 1999 Red Hat, Inc. All rights reserved. Reprinted with permission from Red Hat, Inc.)

stream of improvements for the Linux operating system. The Linux network operating system is very efficient at serving up Web pages and can manage a cluster of up to eight servers.

Solaris

Solaris is the Sun Microsystems variation of the Unix operating system, with total Unix systems representing 19 percent of the 1998 server market share. It has emerged as the server operating system of choice for large Web sites.[8] Solaris is highly reliable and handles the most demanding tasks. It can supervise servers with as many as 64 microprocessors, and eight such computers can be clustered together to work as one. The Solaris operating system can run on Sun's Sparc family of microprocessors as well as computers with Intel microprocessors.[9]

An example of the unique features of Solaris is fault detection and analysis that lets IS administrators establish policies for problematic conditions. For example, if a processor gets too hot, the capability may instruct a system to shut down the processor and reboot. Another feature, the reconfiguration coordination manager, lets administrators write policies that automatically redistribute system capacity. Current processing capacity is a staggering one million concurrent processes.[10]

Apple Computer Operating Systems

While IBM system platforms traditionally use Intel microprocessors and one of the Windows operating systems, Apple computers typically use Motorola microprocessors and a proprietary Apple operating system—the Mac OS. Although IBM and IBM-compatible computers hold the largest share of the business PC market, Apple computers are also quite popular, especially in the fields of publishing, education, and graphic arts. Many people argue that the Mac OS from Apple Computers has always been the easiest and most intuitive of all operating systems. Apple has also been on the leading edge of introducing interesting new features.

Mac OS 9

This operating system began shipping to users in October 1999. A few of the more interesting tools and features are summarized here.

The Sherlock 2 feature of Mac OS 9 is your personal search detective and personal shopper. It searches the Internet to locate multiple sources for what you are shopping for and compares prices and product availability in one convenient window. With the Multiple Users feature, you can safely share your Macintosh computer with other people. Individual users can select their own home page, e-mail settings, bookmarks to Internet Web pages, desktop pattern, system font, and other system and application preferences. Personal files remain private because users each get their own folder. You can log on by typing your password or by using your voice—Mac OS 9 analyzes your voiceprint to prevent others from logging on as you. With the keychain feature of OS 9, the only password you need to remember is the one that gives you access to the Macintosh computer. After you've logged on, it lets you access your e-mail and other accounts without entering additional user IDs or passwords. To keep your computer running smoothly, Auto Updating delivers the latest software updates and system enhancements from Apple.[11]

Mac OS X ("Ten")

This operating system was announced in January 2000 and is expected to be pre-loaded as the standard operating system on all Macintosh computers beginning in early 2001. It is designed to run on all Apple Macintosh computers using PowerPC G3 and G4 processor chips and requires a minimum of 64 MB of memory. Mac OS X is a completely new implementation of the Mac operating system and includes an entirely new user interface called "Aqua," which provides a new visual appearance for users with luminous and semi-transparent elements such as buttons, scroll bars, windows, and fluid animation to enhance the user's experience.

At the core of Mac OS X is Darwin, Apple's advanced operating system kernel. Darwin represents an entirely new foundation for the Mac OS. Mac OS X can run most of the over 13,000 existing Macintosh applications without modification. However, to take full advantage of Mac OS X's new features, developers must "tune up" their applications to use "Carbon," the updated version of APIs used to program Macintosh computers.

The Mac OS X server is the first modern server operating system from Apple Computer. It provides Unix-style process management. Protected memory puts each service in its own well-guarded chunk of dynamically allocated memory, preventing a single process from going awry and bringing down the system or other services. Preemptive multitasking ensures that each process gets the right amount of CPU time and system resources it needs for optimal efficiency and responsiveness.

Workgroup Operating Systems

To keep pace with today's high-tech society, the technology of the future must support a world in which Internet usage doubles every 100 days. It must support a world in which a single on-line service provider can consume 28 TB in 45 days—activity equal to all use on the Internet for one month in 1997. This is massive scaling that pushes the boundaries of computer science and physics. Powerful and sophisticated operating systems are needed to run the servers that meet these business needs for workgroups.

Windows 2000 Server

Although Windows 2000 includes an update to the Windows desktop, the most important changes are in the version for server computers that manage corporate networks, run applications such as accounting and e-mail, and serve up Web pages from company sites. Although Windows 2000 is the next evolution in the Windows NT operating system, the conversion of users in a Windows NT environment to the Windows 2000 environment is complex.[12]

Microsoft designed Windows 2000 to do a host of new tasks that are vital for Web sites and corporate Web applications that run on the Internet. Besides being more reliable than Windows NT, it's capable of handling extremely demanding computer tasks such as order processing. It can be tuned to run on machines with up to 32 microprocessors—satisfying the needs of all but the most demanding of Web operators. Four machines can be clustered together to prevent service interruptions that are disastrous for Web sites.[13]

Microsoft has a long history of creating proprietary implementations that only work—or only work well—with other Microsoft software. With Windows 2000, the company introduced Active Directory, a proprietary application that works only on computers running Microsoft operating systems. Active Directory lets corporations keep track of every employee, computer, software package, and even scrap of data in one place.[14] Security is integrated with Active Directory through log-on authentication and access control to objects in the directory.

Netware

Netware is a network operating system sold by Novell that has 23 percent of the 1998 total server market share.[15] This operating system supports Windows, Macintosh, and Unix (through third-party) clients. Netware from Novell provides directory software to keep track of computers, programs, and people on a network. Novell's directory software (NDS) runs on Windows 2000, Solaris, and Linux servers, as well as Netware, making it easier for large companies to manage complex networks. NDS and another Novell software product, ZENworks, also make network users more productive. People can log on from any computer on the network and still get their own familiar desktop with all their applications, data, and even their desktop preferences. And with Novell's optional Single Sign-on, network administrators can enable users to access all enterprise applications, regardless of network platform, using only their NDS password. They no longer have to remember multiple passwords for multiple systems, reducing frustration and eliminating a major source of help desk calls.

Figure 4.5 summarizes the server market share for 1998, the most current data available.

Enterprise Operating Systems

A few years ago, computer industry pundits were predicting the end of large-scale computing systems. The future of enterprise computing, they claimed, was in networks of distributed Unix and Windows servers that were less expensive to buy, more flexible, and more adaptable to the changing demands facing businesses in today's marketplace. Indeed, Windows NT, Windows 2000, Solaris, and other workgroup operating systems have versions that are designed to operate in an environment of extremely powerful network servers and to meet enterprise computing needs.

In spite of this vision, there has been a renewed appreciation for the traditional strengths of mainframe computers. The new generation of CMOS-based mainframe computers (computers based on complementary metal oxide semiconductor technology to provide smaller, faster, and less expensive systems—see Chapter 3) provides the computing and storage capacity to meet massive data processing requirements and provides a large number of users with high performance and excellent system availability, strong security, and scalability. In addition, a wide range of application software has been developed to run in the mainframe environment, making it possible to purchase software to address almost any business problem. As a result, mainframe computers remain the computing platform of choice for mission-critical business applications for many companies.

OS/390

In the mid-1990s, IBM began transforming its S/390 mainframe computers by introducing parallel servers. Based on innovative CMOS technology, these parallel systems are smaller in size and larger in capacity than their predecessors and deliver greater performance at lower cost. IBM also introduced a new generation of S/390 server operating system, called OS/390, which is an evolution of an early IBM mainframe operating system called MVS first developed in the 1960s. OS/390 will also run on Plug Compatible Mainframes (PCM) from companies like Hitachi, Comparex, and Amdahl.

FIGURE 4.5

1998 Server Market Share
(Data Source: Steve Hamm, Peter Burrows, and Andy Reinhardt, "Is Windows Ready to Run E-Business?" *Business Week*, January 24, 2000, pp. 154–160.)

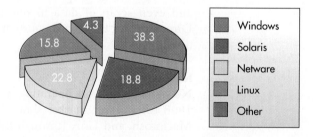

OS/390 evolves the MVS operating system into a complete, network-ready, integrated operational server environment with a full set of modern server functions. OS/390's integrated communications server makes it possible for S/390 computers to manage and share information and transactions across a wide range of platforms and multivendor networks. Within OS/390 is a full implementation of the Unix operating system, so developers can use off-the-shelf Unix packages, develop Unix applications, or port them from other Unix environments. LAN Services, another integrated function of the OS/390 base system, extends S/390's strengths in data and systems management to an enterprise's workstations, whether the operating system they use is Windows, Macintosh, Unix, or AIX (an IBM variant of Unix).

MPE/iX

Multiprogramming Executive with integrated POSIX is the Internet-enabled operating system for the Hewlett-Packard e3000 family of computers using RISC processing. MPE/iX is a robust operating system designed to handle a variety of business tasks, including on-line transaction processing and Web applications. It runs on a broad range of HP e3000 servers—from entry-level to workgroup and enterprise servers within the data centers of large organizations.

Consumer Appliance Operating Systems

New operating systems and other software are changing the way we interact with home entertainment appliances, including TVs, stereos, and other appliances. We look at just two new products here.

Microsoft Pocket PC

Pocket PC is the new name for the Windows CE (Compact Edition) operating system that is preinstalled in read-only memory (ROM) on devices such as digital TV set-top devices, PCs in automobiles, and handheld PCs, which are available in computer and consumer electronics stores and from Microsoft Certified Solution Providers. The standards-based Pocket PC platform makes possible new categories of non-PC business and consumer devices that can communicate with each other, share information with Windows-based PCs, and connect to the Internet. It provides a graphical user interface, incorporating many elements of the familiar Windows user interface, making it easy to use.[16]

Mobil Linux

Mobil Linux is an operating system that runs a new category of tablet-sized Internet-browsing devices called Web pads, which cost under $1,000. This version of Linux has additional driver support for certain video and sound cards, application programs for handwriting recognition, and power management capabilities to minimize battery consumption.[17] The Ericsson Cordless Screen Phone HS210 (see Figure 4.6) allows users to access the Internet, e-mail, and the telephone with a single, contained package. The Screen Phone's interface is a color touch screen, through which the Internet and other features can be quickly and easily accessed. The product is also equipped with a speakerphone to make it more convenient for Web surfers to talk on the phone while they sail through cyberspace. The Screen Phone runs on the Linux operating system.[18]

FIGURE 4.6

The Ericsson Cordless Screen Phone HS210
(Source: Courtesy of Ericcson)

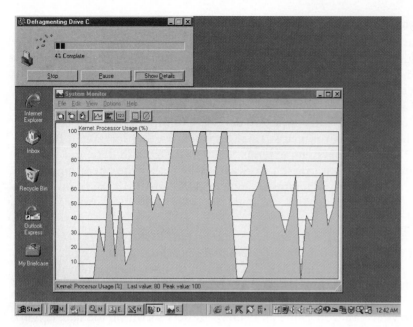

Utility programs can provide detailed information about your entire system and can help improve your computer's performance by scanning for errors and defragmenting your hard drive.

utility programs

programs used to merge and sort sets of data, keep track of computer jobs being run, compress data files before they are stored or transmitted over a network, and perform other important tasks

Utility Programs

Utility programs are used to merge and sort sets of data, keep track of computer jobs being run, compress files of data before they are stored or transmitted over a network (thus saving space and time), and perform other important tasks. Utility programs often come installed on computer systems; a number of utility programs can also be purchased. Norton Utilities, for example, is a set of utility programs that provides a number of capabilities, including checking your hard drive and diskettes for viruses, checking your hard drive for bad storage locations and removing them, and compressing data through a disk compression software package. Data or disk compression allows more data and programs to be stored on the same hard disk by eliminating repeated data and empty storage space. When programs and data are needed, decompression routines make the programs and data available. In some cases, data compression software can almost double the amount of stored data, increasing capacity without high costs. Utility programs can meet the needs of an individual, a workgroup, or an enterprise, as listed in Table 4.3. They perform useful tasks—from tracking jobs to monitoring system integrity.

IBM has created systems management software that allows individual support people to monitor the growing number of desktop computers in a business. With this software, the support people can sit at their personal computers and check or diagnose problems, like a hard disk failure, on individual computer systems on a network. The support people can even repair individual systems anywhere on the organization's network, often without having to leave their desks. The direct benefit is to the system manager, but the business gains from having a smoothly functioning information system.

TestDrive is another example of a utility program. It allows you to try out a software package before you purchase it. Using TestDrive, you can download certain software you are thinking about purchasing from the Internet. TestDrive allows you to use the software for a set length of time. If you like the software, TestDrive enables you to purchase it using a credit card. If you do not like it, TestDrive will automatically delete it from your hard disk.

TABLE 4.3

Examples of Utility Programs

Personal	Workgroup	Enterprise
Software to compress data so that it takes less hard disk space	Software to provide detailed reports of workgroup computer activity and status of user accounts	Software to archive contents of a database by copying data from disk to tape
Screen saver	Software that manages an uninterruptable power source to do a controlled shutdown of the workgroup computer in the event of a loss of power	Software that compares the content of one file with another and identifies any differences
Virus detection software	Software that reports unsuccessful user log-on attempts	Software that reports the status of a particular computer job

APPLICATION SOFTWARE

As discussed earlier in this chapter, the primary function of application software is to apply the power of the computer to give individuals, workgroups, and the entire enterprise the ability to solve problems and perform specific tasks. When you need the computer to do something, you use one or more application programs. The application programs then interact with systems software. Systems software then directs the computer hardware to perform the necessary tasks.

Suppose a manager is concerned that too many employees are getting overtime pay by working more than 40 hours each week, even though many others are working less than 40 hours per week. She would like to have those working below the 40-hour threshold replace those over the threshold, and hence avoid the time-and-a-half overtime pay rate. The manager can enlist a computer to print the names of all employees working significantly more or significantly less than 40 hours per week on average over the last three months.

Programs like this that complete sales orders, control inventory, pay bills, write paychecks to employees, and provide financial and marketing information to managers and executives are examples of application software. Most of the computerized business jobs and activities discussed in this book involve the use of application software.

Types and Functions of Application Software

The key to unlocking the potential of any computer system is application software. A company can either develop a one-of-a-kind program for a specific application (called **proprietary software**) or purchase and use an existing software program (sometimes called **off-the-shelf software**). It is also possible to modify some off-the-shelf programs, giving a blend of off-the-shelf and customized approaches. These different sources of software are represented in Figure 4.7. The relative advantages and disadvantages of proprietary software and off-the-shelf software are summarized in Table 4.4.

Proprietary Application Software

Software to solve a unique or specific problem is called *proprietary application software*. This type of software is usually built, but it can also be purchased from an outside company. If the organization has the time and IS talent, it may opt

proprietary software

a one-of-a-kind program for a specific application

off-the-shelf software

existing software program

FIGURE 4.7

Sources of Software: Proprietary and Off-the-Shelf

Some off-the-shelf software may be modified to allow some customization.

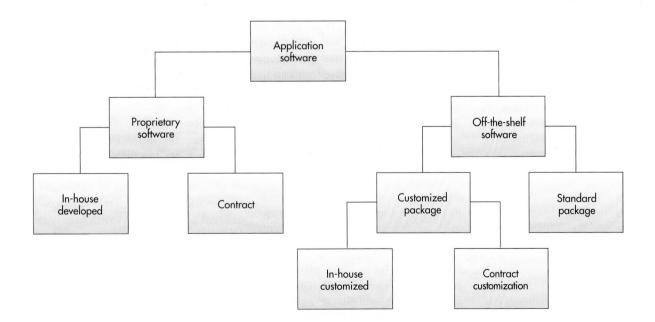

Proprietary Software		Off-the-Shelf Software	
Advantages	**Disadvantages**	**Advantages**	**Disadvantages**
You can get exactly what you need in terms of features, reports, and so on.	It can take a long time and significant resources to develop required features.	The intial cost is lower since the software firm is able to spread the development costs over a large number of customers.	An organization might have to pay for features that are not required and never used.
Being involved in the development offers a further level of control over the results.	In-house system development staff may become hard pressed to provide the required level of ongoing support and maintenance because of pressure to get on to other new projects.	There is a lower risk that the software will fail to meet the basic business needs—you can analyze existing features and the performance of the package.	The software may lack important features, thus requiring future modi-cation or customization. This can be very expensvie because users must adopt future releases of the software.
There is more flexibility in making modifications that may be required to counteract a new initiative by one of your competitors or to meet new supplier and/or customer requirements. A merger with another firm or an acquisition also will necessitate software changes to meet new business need.	There is more risk concerning the features and performance of the software that has yet to be developed.	Package is likely to be of high quality since many customer firms have tested the software and helped identify many of its bugs.	Software may not match current work processes and data standards.

TABLE 4.4

A Comparison of Proprietary and Off-the-Shelf Software

in-house development

development of application software using the company's resources

contract software

software developed for a particular company

for **in-house development** for all aspects of the application programs. Alternatively, an organization may obtain customized software from external vendors. For example, a third-party software firm, often called a value-added software vendor, may develop or modify a software program to meet the needs of a particular industry or company. A specific software program developed for a particular company is called **contract software**.

Off-the-Shelf Application Software

Software can also be purchased, leased, or rented from a software company that develops programs and sells them to many computer users and organizations. Software programs developed for a general market are called off-the-shelf software packages because they can literally be purchased "off the shelf" in a store. Many companies use off-the-shelf software to support business processes.

Customized Package

In some cases, companies use a blend of external and internal software development. That is, off-the-shelf software packages are modified or customized by in-house or external personnel. For example, a software developer may write a collection of programs to be used in an auto body shop that includes such features as generating estimates, ordering parts, and processing insurance. Body shops of all types have these needs. Designed properly—and with provisions for minor tailoring for each user—the same software package can be sold to many users. However, since each body shop has slightly different requirements, some modifications to the software may be needed. As a result, software vendors often provide a wide range of services, including installation of their standard software, modifications to the software required by the customer, training of the end users, and other consulting services.

Some software companies encourage their customers to make changes to their software. In some cases, the software company supplying the necessary software will make the necessary changes for a fee. Other software companies, however, will not allow their software to be modified or changed by those purchasing or leasing it.

application service provider

a company that provides both end user support and the computers on which to run the software from the user's facilities

Another approach to obtaining a customized software package is through the use of an application service provider. An **application service provider** is a company that provides both end user support as well as the computers on which to run the software from the user's facilities. They can also take a complex corporate software package and simplify it for the users so it is easier to set up and manage. They provide contract customization of off-the-shelf software, and they speed deployment of new applications while helping IS managers avoid implementation headaches, reducing the need for many skilled IS staff members and reducing project start-up expenses. Such an approach allows companies to devote more time and resources to more important tasks. The use of an application service provider makes the most sense for relatively small, fast-growing companies with limited information system resources. It is also a good strategy for companies looking to deploy a single, functionally focused application quickly, such as setting up an e-commerce Web site or supporting expense reporting. Contracting with an application service provider may make less sense, however, for larger companies with major systems that have their technical infrastructure already in place.

There is a long tradition of service companies managing a client's computing systems—the so-called outsourcers like IBM and Electronic Data Services. But what's different with application service providers is the use of the Internet. Through application service providers, software can be made available quickly. Users can gather and update information from their applications via simple Web browsers, so even users who recoil from dealing with complicated software can feel comfortable and be trained quickly. Another improvement over the old outsourcing model is that the data traffic is carried over public networks instead of far more expensive private networks. The software that companies are using is changing, too. Application service providers employ standard software packages, replacing, in many cases, custom programs that require small armies of programmers to support. That means application service providers can also devise simple methods for handling all their clients' applications, driving costs even lower.

Using an application service provider is not without risks—sensitive information could be compromised in a number of ways, including unauthorized access by employees or computer hackers, the application service provider being unable to keep its computers and network up and running as consistently as is needed, or a disaster disabling the application service provider's data center, temporarily putting an organization out of business. These are legitimate concerns that the application service provider must address.

Perhaps the biggest advantage of employing an application service provider is that it frees in-house corporate resources from staffing and managing complex computing projects so that they can focus on more important things. Read more about the use of application service providers in the "E-Commerce" special interest box.

Personal Application Software

There are literally hundreds of computer applications that can help individuals at school, home, and work. Personal application software includes general-purpose tools and programs that can support a number of individual needs. For example, a graphics program can be purchased to help a sales

E-COMMERCE

Samsung Employs ASP to Accelerate E-Commerce Initiative

A new breed of company called the application service provider (ASP) is transforming software from expensive, complex products into a more affordable and easy-to-use service. The goal of the ASP is to make sophisticated software readily available. Rather than taking many months, or even years, to install large, complex programs, these service providers can do it in a few weeks or months. As a result, ASPs are making it much easier and cheaper for companies to get into e-commerce.

Samsung Electronics, with 1999 sales revenue of $23 billion, is a world leader in the electronics industry. The Korea-based concern has operations in about 50 countries with 54,000 employees worldwide. The company consists of three main business units: Digital Media Systems, Semiconductors, and Information & Communications Businesses. As the world's largest manufacturer of TFT-LCDs (thin film transistor-liquid crystal displays), Samsung can find its display products in one out of every four notebook PCs or one in every five LCD monitors.

Samsung announced in mid-February 2000 that it had developed the first-ever 64-MB SDRAM with 266-MHz speed expected to lead the formation of a graphics-oriented digital environment. The new SDRAM is 30 percent faster than competing devices and will support three-dimensional moving pictures. Samples will be provided to companies specializing in graphics controller chip sets in June 2000. Samsung expects to earn $300 million from sales of this device alone in 2000.

Samsung planned to build a Web site for selling this chip and other memory chips—a potentially huge advantage in a market dominated by Asian competitors that are reluctant to sell on the Web. But the cost to buy the computers and software to set up the Web site was $2.5 million. Worse, it would take over six months to set up the site and get it running smoothly. Such a delay would evaporate the lead Samsung had over its competitors. The company found a better way—it employed

an application service provider. Instead of Samsung's resources being distracted to manage a half-dozen companies to configure the e-commerce software, set up the network, design business processes, and manage the system day to day, the application service provider did all this. And instead of building or buying the e-commerce software, Samsung rents it. As a result, the Web site was launched in a matter of weeks! Using an ASP put Samsung even further ahead of its competition.

Discussion Questions

1. What advantages did Samsung gain through use of an application service provider?
2. Are there any potential risks with this approach for Samsung?

Critical Thinking Questions

3. Given that Samsung already has a lead on its competitors and that they are reluctant to sell their products over the Web, does the use of an application service provider to build a Web site represent an unfair advantage over the competition?
4. How would you address complaints from members of the information systems organization that use of an application service provider was reducing the amount of work they had to do?

Sources: Adapted from "Samsung Develops High-Speed SDRAM for Graphics," February 14, 2000, accessed at the Samsung Web page March 1, 2000, at http://samsungelectronics.com/news/global/ns_378_en.html; John Madden, "Digex Hosts Apps, Securely," *PC Week*, January 3, 2000, p. 32; Matt Hicks, "Great ASP-irations," *PC Week*, September 8, 1999, pp. 67, 74; Timothy S. Mullaney and Peter Burrows, "Suddenly, Software Isn't a Product, It's a Service," *Business Week*, June 21, 1999, pp. 134–138; and Michael Moeller, "Again, a Trendsetter," *Business Week*, June 21, 1999, p. 138.

manager develop an attractive presentation to give to the salesforce at its annual meeting. A spreadsheet program allows a financial executive to test possible investment outcomes. The primary programs are word processing, spreadsheet analysis, database, graphics, and on-line services. Advanced software tools—like project management, financial management, desktop publishing, and creativity software—are finding more and more use in business. The features of personal application software are summarized in Table 4.5. In addition, there are literally thousands of other personal computer applications to perform specialized tasks: to help you do your taxes, get in shape, lose weight, get medical advice, write wills and other legal documents, make repairs to your computer, fix your car, write music, and edit your pictures and videos (see Figures 4.8 and 4.9). This type of software, often called user software or personal productivity software, includes the general-purpose tools and programs that support individual needs.

Word Processing

If you write reports, letters, or term papers, word processing applications can be indispensable. The majority of personal computers in use today have word processing applications installed. Such applications can be used to create, edit, and print documents. Most come with a vast array of features, including those for checking spelling, creating tables, inserting formulas, creating graphics, and much more. This book (and most like it) was entered into a word processing application using a personal computer. (See Figure 4.10.)

Spreadsheet Analysis

People use spreadsheets to prepare budgets, forecast profits, analyze insurance programs, summarize income tax data, and analyze investments. Whenever numbers and calculations are involved, spreadsheets should be considered. Features of spreadsheets include graphics, limited database capabilities, statistical analysis, built-in business functions, and much more. (See Figure 4.11.)

TABLE 4.5

Examples of Personal Productivity Software

Type of Software	Explanation	Example	Vendor
Word processing	Create, edit, and print text documents	Word WordPerfect	Microsoft Corel
Spreadsheet	Provide a wide range of built-in functions for statistical, financial, logical, database, graphics, and data and time calculations	Excel Lotus 1-2-3 Quattro Pro	Microsoft Lotus/IBM Originally developed by Borland
Database	Store, manipulate, and retrieve data	Access Approach FoxPro dBASE	Microsoft Lotus/IBM Microsoft Borland
On-line information services	Obtain a broad range of information from commercial services	America Online CompuServe Prodigy	America Online CompuServe Prodigy
Graphics	Develop graphs, illustrations, and drawings	Illustrator FreeHand	Adobe Macromedia
Project management	Plan, schedule, allocate, and control people and resources (money, time, and technology) needed to complete a project according to schedule	Project for Windows On Target Project Schedule Time Line	Microsoft Symantec Scitor Symantec
Financial management	Provide income and expense tracking and reporting to monitor and plan budgets (some programs have investment portfolio management features)	Managing Your Money Quicken	Meca Software Intuit
Desktop publishing (DTP)	Works with personal computers and high-resolution printers to create high-quality printed output, including text and graphics; various styles of pages can be laid out; art and text files from other programs can also be integrated into "published" pages	QuarkXPress Publisher PageMaker Ventura Publisher	Quark Microsoft Adobe Corel
Creativity	Helps generate innovative and creative ideas and problem solutions. The software does not propose solutions, but provides a framework conducive to creative thought. The software takes users through a routine, first naming a problem, then organizing ideas and "wishes," and offering new information to suggest different ideas or solutions	Organizer Notes	Macromedia Lotus

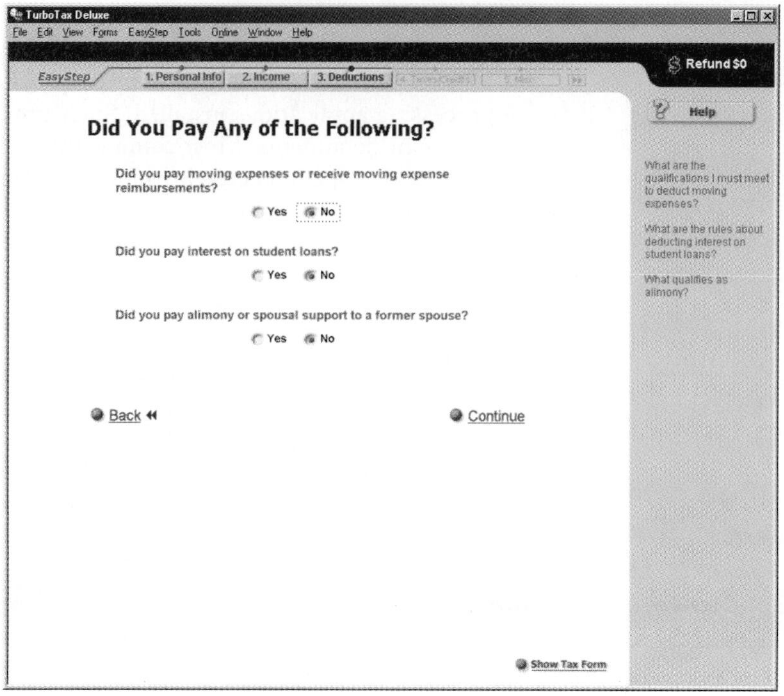

FIGURE 4.8

TurboTax

Tax preparation programs can save hours of work and are typically more accurate than doing a tax return by hand. Programs can check for potential problems and give you help and advice about what you may have forgotten to deduct.
(Source: Courtesy of Intuit)

Database Applications

Database applications are ideal for storing, manipulating, and retrieving data. These applications are particularly useful when you need to manipulate a large amount of data and produce reports and documents. Database manipulations include merging, editing, and sorting data. The uses of a database application are varied. You can keep track of a record or CD collection, the items in your apartment, tax records, and expenses using a database application. A student club can use a database to store names, addresses, phone numbers, and dues paid. In business, a database application can help process sales orders, control inventory, order new supplies, send letters to customers, and pay employees. A database can also be a front end to another application. For example, a database application can be used

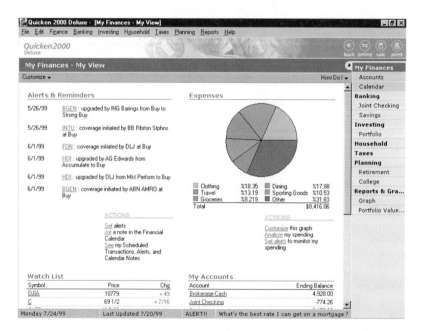

FIGURE 4.9

Quicken

Off-the-shelf financial management programs are useful for paying bills and tracking expenses.
(Source: Courtesy of Intuit)

FIGURE 4.10

Word Processing Program

Word processing applications can be used to write letters, holiday greeting cards, work reports, and term papers.

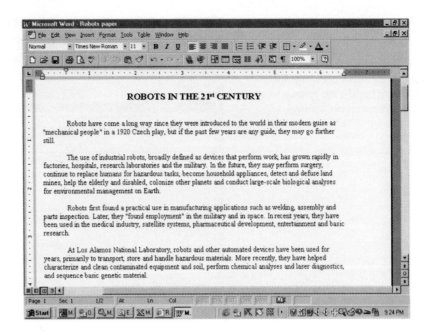

FIGURE 4.10

Word Processing Program

Word processing applications can be used to write letters, holiday greeting cards, work reports, and term papers.

to enter and store income tax information; the stored results can then be exported to other applications, such as a spreadsheet or tax preparation application. (See Figure 4.12.)

Graphics Programs

It is often said that a picture is worth a thousand words. With today's graphics programs, it is easy to develop attractive graphs, illustrations, and drawings. Graphics programs can be used to develop advertising brochures, announcements, and full-color presentations. If you are asked to make a presentation at school or work, you can use a graphics program to develop and display slides while you are making your talk. A graphics program can be used to help you make a presentation, a drawing, or an illustration. (See Figure 4.13.)

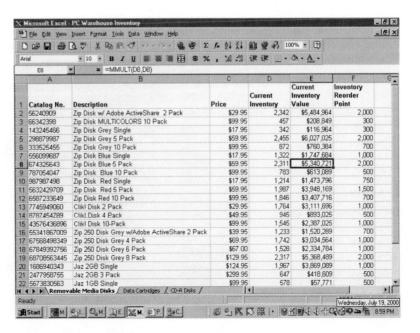

FIGURE 4.11

Spreadsheet Program

Spreadsheet programs should be considered when calculations are required.

FIGURE 4.12

Database Program

Once entered into a database application, information can be manipulated and used to produce reports and documents.

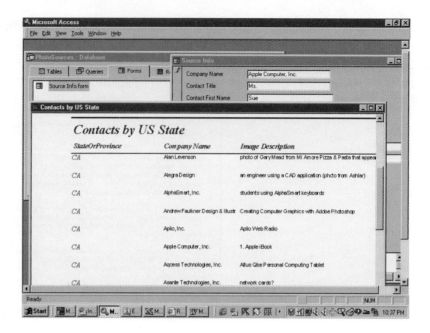

On-Line Information Services

On-line services allow you to connect a personal computer to the outside world through phone lines. Using an on-line service, you can get investment information, make travel plans, and check news from around the world. You can also get prices and features for most consumer items, learn about companies, send e-mail to friends and family, learn about degree programs offered by colleges and universities around the world, and search for job openings in your area. (See Figure 4.14.)

Software Suites

software suite

a collection of single-application software packages in a bundle

A **software suite** is a collection of single-application software packages in a bundle. Software suites can include word processors, spreadsheets, database management systems, graphics programs, communications tools, organizers, and more. There are a number of advantages to using a software suite. The software programs have been designed to work similarly, so once you learn the basics for one application, the other applications are easy to learn and use. Buying software

FIGURE 4.13

Graphics Program

Graphics programs can help you make a presentation at school or work. They can also be used to develop attractive brochures, illustrations, drawings, and maps.

(Source: Courtesy of Adobe Systems Inc.)

FIGURE 4.14

On-Line Services

On-line services provide instant access to information. Prices of cars and trucks, travel discounts, information about companies, stock market data, and much more is available with a few key strokes using an on-line service.

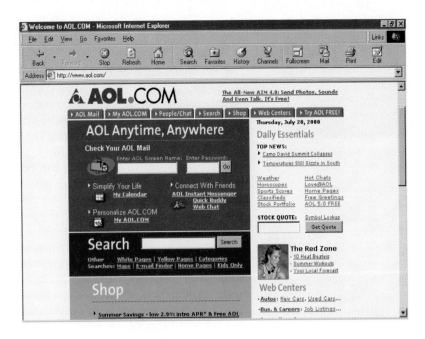

object linking and embedding (OLE)

a software feature that allows you to copy text from one document to another or embed graphics from one program into another program or document

FIGURE 4.15

Software Suite

A software suite, such as Microsoft Office, offers a collection of powerful programs, including word processing, spreadsheet, database, graphics, and other programs. The programs in a software suite are designed to be used together. In addition, the commands, icons, and procedures are the same for all programs in the suite. (Source: Courtesy of Microsoft Corporation.)

in a bundled suite is cost-effective; the programs usually sell for a fraction of what they would cost individually.

Microsoft Office, Corel's WordPerfect Office, and Lotus SmartSuite are examples of popular general-purpose software suites for personal computer users. (See Figure 4.15.) Each of these software suites includes a spreadsheet program, word processor, database program, and graphics package with the ability to move documents, data, and diagrams among them. Thus, a user can create a spreadsheet and then cut and paste that spreadsheet into a document created using the word processing application.

The Microsoft Office 2000 suite includes Excel for spreadsheet, Word for word processing, PowerPoint for presentation preparation, Outlook for messaging and collaboration, PhotoDraw for business graphics (including the ability to edit photos), FrontPage to create Web sites, and Publisher for desktop publishing. Microsoft Office goes beyond its role as a mainstream package of ready-to-run applications with the extensive custom development facilities of Visual Basic for Applications (VBA)—a built-in scripting facility that is part of every Office application. VBA provides a means of enhancing off-the-shelf applications to allow users to tailor the programs with powerful scripting facilities.

Since one or more applications in a suite may not be as desirable as the others, some people still prefer to buy separate packages. Another issue with the use of software suites is the large amount of main memory required to run them effectively. For example, many users find that they must spend hundreds of dollars for additional internal memory to upgrade their personal computer to be able to run a software suite. Continual debates rate one vendor's spreadsheet superior to another vendor's, and yet a third vendor may have the best word processing package. Thus, some users prefer using individual software packages from different vendors rather than a software suite from a single vendor.

Object Linking and Embedding (OLE)

To enable different application programs to work together, many operating systems support **object linking and embedding (OLE)**. With OLE, you can copy text from one document to another or embed graphics from one program into another program or document, such as a word processing report. You can also link many documents and files, such as a table developed in a spreadsheet program that can be linked to a word processing document. When you make a link, you are actually referencing the original document or program. Changes are always made to the source document or program and transferred to the linked documents, such as one or more word processing reports.

If the operating system and the application support OLE, data can be shared among applications in three ways: copying, linking, and embedding. In the following explanation, the data you transfer among applications is called an object. An object can be a picture, graph, text, spreadsheet, or other data. The **server application** is the application that supplies objects that you place into other applications. The **client application** is the application that accepts objects from other applications. An application can be a server, a client, or both. Figure 4.16 shows a graphic image, spreadsheet, and project schedule being incorporated into a word processing document.

Copying

You can copy data from one application and place it in another. Both applications will then display the data. The copy method is used when you do not want to change data shared between applications. To change data, you must first update the data in the client application and then repeat the copy procedure. You can also capture a screen display from one program and copy it into another by pressing the Shift and Print Screen keys at the same time to "capture a copy" of the currently displayed screen, starting the second application (e.g., a word processing program), and finally pressing the Ctrl and V keys at the same time.

Linking

You **link** an object when you want any changes made to the server object to automatically appear in all linked client documents. You must first save an object in the server application before you can link it to any client documents. To link an object, use the Past Link command. Linking an object provides the client documents with a screen image of the object. The real object resides in the server document. Double clicking the object in any client document opens the server application that supplied the object so that you can then edit that object. Any changes you make to the object in the server application automatically appear in all linked client documents.

Embedding

You **embed** an object when you want it to become part of the client document. To embed an object, use the Edit-Paste command. (If the applications you are using support OLE, using the Paste command embeds the object. If the applications do not support OLE, using the Paste command copies the object.) You do not have to save an object in the server application before embedding it in a client document. This makes the document more portable since the server document is not required. Double clicking on the object opens the object in the client server application that supplied the object. Any changes you make to the object appear only in the client documents containing the object. However, to see those changes in any client application, you must use the File-Update command. When you embed an object, each client document receives a copy of the object. The object becomes part of the client document. Documents containing embedded objects require more memory and hard disk storage than documents linked to objects. After you embed an object in the client document, the server document is no longer required.

Workgroup Application Software

The software class known as **groupware** cannot be concretely defined but is, in general, software that helps groups of people work together more efficiently and effectively. **Collaborative computing software**, an only slightly better term, at least conveys the sense that teams are working toward a common goal. Collaborative computing software can support a team of managers working on the

---

server application

the application that supplies objects you place into other applications

client application

the application that accepts objects from other applications

linking

procedure used when you want any changes made to the server object to automatically appear in all linked client objects

embedding

procedure used when you want an object to become part of the client document

groupware

software that helps groups of people work together more efficiently and effectively

collaborative computing software

software that helps teams of people to work together toward a common goal

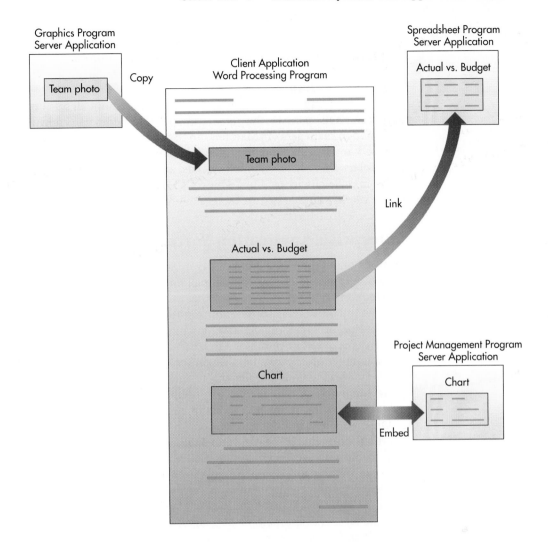

FIGURE 4.16

Object Copying, Linking, and Embedding

A project manager wants to create a project status report using a word processing program. The manager wants to include a copy of a photo of the project team stored in digital format from a graphics program, a linked spreadsheet showing project actual costs vs. budget, and an embedded chart showing the project schedule stored in a project management program. Assuming that the project manager is using personal productivity software in an operating environment that supports object linking and embedding, this figure shows how this can be accomplished.

same production problem, letting them share their ideas and work via connected computer systems. Examples of such software include group scheduling software, electronic mail, and other software that enables people to share ideas. New York–based Swiss Reinsurance America deployed a claims processing application to more than 300 workers. By streamlining its work processes and using workgroup software, it was able to reduce the number of steps needed to process a claim from 18 to 7 and reduce the time it takes to process a claim from three days to one.

Lotus Notes

Lotus has defined knowledge management as the ability to provide individuals and groups of users with a method to find, access, and deliver valuable information in a coherent fashion. Its Lotus Notes product is an attempt to provide this ability. Lotus Notes gives companies the capability of using one software package, and one user interface, to integrate many business processes. For example, it can allow a global team to work together from a common or shared set of documents, have electronic discussions using common threads of discussion, and schedule team meetings. A key design feature is to make Notes 5 very easy to use.

As Lotus Notes matured, Lotus added services to it and renamed it Domino, and now an entire third-party market has emerged around building collaborative software based on Domino. These products, which include Changepoint's Involv, remove the burden of Notes administration and broaden the application's scope to better support the Internet. For example, Domino.Doc is a Domino-based document management application with built-in workflow and archiving capabilities. Its "life cycle" feature tracks a document through the review, approval, publishing, and archiving processes. Similarly, the workflow integration adds support for multiple roles, log tracking, and distributed approval.

Group Scheduling

Group scheduling is another form of groupware, but not all software schedulers approach their tasks in the same way. Some schedulers, known as personal information managers (PIMs), tend to focus on personal schedules and lists, as opposed to coordinating the schedules and meetings of a team or group. Schedulers do not suit everyone's needs, and if they are not truly required, they could impede efficiency. The "Three Cs" rule for successful implementation of groupware is summarized in Table 4.6.

Enterprise Application Software

Software that benefits the entire organization can also be developed or purchased. A fast-food chain, for example, might develop a materials ordering and distribution program to make sure that each fast-food franchise gets the necessary raw materials and supplies during the week. This materials ordering and distribution program can be developed internally using staff and resources in the IS department or purchased from an external software company. Table 4.7 lists a number of applications that can be addressed with enterprise software.

Many organizations are moving to integrated enterprise software that supports supply chain management (movement of raw materials from suppliers through shipment of finished goods to customers), as shown in Figure 4.17.

Organizations can no longer respond to market changes using nonintegrated information systems based on overnight processing of yesterday's business transactions, conflicting data models, and obsolete technology. As a result, many corporations are turning to **enterprise resource planning (ERP)** software, a set of integrated programs that manage a company's vital business operations for an entire multisite, global organization. Thus, an ERP system must be able to support multiple legal entities, multiple languages, and multiple currencies. Although the scope of an ERP system may vary from vendor to vendor, most ERP systems provide integrated software to support manufacturing and finance. In addition to these core business processes, some ERP systems may be capable of supporting additional business functions such as human resources, sales, and distribution.

Successful linking of these business functions means that business requirements must come first and technology constraints second. Software will need to be rethought and reconfigured to reflect that reality. Unfortunately, the effort to implement ERP software is great and the path to success is full of hazards. Software vendors that provide integrated enterprise software are listed in Table 4.8.

Most ERP vendors specialize in software that addresses the needs of well-defined markets, such as automotive, semiconductor, petrochemical, and food/beverage manufacturers with solutions targeted to meet specific needs in those industries. ERP has become one of the most lucrative segments of the software market as companies continue to switch from old information systems to more efficient client/server-based systems.

Increased global competition, new executive management needs for control over the total cost and product flow through their enterprises, and more customer

enterprise resource planning (ERP)

a set of integrated programs that manage a company's vital business operations for an entire multisite, global organization

TABLE 4.6

Ernst & Young's "Three Cs" Rule for Groupware

Convenient	If it's too hard to use, it doesn't get used; it should be as easy to use as the telephone.
Content	It must provide a constant stream of rich, relevant, and personalized content.
Coverage	If it isn't close to everything you need, it may never get used.

TABLE 4.7

Examples of Enterprise
Application Software

Accounts receivable	Sales ordering
Accounts payable	Order entry
Airline industry operations	Payroll
Automatic teller systems	Human resource management
Cash-flow analysis	Check processing
Credit and charge card administration	Tax planning and preparation
Manufacturing control	Receiving
Distribution control	Restaurant management
General ledger	Retail operations
Stock and bond management	Invoicing
Savings and time deposits	Shipping
Inventory control	Fixed asset accounting

interactions are driving the demand for enterprisewide access to current business information. ERP offers integrated software from a single vendor that helps meet those needs. The primary benefits of implementing ERP include eliminating inefficient systems, easing adoption of improved work processes, improving access to data for operational decision making, standardizing technology vendors and equipment, and enabling the implementation of supply chain management. Read the "Ethical and Societal Issues" box to learn how companies must be careful of whom they partner with to build critical enterprise application software.

PROGRAMMING LANGUAGES

programming languages

coding schemes used to write both systems and application software

FIGURE 4.17

Use of Integrated Supply Chain Management Software

Both systems and application software are written in coding schemes called **programming languages**. The primary function of a programming language is to provide instructions to the computer system so that it can perform a processing activity. Specialized IS professionals work with programming languages, which are sets of symbols and rules used to write program code. Programming involves translating what a user wants to accomplish into a code that the computer can understand and execute. Like writing a report or a paper in English, writing a computer program in a programming language requires that the programmer follow a set of

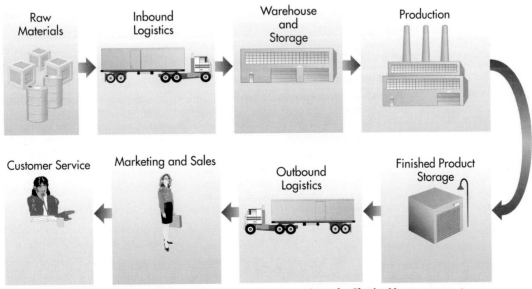

Raw Materials Inbound Logistics Warehouse and Storage Production

Customer Service Marketing and Sales Outbound Logistics Finished Product Storage

Integrated Enterprise Software to Support Supply Chain Management

TABLE 4.8

Selected Enterprise Resource Planning Vendors

SAP	Baan
Oracle	SSA
PeopleSoft	Marcam
Dun & Bradstreet	QAD
JD Edwards	Ross Systems

syntax

a set of rules associated with a programming language

rules. Each programming language uses a set of symbols that have special meaning. Each language also has its own set of rules, called the **syntax** of the language. The language syntax dictates how the symbols should be combined into statements capable of conveying meaningful instructions to the CPU.

Standards and Characteristics

Programming languages, like computers, have evolved over time. For the most part, their evolution has been driven by the desire to efficiently apply the power of information processing to as wide a variety of problem-solving activities as possible. This evolution has been somewhat regulated through standards. Earlier programming languages suffered from a lack of consistency in statements and procedures, resulting in one form of a programming language not working with another program written in a different form of the same language. To avoid this inefficiency, the American National Standards Institute (ANSI) developed standards for popular programming languages.

A program language standard is a set of rules that describe how programming statements and commands should be written. A rule that "variable names must start with a letter" is an example of a standard. (A variable is an item that can take on different values.) Program variable names such as SALES, PAYRATE, and TOTAL follow the standard because they start with a letter, while variables such as %INTEREST, $TOTAL, and #POUNDS do not. By following program language standards, organizations can focus less on code writing and more on efforts to use programming languages to most effectively solve business problems.

Programming languages were developed to help solve a particular type of problem. Since they were each designed for different problems, they contain different attributes. Each of the attributes in Table 4.9 represents two extremes, with most languages falling somewhere between these extremes.

TABLE 4.9

Programming Language Attributes

Extreme 1	Extreme 2
Supports programming of batch processing systems with data collected into a set and processed at one time.	Supports programming of real-time systems with each data transaction processed when it occurs.
Requires programmer to write procedure-oriented code, describing step by step each action the computer must take.	Enables a programmer to write nonprocedure-oriented code, describing the end result desired without having to specify how to accomplish it.
Supports business applications that require the ability to store, retrieve, and manipulate alphanumeric data and process large files.	Supports sophisticated scientific computations.
Programmers write code with a relatively high level of errors.	Programmers write code with a relatively low level of errors.
Programmers are less productive and able to create only a small amount of code per unit time.	Programmers are more productive and are able to create a large amount of code per unit time.

ETHICAL AND SOCIETAL ISSUES
Beware of Your Partners

The U.S. food and beverage industry is a $300 billion-a-year business. One of the biggest challenges food manufacturers face is quickly identifying and acquiring the ingredients they need to make a particular product, because the grade and quality of ingredients can vary widely. Juice manufacturers, for example, may need less sugar if the fruit that is available for purchase is at a certain stage of ripeness. Also, companies need a system to quickly identify customers' changing demands and communicate those needs to suppliers.

Cargill, headquartered in Minneapolis, is an international marketer, processor, and distributor of agricultural, food, financial, and industrial products and services with some 82,000 employees in more than in 59 countries. In North America, Cargill sells over $6 billion of food and beverage ingredients to manufacturers and purchases many packaging components annually.

Cargill and Ariba (a software services provider) announced Novopoint.com, an open, Internet business-to-business exchange for food and beverage manufacturers and their suppliers. Novopoint will be designed by the food and beverage industry to provide a standardized, automated trading and logistics infrastructure for ingredient and materials transaction and supply chain optimization—linking manufacturers and suppliers in a just-in-time fashion. With the new system, food manufacturers can purchase more efficiently and suppliers can reach a broader set of buyers. Transactions that currently generate multiple documents and phone calls will now be captured once in a secure and confidential digital environment. Status of orders will be easily traceable along the delivery chain with the click of the mouse. It is estimated that by using the exchange, the industry will save between $5 billion and $10 billion annually through economies of scale and streamlined processes.

Novopoint will be operated by a neutral, independent company, with the majority of equity ownership to be held by companies recognized as leaders in the food and beverage industry. Discussions have begun with other prospective investors who are recognized world leaders in the food and beverage industry.

Ariba will continually add new functions to Novopoint for users, including logistics and electronic payments. These features will be designed to make jobs easier for the people working in the purchasing, finance, plant operations, sales, and research and development departments.

Although many members of the food and beverage industry were excited by the announcement of Novopoint, Scott Sexton, cofounder of Inc2inc Technologies, was upset. The reason? With Novopoint.com, Ariba has created a direct competitor to Inc2inc's own food industry exchange—which, ironically, Ariba had helped Sexton launch a few weeks prior to the announcement of Novopoint.

As the business-to-business commerce market undergoes a flurry of partnerships and acquisitions, Sexton learned a lesson that many others surely will come to know: be careful with whom you make deals because your partner today could be your competitor tomorrow.

Inc2inc fears losing the competitive edge its partnership with Ariba created. Ariba was helping Inc2inc develop value-added features for the exchange. Now Ariba could transfer that information to Novopoint. Ariba's argument for working with multiple partners is simple: it cannot afford to lock itself into an exclusive agreement with a single customer, since that would take it out of the market.

Discussion Questions

1. What are the benefits that an electronic exchange can bring the food and beverage industry?
2. What sort of transactions must be handled by the electronic exchange?

Critical Thinking Questions

3. Do you think Ariba took unfair advantage of the development work done for Inc2inc? Why or why not?
4. What could Inc2inc have done to protect against this situation?

Sources: Adapted from Lee Pender, "B2B Partners Beware," *PC Week*, March 20, 2000, pp. 1, 20; Julia King, "Cargill Takes Another Stake in Marketplace," *Computerworld*, March 20, 2000, p. 14; and Bill Brady, "Cargill, Ariba Announce Food and Beverage Open Internet B2B Marketplace," Cargill Web site "Press Releases," at http://www.cargill.com/today/releases/00_3_14novopoint.htm March 14, 2000.

The Evolution of Programming Languages

The desire to use the power of information processing efficiently in problem solving has pushed development of newer programming languages. The evolution of programming languages is typically discussed in terms of generations of languages.

First-Generation Languages

machine language

the first generation of programming languages

The first generation of programming languages is **machine languages**. To give you an idea of the complexity of machine language, Figure 4.18 shows how even a simple instruction requires the use of many binary symbols. Machine language is

| 00100101 | 00000010 | 00001101 |

Operation code
(i.e., add,
subtract)

Address location 1
(i.e., first number
to be added)

Address location 2
(i.e., second number
to be added)

FIGURE 4.18

A Simplified Language
Instruction

A machine language instruction
consists of all 0s and 1s. Here,
just a few elements of a single
instruction are presented.

assembly language

second-generation language that
replaced binary digits with symbols
programmers could more easily
understand

considered a low-level language because there is no program coding scheme less sophisticated than that using the binary symbols 1 and 0. ASCII (American Standard Code for Information Interchange) uses all 0s and 1s to represent letters of the alphabet. As this is the language of the CPU, text files translated into ASCII binary sets can be read by almost every computer system platform.

Second-Generation Languages

Developers of programming languages attempted to overcome some of the difficulties inherent in machine language by replacing the binary digits with symbols programmers could more easily understand. Assembly languages use codes like A for add, MVC for move, and so on. This second-generation language was termed **assembly language**, after the system programs, called assemblers, used to translate it into machine code. Systems software programs such as operating systems and utility programs are often written in an assembly language.

Third-Generation Languages

These languages continued the trend toward greater use of symbolic code and away from specifically instructing the computer how to complete an operation. BASIC, COBOL, C, C++, and FORTRAN are examples of third-generation languages that use Englishlike statements and commands. This type of language is easier to learn and use than machine and assembly languages because it more closely resembles everyday human communication and understanding. Third-generation languages use statements such as the following:

PRINT TOTAL_SALES

READ HOURS_WORKED

NORMAL_PAY=HOURS_WORKED*PAYRATE

With third-generation programming languages, each statement in the language translates into several instructions in machine language. In addition, third-generation languages take the programmer one step further from directing the actual operation of the computer. Although easier to program, third-generation languages are not as efficient in terms of operational speed and memory.

The various languages have certain characteristics that make them appropriate for particular types of problems and applications. For example, COBOL has excellent file- and database-handling capabilities for manipulating large volumes of business data, while FORTRAN is better suited for scientific applications. Although other languages are used to write new business applications, there are more lines of code of existing business applications written in COBOL than in any other programming language.

Third-generation languages are relatively independent of computer hardware. This means that the same program can be used on a number of different computers by different computer manufacturers, with only small modifications or no modifications at all. This characteristic is gaining importance as companies connect different computers into distributed processing systems and as changing hardware technologies drive rapid upgrades.

Even though the use of low-level programming languages (machine and assembly) enables a programmer to write efficient code for programs that require the utmost speed, most companies use third-generation languages for the majority of their programs. This is primarily because it takes less effort to write a program using the more intuitive Englishlike third-generation languages. Note that all these programs are general-purpose and procedure-oriented languages.

Fourth-Generation Languages

Fourth-generation programming languages are less procedural and even more Englishlike than third-generation languages. They emphasize what output results are desired rather than how programming statements are to be written. As a result, many managers and executives with little or no training in computers and programming are using fourth-generation languages (4GLs). Some of the features of 4GLs include query and database abilities, code-generation abilities, and graphics abilities. Prime examples include Visual C++, Visual Basic, PowerBuilder, Delphi, Forte, and many others. Visual application development environments are becoming more important than languages because you don't have to be a software engineer to be productive with these "abstract" tool sets.

query languages

programming languages used to ask the computer questions in Englishlike sentences

With some 4GLs, a user can give simple commands or perform simple procedures to retrieve information from a database. These commands are stated in the form of simple questions. With a billing database written in a 4GL, for example, you can click on "AMOUNT" and enter "> $250" to get a listing of all bills that are greater than $250. These languages have been called **query languages** because they ask a computer questions in Englishlike sentences. Many of these languages also operate only on organized databases and are thus called database languages. Some examples of queries in a fourth-generation database or query language are shown here:

PRINT EMP_ NO IF GROSS PAY > 1000

PRINT CUS_NAME IF AMOUNT > 5000 AND IF DUE_DATE > 90

PRINT INV_NOSUB IF ON HAND < 50

Creating, manipulating, and using graphics and illustrations can be easier and simpler with a fourth-generation language than with a lower-level language. For example, fourth-generation languages can be used to develop trend lines and pie charts from data.

structured query language (SQL)

a standardized language often used to perform database queries and manipulations

One popular fourth-generation language is a standardized language called **structured query language (SQL)**, which is often used to perform database queries and manipulations. An example of an SQL statement follows:

SELECT OCCUPANT FROM ROOM_CHG WHERE RM_CHG > 1000

Many other fourth-generation languages—such as FOCUS, Powerhouse, and SAS—are used by end users and programmers alike to develop programs.

Object-Oriented Programming Languages

object-oriented languages

languages that allow interaction of programming objects, including data elements and the actions that will be performed on them

A newer type of programming language—**object-oriented languages**—allows the interaction of programming objects. This approach to programming is called object-oriented programming. Most other programming languages separate data elements from the procedures or actions that will be performed on them, but object-oriented programming languages tie them together into objects. Thus, an object consists of data and the actions that can be performed on the data. For example, an object could be data about an employee and all the operations (such as payroll calculations) that might be performed on the data.

In high-level, query, and database languages, programs consist of procedures to perform actions on each data element. However, in object-oriented systems, programs tell objects to perform actions on themselves. For example, a video display window does not need to be drawn on the screen by a series of instructions. Instead, a window object could be sent a message to open, and the window will appear on the screen. That is because the window object contains the program code for opening itself.

As already stated, in object-oriented programming, data, instructions, and other programming procedures are grouped together into an item called an

encapsulation

the process of grouping items into an object

polymorphism

a process allowing the programmer to develop one routine or set of activities that will operate on multiple objects

inheritance

property used to describe objects in a group of objects taking on characteristics of other objects in the same group or class of objects

reusable code

the instruction code within an object that can be reused in different programs for a variety of applications

FIGURE 4.19

Reusable Code in Object-Oriented Programming

By combining existing program objects with new ones, programmers can easily and efficiently develop new object-oriented programs to accomplish organizational goals. Note that these objects can be either commercially available or designed internally.

object. The process of grouping items into an object is called encapsulation. **Encapsulation** means that functions or tasks are captured (encapsulated) into each object, which keeps them safe from changes because access is protected. Objects often have properties of polymorphism and inheritance. **Polymorphism** allows the programmer to develop one routine or set of activities that will operate or work with multiple (poly) objects. **Inheritance** means that objects in a group of objects can take on, or "inherit," characteristics of other objects in the same group or class of objects. This helps programmers select objects with certain characteristics for other programming tasks or projects.

Building programs and applications using object-oriented programming languages is like constructing a building using prefabricated modules or parts. The object containing the data, instructions, and procedures is a programming building block. Unlike constructing a building, however, the same objects (modules or parts) can be used repeatedly. An object can relate to data on a product, an input routine, or an order-processing routine. An object can even direct a computer to execute other programs or to retrieve and manipulate data. One of the primary advantages of an object is that it contains **reusable code**. In other words, the instruction code within that object can be reused in different programs for a variety of applications, just as the same basic type of prefabricated door can be used in two different houses. Thus, a sorting routine developed for a payroll application could be used in both a billing program and an inventory control program. By reusing program code, programmers are able to write programs for specific application problems more quickly (see Figure 4.19). Object-oriented programming is also an excellent programming approach for database applications.

Object-oriented programming languages offer the potential advantages of reusable code, lower costs, reduced testing, and faster implementation times. Instead of writing detailed lines of code, programmers can combine, modify, and integrate predeveloped modules into a unified program. Even with modifications, using the modular approach can be significantly faster than writing all the statements from scratch. Potential disadvantages of object-oriented programs include slower execution times and higher memory requirements. However, as hardware costs come down and programmer salaries increase, many managers are willing to trade off these disadvantages for much faster program development time.

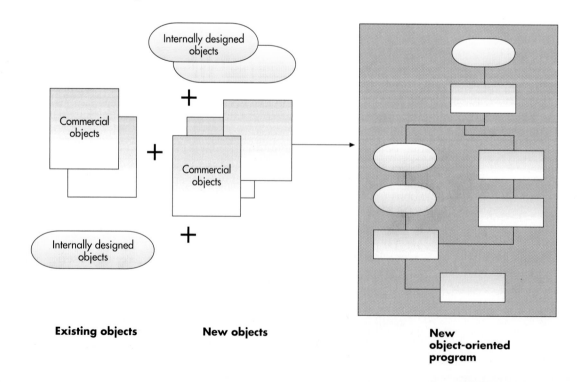

Existing objects   **New objects**   **New object-oriented program**

There are several object-oriented programming languages; some of the most popular include Smalltalk and Java.

Smalltalk

Smalltalk was developed by Alan Kay at Xerox at the Palo Alto Research Center in the 1970s. For more than a decade, the language was not used to a great extent in business settings. With the implementation of the language on desktop computers, this is changing. Some organizations have experienced as much as a three-to-one improvement in application program development productivity using this language versus traditional third-generation languages.

Java

Java
an object-oriented programming language based on C++ that allows small programs—applets—to be embedded within an HTML document

Java is an object-oriented programming language developed by Sun Microsystems. It gives programmers a powerful programming environment and the ability to develop applications that work across the Internet. Java can be used to develop small applications, called applets, which can be embedded into Web pages on the Internet. The Java language includes a debugger, a documentation generator, a utility to convert its source code into executable code, and a viewer for running Java applications without an Internet browser. Java has a structure and uses conventions similar to the C++ programming language. It is a much more complex and abstract language than the third-generation languages. Sun Microsystems' hopes that Java would be the primary language for developing applications on the Internet have been realized—several companies, including Microsoft, IBM, and Netscape, have incorporated Java into their products. In addition, other products such as ActiveX by Microsoft can be used to develop network-ready applets.

One of the key advantages of Java is its cross-platform capabilities. This means that one Java application can be developed to run on a variety of computers and operating systems, including Unix, Windows, and Macintosh operating systems. The Java programming code does not have to be modified for these different systems. Another advantage of Java is its power and functionality. Some developers believe that the language is easier to use than other languages like C++.

Visual Programming Languages

visual programming languages
languages that use a mouse, icons, or symbols on the screen and pull-down menus to develop programs

Many programming languages can be used in a visual or graphical environment. Often called **visual programming languages**, these languages use a mouse, icons, or symbols on the screen and pull-down menus. This visual environment can make programming easier and more intuitive. Visual Basic, PC COBOL, and Visual C++ are examples of visual programming languages. Despite their similar names, programming languages that use a visual or graphic environment are usually substantially different from the languages developed for a strictly character- and text-based environment.

Unfortunately, these programming languages have never been considered appropriate for developing critical enterprise applications. For example, Visual Basic code runs slowly, its screen display performance is considered fair, and it takes major effort to connect to various databases. Also, it lacks the power of C++ and cannot be used for full Web development. Still, developers and novices flocked to Visual Basic because it was easy to use. And the industry that Visual Basic spawned is overtaking it. Visual Basic–style scripting is now integrated into a wide variety of applications. Visio—a drawing package—includes Visual Basic–style scripting. So does SQA TeamTest, a GUI testing tool that ironically is used to make robust and error-free Visual Basic applications. Likewise, all of the applications in Microsoft Office are based on Visual Basic for Applications, a derivative of Visual Basic. On the high end of the spectrum, Microsoft has provided Visual C++ with an integrated development environment that is becoming more like Visual Basic with each version. Eventually, the two environments will be combined, giving Visual C++ the ability to develop enterprise applications rapidly.

Fifth-Generation Languages

knowledge-based programming

an approach to development of computer programs in which you do not tell a computer how to do a job but what you want it to do

Fifth-generation language tool kits appeared around mid-1998. They combine rules-based code generation, component management, visual programming techniques, reuse management, and other advances. In general, software tool vendors call this approach **knowledge-based programming**, which means that you do not tell the computer how to do a job but what you want it to do, and it figures out what you need. SunSoft is currently beta testing Java Studio, a connect-the-dots style of visual development toolset that uses components known as Java Beans. Java Studio is so simple to use that you can easily develop a working program without any previous programming experience.

Rules-based programming may be the best way to develop intelligent, knowledge-based applications because it simulates the human decision-making process and defines clear-cut rules to be followed, thus eliminating the problem of inconsistency. Are artificial intelligence-based development tools and languages likely to become universally accepted? Given the rapid innovation in tools and languages, it appears so. In fact, it is hard to imagine a programmer learning one language and one development environment and sticking with it for the next decade or two.

Selecting a Programming Language

Selecting the best programming language to use for a particular program involves balancing the functional characteristics of the language with cost, control, and complexity issues.

Machine and assembly languages provide the most direct control over computer hardware. For this reason, many vendors of popular application software programs take the time and effort to code portions of their leading programs in assembly language to maximize their speed. When a programmer requires a high degree of control over how various hardware components are used, these languages should be used. In selecting any programming language, the amount of direct control that is needed over the operation of the hardware can be an important factor to consider.

More recent programming languages are typically more complex than earlier programming languages. Although these newer languages appear to be simpler because they are more Englishlike, each command can drive complex routines and functions that operate behind the scenes. It takes less time to develop computer programs using higher-level languages than with lower-level languages. This means that the cost to develop computer programs can be substantially less with these more recent programming languages. Although training programmers to use these higher-level programming languages may produce high up-front costs, using higher-level languages can reduce the total costs to develop computer programs in the long run.

C++ and Java both have advantages and disadvantages, but Java may be the future of programming. Java is far easier to learn and, as a result, people become productive much sooner. Programmers who learn C++ must spend a lot of time debugging rather than learning software engineering techniques. An increasing number of colleges in the United States are using Java as their first programming language. Java is also more portable—with the ability to run on more operating systems and hardware. However, C++ will not disappear anytime soon. A large base of C++ programs is installed, and there are a large number of users because Microsoft uses it for its programming. The ANSI and ISO standards committees have also been working on C and C++ since 1990, and it is apparent that people will continue to develop in C++ in or outside a Microsoft environment.

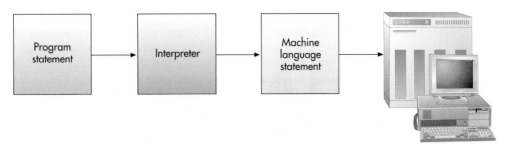

Statement execution

FIGURE 4.20

How An Interpreter Works

An interpreter translates each program statement or instruction in sequence. The CPU then executes the statement, erases it from memory, and translates another statement. An interpreter does not produce a complete machine-readable version of a program.

language translator

systems software that converts a programmer's source code into its equivalent in machine language

source code

high-level program code written by the programmer

object code

machine language code

interpreter

a language translator that translates one program statement at a time into machine code

compiler

a language translator that converts a complete program into a machine language to produce a program that the computer can process in its entirety

Language Translators

Because machine language programming is extremely difficult, very few programs are actually written in machine language. However, machine language is the only language capable of directly instructing the CPU. Thus, every non-machine-language program instruction must be translated into machine language prior to its execution. This is done by systems software called **language translators**. A language translator converts a programmer's source code into its equivalent in machine language. The high-level program code is referred to as the **source code**, whereas the machine language code is referred to as the **object code**. There are two types of language translators—interpreters and compilers.

An **interpreter** translates one program statement at a time, as the program is running. It will display on screen any errors it finds in the statement (Figure 4.20). This line-by-line translation makes interpreters ideal for those who are learning programming, but it does slow down the execution process.

A **compiler** is a language translator that converts a complete program into a machine language to produce a program that the computer can process in its entirety (Figure 4.21). Once the compiler has translated a complete program into machine language, the computer can run the machine language program as many times as needed. A compiler creates a two-stage process for program execution. First, it translates the program into a machine language; second, the CPU executes that program.

Stage 1:   **Convert program**

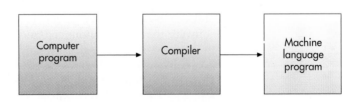

Stage 2:   **Execute program**

FIGURE 4.21

How A Compiler Works

A complier translates a complete program into a complete set of binary data (Stage 1). Once this is done, the CPU can execute the converted program in its entirety (Stage 2).

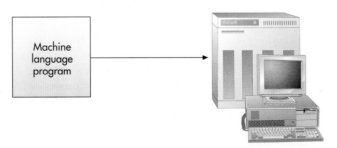

Program execution

Most defects in computer software are unintentional mistakes in syntax or logic. During the course of developing computer programs, the programmer uses an interpreter or a compiler to detect syntax errors. Detecting logic errors usually requires programmers to run through test data and check the results of the program against the results of running the same data by hand or calculator. For example, if you made the logic error $A = B + C$ instead of the correct $A = B - C$, then you tell the computer that B equals 42 and C equals 13 and ask it to find A, you'll quickly find the erroneous + sign when the answer comes out 55 instead of 29.

Unfortunately, few logic errors are so simple or so easily detected. Moreover, most computer programs involve thousands of lines of code. Thus, it can take years, even for teams of programmers, to debug programs such as those used to control emergency shutdown systems on nuclear reactors. Yet, even with the use of sophisticated debugging programs and years of study, experts estimate that every major software program contains dozens of errors. Some programming languages are more prone to the introduction of program defects than others.

SOFTWARE ISSUES AND TRENDS

Because software is such an important part of the computer system, issues such as software bugs, licensing, upgrades, and global software support have received increased attention. We highlight several major software issues and trends here: software bugs, open source software, opensourcing, software licensing, software upgrades, and global software support.

Software Bugs

software bug

a defect in a computer program that keeps it from performing in the manner intended

A **software bug** is a defect in a computer program that keeps it from performing in the manner intended. Some software bugs are obvious and cause the program to terminate unexpectedly. Other bugs are subtler and allow errors to creep into your work. Computer and software vendors tell us that since humans design and program hardware and software, bugs are inevitable. Indeed, according to the Pentagon and the Software Engineering Institute at Carnegie Mellon University, there are typically 5 to 15 bugs in every 1,000 lines of code—the cryptic software instructions that make sense only to computers and programmers.

Most software bugs arise because software manufacturers release new software as early as possible instead of waiting until all software bugs are identified and removed. They are under intense pressure from customers to deliver the software they have announced and from shareholders to begin selling the new product to increase sales. Meanwhile, the software manufacturer's quality-assurance people fight a losing battle for more software testing time to identify and remove bugs. Although the decision of when to release new software is based on a fine line, the industry clearly favors releasing software early and with defects. After all, software companies make money on upgrades, so there is little incentive to achieve a perfect first release.

Software bugs have plagued most new releases of personal computer software. But the most damaging bugs affect software used by major industries, the government, and the military. For example, software bugs have crippled on-line auctions and trading sites. They've delayed product shipments at Hershey and Whirlpool. Faulty programming caused a $250 million Mariner space probe to crash as it entered the Martian atmosphere. A cancer patient received a fatal dose of radiation from a machine designed to be foolproof. The ATMs at a New York bank debited customers twice their actual withdrawals, resulting in a loss of millions of dollars. All told, bad software cost U.S. businesses $85 billion in lost productivity in 1998, according to Jim Johnson, president of market researcher The Standish Group in Dennis, Massachusetts.[19] Table 4.10 summarizes tips for reducing the impact of software bugs.

TABLE 4.10

How to Deal with
Software Bugs

- Register all software so that you receive bug alerts, fixes, and patches.
- Check the manual or read-me files for workarounds.
- Access the support area of the manufacturer's Web site for patches.
- Install the latest software updates.
- Before reporting a bug, make sure that you can re-create the circumstances under which it occurs.
- Once you can re-create the bug, call the manufacturer's tech support line.
- Avoid buying the latest release of software for several months or a year until the software bugs have been discovered and removed.

Open Source Software

open source software

software that is freely available to anyone in a form that can be easily modified

Open source software is software that is freely available to anyone in a form that can be easily modified.[20] Users can download the source code and build the software themselves, or the software's developers can make executable versions available along with the source. Open source software development is a collaborative process employing the Internet to enable open source developers around the world to keep in close touch with each other via e-mail and to download and submit new software. Major software changes can occur in days rather than weeks or months. Open source software is at the heart of many of the Internet's most popular services, including e-mail and the Web. A number of open source software packages are widely used, including the Linux system; Free BSD, another operating system; Apache, the most popular Web server in the world; Sendmail, a program that delivers e-mail for most systems on the Internet; and Perl, a programming language used to develop Internet application software.

Why would an organization run its business using software that's free? How can something that's given away over the Internet be stable or reliable or sufficiently supported to place at the core of a company's day-to-day operations? The answer is surprising—open source software is often *more* reliable than commercial software. How can this be? First, by making a program's source code readily available, users have the opportunity to fix any problems they discover. A fix is often available within hours of the problem's discovery. Second, with the source code for a program accessible to thousands of people, the chances of a bug being discovered and fixed before it does any damage are much greater than with traditional software packages.

The question of software support is the biggest stumbling block to the acceptance of open source software at the corporate level. Getting support for traditional software packages is easy—you call a company's toll-free support number or access its Web site. But how do you get help if an open source package doesn't work as expected? Since the open source community lives on the Internet, you look there for help. Through use of Internet discussion areas, you can communicate with others who use the same software, and you may even reach someone who helped develop it. Users of popular open source packages can get correct answers to their technical questions within a few hours of asking for help on the appropriate Internet forum. Another approach is to contact one of the many companies emerging to support and service such software—for example, Red Hat for Linux, C2Net for Apache, and Sendmail, Inc., for Sendmail. These companies offer high-quality, for-pay technical assistance.

Opensourcing

opensourcing

extending software development beyond a single organization by finding others who share the same problem and involving them in a common development effort

The use of the Internet to spur development of open source software has led to **opensourcing**, or extending software development beyond a single organization by finding others who share the same problem and involving them in a common development effort.[21] Such an approach spreads development costs across multiple organizations and increases the opportunity for people with highly specialized expertise to contribute. It increases the number of developers and users who can identify and eliminate bugs in software. It also fosters the growth of a community of people who can support the software.

Software Licensing

Software is protected by copyright law, which says that the copyright holder has all the rights and that users don't have any, except those that have been specifically granted. As software increases in importance, controlling its use is a concern for some software manufacturers who do not subscribe to the open source software business model. Vendors spend time and resources in software development, so they must try to protect their software from being copied and distributed by individual users and other software companies. Today, companies can copyright software and programs, but this protection is limited. Another approach is to use existing patent laws, but this approach has not been fully accepted by legal experts and the courts.

Software Upgrades

Software companies revise their programs and sell new versions periodically. In some cases, the revised software offers a number of new and valuable enhancements. In other cases, the software uses complex program code that offers little in terms of additional capabilities. Furthermore, revised software can contain bugs or errors. Deciding whether to purchase the newest software can be a problem for corporations and individuals with a large investment in software. Should the newest version be purchased when it is released? Some organizations and individuals do not always get the most current software upgrades or versions, unless there are significant improvements or capabilities. Instead, they may upgrade to newer software only when there are vital new features. Software upgrades usually cost much less than the original purchase price.

Global Software Support

Large, global companies have little trouble persuading vendors to sell them software licenses for even the most far-flung outposts of their company. But can those same vendors provide adequate support for their software customers in all locations? Taking into account the support requirements of local operations is one of the biggest challenges information systems teams face when putting together standardized, companywide systems. They must ensure that the vendor has sufficient support for local operations. In slower technology growth markets, like Eastern Europe and Latin America, there may be no vendor presence at all. Instead, large vendors such as Sybase, IBM, and Hewlett-Packard typically contract out support for their software to local providers.

One approach that has been gaining acceptance in North America is to outsource global support to one or more third-party distributors. The software user company may still negotiate its software license with the software vendor directly, but it then hands over the global support contract to a third-party supplier. The supplier acts as a middleman between software vendor and user, often providing distribution, support, and invoicing. This is how American Home Products Corporation handles global support for both Novell NetWare and Microsoft Office applications in the 145 countries in which it operates. American Home Products, a pharmaceutical and agricultural products company, negotiated the agreements directly with the vendors for both purchasing and maintenance, but fulfillment of the agreement is handled exclusively by Philadelphia-based Softsmart, an international supplier of software and services.

● SUMMARY

PRINCIPLE • When selecting an operating system, you must consider the current and future needs for application software to meet the needs of the organization. In addition, your choice of a particular operating system must be consistent with your choice of hardware.

Software consists of programs that control the workings of the computer hardware. There are two main categories of software: systems software and application software. Systems software is a collection of programs that interacts between hardware and application software. Systems software includes utility programs and operating systems. Application software enables people to solve problems and perform specific tasks. Application software may be proprietary or off-the-shelf. Although there are literally hundreds of computer applications that can help individuals at school, home, and work, the primary ones are word processing, spreadsheet analysis, database, graphics, and on-line services.

All software programs are written in coding schemes called programming languages, which provide instructions to a computer to perform some processing activity. Programming languages provide instructions to the computer system so that it can perform a processing activity.

● ● ●

An operating system (OS) is a set of computer programs that controls the computer hardware to support users' computing needs.

OS hardware functions involve converting an instruction from an application into a set of instructions needed by the hardware. The OS also serves as an intermediary between application programs and hardware, allowing hardware independence. Memory management involves controlling storage access and use by converting logical requests into physical locations and by placing data in the best storage space, perhaps expanded or virtual memory.

Task management allocates computer resources through multitasking and time-sharing. With multitasking, users can run more than one application at a time. Time-sharing allows more than one person to use a computer system at the same time.

● ● ●

An OS also provides a user interface, which allows users to access and command the computer. A command-based user interface requires text commands to send instructions; a graphical user interface (GUI), like Windows, uses icons and menus. Over the years, several popular operating systems have been developed. These include several proprietary operating systems used primarily on mainframes. MS-DOS is an early OS for IBM-compatibles. Older Windows operating systems, such as Windows 3.1, are GUIs used with DOS. Newer versions, like Windows 95, Windows 98, and Windows NT, are fully functional operating systems that do not need DOS. Apple computers use proprietary operating systems like the Mac OS. Unix is the leading portable operating system, usable on many computer system types and platforms.

PRINCIPLE • Do not develop proprietary application software unless doing so will meet a compelling business need that can provide a competitive advantage.

Application software applies the power of the computer to solve problems and perform specific tasks. Application software can support individuals, groups, and organizations. User software, or personal productivity software, includes general-purpose programs that enable users to improve their personal effectiveness, increasing the amount of work that can be done and its quality. Software that helps groups work together is often referred to as groupware. Enterprise software that benefits the entire organization can also be developed or purchased.

● ● ●

Three approaches to developing application software are as follows: build proprietary application software, buy existing programs off the shelf, or use a combination of customized and off-the-shelf application software.

Building proprietary software (in-house or contracting out) has the following advantages: the organization will get software that more closely matches its needs; by being involved with the development, the organization has further control over the results; and the organization has more flexibility in making changes. The disadvantages include the following: it is likely to take longer and cost more to develop; the in-house staff will be hard pressed to provide ongoing support and maintenance; and there is a greater risk that the software features will not work as expected or that other performance problems will occur.

Purchasing off-the-shelf software has its advantages. The initial cost is lower; there is a lower risk that the software will fail to work as expected; and the software is likely to be of higher quality than proprietary software. Some of the disadvantages are that the organization may pay for features in the software it does not need; the software may lack important features requiring expensive customization; and the system may work in such a way that work process reengineering is required.

Some organizations have taken a third approach—customizing software packages. This approach can combine all of the above advantages and disadvantages and must be carefully managed.

PRINCIPLE • Choose a programming language whose functional characteristics are appropriate to the task at hand, taking into consideration the skills and experience of the programming staff.

There are several classes of programming languages, including machine, assembly, high-level, query and database, object-oriented, visual programming, and knowledge-based languages.

Programming languages have gone through changes since their initial development in the early 1950s. In the first generation, computers were programmed in machine language, or binary code, a series of statements written in 0s and 1s. The second generation of languages was termed assembly languages; these languages support the use of symbols and words rather than 0s and 1s. The third generation consists of many high-level programming languages that use Englishlike statements and commands. They also must be converted to machine language by systems software but are easier to write than assembly or machine language code. These languages include BASIC, COBOL, FORTRAN, and others. A fourth-generation language is less procedural and more Englishlike than third-generation languages. The fourth-generation languages include database and query languages like SQL. Fifth-generation programming languages combine rules-based code generation, component management, visual programming techniques, reuse management, and other advances. These languages offer the greatest ease of use yet.

Object-oriented programming languages—like Smalltalk and Java—use groups of related data, instructions, and procedures called objects, which serve as reusable modules in various programs. These languages can reduce program development and testing time. Java can be used to develop applications on the Internet.

● KEY TERMS

application program interface
 (API) 125
application service provider 139
application software 122
assembly language 152
client application 146
collaborative computing software 146
command-based user interface 125
compiler 157
computer programs 122
computer system platform 122
contract software 138
documentation 122
embedding 146
encapsulation 154
enterprise resourcing planning
 (ERP) 148
enterprise sphere of influence 123
graphical user interface (GUI) 125
groupware 146
icon 125

inheritance 154
in-house development 138
interpreter 157
Java 155
kernel 124
knowledge-based programming 156
language translator 157
linking 146
machine language 151
multitasking 127
object code 157
object linking and embedding
 (OLE) 145
object-oriented languages 153
off-the-shelf software 137
open source software 159
opensourcing 159
operating system (OS) 124
paging 127
personal productivity software 123
personal sphere of influence 123

polymorphism 154
programming languages 149
proprietary software 137
query languages 153
reusable code 154
scalability 127
server application 146
software bug 158
software suite 144
source code 157
sphere of influence 123
structured query language (SQL) 153
syntax 150
systems software 122
time-sharing 127
user interface 125
utility programs 136
virtual memory 126
visual programming languages 155
workgroup 123
workgroup sphere of influence 123

● REVIEW QUESTIONS

1. What is a computer system platform? Give two examples.
2. Give four examples of personal productivity software.
3. How do software bugs arise?
4. What is the difference between multitasking and time-sharing?
5. What is meant by scalability? Why is it important?
6. Name four operating systems that support the personal, workgroup, and enterprise spheres of influence.
7. Give three examples of utility software.
8. Identify the two primary sources for acquiring application software.
9. What is an application service provider? What issues arise in considering the use of one?
10. What is open source software? What is the biggest stumbling block with the use of open source software?
11. What is an operating system? What are the key activities that it performs?
12. Discuss the advantages and disadvantages of using a set of individual personal productivity applications from multiple vendors versus an integrated set of applications in a software suite.
13. Describe the term *enterprise resource planning (ERP) system*. What functions does such a system perform?

● DISCUSSION QUESTIONS

1. Assume that you must take a computer programming course next semester. What language do you think would be best for you to study? Why? Do you think that a professional programmer needs to know more than one programming language? Why or why not?
2. What is meant by operating system hardware independence? Why is this important? How do application programming interfaces help meet this goal?
3. Imagine that you have been assigned responsibility to choose the software suite to be adopted by your organization. On what basis would you make this decision? How would you do a comparison of the leading contenders—Microsoft Office, Corel WordPerfect Office, and Lotus SmartSuite? How much consideration would you place on the availability of good user documentation?
4. Explain the difference between a language translator, an interpreter, and a compiler.
5. You are using a new release of an application software package. You think that you have discovered a bug. Outline the approach that you would take to confirm that it is indeed a bug. What actions would you take if it truly were a bug?
6. How can application software improve the effectiveness of a small workgroup? What are some of the benefits associated with implementation of groupware? What are some of the issues that can arise that could keep the use of groupware from being successful?
7. Explain the difference between copying, linking, and embedding.
8. Many organizations have mandated that off-the-shelf application software be the first choice when implementing new software and that the software not be customized at all. What are the advantages and disadvantages of such an approach? What would you recommend for your firm if you were the manager in charge of application development? If you were a senior vice president of a business area?

● PROBLEM-SOLVING EXERCISES

1. Choose a programming language of interest to you and develop a six-slide presentation of its history, current level of usage, typical applications, ease of use, etc.

2. Choose a graphics package to track data of interest to you. Produce a simple line chart that shows at least ten values for each of two variables over a two-week period—high and low stock price of a particular company, high and low temperature at your favorite vacation location, etc. The graph must have a title, and the two line graphs must be labeled with the name of the variables. Use a word processing package to write a paragraph summarizing the data. Cut and paste the graph into the word processing document.

TEAM ACTIVITIES

1. Form a group of three or four classmates. Find articles from business periodicals, search the Internet, or interview people on the topic of software bugs. How frequently do they occur, and how serious are they? What can software users do to encourage defect-free software? Compile your results for an in-class presentation or a written report.

2. Form a group of three or four classmates. Identify and contact an application service provider and request information—via phone interview, Web search, or literature—about its services. Write a brief report summarizing your findings.

WEB EXERCISES

1. A number of companies have developed Web-based tax preparation and filing services for the more than 80 million people who currently use shrink-wrapped software or pencil and paper to file their tax returns. Visit two of the following sites and write a brief summary report comparing the costs and services available at each.

CCH Inc.   www.cch.com
H&R Block   www.hrblock.com

Kiplinger TaxCut Online   www.taxcut.com
Turbotax for the Web   www.turbotax.com
H. D. Vest Inc.   www.hdvest.com

2. Do research on the Web and develop a two-page report on object-oriented programming.
3. Visit the Web site of Microsoft and find out the current status of bug fixes for the most current release of the Windows NT operating system for workstations.

CASES

 A&P Gambles on New Information Systems

Founded in 1859, the Great Atlantic & Pacific Tea Company is the nation's oldest supermarket and one of America's top ten supermarket chains. A&P operates 839 stores in 16 states and Canada under the trade names of A&P, Waldbaum's, Super Fresh, Farmer Jack, Sav-A-Center, Super Foodmart, Food Emporium, Kohl's, Dominion, and Food Basics. It also manufactures and distributes a line of whole-bean coffees under the Eight O'Clock, Bokar, and Royale labels both for sale through its own stores as well as other food and convenience retailers. In 1950, its annual sales were second in the United States only to General Motors.

Today the company is struggling for survival.

In March 2000, A&P announced it would spend $250 million over four years to develop a supply-chain and business operations system. These expenses could cut fiscal 2000 profits by as much as $1.50 per share from about $3 per share. Nonetheless, A&P believes that it will save $325 million over the four years by lowering costs and improving product availability and raise pretax

profits by $100 million per year once it is completed. A&P is hoping that the overhaul will streamline core business processes, support business-to-business e-commerce with its suppliers, and provide store-specific data to tailor purchasing.

The overhaul of its information systems is just the latest move in A&P's campaign to revitalize the troubled chain. In December 1998, A&P announced a series of strategic initiatives designed to improve operating and financial performance. These initiatives, which A&P calls Project Great Renewal Phase I, included accelerating the new store opening and Food Basics conversion programs, undertaking change processes, including leadership development and culture change, realigning distribution functions, closing manufacturing facilities, and closing 132 underperforming stores. Additionally, in the first quarter of 1999 A&P closed the Atlanta, Georgia, market (34 stores). For fiscal 1999, the net pretax costs of Project Great Renewal Phase I were $102.9 million, reducing 1999 reported earnings per share by $1.56. In fiscal 1998, pretax costs of the program were $300.0 million, which reduced reported earnings per share by $3.09.

In spite of these efforts, the A&P turnaround has been slower than hoped. Against growing competition and rapid consolidation within the supermarket industry, many people believe that A&P must find a global partner to survive. When A&P finds that partner, it will have to convert to that partner's information systems and work processes, rendering the planned information systems investment risky and redundant. But increased efficiencies resulting from the information system overhaul could ultimately make A&P more attractive to potential partners.

Discussion Questions

1. According to A&P estimates, how much will the new information systems increase pretax profits over the next five years?

2. Why do some people believe that the investment in new information systems is unwise?

Critical Thinking Questions

3. If you were a member of the A&P board of directors, what assurances would you want that the decision to invest in new information systems was sound?

4. As A&P proceeds with implementing new information systems, should it choose industry standard software or seek to gain a competitive advantage through implementation of unique software? Why?

Sources: Adapted from Great Atlantic & Pacific Tea Company, "Our Company," at http://www.aptea.com/our_company.htm, accessed March 29, 2000; and Sami Lais, "A&P's $250M IT Plan Shunned by Wall Street," *Computerworld*, March 20, 2000, p. 4.

---

2 Cleveland State University Faces Software Problems

Cleveland State University (CSU) is a collection of seven separate colleges offering undergraduate, master's, doctoral, and law degrees. It has an enrollment of 17,000 students and is situated in northeastern Ohio. Its 37 buildings are spread over 70 acres.

PeopleSoft is the vendor for a new information management software system being installed at CSU to support finance, student, and human resources operations. PeopleSoft is a global supplier of enterprise application software for business and government. Company headquarters are in Pleasanton, California, and the firm employs 6,000 people. Six of the top ten Fortune 500 telecommunications companies are users of PeopleSoft applications. In addition, nearly 400 schools nationwide are either currently using or planning to use PeopleSoft. The firm is rated number six on *Fortune* magazine's list of the 100 best companies for which to work. PeopleSoft sales were $815 million in 1997 and increased to $1.314 billion in 1998.

CSU began its implementation of the PeopleSoft software in June 1997 with the initiation of Project PeopleSoft98. The vision was to use the software to provide CSU staff with instant access to information that is accurate and always up-to-date, eliminating the need for duplicate systems and duplicate work processes. Upon successful completion, students, faculty, and staff would be able to initiate many routine tasks, such as address changes, and request reports, thus significantly reducing paper transactions and paper shuffling. The PeopleSoft98 project included the installation of the software on the appropriate computers and servers, the conversion of data from current systems, "turning on" the new software, and training users. Also included in the scope of the PeopleSoft98 project was business process redesign—changing procedures and work tasks to gain the full benefit of the new information technology. Completion of the project would involve hundreds of CSU staff and faculty.

The PeopleSoft software is organized in modules, with one for each of the functional business processes at a university. The units that CSU installed include Admissions, Advising, Budget, Payroll, Financial Aid, Billing, General Ledger, Purchasing, Course Registration, Grades, Benefits, and Accounts Payable. All these CSU business processes would be linked to share common databases, instead of the current separate systems. The goal is to integrate seamlessly the systems for registration, financial aid, bursar's office, and so on.

CSU began its implementation with the installation of the full set of PeopleSoft's student administration applications. This piece is the most complicated of the higher education software suite because it encompasses admissions, financial aid, class registration, and student records. The school was the first to take such an aggressive approach; other schools had implemented software to support other business processes first and then implemented the student administration applications gradually and in a piecemeal fashion. CSU chose to do a fast-track rollout that took less than a year, a schedule thought to be very demanding. And the school stuck to that schedule even after discovering that its mainframe computer couldn't handle the PeopleSoft applications and would have to be replaced by a Unix system.

The modules went live during 1998 and rollout continued throughout the year. Staff working in the central business units were the first to see changes. Eventually, as enhancements were made and as equipment and training extended outside the central business units, everyone was affected—students, faculty, department assistants, deans, deans' staff, and all central administrative staff.

Now, more than a year after it installed the PeopleSoft software, CSU continues to have problems getting the system to work properly. The issues are mostly in financial aid, with many students complaining that they have not been able to receive their money on a timely basis. Nearly 4,000 students have experienced problems with their financial aid. For perspective, some 65,000 students have received over $20 million in aid. Several PeopleSoft modules have been implemented and are running smoothly. Indeed, the PeopleSoft programs in the areas of finance, human resources, and payroll have had no significant problems.

The CSU president appointed a "Just Fix It" task force with a single leader to address the issues that the university is now facing. In a November memo to the board of trustees, the president said, "Given the myriad of problems this fall in the processing of financial aid, it became imperative to initiate extraordinary measures to resolve these problems and to correct both software and related human systems prior to spring 2000 registration." The project team is led by the vice president for human resources and development and labor relations and includes the vice provost of academic affairs, the vice provost of information technology, the dean of enrollment services, the associate director of information services and technology, the director of financial aid, the senior advisor to the president, and the bursar. A faculty member and student will also be named to the team.

The CSU president and the vice provost of information technology and academic innovation spoke with PeopleSoft's vice president of customer services for education and government about the issues. PeopleSoft is working closely with CSU to solve the problems. The company—which has sold its software to more than 400 schools—also said it has more than doubled the number of telephone support workers dedicated to the student administration software and released new documentation and installation guidelines. However, when asked about the situation at CSU, the vice president for college software with PeopleSoft said, "What we see is the normal software implementation process. Cleveland State is not that different from other institutions." However, CSU has identified 35 pieces of the system that either don't work or are missing from the software. Meanwhile, lines of 200 students are common in the financial aid office.

Discussion Questions

1. What are the shortcomings of the PeopleSoft software from the perspective of Cleveland State University?
2. Briefly summarize the actions that PeopleSoft has taken to alleviate the problems. Would you consider these adequate? Why or why not?

Critical Thinking Questions

3. With the benefit of 20-20 hindsight, what could have been done differently during the PeopleSoft98 project that could have alleviated many of the problems CSU is now experiencing? Are these problems due to problems with the PeopleSoft software or are they due to other causes?
4. Imagine that you are the leader of the "Just Fix It" task force. How would you go about making your assessment of the current situation and identifying what needs to be done?

Sources: Adapted from Ray Cooney, "CSU Trustees Contemplate Suit against PeopleSoft," *Cleveland Stater,* November 24, 1999, accessed at http://www.csuohio.edu/clevelandstater February 20, 2000; Becky Muncy, "Kucinich Searching for Financial Aid Solution," *Cleveland Stater,* November 24, 1999, accessed at http://www.csuohio.edu/clevelandstater February 20, 2000; Craig Stedman, "ERP Problems Plague College," *Computerworld,* November 22, 1999, p. 4; PeopleSoft Discussion Groups, accessed at http://www.csuohio.edu/pplsoft/newsgrps.html February 20, 2000.

---

③ Eliminating E-Ticket Hassles

Airline electronic tickets (e-tickets) are stored in the airline's computer rather than printed on paper. Issuing an e-ticket is similar to the traditional paper ticket process, but it eliminates the time and hassle associated with purchasing or exchanging a paper ticket.

Because your e-ticket is held in the airline's computer, you cannot forget it or lose it. More important, your e-ticket cannot be stolen, saving you the cost of a replacement ticket. When you arrive at the airport, you present an official form of identification (driver's license, passport, etc.) to receive your boarding pass. If you used a credit or debit card to pay for the e-ticket, you are also asked to show the credit card you used to purchase your e-ticket. In addition to these advantages for the traveler, airlines have reported savings of $1 to $8 for each e-ticket purchased.

"Ticketing flights on [fill in the name of your favorite air carrier] has never been easier. Our e-ticket program is available on all our flights throughout the United States, Canada, and select destinations in Europe and Asia." So reads the promotional material for every airline's e-ticket program. But what happens if you need to switch carriers after receiving your e-ticket? Currently, you must request the original ticketing airline to issue a paper ticket, find another airline that will honor it, and wait in line to exchange tickets.

In an effort to eliminate this hassle for e-ticket travelers, IBM and the International Air Transport Association (IATA) are developing a global airline information system that will allow carriers to transfer e-tickets among their networks. Rather than have carriers build multiple connections among their proprietary systems, IBM worked with the IATA to create the single-connection Global ET program. Reservation agents can connect to a single powerful computer and transmit the data needed to transfer passengers in real time. This will enable passengers with e-tickets to change airlines by stopping at a single counter rather than shuffling between carriers and waiting in lines for service.

Many carriers think that the entire airline industry needs to adopt such an e-ticketing system. They believe that the ability to transfer e-ticket passengers among airlines should be viewed as a necessary service, not as some sort of competitive advantage. But the carriers are not used to working together, so an industry solution runs counter to the way they've always done business. For example, Delta Air Lines is busy working out e-ticketing agreements with the other carriers in an on-line alliance formed with United, American, Continental, and Northwest.

Discussion Questions

1. What are the advantages and disadvantages of the current e-ticket systems in operation at many airlines today?
2. Why are some of the larger airlines reluctant to accept an industrywide solution by IBM and IATA?

Critical Thinking Questions

3. What are the advantages and disadvantages of a common industrywide solution such as that being developed by IBM and the solution led by the major carriers that evolves and takes on new carriers over time?
4. Must the airline carriers employ common operating systems and on-line reservation application software to use such an industrywide standard solution? Why or why not?

Source: Adapted from Northwest Airlines–Northwest Services Web site, http://www.nwa.com/services/electrav/etick.shtml, accessed May 23, 2000; Delta e-ticket at http://www.delta-air.com/travel/reservations/travel_Info/eticket/index.jsp, accessed May 23, 2000; and Michael Meehan, "Network to Calm E-Ticket Turbulence," *Computerworld*, April 17, 2000, pp. 1, 14.

● NOTES

Sources for the opening vignette on page 121: Adapted from Bryan Larsen, "Home Depot Strives for IT Simplicity," Enterprise Development, March 2000, pp. 10–19; Home Depot Web page, "Company Info and Financial Info," http://homedepot.com, accessed March 28, 2000; and Craig Stedman, "Java Fuels Home Depot Expansion," *Computerworld*, August 23, 1999, p. 34.

1. Mary Jo Foley and Steven J. Vaughan-Nicols, "Microsoft Trims Windows 2000," *PC Week*, September 20, 1999, p. 18.
2. Jason Levett, "Achilles' Heel: Linux Libraries," *Information Week*, January 24, 2000, pp. 65–68.
3. Stephanie Neil, "Testing the Linux Waters," *PC Week*, March 15, 1999, pp. 81–82.
4. Aaron Ricadela, "Linux Comes Alive," *Information Week*, January 24, 2000, pp. 47–63.
5. Aaron Ricadela, "Linux Comes Alive," *Information Week*, January 24, 2000, pp. 47–63.
6. Aaron Ricadela, "Linux Comes Alive," *Information Week*, January 24, 2000, pp. 47–63.
7. Steve Hamm, Peter Burrows, and Andy Reinhardt, "Is Windows Ready to Run E-Business?" *Business Week*, January 24, 2000, pp. 154–160.
8. Steve Hamm, Peter Burrows, and Andy Reinhardt, "Is Windows Ready to Run E-Business?" *Business Week*, January 24, 2000, pp. 154–160.
9. Peter Burrows, "How Sun Became the Eddie and Scottie Show," *Business Week Online*, December 3, 1999, http://www.business-week.com/search.htm, accessed March 5, 2000.
10. "Sun Microsystems Upgrades Solaris," *Information Week Online NewsFlash*, January 24, 2000, http://www.informationweek.com/story/IWK20000124S0001, accessed March 5, 2000.
11. "Mac OS 9 Special Report," *The Macintosh News Network*, November, 17, 1999, at http://www.macnn.com/reports/os9.shtml, accessed March 4, 2000.
12. Kevin Young, "Inside Windows 2000: A Guide to Implementation," *PC Week*, January 31, 2000, p. 47.
13. Steve Hamm, Peter Burrows, and Andy Reinhardt, "Is Windows Ready to Run E-Business?" *Business Week*, January 24, 2000, pp. 154–162.
14. Steve Hamm, Peter Burrows, and Andy Reinhardt, "Is Windows Ready to Run E-Business?" *Business Week*, January 24, 2000, pp. 154–162.
15. Steve Hamm, Peter Burrows, and Andy Reinhardt, "Is Windows Ready to Run E-Business?" *Business Week*, January 24, 2000, pp. 154–160.
16. Russell Kay, "A First Peek at the Newest Windows," *Computerworld*, March 20, 2000, p. 72.
17. Carmen Nobel, "Transmeta's Link to Linux," *PC Week*, January 31, 2000, p. 37.
18. David Penn, "Ericsson's Screen Phone Runs Linux," *Linux Journal*, February 29, 2000, at http://www2.linuxjournal.com/cgi-bin/frames.pl/index.html, accessed March 2, 2000.
19. Neil Gross, "Software Hell," *Business Week*, December 6, 1999, pp. 104–118.
20. Ernie Longmire, "Feeling the Force of Open Source (Why Free Software Suddenly Matters)," *Basis Advantage Magazine*, First Quarter 1999, accessed at http://www.basis.com/advantage/mag-v3n1/opensource.html, February 29, 2000.
21. Tim O'Reilly, "The Open-Source Revolution," Release 1.0 Esther Dyson's Monthly Report, November 1998, accessed at http://www.edventure.com/release1/1198.html, February 29, 2000.

CHAPTER 5

Organizing Data and Information

Organizations that want their employees to make proactive, informed decisions—decisions that provide a real competitive edge—are turning to business intelligence solutions.

— Samuel Greengard, business and technology writer

Principles	Learning Objectives
The database approach to data management provides significant advantages over the traditional file-based approach.	• *Define general data management concepts and terms, highlighting the advantages and disadvantages of the database approach to data management.* • *Name three database models and outline their basic features, advantages, and disadvantages.*
A well-designed and well-managed database is an extremely valuable tool in supporting decision making.	• *Identify the common functions performed by all database management systems and identify three popular end-user database management systems.*
Further improvements in the use of database technology will continue to evolve and yield real business benefits.	• *Identify and briefly discuss recent database developments.*

Catalina Marketing Corporation

Providing Data Services

Catalina Marketing Corporation is a leading supplier of in-store electronic scanner-activated consumer promotions. It provides marketing services to such consumer goods companies as Bristol-Myers Squibb, Campbell, Coca-Cola, and General Mills. Its Catalina Marketing Network provides individually customized communications and promotions based on customer purchases that reach shoppers every week in supermarkets that include Kroger, Meijer, Ralphs, and Winn-Dixie. The network uses actual purchase behavior to target the future buying behavior of more people than newspaper coupons and direct mail.

The shopping patterns of over 150 million shoppers each week in more than 11,000 supermarkets nationwide are captured and stored in an enormous Catalina Marketing database, which has grown in size to more than 2 TB. The database has 18 billion rows of data, encompassing consumer purchases for the past 65 weeks. Information is captured every time there's a transaction at one of the supermarkets that subscribes to the Catalina Network. Buy a gallon of milk, and the store gets a scanned transaction, including what time of day, how it was paid for, and whether you bought bread or cereal with the milk. By itself, this snapshot means little. But combining the millions of transactions at that supermarket and at the entire supermarket chain over a day, a month, or even longer results in a complete profile of customer behavior that consumer goods manufacturers and supermarket retailers can use to better target their promotions and boost sales.

Each retailer that subscribes to the Catalina Network program has a PC in the supermarket. Catalina polls this computer nightly via a data network, and more than 70 million rows of data concerning that day's purchases are loaded into the database. During this polling process, Catalina also downloads to the store PC the information about targeted promotions scheduled for the next day.

With the data from its database, Catalina develops promotions for manufacturers that reach tens of millions of consumers. As groceries are scanned at the checkout counter, the PC flags any item—or the shopper, if the store has a loyalty program—eligible for a promotion. A printer located in each checkout lane prints out a targeted coupon, rebate, in-store game, or other incentive for the consumer. For example, two people buying the same product can be standing behind one another in the checkout line and get different offers. The occasional user might get a coupon, while a user of a competitive product may get a free sample. The result of such a targeted campaign is a much higher redemption rate than can be achieved with traditional coupons. Magazine and newspaper coupon redemption rates are around 0.6 percent, and direct mail's redemption rate is around 4.3 percent. In contrast, Catalina Marketing claims its scanner-based coupon system has an 8.9 percent redemption rate.

The Catalina Marketing Network program also arms consumer-goods manufacturers and supermarket retailers with the latest marketing weapon—stealth campaigns. Stealth marketing lets manufacturers and retailers promote specific items and judge campaign effectiveness without their competitors ever knowing that a promotion took place. Traditional advertising vehicles—newspapers, radio, and television—inform consumers, but they also alert competitors to marketing initiatives.

As you read this chapter, consider the following:

- How can databases be used to support critical business objectives?

- What are some of the issues associated with compiling and managing massive amounts of data?

The bane of modern business is too much data and not enough information. Computers are everywhere, accumulating gigabytes galore. Yet it seems to get harder to find the forest for the trees—that is, to extract significance from the blizzard of numbers, facts, and statistics. Like other components of a computer-based information system, the overall objective of a database is to help an organization achieve its goals. A database can contribute to organizational success in a number of ways, including the ability to provide managers and decision makers with timely, accurate, and relevant information based on data. As we saw in the case of Catalina Marketing, a database can help companies organize data to learn from this valuable resource. Databases also help companies generate information to reduce costs, increase profits, track past business activities, and open new market opportunities. Indeed, the ability of an organization to gather data, interpret it, and act on it quickly can distinguish winners from losers in a highly competitive marketplace. It is critical to the success of the organization that database capabilities be aligned with the company's goals. Because data is so critical to an organization's success, many firms develop databases to help them access data more efficiently and use it more effectively. In this chapter, we will investigate the development and use of different types of databases.

As we saw in Chapter 1, a database is a collection of data organized to meet users' needs. Throughout your career, you will be directly or indirectly accessing a variety of databases, ranging from a simple roster of departmental employees to a fully integrated corporatewide database. You will probably access these databases using software called a **database management system (DBMS)**. A DBMS consists of a group of programs that manipulate the database and provide an interface between the database and the user of the database and other application programs. A database, a DBMS, and the application programs that utilize the data in the database make up a database environment. Understanding basic database system concepts can enhance your ability to use the power of a computerized database system to support IS and organizational goals.

database management system (DBMS)

a group of programs that manipulate the database and provide an interface between the database and the user of the database and other application programs

DATA MANAGEMENT

Without data and the ability to process it, an organization would not be able to successfully complete most business activities. It would not be able to pay employees, send out bills, order new inventory, or produce information to assist managers in decision making. As you recall, data consists of raw facts, like employee numbers and sales figures. For data to be transformed into useful information, it must first be organized in a meaningful way.

The Hierarchy of Data

Data is generally organized in a hierarchy that begins with the smallest piece of data used by computers (a bit) and progresses through the hierarchy to a database. As discussed in Chapter 3, a bit (a binary digit) represents a circuit that is either on or off. Bits can be organized into units called bytes. A byte is typically eight bits. Each byte represents a **character**, which is the basic building block of information. A character may consist of uppercase letters (A, B, C, . . . , Z), lowercase letters (a, b, c, . . . , z), numeric digits (0, 1, 2, . . . , 9), or special symbols (.![+][-]/ . . .).

character

basic building block of information, consisting of uppercase letters, lowercase letters, numeric digits, or special symbols

field

typically a name, number, or combination of characters that describes an aspect of a business object or activity

record

a collection of related data fields

Characters are put together to form a field. A **field** is typically a name, number, or combination of characters that describes an aspect of a business object (e.g., an employee, a location, a truck) or activity (e.g., a sale). A collection of related data fields is a **record**. By combining descriptions of various aspects of an object or activity, a more complete description of the object or activity is obtained. For instance, an employee record is a collection of fields about one employee. One field would be the employee's name, another her address, and

FIGURE 5.1

The Hierarchy of Data

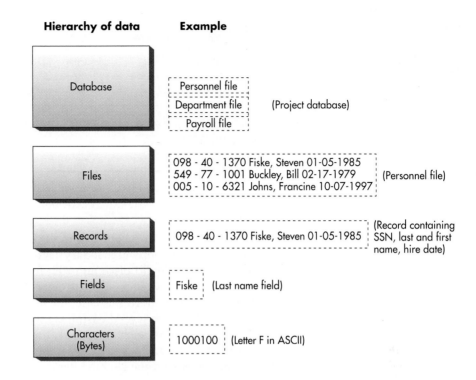

Hierarchy of data   **Example**

Database — Personnel file, Department file, Payroll file (Project database)

Files —
098 - 40 - 1370 Fiske, Steven 01-05-1985
549 - 77 - 1001 Buckley, Bill 02-17-1979
005 - 10 - 6321 Johns, Francine 10-07-1997
(Personnel file)

Records — 098 - 40 - 1370 Fiske, Steven 01-05-1985 (Record containing SSN, last and first name, hire date)

Fields — Fiske (Last name field)

Characters (Bytes) — 1000100 (Letter F in ASCII)

file

a collection of related records

hierarchy of data

bits, characters, fields, records, files, and databases

entity

generalized class of people, places, or things for which data is collected, stored, and maintained

attribute

a characteristic of an entity

data item

the specific value of an attribute

key

a field or set of fields in a record that is used to identify the record

primary key

a field or set of fields that uniquely identifies the record

still others her phone number, pay rate, earnings made to date, and so forth. A collection of related records is a **file**—for example, an employee file is a collection of all company employee records. Likewise, an inventory file is a collection of all inventory records for a particular company or organization. PC database software often refers to files as tables.

At the highest level of this hierarchy is a *database*, a collection of integrated and related files. Together, bits, characters, fields, records, files, and databases form the **hierarchy of data** (Figure 5.1). Characters are combined to make a field, fields are combined to make a record, records are combined to make a file, and files are combined to make a database. A database houses not only all these levels of data but the relationships among them.

Data Entities, Attributes, and Keys

Entities, attributes, and keys are important database concepts. An **entity** is a generalized class of people, places, or things (objects) for which data is collected, stored, and maintained. Examples of entities include employees, inventory, and customers. Most organizations organize and store data as entities.

An **attribute** is a characteristic of an entity. For example, employee number, last name, first name, hire date, and department number are attributes for an employee (Figure 5.2). Inventory number, description, number of units on hand, and the location of the inventory item in the warehouse are examples of attributes for items in inventory. Customer number, name, address, phone number, credit rating, and contact person are examples of attributes for customers. Attributes are usually selected to capture the relevant characteristics of entities like employees or customers. The specific value of an attribute, called a **data item**, can be found in the fields of the record describing an entity.

As discussed, a collection of fields about a specific object is a record. A **key** is a field or set of fields in a record that is used to identify the record. A **primary key** is a field or set of fields that uniquely identifies the record. No other record can have the same primary key. The primary key is used to distinguish records so that they can be accessed, organized, and manipulated. For an employee record such as the one shown in Figure 5.2, the employee number is an example of a primary key.

FIGURE 5.2

Keys and Attributes

The key field is the employee number. The attributes include last name, first name, hire date, and department number.

Employee #	Last name	First name	Hire date	Dept. number
005-10-6321	Johns	Francine	10-07-1997	257
549-77-1001	Buckley	Bill	02-17-1979	632
098-40-1370	Fiske	Steven	01-05-1985	598

Entities (records)

Key field

Attributes (fields)

Locating a particular record that meets a specific set of criteria may require the use of a combination of secondary keys. For example, a customer might call a mail-order company to place an order for clothes. If the customer does not know his primary key (such as a customer number), a secondary key (such as last name) can be used. In this case, the order clerk enters the last name, such as Adams. If there are several customers with a last name of Adams, the clerk can check other fields, such as address, first name, and so on, to find the correct customer record. Once the correct customer record is obtained, the order can be completed and the clothing items shipped to the customer.

The Traditional Approach versus the Database Approach

The Traditional Approach

Organizations are adaptive systems with constantly changing data and information needs. For any growing or changing business, managing data can become quite complicated. One of the most basic ways to manage data is via files. Because a file is a collection of related records, all records associated with a particular application (and therefore related by the application) could be collected and managed together in an application-specific file. At one time, most organizations had numerous application-specific data files; for example, customer records often were maintained in separate files, with each file relating to a specific process completed by the company, such as shipping or billing. This approach to data management, whereby separate data files are created and stored for each application program, is called the **traditional approach**. For each particular application, one or more data files is created (Figure 5.3).

One of the flaws in this traditional file-oriented approach to data management is that much of the data—for example, customer name and address—is duplicated in two or more files. This duplication of data in separate files is known as **data redundancy**. The problem with data redundancy is that changes to the data (e.g., a new customer address) might be made in one file and not another. The order-processing department might have updated its file to the new address, but the billing department is still sending bills to the old address. Data redundancy, therefore, conflicts with **data integrity**—the degree to which the data in any one file is accurate. Data integrity follows from the control or elimination of data redundancy. Keeping a customer's address in only one file decreases the possibility that the customer will have two different addresses stored in different locations. The efficient operation of a business requires a high degree of data integrity.

In many computerized database systems based on the traditional file approach, the data is organized for a particular application program (say, billing). These applications have **program-data dependence**—that is, programs and

traditional approach to data management

an approach whereby separate data files are created and stored for each application program

data redundancy

duplication of data in separate files

data integrity

the degree to which the data in any one file is accurate

program-data dependence

concept according to which programs and data developed and organized for one application are incompatible with programs and data organized differently for another application

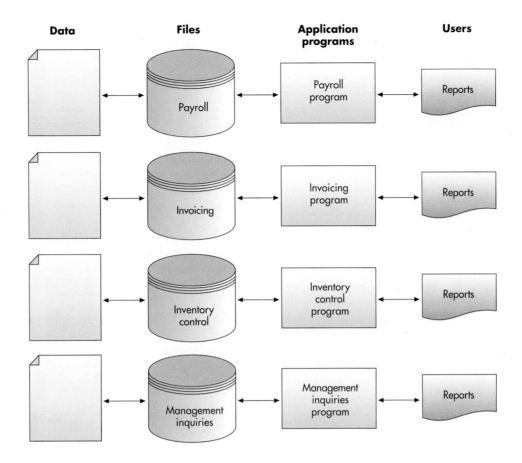

Data	Files	Application programs	Users

Payroll → Payroll program → Reports

Invoicing → Invoicing program → Reports

Inventory control → Inventory control program → Reports

Management inquiries → Management inquiries program → Reports

FIGURE 5.3

The Traditional Approach to Data Management

With the traditional approach, one or more data files is created and used for every application. For example, the inventory control program would have one or more files containing inventory data, such as the inventory item, number on hand, and item description. Likewise, the invoicing program can have files on customers, inventory items being shipped, and so on. With the traditional approach to data management, it is possible to have the same data, such as inventory items, in several different files used by different applications.

database approach to data management

an approach whereby a pool of related data is shared by multiple application programs

data developed and organized for one application are incompatible with programs and data organized differently for another application. For example, one programmer might develop a billing program that stores ZIP code data in one format with five numbers, while another programmer might develop a separate order-processing program that stores ZIP code data in a nine-number format. In a computerized file-based environment, all the programs that access this ZIP code data would need to be changed. Bridging the gap between two files with different program-data dependencies is often difficult and expensive.

Despite the drawbacks of using the traditional file approach in computerized database systems, some organizations continue to use it. For these firms, the cost of converting to another approach is too high.

The Database Approach

Because of the problems associated with the traditional approach to data management, many managers wanted a more efficient and effective means of organizing data. The result was the **database approach** to data management. In a database approach, a pool of related data is shared by multiple application programs. Rather than having separate data files, each application uses a collection of data that is either joined or related in the database.

The database approach offers significant advantages over the traditional file-based approach. For one, by controlling data redundancy, the database approach can use storage space more efficiently and increase data integrity. The database approach can also provide an organization with increased flexibility in the use of data. Because data once kept in two files is now located in the same database, it is easier to locate and request data for many types of processing. A database also offers the ability to share data and information resources. This can be a critical factor in coordinating organizationwide responses across

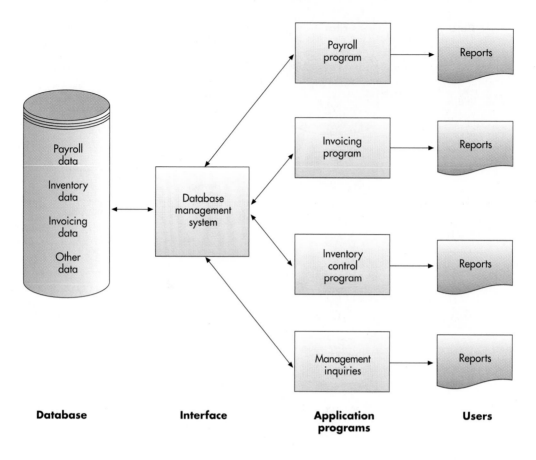

Database   Interface   Application   Users
programs

FIGURE 5.4

The Database Approach to
Data Management

An enterprisewide database
enables Allina Heath System
to reduce healthcare costs
and provide patients with the
best treatment possible. The
database pulls cost information
from its hospital and health
plans, comparing best practices in treatments and matching cost to level of service.
(Source: Stone/Frank Siteman.)

diverse functional areas of a corporation. In sharing data, however, some consistency should exist among software programs.

To use the database approach to data management, additional software—a
database management system (DBMS)—is required. As previously discussed, a
DBMS consists of a group of programs that can be used as an interface
between a database and the user or the database and application programs.
Typically, this software acts as a buffer between the application programs and
the database itself. Figure 5.4 illustrates the database approach.

The database approach to data management involves a combination of hardware and software. Tables 5.1 and 5.2 list some of the primary advantages and
disadvantages of the database approach and explore some of these issues.

Because of the many advantages of the database approach, most businesses
use databases to store data on customers, orders, inventory, employees, and
suppliers. This data is used as the input to the various information systems
throughout an organization. For example, the transaction processing system
can use the data to support daily business processes like billing, inventory tracking, and ordering. This same data can be processed by a management information system to create reports or a decision support system to provide
information to aid managerial decision making.

Many modern databases are enterprisewide, encompassing much of the data
of the entire organization. Often, distinct yet related databases are linked to
provide enterprisewide databases. Much planning and organization go into the
development of such databases. Allina Health System, a healthcare system that
owns and manages 19 hospitals and 50 clinics and provides health plans to over
one million customers, uses an enterprisewide database. Allina must try to
reduce healthcare costs and hold the line on insurance premiums while providing its patients with the best treatment possible. It must also integrate the information systems and business practices of the organizations it has acquired

Advantages	Explanation
Improved strategic use of corporate data	Accurate, complete, up-to-date data can be made available to decision makers where, when, and in the form they need it.
Reduced data redundancy	The database approach can reduce or eliminate data redundancy. Data is organized by the DBMS and stored in only one location. This results in more efficient utilization of system storage space.
Improved data integrity	With the traditional approach, some changes to data were not reflected in all copies of the data kept in separate files. This is prevented with the database approach because there are no separate files that contain copies of the same piece of data.
Easier modification and updating	With the database approach, the DBMS coordinates updates and data modifications. Programmers and users do not have to know where the data is physically stored. Data is stored and modified once. Modification and updating is also easier because the data is stored at only one location in most cases.
Data and program independence	The DBMS organizes the data independently of the application program. With the database approach, the application program is not affected by the location or type of data. Introduction of new data types not relevant to a particular application does not require the rewriting of that application to maintain compatibility with the data file.
Better access to data and information	Most DBMSs have software that makes it easy to access and retrieve data from a database. In most cases, simple commands can be given to get important information. Relationships between records can be more easily investigated and exploited, and applications can be more easily combined.
Standardization of data access	A primary feature of the database approach is a standardized, uniform approach to database access. This means that the same overall procedures are used by all application programs to retrieve data and information.
A framework for program development	Standardized database access procedures can mean more standardization of program development. Because programs go through the DBMS to gain access to data in the database, standardized database access can provide a consistent framework for program development. In addition, each application program need address only the DBMS, not the actual data files, reducing application development time.
Better overall protection of the data	The use of and access to centrally located data are easier to monitor and control. Security codes and passwords can ensure that only authorized people have access to particular data and information in the database, thus ensuring privacy.
Shared data and information resources	The cost of hardware, software, and personnel can be spread over a large number of applications and users. This is a primary feature of a DBMS.

TABLE 5.1

Advantages of the Database Approach

through mergers. To address these issues, management called for an information systems strategy that included the development of enterprisewide databases spanning all its hospitals, clinics, and patients. The goal was to enable Allina to gather information and integrate it in ways never done before—for example, pulling cost information from the hospital and the health plans, comparing best practices in treatments, and matching cost to level of service.[1]

DATA MODELING AND DATABASE MODELS

Because there are so many elements in today's businesses, it is critical to keep data organized so that it can be effectively utilized. A database should be designed to store all data relevant to the business and provide quick access and easy modification. Moreover, it must reflect the business processes of the organization. When building a database, careful consideration must be given to these questions:

• Content: What data should be collected and at what cost?

Disadvantages	Explanation
Relatively high cost of purchasing and operating a DBMS in a mainframe operating environment	Some mainframe DBMSs can cost hundreds of thousands of dollars.
Increased cost of specialized staff	Additional specialized staff and operating personnel may be needed to implement and coordinate the use of the database. However, some organizations have been able to implement the database approach with no additional personnel.
Increased vulnerability	Even though databases offer better security because security measures can be concentrated on one system, they also make more data accessible to the trespasser if security is breached. In addition, if for some reason there is a failure in the DBMS, multiple application programs are affected.

TABLE 5.2

Disadvantages of the Database Approach

- Access: What data should be provided to which users and when?
- Logical structure: How should data be arranged so that it makes sense to a given user?
- Physical organization: Where should data be physically located?

Data Modeling

Key considerations in organizing data in a database include determining what data is to be collected in the database, who will have access to it, and how they might wish to use the data. Based on these determinations, a database can then be created. Building a database requires two different types of designs: a logical design and a physical design. The logical design of a database shows an abstract model of how the data should be structured and arranged to meet an organization's information needs. The logical design of a database involves identifying relationships among the different data items and grouping them in an orderly fashion. Because databases provide both input and output for information systems throughout a business, users from all functional areas should assist in creating the logical design to ensure that their needs are identified and addressed. Physical database design starts from the logical database design and fine-tunes it for performance and cost considerations (e.g., improved response time, reduced storage space, lower operating cost). The person identified to fine-tune the physical design must have an in-depth knowledge of the DBMS to implement the database. For example, the logical database design may need to be altered so that certain data entities are combined, summary totals are carried in the data records rather than calculated from elemental data, and some data attributes are repeated in more than one data entity. These are examples of **planned data redundancy**. It is done to improve the system performance so that user reports or queries can be created more quickly.

One of the tools database designers use to show the logical relationships among data is a data model. A **data model** is a diagram of entities and their relationships. Data modeling usually involves understanding a specific business problem and analyzing the data and information needed to deliver a solution. When done at the level of the entire organization, this is called **enterprise data modeling**. Enterprise data modeling is an approach that starts by investigating the general data and information needs of the organization at the strategic level and then examining more specific data and information needs for the various functional areas and departments within the organization. Various models have been developed to help managers and database designers analyze data and information needs. An entity-relationship diagram is an example of such a data model.

Entity-relationship (ER) diagrams use basic graphical symbols to show the organization of and relationships between data. In most cases, boxes are

planned data redundancy

a way of organizing data in which the logical database design is altered so that certain data entities are combined, summary totals are carried in the data records rather than calculated from elemental data, and some data attributes are repeated in more than one data entity to improve database performance

data model

a diagram of data entities and their relationships

enterprise data modeling

data modeling done at the level of the entire enterprise

entity-relationship (ER) diagrams

a data model that uses basic graphical symbols to show the organization of and relationships between data

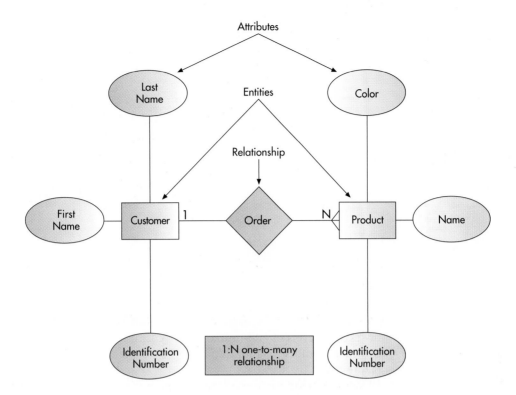

An Entity-Relationship (ER) Diagram for a Customer Ordering Database

Development of this type of diagram helps ensure the logical structuring of application programs that are able to serve users' needs and are consistent with the data relationships in the database.

used in ER diagrams to indicate data items or entities, and diamonds show relationships between data items and entities.

Figure 5.5 shows an ER diagram for a customer who places orders. ER diagrams can show a number of relationships. For example, one customer can place many orders. This is an example of a one-to-many relationship, as shown by the one-to-many symbol (1:N) used in Figure 5.5. Each order can include one or more line items, where a line item specifies the product identification and quantity ordered. One-to-one, many-to-many, and other relationships can also be revealed using ER diagrams. ER diagrams help ensure that the relationships among the data entities in a database are logically structured so that application programs can be developed to serve user needs. In addition, ER diagrams can be used as reference documents once a database is in use. If changes are to be made in the database, ER diagrams can help design them.

Database Models

The structure of the relationships in most databases follows one of three logical database models: hierarchical, network, and relational. Hierarchical and network models are still being used today, but relational models are the most popular. It is important to remember that the records represented in the models are actually linked or related logically to one another. These links dictate the way users can access data with application programs. Because the different models involve different links between data, each model has unique advantages and disadvantages.

Hierarchical (Tree) Models

In many situations, data follows a hierarchical, or treelike, structure. In a **hierarchical database model**, the data is organized in a top-down, or inverted tree, structure. For example, data about a project for a company can follow this type of model, as shown in Figure 5.6. The hierarchical model is best suited to situations in which the logical relationships between data can be properly represented with the one-to-many approach.

hierarchical database model

a data model in which data is organized in a top-down, or inverted tree, structure

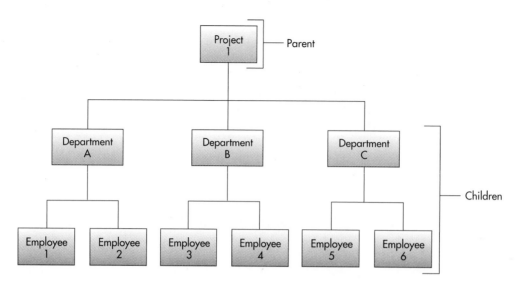

FIGURE 5.6

A Hierarchial Database Model

Project 1 is the top, or root, element. Departments A, B, and C are under this element, with Employees 1 through 6 beneath them as follows: Employees 1 and 2 under Department A, Employees 3 and 4 under Department B, and Employees 5 and 6 under Department C. Thus, there is a one-to-many relationship among the elements of this model.

network model

an expansion of the hierarchical database model with an owner-member relationship in which a member may have many owners

relational model

a database model that describes data in which all data elements are placed in two-dimensional tables, called relations, that are the logical equivalent of files

domain

the allowable values for data attributes

Network Models

A **network model** is an expansion of the hierarchical model. Instead of having only various levels of one-to-many relationships, however, the network model is an owner-member relationship in which a member may have many owners (Figure 5.7).

Databases structured according to either the hierarchical model or the network model suffer from the same deficiency: once the relationships are established between data elements, it is difficult to modify them or to create new relationships.

Relational Models

Relational models have become the most popular database models, and use of these models will increase in the future. The relational model describes data using a standard tabular format. In a database structured according to the **relational model**, all data elements are placed in two-dimensional tables, called relations, that are the logical equivalent of files. The tables in relational databases organize data in rows and columns, simplifying data access and manipulation. It is normally easier for managers to understand the relational model (Figure 5.8) than the hierarchical and network models.

In the relational model, each row of a table represents a data entity, with the columns of the table representing attributes. Each attribute can take on only certain values. The allowable values for these attributes are called the **domain**. The domain for a particular attribute indicates what values can be placed in each of the columns of the relational table. For instance, the domain for an attribute such as gender would be limited to male or female. A domain for pay rate would not include negative numbers. Defining a domain can increase data accuracy. For example, a pay rate of –$5.00 could not be entered into the database because it is a negative number and not in the domain for pay rate.

FIGURE 5.7

A Network Database Model

In this network model, two projects are at the top. Departments A, B, and C are under Project 1; Departments B and C are under Project 2. Thus, the elements of this model represent a many-to-many relationship.

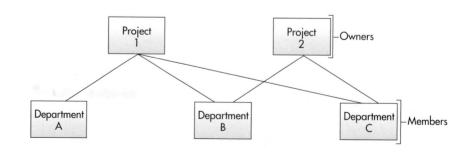

FIGURE 5.8

A Relational Database Model

In the relational model, all data elements are placed in two-dimensional tables, or relations. As long as they share at least one common element, these relations can be linked to output useful information.

Data table 1: Project table

Project number	Description	Dept. number
155	Payroll	257
498	Widgets	632
226	Sales Manual	598

Data table 2: Department table

Dept. number	Dept. name	Manager SSN
257	Accounting	005-10-6321
632	Manufacturing	549-77-1001
598	Marketing	098-40-1370

Data table 3: Manager table

SSN	Last name	First name	Hire date	Dept. number
005-10-6321	Johns	Francine	10-07-1997	257
549-77-1001	Buckley	Bill	02-17-1979	632
098-40-1370	Fiske	Steven	01-05-1985	598

Once data has been placed into a relational database, users can make inquiries and analyze data. Basic data manipulations include selecting, projecting, and joining. **Selecting** involves eliminating rows according to certain criteria. Suppose a project table contains the project number, description, and department number for all projects being performed by a company. The president of the company might want to find the department number for Project 226, a sales manual project. Using selection, the president can eliminate all rows but number 226 and see that the department number for the department completing the sales manual project is 598.

Projecting involves eliminating columns in a table. For example, we might have a department table that contains the department number, department name, and social security number (SSN) of the manager in charge of the project. The sales manager might want to create a new table with only the department number and the social security number of the manager in charge of the sales manual project. Projection can be used to eliminate the department name column and create a new table containing only department number and SSN.

Joining involves combining two or more tables. For example, we can combine the project table and the department table to get a new table with the project number, project description, department number, department name, and social security number for the manager in charge of the project.

As long as the tables share at least one common data attribute, the tables in a relational database can be **linked** to provide useful information and reports (see Figure 5.9). Being able to link tables to each other through common data attributes is one of the keys to the flexibility and power of relational databases.

selecting

data manipulation that eliminates rows according to certain criteria

projecting

data manipulation that eliminates columns in a table

joining

data manipulation that combines two or more tables

linking

data manipulation that combines two or more tables using common data attributes to form a new table with only the unique data attributes

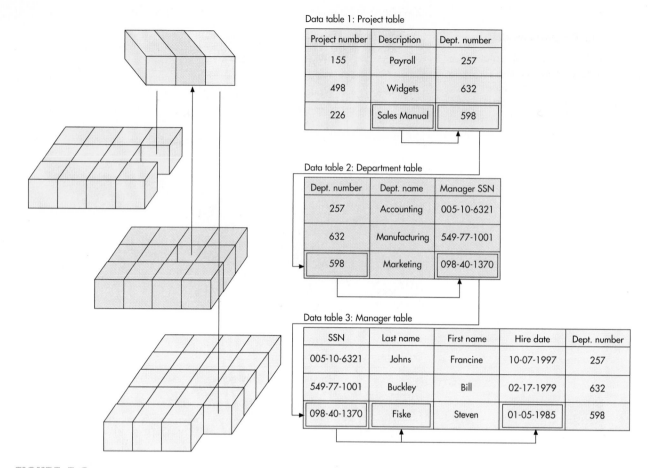

Data table 1: Project table

Project number	Description	Dept. number
155	Payroll	257
498	Widgets	632
226	Sales Manual	598

Data table 2: Department table

Dept. number	Dept. name	Manager SSN
257	Accounting	005-10-6321
632	Manufacturing	549-77-1001
598	Marketing	098-40-1370

Data table 3: Manager table

SSN	Last name	First name	Hire date	Dept. number
005-10-6321	Johns	Francine	10-07-1997	257
549-77-1001	Buckley	Bill	02-17-1979	632
098-40-1370	Fiske	Steven	01-05-1985	598

FIGURE 5.9

Linking Data Tables to Answer an Inquiry

In finding the name and hire date of the manager working on the sales manual project, the president needs three tables; project, department, and manager. The project description (Sales Manual) leads to the department number (598) in the project table, which leads to the manager's SSN (098-40-1370) in the department table, which leads to the manager's name (Fiske) and hire date (01-05-1985) in the manager table.

Suppose the president of a company wants to find out the name of the manager of the sales manual project and how long the manager has been with the company. The president would make the inquiry to the database, perhaps via a desktop personal computer. The DBMS would start with the project description and search the project table to find out the project's department number. It would then use the department number to search the department table for the department manager's social security number. The department number is also in the department table and is the common element that allows the project table and the department table to be linked. The DBMS then uses the manager's social security number to search the manager table for the manager's hire date. The manager's social security number is the common element between the department table and the manager table. The final result: the manager's name and hire date are presented to the president as a response to the inquiry.

One of the primary advantages of a relational database is that it allows tables to be linked, as shown in Figure 5.9. This linkage is especially useful when information is needed from multiple tables, as in our example. The manager's social security number, for example, is maintained in the manager table. If the social security number is needed, it can be obtained by linking to the manager table.

The relational database model is by far the most widely used. It is easier to control, more flexible, and more intuitive than the others because it organizes data in tables. As seen in Figure 5.10, a relational database management system, such as Access, provides a number of tips and tools for building and using database tables. This figure shows the database displaying information about data types and indicating that additional help is available. The ability to link relational tables also allows users to relate data in new ways without having to redefine complex relationships. Because of the advantages of the relational model, many companies use it for large corporate databases, such as in marketing and

Building and Modifying a
Relational Database

Relational databases provide
many tools, tips, and tricks to
simplify the process of creating
and modifying a database.

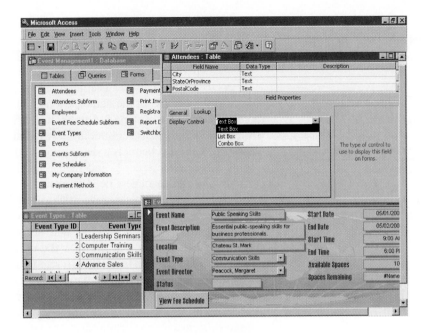

accounting. The relational model can be used with personal computers and
mainframe systems.

Data Cleanup

As discussed in Chapter 1, the characteristics of valuable data include that the
data is accurate, complete, economical, flexible, reliable, relevant, simple,
timely, verifiable, accessible, and secure. The purpose of **data cleanup** is to
develop data with these characteristics. A 4 percent error rate may not sound
like much, but a multibillion-dollar corporation could lose millions of dollars if
bad information caused the firm to bill all but 4 percent of its orders. When a
database is created with data from multiple sources, those disparate sources
may store different values for the same customer due to spelling errors, multiple account numbers, and address variations. The purpose of data cleanup is to
look for and fix these and other inconsistencies that can result in duplicate or
incorrect records ending up in the database.

data cleanup

the process of looking for and
fixing inconsistencies to ensure that
data is accurate and complete

DATABASE MANAGEMENT SYSTEMS (DBMSs)

Creating and implementing the right database system ensures that the database
will support respective business activities and goals. But how do we actually create, implement, use, and update a database? The answer is found in the database
management system. As discussed, a database management system (DBMS) is a
group of programs used as an interface between a database and application programs or a database and the user. DBMSs are classified by the type of database
model they support. For example, a relational database management system follows the relational model. Access by Microsoft is a popular relational DBMS for
personal computers. Popular mainframe relational DBMSs include DB2 by
IBM, Oracle, Sybase, and Informix. All DBMSs share some common functions,
like providing a user view, physically storing and retrieving data in a database,
allowing for database modification, manipulating data, and generating reports.

Providing a User View

Because the DBMS is responsible for access to a database, one of the first steps
in installing and using a database involves telling the DBMS the logical and

schema

a description of the entire database

physical structure of the data and relationships among the data in the database. This description is called a **schema** (as in schematic diagram). A schema can be part of the database or a separate schema file. The DBMS can reference a schema to find where to access the requested data in relation to another piece of data.

A DBMS also acts as a user interface by providing a view of the database. A user view is the portion of the database a user can access. To create different user views, subschemas are developed. A **subschema** is a file that contains a description of a subset of the database and identifies which users can modify the data items in that subset. While a schema is a description of the entire database, a subschema shows only some of the records and their relationships in the database. Normally, programmers and managers need to view or access only a subset of the database. For example, a sales representative might need only data describing customers in her region, not the sales data for the entire nation. A subschema could be used to limit her view to data from her region. With subschemas, the underlying structure of the database can change, but the view the user sees might not change. For example, even if all the data on the southern region changed, the northeast region sales representative's view would not change if she accessed data on her region.

subschema

a file that contains a description of a subset of the database and identifies which users can view and modify the data items in the subset

A number of subschemas can be developed for different managers or users and the various application programs. Typically, the database user or application will access the subschema, which then accesses the schema (Figure 5.11). Subschemas can also provide additional security because programmers, managers, and other users are typically allowed to view only certain parts of the database.

Creating and Modifying the Database

data definition language (DDL)

a collection of instructions and commands used to define and describe data and data relationships in a specific database

Schemas and subschemas are entered into the DBMS (usually by database personnel) via a data definition language. A **data definition language (DDL)** is a collection of instructions and commands used to define and describe data and data relationships in a specific database. A DDL allows the database's creator to describe the data and the data relationships that are to be contained in the schema and the many subschemas. In general, a DDL describes logical access paths and logical records in the database. Figure 5.12 shows a simplified example of a DDL used to develop a general schema. The *Xs* in Figure 5.12 reveal where specific

FIGURE 5.11

The Use of Schemas and Subschemas

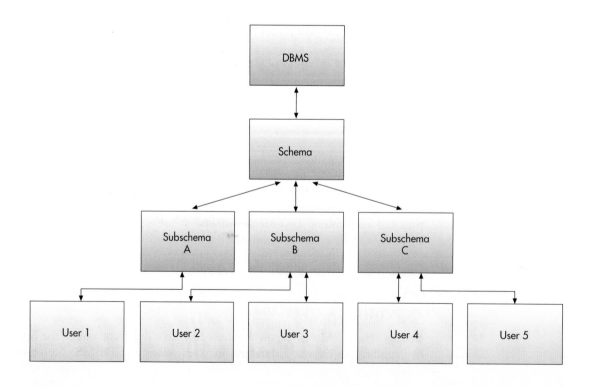

FIGURE 5.12

Using a Data Definition
Language to Define a Schema

```
SCHEMA DESCRIPTION
SCHEMA NAME IS XXXX
AUTHOR        XXXX
DATE          XXXX
FILE DESCRIPTION
      FILE NAME IS XXXX
          ASSIGN XXXX
      FILE NAME IS XXXX
          ASSIGN XXXX
AREA DESCRIPTION
      AREA NAME IS XXXX
RECORD DESCRIPTION
      RECORD NAME IS XXXX
      RECORD ID IS XXXX
      LOCATION MODE IS XXXX
      WITHIN XXXX AREA FROM XXXX THRU XXXX
SET DESCRIPTION
      SET NAME IS XXXX
      ORDER IS XXXX
      MODE IS XXXX
      MEMBER IS XXXX
          .
          .
          .
```

FIGURE 5.12

Using a Data Definition
Language to Define a Schema

information concerning the database is to be entered. File description, area description, record description, and set description are terms the DDL defines and uses in this example. Other terms and commands can be used, depending on the particular DBMS employed.

data dictionary

a detailed description of all the
data used in the database

Another important step in creating a database is to establish a **data dictionary**, a detailed description of all data used in the database. The data dictionary contains the name of the data item, aliases or other names that may be used to describe the item, the range of values that can be used, the type of data (such as alphanumeric or numeric), the amount of storage needed for the item, a notation of the person responsible for updating it and the various users who can access it, and a list of reports that use the data item. Figure 5.13 shows a typical data dictionary entry.

For example, the information in a data dictionary for the part number of an inventory item can include the name of the person who made the data dictionary entry (D. Bordwell), the date the entry was made (August 4, 2000), the name of the person who approved the entry (J. Edwards), the approval date

```
                NORTHWESTERN MANUFACTURING

PREPARED BY:          D. BORDWELL
DATE:                 04 AUGUST 2000
APPROVED BY:          J. EDWARDS
DATE:                 13 OCTOBER 2000
VERSION:              3.1
PAGE:                 1 OF 1

DATA ELEMENT NAME:    PARTNO
DESCRIPTION:          INVENTORY PART NUMBER
OTHER NAMES:          PTNO
VALUE RANGE:          100 TO 5000
DATA TYPE:            NUMERIC
POSITIONS:            4 POSITIONS OR COLUMNS
```

FIGURE 5.13

A Typical Data Dictionary Entry

(October 13, 2000), the version number (3.1), the number of pages used for the entry (1), the part name (PARTNO), other part names that may be used (PTNO), the range of values (part numbers can range from 100 to 5000), the type of data (numeric), and the storage required (four positions are required for the part number). Following are some of the typical uses of a data dictionary.

- *Provide a standard definition of terms and data elements.* This can help in programming by providing consistent terms and variables to be used for all programs. Programmers know what data elements are already "captured" in the database and how they relate to other data elements.
- *Assist programmers in designing and writing programs.* Programmers do not need to know which storage devices are used to store needed data. Using the data dictionary, programmers specify the required data elements. The DBMS locates the necessary data. More important, programmers can use the data dictionary to see which programs already use a piece of data and, if appropriate, copy the relevant section of the program code into their new program, thus eliminating duplicate programming efforts.
- *Simplify database modification.* If for any reason a data element needs to be changed or deleted, the data dictionary would point to specific programs that utilize the data element that may need modification.

A data dictionary helps achieve the advantages of the database approach in these ways:

- *Reduced data redundancy.* By providing standard definitions of all data, it is less likely that the same data item will be stored in different places under different names. For example, a data dictionary would reduce the likelihood that the same part number would be stored as two different items, such as PTNO and PARTNO.
- *Increased data reliability.* A data dictionary and the database approach reduce the chance that data will be destroyed or lost. In addition, it is more difficult for unauthorized people to gain access to sensitive data and information.
- *Faster program development.* With a data dictionary, programmers can develop programs faster. They don't have to develop names for data items because the data dictionary does that for them.
- *Easier modification of data and information.* The data dictionary and the database approach make modifications to data easier because users do not need to know where the data is stored. The person making the change indicates the new value of the variable or item, such as part number, that is to be changed. The database system locates the data and makes the necessary change.

Storing and Retrieving Data

As just described, one function of a DBMS is to be an interface between an application program and the database. When an application program needs data, it requests that data through the DBMS. Suppose that to calculate the total price of a new car, an auto dealer pricing program needs price data on the engine option—six cylinders instead of the standard four cylinders. The application program thus requests this data from the DBMS. In doing so, the application program follows a logical access path. Next, the DBMS, working in conjunction with various system software programs, accesses a storage device, such as disk or tape, where the data is stored. When the DBMS goes to this storage device to retrieve the data, it follows a path to the physical location (physical access path) where the price of this option is stored. In the pricing example, the DBMS might go to a disk drive to retrieve the price data for six-cylinder engines. This relationship is shown in Figure 5.14.

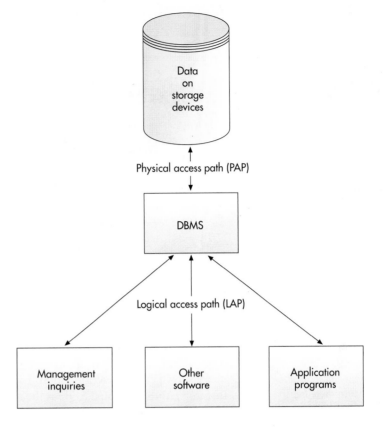

This same process is used if a user wants to get information from the database. First, the user requests the data from the DBMS. For example, a user might give a command, such as LIST ALL OPTIONS FOR WHICH PRICE IS GREATER THAN 200 DOLLARS. This is the logical access path (LAP). Then the DBMS might go to the options price sector of a disk to get the information for the user. This is the physical access path (PAP).

When two or more people or programs attempt to access the same record in the same database at the same time, there can be a problem. For example, an inventory control program might attempt to reduce the inventory level for a product by ten units because ten units were just shipped to a customer. At the same time, a purchasing program might attempt to increase the inventory level for the same product by 200 units because more inventory was just received. Without proper database control, one of the inventory updates may not be correctly made, resulting in an inaccurate inventory level for the product. **Concurrency control** can be used to avoid this potential problem. One approach is to lock out all other application programs from access to a record if the record is being updated or used by another program.

Users increasingly need to be able to access and update databases via the Internet. Many database vendors are incorporating this capability into their products, including Microsoft, Allaire, Inline Internet Systems, Netscape Communications, EveryWhere Development, and StormCloud Development. Such databases allow companies to create an Internet-accessible catalog, which is nothing more than a database of items, descriptions, and prices.

concurrency control

a method of dealing with a situation in which two or more people need to access the same record in a database at the same time

Manipulating Data and Generating Reports

Once a DBMS has been installed, the system can be used by all levels of employees via specific commands in various programming languages. For example, COBOL commands can be used in simple programs that will access

FAO Schwarz uses Allaire Corporation's software to enhance its online store. Enhancements increase the volume of customers the site can handle and enable customers to track their purchases on the site using a UPS tracking number.

data manipulation language (DML)

the commands that are used to manipulate the data in a database

or manipulate certain pieces of data in the database. Here's another example of a DBMS query: SELECT * FROM EMPLOYEE WHERE JOB_CLASSI-FICATION = "C2". The * tells the program to include all columns from the EMPLOYEE table. In general, the commands that are used to manipulate the database are part of the **data manipulation language (DML)**. This specific language, provided with the DBMS, allows managers and other database users to access, modify, and make queries about data contained in the database to generate reports. Again, the application programs go through subschemas, schemas, and the DBMS before actually getting to the physically stored data on a device such as a disk.

In the 1970s, D. D. Chamberlain and others at the IBM Research Laboratory in San Jose, California, developed a standardized data manipulation language called *Structured Query Language (SQL)*, pronounced like the word *sequel* or simply spelled out as *SQL*. The EMPLOYEE query shown earlier is written in SQL. In 1986, the American National Standards Institute (ANSI) adopted SQL as the standard query language for relational databases. Since ANSI's acceptance of SQL, interest in making SQL an integral part of relational databases on both mainframe and personal computers has increased. As discussed in Chapter 4, SQL lets programmers learn one powerful query language and use it on systems ranging from PCs to the largest mainframe computers (Figure 5.15). Programmers and database users also find SQL valuable because SQL statements can be embedded into many programming languages, such as the widely used COBOL. Because SQL uses standardized and simplified procedures for retrieving, storing, and manipulating data in a database system, the popular database query language can be easy to understand and use.

Once a database has been set up and loaded with data, it can produce desired reports, documents, and other outputs (see Figure 5.16). These outputs usually appear in screen displays or hard-copy printouts. The output-control features of a database program allow you to select the records and fields to appear in reports. You can also complete calculations specifically for the report by manipulating database fields. Formatting controls and organization options (like report headings) help you to customize reports and create flexible, convenient, and powerful information-handling tools.

FIGURE 5.15

Structured Query Language

SQL has become an integral part of most relational database packages, as shown by this screen from Microsoft Access.

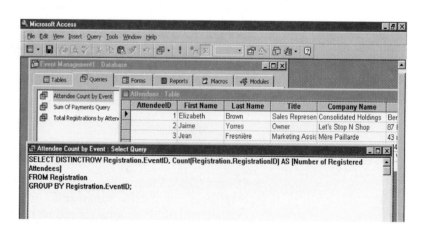

FIGURE 5.16

Database Output

A database application offers sophisticated formatting and organization options to produce the right information in the right format.

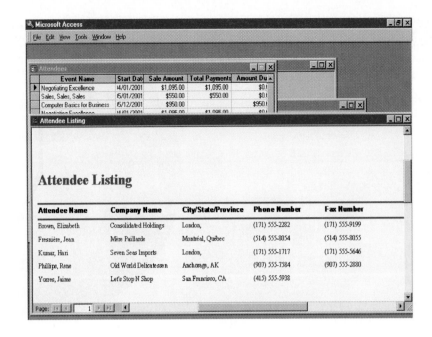

A database program can produce a wide variety of documents, reports, and other outputs that can help organizations achieve their goals. The most common reports select and organize data to present summary information about some aspect of company operations. For example, accounting reports often summarize financial data such as current and past-due accounts. Many companies base their routine operating decisions on regular status reports that show the progress of specific orders toward completion and delivery. Increasingly, companies are using databases to provide improved customer services.

Exception, scheduled, and demand reports, first discussed in Chapter 1, highlight events that require urgent management attention. Database programs can produce literally hundreds of documents and reports. A few examples include these:

• Form letters with address labels
• Payroll checks and reports
• Invoices
• Orders for materials and supplies
• A variety of financial performance reports

Popular Database Management Systems

The latest generation of database management systems makes it possible for end users to build their own database applications. End users are using these tools to address everyday problems like how to manage a mounting pile of information on employees, customers, inventory, or sales and fun stuff like wine lists, CD collections, and video libraries. These database management systems are an important personal productivity tool along with word processing, spreadsheet, and graphics software.

A key to making DBMSs more usable for some databases is the incorporation of "wizards" that walk you through how to build customized databases, modify ready-to-run applications, use existing record templates, and quickly locate the data you want. These applications also include powerful new features such as help systems and Web-publishing capabilities. For example, users can create a complete inventory system and then instantly post it to the Web, where it does double duty as an electronic catalog. Some of the more popular DBMSs for end users include Microsoft Access, Lotus Approach, and Inprise's dBASE.

FIGURE 5.17

Worldwide Database Market Share, 1998
(Source: Data from Julie Pitta, "Squeeze Play: Databases Get Ugly," *Forbes*, February 22, 1999, pp. 50–51.)

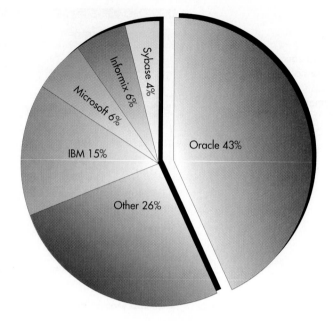

The complete database management software market encompasses software used by professional programmers and that runs on midrange, mainframe, and supercomputers. The entire market generates $10 billion per year in revenue, with Oracle, IBM, Microsoft, Informix, and Sybase the leaders (Figure 5.17).[2] Although Microsoft rules in desktop PC software, its share of database software on bigger computers is small.

Selecting a Database Management System

Selecting the best database management system begins by analyzing database needs and characteristics. The information needs of the organization affect what type of data is collected and what type of database management system is used. Important characteristics of databases include the size of the database, number of concurrent users, performance, the ability of the DBMS to integrate with other systems, the features of the DBMS, vendor considerations, and the cost of the database management system.

Database Size
The database size depends on the number of records or files in the database. The size determines the overall storage requirement for the database. Most database management systems can handle relatively small databases of less than 100 million bytes; fewer can manage terabyte-size databases.

Number of Concurrent Users
The number of simultaneous users that can access the contents of the database is also an important factor. Clearly, a database that is meant to be used in a large workgroup must be able to support a number of concurrent users; if it cannot, then the efficiency of the members of the workgroup will be lowered. The term *scalability* is sometimes used to describe how well a database performs as the size of the database or the number of concurrent users is increased. A highly scalable database management system is desirable to provide flexibility. Unfortunately, many companies make a poor DBMS choice in this regard and then later are forced to convert to a new DBMS when the original does not meet expectations.[3]

Performance

How fast the database is able to update records can be the most important performance criterion for some organizations. Credit card and airline companies, for example, must have database systems that can update customer records and check credit or make a plane reservation in seconds, not minutes. Other applications, such as payroll, can be done once a week or less frequently and do not require immediate processing. If an application demands immediacy, it also demands rapid recovery facilities in the event the computer system shuts down temporarily. Other performance considerations include the number of concurrent users that can be supported and how much main memory is required to execute the database management program.

Integration

A key aspect of any database management system is its ability to integrate with other applications and databases. A key determinant here is what operating systems it can run under—such as Unix, Windows NT, and Windows 2000. Some companies use several databases for different applications at different locations. A manufacturing company with four plants in three different states might have a separate database at each location. The ability of a database program to import data from and export data to other databases and applications can be a critical consideration.

Features

The features of the database management system can also make a big difference. Most database programs come with security procedures, privacy protection, and a variety of tools. Other features can include how easy the database package is to use and the availability of manuals and documentation that can help the organization get the most from the database package. Additional features such as wizards and ready-to-use templates help improve the product ease of use and are very important.

The Vendor

The size, reputation, and financial stability of the vendor should also be considered in making any database purchase. Some vendors are well respected in the information systems industry and have a large staff of support personnel to give assistance, if necessary. A well-established and financially secure database company is more likely to remain in business than others.

Cost

Database packages for personal computers can cost a few hundred dollars, while large database systems for mainframe computers can cost hundreds of thousands of dollars. In addition to the initial cost of the database package, monthly operating costs should be considered. Some database companies rent or lease their database software. Monthly rental or lease costs, maintenance costs, additional hardware and software costs, and personnel costs can be substantial.

DATABASE DEVELOPMENTS

The types of data and information that managers need change as business processes change. A number of developments in the use of databases and database management systems can help managers meet their needs, among them allowing organizations to place data at different locations, setting up data warehouses and marts, using the object-oriented approach in database development, and searching for and using unstructured data such as graphics and video.

Distributed Databases

distributed database

a database in which the data may be spread across several smaller databases connected via telecommunications devices

Distributed processing involves placing processing units at different locations and linking them via telecommunications equipment. A **distributed database**—a database in which the data may be spread across several smaller databases connected via telecommunications devices—works on much the same principle. A user in the Milwaukee branch of a clothing manufacturer, for example, might make a request for data that is physically located at corporate headquarters in Milan. The user does not have to know where the data is physically stored. He or she makes a request for data, and the DBMS determines where the data is physically located and retrieves it (Figure 5.18).

Organizations often find that distributed databases provide some of the same advantages as distributed processing. Distributed databases give corporations more flexibility in how databases are organized and used. Local offices can create, manage, and use their own databases, and people at other offices can access and share the data in the local databases. Giving local sites a more direct way to access highly used data can provide significant organizational effectiveness and efficiency.

Despite its advantages, distributed processing creates additional challenges in maintaining data security, accuracy, timeliness, and conformance to standards. Distributed databases allow more users direct access at different sites; thus, controlling who accesses and changes data is sometimes difficult. Also, because distributed databases rely on telecommunications lines to transport data, access to data can be slower. To reduce telecommunications costs, some organizations will build a replicated database. A **replicated database** holds a duplicate set of frequently used data. At the beginning of the day, the company

replicated database

a database that holds a duplicate set of frequently used data

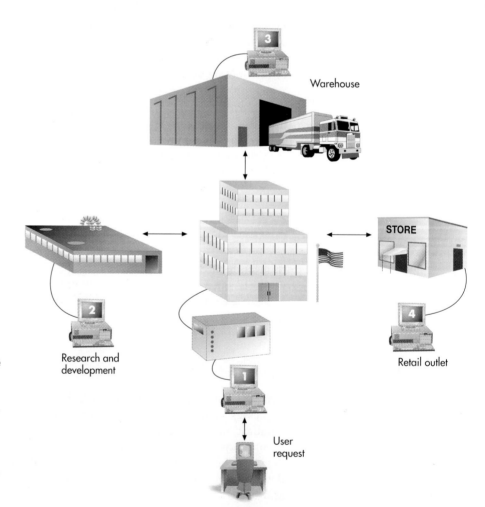

FIGURE 5.18

The Use of a Distributed Database

For a clothing manufacturer, computers may be located at corporate headquarters, in the research and development center, the warehouse, and in a company-owned retail store. Telecommunications systems link the computers so that users at all locations can access the same distributed database no matter where the data is actually stored.

sends a copy of important data to each distributed processing location. At the end of the day, the different sites send the changed data back to be stored in the main database.

Another challenge created by distributed databases involves integrating the various databases. For example, some organizations use many database management systems. A hierarchical database management system may be used at the headquarters of a large manufacturing company, and different relational DBMSs may be used by various regional offices. Businesses must develop a solution that enables them to access data in these different DBMSs.

Data Warehouses, Data Marts, and Data Mining

The raw data necessary to make sound business decisions is stored in a variety of locations and formats—hierarchical databases, network databases, flat files, and spreadsheets, to name a few. This data is initially captured, stored, and managed by transaction processing systems that are designed to support the day-to-day operations of the organization. For decades, organizations have collected operational, sales, and financial data with their on-line transaction processing (OLTP) systems.

Traditional OLTP systems are designed to put data into databases very quickly, reliably, and efficiently. These systems are not good at supporting meaningful analysis of the data. Indeed, tuning a system to provide excellent performance for OLTP often renders rapid data retrieval for data analysis nearly impossible. Furthermore, data stored in OLTP databases is inconsistent and constantly changing. The database contains the current transactions required to operate the business, including errors, duplicate entries, and reverse transactions, which get in the way of a business analyst, who needs stable data. Historical data is missing from the OLTP database, which makes trend analysis impossible. Thus, because of the application orientation of the data, the variety of nonintegrated data sources, and the lack of historical data, companies were limited in their ability to access and use the data for other purposes. So, although the data collected by OLTP systems doubles every two years, it does not meet the needs of the business decision maker—they are data rich but information poor.

Data Warehouses

data warehouse

a database that collects business information from many sources in the enterprise, covering all aspects of the company's processes, products, and customers

A **data warehouse** is a database that collects business information from many sources in the enterprise, covering all aspects of the company's processes, products, and customers.[4] The data warehouse provides business users with a multidimensional view of the data they need to analyze business conditions. Data warehousing is designed specifically to support management decision making, not to meet the needs of transaction processing systems. The data warehouse provides a specialized decision support database that manages the flow of information from existing corporate databases and external sources to end-user decision support applications. A data warehouse stores historical data that has been extracted from operational systems and external data sources (Figure 5.19). This operational and external data is "cleaned up" to remove inconsistencies and integrated to create a new information database that is more suitable for business analysis.

Data warehouses typically start out as very large databases (VLDBs), containing millions and even hundreds of millions of data records. As this data is collected from the various production systems, a historical database is built that business analysts can use. To remain fresh and accurate, the data warehouse receives regular updates. Old data that is no longer needed is purged from the data warehouse. Updating the data warehouse must be fast, efficient, and automated, or the ultimate value of the data warehouse is sacrificed. It is common

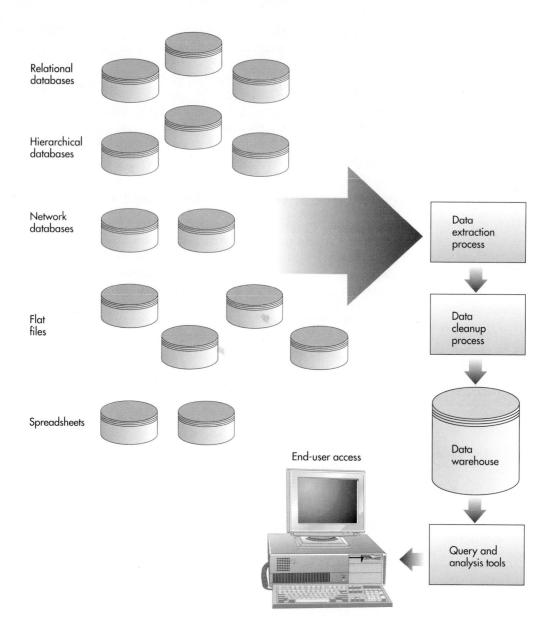

Relational
databases

Hierarchical
databases

Network
databases

Flat
files

Spreadsheets

Data
extraction
process

Data
cleanup
process

Data
warehouse

Query and
analysis tools

End-user access

FIGURE 5.19

Elements of a Data Warehouse

for a data warehouse to contain 3–10 years of current and historical data. Data cleaning tools can merge data from many sources into one database, automate data collection and verification, delete unwanted data, and maintain data in a database management system. Wal-Mart has one of the largest data warehouses in the world—a whopping 101 TB of data representing two years of detailed sales data that goes down to the level of individual receipts.[5] This data is used to make purchasing, pricing, and promotion decisions.

The primary advantage of data warehousing is the ability to relate data in new, innovative ways. However, a data warehouse can be extremely difficult to establish, with the average cost of building a data warehouse estimated at $2.2 million.[6] Table 5.3 provides some advice on how to create a data warehouse.

Data Marts

data mart

a subset of a data warehouse

A **data mart** is a subset of a data warehouse. Data marts bring the data warehouse concept—on-line analysis of sales, inventory, and other vital business data that has been gathered from transaction processing systems—to small and medium-size businesses and to departments within larger companies. Rather than store all enterprise data in one monolithic database, data marts contain a

Wal-Mart's data warehouse contains two years of detailed sales data on which the company bases purchasing, pricing, and promotion decisions.
(Source: Stone/Bruce Forster.)

data mining

an information analysis tool that involves the automated discovery of patterns and relationships in a data warehouse

subset of the data for a single aspect of a company's business—for example, finance, inventory, or personnel. In fact, there may even be more detailed data for a more specific area in a data mart than what a data warehouse would provide.

Data marts are most useful for smaller groups who want to access detailed data. A warehouse is used for summary data that can be used by the rest of the company. Because data marts typically contain tens of gigabytes of data, as opposed to the hundreds of gigabytes in data warehouses, they can be deployed on less powerful hardware with smaller disks, delivering significant savings to an organization. Although any database software can be used to set up a data mart, some vendors deliver specialized software designed and priced specifically for data marts. Already, companies such as Sybase, Software AG, Microsoft, and others have announced products and services that make it easier and cheaper to deploy these scaled-down data warehouses. The selling point: data marts put targeted business information into the hands of more decision makers.

Data Mining

Data mining is an information analysis tool that involves the automated discovery of patterns and relationships in a data warehouse. Data mining represents the next step in the evolution of decision support systems. It makes use of advanced statistical techniques and machine learning to discover facts in a large database, including databases on the Internet. Unlike query tools, which require users to formulate and test a specific hypothesis, data mining uses built-in analysis tools to automatically generate a hypothesis about the patterns and anomalies found in the data and then from the hypothesis predicts future behavior.

Data mining's objective is to extract patterns, trends, and rules from data warehouses to evaluate (i.e., predict or score) proposed business strategies, which in turn will improve competitiveness, improve profits, and transform

Sharply define your goals and objectives before you build the warehouse.
- Are you looking to increase your customer base?
- Do you want to double sales of a product line?
- Would you like to encourage repeat business?

Choose the software that best fits your goals.
- If you need a system that is tightly tied to sales rep activity in the field, choose a contact-management program with database capability.
- If you do a lot of telemarketing, you may need a telemarketing program.

Determine who should be in the database.
- First, figure out what types of customers have the most potential.
- Then construct the database, using everything from salespeople's contacts to lists bought from outside suppliers.

Develop a plan.
- Only after your objectives are laid out and your database is constructed is it time to devise a marketing program.
- Generally, it will fall into one of three categories: direct marketing, rewards for repeat purchases, and relationship-building promotions for long-time customers that generate profits.

Measure results.
- Generate periodic status reports in which you determine items like cost per contact and cost per sale.
- If you want to be really careful, do not launch your program at full tilt. Start with a small prototype; if you like what you see, expand to include the entire database.

TABLE 5.3

How to Design a Customer Data Warehouse
(Source: Adapted from Anne Field, "Precision Marketing," *Inc.*, vol. 18, no. 9, June 18, 1996, p. 54.)

business processes. It is used extensively in marketing to improve customer retention; cross-selling opportunities; campaign management; market, channel, and pricing analysis; and customer segmentation analysis (especially one-to-one marketing). In short, data mining tools help end users find answers to questions they never even thought to ask.

There are thousands of data mining applications. Bell Canada uses data mining to identify patterns and group customers with similar characteristics and create predictive target models that help determine which customers should receive a particular offer.[7] Fingerhut uses data mining to create specialized catalogs and optimize mailings to secure the highest possible revenue from the more than one million catalogs it mails each day.[8] Credit card issuers and insurers mine their data warehouses for subtle patterns within thousands of customer transactions to identify fraud, often just as it happens. One U.S. cellular phone company is using Silicon Graphics MineSet software to dig through mountains of call data and pinpoint illegally cloned cell phone ID numbers. Manufacturers mine data collected from factory floor sensors to identify where an intermittent assembly line error is causing a defect that will show up only months after an appliance goes into use.

E-commerce presents another major opportunity for effective use of data mining. Attracting customers to on-line Web sites is tough; keeping them can be next to impossible. For example, when on-line retail Web sites launch deep-discount sales, they cannot easily figure out how many first-time customers are likely to come back and buy again. Nor do they have a way of understanding which customers acquired during the sale are price sensitive and more likely to jump on future sales. As a result, companies are gathering data on user traffic through their Web sites into databases. This data is then analyzed using data mining techniques to personalize the Web site and develop sales promotions targeted at specific customers.

Faced with nagging questions about how best to push products and understand customers, Reel.com (an on-line provider of movies to rent, buy, and sell; a movie guide; and new-movie information) decided to use data mining techniques to analyze its Web site traffic data to better understand its customers. It employed consumer behavior-tracking software called LifeTime from Verbind. The software crunches through six to twelve months of on-line transaction data to build behavior maps—digital blueprints of a customer's Web site activity. The maps reveal a shopper's "velocity"—what he or she bought, how much was spent, and how often purchases were made. By viewing a customer's past behavior, Reel.com can predict that customer's average transactional amount and develop plans to target offers to that specific customer. Verbind is just one of an array of sophisticated consumer-tracking tools being deployed by e-commerce sites. Analytic profilers such as DataSage, E-piphany, and Personify all mimic Verbind's capabilities.[9]

Traditional DBMS vendors are well aware of the great potential of data mining. Thus, companies like Oracle, Informix Software, Sybase, Tandem, and Red Brick Systems are all incorporating data mining functionality into their products. Table 5.4 summarizes a few of the most frequent applications for data mining.

Reel.com employed data mining to map individual shopper's habits. Armed with a customer's data, Reel.com can develop plans to target offers to that specific customer.

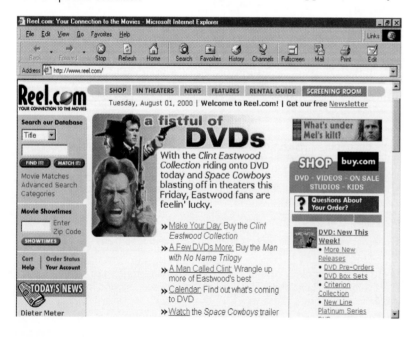

Application	Description
Market segmentation	Identifies the common characteristics of customers who buy the same products from your company.
Customer churn	Predicts which customers are likely to leave your company and go to a competitor.
Fraud detection	Identifies which transactions are most likely to be fraudulent.
Direct marketing	Identifies which prospects should be included in a mailing list to obtain the highest response rate.
Market basket analysis	Identifies what products or services are commonly purchased together (e.g. beer and diapers).
Trend analysis	Reveals the difference between a typical customer this month versus last month.

TABLE 5.4

Common Data Mining Applications

Source: Vance McCarthy, "Strike It Rich," *Datamation*, February 1997, pp. 44–50.

on-line analytical processing (OLAP)

software that allows users to explore data from a number of different perspectives

On-Line Analytical Processing (OLAP)

Most industry surveys today show that the majority of data warehouse users rely on spreadsheets, reporting and analysis tools, or their own custom applications to retrieve data from warehouses and format it into business reports and charts. In general, these approaches work fine for questions that can be answered when the amount of data involved is relatively modest and can be accessed with a simple table lookup.

For nearly two decades, multidimensional databases and their analytical information display systems have provided flashy sales presentations and trade show demonstrations. All you have to do is ask where a certain product is selling well, for example, and a colorful table showing sales performance by region, product type, and time frame automatically pops up on the screen. Called **on-line analytical processing (OLAP)**, these programs are now being used to store and deliver data warehouse information. OLAP allows users to explore corporate data from a number of different perspectives.

OLAP servers and desktop tools support high-speed analysis of data involving complex relationships, such as combinations of a company's products, regions, channels of distribution, reporting units, and time periods. Speed is essential in a booming economy, as businesses grow and accumulate more and more data in their operational systems and data warehouses. Long popular with financial planners, OLAP is now being put in the hands of other professionals. The leading OLAP software vendors include Cognos, Comshare, Hyperion Solutions, Oracle, MineShare, WhiteLight, and Microsoft.

Access to data in multidimensional databases can be very quick because they store the data in structures optimized for speed, and they avoid SQL and index processing. But multidimensional databases can take a great deal of time to update; in very large databases, update times can be so great that they force updates to be made only on weekends. Despite this flaw, multidimensional databases have continued to prosper because of their great retrieval speed. Some software providers are attempting to counteract this flaw through the use of partitioning and calculations-on-the-fly capabilities.

Consumer goods companies use OLAP to analyze the millions of consumer purchase records captured by scanners at the checkout stand. This data is used to spot trends in purchases and to relate sales volume to promotions and store conditions, such as displays, and even the weather. OLAP tools let managers analyze business data using multiple dimensions, such as product, geography, time, and salesperson. The data in these dimensions, called measures, is generally aggregated—for example, total or average sales in dollars or units, or budget dollars or sales forecast numbers. Rarely is the data studied in its raw, unaggregated form. Each dimension also can contain some hierarchy. For example, in the time dimension, users may examine data by year, by quarter,

Characteristic	OLTP Database	Data Warehousing
Purpose	Support transaction processing	Support decision support
Source of data	Business transactions	Multiple files, databases—data internal and external to the firm
Data access allowed users	Read and write	Read only
Primary data access mode	Simple database update and query	Simple and complex database queries with increasing use of data mining to recognize patterns in the data
Primary database model employed	Relational	Relational
Level of detail	Detailed transactions	Often summarized data
Availability of historical data	Very limited—typically a few weeks or months	Multiple years
Update process	On-line, ongoing process as transactions are captured	Periodic process, once per week or once per month
Ease of update	Routine and easy	Complex, must combine data from many sources; data must go through a data cleanup process
Data integrity issues	Each individual transaction must be closely edited	Major effort to "clean" and integrate data from multiple sources

TABLE 5.5

Comparison of OLTP and Data Warehousing

by month, by week and even by day. A geographic dimension may compile data from city, state, region, country, and even hemisphere.

Resort Condominiums International (RCI) is a wholly owned subsidiary of Cendant Corporation and a market leader in vacation exchange services, with more than 3,500 affiliated time-share resorts and over 2.5 million member families worldwide. RCI uses Applix's iTM1 real-time OLAP solution to enhance its budgeting, forecasting, inventory, sales reporting, and analysis for locations in the United States and Europe. Like many companies, this information resides in various locations—such as databases, flat files, and spreadsheets. RCI was able to develop a centralized data warehouse to hold the data. Business analysts use iTM1 to access the data warehouse and quickly develop models incorporating this voluminous and complex data. The OLAP features of iTM1 enable RCI to make better business decisions because they can easily integrate and analyze both operational and financial data.[10]

The value of data ultimately lies in the decisions it enables. Powerful information-analysis tools in areas such as OLAP and data mining, when incorporated into a data warehousing architecture, bring market conditions into sharper focus and help organizations deliver greater competitive value. OLAP provides top-down, query-driven data analysis; data mining provides bottom-up, discovery-driven analysis. OLAP requires repetitive testing of user-originated theories; data mining requires no assumptions and instead identifies facts and conclusions based on patterns discovered. OLAP, or multidimensional analysis, requires a great deal of human ingenuity and interaction with the database to find information in the database. A user of a data mining tool does not need to figure out what questions to ask; instead, the approach is, "Here's the data, tell me what interesting patterns emerge." For example, a data mining tool in a credit card company's customer database can construct a profile of fraudulent activity from historical information. Then, this profile can be applied to all incoming transaction data to identify and stop fraudulent behavior, which may otherwise go undetected.

Table 5.5 compares an on-line transaction processing (OLTP) database with a data warehouse. Table 5.6 compares the OLAP and data mining approaches to data analysis.

Characteristic	OLAP	Data mining
Purpose	Supports data analysis and decision making	Supports data analysis and decision making
Type of analysis supported	Top-down, query-driven data analysis	Bottom-up, discovery-driven data analysis
Skills required of user	Must be very knowledgeable of the data and its business context	Must trust in data mining tools to uncover valid and worthwhile hypothesis

TABLE 5.6

Comparison of OLAP and Data Mining

open database connectivity (ODBC)

standards that ensure that software written to comply with these standards can be used with any ODBC-compliant database

Read the "E-Commerce" special feature to learn about an interesting data warehouse application to introduce state-of-the-art technology into everyday life.

Open Database Connectivity (ODBC)

To help with database integration, many companies rely on **open database connectivity (ODBC)** standards. ODBC standards help ensure that software written to comply with these standards can be used with any ODBC-compliant database, making it easier to transfer and access data among different databases. For example, a manager might want to take several tables from one database and incorporate them into another database that uses a different database management system. In another case, a manager might want to transfer one or more database tables into a spreadsheet program. If all this software meets ODBC standards, the data can be imported, exported, or linked to other applications (see Figure 5.20). For example, a table in an Access database can be exported to a Paradox database or a spreadsheet. Tables and data can also be imported using ODBC. For example, a table in a dBASE database or an Excel spreadsheet can be imported into an Access database. Linking allows an application to use data or an object stored in another application without actually importing the

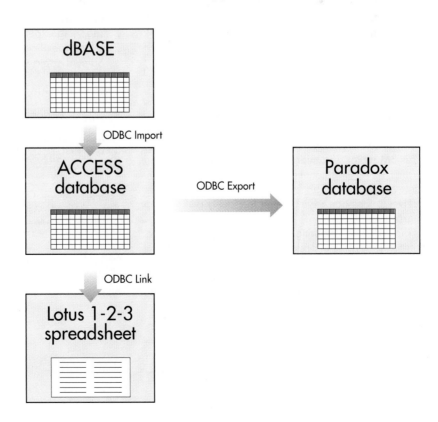

FIGURE 5.20

Advantages of ODBC

ODBC can be used to export, import, or link tables between different applications.

Safeway Customers Involved in E-Commerce Experiment

Basingstoke, England, isn't normally considered a cutting-edge kind of place. The quintessential British suburb lies 60 miles southwest of London, in the Hampshire countryside. "Boringstoke," as it is sometimes called, has long been the butt of jokes for the likes of Gilbert and Sullivan and Monty Python. Lately, however, 500 residents of Basingstoke have been participating in Easi-Order, a shopping experiment at the Safeway supermarket on Worting Road. They use a Palm III PDA equipped with a bar-code scanner, a modem, and software from IBM's Institute for Advanced Commerce to do their shopping from home. The PDA is provided free of charge to the test shoppers.

Here's how it works. Safeway built a data warehouse of every item bought from its inventory of 22,000 products by 10 million British shoppers over the past four years—some 3 TB of grocery-buying intelligence. Data mining software written for Safeway by IBM determines what groceries a family needs based on how long it has been since they visited the store and details from their past orders. The software is programmed to suggest additional items as well. For example, it may suggest that the family try Oracle toothpaste—Safeway's own brand—or promote baby products to new mothers based on birth notices obtained from a government health agency.

A suggested order is transmitted to the family's Palm III PDA whenever they connect by telephone to Safeway. Of course, families sometimes want to buy something they haven't ordered before or something they bought so long ago the Safeway computer figures they've lost interest in it. If the family has an empty box or wrapper for the item, someone swipes the bar code with a scanner built into the PDA, and the item is added to the electronic order. If the family doesn't have one of the items on hand or if it has no bar code, a member just describes the item in a free-format field that turns into e-mail to Safeway: "One quart of strawberries."

When the family finishes editing the order, the person attaches the PDA to the telephone and transmits the order to an IBM server. This midrange computer is a Java-based intranet server that connects to Safeway's S/390 mainframe computer. In addition to the order, the server receives a message saying when the family will pick up its groceries.

The following morning, a Safeway Easi-Order specialist arrives at the Basingstoke store, logs on to the server and prints out all the orders that are scheduled for pickup that day. Then sometime before the scheduled pickup, the Easi-Order specialist goes up and down the store's aisles filling a shopping cart with each order. The specialist logs each item as it is put into the basket by scanning its bar code with a handheld scanner.

When the specialist completes the order, it is brought to a holding area at the front of the store. The scanner is plugged into a docking station that reads the order and holds the information until the family comes. When the family arrives, a member swipes its Safeway account card at the same station, and the system matches the order data with the customer data and sends both to the server, and from there they go to the 3-TB database at Safeway's data center. The order information will rest in the database until the family next connects its PDA to Safeway's computer and obtains a new suggested order.

Most of the test shoppers use the service weekly. Some prefer to buy a month's worth of nonperishable items with Easi-Order and shop weekly for other items in the conventional way. Of course, people have certain items they prefer to pick on their own, such as fresh fruit and vegetables. The big advantage that shoppers appreciate is the time saved using the electronic ordering service—they can go in and out of the store in less than 15 minutes. Some shoppers think they have joined an elite class of shoppers because they no longer have to line up at the checkout counter.

Discussion Questions

1. Identify specific costs and benefits for Safeway associated with setting up this novel approach to shopping. Do you think that Safeway can cost-justify this shopping experiment? Why or why not?
2. What specific pieces of data are contained in the Safeway data warehouse for each family?

Critical Thinking Questions

3. What do you think are some of the potential issues and limitations of this system that could keep it from being successful?
4. Would you consider shopping in this manner? Why or why not?

Sources: Adapted from Gary H. Anthes, "Easi-Order," *Computerworld*, March 30, 2000, pp. 46–48; Chip Bayers, "Capitalist Econstruction," *Wired*, March 2000, accessed at http://www.wired.com/wired/archive/8.03/markets.html; Safeway UK Web site at http://www.safeway.co.uk, accessed April 11, 2000; and Pervasive Computing at IBM Web site at http://www-3.ibm.com/pvc/tech/safeway.shtml, accessed April 11, 2000 (this site includes the capability to demo the Safeway system).

data or object into the application. The Access database, for example, can link to a table in the Lotus 1-2-3 spreadsheet or the Sybase database. Applications that follow the ODBC standard can use these powerful ODBC features to share data between different applications stored in different formats.

ODBC-compliant products do suffer from their all-purpose nature, however. Their overall performance is usually less efficient than that of products designed for use with a specific database. Yet, more and more vendors are building ODBC-compliant products as businesses increasingly use distributed databases. Many organizations are using such tools to allow their workers and managers easier access to a variety of databases and data sources. ODBC standards also make it easier for growing companies to integrate existing databases, to connect more users into the same database, and to move application programs from PC-oriented databases to larger, workstation-based databases, and vice versa.

Object-Relational Database Management Systems

object-relational database management system (ORDBMS)

a DBMS capable of manipulating audio, video, and graphical data

Many of today's newer application programs require the ability to manipulate audio, video, and graphical data. Conventional database management systems are not well suited for this, because these types of data cannot easily be stored in rows or tables. Manipulation of such data requires extensive programming so that the DBMS can translate data relationships. An **object-relational database management system (ORDBMS)** provides a complete set of relational database capabilities plus the ability for third parties to add new data types and operations to the database. These new data types can be audio, images, unstructured text, spatial, or time series data that require new indexing, optimization, and retrieval features.

In such a database, these types of data are stored as objects, which contain both the data and the processing instructions needed to complete the database transaction. The objects can be retrieved and related by an ORDBMS. Businesses can then mix and match these elements in their daily search for clues and information. For example, by clicking on a picture of a red Corvette, a market analyst at General Motors might be able to call up a profile of red Corvette buyers. If he wants to break that up by geographic region, he might circle and click on a map. All in the same motion, he might view a GM sales-training film to see whether the sales pitch is appropriate, given the most recent market trends. Chances are the analyst will do all this using an Internet site tied into a database. As another example, MasterCard is interested in object-oriented technology to combine transactional data with cardholder fingerprints to prevent fraud.

Each of the vendors offering ORDBMS facilities provides a set of application programming interfaces to allow users to attach external data definitions and methods associated with those definitions into the database system. They are essentially offering a standard socket into which users can plug special instructions. DataBlades, Cartridges, and Extenders are the names applied by Informix, Oracle, and IBM to describe the plug-ins to their respective products. Other plug-ins serve as interfaces to Web servers.

Web-based applications increasingly require complex object support to link graphical and other media components back to the database. These systems make sense for developers of systems that are highly dependent on complex data types, particularly Web and multimedia applications. Because it supports so many applications, an ORDBMS is also called a universal database server.

An increasing amount of data that organizations use is in the form of images, which can be stored in object-relational databases. Credit card companies, for example, input pictures of charge slips into an image database using a scanner. The images can be stored in the database and later sorted by customer, printed, and sent to customers along with their monthly statements. Image databases are also used by physicians to store X rays and transmit them to clinics away from the main hospital. Financial services, insurance companies, and government branches are also using image databases to store vital records and replace paper documents.

Image data has some disadvantages, one of which is the increased secondary storage requirements. It is also sometimes difficult to locate the desired data in the database. Retrieval of image data from databases can be made easier by using the object-oriented approach. Other ways to index and cross-reference data are being developed.

Hypertext

The object-relational database provides greater flexibility in defining relationships between data. The previously discussed relational, hierarchical, and network models were designed for data that can be organized into fixed-length data fields and records with structured relationships. With hypertext, users can search and manipulate alphanumeric data in an unstructured way. In an object-relational database, text data is placed in chunks called *nodes*. The user then establishes links between the nodes. The relationships between the data can be created to user specifications, instead of following one of the more structured database models. Suppose a doctor is treating a new patient with symptoms similar to those of two other patients. The notes on each patient are placed in nodes, and the three nodes are linked. The doctor can use hypertext to retrieve and cross-reference these three cases when treating future symptoms of the new patient.

Hypermedia

Hypermedia is an extension of hypertext. Hypermedia allows businesses to search and manipulate multimedia forms of data—graphics, sound, video, and alphanumeric data. A marketing manager, for example, might store notes about the competition and new trends in the marketplace. These notes could include written material on new markets, images of products and advertising brochures, and TV commercials used by competitors. Using a hypermedia database management system, the marketing manager could organize this data into nodes and define the relationships among them. For example, she could link all TV commercials and written brochures about new products from several competitors. With the hypermedia database approach, many types of data can be organized into a web of nodes connected by links established by the user.

Spatial Data Technology

Spatial data technology involves the use of an object-relational database to store and access data according to the locations it describes and to permit spatial queries and analysis. NASA has set up a massive environmental monitoring program that calls for satellites and earth stations to deliver more than a terabyte of data every day to databases. The cumulative data will eventually multiply to a petabyte, which is 1,000 times a terabyte. Once the data is edited, it will be loaded into other databases for analysis. The primary database will offer satellite photo images and measurements. If environmentalists wanted to compare carbon dioxide levels within a certain latitude, longitude, and altitude with another polygon of space, they could easily do so. Implications of spatial data technology are potentially significant. For example, the data derived from the NASA project and others like it could help find oil and mineral reserves or identify sources of pollutants. Builders and insurance companies can use spatial data to make decisions related to natural hazards. Spatial data can even be used to improve financial risk management with information stored by derivative type, currency type, interest rates, and time.

Spatial data technology is used by NASA to store data from satellites and earth stations. Location-specific information can be accessed and compared.
(Source: Courtesy of NASA.)

The painstaking process of ensuring accuracy, completeness, and currency of the data that's been extracted from operational databases and formatted for analysis by a variety of departments is key to the success of any data warehouse. Evolutionary Technologies International, Platinum Technology, Prism Solutions, and other companies sell products for extracting and cleaning data. Also, a new class of tools is emerging for managing **metadata**, the data that describes the contents of a database. Metadata tells users when a piece of data was last updated, its format, where the data originated, and its intended uses. That information can guide users through databases and help them understand the meaning and context of financial data, customer records, and business transactions.[11] But these products address only part of the problem—the harder part is having procedures in place to manage data as it is extracted from operational systems or imported from data providers and transformed into meaningful information. Often data problems can be traced back to a transaction processing error that occurred during update of an operational database. Warehouse implementers should expect to spend four to five times more time and effort on data cleaning activities than they would at first estimate.

metadata

the data that describes the contents of a database

Business Intelligence

Business intelligence is the process of getting enough of the right information in a timely manner and usable form and analyzing it so that it can have a positive impact on business strategy, tactics, or operations. Business intelligence involves turning data into useful information that is then distributed throughout an enterprise. Companies use this information to make improved strategic decisions about which markets to enter, how to select and manage key customer relationships, and how to select and effectively promote products to increase profitability and market share.

business intelligence

the process of getting enough of the right information in a timely manner and usable form and analyzing it so that it can have a positive impact on business strategy, tactics, or operations

Boise Cascade Office Products uses Aonix's NOMAD mainframe reporting software to provide business intelligence to key employees in marketing, accounting, product planning, logistics, and distribution centers across the country. Users can select data from more than 72 databases and, using the power of the IBM S/390 mainframe computer, perform reporting tasks based on any combination of parameters, including market segment, customer size, date range, and product category. Employees can obtain the data needed to zero in on problem areas, obtain a detailed picture of the profitability of any customer, and see which products are selling and which are not. If sales decline, they can track those declines to specific offices or even to individual sales reps to pinpoint problems and take immediate steps to remedy them.[12]

Competitive intelligence is one aspect of business intelligence and is limited to information about competitors and how that knowledge affects strategy, tactics, and operations. Effective **competitive intelligence** is a continuous process involving the legal and ethical collection of information, analysis that doesn't avoid unwelcome conclusions, and controlled dissemination of that information to decision makers. Are you ahead of your competitors or are you racing to catch up? To stay ahead in the marketplace, you must be able to integrate competitive intelligence into your company's strategic plans and decisions. Competitive intelligence is a critical part of your company's ability to see and respond quickly and appropriately to the changing marketplace.

competitive intelligence

a continuous process involving the legal and ethical collection of information, analysis, and controlled dissemination of information to decision makers

Competitive intelligence is not espionage—the use of illegal means to gather information. In fact, almost all the information a competitive intelligence professional needs can be collected by examining published information sources, conducting interviews, and using other legal, ethical methods. Using a variety of analytical tools, a skilled competitive intelligence professional can by deduction fill the gaps in information already gathered. Read the "Ethical and Societal Issues" special interest box to see how one company is using competitive intelligence.

ETHICAL AND SOCIETAL ISSUES

United Technologies Gathers Competitive Intelligence

United Technologies (UTC) provides high-technology products to the aerospace and building systems industries throughout the world. UTC's business units include Pratt & Whitney (aircraft engines), Carrier (air conditioning), Otis (elevators), International Fuel Cells (fuel cell technology), Hamilton Sundstrand (aerospace systems), and Sikorsky (helicopters). Recent annual income was $3 billion from revenues of $24 billion.

UTC gathers a lot of competitive intelligence. Each UTC business unit has long had its own competitive intelligence (CI) function, and UTC recently added a corporate CI unit to offer a big-picture perspective. The unit resides within UTC's Research Center, which monitors and develops technology for the company. In aerospace, consolidation has reduced the number of competitors, while in the worldwide air conditioning and ventilation market, the challenge is to keep an eye on a growing number of rivals. UTC constantly monitors what companies are up to, how they use suppliers, how their distribution chains function, and who their customers are.

UTC's intelligence gathering supports its strong acquisition strategy of buying companies that possess complementary technologies. Three major acquisitions occurred during 1999. Sundstrand joined Hamilton Standard to more than double the size of UTC's aerospace systems activities and presence. International Comfort Products brought $750 million of residential product sales to Carrier's North American cooling and heating businesses. LG Industrial Systems' elevator business added more than $500 million to Otis's sales, mostly in Korea, where LG is the leading elevator company.

To begin researching a competitor, UTC gets as much information as possible from public sources, such as financial reports, Web site information, patents, market activity, and alliances with universities. All that information is brought to UTC's CI group to evaluate the competitor's technology—does it complement UTC's, could UTC improve the competitor's technology, does the competitor seem to be a good value? This intelligence is complemented with information from unconventional sources to complete the picture. For example, UTC staff might question a magazine writer about information on the company that didn't make it into the article, or staffers might contact potential CI sources at a trade association meeting. UTC might also interview customers and suppliers of the company in question.

UTC's CI staff is told to adhere to strict legal and ethical codes in gathering information. They must always identify who they are, where they're from, and their purpose for seeking information.

Discussion Questions

1. Who are the primary sources of CI for UTC? Why? What kinds of information can they provide?
2. Generate a list of three questions you would ask of a supplier to a UTC competitor to gain competitive intelligence.

Critical Thinking Questions

3. Do you think that adherence to its strict legal and ethical code helps or hinders the UTC staff in obtaining information? Why?
4. What additional means of gathering competitive intelligence can you identify? Which of these might be outside the limits of the UTC code of ethics?

Sources: Adapted from "Letter to Shareholders" and "Year in Review," United Technologies 1999 annual report, found at http://www.utc.com/annual99/, accessed April 4, 2000; and Gary Abramson, "All Along the Watchtower," *CIO Enterprise*, July 15, 1999, Section 2, pp. 25–34.

Larger businesses often have staff who can do most of the data gathering and analysis. They will often benefit, though, by having an outside perspective during analysis. Smaller businesses almost never have the internal staff resources to do effective business intelligence. The exception to this is when there is a person (often the CEO) who does the business intelligence work because he or she likes it. Smaller businesses and independent business professionals can usually benefit from outside help with information gathering and with analysis and recommendation.

The term **counterintelligence** describes the steps an organization takes to protect information sought by "hostile" intelligence gatherers. One of the most effective counterintelligence measures is to define "trade secret" information relevant to the company and control its dissemination.

Knowledge management is the process of capturing a company's collective expertise wherever it resides—in computers, on paper, or in people's heads—and distributing it wherever it can help produce the biggest payoff. The goal of knowledge management is to get people to record knowledge (as opposed to data) and then share it. Although a variety of technologies can support it, knowledge management is really about changing people's behavior to

counterintelligence

the steps an organization takes to protect information sought by "hostile" intelligence gatherers

knowledge management

the process of capturing a company's collective expertise wherever it resides—in computers, on paper, in people's heads—and distributing it wherever it can help produce the biggest payoff make their experience and expertise available to others.[13] Knowledge management had its start in large consulting firms and has expanded to nearly every industry. Dow Chemical was one of the first manufacturing companies to employ a disciplined process in managing its knowledge, notably its previously underexploited portfolio of more than 29,000 patents. Intellectual Asset Management is now a formal, six-stage process, whose mission is to capitalize on the revenue potential of Dow's patents through new product development or licensing.[14]

● SUMMARY

PRINCIPLE ● The database approach to data management provides significant advantages over the traditional file-based approach.

Data is one of the most valuable resources a firm possesses. It is organized into a hierarchy that builds from the smallest element to the largest. The smallest element is the bit, a binary digit. A byte (a character such as a letter or numeric digit) is made up of eight bits. A group of characters, such as a name or number, is called a field (an object). A collection of related fields is a record; a collection of related records is called a file. The database, at the top of the hierarchy, is an integrated collection of records and files.

An entity is a generalized class of objects for which data is collected, stored, and maintained. An attribute is a characteristic of an entity. Specific values of attributes—called data items—can be found in the fields of the record describing an entity. A data key is a field within a record that is used to identify the record. A primary key uniquely identifies a record, while a secondary key is a field in a record that does not uniquely identify the record.

The traditional approach to data management has been from a file perspective. Separate files are created for each application. This approach can create problems over time: as more files are created for new applications, data that is common to the individual files becomes redundant. Also, if data is changed in one file, those changes might not be made to other files, reducing data integrity.

Traditional file-oriented applications are often characterized by program-data dependence, meaning that they have data organized in a manner that cannot be read by other programs. To address problems of traditional file-based data management, the database approach was developed. Benefits of this approach include reduced data redundancy, improved data consistency and integrity, easier modification and updating, data and program independence, standardization of data access, and more efficient program development.

Potential disadvantages of the database approach include the relatively high cost of purchasing and operating a DBMS in a mainframe operating environment; specialized staff required to implement and coordinate the use of the database; and increased vulnerability if security is breached and there is a failure in the DBMS.

● ● ● ●

When building a database, careful consideration must be given to content and access, logical structure, and physical organization. One of the tools database designers use to show the relationships among data is a data model. A data model is a map or diagram of entities and their relationships. Enterprise data modeling involves analyzing the data and information needs of the entire organization. Entity-relationship (ER) diagrams can be employed to show the relationships between entities in the organization. Entities may have a one-to-one (1:1), one-to-many (1:N), or many-to-one relationship (N:1).

Databases typically use one of three common models: hierarchical (tree), network, and relational. The hierarchical model has one main record type at the top, with subordinate records below. The network model, an expansion of the hierarchical structure, involves an owner-member relationship in which each member may have more than one owner. The newest, most flexible structure is the relational model. Instead of a hierarchy of predefined relationships, data is set up in two-dimensional tables. Tables can be linked by common data elements, which are used to access data when the database is queried. Each row represents a record. Columns of the tables are called attributes, and allowable values for these attributes are called the domain. Basic data manipulations include selecting, projecting, and joining.

The relational model, the most widely used database model, is easier to control, more flexible, and more intuitive than the other models because it organizes data in tables.

PRINCIPLE ● A well-designed and well-managed database is an extremely valuable tool in supporting decision making.

A DBMS is a group of programs used as an interface between a database and application programs. When an application program requests data from the database, it follows a logical access path. The actual retrieval of the data follows a physical access path. Records can be considered in the same way: a logical record is what the record contains; a physical record is where the record is stored on storage devices. Schemas are used to describe the entire database, its record types, and their relationships to the DBMS.

A database management system provides four basic functions: providing user views, creating and modifying the database, storing and retrieving data, and manipulating data and generating reports.

Subschemas are used to define a user view, the portion of the database a user can access and/or manipulate. Schemas and subschemas are entered into the computer via a data definition language, which describes the data and relationships in a specific database. Another tool used in database management is the data dictionary, which contains detailed descriptions of all data in the database.

Once a DBMS has been installed, the database may be accessed, modified, and queried via a data manipulation language. A more specialized data manipulation language is the query language, the most common being Structured Query Language (SQL). SQL is used in several popular database packages today and can be installed on PCs and mainframes.

Popular end-user DBMSs include Microsoft Access, Lotus Approach, and Inprise's dBASE. These DBMSs provide "wizards" to help end users create a new database and load data and to perform many other functions. Oracle, IBM, Microsoft, Sybase, and Informix are the leading DBMS vendors.

PRINCIPLE ● Further improvements in the use of database technology will continue to evolve and yield real business benefits.

With the increased use of telecommunications and networks, distributed databases, which allow multiple users and different sites access to data that may be stored in different physical locations, are gaining in popularity. To reduce telecommunications costs, some organizations build replicated databases, which hold a duplicate set of frequently used data.

Organizations are also building data warehouses, which are relational database management systems specifically designed to support management decision making.

Multidimensional databases and on-line analytical processing (OLAP) programs are being used to store data and allow users to explore the data from a number of different perspectives.

Data mining, which is the automated discovery of patterns and relationships in a data warehouse, is emerging as a practical approach to generate a hypothesis about the patterns and anomalies in the data that can be used to predict future behavior.

An object-relational database management system (ORDBMS) provides a complete set of relational database capabilities, plus the ability for third parties to add new data types and operations to the database. These new data types can be audio, images, unstructured text, spatial, or time series data that require new indexing, optimization, and retrieval features.

Business intelligence is the process of getting enough of the right information in a timely manner and usable form and analyzing it so that it can have a positive impact on business strategy, tactics, or operations. Competitive intelligence is one aspect of business intelligence limited to information about competitors and how that information affects strategy, tactics, and operations. Competitive intelligence is not espionage—the use of illegal means to gather information. Counterintelligence describes the steps an organization takes to protect information sought by "hostile" intelligence gatherers. One of the most effective counterintelligence measures is to define "trade secret" information relevant to the company and control its dissemination. Knowledge management is the process of capturing a company's collective expertise wherever it resides—in computers, on paper, or in people's heads—and distributing it wherever it can help produce the biggest payoff. The goal of knowledge management is to get people to record knowledge (as opposed to data) and then share it.

KEY TERMS

attribute 171
business intelligence 201
character 170
competitive intelligence 201
concurrency control 185
counterintelligence 202
data cleanup 181
data definition language (DDL) 182
data dictionary 183
data integrity 172
data item 171
data manipulation language (DML) 186
data mart 192
data mining 193
data model 176
data redundancy 172
data warehouse 191
database approach to data
    management 173

database management system
    (DBMS) 170
distributed database 190
domain 178
enterprise data modeling 176
entity 171
entity-relationship (ER) diagrams 176
field 170
file 171
hierarchical database model 177
hierarchy of data 171
joining 179
key 171
knowledge management 202
linking 179
metadata 201
network model 178
object-relational database
    management system (ORDBMS) 199

on-line analytical processing
    (OLAP) 195
open database connectivity
    (ODBC) 197
planned data redundancy 176
primary key 171
program-data dependence 172
projecting 179
record 170
relational model 178
replicated database 190
schema 182
selecting 179
subschema 182
traditional approach to data
    management 172

REVIEW QUESTIONS

1. Describe the hierarchy of data.
2. Define the term *database*. How is it different from a database management system?
3. What is meant by data cleanup?
4. What are the advantages of the database approach to data management, as opposed to the traditional file-based approach?
5. Describe the following three types of database models: hierarchical model, network model, and relational model.
6. How is a database schema used?
7. Identify important characteristics in selecting a database management system.
8. What is the difference between a data definition language (DDL) and a data manipulation language (DML)?
9. What is a distributed database system?
10. What advantages does the open database connectivity (ODBC) standard offer?
11. What is a data warehouse, and how is it different from a traditional database used to support OLTP?
12. What is OLAP? Does OLAP imply the use of a relational database?
13. What is data mining? How is it different from OLAP?
14. What is an ORDBMS? What kind of data can it handle?
15. What is business intelligence? How is it used?
16. What is competitive intelligence?

DISCUSSION QUESTIONS

1. You have been selected to represent the student body on a project to develop a new student database for your school. What actions might you take to fulfill this responsibility to ensure that the project met the needs of students and was successful?
2. Your company is building a new manufacturing facility to handle the increase in sales volume expected from the introduction of a new line of products. Imagine that you are the manager of the new plant. What counterintelligence initiatives might you undertake?
3. What is a data model and what is data modeling? Why is data modeling an important part of strategic planning?
4. You are going to design a database for your wine-tasting club to track its inventory of wines. Identify the database characteristics most important to you in choosing a DBMS to implement this system. Which of the database management systems described in this chapter would you choose? Why? Is it important for you to know what sort of computer the database will run on? Why or why not?

5. How would you distinguish a data mart from a data warehouse? Would you recommend a strategy of developing numerous data marts or a single data warehouse to meet the decision support needs of a large multidivisional organization? Why?

6. Make a list of the databases in which data about you exists. How is the data in each database captured? Who updates each database and how often? Is it possible for you to request a printout of the contents of your data record from each database?

7. Develop a list of conditions under which you would use the OLAP approach to analyze data in a data warehouse. Develop a list of conditions that favor the data mining approach.

8. You are the vice president of information technology for a large commercial bank. You are to make a presentation to the board of directors recommending the investment of $5 million to establish a competitive intelligence organization including people, data gathering services, and software tools. What are your key points in favor of this investment? What counterpoints can you anticipate others might argue?

PROBLEM-SOLVING EXERCISES

1. Use an end-user database management system to design and implement a system to record your personal items and home/apartment furnishings so that you have a log of all valuable items for insurance purposes in case of theft, fire, or natural disaster. For each item, you should record a complete description, date of purchase, cost, and any identifying numbers such as a serial number or registration number. What will be the unique key for the records in your database?

2. A video movie rental store is using a relational database to store information on movie rentals to answer customer questions. Each entry in the database contains the following items: Movie ID No. (primary key), Movie Title, Year Made, Movie Type, MPAARating, Number of Copies on Hand, and Quantity Owned. Movie types are comedy, family, drama, horror, science fiction, and western. MPAA ratings are G, PG, PG-13, R, X, and NR (not rated). Use an end-user database management system to build a data entry screen to enter this data. Build a small database with at least 10 entries.

3. To improve service to their customers, the salespeople at the video rental store have proposed a list of changes being considered for the database in the previous exercise. From this list, choose two database modifications and implement them.

Proposed changes:

a. Add the date that the movie was first available to help locate the newest releases

b. Add the director's name.

c. Add the names of three actors in the movie.

d. Add a rating of one, two, three, or four stars.

e. Add the number of Academy Award nominations.

TEAM ACTIVITIES

1. In a group of three or four classmates, do an analysis of the leading database management system vendors. Which one is currently the market leader in terms of annual revenue? In terms of market share? Which one do you believe has the best products? Why?

2. As a team of three or four classmates, interview business managers from three different businesses that use databases to help them in their work. What data entities and data attributes are contained in each database? How do they access the database to perform analysis? Have they received training in any query or reporting tools? What do they like about their database and what

could be improved? Do any of them use data mining or OLAP techniques? Weighing the information obtained, select one of these databases as being most strategic for the firm and briefly present your selection and the rationale for the selection to the class.

3. Imagine that you and your classmates are a research team developing an improved process for evaluating college applicants. The goal of the research is to predict which students will be most successful in their college career. Those who score well on the profile will be accepted; those who score exceptionally well will be considered for scholarships. Prepare a brief report for your

instructor addressing these questions:

a. What data do you need for each college applicant?

b. What data might you need that is not typically requested on the college application form?

c. From where might you get this data?

Take a first cut at designing a database for this application. Using the chapter material on designing a database, show the logical structure of the relational tables for this proposed database. In your design, include those data attributes you believe are necessary for this database, and show the primary keys in your tables. Keep the size of the fields and tables as small as possible to minimize required disk drive storage space. Fill in the database tables with the sample data for demonstration purposes (10 records). Once your design is complete, implement it for a university database using a relational DBMS.

● WEB EXERCISES

1. Use a Web search engine to find information on one of the following topics: data warehouse, data mining, or OLAP. Find a definition of the term, an example of a company using the technology, and three companies that provide such software. Cut graphics and text material from the Web pages and paste into a word processing document to create a two-page report on your selected topic. At the home page of each software company, request further information from the company about its products.

2. Use a Web search engine to find three companies that provide competitive intelligence services. How are the services that they provide similar? How are they different? Which companies seem to be the most ethical? Why?

● CASES

1 Lockheed Martin Implements OLAP System

Lockheed Martin is a diversified technology company that researches, designs, manufactures, and integrates advanced technology products and services for government and commercial customers. Annual sales exceed $25 billion, and the firm employs 160,000 people worldwide. Its products and services include systems integration, military aircraft, launch vehicles, and defense systems.

The Tactical Aircraft Systems (TAS) division designs, develops, and produces state-of-the-art tactical military aircraft systems. Located in Fort Worth, Texas, it operates the mile-long manufacturing facility designated U.S. Air Force Plant 4. Its current products include the T-16 Fighting Falcon, a small, agile, low-cost, high-capability fighter with outstanding mission success; the F-22 Fighter, designated by the U.S. Air Force as the next-generation air superiority fighter; and the Joint Strike Fighter, the next-generation multirole fighter. TAS is also working with Japan to help design, develop, and manufacture its next fighter, called the F-2 Fighter.

When you design and build high-performance fighter aircraft, gathering information about processes, parts, and procedures is critical to producing a superior airplane while controlling costs. TAS decided to implement a data warehouse and provide OLAP tools to create a single source for one consolidated view of its business information needed for better decision making. The Lockheed division wanted to provide its employees with access to such information as program and business-management metrics, staffing information, risk and technical measurements, sales and cost forecasting, and overhead analysis.

The division began implementing OLAP in 1998 as part of a long-term strategy to bolster business intelligence capabilities. OLAP lets users from engineering, purchasing, manufacturing, and other areas in the company examine data on aircraft design and manufacturing from a multidimensional view. By the end of 1999, the defense contractor had over 20 OLAP applications deployed. One application, for example, is the Manpower Forecast Module used to analyze staffing requirements, employee skills, and skill-set requirements for various contracts. This application allows product team leaders to input their staffing needs and obtain the identity of individuals who match the needs down to the skill level. Another application, Hours Per Unit, measures the cost of aircraft parts in terms of manufacturing hours. This enables analysis of alternative designs and helps evaluate whether specific processes should be outsourced.

In implementing this system, the company needed strong protection for sensitive business data. Considerable effort was devoted to ensuring that security measures were in place, including managing user access permissions and user authentication. As a result, the division's system administrators can control what users are able to do with the OLAP applications and manage how different user groups can interact with the data.

Discussion Questions

1. What benefits does Lockheed Martin gain from the use of OLAP and a data warehouse that are not possible through the use of more traditional database technology?
2. Briefly describe how the Manpower Forecast Module might work.

Critical Thinking Questions

3. Why does Lockheed consider the data in the warehouse to be sensitive? Who else could use the data to their benefit? How might they use this data?
4. Is there a high potential for application of this OLAP solution to other Lockheed Martin divisions? Why or why not?

Sources: James Cope, "New Tools Help Lockheed Martin Prepare for Takeoff," *Computerworld*, March 27, 2000, p. 46; Tatilia Baron, "OLAP Goes Online," *Information Week*, September 20, 1999, pp. 90–92; Lockheed Martin Tactical Aircraft Systems home page, http://www.lmco.com/careers/tactical_aircraft_fortworth_tx.html, accessed December 6, 1999; and "About Lockheed Martin," Lockheed Martin Web site, http://www.lmco.com/about/index.htm, accessed December 6, 1999.

 ## 2   Allina Health System Implements a Data Warehouse

Minneapolis-based Allina Health System is a not-for-profit healthcare system serving one million people living in Minnesota, Wisconsin, North Dakota, and South Dakota. The vertically integrated healthcare system includes 13,000 physicians and 22,000 employees who own and manage 19 hospitals, 57 clinics, and seven nursing homes. Allina provides people with a lifetime of healthcare options and a full continuum of care—from prevention and wellness services, such as health screenings and immunizations, to the highest-quality and technologically advanced inpatient and outpatient services.

As are all healthcare systems, Allina is pressured from all sides to reduce healthcare costs and hold the line on premiums, while providing its patients with the best treatment possible. It must also integrate the information systems and business practices of the multiple organizations it has acquired through various mergers in the recent past.

To meet these challenges, management developed an enterprise data warehouse strategy. The goal is to enable Allina to pull information together and integrate it in ways never done before—for example, extracting cost information from the hospital and the health plans, comparing best practices in treatments, and improved matching of cost to level of service.

Meeting this goal was a data warehouse design challenge. The data modeling effort had to ensure that each data attribute, each data mart, and the data warehouse could be tied together logically and physically. For example, all data attributes are assigned to one of four different data subtypes: people (organizations and patients, members, providers, etc.), places, things, and events.

There were three key elements to the success of this project. First, a multitier database was developed so that both summary and detailed information is available. Second, the large implementation team was divided into three specialized groups, including a data management group in charge of architecture, data modeling, and databases; a tools group focusing on management of data access and data query tools; and an analysis group of business analysts, project managers, and consultants to implement new warehouse subject areas and data marts. Third, the organization used pilot projects, including projects to build a data mart for administrative reporting for one hospital and a large-scale data warehouse on patient histories for all hospitals; when completed, the warehouse size exceeded 70 GB.

Discussion Questions

1. In which group of the implementation team would you feel most comfortable participating? Why?
2. What business and technical challenges made this project difficult?

Critical Thinking Questions

3. Starting with the four data subtypes (people, places, things, and events), sketch an initial draft of a data model to meet the needs of Allina Health System.
4. Imagine that you are on the board of directors for Allina and the IT manager has requested approval to market the data warehouse system to other healthcare systems across the United States. What would be your reaction? Why?

Sources: "About Allina Health System," Allina Health Care System Web site, http://www.allina.com, accessed April 3, 2000; and "Rx for Reducing Paperwork," IBM Web site, http://www.ibm.com/stories, accessed April 4, 2000.

3 Fifth Third Bank Investing in Internet Systems

Fifth Third Bank traces its origins to the Bank of the Ohio Valley, which opened its doors in Cincinnati in 1858. The Third National Bank purchased the bank in 1871. With the turn of the nineteenth century came the union of the Third National Bank and the Fifth National Bank, eventually to become known as the Fifth Third Bank. Today Fifth Third comprises 14 affiliate banks with 650 branches and 12,000 employees located in Ohio, Kentucky, Indiana, Illinois, Florida, Arizona, and Michigan. Fifth Third is headquartered in Cincinnati, Ohio, and has $41 billion in assets and total deposits of $26 billion. Fifth Third is better than most banks at increasing revenues faster than expenses, and that helps it achieve above-average earnings growth—it is coming off its twenty-sixth straight year of record profits.

For the past two years, Fifth Third has invested the majority of its $60 million information systems budget in Internet technology. It has aggressively pushed large corporate customers to give up faxes and move onto the Internet to access and exchange information and even trade stocks on-line. Fifth Third developed tools that allow merchants to access credit card information over the Internet. Using passwords and secure networks, retailers are allowed to view their credit card transactions, refund money to customers, and review information issued by credit card companies. Currently, more than $4 billion of e-commerce transactions is completed every year using Fifth Third's network.

Another of Fifth Third's value-added services is the development of a mobile workforce. The company is testing a program that enables 700 commercial loan officers, armed with laptops, to go to different banks and sell cash management products and credit lines on the spot. Similarly, in the business-to-consumer arena, mortgage loan officers will be able to walk into an open house and help potential home buyers apply for a home mortgage.

Although the industry average for on-line banking hovers at 4 percent of customers, more than 10 percent of Fifth Third's customers rely on the Internet to access its services. The company is successfully attracting more on-line customers by continually improving its ability to serve them over the Internet. For instance, the bank is rolling out a 401(k) retirement planning site that will allow customers to execute trades electronically. Fifth Third will also incorporate the natural language inquiry ability from the popular Web search engine Ask Jeeves. This service will allow customers to use plain-English questions rather than keywords to search for answers regarding retirement planning.

Fifth Third is also using Internet technologies to build its own intranet. By the end of 2000, 95 percent of the company's manuals will be on-line. This move will create a more productive workforce.

Discussion Questions

1. Which of these many information system initiatives will increase bank revenue? Which will decrease bank expenses?

2. Fifth Third has a number of Internet-based services: providing merchants access to credit card information over the Internet; supporting mobile workers in selling cash management, credit lines, and home mortgages; developing a 401(k) retirement planning site; and providing on-line access to company manuals. Briefly summarize the database implications for each one of these applications—the type of database technology required, size of the database, and contents of the database.

Critical Thinking Questions

3. Which one of these new information system services do you think will be the most successful investment for Fifth Third? Why?

4. Is Fifth Third investing too much of its information systems budget in the Internet? If so, what other technologies should it be exploring? If not, why not?

Sources: Adapted from Jeff McKinney, "5/3 Sees Growth Opportunity in Chicago," *Cincinnati Enquirer*, February 13, 2000, pp. E1–E2; Anne Chen, "Fifth Third Places First on *Fast@Track*," *PC Week*, December 15, 1999, p. 90; and "Investor Relations," Fifth Third Web site, http://www.53.com, accessed February 15, 2000.

● NOTES

Sources for the opening vignette on page 169: Adapted from "Catalina Marketing Corporation Reports Record Third-Quarter Results," company press release, *PRNewswire*, January 13, 2000, at http://www.prnewswire.com/cgi-bin/stories; and Catalina Marketing Corporation, "Investor Highlights," fact sheet, *PR Newswire*, http://www.prnewswire.com/cnoc/, accessed April 2, 2000.

1. "About Allina Health System" section of the Allina Health Care System Web site at http://www.allina.com, accessed April 3, 2000.

2. Julie Pitta, "Squeeze Play: Databases Get Ugly," *Forbes*, February 22, 1999, pp. 50–51.

3. Stewart Deck, "SQL Users Turn to Oracle8 for Bulk," *Computerworld*, May 10, 1999, p. 4.

4. Amy Helen Johnson, "Data Warehousing," *Computerworld*, December 6, 1999, p. 75.

5. Craig Stedman, "Wal-Mart CIO Leaves Retailer and IT Leader," *Computerworld*, March 9, 2000, p. 4.

6. Gabrielle Gagnon, "Data Warehousing: An Overview," *PC Magazine*, March 19, 1999, pp. 245–246.

7. Candee Wilde, "Telcos Turn to Analytical Tools to Stay in Touch," *Information Week*, March 13, 2000, pp. 98–102.

8. James M. Connolly, "Mining Your Business," *Computerworld*, May 17, 1999, pp. 94–98.

9. Carol Pickering, "They're Watching You," *Business 2.0*, February 2000, pp. 135–137.

10. "RCI Implements Applix's Real-Time OLAP Solution to Enhance Sales Reporting & Analysis," August 5, 1999, Applix press releases at http://www.applix.com/releases/99-08-05_rci.cfm.

11. Craig Stedman, "Metadata," *Computerworld*, October 18, 1999, p. 74.

12. Samuel Greengard, "Business Intelligence," *Beyond Computing*, January/February 2000, pp. 18–23.

13. Rochelle Garner, "Knowledge Management," *Computerworld*, August 9, 1999, pp. 50–51.

14. James Allen, "The Process of Knowledge Management," presented at the Transforming Knowledge-Intensive Processes Conference, September 28, 1999, Boston, Massachusetts.

CHAPTER 6

Telecommunications and Networks

We want to know what you want, and we want to bring it to you.

— Mike Kelly, chief financial officer of America Online

Principles	Learning Objectives
Effective communications is essential to organizational success.	• *Define the terms* communications *and* telecommunications *and describe the components of a telecommunications system.*
An unmistakable trend of communications technology is that more people are able to send and receive all forms of information over greater distances at a faster rate.	• *Identify three basic types of communications media and discuss the basic characteristics of each.* • *Identify several types of telecommunications hardware devices and discuss the role that each plays.* • *Identify the benefits associated with a telecommunications network.* • *Name three distributed processing alternatives and discuss their basic features.* • *Define the term* network topology *and identify five alternatives.*
The effective use of telecommunications and networks can turn a company into an agile, powerful, and creative organization, giving it a long-term competitive advantage.	• *Identify and briefly discuss several telecommunications applications.*

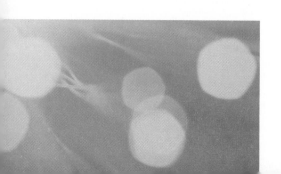

Coca-Cola

Overhauling Its Network for Speedy Communications

Coca-Cola Bottling Consolidated is the second-largest Coke bottler in the United States. The bottler has 6,000 employees at 100 locations in 12 states. Like other Coke bottlers, Consolidated receives concentrate from Coca-Cola. It then adds sweetener and carbonation before it places the finished product into cans, bottles, and other containers that are shipped to stores in its region.

In the past, Consolidated was apprehensive about opening up its network to others. The threat of hackers was a primary concern. Yet, the company realized that it could be more efficient to communicate with suppliers and customers by connecting its network to the Internet. In addition, senior management at Consolidated wanted to increase sales opportunities using the Internet. Increasingly, senior management viewed an effective communications system using the Internet as a competitive advantage.

Driven by the goal to increase communications internally and externally, Consolidated decided to improve its telecommunications system. In the past, the company used a slow and inefficient internal network, which was not able to handle increased demands on communications when suppliers and customers were connected to the network. To increase speed and flexibility, the company has installed a series of high-speed T1 lines and switches. The new system will improve communications internally and will allow Consolidated to achieve speedy access to the Internet. In addition to the new lines, Consolidated has installed software to protect against unauthorized access to its telecommunications system and to give it the ability to monitor all telecommunications activities. The company plans to use Axent Raptor NT for this protection and to monitor and control communications traffic between its network and the Internet. Telemate.net will be used to help Consolidated produce audit logs and reports for Internet and network usage. WebNot will be used to help block viruses and other malicious programs. The plan is to roll out the new telecommunications systems over time. According to one company official, "In the next couple of weeks, we will sign on more people based on business needs."

As you read this chapter, consider the following:

- Why is it so critical for companies like Consolidated to have an efficient and effective telecommunications system?

  - What are the advantages of a T1 line?

In today's high-speed business world, effective communication is critical to organizational success, as it is with Coca-Cola. Often, what separates good management from poor management is the ability to identify problems and solve them with available resources. Efficient communication is one of the most valuable of these resources, because it enables a company to keep in touch with its operating divisions, customers, suppliers, and stockholders. One auto company, for example, is reaching out to its suppliers with a new system, which is like an on-line bazaar of auto parts and supplies.[1] Companies hope to save billions of dollars with these new telecommunications systems.

Memos, notices on bulletin boards, and presentations are all obvious examples of continual communication within a business organization. Other not-so-obvious examples include policy and procedure manuals and even salaries (they communicate the company's perception of the value of the contribution of the person being paid). Communication also exists in other forms—for instance, warning lights from a computer system that monitors manufacturing processes and signals from a building management system that monitors temperature, humidity, lighting, and security of a building. Communication is any process that permits information to pass from a sender to one or more receivers. Communications of all types form a major part of any business system. Therefore, managers must gain an appreciation of communication concepts, media, and devices—as well as an understanding of how these factors may best be employed to develop effective and efficient business systems.

AN OVERVIEW OF COMMUNICATIONS SYSTEMS

Communications

Communications is the transmission of a signal by way of a medium from a sender to a receiver (Figure 6.1). The signal contains a message composed of data and information. The signal goes through some communications medium, which is anything that carries a signal between a sender and receiver. In human speech, the sender transmits a signal through the transmission medium of the air. In telecommunications, the sender transmits a signal through a transmission medium such as a cable.

The components of communication can easily be recognized if you consider human communication (Figure 6.2). When we talk to one another face to face, we send messages to each other. A person may be the sender at one moment and the receiver a few seconds later. The same entity, a person in this case, can be a sender, a receiver, or both. This is typical of two-way communication. The signals we use to convey these messages are our spoken words—our language. For communication to be effective, both sender and receiver must understand the signals and agree on the way they are to be interpreted. For example, if the sender in Figure 6.2 is speaking in a language the receiver does not understand, or if the sender believes a particular word has one meaning and the receiver believes the word has some other meaning, effective communication will not occur.

In addition to the flow of communications, shown in Figure 6.2, communications can be synchronous or asynchronous. With **synchronous communications**, the receiver gets the message instantaneously, when it is sent. Voice and phone communications are examples of synchronous communications. With **asynchronous communications**, the receiver gets the message later—sometimes hours or days after the message is sent. Sending a letter through the post office or e-mail over the Internet are examples of asynchronous communications. Academic researchers are actively investigating the impact of both synchronous and asynchronous communications

synchronous communications

communication in which the receiver gets the message instantaneously

asynchronous communications

communication in which the receiver gets the message minutes, hours, or days after it is sent

FIGURE 6.1

Overview of Communications

The message (data and information) is communicated via the signal. The transmission medium "carries" the signal.

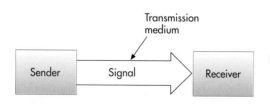

FIGURE 6.2

Communication and
Telecommunication

In human speech, the sender
transmits a signal through the
transmission medium of the air.
In telecommunication, the
sender transmits a signal
through a cable or other
telecommunication medium.

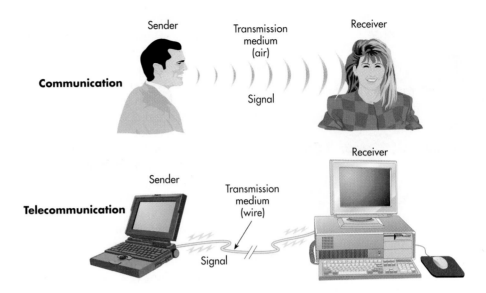

on effectiveness, performance, and other organizational measures. Both types of communications are important in business, regardless of whether the communication is done manually or electronically.

Telecommunications

Telecommunications refers to the electronic transmission of signals for communications, including such means as telephone, radio, and television. Telecommunications has the potential to create profound changes in business because it lessens the barriers of time and distance. Telecommunications not only is changing the way businesses operate but also is altering the nature of commerce itself. As networks are connected with one another and information is transmitted more freely, a competitive marketplace is making excellent quality and service imperative for success. **Data communications**, a specialized subset of telecommunications, refers to the electronic collection, processing, and distribution of data—typically between computer system hardware devices. Data communications is accomplished through the use of telecommunications technology.

Figure 6.3 shows a general model of telecommunications. The model starts with a sending unit (1), such as a person, a computer system, a terminal, or another device, that originates the message. The sending unit transmits a signal (2) to a telecommunications device (3). The telecommunications device performs a number of functions, which can include converting the signal into a different form or from one type to another. A telecommunications device is a hardware component that allows electronic communication to occur or to

data communications

a specialized subset of
telecommunications that refers to the
electronic collection, processing,
and distribution of data—typically
between computer system hardware
devices

FIGURE 6.3

Elements of a
Telecommunications System

Telecommunications devices
relay signals between
computer systems and
transmission media.

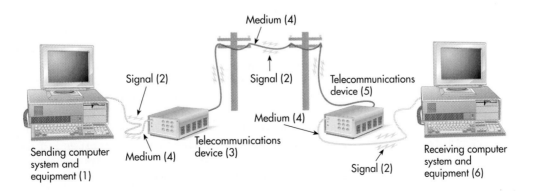

Telecommunications technology enables businesspeople to communicate with coworkers and clients from remote locations.
(Source: Stone/Terry Vine)

telecommunications medium

anything that carries an electronic signal and interfaces between a sending device and a receiving device

computer network

the communications media, devices, and software needed to connect two or more computer systems and/or devices

occur more efficiently. The telecommunications device then sends the signal through a medium (4). A **telecommunications medium** is anything that carries an electronic signal and interfaces between a sending device and a receiving device. The signal is received by another telecommunications device (5) that is connected to the receiving computer (6). The process can then be reversed and another message can go back from the receiving unit (6) to the original sending unit (1). In this chapter, we will explore the components of the telecommunications model shown in Figure 6.3. An important characteristic of telecommunications is the speed at which information is transmitted, measured in bits per second (bps). Common speeds are in the range of thousands of bits per second (Kbps) to millions of bits per second (Mbps).

Advances in telecommunications technology allow us to communicate rapidly with clients and co-workers almost anywhere in the world. Telecommunications also reduces the amount of time needed to transmit information that can drive and conclude business actions. A manufacturing sales representative, for example, can use telecommunications technology to get new product prices from the central sales office while working at a customer's location. This empowers the sales representative and often results in faster, higher-quality customer service. Telecommunications technology also helps businesses coordinate activities and integrate various departments to increase operational efficiency and support effective decision making. The far-reaching developments of telecommunications are having and will continue to have a profound effect on business information systems and on society in general.

Networks

A **computer network** consists of communications media, devices, and software needed to connect two or more computer systems and/or devices. Once connected, computers can share data, information, and processing jobs. More and more businesses are linking computers in networks to streamline work processes and allow employees to collaborate on projects. The effective use of networks can turn a company into an agile, powerful, and creative organization, giving it a long-term competitive advantage. Networks can be used to share hardware, programs, and databases across the organization. They can transmit and receive information to improve organizational effectiveness and efficiency. They enable geographically separated workgroups to share documents and opinions, which fosters teamwork, innovative ideas, and new business strategies.

TELECOMMUNICATIONS

The use of telecommunications can help businesses solve problems and maximize opportunities. Using telecommunications effectively requires careful analysis of telecommunications media, devices, and carriers and services.

Types of Media

Various types of communications media are available. Each type exhibits its own characteristics, including transmission capacity and speed. In developing a telecommunications system, the selection of media depends on the purpose of the overall information and organizational systems, the purpose of the telecommunications subsystems, and the characteristics of the media. As with other system components, the media should be chosen to support the goals of the information and organizational systems at the least cost and to allow for possible modification of system goals over time. The proper media will help a company link subsystems to maximize effectiveness and efficiency.

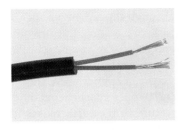

FIGURE 6.4

Twisted-Pair Wire Cable
(Source: Fred Bodin.)

Twisted-Pair Wire Cable

Twisted-pair wire cable is, as you might expect, a cable consisting of pairs of twisted wires (Figure 6.4). A typical cable contains two or more twisted pairs of wire, usually copper. Proper twisting of the wire keeps the signal from "bleeding" into the next pair and creating electrical interference. Because the twisted-pair wires are insulated, they can be placed close together and packaged in one group. Hundreds of wire pairs can be grouped into one large wire cable.

There are two kinds of twisted-pair wire cable: shielded and unshielded. Shielded twisted-pair wire cable has a special conducting layer within the normal insulation. This conducting layer makes the cable less prone to electrical interference, or "noise." Unshielded twisted-pair (UTP) wire cable lacks this special insulation shield. UTP cables have historically been used for telephone service and to connect computer systems and devices. Newer types of cable, however, have begun to replace UTP cable in both businesses and homes.

Coaxial Cable

Figure 6.5 shows a typical coaxial cable, similar to that used in cable television installations. A coaxial cable consists of an inner conductor wire surrounded by insulation, called the dielectric. The dielectric is surrounded by a conductive shield (usually a layer of foil or metal braiding), which is in turn covered by a layer of non-conductive insulation, called the jacket. When used for data transmission, coaxial cable falls in the middle of the cabling spectrum in terms of cost and performance. It is more expensive than twisted-pair wire cable but less so than fiber-optic cable (discussed next). Coaxial cable offers cleaner and crisper data transmission (less noise) than twisted-pair wire cable. It also offers a higher data transmission rate.

Fiber-Optic Cable

Fiber-optic cable, consisting of many extremely thin strands of glass or plastic bound together in a sheathing (a jacket), transmits signals with light beams (Figure 6.6). These high-intensity light beams are generated by lasers and are conducted along the transparent fibers. These fibers have a thin coating, called cladding, which effectively works like a mirror, preventing the light from leaking out of the fiber.

Fiber-optic technology is dramatically improving.[2] In 1995, fiber-optic equipment manufacturers discovered how to send eight signals through a single strand of fiber-optic cable. Each signal is a different color in the light spectrum and could carry 120,000 calls. The latest technology, known as dense wave division multiplexing, can break a fiber into 128 signals, which means the number of calls that can be carried increases from 120,000 to 12 million. Telecommunications companies are now testing fiber-optic systems that may have the potential to carry close to 100 million calls.

The much smaller diameter of fiber-optic cable makes it ideal in situations where there is not room for additional bulky copper wires—for example, in crowded conduits, which can be pipes or spaces used to house and carry the electrical and communications wires. In this case, the use of smaller fiber-optic telecommunications systems is very effective. Because fiber-optic cables are immune to electrical interference, signals can be transmitted over longer distances with fewer expensive repeaters to amplify or rebroadcast the data. A special plus for security-conscious applications is the difficulty of stealing information. With the right equipment installed, it is virtually impossible to tap into fiber-optic cable without being detected. Fiber-optic cable and devices are more expensive to purchase and install than their twisted-pair wire and coaxial counterparts, although the cost is coming down.

Microwave Transmission

Microwave transmissions are sent through the atmosphere and space. Although these transmission media do not entail the expense of laying cable, the transmission devices needed to utilize this medium are quite expensive. Microwave

FIGURE 6.5

Coaxial Cable
(Source: Fred Bodin.)

is a high-frequency radio signal that is sent through the air (Figure 6.7). Microwave transmission is line-of-sight, which means that the straight line between the transmitter and receiver must be unobstructed. Typically, microwave stations are placed in a series—one station will receive a signal, amplify it, and retransmit it to the next microwave transmission tower. Such stations can be up to 30–70 miles apart (depending on the height of the towers) before the curvature of the earth makes it impossible for the towers to "see one another." Microwave signals can carry thousands of channels at the same time.

A communications satellite is basically a microwave station placed in outer space (Figure 6.8). The satellite receives the signal from the earth station, amplifies the relatively weak signal, and then rebroadcasts it at a different frequency. The advantage of satellite communications is the ability to receive and broadcast over large geographic regions. Such problems as the curvature of the earth, mountains, and other structures that block the line-of-sight microwave transmission make satellites an attractive alternative.

Most of today's communications satellites are owned by companies that rent or lease satellite communications capacity to other companies. However, several large companies are now using their own satellites for internal telecommunications. Some large retail chains, for example, use satellite transmission to connect their main offices to retail stores and warehouses throughout the country or the world. General Motors, for example, is planning on having Internet access as a feature on many of its cars and trucks, using existing satellite technology.[3] You will be able to send and receive e-mails, check stock prices, find out what is playing at the local movie theater, shop, make dinner reservations, and of course make phone calls—all from your car. The service is expected to cost about $10 per month.

In addition to standard satellite stations, there are small mobile satellite systems that allow people and businesses to communicate. These portable systems have a dish that is a few feet in diameter and can operate on battery power anywhere in the world. This is important for news organizations that require the ability to transmit news stories from remote locations. Many people are investing in direct satellite dish technology to receive TV and send and receive computer communications. Hughes Electronics, for example, is investing $1.4 billion in a satellite system, called Spaceway, to bring telecommunications and Internet access to anywhere in North America by 2003 through its DirectPC offering.[4]

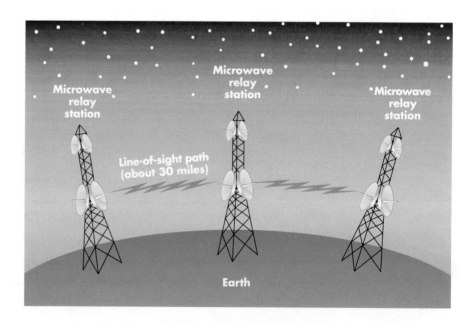

FIGURE 6.8

Satellite Transmission

Communications satellites are relay stations that receive signals from one earth station and rebroadcast them to another.

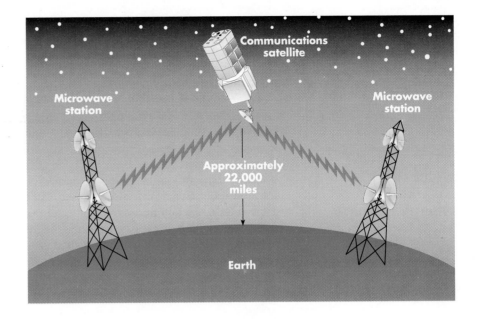

FIGURE 6.9

A Typical Cellular Transmission Scenario

Using a cellular car phone, the caller (1) dials the number. The signal is sent from the car's antenna to the low-powered cellular antenna located in that cell (2). The signal is sent to the regional cellular phone switching office, also called the mobile telephone subscriber office (MTSO) (3). The signal is switched to the local telephone company switching station located nearest the call destination (4). Now integrated into the regular phone system, the call is automatically switched to the number originally dialed (5), all without the need for operator assistance.

Cellular Transmission

With cellular transmission, a local area, such as a city, is divided into cells. As a car or vehicle with a cellular device, such as a mobile phone, moves from one cell to another, the cellular system passes the phone connection from one cell to another (Figure 6.9). The signals from the cells are transmitted to a receiver and integrated into the regular phone system. Cellular phone users can thus connect to anyone that has access to regular phone service, like a child at home or a business associate in London. They can also contact other cellular phone users. Because cellular transmission uses radio waves, it is possible for people with special receivers to listen to cellular phone conversations.

Cellular transmission is used for much more than making phone calls from remote locations. Combining cellular transmission with other devices opens up the power of networks, communications, and the Internet. For example, combining cellular transmission with some phone devices allows people to read and respond to their e-mails, check stock prices, get news on their favorite topics, and more. With cell phone rates dropping, some people are disconnecting their old, wired phones in their homes and offices and using cellular phones for all of their calling needs. New devices, called net phones, combine cellular transmission with access to networks and the Internet. Handheld computers are now using wireless communications to enhance their power. The Palm VII, for

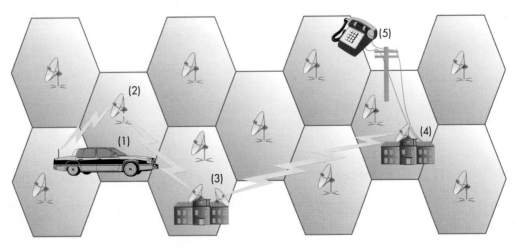

example, can use wireless communications to check appointments from a central computer, download information from corporate computers, and browse the Internet.

Infrared Transmission

Another mode of transmission, called infrared transmission, involves sending signals through the air via light waves. Infrared transmission requires line-of-sight transmission and short distances—under a few hundred yards. Infrared transmission can be used to connect various small devices and computers. For example, infrared transmission has been used to allow handheld computers to transmit data and information to larger computers within the same room. Infrared transmission can also be used to connect a display screen, a printer, and a mouse to a computer. Some special-purpose phones can also use infrared transmission. This means of transmission can be used to establish a wireless network with the advantage that devices can be moved, removed, and installed without expensive wiring and network connections.

Devices

A telecommunications device is one of various hardware devices that allow electronic communication to occur or to occur more efficiently. Almost every telecommunications system uses one or more of these devices to transmit or receive signals.

Modems

In data telecommunications, it is not uncommon to use transmission media of differing types and capacities at various stages of the communications process. If a typical telephone line is used to transfer data, it can only accommodate an **analog signal** (a continuous, curving signal). Because a computer generates a **digital signal** representing bits, a special device is required to convert the digital signal to an analog signal, and vice versa (Figure 6.10). Translating data from digital to analog is called modulation, and translating data from analog to digital is called demodulation. Thus, these devices are modulation/demodulation devices, or **modems**. Penril/Bay Networks, Hayes, Microcom, Motorola, and U.S. Robotics are examples of modem manufacturers.

Modems can automatically dial telephone numbers, originate message sending, and answer incoming calls and messages. Modems can also perform tests and checks on how they are operating. Some modems are able to vary their transmission rates, commonly measured in bits per second. Today, many modems transmit at 56,600 bps. Some of the more expensive modems, called

analog signal

a continuous, curving signal

digital signal

a signal represented by bits

modem

a device that translates data from digital to analog and analog to digital

FIGURE 6.10

How a Modem Works

Digital signals are modulated into analog signals, which can be carried over existing phone lines. The analog signals are then demodulated back into digital signals by the receiving modem.

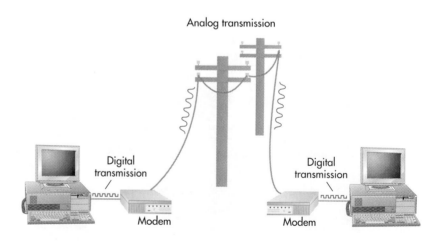

Analog transmission

Digital transmission

Modem

Digital transmission

Modem

PCMCIA modem and desktop
fax modem
(Source: 3Com Corporation.)

multiplexer

a device that allows several
telecommunications signals to
be transmitted over a single
communications medium at the
same time

front-end processor

a special-purpose computer that
manages communications to and
from a computer system

FIGURE 6.11

Use of a Multiplexer to
Consolidate Data
Communications onto a Single
Communications Link

smart modems, contain microprocessors that allow them to operate and function under a variety of circumstances.

Fax Modems

Facsimile devices, commonly called fax devices, allow businesses to transmit text, graphs, photographs, and other digital files via standard telephone lines. A fax modem is a very popular device that combines a fax with a modem, giving users a powerful communications tool.

Special-Purpose Modems

Various types of special-purpose modems are available. Cellular modems are placed in laptop personal computers to allow people on the go to communicate with other computer systems and devices. With a cellular modem, you can connect to other computers while in your car, on a boat, or in any area that has cellular transmission service. Expansion slots used for PC memory cards can also be used for standardized credit-card-size PC modem cards, which work like standard modems. PC modems are becoming increasingly popular with notebook and portable computer users. Cable companies are promoting the cable modem, which has a low initial cost and transmission speeds up to 10 Mbps.[5] A cable modem can deliver network and Internet access up to 500 times faster than a standard modem and phone line. In addition, a cable modem is always on. You can be on the Internet 24 hours a day, 7 days a week. Fees from $30 to $50 per month usually include unlimited service, local news, and an e-mail account. Cable service may cost only an additional $10 per month if you have existing Internet access through an Internet provider and want to upgrade from phone to cable service.

Multiplexers

Because media and channels are expensive, devices that allow several signals to be sent over one channel have been developed. A **multiplexer** is one of these devices. A multiplexer allows several telecommunications signals to be transmitted over a single communications medium at the same time (Figure 6.11).

Front-End Processors

Front-end processors are special-purpose computers that manage communications to and from a computer system. Like a receptionist handling visitors at an office complex, communications processors direct the flow of incoming and outgoing jobs. They connect a midrange or mainframe computer to hundreds or thousands of communications lines. They poll terminals and other devices to see if they have any messages to send. They provide automatic answering and

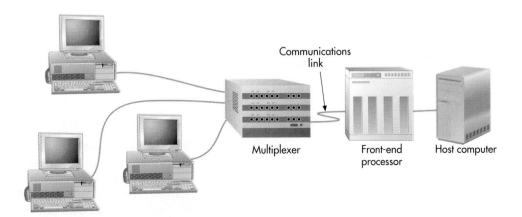

Communications
link

Multiplexer   Front-end   Host computer
processor

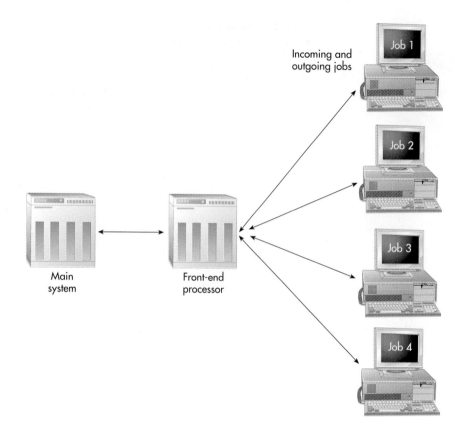

Incoming and
outgoing jobs

Main
system

Front-end
processor

FIGURE 6.12

Front-End Processor

A front-end processor takes the burden of communications management away from the main system processor.

calling, as well as perform circuit checking and error detection. Front-end processors also develop logs or reports of all communications traffic, edit data before it enters the main processor, determine message priority, automatically choose alternative and efficient communications paths over multiple data communications lines, and provide general data security for the main system CPU. Because front-end processors perform all these tasks, the midrange or mainframe computer is able to process more work (Figure 6.12).

Carriers and Services

Telecommunications carriers provide the telephone lines, satellites, modems, and other communications technology used to transmit data from one location to another. They also provide many types of services. Telecommunications carriers are classified as either common carriers or other special-purpose carriers. The **common carriers** are primarily the long-distance telephone companies. American Telephone & Telegraph (AT&T), one of the largest companies providing communications media and services, is a common carrier for long-distance service and a special-purpose carrier for other services. MCI Worldcom, Sprint, and others make up a significant part of the telecommunications industry as well. **Value-added carriers** are companies that have developed private telecommunications systems and offer their services for a fee. Some value-added carriers that offer communications services include SprintNet and Telenet (developed by GTE) and Tymnet.

common carriers

long-distance telephone companies

value-added carriers

companies that have developed private telecommunications systems and offer their services for a fee

switched line

a communications line that uses switching equipment to allow one transmission device to be connected to other transmission devices

Switched and Dedicated Lines

Common carriers typically provide the use of standard telephone lines, called **switched lines**. These lines use switching equipment to allow one transmission device (e.g., your telephone) to be connected to other transmission devices (e.g., the telephones of your friends and relatives). A switch is a special-purpose circuit that directs messages along specific paths in a telecommunications system.

Telecommunications networks require state-of-the-art computer software technology to continuously monitor the flow of voice, data, and image transmission over billions of circuit miles worldwide. (Source: Stone/Roger Tully.)

dedicated line

a communications line that provides a constant connection between two points; no switching or dialing is needed, and the two devices are always connected

private branch exchange (PBX)

a communications system that can manage both voice and data transfer within a building and to outside lines

When you make a phone call, the local telephone service provider's switching equipment connects your phone to the phone of the person you're calling. Fees for a switched business line (versus residential line) can range from $25 to $100 or more per month. A **dedicated line**, also called a leased line, provides a constant connection between two points. No switching or dialing is needed; the two devices are always connected. Many firms with high data transfer requirements between two points—say, an East Coast and a West Coast shared headquarters arrangement—utilize dedicated lines. The high initial cost of purchasing or leasing such a line is offset by eliminating long-distance charges incurred with a switched line. Monthly fees for a dedicated line can range from $100 to $500 or more, but there is no additional charge for use.

Private Branch Exchange

A **private branch exchange (PBX)** is a communications system that can manage both voice and data transfer within a building and to outside lines. In a PBX system, switching equipment routes phone calls and messages within the building. PBXs can be used to connect hundreds of internal phone lines to a few phone company lines. For example, an organization might have five phone lines coming in from the outside phone company. These five lines may be connected to 50 phones within the organization. Any of the 50 phones can use one of the five phone lines to make calls outside the organization. These same five lines may also be used for incoming calls. Furthermore, it is usually possible for any of the 50 phones to connect to another internal phone on an intercom system.

Not only can PBXs store and transfer calls, but they can also serve as connections between different office devices. With PBX technology, a manager could connect his computer to the PBX via a modem and then send instructions to a copy machine through his PC. Another advantage of a PBX system is that it requires a business to have fewer phone lines coming in from the outside. The disadvantage is that the company has to purchase, rent, or lease the PBX equipment. Thus, there is a trade-off between the expense of the PBX equipment and the savings in the reduced number of incoming phone lines.

WATS Service

Wide-area telecommunications service (WATS) is a billing method for heavy users of voice services. When you dial a company at a toll-free 800 or 888 number to place an order or make a query, you are using WATS. The company or organization you call via WATS pays a fee to the phone company, depending on the level of service and usage. The fee varies depending on the caller's geographic location within the United States and the number of incoming and outgoing calls. Companies that rely on phones for customer service typically use WATS services because customers can call the WATS number free of charge. For companies with a high volume of calls, WATS can also be substantially less expensive than a normal billing schedule. It is even possible for individuals to get a personal toll-free number.

Phone and Dialing Services

Common carriers are beginning to provide more and more phone and dialing services to home and business users. Automatic number identification (ANI), or caller ID, equipment can be installed on a phone system to identify and display the number of an incoming call. In a business setting, ANI can be used to identify the caller and link that caller with information stored in a computer. For example, when a customer calls Federal Express, the customer service rep uses the ANI to identify the name and address of the customer, thus saving time when handling a request for a pickup. The ANI can be very useful in helping people screen calls before they are answered. Unwanted phone calls from other people and businesses can be identified before the phone is ever answered. ANI, however, doesn't always work well with different carriers. Common carriers offer even more services to extend the capabilities of the typical phone system. Even with all the advances in computers and telecommunications, services offered by common carriers will remain important. Some of the services are listed here:

- The ability to integrate personal computers so that the telephone number of the caller is automatically captured and used to look up information in a database about this customer
- Access codes to screen out junk calls, wrong numbers, and unwanted phone calls
- Call priorities (e.g., only certain calls would be received during certain times of the day, such as from 10:00 P.M. to 7:00 A.M.)
- The ability to use one number for a business phone, home phone, personal computer, fax, etc.
- Intelligent dialing (when a busy signal is received, the phone redials the number when your line and the line of the party you are trying to reach are both free).

Digital Subscriber Line

digital subscriber line (DSL)

a communications line that uses existing phone wires going into today's homes and businesses to provide transmission speeds exceeding 500 Kbps at a cost of $20 or more per month

A **digital subscriber line (DSL)** uses existing phone wires going into today's homes and businesses to provide transmission speeds exceeding 500 Kbps at a cost of $20 or more per month. This speed means faster Internet access and downloads compared with standard phone lines. A special modem costing a few hundred dollars is also required. DSL lines are not available everywhere. In early 2000, for example, the chairman of US West, a company that offers DSL service, couldn't get his own company to deliver DSL to his home.[6] Even so, DSL is growing rapidly in usage. The number of DSL modems that will be installed in 2000 is expected to be about 6.2 million, which is double the rate of a year earlier.[7] Figure 6.13 shows the expected growth of DSL compared with cable modems.[8] The problem for individuals and businesses is how to choose the best telecommunications option.[9] Each option has its own cost, speed, and reliability to consider. Table 6.1 shows some of the costs, advantages, and disadvantages of different lines and services offered by communications carriers.[10]

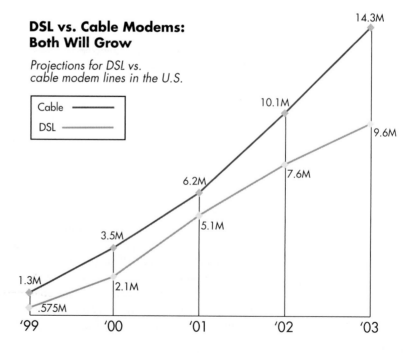

FIGURE 6.13

Growth of DSL versus Cable Modems
(Sources: DSL figures, Telechoice, Inc.; Denver, cable figures, Kenetic Strategies, Inc., Phoenix, found in Matthew Hamblen, "Digital Subscriber Line," *Computerworld*, February 7, 2000, p. 69. Reprinted with permission of *Computerworld*)

DSL vs. Cable Modems: Both Will Grow

Projections for DSL vs. cable modem lines in the U.S.

Cable ——————
DSL ——————

14.3M
10.1M
6.2M
3.5M
1.3M
.575M
2.1M
5.1M
7.6M
9.6M

'99   '00   '01   '02   '03

Line/Service	Speed	Cost per Month	Advantages	Disadvantages
Standard phone service	56 Kbps	$10–$30	Low cost and broadly available	Too slow for video and downloads of large files
ISDN	64–128 Kbps	$70–$120	Fast for video and other applications	Higher costs and not available everywhere
DSL	500 Kbps–1.544 Mbps	$20–$100 in addition to standard phone service	Fast, and the service comes over standard phone lines	Slightly higher costs and not available everywhere
Cable modem	Receive at up to 500 Kbps and send at speeds up to 1.544 Mbps	$20–$100	Fast and uses existing cable that comes into the home	Slightly higher costs and not available everywhere
T1	1.544 Mbps	$1,000	Very fast broadband service, typically used by corporations and universities	Very expensive, high installation fee, and users pay a monthly fee based on distance

TABLE 6.1

Costs, Advantages, and Disadvantages of Several Line and Service Types

integrated services digital network (ISDN)

a technology that uses existing common-carrier lines to simultaneously transmit voice, video, and image data in digital form

ISDN

Many telephone companies now offer **integrated services digital network (ISDN)** service, a technology that uses existing common-carrier lines to simultaneously transmit voice, video, and image data in digital form. ISDN also offers high rates of transmission: the digital service has the capacity to send a 22-page document in about a second! With ISDN, communications devices require a special ISDN board. These data communications systems use an ISDN network switch—a digital switch that allows different communications services to be connected to the system. For example, ISDN allows long-distance services, video and voice services, facsimile devices, telephones, and private branch exchanges to be integrated into one telecommunications system (Figure 6.14). ISDN digital networks are typically faster (from 64 Kbps to 2 Mbps) than standard phone lines and can carry more signals than analog networks. They also allow for easier sharing of image, multimedia, and other complex forms of data across telephone lines.

T1 carrier

a line or channel developed by AT&T and used in North America to increase the number of voice calls that can be handled through existing cables

T1 Carrier

The **T1 carrier** was developed by AT&T to increase the number of voice calls that could be handled through the existing cables. For digital communications, T1 is the carrier used in North America. T1 is also suitable for data and image transmissions. Large companies frequently purchase T1 lines to develop an integrated telecommunications network that can carry voice, data, and images. T1 has a speed of 1.544 Mbps developed from two dozen 64 Kbps channels, together with one 8 Kbps channel for carrying control information. T1 services are quite expensive, with subscribers paying a monthly fee based on the distance (several dollars per mile in some cases), and there is also a high installation fee.

NETWORKS AND DISTRIBUTED PROCESSING

Businesses are linking their personnel and equipment to enable people to work quicker and more efficiently. Computer networks allow organizations flexibility—to accomplish work wherever and whenever it is most beneficial. To take full advantage of networks and distributed processing, it is important to understand strategies, network concepts and considerations, network types, and related topics.[11]

FIGURE 6.14

ISDN Network Switching

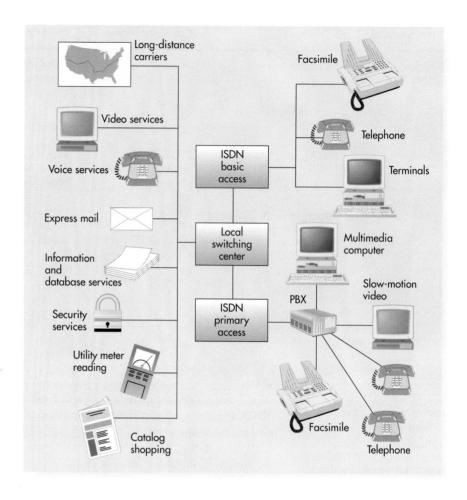

Basic Processing Strategies

When an organization needs to use two or more computer systems, one of three basic processing strategies may be followed: centralized, decentralized, or distributed. With **centralized processing**, all processing occurs in a single location or facility. This approach offers the highest degree of control. For example, centralized processing is useful for financial institutions that require a high degree of security. With **decentralized processing**, processing devices are placed at various remote locations. The individual computer systems are isolated and do not communicate with each other. Decentralized systems are suitable for companies that have independent operating divisions. Some drug store chains, for example, operate each location as a completely separate entity; each store has its own computer system that works independently of the computers at other stores. With **distributed processing**, computers are placed at remote locations but connected to each other via telecommunications devices. Consider a manufacturing company with plants in Milwaukee, Chicago, and Atlanta and a corporate headquarters in New York. Each location has its own computer system. By connecting all the computer systems into a distributed processing system, all the locations can share data and programs. Distributed processing also allows each plant to perform its own processing (say, for example, inventory) while the New York computer system coordinates and processes other applications, like payroll.

One benefit of distributed processing is that processing activity can be allocated to the location(s) where it can most efficiently occur. For example, the New York headquarters may have the largest computer system, but the Atlanta office might have hundreds of employees to input the data. The system's output

centralized processing

processing alternative in which all processing occurs in a single location or facility

decentralized processing

processing alternative in which processing devices are placed at various remote locations

distributed processing

processing alternative in which computers are placed at remote locations but connected to each other via telecommunications devices may be most needed in Chicago, the location of the warehouse. With distributed processing, each of these offices can organize and manipulate the data to meet its specific needs, as well as share its work product with the rest of the organization. The distribution of the processing across the organizational system ensures that the right information is delivered to the right individuals, maximizing the capabilities of the overall information system by balancing the effectiveness and efficiency of each individual computer system.

Network Concepts and Considerations

Networks that link computers and computer devices provide for flexible processing. Building networks involves two types of design: logical and physical. A logical model shows how the network will be organized and arranged. A physical model describes how the hardware and software in the network will be physically and electronically linked.

Network Topology

The number of possible ways to logically arrange the nodes, or computer systems and devices on a network, may seem limitless. Actually, there are only five major types of **network topologies**—logical models that describe how networks are structured or configured. These types are ring, bus, hierarchical, star, and hybrid (Figure 6.15).

The **ring network** contains computers and computer devices placed in a ring, or circle. With a ring network, there is no central coordinating computer. Messages are routed around the ring from one device or computer to another in one direction. A **bus network** consists of computers and computer devices on a single line. Each device is connected directly to the bus and can communicate directly with all other devices on the network. The bus network is one of the most popular types of personal computer networks. The **hierarchical network** uses a treelike structure. Messages are passed along the branches of the hierarchy until they reach their destination. Like a ring network, a hierarchical network does not require a centralized computer to control communications. Hierarchical networks are easier to repair than other topologies because you can isolate and repair one branch without affecting the others. A **star network** has a central hub or computer system. Other computers or computer devices are located at the end of communications lines that originate from the central hub or computer system. The central computer of a star network controls and directs messages. If the central computer breaks down, it results in a breakdown of the entire network. Many organizations use a **hybrid network**, which is simply a combination of two or more of the four topologies just discussed. The exact configuration of the network depends on the needs, goals, and organizational structure of the company involved.

Network Types

Depending on the physical distance between nodes on a network and the communications and services provided by the network, networks can be classified as local area, wide area, or international. Local area networks tie together equipment in a building or local area; international networks are used to communicate between countries. Wide area networks operate over a broad geographic area.

Local Area Networks

A network that connects computer systems and devices within the same geographic area is a **local area network (LAN)**. A local area network can be a ring, star, hierarchical, or hybrid network. Typically, local area networks are

---

network topology

logical model that describes how networks are structured or configured

ring network

a type of topology that contains computers and computer devices placed in a ring, or circle; there is no central coordinating computer; messages are routed around the ring from one device or computer to another

bus network

a type of topology that contains computers and computer devices on a single line; each device is connected directly to the bus and can communicate directly with all other devices on the network; one of the most popular types of personal computer networks

hierarchical network

a type of topology that uses a treelike structure with messages passed along the branches of the hierarchy until they reach their destination

star network

a type of topology that has a central hub or computer system, and other computers or computer devices are located at the end of communications lines that originate from the central hub or computer

hybrid network

a network topology that is a combination of other network types

local area network (LAN)

a network that connects computer systems and devices within the same geographic area

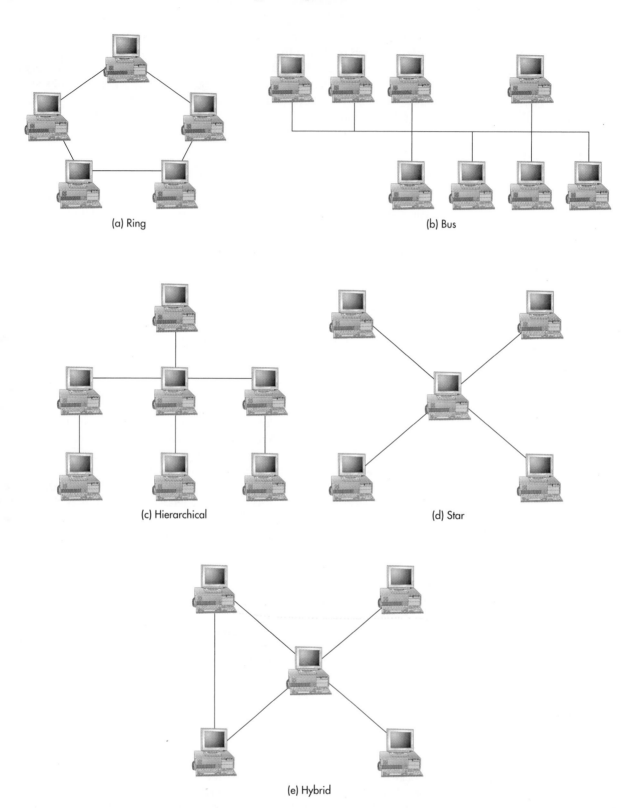

(a) Ring

(b) Bus

(c) Hierarchical

(d) Star

(e) Hybrid

FIGURE 6.15

The four basic types of network topology are (a) ring, (b) bus, (c) hierarchical, and (d) star. In addition, a hybrid configuration (e) can be formed from elements of any of these four topologies.

wired into office buildings and factories (Figure 6.16). While unshielded twisted-pair wire cable is the most widely used media with LANs, other media—including fiber-optic cable—is also popular. They can be built around powerful personal computers, minicomputers, or mainframe computers. When a personal computer is connected to a local area network, a network interface card (NIC) is usually required. A network interface card is a card or board that is placed in a computer's expansion slot to allow it to communicate with the

FIGURE 6.16

A Typical LAN in a Bus Topology

All network users within an office building can connect to each other's devices for rapid communication. For instance, a user in research and development could send a document from her computer to be printed at a printer located in the desktop publishing center.

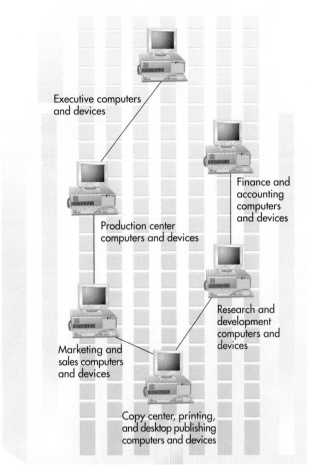

Executive computers and devices

Finance and accounting computers and devices

Production center computers and devices

Research and development computers and devices

Marketing and sales computers and devices

Copy center, printing, and desktop publishing computers and devices

Corporate headquarters

network. A wire or connector from the network is plugged directly into the network interface card. For example, a salesperson whose notebook computer has an interface card can establish a link to the corporate LAN. He or she can access the network while at the office and download data needed for the next sales call.

Another basic LAN is a simple peer-to-peer network that might be used for a very small business to allow the sharing of files and hardware devices such as printers. In a peer-to-peer network, each computer is set up as an independent computer, except that other computers can access specific files on its hard drive or share its printer. These types of networks have no server. Instead, each computer uses a network interface card and cabling to connect it to the next machine. Examples of peer-to-peer networks include Windows for Workgroups, Windows 98, Windows NT, Windows 2000, and AppleShare. Performance of the computers on a peer-to-peer network is usually slower because one computer is actually sharing the resources of another computer. These networks, however, are a good beginning network from which small businesses can grow. The software cost is minimal, and network cards can be used if the company decides to enlarge the system. In addition, peer-to-peer networks are becoming cheaper, faster, and easier to use for home-based businesses.[12]

Wide Area Networks

wide area network (WAN)

a network that ties together large geographic regions using microwave and satellite transmission or telephone lines

A **wide area network (WAN)** ties together large geographic regions using microwave and satellite transmission or telephone lines. When you make a long-distance phone call, you are using a wide area network. AT&T, MCI Worldcom, and others are examples of companies that offer WAN services to the public. Companies also design and implement WANs. These WANs usually consist of

FIGURE 6.17

A Wide Area Network

Wide area networks are the basic long-distance networks used by organizations and individuals around the world. The actual connections between sites, or nodes (shown by dashed lines), may be any combination of satellites, microwave, or cabling. When you make a long-distance telephone call, you are using a WAN.

North America

computer equipment owned by the user, together with data communications equipment provided by a common carrier. (See Figure 6.17.)

International Networks

international network

a network that links systems between countries

Networks that link systems between countries are called **international networks**. However, international telecommunications comes with special problems. In addition to requiring sophisticated equipment and software, global area networks must meet specific national and international laws regulating the electronic flow of data across international boundaries, often called transborder data flow. Some countries have strict laws restricting the use of telecommunications and databases, making normal business transactions such as payroll costly, slow, or even impossible. Other countries have few laws restricting the use of telecommunications and databases. These countries, sometimes called data havens, allow other governments and companies to avoid their country's laws by processing data within their boundaries. International networks in developing countries can have inadequate equipment and infrastructure that can cause problems and limit the usefulness of the network.

Despite the obstacles, numerous private and public international networks exist. United Parcel Service, for example, has invested in an international network, called UPSnet. UPSnet allows drivers to use handheld computers to send real-time information about pickups and deliveries to central data centers. The huge UPS network allows data to be retrieved by customers to track packages or to be used by the company for faster billing, better fleet planning, and improved customer service.[13] (The Internet, which we will discuss in Chapter 7, is the largest public international network.)

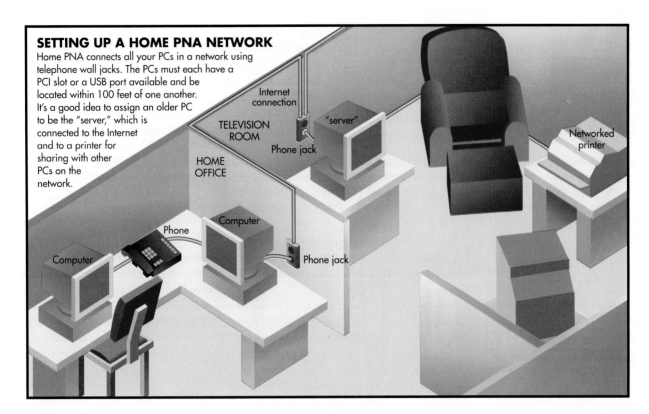

SETTING UP A HOME PNA NETWORK
Home PNA connects all your PCs in a network using telephone wall jacks. The PCs must each have a PCI slot or a USB port available and be located within 100 feet of one another. It's a good idea to assign an older PC to be the "server," which is connected to the Internet and to a printer for sharing with other PCs on the network.

FIGURE 6.18

Connecting Computing Devices Using a Home Network
(Source: Reprinted from "Network Your Home Painlessly," *PC Magazine*, April 4, 2000, p. 108. Copyright © 2000 Ziff Davis Publishing Holdings, Inc. All rights reserved.)

terminal-to-host

an architecture in which the application and database reside on one host computer, and the user interacts with the application and data using a "dumb" terminal

Home and Small Business Networks

With more people working at home, connecting home computing devices and equipment together into a unified network is on the rise. Small businesses are also starting to connect their systems and equipment. With a home or small business network, computers, printers, scanners, and other devices can be connected. A person working on one computer, for example, can use data and programs stored on another computer's hard disk. In addition, a single printer can be shared by several computers on the network.

To make home and small business networking a reality, a number of companies are offering standards, devices, and procedures.[14] Home Phoneline Network Alliance, for example, has developed HomePNA 2. This system allows 10-Mbps speeds over existing phone lines, without interfering with existing phone service. (See Figure 6.18.) Other home networking systems run on electric lines or require new wires and cables.

Terminal-to-Host, File Server, and Client/Server Systems

If an organization is selecting distributed information processing, it can connect computers in several ways. Most common are terminal-to-host, file server, and client/server architecture.

Terminal-to-Host

With **terminal-to-host** architecture, the application and database reside on one host computer, and the user interacts with the application and data using a "dumb" terminal. (Even if you use a PC to access the application, you run terminal emulation software on the PC to make it act as if it were a dumb terminal with no processing capacity.) Since a dumb terminal has no data processing capability, all computations, data accessing and formatting, and data display, are done by an application that runs on the host computer (Figure 6.19).

"Dumb" terminal

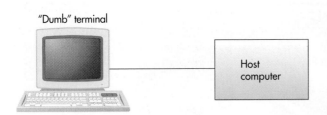

FIGURE 6.19

Terminal-to-Host Connection

file server

an architecture in which the application and database reside on the one host computer, called the file server

client/server

an architecture in which multiple computer platforms are dedicated to special functions such as database management, printing, communications, and program execution

File Server

In **file server** architecture, the application and database reside on the one host computer, called the file server. The database management system runs on the end user's personal computer or workstation. If the user needs even a small subset of the data that resides on the file server, the file server sends the user the entire file that contains the data requested, including a lot of data the user does not want or need. The downloaded data can then be analyzed, manipulated, formatted, and displayed by a program that runs on the user's personal computer (Figure 6.20).

Client/Server

In **client/server** architecture, multiple computer platforms are dedicated to special functions such as database management, printing, communications, and program execution. These platforms are called servers. Each server is accessible by all computers on the network. Servers can be computers of all sizes; they store both application programs and data files and are equipped with operating system software to manage the activities of the network. The server distributes programs and data files to the other computers (clients) on the network as they request them. An application server holds the programs and data files for a particular application, such as an inventory database. Processing can be done at the client or server.

A client is any computer (often an end user's personal computer) that sends messages requesting services from the servers on the network. A client can converse with many servers concurrently. A user at a personal computer initiates a request to extract data that resides in a database somewhere on the network. A data request server intercepts the request and determines on which data server the data resides. The server then formats the user's request into a message that the database server will understand. Upon receipt of the message, the database server extracts and formats the requested data and sends the results to the client. Only the data needed to satisfy a specific query is sent—not the entire file (Figure 6.21). As with the file server approach, once the downloaded data is on the user's machine, it can then be analyzed, manipulated, formatted, and displayed by a program that runs on the user's personal computer.

There are several advantages of the client/server approach over both the terminal-to-host and file server approaches: reduced cost, improved performance, and increased security. For example, the Dominican Republic's first credit bureau has found that using this technology has dramatically cut costs, while increasing reporting and decision support information.[15] According to Asdrubal Pichardo, "Before, it took two hours with the user to install an application. Now, it's only minutes."

Reduced cost potential. The functionality achieved with client/server computing can exceed that provided by a traditional minicomputer or even a mainframe-based computer system at a lower cost. With client/server computing, a

FIGURE 6.20

File Server Connection

The file server sends the user the entire file that contains the data requested. The downloaded data can then be analyzed, manipulated, formatted, and displayed by a program that runs on the user's personal computer.

File downloaded to user

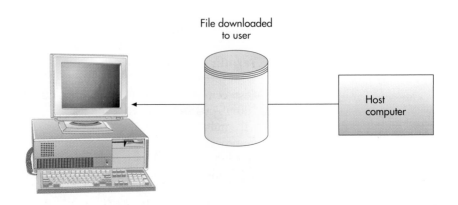

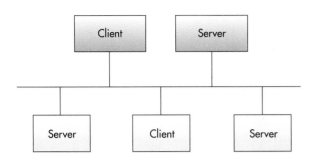

FIGURE 6.21

Client/Server Connection

Multiple computer platforms, called servers, are dedicated to special functions such as database management, data storage, printing, communications, network security, and program execution. Each server is accessible by all computers on the network. A server distributes programs and data files to the other computers (clients) on the network as they request them. The client requests services from the servers, provides a user interface, and presents results to the user. Once data is moved from a server to the client, the data may be processed on the client.

powerful workstation costing less than $25,000 may replace much of the function provided by a midrange computer costing over $100,000. In addition, vendor contracts for workstation software and hardware support are cheaper than for midrange and mainframe computers. Thus, many organizations view the migration of applications from mainframe computers and terminal-to-host architecture to client/server architecture as a significant cost savings opportunity. This downsizing (or, as some call it, "rightsizing") can yield significant savings in reduced hardware and software support costs.

Improved performance. The most important difference between the file server and client/server architecture is that the latter much more efficiently minimizes traffic on the network. With client/server computing, only the data needed to satisfy a user query is moved from the database to the client device, whereas the entire file is sent in file server computing. The smaller amount of data being sent over the network also greatly reduces the amount of time needed for the user to receive a response.

Increased security. Security mechanisms can be implemented directly on the database server through the use of stored procedures. These procedures execute faster than the password protection and data validation rules attached to individual applications on a file server. They can also be shared across multiple applications.

The type of application most appropriate for client/server architecture is one that uses large data files, requires fast response time, and needs strong security and recovery options. All these factors point to the kind of applications that are central to the operation and management of the business. On-line transaction processing and decision support applications are particularly good candidates for client/server computing.

Although client/server systems have much to offer in terms of practical benefits, some problems are associated with such systems: increased cost, loss of control, and complexity of vendor environment.

Increased cost potential. If all costs associated with client/server computing are accounted for, expected savings may fail to materialize. Moving to client/server architecture is a major two- to five-year conversion process. Over that time period, considerable costs are incurred for hardware, software, communications equipment and links, data conversion, and training. Costs are even higher for multiple-site companies converting to client/server computing. These expenses are difficult for the IS organization to track because they are often paid by the end users directly. Thus, the move to client/server architecture may be much more expensive than the IS organization realizes.

Loss of control. Controlling the client/server environment to prevent unauthorized use, invasion of privacy, and viruses is also difficult. Despite these concerns, many companies expect to gain long-term efficiency and effectiveness by moving away from large mainframe systems. The overall use of mainframe computers may not decrease, however, because mainframes are often reconfigured to become primary servers for large-scale client/server systems.

Complex multivendor environment. Implementation of client/server architecture leads to operating in a multivendor environment with, in many cases, relatively new and immature products. Situations such as these make it likely that problems will arise. Often such problems are difficult to identify and isolate to the appropriate vendor.

Nevertheless, the dominance of single-vendor environments and terminal-to-host architecture is fading fast as corporations move into the much more complex client/server environment with multiple vendors for networks, hardware, and software. Open systems are essential to implementing a client/server architecture so that managers are free to choose clients and servers and be assured that their combinations will be able to communicate with one another.

Communications Software and Protocols

Communications Software

communications software

software that provides a number of important functions in a network, such as error checking and data security

Communications software provides a number of important functions in a network. Most communications software packages provide error checking and message formatting. In some cases, when there is a problem, the software can indicate what is wrong and suggest possible solutions. Communications software can also maintain a log listing all jobs and communications that have taken place over a specified period of time. In addition, data security and privacy techniques are built into most packages.

In Chapter 4, you learned that all computers have operating systems that control many functions. When an application program requires data from a disk drive, it goes through the operating system. Now consider a situation in which a computer is attached to a network that connects large disk drives, printers, and other equipment and devices. How does an application program request data from a disk drive on the network? The answer is through the network operating system.

network operating system (NOS)

systems software that controls the computer systems and devices on a network and allows them to communicate with each other

A **network operating system (NOS)** is systems software that controls the computer systems and devices on a network and allows them to communicate with each other. An NOS performs the same types of functions for the network as operating system software does for a computer, such as memory and task management and coordination of hardware. When network equipment (such as printers, plotters, and disk drives) is required, the network operating system makes sure that these resources are correctly used. In most cases, companies that produce and sell networks provide the NOS. For example, NetWare is the NOS from Novell, a popular network environment for personal computer systems and equipment. Windows NT and Windows 2000 are other commonly used network operating systems.

network management software

software that enables a manager on a networked desktop to monitor the use of individual computers and shared hardware (like printers), scan for viruses, and ensure compliance with software licenses

Software tools and utilities are available for managing networks. With **network management software**, a manager on a networked desktop can monitor the use of individual computers and shared hardware (like printers), scan for viruses, and ensure compliance with software licenses. Network management software also simplifies the process of updating files and programs on computers on the network—changes can be made through a communications server instead of being made on individual computers. Network management software also protects software from being copied, modified, or downloaded illegally and performs error control to locate telecommunications errors and potential network problems. Some of the many benefits of network management software include fewer hours spent on routine tasks (like installing new software), faster response to problems, and greater overall network control.

Communications Protocols

protocols

rules that ensure communications among computers of different types and from different manufacturers

Open Systems Interconnection (OSI) model

a standard model for network architectures that divides data communications functions into seven distinct layers to promote the development of modular networks that simplify the development, operation, and maintenance of complex telecommunications networks

Communications protocols make communications possible. A number of communications **protocols** are used by companies and organizations of all sizes. Just as standards are important in building computer and database systems, established protocols help ensure communications among computers of different types and from different manufacturers.

Many protocols have layers of standards and procedures. The **Open Systems Interconnection (OSI) model** serves as a standard model for network architectures and is endorsed by the International Standards Committee. The OSI model divides data communications functions into seven distinct layers to promote the development of modular networks that simplify the development, operation, and maintenance of complex telecommunications networks. These layers are described in Figure 6.22.

FIGURE 6.22

The Seven Layers of the OSI Model

This Open Systems Interconnection (OSI) model is designed to permit communication among different computers from different manufacturers using different operating systems—as long as each conforms to the OSI model.

(Source: *Information Systems for Managers 3/e*, pp. 134–135, by George Reynolds, 1995, West Publishing. Reprinted with permission from Course Technology.)

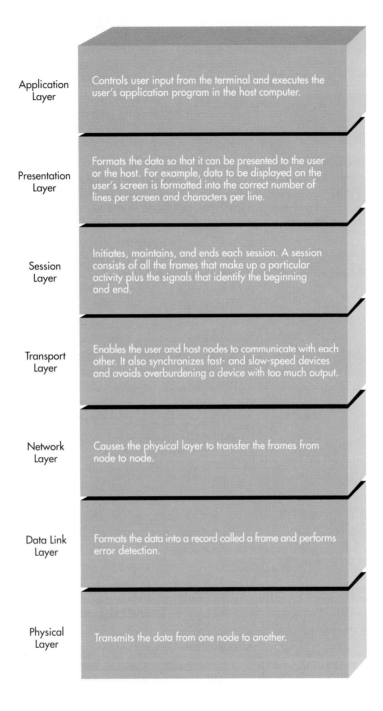

Application Layer	Controls user input from the terminal and executes the user's application program in the host computer.
Presentation Layer	Formats the data so that it can be presented to the user or the host. For example, data to be displayed on the user's screen is formatted into the correct number of lines per screen and characters per line.
Session Layer	Initiates, maintains, and ends each session. A session consists of all the frames that make up a particular activity plus the signals that identify the beginning and end.
Transport Layer	Enables the user and host nodes to communicate with each other. It also synchronizes fast- and slow-speed devices and avoids overburdening a device with too much output.
Network Layer	Causes the physical layer to transfer the frames from node to node.
Data Link Layer	Formats the data into a record called a frame and performs error detection.
Physical Layer	Transmits the data from one node to another.

Transmission Control Protocol/Internet Protocol (TCP/IP)

the primary communications protocol of the Internet, originally developed to link defense research agencies

TCP/IP

In the 1970s, the U.S. government pioneered development of the **Transmission Control Protocol/Internet Protocol (TCP/IP)** to link its defense research agencies. The government has adopted OSI standards to replace TCP/IP, but TCP/IP remains the major network protocol used by schools and businesses. It is the primary communications protocol of the Internet. The most recent version before publication of this book was TCP/IP 4.0.[16]

SNA

IBM has also developed a communications protocol, called Systems Network Architecture (SNA), which is a protocol used for IBM systems. Because of the popularity of IBM systems, many other computer manufacturers and communications companies have made their systems compatible with the SNA protocol.

Ethernet

Ethernet is a popular communications protocol often used with local area networks. The ethernet standard is designed for LANs that use a bus topology; the standard helps ensure compatibility among devices so that many people can attach to a common cable to share network facilities and resources. Some ethernet switches can operate at a gigabit per second, giving companies fast data transfer within an ethernet network.[17]

X.400 and X.500

The X.400 and X.500 protocols are also used in many organizations. Many international companies have adopted one of these protocols as their standard. With more international acceptance of these protocols, telecommunications among countries will become simpler. As businesses continue to move toward global operations, the importance of adopting standard international protocols will increase.

X.400 is a set of messaging handling standards ranging from X.400 to X.440 that define a Message Handling System and a Message Transfer Service. They are often used by businesses to process transactions electronically. For example, a company can use these standards to purchase parts from a supplier.

X.500 is a set of standards dictating the design of networking directories, which contain information on users—from names and e-mail addresses to job titles and resource-access privileges. Since X.500 is standards based, it provides a technological foundation for connecting proprietary and nonproprietary directories, such as those used in Windows NT and Unix networks, as well as in Internet-based networks. One large company that manufactures and sells more than $50 billion a year worldwide in packaged goods built an X.500-based directory to streamline the maintenance of its information system and virtually eliminate e-mail addressing errors.

Bridges, Routers, Gateways, and Switches

Many LANs have hardware and software devices that allow them to communicate with other networks that employ different transmission media or protocols. (See Figure 6.23.)

Bridge

bridge

connection between two or more networks at the media access control portion of the data link layer; the two networks must use the same communications protocol

A **bridge** connects two or more networks at the media access control portion of the data link layer. The two networks must use the same communications protocol.

Router

router

connection that operates at the network level of the OSI model and features more sophisticated addressing software than bridges; whereas bridges simply pass along everything that comes to them, routers can determine preferred paths to a final destination

A **router** operates at the network level of the OSI model and features more sophisticated addressing software than bridges. Whereas bridges simply pass along everything that comes to them, routers can determine preferred paths to a final destination. Routers also perform useful network management functions. They can break a network into subnets to create separate administrative network domains, thus helping to manage network distribution. Lucent Technology has developed an all-optical router that uses mirrors.[18] The new technology can speed data transfer through the router by a factor of 16.

Routers are also used as security firewalls (discussed in Chapter 7) between networks and the public Internet. The firewall keeps unwanted messages and users out of the organization's network. A specific router works with only one particular protocol.

Gateway

gateway

connection that operates at or above the OSI transport layer and links LANs or networks that employ different, higher-level protocols and allows networks with very different architectures and using dissimilar protocols to communicate

A **gateway** operates at or above the OSI transport layer and links LANs or networks that employ different higher-level protocols. Data received by a gateway must be restructured into a format understandable by the destination network.

FIGURE 6.23

Bridges, Routers, and Gateways

A switch can perform the role of either a router or a gateway. (Source: *Information Systems for Managers 3/e*, pp. 134–135, by George Reynolds, 1995, West Publishing. Reprinted with permission from Course Technology).

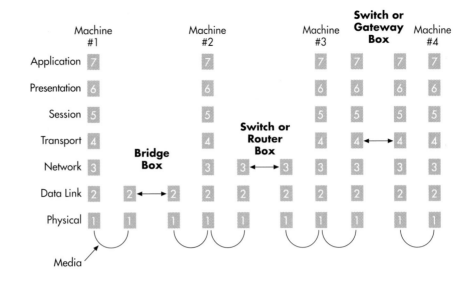

Thus, a gateway allows networks with very different architectures and using dissimilar protocols to communicate. Corporations often have a gateway to other types of networks so that workers can access the programs and data contained on networks outside their geographic region. In most cases, a user can click an icon or give a few simple commands to access these other networks. The user's computer and the server on the local area network will then automatically perform the tasks needed to link to the other network.

Switch

switch

a device that routes or switches data to its destination

A data **switch** is a device by which data can be routed or switched to its destination. A switch needs to be able to establish the desired connection. There are two main ways of doing this. In a matrix approach, each input channel has a predefined connection with each output channel. To pass something from an input channel to an output channel is merely a matter of following the connection. In a shared memory approach, the input controller writes material to a reserved area of memory and the specified output channel reads the material from this memory area.

A switch may also need to translate the input before sending it to an output channel. Switches are now generally replacing earlier, less flexible devices such as bridges and gateways. For example, a gateway may be able to connect two different architectures, but a switch may be able to connect several.

TELECOMMUNICATIONS APPLICATIONS

It is easy to see how telecommunications and networks can be applied to support information systems and organizational goals. For example, suppose a business needs to develop an accurate monthly production forecast. This can require a manager to download data from databases of sales forecasts from its customers. Telecommunications can provide a network link so that the manager can access the data needed for the production forecast report, which in turn supports the company's objective of better financial planning.

The consumer goods giant Procter & Gamble uses local area networks in all its plants to link office and plant workers to common software and shared databases and to provide e-mail services. The result is faster, more cost-effective, higher-quality product manufacturing. Other organizations transfer millions of important and strategic messages from one location to another every day. Telecommunications has become a critical component of information systems.

In some industries it is almost a requirement for doing business; most companies could not survive without it. This section will look at some significant business applications of networks.

Linking Personal Computers to Mainframes and Networks

One of the most basic ways telecommunications connect an individual to information systems is by connecting personal computers to mainframe computers so data can be downloaded or uploaded. For example, a data file or document file from a database can be downloaded to a personal computer for an individual to use. Some communication software programs will instruct the computer to connect to another computer on the network, download or send information, and then disconnect from the telecommunications line. These are called unattended systems because they perform these functions automatically, without user intervention.

Voice and Electronic Mail

voice mail

technology that enables users to leave, receive, and store verbal messages for and from other people around the world

With **voice mail**, users can leave, receive, and store verbal messages for and from other people around the world. Suppose Leslie May calls Paul Davis, who is out of his office. A digitized voice tells her to enter a code for Paul (perhaps his extension number). Leslie will then hear a voice message from Paul and can leave a message, just as she would if she reached an answering machine. When Paul Davis calls the voice mail system and enters his access code, he will hear his messages, including the one Leslie left. In some voice mail systems, a code can be assigned to a group of people instead of an individual. Suppose the code 100 stands for all 250 sales representatives in a company. If Leslie calls the voice mail system, enters the number 100, and leaves her message, all 250 sales representatives will receive Leslie's message.

People can also send messages to others via electronic mail, also called e-mail. With the right hardware and software, a sender can connect his or her computer to a network, type in a message, and send it to another person on the network. The receiving person is informed that there is a message waiting and can access it through communications software on his or her computer. It is also possible to attach files (spreadsheets, word processing documents, etc.) to an e-mail message. E-mail messages can be forwarded to other people. It is also possible to have e-mail make an automatic reply. A manager, for example, might be traveling without access to e-mail. The automatic reply feature allows the manager to reply to any incoming e-mail messages about when the manager will return or how he or she can be reached.

E-mail services are typically included with an Internet connection, through a company like America Online or Prodigy. In addition, there are a number of free Internet providers with excellent features. Yahoo!, Hotmail, and Eudora, for example, offer free e-mail.[19] In some cases, you have to watch ads with these services to be able to use the e-mail features.

While e-mail has become an indispensable part of business, there are also potential problems. E-mail can contain viruses that infect entire computer systems. E-mail can also be used to threaten people and organizations. Read the "Ethical and Societal Issues" box to see how it has been used to make threats recently. From an employee's perspective, organizations can also use e-mails sent at work in lawsuits against employees. Some employees have been disciplined or fired because of inappropriate e-mail messages they sent from the office. In addition, some e-mails may contain confidential corporate information that should not be shared with everyone. E-mail can also be used to invade personal privacy. To combat some of these problems, software programs that

ETHICAL AND SOCIETAL ISSUES
Threatening E-mail Goes to Trial

One of the biggest advantages of telecommunications and the Internet is the open communications it provides. But this advantage is also one of the biggest potential problems. People can send misinformation and malicious messages without carefully considering the consequences. This was the case with an 18-year-old man from Florida.

After the Columbine High School tragedy, where two students shot and killed 13 of their fellow students, the community of Littleton, Colorado, was in shock. How could this happen? Could it have been prevented? It was a parent's worst nightmare, and it was about to get worse. The Florida man sent an e-mail message to a Columbine student and said that he would "finish what had begun." The e-mail also threatened the Columbine student that she "would be the first to go" if she told anyone else. As a result of the message, Columbine High School closed for two days.

The young Florida man was charged with threatening a Columbine student on the Internet. Originally pleading

not guilty, he later changed his plea to guilty. In a press conference, he admitted, "I am horrified at what I have done and its consequences. . . . I hope this is a lesson to kids out there not to make a stupid mistake like I did."

Discussion Questions

1. How would you define appropriate e-mail?
2. What is inappropriate or malicious e-mail?

Critical Thinking Questions

3. Can anything be done to prevent this type of malicious e-mail message?
4. What new laws, if any, would you suggest to prevent or punish this type of behavior?

Sources: Adapted from Karen Abbot, "Teen Can't Get Fair Trial," *Denver Rocky Mountain News*, January 12, 2000, p. 4A; and "Florida Teen to Be Tried in Colorado," *ABCNews Online*, December 12, 1999, accessed on December 12, 1999.

provide security for e-mail transmissions have been developed. For example, Qvtech, a Colorado Springs company, is developing a program that will make e-mail self-destruct after a certain length of time.[20] To protect privacy, public-key infrastructure (PKI) can be used to help keep e-mails, such as those used to send medical and sensitive corporate information, private and secure.[21]

Electronic Software and Document Distribution

electronic software distribution

process that involves installing software on a file server for users to share by signing onto the network and requesting that the software be downloaded onto their computers over a network

Electronic software distribution involves installing software on a file server for users to share by signing onto the network and requesting that the software be downloaded onto their computers over a network. Electronic software distribution is quicker and more convenient than traditional ways of acquiring software and significantly less costly and more efficient than having a network administrator constantly install upgrades. Ingram Micro, for example, has developed digital technology to electronically send software, music, video, and other digital content across the Internet.[22] "We've spent close to two years making sure the technology works," says Guy Abramo, senior vice president of Ingram Micro. The technology also allows people to rent software or try it before they purchase it (try-before-you-buy). Companies are also developing services that will allow buyers to order software and have it delivered and installed, without ever leaving their computer. One problem with electronic software distribution is the size of software programs—downloading large programs requires high-capacity telecommunications media and takes a lot of time. Controlling software piracy is another issue that must be addressed.

electronic document distribution

process that involves transporting documents—such as sales reports, policy manuals, and advertising brochures—over communications lines and networks

Networks also allow organizations to transmit documents without using paper at all. It is not known how many millions of dollars companies spend on printing, distributing, and storing documents of all types, but the amount is staggering. **Electronic document distribution** involves transporting documents—such as sales reports, policy manuals, and advertising brochures—over communications lines and networks. Electronic document distribution software allows word processing and graphics documents to be converted into binary code and sent over networks. Acrobat from Adobe, for example, is a software package that allows documents to be transmitted between different types of computer system platforms. For example, a color advertising pamphlet can be created by an Apple G4 and sent electronically to an IBM personal computer. Electronic document distribution can save a substantial amount of time and money versus standard mailing and storage of hard-copy documents.

Telecommuting

telecommuting

a work arrangement whereby employees work away from the office using personal computers and networks to communicate via e-mail with other workers and to pick up and deliver results

More and more work is being done away from the traditional office setting. Many enterprises have adopted policies for **telecommuting** that enable employees to work away from the office using personal computers and networks. According to one study sponsored by AT&T, a company can save about $10,000 per year for each employee it allows to telecommute.[23]

There are several reasons why telecommuting is popular among workers. Single parents find that it helps balance family and work responsibilities because it eliminates the daily commute. It also enables qualified workers who may be unable to participate in the normal workforce (e.g., those who are physically challenged or who live in rural areas too far from the city office to commute on a regular basis) to become productive workers. Extensive use of telecommuting can lead to decreased need for office space, potentially saving a large company millions of dollars. Corporations are also being encouraged by public policy to try telecommuting as a means of reducing traffic congestion and air pollution.

Some types of jobs are better suited for telecommuting than others. These include jobs held by salespeople, secretaries, real estate agents, computer programmers, and legal assistants, to name a few. It also takes a special personality type to be effective while telecommuting. Telecommuters need to be strongly self-motivated, organized, able to stay on track with minimal supervision, and have a low need for social interaction. Jobs not good for telecommuting include those that require frequent face-to-face interaction, need much supervision, and have lots of short-term deadlines. Employees who choose to work at home must be able to work independently, manage their time well, and balance work and home life.

videoconferencing

a telecommunication system that combines video and phone call capabilities with data or document conferencing

Videoconferencing

Videoconferencing enables people to have a conference by combining voice, video, and audio transmission. Not only are travel expenses and time reduced, but managerial effectiveness is increased through faster response to problems, access to more people, and less duplication of effort by geographically dispersed sites.[24] Almost all **videoconferencing** (Figure 6.24) systems combine video and phone call capabilities with data or document conferencing. You can see the other person's face, view the same documents, and swap notes and drawings. With some of the systems, callers can make changes to live documents in real time. Many businesses find that the document and application sharing feature of the videoconference enhances group productivity and efficiency. Meeting over phone lines also fosters teamwork and can save corporate travel time and expense. Group videoconferencing is used daily in a variety of businesses as an easy way to connect work teams. Members of a team go to a specially prepared videoconference room equipped with sound-sensitive cameras that automatically focus on the person speaking, large TV-like monitors for viewing the participants at the remote location, and high-quality speakers and microphones. It costs around $60,000 to set up a typical group videoconferencing room. There are additional expenses associated with use of the telecommunications network to relay voice, video, data, and images.

Many companies have adopted policies for telecommuting that enable employees to work away from the office using personal computers and networks.
(Source: Stone/Ian Shaw.)

The Royal Bank of Scotland, for example, uses videoconferencing to save the time and hassle of traveling between Scotland and England for its managers and executives.[25] The system has been in place for more than 10 years and is constantly being updated and improved. Today, the videoconferencing system has 13 studios for general use, 6 facilities in directors' offices, and another 30 desktop videoconferencing devices. "We do about 16 or 17 videoconferences a day between the studios," says George Clark, senior manager of telecommunications.

Electronic Data Interchange

Electronic data interchange (EDI) is an intercompany, application-to-application communication of data in standard format, permitting the recipient to perform the functions of a standard business transaction, such as processing purchase orders. Connecting corporate computers among organizations is the idea behind EDI, which uses network systems and follows standards and procedures that allow output from one system to be processed directly as input to other systems, without human intervention. With EDI, the computers of customers, manufacturers, and suppliers can be linked (Figure 6.25). This technology eliminates the need for paper documents and substantially cuts down on costly errors. Customer orders and inquiries are transmitted from the customer's computer to the manufacturer's computer. The manufacturer's computer can then determine when new supplies are needed and can automatically place orders by connecting with the supplier's computer.

FIGURE 6.24

Videoconferencing

Videoconferencing allows participants to conduct long-distance meetings "face to face" while eliminating the need for costly travel.

(Source: Courtesy of Zydacron.)

electronic data interchange (EDI)

an intercompany, application-to-application communication of data in standard format, permitting the recipient to perform the functions of a standard business transaction

FIGURE 6.25

Two Approaches to Electronic Data Interchange

EDI is no longer just the wave of the future; many organizations now insist that their suppliers operate using EDI systems. Often the EDI connection is made directly between vendor and customer (a); alternatively, the link may be provided by a third-party clearinghouse, which provides data conversion and other services for the participants (b).

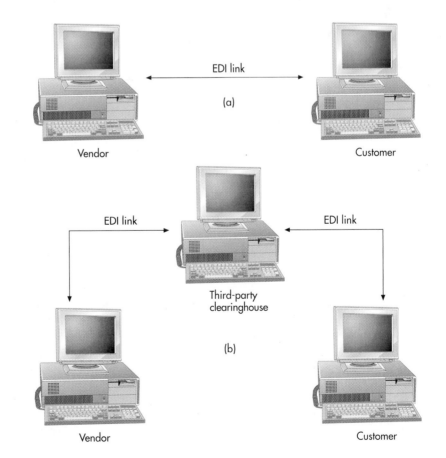

As discussed in this chapter, EDI allows companies to transact business between suppliers and customers. Although the Internet is an important e-commerce tool, EDI remains popular and useful. In fact, some companies require the use of EDI for all their suppliers.

Medoutlook.com, an on-line medical support company, helps coordinate communications between healthcare providers and their suppliers. According to Mark Kingston, CEO of Medoutlook.com, "When you look at healthcare, what's apparent is the lack of electronic streamlining of business services." Medoutlook.com is trying to solve this problem. Concentrating on medical suppliers, the company provides storefronts for products and services offered to the medical community. The system uses electronic data interchange as the primary technology, and it has wide appeal. Johnson & Johnson, for example, bills more than $500,000 monthly using EDI, but more than half of its business partners, mostly the medical community, have not been able or prepared to use the EDI system. Major manufacturing companies will be charged from $75,000 to $150,000 to set up a trading

network that uses EDI under the Medoutlook.com system. Each trading partner of the manufacturing company will pay only $50 to $300 per year. Buyers may also pay a low transaction fee for using the EDI system. "It's an interesting approach that should have some appeal," said Christopher Selland, an analyst at the Yankee Group.

Discussion Questions

1. Why is Medoutlook important for healthcare providers and suppliers?
2. Would Medoutlook be useful to allow patients to communicate with their healthcare providers?

Critical Thinking Questions

3. Why is a technology like EDI important to the health care field?
4. In what other industries could this approach be used?

Sources: Adapted from Molly Tschida, "E-fficiency," *Modern Physician,* February 1, 2000, p. 8; and Richard Karpinske, "Health Care Portal Launched," *Internet Week,* May 24, 1999, p. 18.

For some industries, EDI is a necessity. Read the "E-Commerce" box to see how Medoutlook uses EDI to set up medical storefronts. For many large companies, including General Motors and Dow Chemical, it is not uncommon for most computer input to originate as output from another computer system. Some companies will do business only with suppliers and vendors using compatible EDI systems, regardless of the expense or effort involved. For small companies that may not be able to afford their own EDI system, there are vendors that provide EDI services, such as St. Paul Software.[26] For a fee that can average $1.50 per transaction and a minimum of $50 per month, these companies allow smaller companies to provide products to larger corporations that require EDI capabilities. As more industries demand that businesses have this ability to stay competitive, EDI will cause massive changes in the work activities of companies. Companies will have to change the way they deal with processes as simple as billing and ordering, while new industries will emerge to help build the networks needed to support EDI.

Public Network Services

public network services

systems that give personal computer users access to vast databases and other services, usually for an initial fee plus usage fees

Public network services give personal computer users access to vast databases, the Internet, and other services, usually for an initial fee plus usage fees. Public network services allow customers to book airline reservations, check weather forecasts, get information on TV programs, analyze stock prices and investment information, communicate with others on the network, play games, and receive articles and government publications. Fees, based on the services used, can range from under $15 to over $500 per month (Figure 6.26). Providers of public network services include Microsoft, America Online, and Prodigy. These companies provide a vast array of services, including news, electronic mail, and investment information. AOL is the number one provider of public network services in terms of size.

FIGURE 6.26

Public network services provide users with the latest information required to remain competitive. AOL, for example, enables subscribers to obtain up-to-the-minute stock quotes.

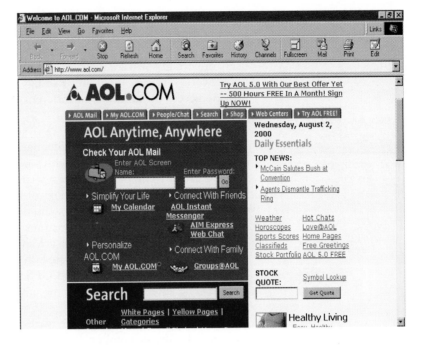

Distance Learning

Telecommunications can be used to extend the classroom. General Motors, for example, has developed an interactive system to reach its 175,000 employees at over 7,000 dealerships.[27] Before the system was installed, corporate training was a nightmare. The new system allows employees to view a live course without leaving their dealership. Travel costs are being dramatically cut with the new system, and the quality of the training is expected to greatly improve. Often called **distance learning** or cyberclass, these electronic classes are likely to thrive in the future.

distance learning

the use of telecommunications to extend the classroom

With distance learning software and systems, instructors can easily create course home pages on the Internet. Students can access the course syllabus and instructor notes on the Web page. E-mail mailing lists can be established so students and the instructor can easily mail one another as a means of turning in homework assignments or commenting and asking questions about material presented in the course. It is also possible to form chat groups so that students can work together as a "virtual team" that meets electronically to complete a group project.

Specialized Systems and Services

In addition to the telecommunications applications just discussed, there are a number of other specialized telecommunications systems and services. For example, with millions of personal computers in businesses across the country, interest in specialized and regional information services is increasing. Specialized services, which can be expensive, include professional legal, patent, and technical information. For example, investment companies can use systems such as Quotron and Shark to get up-to-the-minute information on stocks, bonds, and other investments.

Regional services, also called metropolitan services, include local electronic bulletin boards and electronic mail facilities that offer information regarding local club, school, and government activities. An electronic bulletin board is a message center that displays messages in electronic form, much like a bulletin board displays paper messages in schools and offices. An electronic bulletin board can be accessed by subscribers with personal computers, network equipment, and software. In addition to regional bulletin boards, national and international bulletin boards are available for people and groups with special interests or needs. These types of bulletin boards exist for many users, such as users of certain software packages and users with certain hobbies. Many public network services, including Prodigy, America Online, and CompuServe, provide access to hundreds of different bulletin boards on a variety of topics and interest areas.

Global positioning systems (GPSs) are other types of specialized telecommunications services. They have long been used by the military to find locations of troops, equipment, and the enemy—within yards in some cases. Today, GPS is being used by companies to survey land and buildings and by individuals to locate their positions while camping or exploring. The Garmin GPS III Plus, for example, is a nine-ounce system that displays your location on built-in maps.[28] The system can also determine speed and altitude. The Axiom Sports Tracker can be used on the slopes at Vail and Copper Mountain. After a day of skiing, you can download and print the runs you completed during the day. Some auto companies have placed GPS systems in their cars to assist travelers in need. The auto systems and some advanced GPSs combine basic GPS features with a cellular phone.

The use of pagers is also on the increase, with many features, including two-way paging. Airline companies have placed phone and Internet services on many of their aircraft, allowing people to stay in touch at 30,000 feet. With all these telecommunications systems and services, it is no wonder that managers and workers are able to conduct business from remote locations. Often called virtual workers, these managers and employees can conduct business at any time and at any place. Today, new telecommunications systems and services are being introduced every month. In the future, we can expect even more innovations in telecommunications that will dramatically alter how businesses and individuals stay connected and in touch.

● SUMMARY

PRINCIPLE ● Effective communications is essential to organizational success.

Communications is any process that permits information to pass from a sender to one or more receivers. Communications of all types form a major part of any business system. Telecommunications refers to the electronic transmission of signals for communications, including such means as telephone, radio, and television. Telecommunications has the potential to create profound changes in business because it lessens the barriers of time and distance.

The elements of a telecommunications system start with a sending unit, such as a person, a computer system, a terminal, or another device, that originates the message. The sending unit transmits a signal to a telecommunications device. The telecommunications device performs a number of functions, which can include converting the signal into a different form or from one type to another. A telecommunications device is a hardware component that allows electronic communication to occur or to occur more efficiently. The telecommunications device then

sends the signal through a medium. A telecommunications medium is anything that carries an electronic signal and interfaces between a sending device and a receiving device. The signal is received by another telecommunications device that is connected to the receiving computer. The process can then be reversed, and another message can go back from the receiving unit to the original sending unit. With synchronous communications, the receiver gets the message instantaneously, when it is sent. Voice and phone communications are examples. With asynchronous communications, the receiver gets the message hours or days after the message is sent.

PRINCIPLE ● An unmistakable trend of communications technology is that more people are able to send and receive all forms of information over greater distances at a faster rate.

The telecommunications media that physically connect data communications devices are twisted-pair wire cable, coaxial cable, and fiber-optic cable. Twisted-pair cable consists of pairs of twisted wires, either shielded or unshielded. A coaxial cable consists of an inner conductor wire surrounded by insulation (the dielectric) and a nonconductive insulating shield (the jacket). Fiber-optic cable consists of thousands of extremely thin glass or plastic strands bound together in a sheathing (a jacket) for transmitting signals via light. Fiber-optic cables are faster and more reliable than other types, but they are expensive to install. Microwave transmission consists of a high-frequency radio signal sent through the air. Other transmission options that give organizations portability and flexibility in transmitting data are cellular and infrared transmission.

· · ·

There are several types of telecommunications hardware devices. Four types of modems are internal modems, external modems, fax modems, and special-purpose modems. Two additional hardware devices include multiplexers and communications processors.

Modems convert signals from digital to analog for transmission, then back to digital. Modems can be either internal or external. A fax modem allows sending and receiving of faxes on a PC while working on other applications. Special-purpose modems include cellular modems placed in portable computers to allow communication with other computer systems and devices and cable modems designed for use with coaxial cable media. A multiplexer allows several signals to be transmitted over a single communications medium at the same time. A communications processor connects to a large number of communications lines and performs a number of tasks, including polling, providing automatic answering and calling, performing circuit checking and error detection, developing logs or reports of all communications traffic, editing basic data entering the main processor, determining message priority, choosing alternative and efficient communications paths over multiple data communications lines, and providing general data security for the main system.

· · ·

The effective use of networks can turn a company into an agile, powerful, and creative organization, giving it a long-term competitive advantage. Networks can be used to share hardware, programs, and databases across the organization. They can transmit and receive information to improve organizational effectiveness and efficiency. They enable geographically separated workgroups to share documents and opinions, which fosters teamwork, innovative ideas, and new business strategies.

· · ·

When an organization needs to use two or more computer systems, one of three basic data processing strategies may be followed: centralized, decentralized, or distributed. With centralized processing, all processing occurs in a single location or facility. This approach offers the highest degree of control. With decentralized processing, processing devices are placed at various remote locations. The individual computer systems are isolated and do not communicate with each other. With distributed processing, computers are placed at remote locations but connected to each other via telecommunications devices. Three distributed processing alternatives include terminal-to-host, file server, and client/server.

With terminal-to-host architecture, the application and database reside on the same host computer, and the user interacts with the application and data using a "dumb" terminal. Since a dumb terminal has no data processing capability, all computations, data accessing and formatting, as well as data display, are done by an application that runs on the host computer.

In the file server approach, the application and database reside on the same host computer, called the file server. The database management system runs on the end user's personal computer or workstation. If the user needs even a small subset of the data that resides on the file server, the file server sends the user the entire file that contains the data requested, including a lot of data the user does not want or need. The downloaded data can then be analyzed, manipulated, formatted, and displayed by a program that runs on the user's personal computer.

A client/server system is a network that connects a user's computer (a client) to one or more host computers (servers). A client is often a PC that requests services from the server, shares processing tasks with the server, and displays the results. Many companies have reduced their use of mainframe computers in favor of client/server systems using midrange or personal computers to achieve cost savings, provide more control over the desktop, increase flexibility, and become more responsive to business changes. The start-up costs of these systems can be high, and the systems are more complex than a centralized mainframe computer.

· · ·

Network topology refers to the manner in which devices on the network are physically arranged. Communications networks can be configured in numerous ways, but five designs are most prevalent: bus, hierarchical, star, ring, and hybrid (hybrid networks combine the basic designs of the four other topologies to suit the specific communication needs of an organization).

The physical distance between nodes on the network determines whether it is called a local area network (LAN), a wide area network (WAN), or an international network. The major components in a LAN are a network interface card, a file server, and a bridge and/or gateway. WANs tie together large geographic regions using microwave and satellite transmission or telephone lines. Value-added networks (VANs) are special WANs with additional services that permit more economical and faster communications. International networks involve communications between countries, linking systems together from around the world. These networks are also called global area networks. The electronic flow of data across international and global boundaries is often called transborder data flow.

PRINCIPLE
The effective use of telecommunications and networks can turn a company into an agile, powerful, and creative organization, giving it a long-term competitive advantage.

There are many applications of telecommunications, including the following: personal computer to mainframe links, voice and electronic mail, electronic document distribution, electronic software distribution, telecommuting, videoconferencing, electronic data interchange, public network services, specialized and regional information services, and distance learning. Personal computer to mainframe links enable people to upload and download data. Voice and electronic mail users can leave, receive, and store messages from other people around the world. Electronic document distribution allows organizations to transmit documents without the use of paper, thus cutting costs and saving time. Electronic software distribution involves installing software on a computer by sending programs over a network so they can be downloaded into individual computers. Telecommuting employs information technology to enable workers to work away from the office. Videoconferencing brings groups together in voice, video, and audio calls. Electronic data interchange (EDI), another rapidly growing area, enables customers, suppliers, and manufacturers to electronically exchange data. EDI reduces the need for manual paper systems while speeding up the rate at which business can be transacted. Public network services give users access to vast databases and services, usually for an initial fee plus usage fees. Distance learning is a way to support education of students who are unable to meet frequently with their instructor. Specialized services, which are more expensive, include legal, patent, and technical information. Regional services include local electronic bulletin boards that offer e-mail facilities and information regarding local activities.

● KEY TERMS

analog signal 218
asynchronous communications 212
bridge 234
bus network 225
centralized processing 224
client/server 230
common carriers 220
communications software 232
computer network 214
data communications 213
decentralized processing 224
dedicated line 221
digital signal 218
digital subscriber line (DSL) 222
distance learning 241
distributed processing 224
electronic data interchange (EDI) 239
electronic document distribution 237
electronic software distribution 237

file server 230
front-end processor 219
gateway 234
hierarchical network 225
hybrid network 225
integrated services digital network (ISDN) 223
international network 228
local area network (LAN) 225
modem 218
multiplexer 219
network management software 232
network operating system (NOS) 232
network topology 225
Open Systems Interconnection (OSI) model 232
private branch exchange (PBX) 221
protocols 232
public network services 240

ring network 225
router 234
star network 225
switch 235
switched line 220
synchronous communications 212
T1 carrier 223
telecommunications medium 214
telecommuting 238
terminal-to-host 229
Transmission Control Protocol/Internet Protocol (TCP/IP) 233
value-added carriers 220
videoconferencing 238
voice mail 236
wide area network (WAN) 227

● REVIEW QUESTIONS

1. Define the term *communications.* How does telecommunications differ from data communications?
2. Describe the steps involved in the communications process.
3. What is a telecommunications medium? List three media in common use.
4. What is the difference between synchronous and asynchronous communications?
5. Discuss the function and use of a modem and a multiplexer.
6. What is the difference between a switched and a dedicated line?
7. Define the term *computer network.*
8. Identify three distributed data processing alternatives.
9. What advantages and disadvantages are associated with the use of client/server computing?
10. What is a T1 line? How might it be used?
11. What is a network operating system? What is network management software?
12. Identify four basic types of network topologies.
13. What role do the bridge, router, gateway, and switch play in a network?
14. Describe a local area network and its various associated components.
15. What is a wide area network? What is a value-added network?
16. What is EDI? Why are companies using it?
17. What are the key elements of an effective distance learning course?

● DISCUSSION QUESTIONS

1. Why is effective communications critical to organizational success?
2. Discuss the human communications process of a message going from a sending unit to a receiving unit. What are some of the problems that may cause people to have a miscommunication?
3. What sort of issues would you expect to encounter in establishing an international network for a large, multinational company?
4. A company can use the Internet or EDI to connect suppliers to a company. Describe the advantages and disadvantages of each approach. Under what circumstances would you use EDI instead of the Internet?
5. Client/server computing should always be employed versus other forms of distributed data processing. Do you agree or disagree? Why?
6. If it were available, which would you rather have in your home—T1, ISDN, or DSL? Why?
7. What factors are limiting the competitiveness for local telephone service?
8. Consider an industry that you are familiar with through work experience, coursework, or a study of industry performance. How could electronic data interchange be used in this industry? What limitations would EDI have in this industry?
9. What is telecommuting? What are the advantages and disadvantages of telecommuting? Do you anticipate that you will telecommute in your future career?
10. Discuss the pros and cons of conducting this course as a distance learning course.

● PROBLEM-SOLVING EXERCISES

1. You have been hired as a telecommunications consultant for a small but growing consumer electronics manufacturer. The company wishes to develop a telecommunications system to link the company to its suppliers. There are many options, including EDI and the Internet. The company has hired you to review its needs for the new system and to select a telecommunications solution. How would you proceed? What questions need to be asked? Use word processing software to prepare a list of at least 10 questions that you need to answer to evaluate this project. Make some assumptions about the answers and write your opinion of this project. Embed a spreadsheet that details the approximate cost to set up the telecommunications connection between the company and its suppliers.

2. You work in the Los Angeles national headquarters of a large multinational company. After years of fighting the smog and traffic, you and your fellow workers have decided to do something about it. Your co-workers have elected you as spokesperson to develop a recommendation for management to implement a

telecommuting program. Use PowerPoint or similar software to make a convincing presentation to management to adopt such a program. Your presentation must address such questions as what the benefits are to the company, what the costs of the hardware and software are, how individuals will be selected for participation, and what benefits employees will receive.

● TEAM ACTIVITIES

1. With a group of your classmates, visit one of the following: a cellular phone company, the college computing center, a phone or cable company, the police department, or another interesting organization that uses telecommunications. Prepare a report on how the organization is planning to use communications to enhance access to information—both yours and its. Find out what kind of telecommunications media and devices it uses currently and what changes the organization might make to improve data and information access.

2. Your team should carefully analyze the different options to connect a PC to the Internet, including a standard phone line, DSL, cable, and satellite. Carefully summarize the costs, advantages, and disadvantages of these options. Investigate what options are available in your area. Discuss your findings with the class. How would you evaluate the competitiveness of the local service providers in your area?

● WEB EXERCISES

1. Digital subscriber lines (DSLs) and asynchronous digital subscriber lines (ADSLs) offer faster access to the Internet. Search the Internet to get additional information about this communications technology. Describe a few companies that manufacture this type of equipment and their products. You may be asked to develop a report or send an e-mail to your instructor about what you found.

2. There are a number of on-line job-search companies, including Monster.com. Investigate one or more of these companies and research the positions available in the telecommunications industry, including the Internet. You may be asked to summarize your findings for your class in a written or verbal report.

● CASES

 Increasing Speed, Reducing Costs

DSW Shoe Warehouse has 50 sales offices throughout the country. To streamline communications and increase profitability, the company decided to develop a network to connect all of its stores. Like other companies with a large number of remote locations, DSW Shoe Warehouse had a number of options, including an approach called frame-relay. Fred Brunel, the MIS director for DSW, carefully considered this approach. It would have guaranteed telecommunications speeds ranging from 56 Kbps to 64 Kbps. Although the speed was more than adequate, the cost was estimated to be from $25,000 to $30,000 per month. A less expensive dial-up frame-relay system was expected to cost $13,000 to $15,000 per month. Because of the high expense, Brunell decided to look at other options.

Using AT&T's telecommunications backbone, DSW was able to develop a network for all of its locations for less. According to Brunell, "It's very inexpensive, offers a great degree of flexibility, and provides a foundation for us to deploy new systems for our stores." The new AT&T backbone also provided excellent security from outside hackers and intruders. Using the AT&T telecommunications system to create a network, DSW was able to achieve speeds up to 56 Kbps. The big advantage was cost. Using the AT&T telecommunications system was expected to cost DSW only $1,200 to $1,500 per month. According to Fred Brunell, "What we're creating is a pseudo wide area network that's available around the clock for our stores." With the new system, a DSW employee is first authenticated against a list of authorized users. If the employee is approved, a fast speed modem connects him to the closest AT&T entry point on the network. A 3Com Corporation router is used to encrypt the information and transport it to the desired location. Another 3Com router is used at the receiving site to decrypt the information.

Discussion Questions

1. What alternatives did DSW consider in developing a network for its company?
2. What were some of the factors that led DSW to choose the AT&T backbone for its pseudo wide area network?

Critical Thinking Questions

3. Discuss the advantages and disadvantages of a wide area network versus the approach that DSW used.
4. Do you think that the speed of DSW's network could be a problem in the future?

Sources: Adapted from Bob Wallace, "Shoe Chain Likes Fit," *Computerworld*, May 3, 1999, p. 66; and "Shoe Chain on the March," *Retail Week*, January 14, 2000, p. 18.

 Using Networks to Increase Security

Crown Central is a large refinery with annual sales averaging $1.6 billion. The company has refinery, distribution, and retail facilities in the U.S. Southeast. The company has approximately 150 remote users that need to be tied into Crown Central's computer system. In the past, Crown used toll-free numbers to connect its remote users. The monthly cost ranged from about $3,000 to $4,000. To reduce costs and still provide needed communications, Crown Central decided to institute a formal bidding process to determine the best telecommunications solution. As a result of the bidding process, Crown decided to turn to GTE to provide basic telecommunications and network capabilities.

GTE Internetworking, a division of GTE, offers voice and data service in almost every city where there is a Crown Central office. With the new system, workers in refineries and distribution centers connect to the network by making a local call to a GTE remote access server. The server gives them a secure connection across the GTE network, the Internet, and the Crown network. Communications security is achieved through GTE's network operations center in Burlington, Massachusetts. The security system uses encryption and X.509 certificates.

Crown hopes to save at least 10 percent of its current costs. GTE Internetworking guarantees 99.9 percent availability for companies and users of its backbone network. This was an important consideration for Crown. According to Miguel Montanez, Crown's group manager of information systems, "It will achieve savings and also give us the opportunity to outsource our remote access in a way that is secure. GTE has the experience to provide that service, and Crown won't have to worry about growing experts in this very complex security technology." The GTE system also includes data encryption capabilities. Its reliability and encryption abilities make GTE one of the industry leaders in telecommunications.

Discussion Questions

1. What needs did Crown try to satisfy using its telecommunications approach?
2. Why did Crown turn to GTE for its telecommunications needs?

Critical Thinking Questions

3. What are the advantages to a company of using a common carrier, like GTE, compared with developing its own network?
4. In general, when should a company outsource its telecommunications system?

Sources: Adapted from Matt Hamblen, "Virtual Net Increases Security," *Computerworld,* January 14, 1999, p. 37; and "Apex Oil Seeks to Acquire 35% Stake in Crown Central Petroleum," *Business and Industry,* January 2000, p. 48.

---

2 Mirage on the Net

When it comes to building a top-notch network, Mirage decided not to gamble, especially with its newest facility, the Bellagio. This new 3,000-room luxury hotel and casino has original art, a spacious decor, and a complex water fountain that moves to music, ranging from classic to rock. "When the president of Bellagio looks me in the eye and says, 'Glenn, is this going to work?' I have to be able to guarantee success," says Glenn Bonner, chief information officer at Mirage Resorts. "This industry is littered with careers that chose new systems for opening day."

Bonner's conservative approach to developing a telecommunications system for Mirage appears to have succeeded. First, Bonner developed a regional area network, also called a metropolitan-area network, to tie the four casinos in Las Vegas together. The network uses 3Com CoreBuilder 7000 backbone switches in the four hotels and the data center in Las Vegas. The links between the hotels and the data center use a 155-Mbps mesh. "The mesh provides the redundancy and resiliency we needed," Bonner said.

According to some, however, the jewel in the crown for Mirage is the local area network (LAN) at the Bellagio. The ethernet LAN consists of servers

and high-speed switches, which can provide a speed of 10 Mbps. According to Bonner, "Networking is not an exact science; it's an art. So you go and architect for scalability and capacity, and when you need it, you turn it on." Because a casino never closes, it is hard to shut down the network for maintenance or updates. Thus, it is critical to build the best network possible at the start. And because of the nature of the business, the network must be secure.

Discussion Questions

1. Briefly describe the network system used by Mirage.
2. What are the advantages and disadvantages of this approach?

Critical Thinking Questions

3. What special network issues would a casino company have to consider versus a more traditional company?
4. What other types of computer technology would be useful for a casino company?

Sources: Adapted from Jon Fontana, "Mirage Makes Broadband Bet," *Computerworld,* January 25, 1999, p. 23; and Richard Siklos, "Weaving Yet Another Web for Women," *Business Week,* January 17, 2000, p. 101.

NOTES

Sources for the opening vignette on page 211: Adapted from Dean Foust, "Doug Daft Isn't Sugarcoating Things," *Business Week,* February 7, 2000, p. 36; and Rutrell Yasin, "Coke Unbottles Web Potential," *Internet Week,* March 22, 1999, p. 9.

1. Jennifer Reingold et al., "Why the Productivity Revolution Will Spread," *Business Week,* February 14, 2000, p. 112.

2. Steve Rosen, "Charge of the Light Brigade," *Business Week,* January 31, 2000, p. 62.

3. David Welch et al., "Now You Can Cruise and Surf," *Business Week,* January 24, 2000, p. 54.

4. Peter Speigel, "Dishing Out Data," *Forbes,* January 24, 2000, p. 110.

5. Steve Kichen, "Cable Guys," *Forbes,* May 3, 1999, p. 230.

6. Leslie Cauley, "Even US West's Chairman Can't Get DSL," *The Wall Street Journal,* February 17, 2000, p. B1.

7. Eric Brown, "Broadband Blues," *PC World,* February 2000, p. 57.

8. Matthew Hamblen, "Digital Subscriber Line," *Computerworld,* February 7, 2000, p. 69.

9. Kate Murphy, "Cruising the New-In Hyperdrive," *Business Week,* January 24, 2000, p. 170.

10. Scott Wooley, "Kiss That Duopoly Good-Bye," *Forbes,* February 21, 2000, p. 135.

11. Young Myung et al., "Design of Communication Networks with Survivability Constraints," *Management Science,* February 1999, p. 238.

12. Stephen Wildstrom, "The Network Comes Home," *Business Week,* January 31, 2000, p. 28.

13. UPS Web site at http://www.upsnet.com, accessed February 18, 2000.

14. Neil Rubenking, "Network Your Home Painlessly," *PC Magazine,* April 4, 2000, p. 108.

15. Robin Robinson, "Thin-Client Apps Give Credit Where Credit's Due," *Computerworld,* January 17, 2000, p. 64.

16. Lee Copeland, "TCP/IP," *Computerworld,* January 17, 2000, p. 72.

17. Paul Korzeniowski, "Gigabit Ethernet Gains Momentum," *Internet Week,* May 10, 1999, p. 49.

18. James Cope, "Doing It with Mirrors: Lucent to Unveil All-Optical Router," *Computerworld,* November 22, 1999, p. 59.

19. Oliver Rist, "Eudora Now Free (If You Can Stand the Ads)," *PC World,* March 2000, p. 90.

20. Neil Gross, "E-mail That Won't Come Back to Haunt You," *Business Week,* November 15, 1999, p. 136.

21. Roberta Fusary, "Digital Protection," *Computerworld,* March 1, 1999, p. 14.

22. Karen Franse, "Getting Software from Here to There," *VAR Business,* December 6, 1999, p. 49.

23. Sonya Donaldson, "Teleworker," *Home Office Computing,* February 2000, p. 101.

24. Devin Burden, "Companies Now Find Videoconferencing a More Viable Alternative to Meeting in Person," *Computerworld,* January 11, 1999, p. 91.

25. "Videoconferencing," *Management Consultancy,* February 4, 2000, p. 16.

26. Richard Karpinski, "EDI Developer Extends E-Trading Boundaries," *Internet Week,* May 3, 1999, p. 10.

27. "Log on for Company Training," *Business Week,* January 10, 2000, p. 140.

28. Gordon Bass, "On a Road to Nowhere?" *PC Computing,* December 1999, p. 308.

CHAPTER 7

The Internet, Intranets, and Extranets

The Internet changes everything.

— Larry Ellison, CEO of
Oracle Corporation

Principles	Learning Objectives
The Internet is like many other new technologies—it provides a wide range of services, some of which are effective and practical for use today, others of which are still evolving, and still others of which will fade away from lack of use.	• *Briefly describe how the Internet works, including alternatives for connecting to it and the role of Internet service providers.* • *Identify and briefly describe the services associated with the Internet.* • *Describe the World Wide Web, including tools to view and search the Web.*
Before the Internet becomes universally used and accepted for business use, management issues, service bottlenecks, and privacy and security issues must be addressed and solved.	• *Identify who is using the Web to conduct business and discuss some of the pros and cons of Web shopping.* • *Outline a process for creating Web content.* • *Describe Java and discuss its potential impact on the software world.* • *Define the terms* intranet *and* extranet *and discuss how organizations are using them.* • *Identify several issues associated with the use of networks.*

FedEx

Sparring with UPS on the Internet

As more companies are setting up storefronts on the Internet, there is a bigger need for them to find reliable and cost-effective ways to ship their products to consumers. There are a number of shipping companies to choose from, but UPS and FedEx are the largest. UPS ships about 12 million packages daily. Its annual revenues are estimated to be about $26 billion. Of this total, $14.4 billion is for U.S. ground shipments, $5.2 billion is for U.S. overnight deliveries, $2.7 billion is for U.S. two-day deliveries, and $3.6 billion is for international deliveries. UPS currently has about 326,000 employees. FedEx, on the other hand, has annual revenues of about $16.5 billion. Unlike UPS, however, FedEx's largest revenue component is U.S. overnight shipments, worth $7.4 billion annually. International shipments are the next largest contributor to total revenues, worth $3.8 billion. U.S. ground shipments represent $3 billion, and U.S. two-day air represents $2.3 billion of FedEx's total revenues. FedEx has about 141,000 employees. Clearly, UPS is doing more with ground transportation. The company has about 150,000 trucks, compared with 43,500 trucks for FedEx. The two companies are similar in terms of their air shipments. FedEx has a fleet of 637 planes, which is slightly higher than the 610-plane fleet for UPS. The impressive size and success of these two companies can be directly traced to their Internet strategies.

FedEx was one of the first shipping companies to totally integrate use of the Internet into its operations. According to Fred Smith, CEO of FedEx, "The information about the package is almost as important as the package itself." This total information approach was implemented using the convenience and power of the Internet. Being one of the first shipping companies to install elaborate scanning and tracking equipment, FedEx was able to tell its customers the location of their packages and shipments from the moment the deliveries left the customer's driveway. FedEx customers started to rely on FedEx when the package "absolutely, positively" had to be there on time. More recently, FedEx decided to invest an additional $100 million in technology and equipment that will allow it to deliver packages faster and with greater control.

Although UPS was late to implement sophisticated scanning and tracking systems, the company has now listed technology and the Internet as one of its primary strategic initiatives. During the last several years, UPS has invested about $1 million each year on its Internet site and technology in general. In 1999, UPS did about $20 million of business from its Internet site. This amount is expected to increase to $180,000 million within five years. This investment has allowed UPS to compete with FedEx on reliable, overnight deliveries. Both FedEx and UPS are integrating their Web sites with back-end computer systems and operations. In addition, both companies are looking to customize their shipping operations. Corporations such as Nike, Hewlett-Packard, and Cisco are using FedEx and UPS to streamline their shipping operations.

Even with huge investments in the Internet, these two companies are still locked in fierce competition to be the best. According to James Tompkins, head of a shipping and logistics consulting company in Raleigh, "There's no difference." Clad in purple uniforms, FedEx employees are in a close race with the coffee-colored-uniformed UPS employees. Fortunately for both companies, the Internet has dramatically expanded the need for shipping. While these two companies are fighting it out to get a larger slice of the shipping pie, the pie's size is ever increasing.

As you read this chapter, consider the following:

- What impact has the Internet had on FedEx and UPS?

- In the next five years, what impact will the Internet have on businesses and the economy in general?

To speed communications and share information, businesses are linking employees, branch offices, and global operations via networks, whether they set up their own or use outside services. As seen with FedEx and UPS in the opening vignette, companies are using the Internet not only to speed communications but to compete for market share and profits. Everyone is talking about the Internet, and companies are increasingly using it for competitive advantage, but what exactly is it? The Internet is the world's largest computer network. Actually, the Internet is a collection of interconnected networks, all freely exchanging information (see Figure 7.1). Research firms, colleges, and universities have long been part of the Internet, and now businesses, high schools, elementary schools, and other organizations are joining up as well. Nobody knows exactly how big the Internet is because it is a collection of separately run smaller computer networks with no single place where all the connections are registered.

USE AND FUNCTIONING OF THE INTERNET

The Internet is truly international in scope, with users on every continent—including Antarctica. However, the United States has the most usage, by far. Although the United States still claims more Web activity than other countries, the Internet is expanding around the globe.[1] In Japan, for example, approximately 18 million people use the Internet.[2] In a few short years, the numbers are expected to swell to over 60 million Internet users. In Brazil, Latin America's biggest Internet market, a number of Internet companies are about to launch free Internet service.[3] These companies hope to make a profit from sales transactions and advertising fees. If successful, free Web access may become more common worldwide.

China is also getting into the Internet to a greater extent with the help of American companies like Dell Computer Corporation and International Data Group.[4] International Data Group has already invested approximately $100 million into China's Internet infrastructure. The use of the Internet in Europe lags that in the United States by about two years, and Canada is moving slower in its use of the Internet because of high communications costs and strict government control over its growth. In Africa, Internet connectivity is limited in every country

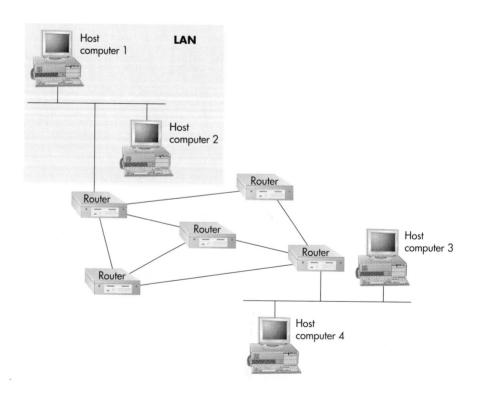

FIGURE 7.1

Routing Messages over the Internet

except South Africa. But even there, its use is very limited, since even a slow 14.4-Kbps modem costs more than a month's salary for most people. In Russia, using the Internet's e-mail capabilities provides a timely mail service, whereas it may take weeks for an airmail letter to reach the United States. Although technically the Internet is global, outside the United States it is seen as overwhelmingly U.S.-centric, inundated with English-language sites and U.S.-generated content.

Use of the Internet is not a male-dominated activity, as some still believe. One study predicted that women would outnumber men in the use of the Internet in the near future.[5] "Women seem well on their way to outnumbering men in a medium that until recently was almost exclusively male," said Charles Buchwaler, a vice president for AdRelevance, a company that studies Internet trends.

ARPANET

project started by the U.S. Department of Defense (DOD) in 1969 as both an experiment in reliable networking and a means to link DOD and military research contractors, including a large number of universities doing military-funded research

The ancestor of the Internet was the **ARPANET**, a project started by the U.S. Department of Defense (DOD) in 1969. The ARPANET was both an experiment in reliable networking and a means to link DOD and military research contractors, including a large number of universities doing military-funded research. (*ARPA* stands for the Advanced Research Projects Agency, the branch of the DOD in charge of awarding grant money. The agency is now known as DARPA—the added *D* is for *Defense*.)

The ARPANET was highly successful, and every university in the country wanted to sign up. This wildfire growth made it difficult to manage the ARPANET, particularly the large and rapidly growing number of university sites on it. It was decided to break the ARPANET into two networks: MILNET, which included all military sites, and a new, smaller ARPANET, which included all the nonmilitary sites. The two networks remained connected, however, through use of the **Internet protocol (IP)**, which enabled traffic to be routed from one network to another as needed. All the networks connected to the Internet speak IP, so they all can exchange messages.

Internet protocol (IP)

communication standard that enables traffic to be routed from one network to another as needed

Unlike a corporate network with a centralized infrastructure, the Internet is nothing more than an ad hoc linkage of many networks that adhere to basic standards. Since these networks are constantly changing and being improved, the Internet itself is in a perpetual state of evolution. However, since the Internet is such a loose collection of networks, there is nothing to prevent some participants from using outdated or slow equipment.

To speed Internet access, a group of corporations and universities, called the University Corporation for Advanced Internet Development (UCAID), is working on a faster, new Internet.[6] Called Internet2 (I2), Next Generation Internet (NGI), and Abilene, depending on the universities or corporations involved, the new Internet offers the potential of faster Internet speeds, up to 2 Gbits per second or more. Although not related to the efforts of UCAID, Project Oxygen could also improve the speed of the Internet. The project is a global fiber-optic, undersea network that will connect almost 80 countries. The first phase of the project should be completed by 2003.

A number of companies are involved in developing NGI and systems to take advantage of the promise of faster speeds. Qwest Communications International received a $50 million Department of Energy contract to supply very fast links between national laboratories and various research facilities.[7] The new links have the potential to be 500 times faster than what is available today. Qwest operates the Abilene backbone. FedEx and other companies are starting to build global systems to take advantage of I2 when it arrives.[8] The initial system should connect 100,000 FedEx employees in 90 countries.

How the Internet Works

The Internet transmits data from one computer (called a host) to another (see Figure 7.1). If the receiving computer is on a network to which the first computer is directly connected, it can send the message directly. If the receiving computer is not on a network to which the sending computer is connected, the sending computer relays the message to another computer that can forward it. The message may be sent through a router (see Chapter 6) to reach the forwarding computer. The forwarding host, which presumably is attached to at least one other network, in turn delivers the message directly if it can or passes it to yet another forwarding host. It is quite common for a message to pass through a dozen or more forwarders on its way from one part of the Internet to another.

The various networks that are linked to form the Internet work pretty much the same way—they pass data around in chunks called *packets,* each of which carries the addresses of its sender and its receiver. The set of conventions used to pass packets from one host to another is known as the Internet protocol (IP), which operates at the network layer of the seven-layer OSI model discussed in Chapter 6. Many other protocols are used in connection with IP. The best known is the **transport control protocol (TCP),** which operates at the transport layer. TCP is so widely used as the transport layer protocol that many people refer to TCP/IP, the combination of TCP and IP used by most Internet applications. Adhering to the same technical standards allows the more than 100,000 individual computer networks owned by governments, universities, nonprofit groups, and companies to constitute the Internet. Once a network following these standards links to a **backbone**—one of the Internet's high-speed, long-distance communications links—it becomes part of the worldwide Internet community.

Each computer on the Internet has an assigned address called its **uniform resource locator,** or **URL,** to identify it from other hosts. The URL gives those who provide information over the Internet a standard way to designate where Internet elements such as servers, documents, newsgroups, and so on can be found. Let's look at the URL for Course Technology, http://www.course.com/home.cfm.

The "http" specifies the access method and tells your software to access this particular file using the HyperText Transfer Protocol. This is the primary method for interacting with the Internet. The "www" part of the address signifies that the address is associated with the World Wide Web service (discussed later). The "course.com/home.cfm" part of the address is the domain name that identifies the Internet host site, and domain names must adhere to strict rules. They always have at least two parts separated by dots (periods). For all countries except the United States, the rightmost part of the domain name is the country code (au for Australia, ca for Canada, dk for Denmark, fr for France, jp for Japan, etc.). Within the United States, the country code is replaced with a code denoting affiliation categories (see Table 7.1). The leftmost part of the domain name identifies the host network or host provider, which might be the name of a university or business.

Herndon, Virginia–based Network Solutions Inc. (NSI) was once the sole company in the world with the direct power to register addresses using .com, .net, or .org domain names. But this government contract ended in October 1998, as part of the U.S. government's move to turn management of the Web's address system over to the private sector.[9] Today, 20 companies, called *registrars,* can register domain names, and an additional 33 companies are accredited to register domain names. Another 30 companies are seeking accreditation to register domain names from the Internet Corporation for Assigned Names and Numbers (ICANN).[10] Some registrars are concentrating on large corporations, where the profit margins may be higher, compared with small businesses or individuals.

It has been estimated that there are over 270,000 registered domain names.[11] Some people, called cybersquatters, have registered domain names in the hope of selling the names to corporations at a later date. The domain name Business.com, for example, sold for $7.5 million. But some companies are

transport control protocol (TCP)

widely used transport layer protocol that is used in combination with IP by most Internet applications

backbone

one of the Internet's high-speed, long-distance communications links

uniform resource locator (URL)

an assigned address on the Internet for each computer

TABLE 7.1

U.S. Top-Level Domain Affiliations

Affiliation ID	Affiliation
arts	cultural and entertainment activities
com	business organizations
edu	educational sites
firm	businesses and firms
gov	government sites
info	information service providers
mil	military sites
nom	individuals
net	networking organizations
org	organizations
rec	recreational activities
store	businesses offering goods for purchase
web	entities related to World Wide Web activities

fighting back. Ford Motor Company, for example, sued a person who tried to sell the domain names Ford-quality.com and Lincoln-quality.com on an Internet auction site.[12]

Accessing the Internet

There are three ways to connect to the Internet (Figure 7.2). Which method is chosen is determined by the size and capability of the organization or individual.

Connect via LAN Server

This approach requires the user to install on his or her PC a network adapter card and Open Datalink Interface (ODI) or Network Driver Interface Specification

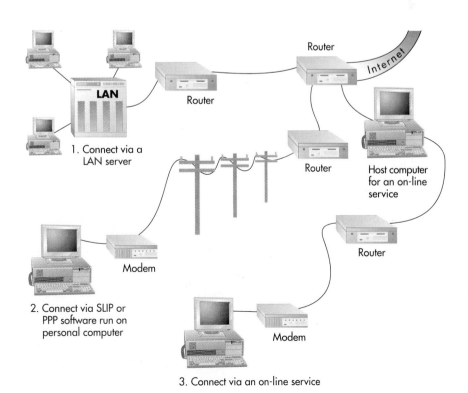

FIGURE 7.2

Three Ways to Access the Internet

There are three ways to access the Internet—using a LAN server, dialing into the Internet using SLIP or PPP, or using an on-line service with Internet access.

(NDIS) packet drivers. These drivers allow multiple transport protocols to run on one network card simultaneously. LAN servers are typically connected to the Internet at 56 Kbps or faster. Such speed makes for an exciting trip on the Internet but is also very expensive—$2,000 or so a month! However, the cost of this connection can be shared among several dozen LAN users to get a reasonable cost per user. Additional costs associated with a LAN connection to the Internet include the cost of the software mentioned at the beginning of this section.

Connect via SLIP/PPP

This approach requires a modem and the TCP/IP protocol software plus **serial line internet protocol (SLIP)** or **point-to-point protocol (PPP)** software. SLIP and PPP are two communications protocols that transmit packets over telephone lines, allowing dial-up access to the Internet. If you are running Windows, you will also need Winsock. Users must also have an Internet service provider that lets them dial into a SLIP/PPP server. SLIP/PPP accounts can be purchased for $30 a month or less from regional providers. With all this in place, a modem is used to call into the SLIP/PPP server. Once the connection is made, you are on the Internet and can access any of its resources. The costs include the cost of the modem and software, plus the service provider's charges for access to the SLIP/PPP server. The speed of this Internet connection is limited to the power of your computer's modem and the speed of the modem of the SLIP/PPP server to which you connect.

Connect via an On-Line Service

This approach requires nothing more than what is required to connect to any of the on-line information services—a modem, standard communications software, and an on-line information service account. There is normally a fixed monthly cost for basic services, including e-mail. The on-line information providers offer a wide range of services, including e-mail and access to the World Wide Web. America Online, Microsoft Network, and Prodigy are examples of such services.

Internet Service Providers

An **Internet service provider (ISP)** is any company that provides individuals and organizations with access to the Internet. ISPs do not offer the extended informational services offered by commercial on-line services such as America Online and Prodigy. There are literally thousands of Internet service providers, ranging from universities making unused communications line capacity available to students and faculty to major communications giants such as AT&T and MCI. To use this type of connection, you must have an account with the service provider and software that allows a direct link via TCP/IP.

In choosing an Internet service provider, the important criteria are cost, reliability, security, the availability of enhanced features, and the service provider's general reputation. Reliability is critical because if your connection to the ISP fails, it interrupts your communications with customers and suppliers. Among the value-added services ISPs

serial line Internet protocol (SLIP)

a communications protocol that transmits packets over telephone lines

point-to-point protocol (PPP)

a communications protocol that transmits packets over telephone lines

Internet service provider (ISP)

any company that provides individuals and organizations with access to the Internet

To use an ISP like AT&T, you must have an account with the service provider and software that allows a direct link via TCP/IP.

Internet Service Provider	Web Address
AT&T's WorldNet Service	www.att.com
BellSouth	www.bellsouth.com
Digex, Inc.	www.digex.net
Earthlink/Sprint	www.earthlink.net
GTE Internetworking	www.gte.net
IBM Internet Connection	www.ibm.net
WorldCom	www.wcom.com
Sprintlink	www.sprint.net
UUnet Technologies	www.uu.net

provide are electronic commerce, networks to connect employees, networks to connect with business partners, host computers to establish your own Web site, Web transaction processing, network security and administration, and integration services. Many corporate IS managers welcome the chance to turn to ISPs for this wide range of services because they do not have the in-house expertise and cannot afford the time to develop such services from scratch. In addition, when organizations go with an ISP-hosted network, they can also tap the ISP's national infrastructure at minimum cost. That's important when a company has offices spread across the country.

In most cases, ISPs charge a monthly fee that can range from $15 to $30 for unlimited Internet connection. The fee normally includes use of e-mail. Some ISPs, however, are experimenting with no-fee Internet access, as mentioned earlier.[13] But there are strings attached to the no-fee offers in most cases. Some free ISPs require that customers provide detailed demographic and personal information. In other cases, customers must put up with extra advertising banners on every Web site. Free net ISPs include NetZero, with approximately three million subscribers, and AltaVista. AltaVista's free net service obtained about 1.5 million customers in its first five months of operation. Table 7.2 identifies the major corporate Internet service providers.[14]

An increasing number of ISPs are offering Internet connection via satellite. Hughes Network Systems began its DirectPC service in October 1996 as a means to provide a high-speed connection to the Internet without the installation delays and cost of land lines. The cost of the service varies greatly depending on the type of service the customer chooses. For unlimited access at rates up to 200 Kbps from 6 P.M. to 6 A.M. on weekdays and unlimited access on weekends and holidays, you pay $19.95 per month. Other options offered by some ISPs include DSL, ISDN, and cable modems.[15] See Chapter 6 for a description of these connections.

INTERNET SERVICES

The types of Internet services available are vast and ever expanding. Some commonly used Internet services include e-mail; telnet; FTP; Usenet and newsgroups; chat rooms; Internet phone service; Internet video conferencing; content streaming; instant messaging; shopping on the Web; Web auctions; music, radio, and video on the Web; Office on the Web; three-dimensional Internet sites; and a vast array of free software and services. Other commonly used services include connecting to other computers, finding and downloading files, and accessing different kinds of servers for information. These services are discussed next and summarized in Table 7.3.

Service	Description
E-mail	Enables you to send text, binary files, sound, and images to others
Telnet	Enables you to log on to another computer and access its public files; users can log on to a work computer from an offsite location
FTP	Enables you to copy a file from another computer to your computer
Usenet and newsgroups	Focuses on a particular topic in an on-line discussion group format
Chat rooms	Enables two or more people to carry on on-line text conversations in real time
Internet phone	Enables you to communicate with other Internet users around the world who have equipment and software compatible to yours
Internet videoconferencing	Supports simultaneous voice and visual communications
Content streaming	Enables you to transfer multimedia files over the Internet so that the data stream of voice and pictures plays more or less continuously
Instant Messaging	Allows two or more people to communicate instantly on the Internet
Shopping on the Web	Allows people to purchase products and services on the Internet
Web auctions	Lets people bid on products and services
Music, radio, and video	Lets users play or download music, radio, and video
Office on the Web	Allows people to have access to important files and information through a Web site
Internet sites in 3-D	Allow people to view products and images at different angles in what appears to be three dimensions
Free software and services	Allows people to obtain a wealth of free software, advice, and information on the Internet; unwanted advertising and false information are potential drawbacks

TABLE 7.3

Summary of Internet Services

E-Mail

Electronic mail, or e-mail, has been used in business networks for years, but with the spread of Internet use, it is now commonly used for national and international communications. E-mail is no longer limited to simple text messages. Depending on your hardware and software and the hardware and software of the recipient, you can embed sound and images in your message and attach files that contain text documents, spreadsheets, graphs, or executable programs. E-mail travels through the systems and networks that make up the Internet. Gateways can receive e-mail messages from the Internet and deliver them to users on other networks.

Many of these networks have agreements with the Internet and with each other to exchange e-mail, just as countries exchange regular mail across their borders. Similarly, an e-mail message may pass through a series of intermediate networks to reach the destination address. Since not all networks use the same e-mail format, a gateway translates the format of the e-mail message into one that the next network can understand. Each gateway reads the "To" line of the e-mail message and routes the message closer to the destination mailbox. Thus, you can send e-mail messages to literally anyone in the world if you know that person's e-mail address and if you have access to the Internet or another system that can send e-mail.

E-mail has changed the way people communicate. It improves the efficiency of communications by reducing interruptions from the telephone and unscheduled personal contacts. Furthermore, messages can be distributed to multiple recipients easily and quickly without the inconvenience and delay of scheduling meetings. Because past messages can be saved, they can be reviewed, if necessary. And because messages are received at a time convenient to the recipient, the recipient has time to respond more clearly and to the point. Opinions and feedback from remote experts and other interested people are easy to obtain, thus improving the quality of decisions and the probability of acceptance. For large organizations whose operations span a country or the world, e-mail allows people to work around the time zone changes. Some users of e-mail estimate that they eliminate two hours of verbal communications for every hour of e-mail use. But the person at the other end still must check the mailbox to receive messages.

Telnet and FTP

Telnet

a terminal emulation protocol that enables users to log on to other computers on the Internet to gain access to public files

file transfer protocol (FTP)

a protocol that describes a file transfer process between a host and a remote computer and allows users to copy files from one computer to another

Telnet is a terminal emulation protocol that enables you to log on to other computers on the Internet to gain access to their publicly available files. Telnet is particularly useful for perusing library card files and large databases. It is also called *remote logon*.

File transfer protocol (FTP) is a protocol that describes a file transfer process between a host and a remote computer. Using FTP, you can copy a file from another computer to your computer. FTP is often used to gain access to a wealth of free software on the Internet.

Usenet and Newsgroups

Usenet

a system closely allied with the Internet that uses e-mail to provide a centralized news service; a protocol that describes how groups of messages can be stored on and sent between computers

Usenet is a system closely allied with the Internet that uses e-mail to provide a centralized news service. It is actually a protocol that describes how groups of messages can be stored on and sent between computers. Following the Usenet protocol, e-mail messages are sent to a host computer that acts as a Usenet server. This server gathers information about a single topic into a central place for messages. A user sends e-mail to the server, which stores the messages. The user can then log on to the server to read these messages or have software on the computer log on and automatically download the latest messages to be read at leisure. Thus, Usenet forms a virtual forum for the electronic community, and this forum is divided into newsgroups.

newsgroups

on-line discussion groups that focus on specific topics

Newsgroups are what make up Usenet, a worldwide discussion system classified by subject. Articles and messages are posted to newsgroups using newsreader software and are then broadcast to other interconnected computer systems via a wide variety of networks. A newsgroup is essentially an on-line discussion group that focuses on a particular topic. Newsgroups are organized into various hierarchies by general topic, and within each topic there can be many subtopics. On the Internet, there are tens of thousands of newsgroups, covering topics from astrology to zoology (see Table 7.4). Discussions take place via e-mail, which is sent to the newsgroup's address. A newsgroup may be moderated or unmoderated. If a newsgroup is moderated, e-mail is automatically routed to the moderator, a person who screens all incoming e-mail to make sure it is appropriate before posting it to the newsgroup. Some people who are frequent newsgroup users invest in a newsgroup reader that makes reading and posting messages easier.[16]

TABLE 7.4

Selected Usenet Newsgroups

alt.airline	alt.college
alt.aol	alt.current-events
alt.art	alt.fan.leonardo-dicaprio
alt.books	alt.history
alt.fan	alt.elvis
alt.sports.baseball.cinci-red	alt.music
alt.sports.basketball.nba.la-lakers	alt.politics
alt.sports.college.sec	alt.hackers
alt.sports.clothing	gov.us.fed.congress.record.senate

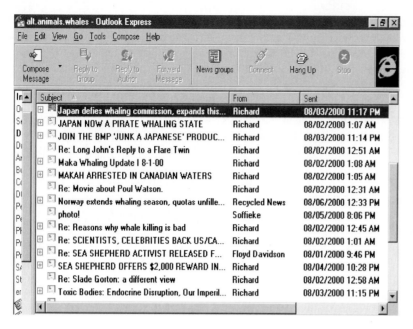

The participants of this newsgroup focus their discussion on the whaling industry.

Newsgroup servers around the world host newsgroups that share information and commentary on predefined topics. Each group takes the form of a large bulletin board where members post and reply to messages, creating what is called a message thread. The open nature of newsgroups encourages participation, but the discussions often become rambling and unfocused. As a result of so much active participation, newsgroups can evolve into tight communities where certain members tend to dominate the discussions.

Here are some tips to consider when accessing newsgroups. When you join a newsgroup, first check its list of Frequently Asked Questions, or FAQs (pronounced "facks"), before submitting any questions to the newsgroup. The FAQ list will have answers to common questions the group receives. It is considered impolite to waste the group's time by asking common questions when FAQs are available. Most new users just read messages without responding at first. Many newsgroups include members from around the world, and in the interest of courtesy, you should pick up some sense of the audience and its culture before jumping in with questions and opinions. It is impolite to jump in on the middle of a conversation (also called a thread). You may raise points and issues long since discussed and abandoned. Be concerned about what you say and the feelings of others. Remember, a person is receiving your messages. Do not use extreme words or repeat rumors (you could risk libel or defamation lawsuits). Do not post copyrighted material, and be careful how you use copyrighted material downloaded to your computer. Protect yourself by not offering personal information such as home address, employer, or phone number. Remember that this global on-line community has fragmented into thousands of different groups for a reason—to maintain the focus of each conference. Respect the specific subject matter of the group.

Chat Rooms

chat room

a facility that enables two or more people to engage in interactive "conversations" over the Internet

A **chat room** is a facility that enables two or more people to engage in interactive "conversations" over the Internet. Indeed, when you participate in a chat room there may be dozens of participants from around the world. Multiperson chats are usually organized around specific topics, and participants often adopt nicknames to maintain anonymity. One form of chat room, Internet Relay Chat (IRC), requires participants to type their conversation rather than speak. Voice chat is also an option, but you must have a microphone, sound card and speakers, a fast modem, and voice-chat software that's compatible with that of the other participants.

Internet Phone and Videoconferencing Services

Internet phone service enables you to communicate with other Internet users around the world who have equipment and software compatible to yours.[17] This service is relatively inexpensive and can make sense for international calls. With some services, like Net2Phone, it is now possible for someone using the Internet to make a call to someone using a standard phone. The cost to make a call can be as low as 1 cent a minute for calls in the United States. Low rates are available for calling outside the United States. Voice mail and fax capabilities are also available.

Using **voice-over-IP (VOIP)** technology, network managers can route phone calls and fax transmissions over the same network they use for data—which means no more phone bills. Gateways installed at both ends of the communications link convert voice to IP packets and back. With the advent of widespread, low-cost Internet telephony services, traditional long-distance providers are being pushed to either respond in kind or trim their own long-distance rates.

Here's how VOIP works (see Figure 7.3). Voice travels over the corporate intranet or Internet rather than the circuit-switched public network. Most corporate-class IP telephony applications use gateways that sit between the PBX and a router to convert calls into IP packets and shunt them onto the network. Using packets allows multiple parties to share digital lines, so data transmission is much more efficient than traditional phone conversations, each of which requires a line. When the packets hit the destination gateway, the message is depacketized, converted back into voice, and sent out via local phone lines. With newer, multiVOIP, the PBX (seen in Figure 7.3), is not needed. Phones are directly connected to a multiVOIP box.[18] This arrangement can be cheaper than standard VOIP, making the technology more attractive to small businesses.

Net telephony first hit the Web in 1995—only to be quickly derided for poor voice quality and annoying delays. One big problem has been that data networks break speech into little packets, so it's possible for some packets to arrive out of order or too late to be included in the conversation. Another is the lag

voice-over-IP (VOIP)

technology that enables network managers to route phone calls and fax transmissions over the same network they use for data

FIGURE 7.3

How Voice-Over-IP (VOIP) Works

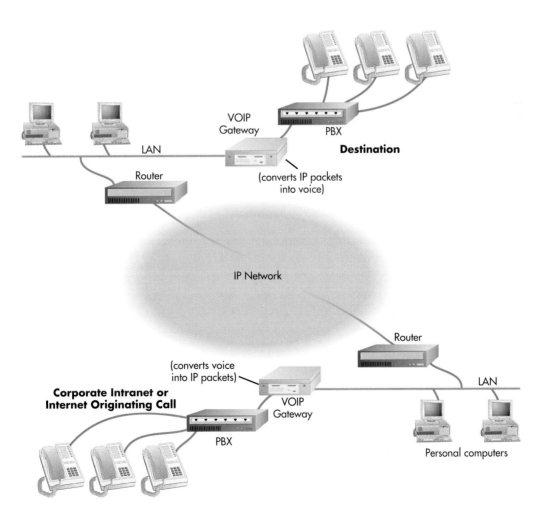

time inherent in the Net—speech packets have to travel through a dozen or more routers, and each router takes a split second to do its job. However, improvements can be expected. One major breakthrough was the development of a gateway server that connects data networks—such as the Internet and internal business networks—to the public telephone networks. This technology allows Internet calls from computer to phone or, with gateways on both ends, from phone to phone.

What is especially interesting about VOIP is the promise of new ways for merging voice with video and data communications over the Web or a company's data network. In the long run, it's not the cost savings that will boost the market, it's the multimedia capabilities it gives us and the smart call-management capabilities. Travel agents could use voice and video over the Internet to discuss travel plans; Web merchants could use it to show merchandise and take orders; customers could show suppliers problems with their products. Table 7.5 lists some current suppliers of Internet phone services.

A codec (*co*mpression-*dec*ompression) device prepares the voice transmission by squeezing the recorded sound data and slicing it into packets for transfer over the Internet. On the receiving end, a codec reassembles and decompresses the data for playback. Different codecs are optimized for different uses and conditions, and the characteristics of a specific codec can affect voice quality.

Internet videoconferencing, which supports both voice and visual communications, is another emerging service. Hardware and software are available to support a two-party conferencing system. The key here is a video codec to convert visual images into a stream of digital bits and translate them back again. However, the ideal video product will support multipoint conferencing in which multiple users appear simultaneously on the multiple screens.

Content Streaming

content streaming
a method for transferring multimedia files over the Internet so that the data stream of voice and pictures plays more or less continuously, without a break, or very few of them; enables users to browse large files in real time

Content streaming is a method for transferring multimedia files over the Internet so that the data stream of voice and pictures plays more or less continuously, without a break, or very few of them. It also enables users to browse large files in real time. For example, rather than wait the half-hour it might take for an entire 5-MB video clip to download before they can play it, users can begin viewing a streamed video as it is being received.

Instant Messaging

instant messaging
a method that allows two or more individuals to communicate on-line using the Internet

Instant messaging is on-line, real-time communication between two or more people who are connected to the Internet.[19] With instant messaging, two or more screens open up. Each screen displays what one person is typing. Because

TABLE 7.5

Current IP Telephone Services

Company	Service	Availability	Per-Minute Rate
AT&T	AT&T WorldNet Voice	Three cities yet to be named, expanding to 16 cities by the end of 1998	$.075–$.09
Delta Three	PC-to-Phone	16 countries, including the United States	$.30–$.40 for international calls
IDT	Net2Phone Direct	Nationwide using "800" or local access	$.05–$.08 for domestic calls
Qwest Communications International	Q.talk	Nine U.S. cities	$.075

ICQ is a client program that informs you who's on-line and enables you to contact them and chat with them in real time.

the typing is displayed on the screen in real time, it is like talking to someone using the keyboard.

A number of companies offer instant messaging, including America Online, Yahoo!, and Microsoft. America Online is one of the leaders in instant messaging, with about 40 million users of its Instant Messenger and about 50 million people using its client program ICQ. In addition to being able to type messages on a keyboard and have the information instantly displayed on the other person's screen, some instant messaging programs are allowing voice communication or connection to cell phones. One wireless service provider announced that it has developed a technology that can detect when a person's cell phone is turned on.[20] With this technology, it will be possible for someone on the Internet using instant messaging to communicate with someone on a cell phone anywhere in the world.

Shopping on the Web

Shopping on the Web for books, clothes, cars, drugs, and even medical advice can be convenient and easy. Read the "E-Commerce" box to see how builders are now shopping for materials on the Web. For some items, Web sales are booming. It has been reported that most people shopping for books over the 1999–2000 holiday season got them on-line.[21] In total, about 47 percent of all shoppers purchased their books online. This compares with 37 percent from traditional retail stores. The remaining customers purchased their books through catalogs and other sources. Items that require fit and style are less likely to be purchased over the Web. Only 29 percent of people shopping for clothes purchased them through the Internet. Postage stamps can also be purchased on the Internet at sites like Stamps.com. Simply Postage uses a small postage meter that connects to the Internet and prints the correct postage on special labels on demand; a PC is not needed. The surge in on-line shopping is one reason *Time* magazine named Jeff Bezos, founder of Amazon.com, its Person of the Year for 1999. Amazon.com sells books, videos, and other products on the Internet.

bot

a software tool that searches the Web for information, products, prices, and so on

Increasingly, people are using bots to help them shop on the Internet. A **bot** is a software tool that searches the Web for information, products, prices, and so on. A bot, short for "knowledge robot," can find the best prices or features from multiple Web sites. MortgageMarvel, for example, is a bot that helps people search a number of Internet sites for the best home loan or mortgage. BargainBot searches the Internet for the best prices for books and related items. In addition to shopping, bots are also useful in other areas. AdHound, for example, is a news bot that searches Internet classified ads of hundreds of newspapers across America. AdHound will send you an e-mail with the classified ads that match your search parameters.

Web auction

an Internet site that matches people who want to sell products and services with people who want to purchase these products and services

Web Auctions

A **Web auction** is a way to match people and companies that want to sell products and services with people who want to buy products and services. In addition to typical products and services, Internet auction sites excel at offering

E-COMMERCE
Home Builders Build Sites on the Internet

The Internet has already transformed book sales, travel, and many other industries. Customers are flocking to the Internet for information and deals in record numbers. Today, the Internet is starting to have a dramatic impact even on older, more traditional businesses, such as home building.

Weather-Tek Design Center, a building company in Waukesha, Wisconsin, recently discovered the power of the Internet. The design center, located in a community close to Milwaukee and Lake Michigan, had inventory problems. The company had amassed too much inventory of custom door and window materials. Inventory that sits for months or years can be a huge expense. In addition to the money invested that can't be used for other purposes, excess inventory usually depreciates in value as new products are introduced, and there is always the threat that the inventory will be stolen or damaged. To solve its inventory problem, Weather-Tek turned to the Internet.

BuildersExpress.com is a Web site for builders looking for hard-to-find items. Builders can advertise what they want to unload, and other builders can search the BuildersExpress Web site to find what building materials they need. After stumbling on the Chicago-based BuildersExpress Web site, Weather-Tek decided to list all 140 inventory items it no longer needed, instead of letting the items collect dust. In 90 days, all 140 items were sold through the BuildersExpress Web site. According to Michael Gavin, founder of BuildersExpress, "This really wouldn't have been possible without the Internet."

Although there are a few big builders in the United States, most of the industry consists of tens of thousands of small builders. Many of them are traditional mom-and-pop operations. Because the industry is not organized or centralized, there are inefficiencies. Prices vary, materials can be scarce, and there can be a long wait for needed supplies. Building companies, such as Building Materials Holding Corporation, are seeing dramatic results from

putting lists of all their building materials on the Internet. Building Materials has also been able to use the Internet to receive payments faster from customers. "This is a way for us to add profits to a low-margin business," says Robert Mellor, CEO of Building Materials.

Other building companies are also starting to venture on-line. USBuild.com, for example, is developing software to help streamline pricing practices. The old system relied on subcontractors to bid on various projects, which was inefficient and not always cost effective. With USBuild.com, companies will be able to put jobs on the Internet for bid instead of relying on a handful of local subcontractors. The result could be lower total building costs. Equalfooting.com, another building Web site, offers price discounts to small builders. These discounts were typically only available to larger, high-volume builders. The Equalfooting.com Web site is backed by a number of influential individuals, including Steve Case, chairman of America Online, and Marc Andreessen, founder of Netscape Communications.

Discussion Questions

1. How was Weather-Tek able to use the Internet to its benefit?
2. Explore the Internet and describe other building sites you find.

Critical Thinking Questions

3. How was the Internet able to help small builders become more efficient and profitable?
4. What other industries could benefit from the Internet?

Sources: Adapted from Jim Carlton, "Home Builders Learn to Love the Internet," *The Wall Street Journal*, February 7, 2000, p. A2; and Brae Canlen, "Pro Dealers Seek E-Link to Builders," *National Home Center News*, January 24, 2000, p. 1.

unique and hard-to-find items. Finding these items without an Internet auction site is often difficult, time consuming, and expensive. Many businesses are now using auction sites to buy and sell products. This business-to-business application of auction sites is expected to continue. Almost anything you may want to buy or sell can be found on auction sites.

One of the most popular auction sites is eBay. The eBay site is easy to use, and a large number of products and services can be found there. Like some other Web sites that have realized stunning growth, eBay has experienced some technical problems with its Web site in the past. The company is working hard to prevent such service interruptions from happening in the future. In addition to eBay, there are a number of other auction sites on the Web. Traditional companies are even starting their own auction sites.[22]

Although auction Web sites are excellent for matching buyers and sellers, there are potential problems. Auction sites on the Web are not always able to determine whether products and services listed by people and companies are legitimate. In addition, some Web sites have had illegal or questionable items offered. Even with these potential problems, the use of Web auction sites is expected to grow rapidly in the future.

Music, Radio, and Video on the Web

Music, radio, and video are now available on the Web. In fact, music is one of the hottest growth areas. Audio and video programs can be played on the Internet, or files can be downloaded. Using music players and music formats like MP3, discussed in Chapter 3, it is possible to download music from the Internet and listen to it anywhere using small, portable music players. A number of companies are now offering music over the Internet.[23] A few weeks after the announcement of the huge America Online (AOL)–Time Warner merger, the company also announced that the new merged company would acquire EMI Group for $20 billion to allow AOL–Time Warner to offer music over the Internet. According to AOL's CEO, Steve Case, "We have the opportunity to create a personal jukebox in the house and the car." Another key executive of AOL said, "This is the take-off point for the music business." Music on the Internet, however, is not without controversy. One Internet site that allows people to share perfect digital copies of songs and music has been sued.[24] In Germany, a court ruled that a major Internet service provider may be liable for music copyright violations, and in China, imported music and videos have been banned.

It is also possible to listen to radio broadcasts over the Internet and to download radio programs. Entire audio books can also be downloaded for later listening, using devices like the Audible Mobile Player. This technology is similar to the popular books-on-tape media, except you don't need a cassette tape or a tape player. Worldstream Communications is now offering interactive talk shows on the Internet, with a format that resembles the popular TV talk shows. Topics range from politics, to economics, to news.[25] A typical program has a reporter interviewing a guest. On the Internet, you can see pictures of the guest and hear the live talk and audience reactions.

Some corporations have also started to use Internet video to broadcast corporate messages or to advertise on the Web.[26] Victoria's Secret, for example, used Internet video to advertise its lingerie line. The video was so popular that the 1.5 million viewers jammed the Internet site. In addition to advertising, some companies are now investigating the use of Internet video to broadcast stockholder meetings, statements to the public from top-level executives, and other messages. Doctors can also use Internet video to monitor and even control surgical operations that take place thousands of miles from them. As mentioned in Chapter 6, Internet video is also being used successfully for teleconferencing, which can connect employees, managers, and corporate executives around the world in private conversations. Using Internet video, it is also possible to receive TV programs from an Internet site.

Office on the Web

With people traveling more for pleasure and business, it can be difficult to keep connected or conduct business. When traveling for pleasure, for example, you may want to call, e-mail, or write friends or family members. But if you didn't remember to bring the needed phone numbers or addresses, you could be out of luck. If you are a business traveler, you could encounter even more potential problems. Perhaps you need to get an important sales report completed before you return from your business trip. Or you may have received a phone call from your boss, who wants you to send a financial document to a co-worker immediately. You may also want to set up conference calls or track upcoming appointments. It can be difficult for business travelers to bring all important files or information with them, and they cannot anticipate all needs and emergencies.

To help solve these problems, you can set up an Internet office, which is a Web site that contains files, phone numbers, e-mail addresses, an appointment calendar, and more.[27] Using a standard Web browser, you can access important files and information. You can download a draft of a sales report to your laptop computer or send documents from the Web site to co-workers. You can also search your online appointment book.

An Internet office makes it possible to have the capabilities of your desktop computer, phone books, appointment schedulers, and other important information with you at all times, regardless of where you are in the world. With products such as HotOffice, eRoom, and QuickPlace, you can travel light and never leave important information behind. For as little as $13 per month, a company can set up a virtual office on the Internet for up to 20 employees. For individuals and employees who travel, these services can be invaluable.

Internet Sites in Three Dimensions

Using new technology by MetaCreations, HotMedia, and others, some Web sites offer three-dimensional views of places and products.[28] For example, a 3-D Internet auto showroom allows people to get different views of a car, simulating the experience of walking around in a real auto showroom. When looking at a 3-D real estate site on the Web, people can tour the property, go into different rooms, look at the kitchen appliances, and even take a virtual walk in the garden. Some 3-D sites allow people to interact with the product they are viewing on the Web. Sony, for example, is experimenting with this technology to allow people to open and close the lid of a Sony laptop using a mouse. Although still under development, 3-D Internet sites may become common in the next few years.

Free Software and Services

The Internet has always been known as a source for free software, advice, and services. The software for many of the services just discussed can be downloaded from the Internet free of charge. ICQ, the instant messaging system discussed earlier, for example, can be downloaded at no charge. Some e-mail services, such as those offered by Yahoo! and Hotmail, are also free. In addition, there is a wealth of information and advice on the Internet. Using a search engine, it is possible to obtain free information on almost any topic, ranging from investments to dating.

There can be disadvantages to free services, however. There is a wealth of free software and services on the Internet, but many of the sites bombard the user with annoying advertising. In addition, the information and advice may not always be truthful or helpful. For example, some people have posted false information on investment chat rooms, hoping to manipulate the price of a stock and make a profit. In some cases, groups or sites on the Internet have a bias. They can appear to be helpful but are actually trying to advance their own agenda by posting false or misleading information on the Internet. Thus, great care must be exercised when obtaining free software or services from the Internet.

THE WORLD WIDE WEB

World Wide Web (WWW, or W3)

a collection of tens of thousands of independently owned computers that work together as one in an Internet service

The World Wide Web was developed by Tim Berners-Lee at CERN, the European Organization for Nuclear Research in Geneva. He originally conceived of it as an internal document-management system. This server can be located at http://www.cern.ch. From this modest beginning, the **World Wide Web** (the Web, WWW, or W3) has grown to a collection of tens of thousands of independently owned computers that work together as one in an Internet service. These computers, called Web servers, are scattered all over the world and contain every imaginable type of data. Thanks to the high-speed Internet circuits connecting them and some clever cross-indexing software, users can jump from one Web computer to another effortlessly—creating the illusion of using one big computer. Because of its ability to handle multimedia objects, including linking multimedia objects distributed on Web servers around the world, the Web is emerging as the most popular means of information access on the Internet today.

The Web is a menu-based system that uses the client/server model. It organizes Internet resources throughout the world into a series of menu pages, or screens, that appear on your computer. Each Web server maintains pointers, or links, to data on the Internet and can retrieve that data. However, you need the right hardware and telecommunications connections, or the Web can be painfully slow. Traditionally, graphics and photos have taken a long time to materialize on the screen, and an ordinary phone line connection may not always provide sufficient speed to use the Web effectively. Serious Web users need to connect via the LAN server or SLIP/PPP approaches discussed earlier. Thankfully, some phone companies are offering new high-speed Internet access services, such as DSL, discussed in Chapter 6, based on new technology that allows traditional copper phone lines to connect to the Web. As discussed in Chapter 6, cable companies also offer on-line services using cable modems with faster data speeds compared with traditional phone lines.

Data can exist on the Web as ASCII characters, word processing files, audio files, graphic and video images, or any other sort of data that can be stored in a computer file. A Web site is like a magazine, with a cover page called a **home page** that has color graphics, titles, and text. All the type that is underlined is hypertext, which links the on-screen page to other documents or Web sites. **Hypermedia** connects the data on pages, allowing users to access topics in whatever order they wish. As opposed to a regular document that you read linearly, hypermedia documents are more flexible, letting you explore related documents at your own pace and navigate in any direction. For example, if a document mentions the Egyptian pharaohs, you can choose to see a picture of the pyramids, jump into a description of the building of the pyramids, and then jump back to the original document. Hypertext links are maintained using URLs. Table 7.6 lists some interesting Web sites. Many PC and business magazines also

home page

a cover page for a Web site that has titles, graphics, and text

hypermedia

tools that connect the data on Web pages, allowing users to access topics in whatever order they wish

TABLE 7.6

Some Interesting Web Sites

Site	Description	URL
Monster	This is a job-hunting site. You can search for a job by type or company, list your résumé, and perform basic company research. One feature, Talent Market, allows people to put their skills up for bid.	www.monster.com
AskMe	This site offers expert advice on a variety of topics.	www.askme.com
ICQ	This is a chat facility that offers free chat services for two or more people.	web.icq.com
MyHelpDesk	This site offers information and help on over 1,500 products. The site includes vendor phone numbers, chat room addresses, and message boards.	www.myhelpdesk.com
MSN MoneyCentral	This Microsoft site offers a large range of financial and investment information.	moneycentral.msn.com
Britannica	This site provides the popular encyclopedia on-line.	www.britannica.com
DealTime	This site helps you find what you want on the Internet.	www.dealtime.com
eBay	This is a popular auction site on the Internet.	www.ebay.com
Amazon.com	This popular site sells books, videos, music, furniture, and much more.	www.amazon.com
Travelocity	This large site offers travel information and bargains.	www.travelocity.com
WebMD	This site provides medical information and advice.	www.webmd.com

hypertext markup language (HTML)

the standard page description language for Web pages

HTML tags

codes that let the Web browser know how to format text: as a heading, as a list, or as body text and whether images, sound, and other elements should be inserted

Extensible Markup Language (XML)

markup language for Web documents containing structured information, including words, pictures, and other elements

Web browser

software that creates a unique, hypermedia-based menu on your computer screen that provides a graphical interface to the Web

FIGURE 7.4

Sample Hypertext Markup Language

Shown at the left on the screen is a document, and at the right are the corresponding HTML tags.

publish interesting and useful Web sites, and they are often evaluated and reviewed in print media and on-line.[29]

Hypertext markup language (HTML) is the standard page description language for Web pages. One way to think about HTML is as a set of highlighter pens in different colors that you use to mark up plain text to make it a Web page—red for the headings, yellow for bold, and so on. The **HTML tags** let the browser know how to format the text: as a heading, as a list, or as body text. HTML also tells whether images, sound, and other elements should be inserted. Users mark up a page by placing HTML tags before and after a word or words. For example, to turn a sentence into a heading, you place the <H1> tag at the start of the sentence. At the end of the sentence, you place the closing tag </H1>. When you view this page in your browser, the sentence will be displayed as a heading. This means that a Web page is made up of two things: text and tags. The text is your message, and the tags are codes that mark the way words will be displayed. All HTML tags are encased in a set of less than (>) and greater than (<) arrows, such as <H2>. The closing tag has a forward slash in it, such as </B> for closing bold. Figure 7.4 shows a simple document and its corresponding HTML tags.

A number of new Web standards are undergoing definition and early use. These include Extensible Markup Language (XML), cascading style sheets (CSSs), and Dynamic HTML (DHMTL). **Extensible Markup Language (XML)** is markup language for Web documents containing structured information, including words, pictures, and other elements. With XML, there is no predefined tag set. With HTML, for example, the <H1> tag always means a first-level heading. With XML, tags and relationships between them can be defined. XML Web documents consist of content and markup. For example, <?xml version="1.0"?> is a processing markup that identifies a document as an XML document. <!ELEMENT oldjoke (burns +, allen, applause?)> is an example of an element declaration markup. XML includes the capabilities to define and share document information over the Web. CSSs improve Web page inheritance and presentation, and DHTML provides dynamic presentation of Web content. These standards move more of the processing for animation and dynamic content to the browser and will provide quicker access and displays. XML documents contain tags that pertain specifically to the information you requested. Thus, you can navigate to desired data more quickly. However, entire industries will need to agree on sets of tags to achieve this goal. Fortunately, software is available to help people interested in Web authoring in HTML, XML, and other markup languages.

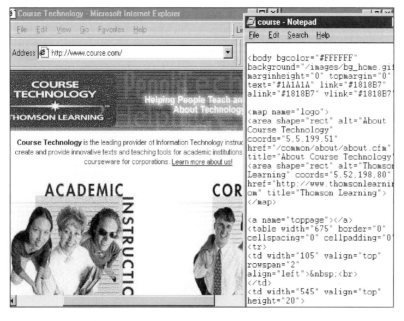

Web Browsers

A **Web browser** creates a unique, hypermedia-based menu on your computer screen that provides a graphical interface to the Web. The menu consists of graphics, titles, and text with hypertext links. The hypermedia menu links you to Internet resources, including text documents, graphics, sound files, and newsgroup servers. As you choose an item or resource, or move from one document to another, you may be jumping between computers on the Internet without knowing it, while the Web handles all the connections. The beauty of Web browsers and the Web is that they make surfing the Internet fun. Just clicking with a mouse on a highlighted word or a graphical button whisks you effortlessly to computers halfway around the world. Most browsers offer basic features such as support for backgrounds and tables, the ability to view a Web page's HTML source code, and a way to create hot lists of your favorite sites.

By 1996, Netscape Communications' Navigator had become the most widely used Web browser. It was setting the pace of browser development by embracing the HTML 3.0 standard and other hot new technologies. Navigator enables net surfers to view more complex graphics and 3-D models as well as audio and video material and runs small programs embedded in Web pages called **applets**. Many Web site developers boast that their sites use the latest Netscape innovations (such as frames that let the user split a screen into multiple, independent frames) and that their Web pages are best viewed with Navigator. The latest version of Netscape, Netscape 6, was under testing in early 2000 to be released later in the year.

applet

small program embedded in Web pages

Microsoft released Internet Explorer in the summer of 1995 to compete with Netscape. As of this writing, the U.S. Department of Justice and Microsoft are discussing whether it is appropriate for Explorer to come bundled with Windows 98 and Windows 2000. This type of giveaway is standard procedure—that's how Navigator became the number one browser. In addition to voice, chat, and whiteboard sharing, Microsoft's NetMeeting lets two or more people on the Internet work together in a Windows application such as Microsoft Excel. Microsoft developed and is marketing ActiveX as a method for creating network-ready applets.

Search Engines

Looking for information on the Web is a little like browsing in a library—without the card catalog, it is extremely difficult to find information. Web search tools—called **search engines**—take the place of the card catalog. Most search engines, like Yahoo.com, are free. They make money by charging advertisers to put ad banners on their sites. In one case, a search engine company (iwon.com) offers thousands of dollars in prize money for users who use its search engine at certain specified time intervals. This promotion is intended to generate traffic to its site.

search engine

a Web search tool

meta-search engine

a tool that submits keywords to several individual search engines and returns the results from all search engines queried

The Web is a huge place, and it gets bigger with each passing day, so even the largest search engines do not index all Internet pages. Even if you do find a search site that suits you, your query might still miss the mark. So when searching the Web, you may wish to try more than one search engine to expand the total number of potential Web sites of interest. Another option is to use a meta-search engine. A **meta-search engine** submits keywords to several individual search engines and returns the results from all search engines queried. MetaCrawler (*www.metacrawler.com*), for example, is a meta-search engine that searches AltaVista, Excite, Google, Lycos, and other search engines. Once you find a document that comes close to your goal, you can usually find related material by following the highlighted entries that take you to other Web pages when you click on them. And if you come across something you think you'll want to return to, you can add it to the "hot list" or "favorites" list on your Web browser to save time in the future.

Yahoo! Is one of the most popular search engines on the Web.

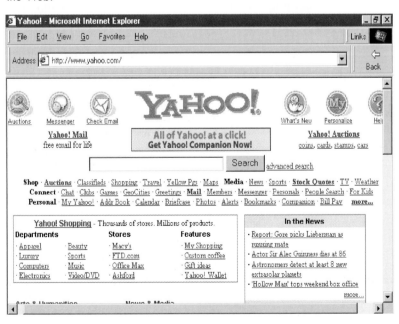

Search engines that use keyword indexes produce an index of all the text on the sites they examine. Typically, the

engine reads at least the first few hundred words on a page, including the title, the HTML "alt text" coded into Web-page images, and any keywords or descriptions that the author has built into the page structure. The engine throws out words such as *and, the, by* and *for*. The engine assumes whatever words are left are valid page content; it then alphabetizes these words (with their associated sites) and places them in an index where they can be searched and retrieved.

This type of search engine usually does no content analysis per se but uses word placement and frequency to determine how a page ranks among other pages containing the same or similar words. For example, when someone searches for the word *alien*, a page with "alien" in its title will appear higher in the search results than a site that doesn't mention "alien" in the title. Likewise, a page with 20 mentions of "alien" in the body text will rank higher than a page with one instance of the word.

Keyword indexes tend to be fast and broad; you'll typically get search results in seconds (faster than other kinds of engines). But unless you're careful about how you construct your query, you're likely to be overwhelmed with data.

Subject directories operate like a card catalog: They assign sites to specific topic categories based on the site's content and human judgment. The advantage of this approach is that sites are pregrouped and easier to browse than those in a raw keyword index. A human-generated subject directory also allows more nuance and subtlety than machine-generated keyword indexes and can offer advice on not only where the content is but how good or bad it is. However, human-generated directories can never be as comprehensive or up-to-date as machine-generated sites. Some sites use both keyword and subject directories. There are a number of Web search tools to choose from, as summarized in Table 7.7.

Java

Java
an object-oriented programming language from Sun Microsystems based on C++ that allows small programs (applets) to be embedded within an HTML document

Java is an object-oriented programming language from Sun Microsystems based on C++ that allows small programs—the applets mentioned earlier—to be embedded within an HTML document. When the user clicks on the appropriate part of the HTML page to retrieve it from a Web server, the applet is downloaded onto the client workstation environment, where it begins executing.

Java lets software writers create compact "just-in-time" programs that can be dispatched across a network such as the Internet. On arrival, the applet automatically loads itself on a personal computer and runs—reducing the need for computer owners to install huge programs anytime they need a new function. And unlike other programs, Java software can run on any type of computer. So far, Java is used mainly by programmers to make Web pages come alive, adding splashy graphics, animation, and real-time updates. Java-enabled Web pages are more interesting than plain Web pages.

Search Engine	Web Address
AltaVista	http://www.altavista.com
Ask Jeeves	http://www.ask.com
Excite	http://www.excite.com
Galaxy	http://www2.galaxy.com
GO Network	http://www.go.com
Google	http://www.google.com
Lycos	http://www.lycos.com
WebCrawler	http://www.webcrawler.com
Yahoo!	http://www.yahoo.com

TABLE 7.7

Popular Search Engines

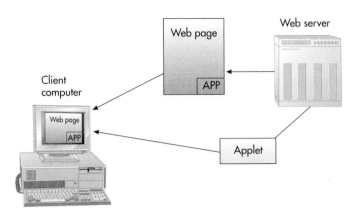

FIGURE 7.5

Downloading an Applet from a Web Server

The user accesses the Web page from a Web server. If the user clicks on the APP rectangle to execute the Java application, the client's computer checks for a copy on its local hard drive. If the applet is not present, the client requests that the applet be downloaded.

The relationship among Java applets, a Java-enabled browser, and the Web is shown in Figure 7.5 and explained here. To develop a Java applet, the author writes the code for the client side and installs that on the Web server. The user accesses the Web page and pulls it down to his personal computer, which serves as a client. The Web page contains an additional HTML tag called APP, which refers to the Java applet. A rectangle on the page is occupied by the Java application. If the user clicks on the rectangle to execute the Java application, the client computer checks to see if a copy of the applet is already stored locally on the computer's hard drive. If it is not, the computer accesses the Web server and requests that the applet be downloaded. The applet can be located anywhere on the Web. If the user's Web browser is Java-enabled (e.g., with Sun's HotJava browser or Netscape's Navigator product), then the applet is pulled down into the user's computer and executed within the browser environment.

The Web server that delivers the Java applet to the Web client is not capable of determining what kind of hardware or software environment the client is running on, and the developer who creates the Java applet does not want to worry about whether it will work correctly on OS/2, Windows, Unix, and MacOS. Java is thus often described as a "cross-platform" programming language.

The development of Java has had a major impact on the software industry. Sun Microsystems' strategy is to open up Java to any and all. Any software vendors and individual developers—from development tool vendors, language compiler developers, database management system vendors, and client/server application vendors to small businesses—can then use Java to create Internet-capable, run-anywhere applications and services (Figure 7.6). As a result, the Java community is becoming broader every day, encompassing some of the world's biggest independent software vendors, as well as users ranging from corporate CIOs, programmers, multimedia designers, and marketing professionals to educators, managers, film and video producers, and hobbyists.

FIGURE 7.6

Web Page with a Java Applet

Free Java Applets can be downloaded from this site for use on your own Web site.

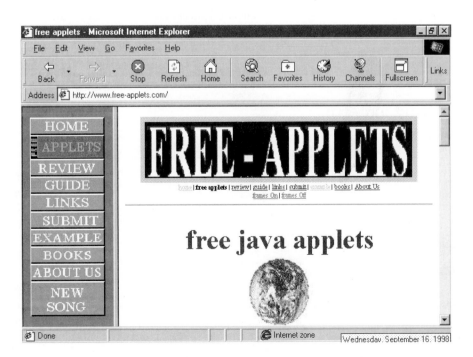

Java could change the economics of paying for software—the software industry today is based on the concept of entire applications delivered to the marketplace in shrink-wrapped boxes for a fixed, one-time cost, for which the customer is given a license that allows him or her to use the software forever on a single computer platform. Java makes it possible to sell one-time usage of a piece of software. This usage could be defined for a single transaction or for a single session during which the user is connected to a Web server.

Push Technology

push technology

automatic transmission of information over the Internet rather than making users search for it with their browsers

Push technology is used to send information automatically over the Internet rather than make users search for it with their browsers. Frequently the information, or "content," is customized to match an individual's needs or profile. The use of push technology is also frequently referred to as "Webcasting." Most push systems rely on HTTP (HyperText Transfer Protocol) or Java technology to collect content from Web sites and deliver it to users' desktops. Before they can be "pushed," users must download and install software that acts like a TV antenna, capturing transmitted content. As with any new technology, the people paying for push have yet to venture beyond rudimentary applications. Most are focusing on improving communications with employees, customers, and business partners.

A number of companies are using push technology to deliver critical information over the Internet.[30] SAP, for example, uses push technology to deliver critical enterprise resource planning software over the Internet. Enterprise resource planning software is used by many large corporations to streamline operations. SAP uses software from BackWeb. Other companies, including oil service supplier Schlumberger and Carlson Wagonlit Travel, also use push technology to make priority deliveries of information over the Internet.

There are drawbacks to the use of push technology. One issue, of course, is information overload. Another is that the volume of data being broadcast is so great that push technology can clog up the Internet communications links with traffic.

Business Uses of the Web

Drugstore.com is a Web shopping site that includes both a pharmacy and a retail store with health and beauty supplies.

In 1991, the Commercial Internet Exchange (CIX) Association was established to allow businesses to connect to the Internet. Since then, businesses have been using the Internet for a number of applications. Electronic mail is a major application for most companies. Many companies display products over the Internet, including catalogs and sample texts. Customers can place orders by keying in payment information and shipping addresses.

When the technology was still new, on-line shopping had been slowly adopted. Lack of commerce infrastructure, fears about the security of on-line financial transactions, and poor execution severely hampered early on-line selling efforts. However, as software and procedures are perfected to eliminate these problems, use of the Web to support home shopping, pay-per-use entertainment, and other commercial ventures has grown rapidly.

By linking buyers and sellers electronically on the Web, businesses are able to establish new and ongoing relationships with customers, allowing them access to information and products whenever it suits them. Businesses can use the Web as a tool for marketing, sales, and customer support. The Web can also serve as a low-cost alternative to fax, express mail, and other communications channels. It also can eliminate paperwork and drive down the cost per business transaction.

The Internet's business potential has just begun to be tapped. As more and more people gain access to the World Wide Web, its functions are changing drastically. We discuss a couple of these applications, corporate intranets and extranets, shortly. Chapter 8 will cover e-commerce in more detail.

Developing Web Content

Web authors work with several standards to create their pages. Unfortunately, current Web technical limitations make it difficult to create a "singing and dancing" Web site with beautiful text layout; large, photographic-quality images; voice-over; music; and video clips. The two main problems—confusing HTML standards and slow communications speeds—can limit creativity.

The HTML standards are created by a committee of various people involved in the Web. Anyone can create tags, and others may adopt them, modify them, or reject them. Thus, the HTML standards are evolving. HTML 1.0 was the standard in 1994 and is now obsolete. HTML 2.0 introduced a forms feature for allowing users to enter data and became the standard in 1995. HTML 3.0 allows banners, centering, right text alignment, tables, mathematical formulas, and image alignment. Netscape has added a number of new tags to this standard, such as <BLINK>, which causes blinking words that work only within Netscape. Microsoft wants to develop a new set of standards. Thus, not all browsers will work the same way when used to view the same Web page. For some browsers, a tag will work wonderfully. On others, that tag won't do anything at all, or, worse, it may cause problems. Some browsers are strictly HTML 2.0 and therefore ignore the Netscape extensions or newer HTML 3.0 features, some browsers use the Netscape extensions, and others are already using unapproved parts of the new HTML standards. Web authors need to keep these inconsistencies in mind when they develop pages. The art of Web design involves getting around the technical limitations of the Web and using a limited set of tools to make appealing designs. Following are tips for creating a Web page.

1. Your computer must be linked to a Web server, which can deliver Web pages to other browsers.
2. You will need a Web browser program to look at HTML pages you create.
3. The actual design can take one of the following approaches: (a) Write your copy with a word processor, then use an HTML converter to convert the page into HTML format complete with tags so the browser knows how it should format the page. (b) Use an HTML editor to write text and add HTML tags at the same time. (c) Edit an existing HTML template (with all the tags ready to use) to meet your needs. (d) Use an ordinary text editor and type in the start and end tags for each item.
4. Open the page with the browser and see the result. You can correct mistakes by correcting the tags.
5. Add links to your home page to allow your readers to click on a word and be taken to a related home page. The new page may be either a part of your Web site or a home page on a different Web site.
6. To add pictures, you must first store them as a file on your hard drive. This can be done in one of several ways: draw them yourself using a graphics software package, copy pictures from other Web pages, buy a disk of clip art, scan photos, or use a digital camera.

7. You can add sound by using a microphone connected to your computer to record a sound file; adding links to the page will enable those who access your Web page to hear it.
8. Upload the HTML file to your Web site using e-mail or FTP.
9. Review the Web page to make sure that all links are correctly established to other Web sites.
10. Advertise your Web page to others and encourage them to stop, take a look, and send feedback by e-mail.

After Web content development, the next step is to place the content on a Web site or home page.[31] Popular options include ISPs, free sites, and Web hosting. Some Internet service providers include limited Web space, typically 1–6 MB, as part of their monthly fee. If more disk space is needed, there are additional charges. Free sites, such as Angelfire.com and GeoCities.com, offer limited space for an Internet site. In return, free sites often require the user to view advertising or agree to other terms and conditions. Using a Web host is another option. A Web host can charge $15 or more per month, depending on services. Some Web hosting sites include domain name registration, Web authoring software, and activity reporting and monitoring of the Web site.

INTRANETS AND EXTRANETS

An intranet is an internal corporate network built using Internet and World Wide Web standards and products. It is used by the employees of the organization to gain access to corporate information. After getting their feet wet with public Web sites that promote company products and services, corporations are seizing the Web as a swift way to streamline—even transform—their organizations. These private networks use the infrastructure and standards of the Internet and the World Wide Web. A big advantage of using an intranet is that many people are already familiar with the Internet and Web, so they need little training to make effective use of their corporate intranet.

Most companies already have the foundation for an intranet—a network that uses the Internet's TCP/IP protocol. Computers using Web server software can store and manage documents built on the Web's HTML format. With a Web browser on your PC, you can call up any Web document—no matter what kind of computer it is on.

An intranet is an inexpensive yet powerful alternative to other forms of internal communications, including conventional computer setups. One of an intranet's most obvious virtues is its ability to slash the need for paper. Because Web browsers run on any type of computer, the same electronic information can be viewed by any employee. That means that all sorts of documents (such as internal phone books, procedure manuals, training manuals, and requisition forms) can be inexpensively converted to electronic form on the Web and be constantly updated. An intranet provides employees with an easy and intuitive approach to access information that was previously difficult to obtain. For example, it is an ideal solution to providing information to a mobile sales force that needs access to a lot of rapidly changing information. Intranets can also do something far more

An intranet is an internal corporate network used by employees to gain access to company information.

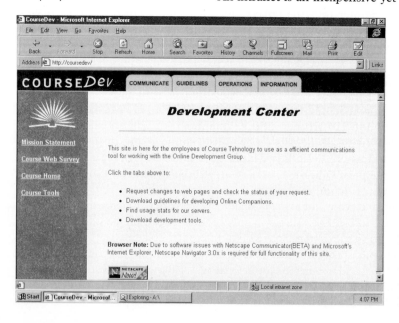

important. By presenting information in the same way to every computer, they can do what computer and software makers have frequently promised but never actually delivered: pull all the computers, software, and databases that dot the corporate landscape into a single system that enables employees to find information wherever it resides.

Universal reach is what made the Internet grow so rapidly. But Internet enthusiasts tended to focus on how to link far-flung people and businesses. When the Internet caught on, people were not considering it as a tool for running their business—but that is what is happening, with amazing speed. Just as the simple act of putting millions of computers around the world on speaking terms initiated the Internet revolution, so connecting all the islands of information in a corporation is sparking unprecedented collaboration. Corporate intranets are breaking down the walls within corporations.

Once a company sees the value of intranet access, it will want to move to the next stage of intranet usage—interactive transaction-based applications. At this stage, employees can query corporate databases to see the status of a customer order, a raw material shipment, or a manufacturing production run of a finished product.

More advanced use of the corporate intranet supports what has come to be known as workgroup computing. Workgroup computing involves many aspects, but basically it is an approach to supporting people working together in teams. One of the key aspects is the ability to store and share information in any form—text, video, sound, graphics, handwritten memos, or hand-drawn figures—which is often called a *knowledge base*. The key feature of workgroup computing is being able to organize and retrieve all this data simply. Group calendaring and scheduling allows an employee to check others' schedules and set up meetings. Another advantage of intranets is support for real-time meetings with people linked over networks, instead of making them travel to one place. Workgroup computing also supports work flow processes, tracking the status of documents—who has them, who is behind or ahead of schedule, and who gets them next.

A rapidly growing number of companies have advanced beyond the workgroup stage to offer limited network access to selected customers and suppliers. Such networks are referred to as extranets, which connect people who are external to the company. An extranet is a network that links selected resources of the intranet of a company with its customers, suppliers, or other business partners. Again, an extranet is built based on Web technologies.

Security and performance concerns are different for an extranet than for a Web site or network-based intranet. Authentication and privacy are critical on an extranet so that information is protected. Obviously, performance must be good to provide quick response to customers and suppliers. Table 7.8 summarizes the differences between users of the Internet, intranets, and extranets.

Secured intranet and extranet access applications usually require the use of a virtual private network (VPN). A **virtual private network (VPN)** is a secure connection between two points across the Internet.[32] VPNs transfer information by encapsulating traffic in IP packets and sending the packets over the Internet, a practice called **tunneling**. Most VPNs are built and run by Internet service providers. Companies that use a VPN from an Internet service provider

virtual private network (VPN)

a secure connection between two points across the Internet

tunneling

the process by which VPNs transfer information by encapsulating traffic in IP packets over the Internet

TABLE 7.8

Summary of Internet, Intranet, and Extranet Users

Type	Users	Need for User ID and Password
Internet	Anyone	No
Intranet	Employees	Yes
Extranet	Business partners	Yes

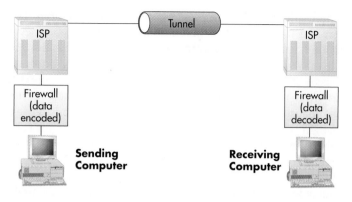

FIGURE 7.7

Virtual Private Network (VPN)

have essentially outsourced their networks to save money on wide-area network equipment and personnel. In using a VPN, a user sends data from his or her personal computer to the company's security device (a *firewall,* discussed later), which converts the data into a coded form that cannot be easily read by an interceptor. The coded data is then sent via an access line to the company's Internet service provider. From here, the data is transmitted through tunnels across the Internet to the recipient's Internet service provider and then over an access line to the receiving company's security device, where it is decoded and sent to the receiver's personal computer (Figure 7.7).

Companies and government agencies are big users of VPNs. NASA, for example, uses a VPN to transfer data between the space shuttles and earth.[33] The technology makes the shuttle appear as a node on the Internet. This allows NASA specialists on earth to control experiments being done in space. It also allows people to monitor shuttle operations using a standard Web browser.

NET ISSUES

The topics raised in this chapter apply not only to the Internet and intranets but also to LANs, private WANS, and every type of network. Control, access, hardware, and security issues affect all networks, so it is important to mention some of these management issues.

Management Issues

Although the Internet is a huge, global network, it is managed at the local level; no centralized governing body controls the Internet. Although the U.S. federal government provided much of the early direction and funding for the Internet, the government does not own or manage the Internet. The Internet Society and the Internet Activities Board (IAB) are the closest the Internet has to centralized governing bodies. These societies were formed to foster the continued growth of the Internet. The IAB oversees a number of task forces and committees that deal with Internet issues. One of the main functions of the IAB is to manage the network protocols used by the Internet, including TCP/IP. Some universities and government agencies are investigating how the Internet can be controlled to prevent sensitive information and pornographic material from being placed on the Internet.

Service Bottlenecks

The primary cause of service bottlenecks is simply the phenomenal growth in traffic. Traffic volume on company intranets is growing even faster than the Internet. Companies setting up an Internet or intranet Web site often underestimate the amount of computing power and communications capacity they need to service all the "hits" (requests for pages) they get from Web cruisers. Web server computers can be overwhelmed with thousands of hits per hour.

Slow modems and the copper-based telephone wire system that carries the signal into an office or home are the two current primary Internet bottlenecks. For most users, these two limit maximum access speed to around 56 Kbps, which is still too slow for 16-bit stereo sound and smooth, full-screen video.

Connection agreements exist among the backbone companies to accept one another's traffic and provide a certain level of service. Some Internet providers do a good job and provide a high level of quality and service. Others do not do such a good job, creating wide variations in the quality of the Internet. At some interconnect points where major Internet operators hand off to one another, one operator may not be able to accept incoming traffic fast enough because its lines are overloaded with traffic. For example, linking with Pacific Rim nations is especially difficult. Most providers lease their lines from phone companies, and the cost of even a medium-speed line to connect California with Australia is over $1.2 million per year—ten times the cost of a New York-to-San Francisco link, so some providers scrimp. This leads to inadequate capacity, which slows transmissions.

Routers, the specialized computers that send packets down the right network pathways, can also become bottlenecked. For each packet, every router along the way must scan a massive address book of about 40,000 area destinations (akin to Internet ZIP codes) to pick the right one. These routers can get overloaded and lose packets. The TCP/IP protocol compensates for this by detecting a missing packet and requesting the sending device to resend the packet. However, this leads to a vicious circle, as the network devices continually try to resend lost packets, further taxing the already overworked routers. This leads to long response times or loss of the connection to the network.

Several actions are being taken to open up the bottleneck. One solution involves the various backbone providers upgrading their backbone links. In some cases they are installing bigger, faster "pipes," and in others they are converting to newer transmission technology, such as asynchronous transfer mode (ATM), which can send a message down the right path more quickly than standard packet-switching technology. Each ATM transmission is preaddressed with its own route, so routing addresses do not have to be looked up and the packet can zip right through an ATM switch. A second solution to the bottleneck problem is provided by router manufacturers, who are working to develop improved models with increases in hardware capacity and more efficient software to provide quick access to addresses. Yet a third solution is to prioritize traffic. Today, all network traffic travels through the same big backbone pipes. There is no way to make sure that your urgent message is not stalled behind someone downloading a magazine page. With prioritized service, customers could pay more for guaranteed delivery speed, much like an overnight package costs more than second-day delivery. If implemented, this solution could also affect the cost of network services that generate a lot of traffic, such as Internet phone and videoconference services. DSL and cable modems, discussed in Chapter 6, can also be used to speed Internet access.[34]

Privacy, Security, and Unauthorized Internet Sites

As use of the Internet grows, privacy and security issues become even more important. People and companies can be reluctant to embrace the Internet unless these issues are successfully addressed. From a consumer perspective, the protection of individual privacy is essential. Yet, many people use the Internet without realizing that their privacy may be in jeopardy. For instance, many Internet sites use cookies to gather information about people who visit their sites. A **cookie** is a text file that an Internet company can place on the hard disk of a computer system. These text files keep track of visits to the site and the actions people take. To help prevent this potential problem, some companies are developing software to prevent these files from being placed on computer systems. CookieCop, for example, allows Internet users to accept or reject cookies by Internet site.[35] Read the "Ethical and Societal Issues" box, which discusses privacy issues.

cookie

a text file that an Internet company can place on the hard disk of a computer system

ETHICAL AND SOCIETAL ISSUES
Outrage on the Web

Advertising on the Web has always been a way to generate revenues for companies. Instead of charging users when they visit a Web site, companies often place banners on their sites that advertise the products and services of other companies. Some Internet companies specialize in using Internet banner ads.

One of the top ad-server companies on the Internet has its banner ads on about 1,500 Web sites. Many of these ads are aimed at specific customer types by using "cookies." A cookie is a text file that collects information about people's behaviors and viewing patterns on the Web. Some estimate that these types of companies have amassed about 100 million profiles, each one containing information about a person's browsing and Internet viewing habits. Once collected, this information can be combined with other data. One company, for example, recently paid over $1 billion for a data-warehouse company that has detailed records of customer catalog purchases and related activities. By combining personal data from several sources, companies can put together comprehensive and detailed profiles on people. These profiles not only contain shopping information but can contain people's home addresses, their places of employment, phone numbers, and much more.

Marketing and advertising companies can greatly benefit from this detailed information about people. Some advertising companies, for example, are willing to pay 10 to 20 percent more for this targeted customer information. According to Susan Nathan, a senior vice president of an advertising company, "The Internet is the first medium that offers advertisers the ability to speak to your customers." While advertising companies are embracing the wealth of personal information that comes from combining databases from the Internet and other sources, many people are very concerned about their privacy. Will highly personal information be for sale to the highest bidder?

Privacy lawsuits are now being filed against some Internet companies. Harriet Judnick, for example, recently sued an Internet company for privacy invasion. After visiting a few Web sites, she started to get phone calls and e-mails from insurance companies, loan companies, and other companies trying to sell her something. She even found her medical insurance information on the Internet.

Discussion Questions

1. Describe banner ad companies discussed in this box.
2. What is their primary business?

Critical Thinking Questions

3. How can people protect their privacy when using the Internet?
4. What would you like to see in a corporate privacy statement to guarantee your privacy?

Sources: Adapted from Heather Green, "Privacy: Outrage on the Web," *Business Week*, February 14, 2000, p. 38; and Thomas Weber, "Can Your Complaints Build a Web Business," *The Wall Street Journal*, January 10, 2000, p. B1.

---

cryptography

the process of converting a message into a secret code and changing the encoded message back to regular text

encryption

the conversion of a message into a secret code

From a corporate strategy perspective, security of data is essential. Such approaches as cryptography can help. **Cryptography** is the process of converting a message into a secret code and changing the encoded message back to regular text. The original conversion is called **encryption**. The unencoded message is called *plaintext;* the encoded message is called *ciphertext*. Decryption converts ciphertext back into plaintext (see Figure 7.8). For much of the Cold War era, cryptography was the province of military and intelligence agencies; uncrackable codes were reserved for people with security clearance only.

Widespread deployment of cryptography requires additional hardware and software but is becoming increasingly necessary to support electronic commerce, copyright management, and electronic delivery of services. Without cryptography, people will not trust that electronic financial transactions, secret or private data, and valuable intellectual property will remain confidential across networks.

A cryptosystem is a software package that uses an algorithm, or mathematical formula, plus a key to encrypt and decrypt messages. The algorithm is calculated with the key and converts every character of the plaintext into other coded characters, thus creating the ciphertext. Only someone with the correct key should be able to decode the ciphertext. Good ciphertext appears to be nothing more than random characters. Encryption makes information useless to hackers and thieves.

FIGURE 7.8

Cryptography is the process of converting a message into a secret code and changing the encoded message back into regular text.

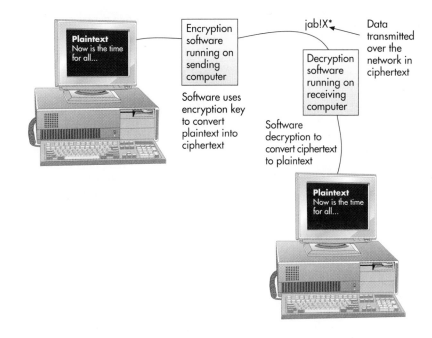

The Data Encryption Standard (DES), adopted as a federal standard in 1977 to protect unclassified communications and data, was designed by IBM and modified by the National Security Agency. It uses 56-bit keys, meaning that a user must employ precisely the right combination of 56 1s and 0s to decode information correctly. Other technologies offer a range of key lengths, up to 2,048 bits in the case of RC5, for instance. The RSA protocol, meanwhile, has no limit on key length, but it can slow things down, since it uses separate keys for encryption and decryption. Many products mix technologies—they use a fast algorithm like DES for the actual encryption but send the DES key through a more secure method like RSA.

U.S. banks and brokerage houses use the federal government's DES algorithm to protect the integrity and confidentiality of fund transfers totaling some $2.3 trillion a day worldwide. Organizations encrypt the words and videos of their teleconferencing sessions. Individuals encode their electronic mail. And researchers use encryption to hide information about new discoveries from prying eyes.

Encryption is not just for keeping secrets. It can also be used to verify who sent a message and to tell whether the message was tampered with en route. A **digital signature** is a technique used to meet these critical needs for processing on-line financial transactions. Digital signatures involve a complicated technique that combines the public-key encryption method with a "hashing" algorithm that prevents reconstructing the original message. The hashing algorithm provides further encoding by using rules to convert one set of characters to another set (e.g., the letter *s* is converted to a *v*, *2* is converted to *7*). Thus, encryption also can prevent electronic fraud by authenticating senders' identities with digital signatures.

Unauthorized and unwanted Internet sites are problems faced by some companies. A competitor or an unhappy employee can create an Internet site with an address that is very similar to the company's Internet address. When someone goes to the Internet to search for information about the company, he or she may find an unauthorized or unwanted site instead. In some cases, the unauthorized or unwanted site may appear to be the legitimate, official corporate site. In other cases, it is obvious that the site is not sponsored or authorized by the company.

Some unauthorized or unwanted sites can contain some very damaging information about the company. In some cases, the information is true, but in other cases, the information is false or misleading. A fired employee, for example, might post stories about his boss and the company that may not be entirely true. A competitor may post information about why a customer should not do business with

digital signature

encryption technique used to verify the identity of a message sender for the processing of on-line financial transactions the company. In still other cases, an environmental group may post why it feels the company may be damaging the environment. Unauthorized and unwanted Internet sites can be very troublesome, and it is not unusual for companies to sue those who post these sites.

Firewalls

When it comes to security on the Internet, intranets, or extranets, it is essential to remember two things. First, there is no such thing as absolute security. Second, plenty of clever people consider it great sport to try to breach any security measures—the better your security, the greater the challenge to them.

firewall

a device that sits between your internal network and the outside Internet and limits access into and out of your network based on your organization's access policy

The most popular method of preventing unauthorized access to corporate computer data is to construct what is known as a firewall between your company computers and the Internet. An Internet **firewall** is a device that sits between your internal network and the outside Internet. Its purpose is to limit access into and out of your network based on your organization's access policy. A firewall can be anything from a set of filtering rules set up on the router between you and the Internet to an elaborate application gateway consisting of one or more specially configured computers that control access. Firewalls permit desired services on the outside, such as Internet e-mail, to pass. In addition, most firewalls now allow access to the Web from inside protected networks. The idea is to allow some services to pass but deny others. For example, you may be able to use the Telnet utility to log into systems on the Internet, but users on remote systems cannot use it to log into your local system because of the firewall.

A firewall can be set up to allow access from only specific hosts and networks or to prevent access from specific hosts. In addition, you can give different levels of access to various hosts; a preferred host may have full access, whereas a secondary host may have access to only certain portions of your host's directory structure.

For a higher level of security, you can have an assured pipeline, which uses more sophisticated methods to prevent access. A firewall looks at only the header of a packet, but an assured pipeline looks at the entire request for data and then determines whether the request is valid. Inappropriate requests can be routed away from the Internet, whereas files that meet specific criteria (such as those that contain the word *confidential*) can never be sent over the Internet.

Government agencies use firewalls to help secure their Internet communications. NASA, for example, is working with Veridian Trident Data Systems to develop a firewall for space.[36] As discussed earlier, NASA is also developing a VPN for space shuttles. The firewall would ensure that communications between earth and space shuttles are secure and private.

The Federal Computer Incident Response Capability assists civilian government agencies with computer security incidents.

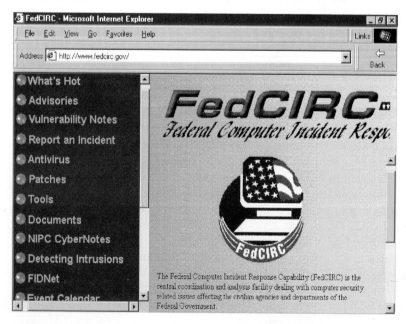

The Federal Computer Incident Response Capability (FedCIRC) responds to virus attacks, network intrusions, and other threats. It also provides security training and consulting services to individual agencies. Vulnerability and threat information is shared with the public at this group's Web site (http://www.fedcirc.gov).

Network management issues will take an increasing amount of time for IS personnel, but any user needs to be aware of the basics to function effectively in business. Communications, service, and daily work are all at stake.

● SUMMARY

PRINCIPLE • The Internet is like many other new technologies—it provides a wide range of services, some of which are effective and practical for use today, others of which are still evolving, and still others of which will fade away from lack of use.

The Internet is the world's largest computer network. Actually, it is a collection of interconnected networks, all freely exchanging information. The Internet transmits data from one computer (called a host) to another. The set of conventions used to pass packets from one host to another is known as the Internet protocol (IP). Many other protocols are used in connection with IP. The best known is the transport control protocol (TCP). TCP is so widely used that many people refer to TCP/IP, the combination of TCP and IP used by most Internet applications. Each computer on the Internet has an assigned address to identify it from other hosts. There are three ways to connect to the Internet: via a LAN whose server is an Internet host, via SLIP or PPP, and via an on-line service that provides Internet access.

An Internet service provider is any company that provides individuals or organizations with access to the Internet. To use this type of connection, you must have an account with the service provider and software that allows a direct link via TCP/IP. Among the value-added services ISPs provide are electronic commerce, intranets and extranets, Web site hosting, Web transaction processing, network security and administration, and integration services.

● ● ●

Internet services include e-mail; Telnet; FTP; Usenet and newsgroups; chat rooms; Internet phone and videoconferencing; content streaming; instant messaging; shopping on the Web; Web auctions; music, radio, and video; office on the Web; 3-D Internet sites; and free software and services. E-mail is used to send messages. Telnet enables you to log on to remote computers. FTP is used to transfer a file from another computer to your computer. Usenet supports newsgroups, which are on-line discussion groups focused on a particular topic. Chat rooms let you talk to dozens of people, who can be located all over the world, at one time. Internet phone service enables you to communicate with other Internet users around the world who have equipment and software compatible with yours. Internet videoconferencing enables people to conduct virtual meetings. Content streaming is a method of transferring multimedia files over the Internet so that the data stream of voice and pictures plays continuously. Instant messaging allows people to communicate in real time using the Internet. Shopping on the Web is popular for books, videos, music, and a host of other items and services. Items that may require style and fit, such as clothes, are not as popular on the Web. Web auctions are a way to match people looking for products and services with people selling these products and services. The Web can also be used to download and play music, radio, and video programs. With office on the Web, it is possible to store important files and information on the Internet; when traveling, a user can download these files and information or send them to other people. Some Internet sites are three dimensional, allowing people to manipulate the sites to see different views of products and images. A wealth of free software and services are available through the Internet. Some of the free information, however, may be misleading or even false.

● ● ●

The Web is a collection of over 30,000 independently owned computers that work together as one in an Internet service. High-speed Internet circuits connect these computers, and cross-indexing software is employed to enable users to jump from one Web computer to another effortlessly. Because of its ability to handle multimedia objects and hypertext links between distributed objects, the Web is emerging as the most popular means of information access on the Internet today.

A Web site is like a magazine, with a cover page called a home page that has graphics, titles, and text. Hypertext links are maintained using URLs (uniform resource locators), a standard way of coding the locations of the HTML (hypertext markup language) documents. Web pages are loosely analogous to chapters in a book.

The client communicates with the server according to a set of rules called HTTP (hypertext transfer protocol), which retrieves the document and presents it to the users. HTML is the standard page description language for Web pages. The HTML tags let the browser know how to format the text: as a heading, as a list, or as body text. HTML also tells whether images, sound, and other elements should be inserted.

A Web browser reads HTML and creates a unique, hypermedia-based menu on your computer screen that provides a graphical interface to the Web. The browser uses data about links to accomplish this; the data is stored on the Web server. The hypermedia menu links you to other Internet resources, not just text documents, graphics, and sound files.

A rapidly growing number of companies are doing business on the Web and enabling shoppers to search for and buy products on-line. The travel, entertainment, gift, greetings, book, and music businesses are experiencing the fastest growth on the Web. For many people, it is easier to shop on the Web than search through catalogs or trek to the shopping mall. However, some shoppers are concerned about the potential for credit card numbers to be stolen over the Internet.

• • •

Web authors work with several standards to create their pages. Unfortunately, current Web technical limitations make it difficult to create a "singing and dancing" Web site with beautiful text layout; large, photographic-quality images; voice-over; music; and video clips. The two main problems—confusing HTML standards and slow communications speeds—can limit creativity.

The steps to creating a Web page include getting space on a Web server; getting a Web browser program; writing your copy with a word processor, using an HTML editor, editing an existing HTML document, or using an ordinary text editor to create your page; opening the page using a browser, viewing the result, and correcting any tags; adding links to your home page to take viewers to another home page; adding pictures and sound; uploading the HTML file to your Web site; reviewing the Web page to make sure that all links are working correctly; and advertising your Web page.

• • •

Java is an object-oriented programming language from Sun Microsystems based on C++ that allows small programs—applets—to be embedded within an HTML document. When the user clicks on the appropriate part of the HTML page to retrieve it from a Web server, the applet is downloaded onto the client workstation environment, where it begins executing. The development of Java has had a major impact on the software industry and could change the economics of paying for software. Java makes it possible to sell one-time usage of a piece of software.

• • •

An intranet is an internal corporate network built using Internet and World Wide Web standards and products. It is used by the employees of the organization to gain access to corporate information. Computers using Web server software store and manage documents built on the Web's HTML format. With a Web browser on your PC, you can call up any Web document—no matter what kind of computer it is on. Because Web browsers run on any type of computer, the same electronic information can be viewed by any employee. That means that all sorts of documents can be converted to electronic form on the Web and constantly updated.

An extranet is a network that links selected resources of the intranet of a company with its customers, suppliers, or other business partners. It is built based on Web technologies. Security and performance concerns are different for an extranet than for a Web site or network-based intranet. Authentication and privacy are critical on an extranet. Obviously, performance must be good to provide quick response to customers and suppliers.

• • •

Management issues, service bottlenecks, privacy and security, and firewalls are issues that affect all networks. No centralized governing body controls the Internet. Also, because the amount of Internet traffic is so large, service bottlenecks often occur. Before the use of the Internet will be even more common, privacy and security issues must be addressed and solved. Cryptography techniques and firewalls are required to combat information thieves and provide as much security as possible. Some companies have unauthorized and unwanted Internet sites that can contain damaging information about the company. These sites can be placed on the Internet by unhappy employees, competitors, or other individuals and groups.

KEY TERMS

applet 269
ARPANET 253
backbone 254
bot 263
chat room 260
content streaming 262
cookie 277
cryptography 278
digital signature 279
encryption 278
Extensible Markup Language (XML) 268
file transfer protocol (FTP) 259
firewall 280

home page 267
HTML tags 268
hypermedia 267
hypertext markup language (HTML) 268
instant messaging 262
Internet protocol (IP) 253
Internet service provider (ISP) 256
Java 270
meta-search engine 269
newsgroups 259
point-to-point protocol (PPP) 256
push technology 272
search engine 269

serial line Internet protocol (SLIP) 256
Telnet 259
transport control protocol (TCP) 254
tunneling 275
uniform resource locator (URL) 254
Usenet 259
virtual private network (VPN) 275
voice-over-IP (VOIP) 261
Web auction 263
Web browser 268
World Wide Web (WWW, or W3) 266

REVIEW QUESTIONS

1. What is the Internet? Who uses it and why?
2. What is the TCP/IP protocol? How does it work?
3. Explain the naming conventions used to identify Internet host computers.
4. Briefly describe three different ways to connect to the Internet. What are the advantages and disadvantages of each approach?
5. What are Internet service providers? What services do they provide?
6. What is a newsgroup? How would you use one?
7. What are Telnet and FTP used for?
8. How does an Internet phone work? What are some of its advantages and disadvantages?
9. How would you use Internet videoconferencing?
10. What is content streaming?
11. What is instant messaging?
12. Briefly describe a Web auction.

13. What is the Web? Is it another network like the Internet, or a service that runs on the Internet?
14. What is hypermedia?
15. What is a URL and how is it used?
16. What is HTML and how is it used?
17. What is a Web browser? How is it different from a Web search engine?
18. What is push technology?
19. What is an intranet? Provide three examples of the use of an intranet.
20. What is an extranet? How is it different from an intranet?
21. What is a virtual private network? Why might an organization use one?
22. What is cryptography?
23. What are firewalls? How are they used?

DISCUSSION QUESTIONS

1. Instant messaging is being used to a greater extent today. Describe how this technology could be used in a business setting. Are there any drawbacks or limitations in using instant messaging in a business setting?
2. Briefly describe how the Internet phone service operates. Discuss the potential for this service to impact traditional telephone services and carriers.
3. The U.S. federal government is against the export of strong cryptography software. Discuss why this may be so. What are some of the pros and cons of this policy?

4. Identify three companies with which you are familiar who are using the Web to conduct business. Describe their use of the Web.
5. Outline a process to create a Web page. What computer hardware and software do you need if you wish to create a Web home page containing both sound and pictures?
6. One of the key issues associated with the development of a Web site is getting people to visit it. If you were developing a Web site, how would you inform others about it and make it interesting enough that they would return and also tell others about it?

7. Getting music, radio, and video programs from the Internet is getting easier, but some companies are worried that people will illegally obtain copies of this programming without paying the artists and producers royalties. If you were an artist or producer of this type of programming, what would you do?

8. How has the Java programming language changed the software industry?

9. Briefly summarize the differences in how the Internet, a company intranet, and an extranet are accessed and used.

10. Can you envision this course being delivered more effectively over the Internet than with the use of a textbook? Or would this lessen the value of the course? Explain your position.

● PROBLEM-SOLVING EXERCISES

1. Do research on the Web to find several popular Web auction sites. After researching these sites, use a word processor to write a report on the advantages and potential problems of using a Web auction site to purchase a product or service. Also discuss the advantages and potential problems of selling a product or service on a Web auction site.

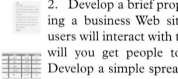

2. Develop a brief proposal for developing a business Web site. Describe how users will interact with the Web site. How will you get people to visit your site? Develop a simple spreadsheet to analyze income and expenses.

3. You are a manager in a Fortune 1000 company. You have been appointed to lead a project to develop an extranet linking your company to key suppliers. The extranet will provide suppliers with access to production planning and inventory data so that the suppliers can ship raw materials, packing materials, and supplies to your plant just-in-time. The goal is to reduce the level of inventory you must carry while not adversely affecting production. Develop a one-page project charter that defines the scope and purpose of the project, identifies the key members of the project team (by title), outlines the key technical and nontechnical issues you will face, and defines the key project success criteria.

● TEAM ACTIVITIES

1. Identify a company that is making effective use of a company extranet. Find out all you can about its extranet. Try to speak with one or more of the customers or suppliers who use the extranet and ask what benefits it provides from their perspective.

2. Work with your team to find several sites that can be used to download music from the Internet. Compare the costs of downloading a piece of music from several sites. Describe which site you would use if you had a music player.

● WEB EXERCISES

1. This chapter covers a number of powerful Internet tools, including Internet phones, search engines, browsers, e-mail, newsgroups, Java, intranets, and much more. Pick one of these topics and get more information about it from the Internet. You may be asked to develop a report or send an e-mail message to your instructor about what you found.

2. Shopping on the Internet is becoming more popular. Using a search engine, locate several Internet sites that sell new cars. Compare the costs and features of several cars you find interesting. Would you purchase a car on-line? What are the advantages and disadvantages of buying a car on the Internet?

● CASES

 JCPenney Auctions Its Overstock Inventory

Web auction sites, such as eBay, have become tremendously successful, matching buyers and sellers. Although there are potential problems with on-line auction sites, they provide an efficient marketplace for unusual and hard-to-find items. Sayings such as "one person's junk is another person's treasure" are often used to describe on-line auction operations. On-line auctions can be like gigantic flea markets, allowing people to sell things they don't want. The success of on-line auction sites has not gone unnoticed by companies. After seeing the success of auction sites such as eBay, JCPenney decided to investigate the possibility of using this approach to sell overstocked merchandise from its catalog and retail outlets. "The whole idea of auction came about just from looking at how successful the auction concept has been on-line," says Richard Last, JCPenney's vice president of e-commerce. The company's on-line presence is minimal at this time. On-line sales were $15 million in 1998 and $100 million in 1999, compared with catalog sales of $4 billion. Yet the Internet has great promise for companies like JCPenney. To make its Web site more attractive to customers, the company is planning on adding a number of on-line specialty shops. According to Last, "We saw it as a way of keeping the site interesting."

If JCPenney is successful, other retailers will no doubt jump on the on-line auction idea. Rebecca Nidositko, an analyst for The Yankee Group, predicts that "a lot of retailers on- and off-line will be adopting auctions this year." The Yankee Group predicts that retail auctions will generate $200 million in sales in 2000 and over $2 billion by 2003.

Discussion Questions

1. How should JCPenney organize its Web site to make it efficient and easy to use?
2. Should JCPenney auction inventory that is not overstock?

Critical Thinking Questions

3. What are the advantages and disadvantages of a retail auction site?
4. If you were a consultant, what products and services would you recommend for an auction site? What products and services should be sold using normal channels?

Sources: Adapted from Stacy Collett, "J.C. Penney to Add Online Auction Service," *Computerworld*, January 24, 2000, p. 18; and "A Conversation with Vanessa Castagna," *Chain Store Age*, January 1, 2000, p. 60.

2 Kaiser Permanente Looks for a Cure on the Web

Many healthcare companies have been on the cutting edge of medicine but lag behind in the use of information systems and the Internet. Loading old steel filing cabinets with tons of medical and insurance records was typical. These massive paper documents were stored in huge rooms and shipped around the country by the truckload. Today, some healthcare providers are starting to take a more healthy approach to their record keeping by turning to technology and the Internet.

In a $2 billion project, Kaiser Permanente is launching a massive technology initiative. The objective is to move all its operations to the Internet. The health-care provider plans to develop a digital medical record of all its nine million members. The record will be linked to the company's 361 hospitals and 10,000 doctors, nurses, and other healthcare providers. The heart of the system will be an Internet site. The current plans are to give companies, such as General Motors and Wells Fargo, access to coverage rates and related information. Separate Internet sites will allow doctors and healthcare administrators the ability to order a wide array of supplies and equipment, from inexpensive bandages to costly CAT scan devices. Kaiser's CEO, David Lawrence, believes that the Internet "will be the central nervous system for tying together all of the elements needed to care for patients better—and it will do so in ways now unimaginable."

The new Internet strategy is aimed not only at improving patient care but also at improving the bottom line for Kaiser. In 1998, Kaiser had losses of $288 million on revenues of $15.5 billion. Some blame the disappointing loss to underestimating the amount of care the company needed to give patients. The new Internet system could help reduce costs. For example, prescription drug costs are rising 16 to 18 percent annually. Kaiser hopes that its investment in technology will help the company locate alternative drugs that are less expensive and less dangerous to patients. Kaiser also hopes that the vast Internet system will allow it to compare doctor's bills that can vary substantially. The hope is to find standard treatments that can reduce costs. Kaiser also hopes that the new system will cut paperwork and record-keeping costs. According to Lawrence, "Companies have already squeezed what savings they can out of managed care, so HMOs like Kaiser must find new improvements to compete, or falter."

Discussion Questions

1. Describe Kaiser's move to the Internet.
2. Do you think Kaiser's Internet initiative was motivated primarily to improve patient care or reduce costs?

Critical Thinking Questions

3. In general, what healthcare systems and services would you recommend for the Internet?
4. Which company operations should be moved to the Internet? Are there operations that should not be moved to the Internet?

Sources: Adapted from Douglas Gantenbein et al., "Kaiser Takes the Cyber Cure," *Business Week*, February 7, 2000, p. 80; and Barbara Bowers, "Driven to the Top," *Best's Review Life/Health Edition*, April 1999, p. 32.

The Washington Post Tries the Web

Today, more people are using the Internet to get their local, regional, and national news. Newspapers across the country are placing content from their publications on the Internet. There are many benefits to consumers. Most of the sites do not charge a subscription fee; depending on the publication, this can save some subscribers hundreds of dollars annually. When traveling, people can easily keep up with local news on-line. Many news sites also provide past issues, and most have powerful search engines. If you want to get news on a particular topic or if you want to search the classified ads for a used car, the search engine can quickly return the desired information. Some sites also allow customers to tailor their news. By filling out a short questionnaire, the on-line news service will give you only the news you really want. In addition, online newspapers also help reduce the need for paper and trees to make paper.

The Washington Post has developed an Internet site that has been extremely successful. The well-respected on-line newspaper is read by 20 percent of residents in the Washington, D.C., area. According to Media Metrix, a Web-traffic tracking company, *The Washington Post* Internet site is rated thirteenth in the region. But although on-line newspapers like *The Washington Post* provide useful information for people, are they profitable?

On-line newspapers generate revenues from advertisements, but there are also costs associated with developing and running any Internet site. *The Washington Post*, for example, received about $17 million in additional advertising revenues from its Internet site. But it cost the newspaper approximately $65 million to develop and operate the site. Why would any company spend more on an Internet site than it returns in revenue? Many newspapers with well-established brand names are worried that other start-up, on-line-only news sites will eventually take away their core business of subscriptions, classified ad revenues, and general advertising revenues. A wrong move by these established news organizations could spell disaster. According to Patrick Heane, a senior analyst at Jupiter Communications, "It's the newspapers' game to lose." But how long will newspapers be willing to take losses on their Internet sites? According to John Borse, chief financial officer for *The Washington Post*, "If we knew that, we'd all feel a lot better."

Discussion Questions

1. Why is it important for a company like *The Washington Post* to have a site on the Internet?
2. If you were the CEO of *The Washington Post*, would you continue to operate an Internet site?

Critical Thinking Questions

3. What types of companies need an Internet site? What types of companies do not?
4. Should standard performance measures (discussed in Chapter 2), such as return on investment, be used to justify an Internet site like *The Washington Post* site?

Sources: Adapted from Erin White, "Washington Post Stays the Course with Web Operations," *The Wall Street Journal*, February 2, 2000, p. B4; and "The Emergence of Convergence," *American Journalism Review*, January 2000, p. 88.

● NOTES

Sources for opening vignette on page 251: Sources: Adapted from David Rynecki, "They've Got Mail," *Fortune,* February 7, 2000, p. 101; and David Rocks, "Going Nowhere Fast in Cyberspace," *Business Week,* January 31, 2000, p. 58.

1. Kerry Capell, "Europe's Top e.Business Leaders," *Business Week,* February 7, 2000, p. 58.
2. Jim Rohwer, "Japan Goes Web Crazy," *Forbes,* February 7, 2000, p. 115.
3. Pamela Druckerman, "Brazil Is Test Case for Free Web Access," *The Wall Street Journal,* February 7, 2000, p. A28.
4. Chikako Mogi, "One U.S. Investor Leads Pack in Chinese Internet Deals," *The Wall Street Journal,* February 1, 2000, p. A18.
5. Lisa Hoffman, "Women on Course to Pass Men on the Web," *Denver Rocky Mountain News,* January 24, 2000, p. 3B.
6. John Gantz, "Internet2 Is on the Way," *Computerworld,* March 1, 1999, p. 34.
7. "Qwest Wins Big," *Washington Technology,* January 10, 2000, p. 19.
8. Andrew Nachison, FedEx Bets on Internet2," *Computerworld,* May 17, 1999, p. 38.
9. "Do-It-Yourself Domains," *Tele.com,* January 24, 2000, p. 28.
10. Kathleen Murphy, "To Challenge the Empire," *Internet World,* January 15, 2000.
11. Jane Weaver, "The Name Game," *PC Computing,* April 2000, p. 108.
12. Lee Copeland, "Automakers Race After Alleged Cybersquatters," *Computerworld,* February 28, 2000, p. 12.
13. Larry Armstrong, "No-Fee Net Access Is Making the Old Profit Models Obsolete," *Business Week,* February 7, 2000, p. 46.
14. Roberta Furger, "The Best ISPs," *PC World,* March 1999, p. 124.
15. Les Freed et al., The Faster Web," *PC Magazine,* April 20, 1999, p. 158.
16. "Get the Software," *PC World,* January 2000.
17. Rebecca Blumenstein, "The Next Web Battle: Phone Calls," *The Wall Street Journal,* December 27, 1999, p. B1.
18. Les Freed, "Talk Gets Cheaper," *PC Magazine,* February 9, 1999, p. 77.

19. Frank Deffler, "Instant Messaging to Work," *PC Magazine,* January 18, 2000, p. 82.
20. Stan Miastkowski, "Instant Messaging Heads for Cell Phones," *CNN Online,* December 13, 1999.
21. Doreen Carajal, "More Books Bought Online than in Stores," *The New York Times,* reprinted in the *Denver Rocky Mountain News,* January 17, 2000, p. 4B.
22. Sebastian Rupley, "Biz-to-Biz Auctions," *PC Magazine,* January 4, 2000, p. 34.
23. Ronald Grover et al., "A Little Net Music," *Business Week,* February 7, 2000, p. 34.
24. Walter Mossberg, "Behind the Lawsuit: Napster Offers Model for Music Distribution," *The Wall Street Journal,* May 11, 2000, p. B1.
25. Nikhil Jutheesing, "Webcasting," *Forbes,* February 8, 1999, p. 104.
26. Adam Penenberg, "Informerical.com," *Forbes,* March 8, 1999, p. 116.
27. Jason Compton, "The Anywhere Office," *PC Computing,* March 2000, p. 146.
28. "E-Commerce in 3-D," *PC Magazine,* March 7, 2000, p. 94.
29. "Best of the Web," *Forbes,* May 22, 2000, p. 60; and Don Willmott, "The Top 100 Web Sites," *PC Magazine,* February 8, 2000, p. 144.
30. "SAP Opts for Push Technology," *Computing,* May 13, 1999, p. 8.
31. Doug Beizer, "A Web Site of Your Own," *PC Magazine,* April 6, 1999, p. 148.
32. Frank Derfler, "Virtual Private Networks," *PC Magazine,* January 4, 2000, p. 146.
33. "Virtual Private Networks," *Network News,* November 24, 1999, p. 4.
34. Eric Brown, "Broadband, Narrow Choices," *PC World,* February 2000, p. 139.
35. Mark Sweeney, "Accept Cookies by Site," *PC Magazine,* February 22, 2000, p. 99.
36. Michael Dornheim, "Firewall in Space," *Aviation Week & Space Technology,* December 13, 1999, p. 29.

PART III

Business Information Systems

CHAPTER 8

Electronic Commerce

*W*hat's happening is no less than a transformation of the most fundamental business precepts: how a company does business, how it enters new markets, how it communicates across the enterprise, and how it deals with suppliers.

— Clinton Wilder, technology writer discussing e-transformation in the September 13, 1999, issue of *Information Week*

Principles	Learning Objectives
E-commerce is a new way of conducting business, and as with any other new application of technology, it presents both opportunities for improvement and potential problems.	• *Identify several advantages of e-commerce.* • *Outline a multistage model that describes how e-commerce works.* • *Identify some of the major challenges companies must overcome to succeed in e-commerce.* • *Identify several e-commerce applications.*
E-commerce requires the careful planning and integration of a number of technology infrastructure components.	• *Outline the key components of technology infrastructure that must be in place for e-commerce to succeed.* • *Discuss the key features of the electronic payments systems needed to support e-commerce.*
Users of the new e-commerce technology must take safeguards to protect themselves.	• *Identify the major issues that represent significant threats to the continued growth of e-commerce.*
Organizations must define and execute a strategy to be successful in e-commerce.	• *Outline the key components of a successful e-commerce strategy.*

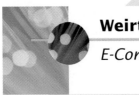

Weirton Steel

E-Commerce Pioneer

Weirton Steel Corporation is the nation's eighth largest steel company and the largest producer of tin-coated steel. Founded in 1910, Weirton Steel has survived by embracing change and innovation. Surprising as it may seem, this old-line company is recognized as an e-commerce pioneer.

Metals companies have increasingly found themselves squeezed by aggressive price competition, particularly from imports, substitution with other materials, and increasing customer demands for higher quality and faster, more reliable delivery times. Furthermore, a large portion of the metals industry is a commodity market, whereby the customer can buy the same or functionally equivalent products from multiple sources at about the same price. So, metals companies must somehow differentiate themselves in this crowded market. In such a tough economic climate, Weirton had to learn new ways to remain profitable to survive.

In the summer of 1998, Weirton announced the formation of an independent company called MetalSite to offer a secure Web-based marketplace for on-line purchases of metals products from various U.S. suppliers. The new venture was launched in phases beginning in the fall of 1998. Weirton spent more than $3 million and two years of research and validation of the site's concept and spent an additional $2 million to launch the site. Currently, 15 suppliers, including steel producers and service centers, offer a wide range of products through the MetalSite marketplace.

For metals buyers, MetalSite is a one-stop, up-to-date resource for comprehensive information about its industry and the products that manufacturers are offering. For sellers, it provides an opportunity to expand their customer base, increase efficiencies in the selling process, turn inventory more quickly, and free up sales staff from non-value-added tasks. Sellers provide MetalSite with daily product availability and pricing information independently of each other. In effect, MetalSite serves as an independent clearinghouse for metals industry information, product availability, and on-line purchases. The individual sellers, not MetalSite, are responsible for pricing, order fulfillment, and payment processing through traditional channels.

Buyers can access and use the site for free. Sellers are assessed a transaction fee for each on-line sale and an annual service fee for consulting and development of an individual market center, which basically is a mini Web site for each seller within MetalSite. MetalSite also offers users numerous ways to communicate with their peers throughout the world—much like they do at industry trade shows. The on-line tools they can use include bulletin boards, presentations, and forums.

The site features Internet security technology that is the most widely used by businesses throughout the world. Arthur Andersen LLP conducts ongoing security audits, as well as periodic audits of MetalSite's information technology, processes, and procedures.

As you read this chapter, consider the following:

- How are organizations using e-commerce to provide improved customer service, become more productive, and remain competitive?

- What are the key issues that must be addressed to ensure effective implementation of e-commerce?

AN INTRODUCTION TO ELECTRONIC COMMERCE

business-to-business (B2B) e-commerce

a form of e-commerce in which the participants are organizations

business-to-consumer (B2C) e-commerce

a form of e-commerce in which customers deal directly with the organization, avoiding any intermediaries

E-commerce involves conducting business activities using electronic data transmission involving computers, telecommunications networks, and streamlined work processes. Weirton Steel (see the opening vignette) provides an example of **business-to-business (B2B) e-commerce**, in which the participants are organizations. There is also **business-to-consumer (B2C) e-commerce**, in which customers deal directly with the organization and avoid any intermediaries. Worldwide business-to-business transactions over the Internet are projected to grow eighteenfold, from $403 billion in 2000 to over $7,300 billion in 2004.[1] The dollar volume of business-to-business e-commerce is 10 to 15 times greater than that of business-to-consumer e-commerce. Even though on-line purchases accounted for less than 1 percent of total sales dollar volume as recently as 1999, traditional retailers view e-commerce as a potent threat to brick-and-mortar stores because of the rapid growth of this component of the world's economy. As a result, they are beginning to make e-commerce part of their brick-and-mortar operations. For example, Borders Books & Music offers in-store kiosks, allowing shoppers access to its immense on-line inventory.

Businesses and individuals use e-commerce to reduce transaction costs, speed the flow of goods and information, improve the level of customer service, and enable close coordination of actions among manufacturers, suppliers, and customers. E-commerce also enables consumers and companies to gain access to worldwide markets.

Business processes that are strong candidates for conversion to e-commerce are those that are paper based and time consuming and those that can make business more convenient for customers. Thus, it comes as no surprise that the first business processes that companies converted to an e-commerce model were those related to buying and selling. For example, after Cisco Systems, the maker of Internet routers and other telecommunications equipment, put its procurement operation on-line in 1998, the company said it saved $350 million in paperwork and transaction costs in the first year.[2]

Some companies, such as those in the automotive and aerospace industries, have been conducting e-commerce for decades through the use of electronic data interchange (EDI), which involves application-to-application communications of business data (invoices, purchase orders, etc.) between companies in a standard data format. Many companies have now gone beyond simple EDI-based applications to launch e-commerce initiatives with suppliers, customers, and employees to address business needs in new areas.

Multistage Model for E-Commerce

A successful e-commerce system must address the many stages consumers experience in the sales life cycle. At the heart of any e-commerce system is the need for the user to search for and identify items for sale; select those items and negotiate prices, terms of payment, and delivery date; send an order to the vendor to purchase the items; pay for the product or service; obtain product delivery; and receive after-sales support. Figure 8.1 shows how e-commerce can support each of these stages. Product delivery may involve tangible goods delivered in a traditional form (e.g., clothing delivered via a package service) or goods and services delivered electronically (e.g., software downloaded over the Internet).

A multi-stage model for purchasing over the Internet includes search and identification, selection and negotiation, purchasing, product or service delivery, and after-sales support, as shown in Figure 8.1.

FIGURE 8.1

Multistage Model for
E-Commerce (B2B and B2C)

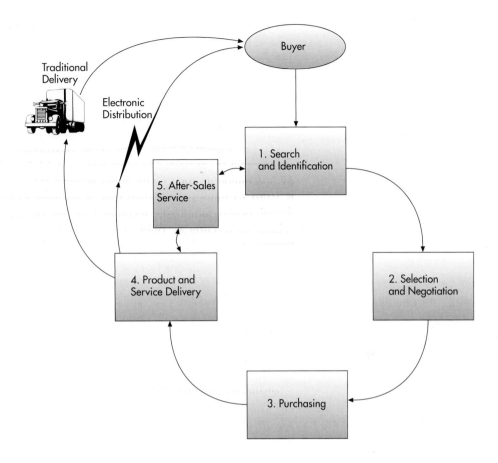

Search and Identification

An employee ordering parts for a storeroom at a manufacturing plant would follow the steps shown in Figure 8.1. Such a storeroom stocks a wide range of office supplies, spare parts, and maintenance supplies. The employee prepares a list of needed items—for example, fasteners, piping, and plastic tubing. Typically, for each item carried in the storeroom, a corporate buyer has already identified a preferred supplier based on the vendor's price competitiveness, level of service, quality of products, and speed of delivery. The employee then logs on to the Internet and goes to the Web site of the preferred supplier.

From the supplier's home page, the employee can access a product catalog and browse until finding the items that meet the storeroom's specifications. The employee fills out a request-for-quotation form by entering the item codes and quantities of the items needed. When the employee completes the quotation form, the supplier's Web application prices the order with the most current prices and shows the additional cost for various forms of delivery—overnight, within two working days, or next week. The employee may elect to visit other suppliers' Web home pages and repeat this process to search for additional items or obtain competing prices for the same items. As mentioned in Chapter 7, bots are software programs that can follow a user's instructions; they can also be used for search and identification.

Selection and Negotiation

Once the price quotations have been received from each supplier, the employee examines them and checks off, on the request-for-quotation form, which of the items, if any, will be ordered from a given supplier. The employee also specifies the desired delivery date. This data is used as input into the supplier's order processing TPS. In addition to price, an item's quality and the supplier's service and speed of delivery can be important in selection and negotiation.

Purchasing Products and Services Electronically

The employee completes the purchase order by sending a completed electronic form to the supplier. Complications may arise in paying for the products. Typically, a corporate buyer who makes several purchases from the supplier each year has established credit with the supplier in advance, and all purchases are billed to a corporate account. But when individuals make their first, and perhaps only, purchase from the supplier, additional safeguards and measures are required. Part of the purchase transaction can involve the customer providing a credit card number. However, computer criminals could capture this data and use the information to make their own purchases. To avoid this problem, some companies have developed security programs and procedures. For example, Secure Electronic Transactions (SET) is endorsed by IBM, Microsoft, MasterCard, and others. Another approach to paying for goods and services purchased over the Internet is using electronic money, which can be exchanged for hard cash; CyberCash is one example. These and many other security procedures make buying products and services over the Internet easier and safer.

Product and Service Delivery

The Internet can also be used to deliver products and services, primarily software and written material. You can download software, reports on the stock market, information on individual companies, and a variety of other written reports and documents directly from the Internet. Often called electronic distribution, sending software, music, pictures, and written material through the Internet is faster and can be less expensive than with regular order processing. Electronic distribution can also eliminate inventory problems for manufacturers, who do not have to stock hundreds or thousands of copies of the software, reports, or documents; one copy can be downloaded to customers' computers when needed.

As more and more people start using the Internet, the electronic distribution of products and services could be a major revenue source for software and publishing companies. Yet, most products cannot be delivered over the Internet. So they are delivered in a variety of ways: overnight carrier, regular mail service, truck, or rail. In some cases, the customer may elect to drive to the supplier and pick up the product.

After-Sales Service

In addition to capturing the information to complete the order, comprehensive customer information is captured from the order and stored in the supplier's customer database. This information can include customer name, address, telephone numbers, contact person, credit history, and some order details. If, for example, the customer later telephones the supplier to complain that not all items were received, that some arrived damaged, or even that product use instructions are not clear, all customer service representatives can retrieve the order information from the database via a personal computer on their desks. Companies are adding the capability to answer many after-sales questions to their Web sites—how to maintain a piece of equipment, how to effectively use the product, how to receive repairs under warranty, and so on.

E-Commerce Challenges

A number of challenges must be overcome for a company to convert its business processes from the traditional form to e-commerce processes. This section summarizes a few.

The first major challenge is for the company to define an effective e-commerce model and strategy. Although a number of different approaches can be used, the most successful e-commerce models include three basic components: community,

Three Basic Components of a
Successful E-Commerce Model

content, and commerce, as shown in Figure 8.2.[3] Message boards and chat rooms
are used to build a loyal *community* of people who are interested in and enthusias-
tic about the company and its products and services. Providing useful, accurate,
and timely *content*—such as industry and economic news and stock quotes—is a
sound approach to get people to return to your Web site time and again. *Commerce*
involves consumers and businesses paying to purchase physical goods, information,
or services that are posted or advertised on-line.

Weirton Steel designed its MetalSite Web site with these three components
in mind. For community, MetalSite offers bulletin boards, presentations, and
forums in which users can communicate with people in the metal industry
around the world. For content, it provides comprehensive information about
the industry and the products being offered by the manufacturers. For commerce, MetalSite provides a means for buyers and sellers to transact business.

Another major challenge for companies moving to business-to-consumer
e-commerce is the need to change distribution systems and work processes to
be able to manage shipments of individual units directly to consumers.
Traditional distribution systems send complete cases of a product to a store.
The store opens the cases, takes the individual units out, and stacks them on
a shelf. Then consumers walk through the aisles and pick up what they need.
In business-to-consumer e-commerce, companies need a distribution system
that can manage **split-case distribution**, in which cases of goods are split
open on the receiving dock and the individual items are stored on shelves or
in bins in the warehouse. The distribution system must also be able ship and
track individual items. The demands of business-to-consumer e-commerce
fulfillment are so great that many on-line vendors outsource the function to
companies like FedEx and UPS.[4]

Another tough challenge for e-commerce is the integration of new Web-
based order processing systems with traditional mainframe computer–based
inventory control and production planning systems. It's one thing to allow sales
reps to place orders over the Web, but they must also be able to find out what
products are in (or soon to be in) inventory and available for sale. That means
that front-end Web-enabled applications such as order taking need to be tightly
integrated to traditional back-end applications such as inventory control and
production planning, as shown in Figure 8.3.

split-case distribution

a distribution system that requires
cases of goods to be opened
on the receiving dock and the
individual items from the cases are
stored in the manufacturer's
warehouse

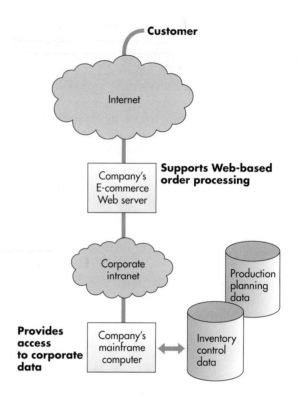

The E-Commerce Supply Chain

As discussed in Chapter 2, all business organizations contain a number of value-added processes. The supply chain management process is a key value chain that, for most companies, offers tremendous business opportunities if converted to e-commerce. **Supply chain management** is composed of three subprocesses: demand planning to anticipate market demand, supply planning to allocate the right amount of enterprise resources to meet demand, and demand fulfillment to fulfill demand quickly and efficiently (Figure 8.4). The objective of demand planning is to understand customers' buying patterns and develop aggregate, collaborative long-term, intermediate-term, and short-term

supply chain management

a key value chain composed of demand planning, supply planning, and demand fulfillment

FIGURE 8.4

Supply Chain Management

Demand Planning

Analyzing buying patterns

Developing customer demand forecasts

Supply Planning

Strategic planning

Inventory planning

Distribution planning

Procurement planning

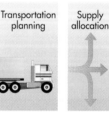
Transportation planning

Supply allocation

Demand Fulfillment

Order fulfillment

Backlog management

Order promising

Customer verification

Order capture

forecasts of customer demand. Supply planning includes strategic planning, inventory planning, distribution planning, procurement planning, transportation planning, and supply allocation. The goal of demand fulfillment is to provide fast, accurate, and reliable delivery for customer orders. Demand fulfillment includes order capturing, customer verification, order promising, backlog management, and order fulfillment.

Conversion to e-commerce supply chain management provides businesses an opportunity to achieve operational excellence throughout the extended supply chain by increasing revenues, decreasing costs, improving customer satisfaction, and reducing inventory. But to achieve this goal requires integrating all subprocesses that exchange information and move goods between suppliers and customers, including manufacturers, distributors, retailers, and any other enterprise within the extended supply chain.

Increased Revenues and Decreased Costs
By eliminating or reducing time-consuming and labor-intensive steps throughout the order and delivery process, more sales can be completed in the same time period and with increased accuracy.

Improved Customer Satisfaction
Increased and more detailed information about delivery dates and current status can increase customer loyalty.

Inventory Reduction across the Supply Chain
With increased speed and accuracy of customer order information, companies can reduce the need for inventory—from raw materials, to safety stocks, to finished goods—at all the intermediate manufacturing, storage, and transportation points.

Through improved supply chain management, companies can set their sights not only on improving their profitability and service but also on transforming entire industries. For example, Weirton Steel wants to increase the rate of annual inventory turns (a measure of inventory effectiveness calculated by dividing the cost of goods sold from stock over the past year by the average inventory level for the past year) in its tin plate business, the material in food cans that accounts for about 45 percent of Weirton's revenue. The current rate of inventory turns in the entire supply chain—from the raw-material producer to the can on the grocery shelf—is only 1.2 per year. If Weirton could double that, it could save millions. So, Weirton is encouraging food-processing customers to share three years of production data via the Internet. Weirton will then analyze that data with its supply-chain forecasting software to coordinate production scheduling throughout the value chain.[5]

Business to Business (B2B)

Although the business-to-consumer (B2C) market grabs more of the news headlines, the business-to-business (B2B) market is considerably larger and expected to grow much more rapidly. Business-to-business e-commerce offers enormous opportunities. It allows manufacturers to buy at a low cost worldwide, and it offers enterprises the chance to sell to a global market right from the start. Moreover, e-commerce offers great promise for developing countries, helping them to enter the prosperous global marketplace, and hence helping reduce the gap between rich and poor countries.

The rapid development of e-commerce presents great challenges to society. Even though e-commerce is creating new job opportunities, it could also cause a loss of employment in traditional job sectors. Many companies may fail in the intense competitive environment of e-commerce and find themselves out of business. Therefore, it is vital that the opportunities and implications of e-commerce be understood.

Business to Consumer (B2C) Internet Shopping

E-commerce for consumers is still in its early stages. Most shoppers are not yet convinced that it is worthwhile to connect to the Internet, search for shopping sites, wait for the images to download, try to figure out the ordering process, and then worry about whether their credit card numbers will be stolen by a hacker. But attitudes are changing, and an increasing number of shoppers are beginning to appreciate the importance of e-commerce. For time-strapped households, consumers are asking themselves, Why waste time fighting crowds in shopping malls when from the comfort of home I can shop on-line anytime and have the goods delivered directly? These shoppers have found that many goods and services are cheaper when purchased via the Web—for example, stocks, books, newspapers, airline tickets, and hotel rooms. They can also get information about automobiles, cruises, and homes to cut better deals. More than a new tool for placing orders, the Internet is emerging as a paradise for comparison shoppers. Internet shoppers can, for example, unleash shopping bots or access sites such as www.jango.com (a shopping resource of Excite) to browse the Internet and obtain lists of items, prices, and merchants.

E-COMMERCE APPLICATIONS

electronic retailing (e-tailing)

the direct sale from business to consumer through electronic storefronts, typically designed around an electronic catalog and shopping cart model

cybermall

a single Web site that offers many products and services at one Internet location

AutoNation.com is the largest auto dealership network, with 378 dealerships and 37 used-vehicle megastores.

Since B2B and B2C e-commerce use is spreading, it's important to examine some of the most common current uses. E-commerce is being applied to retail and wholesale, manufacturing, marketing, investment and finance, and auctions.

Retail and Wholesale

There are numerous examples of e-commerce in retail and wholesale. **Electronic retailing**, sometimes called e-tailing, is the direct sale from business to consumer through electronic storefronts, which are typically designed around the familiar electronic catalog and shopping cart model. Companies such as Office Depot, Wal-Mart, and many others have used the same model to sell wholesale to employees of corporations. There are tens of thousands of electronic retail Web sites—selling literally everything from soup to nuts. In addition, cybermalls are another means to support retail shopping. A **cybermall** is a single Web site that offers many products and services at one Internet location—the basic idea of a regular shopping mall. The cybermall employs the Internet's ubiquity to pull together multiple buyers and sellers into one virtual place, easily reachable through a Web browser.[6]

Auto retailing provides an excellent example of how e-commerce is transforming retail selling. AutoNation, with $17 billion in sales, 376 dealerships, and 37 used-vehicle megastores, is the largest auto dealership network. Its goal is to make every step of the purchase process available on-line—that is, everything but the test drive and handing over the keys. The e-commerce unit has only 22 employees, but it expects to generate three-quarters of a billion dollars in on-line sales using Compass, AutoNation's dealer extranet built with tools and services from Fusion.com and an Oracle8 database. The

system delivers and manages on-line sales leads. Now sales reps, instead of sitting and waiting to pounce on the next customer who walks in, use the computer to e-mail interested customers with information or to set up appointments.[7]

Spending on manufacturing, repair, and operations (MRO) goods and services—from simple office supplies to mission-critical equipment, such as the motors, pumps, compressors, and instruments that keep manufacturing facilities up and running smoothly—provides an example of e-commerce applied to wholesale selling. Spending on MRO goods and services often approaches 40 percent of a manufacturing company's total revenues. Despite this, spending can be haphazard, without automated controls for purchasing materials. In addition to these external purchase costs, companies face significant internal costs resulting from ineffective and cumbersome MRO management processes. The result is lost productivity and capacity. Estimates show that a high percentage of manufacturing downtime is often a result of not having the right part at the right time in the right place. E-commerce software for plant operations provides powerful comparative searching capabilities to enable identification of duplicate and functionally equivalent items, giving the insight to spot opportunities for cost savings. Comparing various suppliers, coupled with consolidating more spending with fewer suppliers, leads to decreased profits. In addition, automated workflows enable the implementation of best practices, such as requesting and approving part additions and deletions in a master record in the plant maintenance software.

Manufacturing

electronic exchange

an electronic forum where manufacturers, suppliers, and competitors buy and sell goods, trade market information, and run back-office operations

One approach taken by many manufacturers to raise profitability and improve customer service is to move their supply chain operations onto the Internet. Here they can form an **electronic exchange** to join with competitors and suppliers alike using computers and Web sites to buy and sell goods, trade market information, and run back-office operations, such as inventory control, as shown in Figure 8.5. With such an exchange, the business center is not a physical building but a network-based location where business interactions occur. This approach has greatly speeded the movement of raw materials and finished products among all members of the business community, thus reducing the amount of inventory that must be maintained. It has also led to a much more competitive marketplace and lower prices.

By mid-2000, nearly 1,000 on-line marketplaces in 70 industries had been announced by Internet and brick-and-mortar companies, including the following: the Worldwide Retail Exchange led by Safeway and Kmart; a giant marketplace for high-tech firms overseen by Hewlett-Packard and Compaq Computer; a global exchange for the auto industry run by automakers Ford, General Motors, and DaimlerChrysler; Chemdex in the chemical industry; and eSteel in the steel industry. It is predicted that more than 10,000 on-line exchanges will sweep the business world between 2000 and 2003; however, Arthur Andersen estimates barely one in four will survive.[8] To date, only a few exchanges are seeing trickles of revenue from transaction fees, software licensing, and other charges; none is believed to be profitable yet.

There are several strategic and competitive issues associated with the use of exchanges. Many companies distrust their corporate rivals and fear they may lose trade secrets through participation in such exchanges. Suppliers worry that the on-line marketplaces and their auctions will drive down the prices of goods and favor buyers. Suppliers also can spend a great deal of money in the setup to participate in multiple exchanges. For example, more than a dozen new exchanges have appeared in the oil industry, and the printing industry is up to more than 20 on-line marketplaces. Until a clear winner emerges in particular industries, suppliers are more or less forced to sign on to several or all of them.[9]

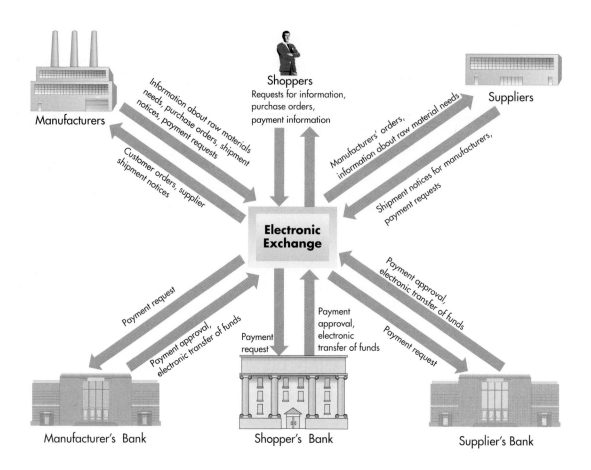

FIGURE 8.5

Model of an Electronic
Exchange

Yet another issue is potential government scrutiny of exchange participants—anytime competitors get together to share information, it raises questions of collusion or antitrust behavior.

Many companies that already use the Internet for their private exchanges have no desire to share their on-line expertise with competitors. At Wal-Mart, the world's number one retail chain, executives have turned down several invitations to join exchanges in the retail and consumer goods industries. Wal-Mart is pleased with its in-house exchange, Retail Link, which connects the company to 7,000 worldwide suppliers that sell everything from toothpaste to furniture.

In addition to formal exchanges, manufacturers are using e-commerce to improve the efficiency of the selling process. For example, with annual revenue of $1.7 billion, Timken is the world's largest manufacturer of tapered roller bearings, a product used in everything from automobile transmissions to paper mills. Timken is over 100 years old and had established ways of doing business. It sells primarily to distributors and for years has used EDI networks to exchange purchase orders and invoices with its business partners. But globalization, consolidation, an emphasis on one-stop shopping, and comprehensive contracts put increased pressure on the company to change its ways. Timken developed Timken Direct to move customer queries on-line—and to improve customer relationships. Previously, to maintain good customer relationships, Timken's sales associates had to travel with four-wheel suitcases to carry all the sales information and sample products. Not any more. With links to Timken's back-end inventory databases, the secure Timken Direct Web site handles the two customer concerns that occupied most of the service representatives' time: product availability and price. Distributors can go to the Web site and get this information for themselves. As a result, Timken has been able to free up 15 percent of its service reps and convert them to sales reps, so they can make sales calls instead of fielding availability and cost queries.[10]

Marketing

The nature of the Web allows firms to gather much more information about customer behavior and preferences than they could using other marketing approaches. Marketing organizations can measure a large number of activities as customers and potential customers gather information and make their purchase decisions. Analysis of this data is complicated because of the Web's interactivity and because each visitor voluntarily provides or refuses to provide personal data such as name, address, e-mail address, telephone number, and demographic data. Internet advertisers use the data they gather to identify specific portions of their markets and target them with tailored advertising messages. This practice, called **market segmentation**, divides the pool of potential customers into segments, which are usually defined in terms of demographic characteristics such as age, gender, marital status, income level, and geographic location.

The idea of technology-enabled relationship management has become possible when promoting and selling on the Web. **Technology-enabled relationship management** occurs when a firm obtains detailed information about a customer's behavior, preferences, needs, and buying patterns and uses that information to set prices, negotiate terms, tailor promotions, add product features, and otherwise customize its entire relationship with that customer.

DoubleClick is a leading global Internet advertising company that leverages technology and media expertise to help advertisers use the power of the Web to build relationships with customers. The DoubleClick Network is its flagship product, a collection of high-traffic and well-recognized sites on the Web (AltaVista, Dilbert, US News, Macromedia, and more than 1,500 others). This network of sites is coupled with DoubleClick's proprietary DART targeting technology, which allows advertisers to target their best prospects based on the most precise profiling criteria available. DoubleClick then places a company's ad in front of those best prospects. Comprehensive on-line reporting lets advertisers know how their campaign is performing and what type of users are seeing and clicking on their ads. This high-level targeting and real-time reporting provide speed and efficiency not available in any other medium. The system is also designed to track advertising transactions, such as impressions and clicks, to summarize these transactions in the form of reports, and to compute DoubleClick Network member compensation.

market segmentation

the identification of specific markets and targeting them with advertising messages

technology-enabled relationship management

the use of detailed information about a customer's behavior, preferences, needs, and buying patterns to set prices, negotiate terms, tailor promotions, add product features, and otherwise customize the entire relationship with that customer

DoubleClick is a leading global Internet advertising company that leverages technology and media expertise to help advertisers use the power of the Web to build relationships with customers.

Investment and Finance

The Internet has revolutionized the world of investment and finance. Perhaps the changes have been so great because this industry had so many built-in inefficiencies and so much opportunity for improvement.

On-line Stock Trading

Before the World Wide Web, if you wanted to invest in stocks, you called your broker and asked what looked promising. He'd tell you about two or three companies and then would try to sell you shares of a stock or perhaps a mutual fund. The sales commission was well over $100 for the stock (depending on the price of the stock and the number of shares purchased) or as much as an 8 percent sales charge on the mutual fund. If you wanted information about the company before you invested, you would have to wait two or three days for a one-page Standard and Poor's stock report providing summary information and a chart of the stock price for the past two years. Once you purchased or sold the stock, it would take two days to get an order confirmation in the mail, detailing what you paid or received for the stock.

Today, more than 14 percent of all stock trades are made on-line, and the average commission charged by the top ten on-line brokers was approximately $15 in 1999.[11] Now all you need to do is log on to the Web site of your on-line broker and, with a few keystrokes and a few clicks of your mouse to identify the stock and number of shares involved in the transaction, you can buy and sell securities in seconds. In addition, an overwhelming amount of free information is available to on-line investors—from the latest Securities & Exchange filings to the rumors spread in chat rooms. See Table 8.1 for a short list of the more valuable sites.

One indispensable tool of the on-line investor is a portfolio tracker. This tool allows you to enter information about the securities you own—ticker symbol, number of shares, price paid, and date purchased—at a tracker Web site. You can then access the tracker site to see how your stocks are doing. (There is typically a 15- to 20-minute delay between the price displayed at the site and the price at which the stock is actually being sold.) In addition to reporting the current value of your portfolio, most sites provide access to news, charts, company profiles, and analyst ratings on each of your stocks. You can also program many of the trackers to watch for certain events (e.g., stock price change of more than +/– 3 percent in a single day). When one of the events you specified occurs, an "alert" symbol is posted next to the affected stock. Table 8.2 lists a number of the more popular tracker Web sites.

TABLE 8.1

Web Sites Useful to Investors

Name of Site	URL	Description
411Stocks	www.411stocks.com	One-stop location to get lots of information about a stock—price data, news, discussion groups, charts, basic data, financial statements, and delayed quotes
MarketReporter	www.marketreporter.com	Provides financial news, recommendations, upgrades, downgrades, message boards, stock market simulation game
Thomson Investors Network	www.thomsoninvest.com	Financial commentary from a number of stock market publications, including *First Watch* and *Stocks to Watch*
Elite Trader	www.elitetrader.com	Virtual gathering place for day traders with bulletin boards and chat rooms
Dayinvestor.com	www.dayinvestor.com	News and stock alerts with frequent briefs on market activity and rumors
DRIP Advisor	www.dripadvisor.com	Covers the basics of dividend reinvestment programs (DRIPs), what companies offer DRIPs, and how to start a DRIP
FertileMind	www.fertilemind.net	Members share information and insights on stocks and the market
The Raging Bull	www.ragingbull.com	Contains lots of message boards; guest experts produce news, commentary, and analysis
EDGAR Online	www.edgar-online.com	Provides access to company filings with the Securities and Exchange Commission (SEC)
Federal Filings Online	www.fedfil.com	Dow Jones directory of documents filed with the federal government, including bankruptcy proceedings, initial public offering (IPO) filings, SEC reports, and court cases

TABLE 8.2

Popular Stock Tracker
Web Sites

Name of Web Stock Tracker Site	URL
MSN MoneyCentral	moneycentral.msn.com/investor
Quicken.com	www.quicken.com
The Motley Fool	www.fool.com
Yahoo!	quote.yahoo.com
Morningstar	www.morningstar.com

On-line Banking

On-line banking customers can check balances of their savings, checking, and loan accounts; transfer money among accounts; and pay their bills. With on-line banking, these customers think they can gain a better current knowledge of how much they have in the bank, eliminate the need to write checks in long-hand, and reduce how much they spend on envelopes and stamps. All of the nation's major banks and many of the smaller banks enable their customers to pay bills on-line. Fewer than a million consumers banked on-line in 1997, but nearly 7.5 million will by the end of 2000, and that number may double to 15.7 million by 2003.[12]

Here's how electronic bill payment works. You first set up a list of frequent payees, along with their addresses and a code describing the type of payment, such as "home mortgage." Then when you go on-line to pay your bills, you simply enter the code or name assigned to the check recipient, the amount of the check, and the date you want it paid. In many cases, the bank still prints and mails a check, so you have to time your on-line transactions to allow for bank processing and mail delays. But most bill-paying programs allow you to schedule recurring payments for every week, month, or quarter, which you might want to do for your auto loan or health insurance bill.

electronic bill presentment

a method of billing whereby the biller posts an image of your statement on the Internet and alerts you by e-mail that your bill has arrived

The next advance in on-line bill paying is **electronic bill presentment**, which will eliminate all paper, right down to the bill itself. Under this process, the biller posts an image of your statement on the Internet and alerts you by e-mail that your bill has arrived. You then direct your bank to pay it. As of May 2000, CheckFree (http://www.checkfree.com) offers such a service from 76 companies, including major utility companies, Amoco, Phillips 66, and Sears. Many major banks, including Key, Norwest, and BancOne, are planning to offer a similar service that will be run by Microsoft and First Data Corp. directly at the banks' Web sites in late 2000.

Fool.com is a Web site that allows users to track the performance of their stock portfolios.

Auctions

As discussed in Chapter 7, the Internet has created many new options for electronic auctions, where geographically dispersed buyers and sellers can come together. A special type of auction called *bidding* allows a prospective buyer to place only one bid for an item or a service. Priceline.com is the patented Internet bidding system that enables consumers to achieve significant savings by naming their own price for goods and services. Priceline.com takes these consumer

Priceline.com is a patented Internet bidding system that enables consumers to save money by naming their own price for goods and services.

offers and then presents them to sellers, who can fill as much of that guaranteed demand as they wish at price points determined by buyers.

Now that we've examined some of the applications of e-commerce, let's look at some of the more technical issues related to information systems and technology that make it possible.

TECHNOLOGY INFRASTRUCTURE

For e-commerce to succeed, a complete technology infrastructure must be in place. These infrastructure components must be chosen carefully and integrated to be capable of supporting a large volume of transactions with customers, suppliers, and other business partners worldwide. Indeed, on-line consumers complain that poor Web site performance (e.g., slow response time and "lost" orders) drives them to abandon some e-commerce sites in favor of those with better, more reliable performance. This section provides a brief overview of the key technology infrastructure components (Figure 8.6).

Hardware

A Web server hardware platform complete with the appropriate software is a key e-commerce infrastructure ingredient. The amount of storage capacity and computing power required of the Web server depends primarily on two things—what software must run on the server and the volume of e-commerce transactions that must be processed. Although it is possible for information systems staff to define the software to be used, sometimes there is much guesswork in

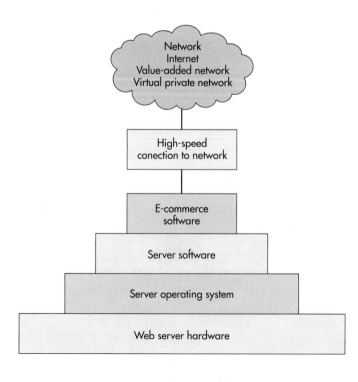

FIGURE 8.6

Key Technology Infrastructure Components

estimating how much traffic the site will generate. As a result, the most successful e-commerce solutions are designed to be highly scalable so that they can be upgraded to meet unexpected user traffic.

A key decision facing new e-commerce companies is whether to host their own Web site or to let someone else do it. Many companies decide that using a third-party Web service provider is the best way to meet initial e-commerce needs. The third-party company rents space on its computer system and provides a high-speed connection to the Internet, which minimizes the initial out-of-pocket costs for e-commerce start-up. The third-party service provider can also provide personnel trained to operate, troubleshoot, and manage the Web server. Of course, a number of companies decide to take full responsibility for acquiring, operating, and supporting the Web server hardware and software, but this approach requires considerable up-front capital and a set of skilled and trained individuals. Whichever approach is taken, there must be adequate hardware backup to avoid a major business disruption in case of a failure of the primary Web server.

Web Server Software

In addition to the Web server operating system, each e-commerce Web site must have Web server software to perform a number of fundamental services, including security and identification, retrieval and sending of Web pages, Web site tracking, Web site development, and Web page development. The two most popular Web server software packages are Apache HTTP Server and Microsoft Internet Information Server.

Security and Identification
Security and identification services are essential for intranet Web server implementation to identify and verify employees coming into the server from the Internet. Access controls provide or deny access to files based on the user name or URL. Web servers support encryption processes for transmitting private information securely over the public Internet.

Retrieving and Sending Web Pages
The fundamental purpose of a Web server is to process and respond to client requests that are sent using the HTTP protocol. In response to such a request, the Web server program locates and fetches the appropriate Web page, creates an HTTP header, and appends the HTML document to it. For dynamic pages, the server involves other programs, retrieves the results from the backend process, formats the response, and sends the pages and other objects to the requesting client program.

Web Site Tracking
Web servers capture visitor information, including who is visiting the Web site (the visitor's URL), what search engines and key words they used to find the site, how long their Web browser viewed the site, the date and time of each visit, and which pages were displayed. This data is placed into a **Web log file** for future analysis.

Web Site Development
Web site development tools include features such as an HTML/visual Web page editor (e.g., Microsoft's FrontPage, NetStudio's NetStudio, SoftQuad's HoTMetal), software development kits that include sample code and code development instructions for languages such as Java or Visual Basic, and Web page upload support to move Web pages from a development PC to the Web

Web log file

a file that contains information about visitors to a Web site

Web site development tools

tools used to develop a Web site, including HTML or visual Web page editor, software development kits, and Web page upload support site. Which tools are bundled with the Web server software depends on which Web server software you select.

Web Page Construction

Web page construction software

software that uses Web editors and extensions to produce both static and dynamic Web pages

static Web pages

Web pages that always contain the same information

dynamic Web pages

Web pages that contain variable information built in response to a specific Web visitor's request

Web page construction software uses Web editors and extensions to produce Web pages—either static or dynamic. **Static Web pages** always contain the same information—for example, a page that provides text about the history of the company or a photo of corporate headquarters. **Dynamic Web pages** contain variable information and are built in response to a specific Web visitor's request. For example, if a Web site visitor inquires about the availability of a certain product by entering a product identification number, the Web server will search the product inventory database and generate a dynamic Web page based on the current product information it found, thus fulfilling the visitor's request. This same request by another visitor in the day may yield different results due to ongoing changes in product inventory. A server that handles dynamic content must be able to access information from a variety of databases. The use of open database connectivity enables the Web server to assemble information from different database management systems, such as SQL Server, Oracle, and Informix.

E-Commerce Software

e-commerce software

software that supports catalog management, product configuration, shopping cart facilities, e-commerce transaction processing, and Web traffic data analysis

Once you have located or built a host server, including the hardware, operating system, and Web server software, you can begin to investigate and install e-commerce software. There are five core tasks that **e-commerce software** must support: catalog management, product configuration, shopping cart facilities, e-commerce transaction processing, and Web traffic data analysis.

The specific e-commerce software you choose to purchase or install depends on whether you are setting up for B2B or B2C. For example, B2B transactions do not include sales tax calculations, and software to support B2B must incorporate electronic data transfers between business partners, such as purchase orders, shipping notices, and invoices. B2C software, on the other hand, must handle the complication of accounting for sales tax based on the current laws and rules in effect in the various states.

Catalog Management

catalog management software

software that automates the process of creating a real-time interactive catalog and delivering customized content to a user's screen

Any company that offers a wide range of product offerings requires a real-time interactive catalog to deliver customized content to a user's screen. **Catalog management software** combines different product data formats into a standard format for uniform viewing, aggregating, and integrating catalog data into a central repository for easy access, retrieval, and updating of pricing and availability changes. The data required to support large catalogs is almost always stored in a database on a computer that is separate from, but accessible to, the e-commerce server machine

Product Configuration

product configuration software

software used by buyers to build the product they need on-line

Customers need help when an item they are purchasing has many components and options. Product configuration software tools were originally developed in the 1980s to assist B2B salespeople to match their company's products to customer needs. Buyers use the new Web-based **product configuration software** to build the product they need on-line with little or no help from salespeople. For example, Dell customers use product configuration software to build the computer of their dreams. Use of such software can expand into the service arena as well, with consumer loans and financial services to help people decide what sort of loan or insurance is best for them.

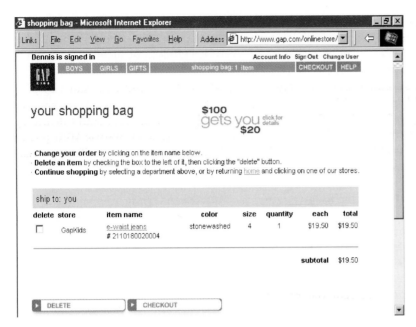

FIGURE 8.7

Electronic Shopping Cart

An electronic shopping cart (or bag) allows online shoppers to view their selections and add or remove items.

electronic shopping cart

a model commonly used by many e-commerce sites to track the items selected for purchase, allowing shoppers to view what is in their cart, add new items to it, and remove items from it

e-commerce transaction processing software

software that provides the basic connection between participants in the e-commerce economy, enabling communications between trading partners, regardless of their technical infrastructure

Web site traffic data analysis software

software that processes and analyzes data from the Web log file to provide useful information to improve Web site performance

Shopping Cart

Today many e-commerce sites use an **electronic shopping cart** to track the items selected for purchase, allowing shoppers to view what is in their cart, add new items to it, or remove items from it, as shown in Figure 8.7. To order an item, the shopper simply clicks that item. All the details about it—including its price, product number, and other identifying information—are stored automatically. If the shopper later decides to remove one or more items from the cart, he or she can do so by viewing the cart's contents and removing any unwanted items. When the shopper is ready to pay for the items, he or she clicks a button (usually labeled "proceed to checkout") and begins a purchase transaction. Clicking the "Checkout" button displays another screen that usually asks the shopper to fill out billing, shipping, and payment method information and to confirm the order.

E-Commerce Transaction Processing

E-commerce transaction processing software connects participants in the e-commerce economy and enables communication between trading partners, regardless of their technical infrastructure. This software fully automates transaction processes from order placement to reconciliation.

Basic transaction processing software takes data from the shopping cart and calculates volume discounts, sales tax, and shipping costs to arrive at the total cost. In some cases, the software determines shipping costs by connecting directly to shipping companies such as UPS, FedEx, and Airborne. In other cases, shipping cost may be a predetermined amount for each item ordered.

More and more e-commerce companies are outsourcing the actual inventory management and order fulfillment process to a third party. In this situation, the e-commerce transaction processing software must actually route order information to one of the shipping companies to ship from inventory under their management. For example, Hewlett-Packard has an arrangement with FedEx to manage orders for HP printers.

Web Traffic Data Analysis

It is necessary to run third-party **Web site traffic data analysis software** to make sense of all the data captured in the Web log file—to turn it into useful information to improve Web site performance. For example, when someone queries a search engine for a keyword related to your site's products or services, does your page appear in the top ten matches, or does your competitor's? If you're listed but not within the first two or three pages of results, you lose, no matter how many engines you submitted your site to.

Network and Packet Switching

Several technologies must be in place for e-commerce to work. The most obvious is that some sort of network must be in place since e-commerce depends on the secure transmission of data over networks, whether the Internet, corporate extranets, value-added networks (VANs), or virtual private networks (VPNs, discussed in Chapter 7). All these approaches rely on basic packet switching technology and the use of routers to help each packet arrive at its destination in the quickest and most economical manner. In choosing among the options, the information systems staff must weigh cost, availability, reliability, security, and redundancy issues. Cost includes the initial development cost as well as the ongoing operational and support costs. Availability concerns the hours the network is *scheduled* to be available for normal use. Reliability is the percentage of available time the network must actually be fully operational, typically 99 percent or more. Security is the ability to keep messages from being intercepted. And redundancy is the ability of the network to keep operating if key elements fail. All are important factors that must be considered.

ELECTRONIC PAYMENT SYSTEMS

digital certificate

an attachment to an e-mail message or data embedded in a Web page that verifies the identity of a sender or a Web site

certificate authority (CA)

a trusted third party that issues digital certificates

secure sockets layer (SSL)

a communications protocol used to secure all sensitive data

The SSL (Secure Sockets Layer encryption) communications protocol assures customers that information they provide to retailers, such as credit card numbers, cannot be viewed by anyone else on the Web.

Electronic payment systems are a key component of the e-commerce infrastructure. Electronic payments make up only about 1 percent of all consumer settlements now, but they should reach 5 percent by 2005. That represents growth from $43 billion to $325 billion, or nearly 800 percent, in just six years.[13] Current e-commerce technology relies on the use of user identification and encryption to safeguard business transactions. Actual payments are made in a variety of ways, including electronic cash; electronic wallets; and smart, credit, charge, and debit cards.

As discussed in Chapter 7, authentication technologies are used by organizations to confirm the identity of a user requesting access to information or assets. A **digital certificate** is an attachment to an e-mail message or data embedded in a Web site that verifies the identity of a sender or Web site. A **certificate authority (CA)** is a trusted third-party organization or company that issues digital certificates. The CA is responsible for guaranteeing that the individuals or organizations granted these unique certificates are, in fact, who they claim to be. Digital certificates thus create a trust chain throughout the transaction, verifying both purchaser and supplier identities.

Secure Sockets Layer

All on-line shoppers fear the theft of credit card numbers and banking information. To help prevent this from happening, the **secure sockets layer (SSL)** communications protocol is used to secure sensitive data. The SSL communications protocol sits above the TCP layer of the OSI model discussed in Chapter 7, and other protocols such as Telnet and HTTP can be layered on top of it. SSL includes a handshake stage, which authenticates the server (and the client, if needed), determines the encryption and hashing algorithms to be used, and exchanges encryption keys. The handshake may use public key encryption. Following the handshake stage, data may be transferred. The data is always encrypted. This ensures that your transactions are not subject to interception or "sniffing" by a third party. For companies wishing to conduct serious e-commerce, such as receiving credit card numbers or other sensitive information, SSL is a must. Although SSL handles the encryption

part of a secure e-commerce transaction, a digital certificate is necessary to provide server identification.

One tip to the security of a transaction is visible on-screen. Look at the bottom left corner of your browser before sending your credit card number to an e-commerce vendor. If you use Netscape Navigator, make sure you see a solid key in a small blue rectangle. If you use Microsoft Explorer, the words "Secure Web site" appear near a little gold lock. And, if you're worried about how secure a secure connection is, visit Netcraft at http://www.netcraft.com/security/. At this site you can type in any Web site address and determine the equipment being used for secure transactions. One more tip: to ensure security, you should always use the newest browser available, the newer the browser, the better the security.

Electronic Cash

electronic cash

an amount of money that is computerized, stored, and used as cash for e-commerce transactions

Electronic cash is an amount of money that is computerized, stored, and used as cash for e-commerce transactions. A consumer must open an account with a bank and show some identification to establish identity to obtain electronic cash. Then whenever the consumer wants to withdraw electronic cash to make a purchase, he or she accesses the bank via the Internet and presents proof of identity—typically a digital certificate issued by a certification authority. After the bank verifies the consumer's identity, it issues the consumer the requested amount of electronic cash and deducts the same amount from the consumer's account. The electronic cash is stored in the consumer's electronic wallet on his or her computer's hard drive, or on a smart card (both are discussed later).

Consumers can spend their electronic cash when they locate e-commerce sites that accept electronic cash for payment. The consumer sends electronic cash to the merchant for the specified cost of the goods or services. The merchant validates the electronic cash to be certain it is not forged and belongs to the customer. Once the goods or services are shipped to the consumer, the merchant presents the electronic cash to the issuing bank for deposit. The bank then credits the merchant's account for the transaction amount, minus a small service charge.

There are two distinct types of electronic cash: identified electronic cash and anonymous electronic cash (also known as *digital cash*). Identified electronic cash contains information revealing the identity of the person who originally withdrew the money from the bank. Also, in much the same manner as credit cards, identified electronic cash enables the bank to track the money as it moves through the economy. The bank can determine what people or organizations bought, where they bought it, when they bought it, and how much they paid. Anonymous electronic cash works just like real paper cash. Once anonymous electronic cash is withdrawn from an account, it can be spent or given away without leaving a transaction trail. A few of the companies that provide electronic cash mechanisms include CyberCash, Mondex, SpendCash, and Visa.

Electronic Wallets

electronic wallet

a computerized stored value that holds credit card information, electronic cash, owner identification, and address information

On-line shoppers quickly tire of repeatedly entering their shipment and payment information each time they make a purchase. An **electronic wallet** holds credit card information, electronic cash, owner identification, and address information. It provides this information at an e-commerce site's checkout counter. When consumers click on items to purchase, they can then click on their electronic wallet to order the item, thus making on-line shopping much faster and easier.

Household International, a leading provider of consumer finance, credit card, auto finance, and credit insurance products, in partnership with General

Motors Corporation and CyberCash, a provider of e-commerce technologies and services for merchants, introduced the GM Card easyPay electronic wallet marketed to a large portion of GM card members. The GM Card easyPay Wallet stores a shopper's name, credit card information, shipping details, and other pertinent facts that can be called up to make an on-line purchase with a single click of a computer mouse. The wallet is available to GM card members and interested consumers at www.GMCard.com.[14]

Smart, Credit, Charge, and Debit Cards

On-line shoppers use credit and charge cards for the majority of their Internet purchases. A credit card, such as Visa or MasterCard, has a preset spending limit based on the user's credit limit, and each month the user can pay off a portion of the amount owed or the entire credit card balance. Interest is charged on the unpaid amount. A charge card, such as American Express, carries no preset spending limit, and the entire amount charged to the card is due at the end of the billing period. Charge cards do not involve lines of credit and do not accumulate interest charges.

Debit cards look like credit cards or automated teller machine (ATM) cards, but they operate like cash or a personal check. While a credit card is a way to "buy now, pay later," a debit card is a way to "buy now, pay now." Debit cards allow you to spend only what is in your bank account. It is a quick transaction between the merchant and your personal bank account. When you use a debit card, your money is quickly deducted from your checking or savings account. Credit, charge, and debit cards currently store limited information about you on a magnetic stripe. This information is read each time the card is swiped to make a purchase. All credit card customers are protected by law from paying any more than $50 for fraudulent transactions.

smart card

a credit card–sized device with an embedded microchip to provide electronic memory and processing capability

The **smart card** is a credit card–sized device with an embedded microchip to provide electronic memory and processing capability. Smart cards can be used for a variety of purposes, including storing a user's financial facts, health insurance data, credit card numbers, and network identification codes and passwords. They can also store monetary values for spending.

Smart cards are better protected from misuse than conventional credit, charge, and debit cards because the smart card information is encrypted. Conventional credit, charge, and debit cards clearly show your account number on the face of the card. The card number, along with a forged signature, is all that a thief needs to purchase items and charge them against your card. A smart card makes credit theft practically impossible because a key to unlock the encrypted information is required, and there is no external number that a thief can identify and no physical signature a thief can forge.

Smart cards have been around for over a decade and are widely used in Europe, Australia, and Japan, but they have not caught on in the United States. Use has been limited because there are so few smart card readers to record payments, and U.S. banking regulations have slowed smart card marketing and acceptance as well. American Express launched its Blue card smart card in 1999. The card comes with a "reader" that attaches to your PC monitor with sticky tape. You must visit the American Express Web site to get an electronic wallet to store your credit card information and shipping address. When you want to buy something on-line, you go to the checkout screen of a Web merchant, swipe your Blue card through the reader, type in a password, and you're done. The digital wallet automatically tells the vendor your credit card number, its expiration date, and your shipping information.[15]

The "E-Commerce" special interest box outlines why companies are preparing their Web sites to handle customers from outside the United States and what steps they must take to do so.

E-commerce sites need to shift their focus from North American consumers as the use of the Internet grows rapidly in markets throughout Europe, Asia, and Latin America. The majority of Internet users will live outside the United States by 2003, and the U.S. share of all e-commerce revenues is projected to shrink from 69 percent in 2000 to 59 percent by 2003. On-line retail sales in Europe will grow at a rate of 98 percent annually over the next five years, soaring from 2.9 billion euros in 1999 to 175 billion euros in 2005. Companies that want to succeed on the Web cannot ignore the global shift. Developing a sound strategy is critical for ensuring that a global e-commerce effort leads to Web sites that are relevant to the consumers and businesses the company wants to reach, whether those customers are in Cleveland, Singapore, or Frankfurt.

The first step in developing a global e-commerce strategy is to determine which global markets make the most sense for selling products or services on-line. One approach is to target regions and countries in which a company already has on-line customers. Companies can track the country domains from which current users of a U.S.-centric site are visiting, and established global companies can look to their overseas offices to help determine the languages and countries to target for their Web sites.

Once the company decides which global markets it wants to reach with its Web site, it must adapt the existing U.S.-centric site to another language and culture—a process called *localization*. Localization requires companies to have a deep understanding of the country, its people, and the market, which means either building a physical presence in the country or forming partnerships so that detailed knowledge can be gathered. Companies must be prepared to take painstaking steps to ensure that e-commerce customers have a local experience even though they're shopping at the Web site of an American company.

Some of the steps involved in localization require recognizing and conforming to the nuances, subtleties, and tastes of local cultures, as well as supporting basic trade laws and technological capabilities such as each country's currency, local connection speeds, payment preferences, laws, taxes, and tariffs. For example, when Dell Computer launched an e-commerce site to sell PCs to consumers in Japan, it made the mistake of surrounding most of the site's content with black borders, a negative sign in Japanese culture. Japanese Web shoppers took one look at the site and fled. Support for Asian languages is difficult because Asian alphabets are more complex and not all Web development tools are capable of handling them. As a result, many companies choose to tackle Asian markets last. In addition, great care must be taken to choose icons that are relevant to a country. For example, the use

of mailboxes and shopping carts may not be familiar to global consumers. Users in European countries don't take their mail from large, tubular receptacles, nor do many of them shop in stores large enough for wheeled carts.

One of the most important and most difficult decisions in a company's global Web strategy is whether Web content should be generated and updated centrally or locally. Companies that expand through international partnerships may be tempted to hand control to the new international entities to take the greatest advantage of the expertise of employees in the new markets. But turning over too much control can lead to a muddle of country-specific sites with no consistency and a scattered corporate message. A mixed model of control may be best. Decisions about corporate identity, brand representation, and the technology used for the Web sites are made centrally to minimize Web development and support effort as well as to present a consistent corporate and brand message. But a local authority decides on content and services best tailored for given markets.

Companies must also be aware that consumers outside the United States will access sites with different devices and modify their site design accordingly. In Europe, for example, closed-system iDTVs (interactive digital televisions) are becoming a popular way to access on-line content, with iDTVs projected to reach 80 million European households by 2005. Such devices have better resolution and more screen space than PC monitors U.S. consumers use to access the Internet. So users of iDTVs expect more ambitious graphics.

A new group of software and service vendors has emerged to address Web globalization issues. The group includes companies such as Idiom, GlobalSight, and Uniscape.com. Their software can integrate with popular e-commerce and Web content management software from vendors such as Vignette, BroadVision, and Interwoven. The multilingual Web site management software can work especially well for global sites with central management.

On the Web, ultimately the only way to fight global companies is to be a global company. Successful firms operate with a portfolio of storefronts designed for each target market, with shared sourcing and infrastructure to support the network of stores, and with local marketing and business development teams to take advantage of local opportunities. Service providers continue to emerge to solve the cross-border logistics, payments, and customer service needs of these pan-European retailers.

Discussion Questions

1. Outline the major steps in taking a U.S. Web site global.
2. What is meant by central control versus local control of the Web site content? Which approach is better? Why?

E-COMMERCE

Get Ready for Global E-Commerce (continued)

Critical Thinking Questions

3. Which approach is better to gain a deep understanding of a country, its people, and the market: form a partnership with a company in the country, or hire a software vendor familiar with Web globalization issues? Why?

4. Your company has just completed globalization of its Web site to address the needs of customers in seven countries in Latin America. How would you go about evaluating the success of this effort?

Sources: Adapted from Matt Hicks and Anne Chen, "Dress for Global Success," *PC Week*, April 3, 2000, pp. 65–73; Anne Chen and Matt Hicks, "Going Global? Avoid Cultural Clashes," *PC Week*, April 3, 2000, p. 6; Idiom Web, "E-Business Globalization" page for global consulting, http://www.idiominc.com, accessed May 20, 2000; "European Online Retail Will Soar to 175 Billion Euros by 2005," press release, accessed March 28, 2000, http://www.forrester.com/ER/Press/Release/0,1769,266,FF.html; and Clay Shirky, "Go Global or Bust," *Business 2.0*, March 2000, http://www.business2.com/articles/2000/03/content/break_3.html, accessed May 20, 2000.

THREATS TO E-COMMERCE

As with any revolutionary change, a host of issues must be dealt with to ensure that e-commerce transactions are safe and consumers are protected. Many represent significant threats to its continued growth. The following sections summarize a number of these threats and present practical ideas on how to minimize their impact.

Security

As mentioned previously, organizations use identification technologies to confirm the identity of a user requesting access to information or assets.

Just as shoppers are concerned with the legitimacy of sites, e-businesses are concerned with the legitimacy of their customers. As a result, an increasing number of companies are investing in biometric technology to protect both sides. Companies such as Drug Emporium, ING Direct Canada, and Election.com have begun testing fingerprint scans, among other things, to verify users' identities when prescribing drugs, accessing bank accounts, and voting on-line.[16] Biometric technology, which digitally encodes physical attributes of a person's voice, eye, face, or hand and associates them with biological attributes stored in a file, are commonly used in organizations such as the FBI to allow clearance into a building, for instance. Currently, using the technology to secure on-line transactions is rare for both cost and privacy reasons. It can be expensive to outfit every customer with a biometric scanner, and it is difficult to convince consumers to supply something as personal and distinguishing as a fingerprint.

Yet, biometrics as an e-commerce authentication tool is being pushed into the limelight by lawmakers. In California and Ohio, state legislators are requiring that physicians who want to prescribe drugs on-line must be authenticated using biometrics. Using software and services from BioNetrix, Drug Emporium is testing an application that allows physicians to order prescriptions over the Internet with whatever method of biometric identification the doctor prefers. Drug Emporium is paying for all the doctors' equipment and sending the BioNetrix consultants on-site to each physician's location to install hardware and conduct the initial biometric scan. Then, BioNetrix's Authentication Suite will

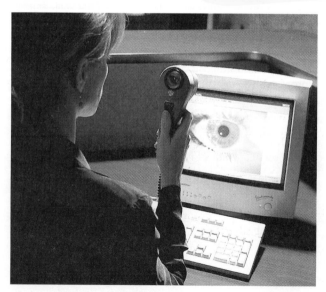

Biometric technology such as iris scanning can provide security for accessing files or obtaining clearance to enter a building.

(Source: Courtesy of IriScan, Inc.)

manage all authentication processes, including password, tokens, fingerprints, retinal scans, and voice recognition, from one console through its Authentication Management Infrastructure software. All data is encrypted using SSL encryption. The system allows Drug Emporium to identify that the prescription was received by a licensed pharmacy and sent by the doctor who has been authenticated.[17]

Intellectual Property

intellectual property

music, books, inventions, paintings, and other special items protected by patents, copyrights, or trademarks

Lawsuits over **intellectual property** (music, books, inventions, paintings, and other special items protected by patents, copyrights, or trademarks) have created a virtual e-commerce war zone. From music files to patented on-line coupon schemes, hardly a day goes by without some new suit being filed. Since 1995, federal lawsuits over patents, copyrights, and other intellectual property have risen ten times faster than other cases brought under federal law, topping 8,200 cases in 1999. Sometimes whole industries, such as entertainment and travel, have their futures riding on the outcomes of these intellectual property battles.[18]

Some argue that unscrupulous e-commerce companies threaten creativity by enabling people to steal the original works of innovators and artists. For example, Napster, a maker of music-sharing software, is seeing its very existence threatened by lawsuits from the music industry and individual artists. MP3.com lost a copyright infringement case brought by a record industry trade group and is facing possible damages that could exceed $6 billion. In response to all these disputes, the life of patents has been increased (now lasting 20 years instead of 17), copyrights now last 70 years after the artist dies (up from 50), and more kinds of copyright infringements are now criminal offenses.

Another class of intellectual property lawsuits involves patents on business processes. For example, even though Amazon.com believes that the trend toward business process patents may hurt e-commerce, it went to court for an injunction to stop rival bookstore Barnes and Noble from using "one-click" shopping during the Christmas 1999 shopping season. Critics say the process was too obvious to be patented, but Amazon disagrees. Other companies have been sued because their Web site copies the "look and feel" of another company's site. Such lawsuits highlight a potentially debilitating problem. What is the proper extent of property rights in an information-based economy? The flurry of e-commerce patent lawsuits raises worries of overly aggressive protection of intellectual property rights and of stalling economic activity.

Fraud

As more people use the Internet for e-commerce, they need to know that the merchant with whom they're dealing is legitimate. The Better Business Bureau's OnLine Reliability seal helps distinguish trustworthy companies among the thousands of businesses marketing on-line, allowing consumers to easily identify them.

Consumer fraud may be bigger than the estimated $5 billion in legitimate business-to-consumer retail Web sales.[19] On-line swindlers include everyone from whiz-kid vandals to the equivalent of the old boiler-room brokers who hawk bogus offshore tax shelters. The good news is that you can enjoy the advantages of e-commerce and still keep yourself largely protected. Following are descriptions of the most common scams and some advice to help protect you.

On-line Auction Fraud

On-line auctions brought in about $1 billion in sales in a recent year, and they represent the number one Internet fraud, according to the National Consumers League. The majority of that fraud comes from so-called person-to-person auctions, which account for roughly half the auction sites. On these sites, it is up to the buyer and seller to resolve details of payment and delivery; the auction sites offer no guarantees. Sticking with auction sites like Onsale (www.onsale.com) and First Auction (www.firstauction.com) that ensure the delivery and quality of all the items up for bidding can help buyers avoid trouble.

Spam

America Online estimates that unsolicited e-mail offering bargains and get-rich schemes represents up to 30 percent of its 28 million e-mail messages sent each day. Such e-mail that is sent to a wide range of people and Usenet groups indiscriminately is called **spam**. Spam allows peddlers to hawk their products instantly to thousands of people at virtually no cost. And obtaining e-mail addresses to spam is now a snap, thanks to so-called *harvester programs* that snoop Usenet chat groups and collect thousands of e-mail addresses in a single day. Do not respond to spam; responding will only confirm to the spammer that your email address is accurate and active. Instead, simply delete spam as soon as you get it. If the spam is truly offensive or obviously fraudulent, forward the entire message to your Internet service provider, the FTC (www.ftc.gov), or the National Fraud Information Center (www.fraud.org) and ask that the sender be barred from sending additional messages.

Pyramid Schemes

Traditional pyramid schemes work by getting new investors to pay their recruiters to join the pyramid at the bottom. The new investors then (theoretically) get rich by recruiting additional new investors who will funnel money up the pyramid. Of course, eventually the people on the bottom of the pyramid have trouble finding new recruits, and it collapses. For that reason, pyramids are illegal. On-line pyramids often bundle the sale of a product or service like vitamins, credit cards, or even electricity to justify downstream recruitment fees. Usually, the product that these supposed multilevel-marketing (MLM) companies hawk is so overpriced or unwanted that the company relies mainly on recruiting fees to provide its cash flow. Be wary of any company that depends on recruitment fees to pay you. Also, avoid "opportunities" that force you to buy costly inventory. Finally, check with the Better Business Bureau (www.bbb.com) and your state attorney general's office before joining any company you're unsure about.

Investment Fraud

The North American Securities Administrators Association estimates that up to $10 billion will be lost in a year to investment fraud, primarily through the sale of bogus investments. Not all this fraud is committed on-line. Some of the more colorful scams involve eel farms, imported diamonds, and super-fast-growing trees that can reach 80 feet in three years. Call your state securities board to make sure the seller and its "security" are registered. Avoid dealing with unregistered securities and brokers.

Stock Scams

Before the Internet, most individuals had a tough time spreading rumors about a stock, but now anybody can do it. Net chat rooms like Usenet's misc.invest.stocks make it cheap and easy for scammers to talk up (or down) a stock based on false information. The scammer buys the stock at a low price, spreads false rumors that help drive the stock price up, and then sells at an artificially high price before the bottom falls out. Always verify tips and information you get on-line and in chat rooms with the company directly or with reputable financial advisers. To be truly safe, avoid thinly traded stocks or penny stocks that sell for less than $1 a share, because they are the most easily manipulated.

Privacy

On-line consumers are more at risk today than ever before. One of the primary factors causing higher risk is *on-line profiling*—the practice of Web advertisers' recording on-line behavior for the purpose of producing targeted advertising. **Clickstream data** is the data gathered based on the Web sites you visit and what items you click on. From the marketers' perspective, the use of on-line profiling allows businesses to market to customers electronically one to one. The benefit to customers is better, more effective service; the benefit to providers is the increased business that comes from building relationships and encouraging customers to return for subsequent purchases. From the consumers' perspective, on-line profiling squeezes out anonymity, which remains crucial to privacy on the Internet. And consumers also fear that the personal information thus gathered will be shared with others without their knowledge. For example, a number of lawsuits have been filed against on-line marketers who intended to create massive data warehouses of consumer information by linking clickstream data with data (such as automobile registrations and product warranty registrations) from other sources.

The U.S. government has not implemented a federal privacy policy and instead relies on e-commerce self-regulation in matters of data privacy. The Better Business Bureau Online and TRUSTe are independent, nonprofit privacy initiatives dedicated to building users' trust and confidence on the Internet and supporting the accelerating growth of e-commerce. The BBB Online Privacy seal program, which helps consumers identify online businesses that honor privacy protection policies, now covers 5,000 Web sites and has hundreds of applications in process. Figures 8.8 and 8.9 show the TRUSTe and BBB Online Privacy seals, respectively. The TRUSTe program is based on a multifaceted assurance process that establishes Web site credibility, thereby making users more comfortable when making on-line purchases and providing personal information.

A set of **safe harbor principles** has been established that address the issues of notice, choice, and access associated with data privacy. A company following these principles must notify consumers of the purpose of data collection, allow consumers to opt out of having their data shared with third parties, and provide users with access to their personal information to review and possibly correct it. With an opt-out policy, the Web site is free to gather and sell information

clickstream data

data gathered based on the Web sites you visit and what items you click on

safe harbor principles

a set of principles that address the e-commerce data privacy issues of notice, choice, and access

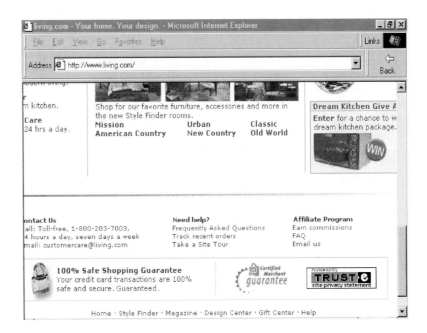

FIGURE 8.8

TRUSTe Seal

FIGURE 8.9

BBB Online Privacy Seal

about you unless you specifically tell it that it cannot, which typically involves clicking on a button. With an opt-in policy, the gathering or selling of your data is forbidden unless you specifically give the Web site permission. The safe harbor principles also require organizations using personal information to "take reasonable precautions to protect it from loss, misuse, and unauthorized access, disclosure, alteration, and destruction."

Table 8.3 lists several tips to help safeguard your privacy. Also, read the "Ethical and Societal Issues" special interest box to see how the lack of data privacy regulations in the United States is affecting our international e-commerce.

STRATEGIES FOR SUCCESSFUL E-COMMERCE

With all the constraints to e-commerce just covered, it is important for a company to develop an effective Web site—one that is easy to use and accomplishes the goals of the company, yet is affordable to set up and maintain. We cover several issues for a successful e-commerce site here.

TABLE 8.3

How to Protect Your Privacy
While On-line

Tip	Rationale
Never give out personal information, especially your social security number.	Assume that any information you give out about yourself will appear on a database somewhere. Even something as innocuous as a warranty card may ask for income and phone numbers—don't provide such data.
Before registering with a Web site, check out its privacy policy.	Look for the BBB or TRUSTe seal. If a site doesn't explain how it plans to use your information, beware!
Be discreet in sending e-mail, posting messages to a Usenet newsgroup, and "talking" in chat rooms.	Anything you send in these forums can be viewed and used by a marketer to update your personal profile.
Update your browser software.	The newest versions of Netscape Navigator and Microsoft Internet Explorer, by far the two most popular Web browsers, have the best encryption technology available. Generally speaking, if these companies release an update, someone has found a hole in their security.
Consider the purchase of a digital ID to use in place of your real name.	This permits users to retrieve Web pages anonymously.

ETHICAL AND SOCIETAL ISSUES
Maintaining the Flow of Global E-Commerce Data

In October 1998, the European Union (EU) issued a directive to ensure the free flow of data across the 15 EU member countries by establishing strict standards for data privacy. The directive mandates that member states must ensure that data transferred outside the EU is protected and bars the export of data to countries that do not have comparable data privacy protection. Failure to meet this requirement could lead to the EU's blocking e-commerce data flows.

Negotiations regarding data flows with the United States have been plagued by EU concerns that the largely voluntary system of data privacy in the United States does not meet the EU directive's stringent standards. Each side has its own privacy philosophy, with the key difference being enforcement. Europe favors strict government regulation of data privacy, and the enforcement is done by a set of commissioners. The U.S. government follows a more laissez-faire approach, with no federal privacy policy, and it relies on self-regulation overseen by such watchdog organizations as the U.S. Department of Commerce, Better Business Bureau Online, and TRUSTe.

In May 2000, the EU agreed that U.S. companies that voluntarily enter safe harbor will be deemed as providing adequate privacy protection and that data transfers to those firms will continue. However, this decision is reversible if the EU determines that the safe harbor participants are not living up to their commitments or that the U.S. Department of Commerce is not overseeing the process with a high level of enforcement. Should this happen, the EU could withdraw the classification and possibly block data flows to the United States. European governments are particularly concerned that safe harbor does not provide for individual redress of data privacy violations.

Prior to this agreement, U.S. firms doing business with European citizens or their overseas subsidiaries needed a signed contract each time a transaction called for personal data. Requiring contractual agreements every time data is exchanged is very cumbersome and administratively prohibitive.

If European data privacy commissioners and other privacy advocates had their way, the United States would pass federal laws and create a European-style data-protection authority, eliminating the need for U.S. exceptions. They feel it will become clearer and clearer, as more countries adopt privacy laws, that the United States is the exception in failing to protect its citizens. The shocking result of safe harbor is that Americans could become second-class citizens in their own country—having fewer safeguards than foreign citizens.

Discussion Questions

1. What is the primary difference in philosophy between the European Union and the United States with regard to data privacy policy?
2. Do a Web search on safe harbor and data privacy. Summarize the current state of affairs.

Critical Thinking Questions

3. What would be the impact on e-commerce if the EU and United States could not agree on standards for ensuring data privacy of e-commerce data flows? How might this affect the economies of the countries involved?
4. In what way, relative to data privacy, could Americans become second-class citizens in their own country?

Sources: Adapted from Elizabeth De Bony, "U.S. Data-Privacy Rules Pass First EU Hurdle," *Industry Standard*, March 2000, http://www.thestandard.com/articles/display, accessed May 20, 2000; Jonathan Koppell, "Big Government," *Industry Standard*, February 7, 2000, http://www.thestandard.com/article/display/0,1151,9516,00.html, accessed May 22, 2000; Declan McCullagh, "U.S., E.U. Approach Safe Harbor," *Wired News*, May 9, 2000, http://www.wired.com/news/print/0,1294,36235,00.html; Ayla Jean Yackley, "Safe Harbor Vote Delayed," *Wired News*, April 17, 2000, http://www.wired.com/news/politics/0,1283,35406,00.html; "BBB Online Reliability Program Reaches 5,000 Businesses," May 16, 2000, Better Business Bureau Web site, http://www.bbb.org/alerts/BBBOn5000.asp.

Developing an Effective Web Presence

Most companies are still developing a Web presence. According to International Data Corporation, as of June 2000, 60 percent of U.S. companies have in-house Web sites but only about 20 percent of them actually sell anything over the Web.[20] When building a Web site, the first thing to decide is which tasks the site must accomplish. Most people agree that an effective Web site is one that creates an attractive presence and that meets the needs of its visitors, including the following:

- Obtaining general information about the organization
- Obtaining financial information for making an investment decision
- Learning the organization's position on social issues

- Learning about the products or services that the organization sells
- Buying the products or services that the company offers
- Checking the status of an order
- Getting advice or help on effective use of the products
- Registering a complaint about the organization's products
- Registering a complaint concerning the organization's position on social issues
- Providing a product testimonial or idea for a product improvement or new product
- Obtaining information about warranties or service and repair policies for products
- Obtaining contact information for a person or department in the organization

Once a company determines which objectives its site should accomplish, it can proceed to the details of actually putting up a site.

Putting Up a Web Site

Companies large and small can establish Web sites. Previously, companies had to develop their sites in-house or find and hire contractors to develop their sites. But no longer must you learn the intricacies of HTML or Java, master Web design software, or hire someone to build your site. Web site hosting services and the use of storefront brokers are two options to designing, building, operating, and maintaining a Web site. Both options offer the advantage of getting the Web site running faster and cheaper than doing it yourself, especially for a firm with few or no experienced Web developers.

Web Site Hosting Services

Web site hosting companies

companies that provide the tools and services required to set up a Web page and conduct e-commerce within a matter of days and with little up-front cost

Web site hosting companies such as Bigstep.com, eCongo.com, and freemerchant.com have made it possible to set up a Web page and conduct e-commerce within a matter of days and with little up-front cost. Such companies have packaged all the basic development tools and made them available for free. Using only a browser, you can choose a Web site template suited to your specific type of business, whatever it may be—clothing, collectibles, or sports equipment. Using the tools provided, you can write the descriptive text and add images of the products you want to sell. Within minutes, your site is up on the Web and accessible to millions.

These companies can also provide free hosting for your store; but to allow visitors to pay for merchandise with credit cards, you need a merchant account with a bank. If your company doesn't already have one, then it must establish one. It costs $14.95 per month plus $.20 per transaction at Bigstep.com and takes at least a few days to establish a merchant account. Without such an account, you must accept payment by check or money order. In many cases, your site won't have its own distinct Web address—www.wearmyhats.com, for instance. Instead, your store will appear as a subsection of your service provider's URL.[21]

Storefront Brokers

storefront broker

companies that act as middlemen between your Web site and on-line merchants that have the products and retail expertise

Another model for setting up a Web site is the use of a **storefront broker**, which serves as a middleman between your Web site and on-line merchants that have the actual products and retail expertise. At sites such as Pop2it (www.pop2it.com), and Vstore (www.vstore.com), you pick and choose what to sell to match the themes of your site and the products you think might interest your visitors. The products are displayed on your own Web pages, creating, in effect, a virtual storefront stocked with merchandise that is actually handled by another on-line merchant.

The storefront broker deals with the details of the transactions, including who gets paid for what, and is responsible for bringing together merchants and reseller sites. The storefront broker is similar to a distributor in standard retail operations, but in this case no product moves—only electronic data flows back and forth. Products are ordered by a customer at your site, transacted through a user interface provided by the storefront broker, and shipped by the merchant.

Storefront brokers make their money by taking a commission from the merchant—anywhere from 5 percent to 25 percent—or by collecting a finder's fee per customer from the merchant. The storefront operator usually takes a commission, again in the range of 5 to 25 percent. Although these brokered sales represent a loss of margin for the original merchant, the increased volume can often make up for lost margins. By multiplying the number of outlets for the products—sometimes by the thousands—merchants stand to make more money.[22]

Consider the case of Fruit of the Loom. The underwear manufacturer supplies plain white T-shirts to distributors who, in turn, sell the shirts to designers who add logos touting colleges and other organizations. Fruit of the Loom's e-commerce system automatically ships T-shirts to distributors at the negotiated price whenever stocks run low.

Building Traffic to Your Web Site

The Internet includes hundreds of thousands of e-commerce Web sites. With all those potential competitors, a company must take strong measures to ensure that the customers it wants to attract can find its Web site. The first step is to obtain and register a domain name. It helps if your domain name says something about your business. For instance, www.lookiehere.com might seem to be a good catchall, but it doesn't describe the nature of the business—it could be anything. If you want to sell soccer uniforms and equipment, then you'd try to get a domain name like www.soccerstuff.com, www.soccerworld.com, or www.soccervillage.com. The more specific the Web address, the better.

meta tag

a special HTML tag, not visible on the displayed Web page, that contains keywords representing your site's content, which search engines use to build indexes pointing to your Web site

The next step to attracting customers is to make your site search-engine-friendly by including a meta tag in your store's home page. A **meta tag** is a special HTML tag, not visible on the displayed Web page, that contains keywords representing your site's content, which search engines use to build indexes pointing to your Web site. Again, the selection of keywords is critical to attracting customers, so they should be chosen carefully.

You can also use Web site traffic data analysis software to turn the data captured in the Web log file into useful information. This data can tell you the URLs from which your site is being accessed, the search engines and keywords that find your site, and other useful information. Using this data can help you identify search engines to which you need to market your Web site, allowing you to submit your Web pages to them for inclusion in the search engine's index.

These tips and suggestions are only a few ideas that can help a company set up and maintain an effective e-commerce site. With technology and competition changing continually, managers should read articles in print and on the Web to keep up to date on ever-evolving issues.

SUMMARY

PRINCIPLE • E-commerce is a new way of conducting business, and as with any other new application of technology, it presents both opportunities for improvement and potential problems.

Businesses and individuals use e-commerce to reduce transaction costs, speed the flow of goods and information, improve the level of customer service, and enable the close coordination of actions among manufacturers, suppliers, and customers. E-commerce also enables consumers and companies to gain access to worldwide markets.

. . . .

A successful e-commerce system must address the many stages consumers experience in the sales life cycle. At the heart of any e-commerce system is the ability of the user to search for and identify items for sale; select those items; negotiate prices, terms of payment, and delivery date; send an order to the vendor to purchase the items; pay for the product or service; obtain product delivery; and receive after-sales support.

. . . .

The first major challenge is for the company to define an effective e-commerce strategy. Although there are a number of different approaches to e-commerce, the most successful models include three basic components: community, content, and commerce. Another major challenge for companies moving to business-to-consumer e-commerce is the need to change their distribution systems and work processes to be able to manage shipments of individual units directly to consumers and deal with split-case distribution. A third tough challenge for e-commerce is the integration of new Web-based order processing systems with traditional mainframe computer-based inventory control and production planning systems.

. . . .

Electronic retailing (e-tailing) is the direct sale from business-to-consumer through electronic storefronts designed around an electronic catalog and shopping cart model. A cybermall is a single Web site that offers many products and services at one Internet location. Manufacturers are joining electronic exchanges where they can join with competitors and suppliers to use computers and Web sites to buy and sell goods, trade market information, and run back-office operations such as inventory control. They are also using e-commerce to improve the efficiency of the selling process by moving customer queries about product availability and prices online. The nature of the Web allows firms to gather much more information about customer behavior and preferences than they could using other marketing approaches. This new technology has greatly enhanced the practice of market segmentation and enabled companies to establish closer relationships with their customers. Detailed information about a customer's behavior, preferences, needs, and buying patterns allow companies to set prices, negotiate terms, tailor promotions, add product features, and otherwise customize a relationship with a customer. The Internet has also revolutionized the world of investment and finance, especially online stock trading and online banking. The Internet has also created many options for electronic auctions, where geographically dispersed buyers and sellers can come together.

PRINCIPLE • E-commerce requires the careful planning and integration of a number of technology infrastructure components.

A number of infrastructure components must be chosen and integrated to support a large volume of transactions with customers, suppliers, and other business partners worldwide. These components include hardware, Web server software, e-commerce software, and network and packet switching.

. . . .

Current e-commerce technology relies on the use of identification and encryption to safeguard business transactions. A digital certificate is an attachment to an e-mail message or data embedded in a Web page that verifies the identity of a sender or a Web site. To help prevent the theft of credit card numbers and banking information, the Secure Sockets Layer communications protocol is used to secure all sensitive data. Actual payments are made in a variety of ways including electronic cash, electronic wallets, and smart, credit, charge, and debit cards.

PRINCIPLE • Users of the new e-commerce technology must take safeguards to protect themselves.

As with any new revolutionary change, there are a host of issues associated with e-commerce. Among these issues are security, intellectual property rights, fraud, and privacy. e-commerce shoppers must be on constant guard to protect their rights, security, and personal privacy.

PRINCIPLE • Organizations must define and execute a strategy to be successful in e-commerce.

Most people agree that an effective Web site is one that creates an attractive presence and meets the needs of its visitors. These are many and varied.

E-commerce start-ups must choose whether they will build and operate the Web site themselves or outsource this function. Web site hosting services and storefront brokers provide options to building your own Web site. It is also critical to build traffic to your Web site by registering a domain name that is relevant to your business, making your site search-engine-friendly by including a meta tag in your home page, and using Web site traffic data analysis software to attract additional customers.

● KEY TERMS

business-to-business (B2B) e-commerce 292
business-to-consumer (B2C) e-commerce 292
catalog management software 306
certificate authority (CA) 308
clickstream data 315
cybermall 298
digital certificates 308
dynamic Web pages 306
e-commerce software 306
e-commerce transaction processing software 307

electronic bill presentment 303
electronic cash 309
electronic exchange 299
electronic retailing (e-tailing) 298
electronic shopping cart 307
electronic wallet 309
intellectual property 313
market segmentation 301
meta tag 319
product configuration software 306
safe harbor principles 315
secure sockets layer (SSL) 308
smart card 310

spam 314
split-case distribution 295
static Web pages 306
storefront broker 318
supply chain management 296
technology-enabled relationship management 301
Web log file 305
Web page construction software 306
Web site development tools 305
Web site hosting companies 318
Web site traffic data analysis software 307

● REVIEW QUESTIONS

1. Define the term *e-commerce*. What is the expected size of the worldwide e-commerce market?
2. What sort of business processes are good candidates for conversion to e-commerce?
3. What is meant by e-commerce supply chain management, and what is its goal?
4. Briefly describe the three subprocesses that make up supply chain management.
5. What are some of the business opportunities presented by business-to-business e-commerce? What are some of the key issues?
6. Briefly describe the multistage model for purchasing over the Internet.
7. Which is bigger today—business-to-business or business-to-consumer e-commerce? Which will be bigger by 2003?
8. What is a portfolio tracker? What services does it provide?
9. Briefly describe how on-line banking works.

10. What is an electronic exchange? How does it work?
11. What are some of the issues associated with the use of electronic exchanges?
12. What are some of the advantages of using a service like Bigstep.com or eCongo.com to set up a Web site? What are some of the disadvantages?
13. Outline the steps a firm needs to take to adapt an existing U.S.-centric Web site to another language and culture.
14. What role do digital certificates and certificate authorities play in e-commerce?
15. What is the secure sockets layer and how does it support e-commerce?
16. Briefly explain the differences among smart, credit, charge, and debit cards.
17. What actions can you take to minimize the risk of being a victim of e-commerce fraud?
18. What actions can you take to safeguard your individual privacy as you surf the Web?

● DISCUSSION QUESTIONS

1. Why is it important in effective e-commerce for front-end Web-enabled applications like order taking to be tightly integrated to back-end applications like inventory control and production planning?
2. How does e-commerce support the build-to-order manufacturing model?
3. Why hasn't on-line banking caught on more quickly than it has?
4. Wal-Mart, the world's number one retail chain, has turned down several invitations to join exchanges in the retail and consumer goods industries. Is this good or bad for the overall U.S. economy? Why?
5. How would you decide whether to use a service like Bigstep.com or a storefront broker like Pop2it

to set up a Web site for a small business you want to run out of your home?
6. What are some of the ethical and societal issues associated with integrating clickstream, demographic, and transactional data into one comprehensive data warehouse? What actions can you take to minimize these issues?
7. Briefly discuss actions e-commerce companies have taken to ensure the legitimacy of their customers. Which of these approaches do you think is best? Why?
8. Discuss the pros and cons of e-commerce companies capturing data about you as you visit their sites.

● PROBLEM-SOLVING EXERCISES

1. Imagine that you work for a company like DoubleClick that captures information about e-commerce shoppers for use in marketing. Develop a logical data model that represents the data you would ideally like to capture for each e-commerce shopper. Develop a second model of the data you would like to capture about each visit of an e-commerce shopper to a specific Web site. Use a database management software package to implement these models.

 2. Do research to get current data about the growth of B2B or B2C e-commerce—either in the United States or worldwide. Use a graphics software package to create a line graph representing this growth. Extend the growth line ten years beyond the available data using two different modeling tools available with the software package. Write a paragraph discussing the accuracy of your ten-year projection and the likelihood that e-commerce will achieve this forecast.

● TEAM ACTIVITIES

1. Do research on TRUSTe and develop a position paper on how well this organization safeguards individual data privacy. Identify additional safeguards you think need to be implemented.
2. As a team, choose an idea for a Web site—products or services you would provide. Develop an implementation plan that outlines the steps you need to take and the decisions you must make to

set up the Web site and make it operational.
3. As a team, choose an idea for forming a company to provide services to companies that want to or are operating e-commerce Web sites. Where do you think the greatest needs and opportunities are? What sort of investments would you need in terms of capital, equipment, and people to get started in your new business?

● WEB EXERCISES

1. Some of the most private areas of your life are those surrounding the healthcare of you and your family. Details such as medical history, prescriptions, insurance—even your basic health and beauty aid purchases—are no one's business but your own. Access the Web sites of several on-line drugstores, such as Drug Emporium, Walgreens, and CVS, and read their privacy policies and perform an assessment of their security. How do you rate these sites?

2. CyberCash provides e-commerce technologies and services, enabling commerce across the entire spectrum from electronic retailing to the Internet. CyberCash provides a complete line of software products and services allowing merchants, financial institutions, and consumers to conduct secure transactions and other e-commerce functions using a broad array of popular payment forms. Visit the CyberCash Web site at http://www.cybercash.com and become familiar with its products and services. Find the Web site of a CyberCash competitor and compare and contrast the products and services of the two companies.

3. The Electronic Frontier Foundation (EFF), one of TRUSTe's founders, is a nonprofit organization working to promote privacy, free expression, and social responsibility in new media. EFF believes that the TRUSTe program educates users about electronic privacy and provides a positive means for companies to build informed consent and trust into their on-line business models. Visit the EFF Web site (http://www.eff.org) and write a brief report summarizing the scope of this site.

● CASES

 Drug Emporium Tests Biometrics

Drug Emporium, a drugstore chain, created its Web site, DrugEmporium.com, in 1997 to offer on-line prescriptions and advice from pharmacists. But because the parent company operates 190 stores in 26 states, complying with state regulations governing secure access to the site proved challenging. States require different combinations of password, token, fingerprint, or retinal scans to authenticate the identity of doctors and pharmacists on the site. For example, California and Ohio state legislators require that physicians who want to prescribe drugs on-line must be authenticated using biometrics.

Drug Emporium needed a solution to accommodate all the different rules and regulations it had to meet across the United States. Using software and services from BioNetrix, Drug Emporium is testing an application that allows physicians to order prescriptions over the Internet with whatever method of biometric identification the doctor prefers. Drug Emporium is paying for all the doctors' equipment and sending the BioNetrix consultants to each physician's location to install hardware and conduct the initial biometric scan. Then, BioNetrix's software manages all authentication processes, including password, tokens, fingerprints, retinal scans, and voice recognition. All data

is encrypted with 40-bit and higher SSL encryption and sent to a single server for authentication by comparing the received data with data on file for that individual. The system allows Drug Emporium to confirm that the prescription was received by a licensed pharmacy and that the doctor is who he says he is.

Discussion Questions

1. Why is there a need to encrypt authentication data?
2. What are the advantages of biometric authentication versus the use of log-on identification codes and passwords? What are the disadvantages?

Critical Thinking Questions

3. Imagine that you are the Drug Emporium employee responsible for evaluating the results of this security system. What key factors would you use to determine the success or failure of this system? What specific tests would you conduct?
4. Which forms of biometric authentication do you think are the most and least reliable? Why?

Sources: Adapted from Anne Chen, "Buying Online: May We Have Your Fingerprint?" *PC Week*, February 28, 2000, pp. 55, 58; and Anne Harrison, "Online Pharmacy Tests Biometrics," *Computerworld*, November 8, 1999, p. 12.

2 On-line Pet Stores

The pet supply market is bigger than toys or music, with Americans spending $23 billion on pet food and supplies in 1999. Although only $299 million of that was spent on-line, the Web market for pet supplies is expected to hit $2.5 billion by 2000—nearly a tenfold increase—according to Forester Research. So it was no surprise that during 1999 more than a dozen major pet supply Web sites were established. Indeed, the Net seems to be a natural for pet owners. Pet owners can check out all the Web sites for specials on food, vitamins, and toys. Most tend to buy the same supplies regularly and welcome the convenience of shopping on-line and having supplies shipped directly to their home within a few days.

Most pet supply Web sites provide a wealth of information, including expert advice from a staff of pet industry experts and veterinarians. Their goal is to give consumers the confidence that they are providing their pets with the best possible care. However, with operating margins a razor-thin 2 percent at leading retailer PETsMART and 4.5 percent at rival Petco, pet supplies have never been a lucrative business. Between start of operation in February 1999 to the end of 1999, Pets.com lost $62 million on a meager $6 million in sales. PETsMART.com fared only slightly better during 1999, losing $48 million on sales of $10 million.

With such thin margins, companies need to sell more than just dog food. Where the companies can improve profitability is through the sale of extras, such as designer collars, grooming supplies, and beds. In a brick-and-mortar location, the 20-pound bags of dog food can be stored in the back, so people have to walk past the colored leashes, but how do you showcase those items on-line? Pet sites are trying. Most send e-mail to customers regularly, pointing out whatever they have on sale.

Pet site executives say that things aren't as dire as they seem. Their goal is to grab market share now so that they can profit in the future while cutting e-commerce-related expenses. For example, one big expense for Pets.com comes from shipping goods by air to the East Coast so that customers get their orders in three days. Those costs will ease once Pets.com builds its distribution center in Indianapolis. Pets.com also expects a big payoff from the $21 million it invested in marketing and advertising trying to create a unique identity. Its branding campaign centers on a sock puppet that looks like a white dog with black patches. It made an appearance in Macy's Thanksgiving Day parade in New York as a 36-foot-high balloon. The singing mascot was also featured in a Super Bowl ad that cost Pets.com more than $2 million.

While Pets.com is a pure Net company, PETsMART.com, the on-line arm of the 480-store chain, is showing the advantage of a clicks-and-mortar strategy. Founded in 1987 in Phoenix, Arizona, PETsMART is a leading worldwide operator of superstores specializing in pet food, pet supplies and pet services. The average domestic superstore stocks more than 12,000 items and offers the most complete assortment of products and services available for household pets and companion animals. PETsMART, which employs approximately 21,600 associates in North America, guarantees its customers the lowest price on each product it sells. PETsMART.com has somewhat lower expenses than Pets.com because its stores help promote its Web site at a relatively low cost. Also, PETsMART.com uses existing PETsMART warehouses to fulfill its orders. Another advantage is its use of idealab!, the Los Angeles–based Internet Web site innovator that spawned successes such as eToys, to build its Web site.

On-line since August 1999, Petopia.com offers its fellow animal lovers everything from expert advice and near-obsessive pet-loving opinion to a full range of food, care products, toys, and enthusiast materials. Petopia.com is truly a place for pets and their people. Utilizing message boards, chat groups, and regular customer feedback, Petopia.com is a forum for pet owners to share and gain knowledge, insight, and maybe even a few housebreaking tips. Petopia.com also has partnerships with several leading companies in the pet care and Internet industries. It calls companies and organizations like the ASPCA, Yahoo!, AOL, and its suppliers partners in the industry of pet loving.

Discovery.com, a sister company to the Animal Planet cable channel, invested $97 million in Petstore.com. To date, with two rounds of investments from Discovery, Advanced Technology, and Battery Ventures, total funding exceeds $150 million. Today, Petstore.com is composed of a group of 200-plus pet lovers who are dedicated to providing expert advice and solutions so that users can get the information needed to choose the right products for their pets. Petstore.com has created an environment resembling favorite neighborhood pet stores—a place where people take the time to help you choose the best food for your new dog, cat, bird, or fish and discuss the advantages of each product.

Because so many pet stores are rushing to the Web, the market is headed for a messy shakeout. On-line pet shops are headed for brutal competition over the next few years, but customers will make winners out of a few.

Discussion Questions

1. What are the advantages of on-line shopping for pet owners? What are some of the issues?
2. What are the key issues faced by on-line pet stores?

Critical Thinking Questions

3. Search for the Web site of the companies mentioned in the case and write a brief summary of their status. Which sites are still operating and profitable? Can you find any new sites?
4. Which component of the e-commerce model—community, content, or commerce—do you think is most important in drawing a pet owner to a particular site? Why?

Sources: Adapted from Arlene Weintraub and Robert D. Hof, "For Online Pet Stores, It's Dog-Eat-Dog," *Business Week*, March 6, 2000, pp. 78–80; "Who Are We?" Pets.com Web site, http://www.pets.com, accessed May 18, 2000; "Corporate Profile," PETsMART.com Web site, http://www.petsmart.com, accessed May 18, 2000; "Company Information—Who Are We?" Petopia Web site, http://www.petopia.com, accessed May 18, 2000; "About Us," Petstore.com Web site, http://www. petstore.com, accessed May 18, 2000.

3 Reapplication of General Electric Web Site

General Electric (GE) is a diversified services, technology, and manufacturing company that operates in more than 100 countries and employs nearly 340,000 people worldwide, including 197,000 in the United States. GE Aircraft Engines (GEAE) is a division of GE, with annual sales in excess of $10 billion and 33,000 employees worldwide. It is the world's largest producer of large and small jet engines for commercial and military aircraft. GEAE also supplies engines for boats and provides aviation services. Throughout the 1990s, more than 50 percent of the world's large commercial jet engine orders were awarded to GE or CFM International, a joint company of GE and Snecma of France. W. James McNerney is the president and CEO of GEAE, which is headquartered in Cincinnati.

Under the leadership of its highly respected chairman and CEO, John F. Welch, GE has earned a number of awards as the World's Most Admired Company (by *Fortune* magazine in 1998 and 1999) and the World's Most Respected Company (by the *Financial Times* in 1998 and 1999). However, in May 1999, it became clear that a key e-commerce project was floundering. To fix the situation,

McNerney recruited a former Green Beret, John Rosenfeld, to take on part of the failed project, a customer Web center for the complex spare-parts business, and build it at lightning speed. The mandate was to be operational within seven months. This challenging goal was met, and today GEAE's Customer Web Center has the following functions:

- **Spare-parts order management.** Checks parts availability, finds alternative parts, places orders, builds automated parts lists (for parts usually ordered together), checks status of orders, and tracks orders via UPS and FedEx.
- **Component repair and engine overhaul.** Checks status of repair and overhaul work in shops around the world and accesses initial findings reports and cost estimates, including corresponding high-resolution digital photos of damaged parts.
- **Spare-parts warranty.** Submits warranty claims, reviews status of claims, and generates reports.
- **Technical publications.** Finds information from illustrated parts catalogs, engine service manuals, standard practice manuals, fleet highlights, and service bulletins.
- **On-line wizards.** Will determine return on investment for engine upgrades (e.g., part X will save Y dollars in maintenance over time frame Z).

- **On-line video training.** Will help users brush up on how to remove, install, or service a part.

Discussion Questions

Assume you are the manager of a critical e-commerce project in the Consumer Appliances division of GE that must be completed as soon as possible. You have made a visit to Cincinnati to meet with Jim McNerney and John Rosenfeld to see if you can reapply the GEAE Web site for Consumer Appliances.

1. What questions will you ask about the technical infrastructure and capabilities of the Web site to assess whether it will meet the needs of the Consumer Appliances division?
2. To what degree will you involve the customers

of the Consumer Appliances division in the assessment of the ability of the GEAE Web site to meet their needs?

Critical Thinking Questions

3. What similarities are there between the spare-parts business for aircraft engines and consumer appliances? What are some of the differences?
4. How might you assess whether the GEAE Web site has the functionality needed to support the Consumer Appliances division?

Sources: Adapted from Marcia Stepanek, "How to Jump-Start Your E-Strategy," *Business Week,* June 5, 2000, pp. 96–100; "The $11 Billion Web Start-Up," *Computerworld,* May 1, 2000, pp. 56–60; and "GE Fact Sheet," General Electric Web site, http://www.ge.com/factsheet.html, accessed May 26, 2000.

● NOTES

Sources for the opening vignette on p. 291: Adapted from "Investor Relations" and "About Us," the Weirton Steel Corporation Web site, http://www.weirton.com, accessed May 4, 2000; MetalSite Web page, http://www.metalsite.com, press releases, accessed May 4, 2000; and Clinton Wilder, "E-Transformation," *Information Week,* September 13, 1999, pp. 44–62.

1. Edward Iwata, "Despite the Hype, B2B Marketplaces Struggle," *USA Today,* May 10, 2000, pp. 1B–2B.
2. David Judson, "Net Links Cut Costs for Firms," *The Cincinnati Enquirer,* June 4, 2000, p. F4.
3. Jacqueline Emigh, "E-Commerce Strategies," *Computerworld,* August 16, 1999, p. 53.
4. Steve Alexander, "E-Commerce Distribution," *Computerworld,* March 26, 2000, p. 58.
5. Clinton Wilder, "E-Transformation," *Information Week,* September 13, 2000, pp. 43–62.
6. Robert Hof, "E-Malls for Business," *Business Week,* March 3, 2000, pp. 32–34.
7. Clinton Wilder, "AutoNation: A Different Style of Sales Pitch," *Information Week,* September 13, 1999, p. 48.
8. Edward Iwata, "Despite the Hype, B2B Marketplaces Struggle," *USA Today,* May 10, 2000, pp. 1B–2B.
9. Julia King and Jaikumar Vijayan, "It's 'Supplier Beware' Online," *Computerworld,* March 27, 2000, p. 6.
10. Clinton Wilder, "Timken: A Big Step for an Old-Line Industry," *Information Week,* September 13, 1999, p. 50.
11. William Glassgall, "The Investor Revolution," *Business Week*

Online, February 22, 1999, accessed at http://www.business-week.com/search.htm.
12. Ellen Stark, "Banking from Home," Money.com Guide to Online Investing, accessed May 30, 2000, at http://www.money.com/money/onlineinvesting/ banking.
13. Bethany McLean, "The Fortune 50 Index, More Than Just Dot-Coms," *Fortune,* December 6, 1999, pp. 130–138.
14. "The GM Card to Drive E-Commerce with Introduction of the GM Card easyPay Electronic Wallet," News Releases section of the Cybercash Web site, http://www.cybercash.com/cybercash/company/, accessed May 13, 2000.
15. Geoffrey Smith, "New Money for the Net," *Business Week,* June 5, 2000, p. EB 14.
16. Anne Chen, "Buying Online: May We Have Your Fingerprint?" *PC Week,* February 28, 2000, pp. 55, 58.
17. Anne Chen, "Buying Online: May We Have Your Fingerprint?" *PC Week,* February 28, 2000, pp. 55, 58.
18. Timothy J. Mullaney and Spencer E. Ante, "Info Wars," *Business Week,* June 5, 2000, pp. 107–116.
19. Malcolm Fitch, "The New Faces of Cybercrime," On Line Investing, accessed May 30, 2000, http://www.money.com/money/onlineinvesting/privacy/privacy1d.html.
20. Marcia Stepanek, "How to Jump-Start Your E-Strategy," *Business Week,* June 5, 2000, pp. 96–100.
21. "Build Your E-Business," *Fortune,* December 1, 1999, pp. 187–195.
22. Neil Randall, "Setting Up Shop Online," *PC Magazine,* November 16, 1999, pp. 137–154.

CHAPTER 9

Transaction Processing and Enterprise Resource Planning Systems

Until recently, order fulfillment hasn't been an overriding concern of "e-tailers," but it's starting to be viewed as a significant differentiator among customers. . . . It represents a job opportunity for IT people.

— Greg Runyan, senior analyst of the Yankee Group

Principles	Learning Objectives
An organization's TPS must support the routine, day-to-day activities that occur in the normal course of business and help a company add value to its products and services.	• *Identify the basic activities and business objectives common to all transaction processing systems.* • *Describe the inputs, processing, and outputs for the transaction processing systems associated with order processing.* • *Describe the inputs, processing, and outputs for the transaction processing systems associated with purchasing.* • *Describe the inputs, processing, and outputs for the transaction processing systems associated with accounting business processes.*
Implementation of an enterprise resource planning system enables a company to achieve numerous business benefits through the creation of a highly integrated set of systems.	• *Define the term* enterprise resource planning system *and discuss the advantages and disadvantages associated with the implementation of such a system.*

Steelcase

Struggling to Redesign Business Processes with ERP

Enterprise resource planning (ERP) offers the potential to integrate a company's entire transaction processing system into a powerful business tool. ERP can tie together order processing, invoicing, inventory control, billing, and other critical business functions. In addition, ERP can help a company find better ways to conduct business. Often called "best practices," sound business procedures and approaches are built into most ERP packages. Thus, companies using the ERP approach often can improve their day-to-day operations. These potential advantages were attractive to Steelcase, which decided to implement the R/3 ERP system by SAP AG, a German software firm.

Steelcase, a large furniture and office equipment manufacturer, realizes about $3 billion in sales every year. The Grand Rapids, Michigan, firm purchased its first version of SAP's ERP software more than six years ago. Steelcase has more than 60,000 different products, and their identification numbers all had to be placed into the ERP system. Entering that many product numbers caused the firm to expend a great amount of time and effort to implement its enterprise resource planning system. In addition, there were rules and relationships between many of the products that needed to be accounted for in the new system. Certain furniture items, for example, required certain parts. According to Dick Reimink, the technology manager for Steelcase, "There's a lot of data in our legacy systems."

Steelcase was also the first company to install Internet Pricing Configurator. This innovative product allows dealers to design offices for a new product line, called Pathways. The new Internet Pricing Configurator, however, also had to be integrated into the company's ERP system—no easy feat. After six years, Steelcase is still struggling to get its ERP system up and running and to realize its benefits. According to one expert, "I don't care what [ERP system] Steelcase picked, they still would have had these issues." And it appears that other companies are also having difficulties implementing their ERP systems.

Crown Crafts had more than a 20 percent drop in its sales in one quarter because of problems implementing its ERP system. Crown Crafts has sales of about $350 million annually. Like Steelcase, the company is using SAP's ERP program. Tri Valley Growers, a farming cooperative in San Francisco that generates $800 million in sales annually, also had problems with its ERP system. As a result, it is suing Oracle, another ERP vendor, for $20 million. Other companies have also had problems implementing and using their ERP systems. Some companies, however, are finding solutions to their ERP problems. To solve ERP implementation and operations problems, companies are turning to experienced application service providers (ASPs). Experienced ASPs offer a number of advantages, including skilled personnel. Avoiding the time and costs of hiring in-house IS personnel has the potential to save companies in the long run. According to Craig Kinyon, chief financial officer for Ried Hospital, "The ASP model takes care of what I call the 'Tylenol Factors' of applications—the maintenance, support, upgrades, and hardware." Today, a number of companies are getting help from others to reduce their ERP headaches, while capitalizing on the many benefits of the ERP approach.

As you read this chapter, consider the following:

- What are the advantages and disadvantages of an ERP system?

- How can some of the problems of an ERP system be solved to help a company achieve the benefits of the ERP approach?

As you can see in the opening vignette, businesses have turned increasingly to information systems to integrate their daily transaction activities. The many business activities associated with supply, distribution, sales, marketing, accounting, and taxation can be performed quickly while avoiding waste and mistakes. The goal of this computerization is ultimately to satisfy a business's customers and provide a competitive advantage by reducing costs and improving service.

Transaction processing was one of the first business processes to be computerized, and without information systems, recording and processing business transactions would consume huge amounts of an organization's resources. The transaction processing system (TPS) also provides employees involved in other business processes—the management information system/decision support system (MIS/DSS) and the artificial intelligence/expert systems (AI/ES)—with data to help them achieve their goals. A transaction processing system serves as the foundation for the other systems (Figure 9.1). Transaction processing systems perform routine operations such as sales ordering and billing, often performing the same operations daily or weekly. The amount of support for decision making that a TPS directly provides managers and workers is low.

These systems require a large amount of input data and produce a large amount of output without requiring sophisticated or complex processing. As we move from transaction processing to management information/decision support, and artificial intelligence/expert systems, we see less routine, more decision support, less input and output, and more sophisticated and complex processing and analysis. But the increase in sophistication and complexity in moving from transaction processing does not mean that it is less important to a business. In most cases, all these systems start as a result of one or more business transactions.

AN OVERVIEW OF TRANSACTION PROCESSING SYSTEMS

Every organization has manual and automated *transaction processing systems (TPSs)*, which process the detailed data necessary to update records about the fundamental business operations of the organization. These systems include order entry, inventory control, payroll, accounts payable, accounts receivable, and general ledger, to name just a few. The input to these systems includes basic business transactions such as customer orders, purchase orders, receipts, time cards, invoices, and payroll checks. The result of processing business transactions is that the organization's records are updated to reflect the status of the operation at the time of the last processed transaction. Automated TPSs consist of all the components of a CBIS, including databases, telecommunications, people, procedures, software, and hardware devices used to process transactions. The processing activities include data collection, data edit, data correction, data manipulation, data storage, and document production.

For most organizations, TPSs support the routine, day-to-day activities that occur in the normal course of business that help a company add value to its

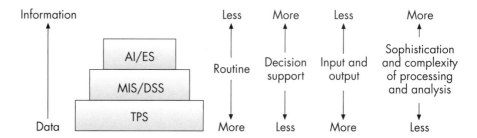

FIGURE 9.1

TPS, MIS/DSS, and AI/ES in Perspective

UPS adds value to its service by providing timely and accurate data online on the exact location of a package.

products and services. Depending on the customer, value may mean lower price, better service, higher quality, or uniqueness of product. By adding a significant amount of value to their products and services, companies ensure further organizational success. Because the TPSs often perform activities related to customer contacts—like order processing and invoicing—these information systems play a critical role in providing value to the customer. For example, by capturing and tracking the movement of each package, United Parcel Service (UPS) is able to provide timely and accurate data on the exact location of a package. Shippers and receivers can access an on-line database and, by providing the airbill number of a package, find the package's current location. Such a system provides the basis for added value through improved customer service.

Traditional Transaction Processing Methods and Objectives

When computerized transaction processing systems first evolved, only one method of processing was available. All transactions were collected in groups, called *batches,* and processed together. With **batch processing systems,** business transactions are accumulated over a period of time and prepared for processing as a single unit or batch (Figure 9.2a). The time period during which transactions are accumulated is whatever length of time is needed to meet the needs of the users of that system. For example, it may be important to process invoices and customer payments for the accounts receivable system daily. On the other hand, the payroll system may receive time cards and process them biweekly to create checks and update employee earnings records as well as to distribute labor costs. The essential characteristic of a batch processing system is that there is some delay between the occurrence of the event and the eventual processing of the related transaction to update the organization's records.

Today's computer technology allows another processing method, called *on-line, real-time,* or **on-line transaction processing (OLTP).** With this form of data processing, each transaction is processed immediately, without the delay of accumulating transactions into a batch (Figure 9.2b). As soon as the input is available, a computer program performs the necessary processing and updates the records affected by that single transaction. Consequently, at any time, the data in an on-line system always reflects the current status. When you make an airline reservation, for instance, the transaction is processed and all databases, such as seat occupancy and accounts receivable, are updated immediately. This type of processing is absolutely essential for businesses that require data quickly and update it often, such as airlines, ticket agencies, and stock investment firms. Many companies have found that OLTP helps them provide faster, more efficient service—one way to add value to their activities in the eyes of the customer. Increasingly, companies are using the Internet to perform many OLTP functions.[1]

A third type of transaction processing, called *on-line entry with delayed processing,* is a compromise between batch and on-line processing. With this type of system, orders or transactions are entered into the computer system when they occur, but they are not processed immediately. For example, when you call

batch processing system

method of computerized processing in which business transactions are accumulated over a period of time and prepared for processing as a single unit or batch

on-line transaction processing (OLTP)

computerized processing in which each transaction is processed immediately, without the delay of accumulating transactions into a batch

Batch versus On-Line
Transaction Processing

Batch processing (a) inputs
and processes data in groups.
In on-line processing (b),
transactions are completed
as they occur.

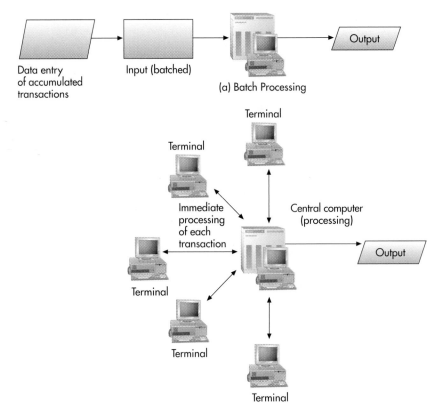

(a) Batch Processing

(b) On-Line Transaction Processing

a toll-free number and order a product, your order is typically entered into the computer when you make the call. However, the order may not be processed until that evening after business hours.

Even though the technology exists to run TPS applications using on-line processing, it is not done for all applications. For many applications, batch processing is more appropriate and cost-effective. Payroll transactions and billing are typically done via batch processing. Specific goals of the organization define the method of transaction processing best suited for the various applications of the company. Figure 9.3 shows the total integration of a firm's transaction processing systems.

Because of the importance of transaction processing, organizations expect their TPSs to accomplish a number of specific objectives, including the following:

Process data generated by and about transactions. The primary objective of any TPS is to capture, process, and store transactions and to produce a variety of documents related to routine business activities. These business activities can be directly or indirectly related to selling products and services to customers. Processing orders, purchasing materials, controlling inventory, billing customers, and paying suppliers and employees are all business activities that result from customer orders. These activities result in transactions that are processed by the TPS.

Maintain a high degree of accuracy and integrity. One objective of any TPS is error-free data input and processing. Even before the introduction of computer technology,

When you order a product
over the phone, the vendor
may use on-line entry with
delayed processing. Your order
is entered into the computer at
the time of the call but is not
processed immediately.
(Source: © 2000 Photodisc)

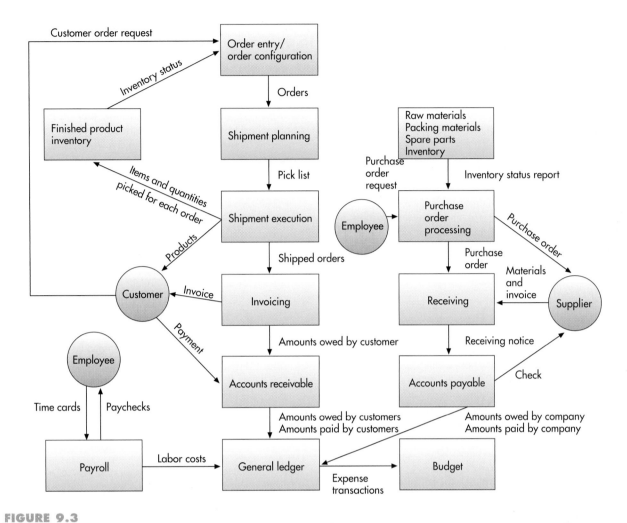

FIGURE 9.3

Integration of a Firm's TPSs

employees visually inspected all documents and reports introduced into or produced by the TPS. Because humans are fallible, the transactions were often inaccurate, resulting in wasted time and effort and requiring resources to correct them. An editing program, for example, should have the ability to determine that an entry that should read "40 hours" is not entered as "400 hours" or "4000 hours" because of a data entry error. As seen in the "Ethical and Societal Issues" box, preventing transaction processing fraud is also an important part of maintaining a high degree of accuracy.

An important component of data integrity is to avoid fraudulent transactions. E-commerce companies face this problem when accepting credit or debit card information over the Internet. How can these companies make sure that the people making the purchases are who they say they are? One approach is to use a digital certificate.[2] A digital certificate is a small computer file that serves as both an ID card and a signature. Some believe that digital certificates, which use complex mathematical codes, are almost fraud proof.

As the volume of data being processed and stored increases, it becomes more difficult for individuals and machines to review all input data. Doing so is critical, however, because data and information generated by the TPS are often used by other information systems in an organization. So, a company must ensure both data integrity and accuracy.

Produce timely documents and reports. Manual transaction processing systems can take days to produce routine documents. Fortunately, the use of computerized TPSs significantly reduces this response time.[3] Improvements in information technology, especially hardware and telecommunications links,

ETHICAL AND SOCIETAL ISSUES
Credit Card Fraud and Transaction Processing

Some believe that companies that do not ride the wave of the Internet era will not survive in the long term. Companies—from retail firms to job-search services—are going on-line in record numbers. For most companies, this means putting most or all of their transaction processing systems, including sales, order processing, inventory control, and billing, on the Internet. Although the specific products and Internet approaches vary tremendously from one firm to the next, one transaction processing activity seems to be similar for all Internet companies: the billing application of most companies operating on the Internet involves collecting fees and funds from credit cards. For most of these companies, there is a fear of credit card fraud.

Although customers and banks feel the impact of credit card fraud, the companies operating on the Internet often are dealt the biggest blow. First, fraud is one of the biggest reasons for consumers and merchants alike to avoid e-commerce. According to the technology officer of one company, "There were days when we had more fraud than legitimate sales." One fraud scam ripped off hundreds of thousands of dollars in merchandise from such companies as Amazon.com, Cyberian Outpost, and others. The merchandise included computers, software, books, and other items.

Law enforcement agencies are striving to protect companies and consumers from credit card fraud, but many companies think that these efforts are too little, too late. As a result, companies are increasingly looking to other approaches to stop credit card fraud, including fraud-busting software. With names like SecurePay and Prism CardAlert, these software products are helping companies and customers gain more confidence in buying and selling on-line.

Discussion Questions

1. Why is bill paying such an important part of a company's transaction processing system?
2. What other transaction processing applications are critical for successfully completing transactions on the Internet?

Critical Thinking Questions

3. If you were starting to sell products on the Internet, what steps would you take to help reduce credit card fraud?
4. What new laws, if any, would you recommend to help curb credit card fraud?

Sources: Adapted from Margaret Mannix, "High Tech Card Fraud Goes on Right Behind You," *US News & World Report*, February 14, 2000, p. 54; Cynthia Morgan, "Ripped Off," *Computerworld*, March 8, 1999, p. 1; and Lucy Dixon, "Unibanco Offers Virtual Credit Card," *Precision Marketing*, September 20, 1999, p. 10.

---

allow transactions to be processed in a matter of seconds. The ability to conduct business transactions in a timely way can be very important for the profitable operation of the organization. For instance, if bills (invoices) are sent out to customers a few days earlier than usual, payment may be received earlier. A number of transaction processing systems have built-in capabilities to monitor how timely a company is when processing transactions and producing reports and documents.[4] Some monitoring software packages can compare actual performance with corporate goals and objectives.

Timing is also crucial for related applications such as order processing, invoicing, accounts receivable, inventory control, and accounts payable. Because of electronic recording and transmission of sales information, transactions can be processed in seconds rather than overnight, thus improving companies' cash flow. Customers find credit card charges they made on the final day of the billing period on their current monthly bill.

Increase labor efficiency. Before computers existed, manual business processes often required rooms full of clerks and equipment to process the necessary business transactions. Today, transaction processing systems can substantially reduce clerical and other labor requirements. A small minicomputer linked to a company's cash registers replaces a room full of clerks, typewriters, and filing cabinets. Many TPSs are cost-justified by labor savings.

Help provide increased service. Without question, we are quickly becoming a service-oriented economy. Even strong manufacturing companies, including household appliance makers and automobile manufacturers, realize the importance of providing superior customer service. A transaction processing system for tickets from Ticketmaster, for example, allows concert enthusiasts to order tickets over the Internet instead of standing in line for hours, or even days in some cases.[5] According to an Auburn University business student looking for tickets to a Fiona Apple concert, "The seats are great, and ordering on-line was quick and easy." One objective of any TPS is to assist the organization in providing this type of service. For example, some companies have EDI systems (see Chapter 6) that allow customers to place orders electronically, thus bypassing slower and more error-prone methods of written or oral communication. A golf club manufacturer in Phoenix, Arizona, for example, was receiving over 4,000 e-mail messages each month from customers asking questions about previous transactions or products.[6] The company had only two people responding to the e-mails, which was frustrating for customers and expensive for the company. To improve service, the company started using software that could analyze the incoming e-mails and automatically send responses in many cases. When the software couldn't understand the e-mail, the e-mail was forwarded to a customer service representative.

Help build and maintain customer loyalty. A firm's transaction processing systems are often the means for customers to communicate. It is important that the customer interaction with these systems keeps customers satisfied and returning.

Achieve competitive advantage. A goal common to almost all organizations is to gain and maintain a competitive advantage. As discussed in Chapter 2, a competitive advantage provides a significant and long-term benefit for the organization. When a TPS is developed or modified, the personnel involved should carefully consider how the new or modified system might provide a significant and long-term benefit. Some of the ways that companies can use transaction processing systems to achieve competitive advantage are summarized in Table 9.1.

Depending on the specific nature and goals of the organization, any of these objectives may be more important than others. By meeting these objectives, transaction processing systems can support corporate goals such as reducing costs; increasing productivity, quality, and customer satisfaction; and running more efficient and effective operations.

For example, overnight delivery companies such as FedEx expect their transaction processing systems to increase customer service. These systems can locate a client's package at any time—from initial pickup to final delivery. This improved customer information allows companies to produce timely information and be more responsive to customer needs and queries.

Brokerage firms employ specialized transaction processing systems that process orders, provide customer statements, and produce managerial reports. Increasingly, brokerage firms offer their clients software to monitor their own

TABLE 9.1

Examples of Transaction Processing Systems for Competitive Advantage

Competitive Advantage	Example
Customer loyalty increased	Use of customer interaction system to monitor and track each customer interaction with the company
Superior service provided to customers	Use of tracking systems that are accessible by customers to determine shipping status
Better relationship with suppliers	Use of an Internet marketplace to allow the company to purchase products from suppliers at discounted prices
Superior information gathering	Use of order configuration system to ensure that products ordered will meet customer's objectives
Costs dramatically reduced	Use of warehouse management system employing scanners and bar-coded product to reduce labor hours and improve inventory accuracy

accounts on-line. These systems indicate all buy and sell orders and associated commissions, the securities held, the purchase price, the current market price, the dividends or interest per period, the dividends per year, and the anticipated yield on each security as well as the anticipated yield of each client's portfolio of securities. The objectives are to increase customer service and provide timely reports. Obviously, maintaining a high degree of data accuracy and integrity is also important. Discount brokerage firms are even offering software that allows clients to enter their own buy and sell requests without first speaking with a broker. This increases labor efficiency but raises issues about the level of commissions that should be charged.

Transaction Processing Activities

Along with having common characteristics, all transaction processing systems perform a common set of basic data processing activities. TPSs capture and process data that describes fundamental business transactions. This data is used to update databases and to produce a variety of reports for use by people both within and outside the enterprise (Figure 9.4). The business data goes through a **transaction processing cycle** that includes data collection, data editing, data correction, data manipulation, data storage, and document production (Figure 9.5).

Data Collection
The process of capturing and gathering all data necessary to complete transactions is called **data collection**. In some cases this can be done manually, such as by collecting handwritten sales orders or changes to inventory. In other cases, data collection is automated via special input devices like scanners, point-of-sale devices, and terminals.

Data collection begins with a transaction (e.g., taking a customer order) and results in the origination of data that is input to the transaction processing system. Data should be captured at its source, and it should be recorded accurately, in a timely fashion, with minimal manual effort, and in a form that can be directly entered into the computer rather than keying the data from some type of document. This approach is called source data automation. An example of source data automation is the use of scanning devices at the grocery checkout to

transaction processing cycle

the process of data collection, data editing, data correction, data manipulation, data storage, and document production

data collection

the process of capturing and gathering all data necessary to complete transactions

FIGURE 9.4

A Simplified Overview of a Transaction Processing System

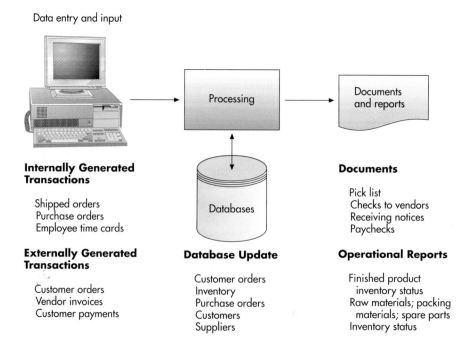

Data entry and input

Processing → Documents and reports

Databases

Internally Generated Transactions

Shipped orders
Purchase orders
Employee time cards

Externally Generated Transactions

Customer orders
Vendor invoices
Customer payments

Database Update

Customer orders
Inventory
Purchase orders
Customers
Suppliers

Documents

Pick list
Checks to vendors
Receiving notices
Paychecks

Operational Reports

Finished product
inventory status
Raw materials; packing
materials; spare parts
Inventory status

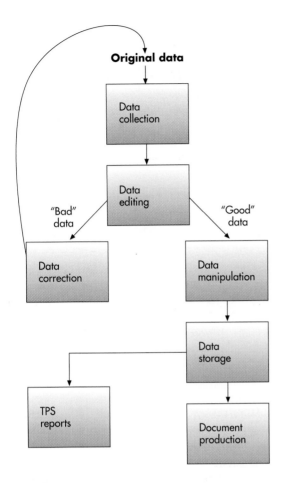

Original data

Data collection

Data editing

"Bad" data

"Good" data

Data correction

Data manipulation

Data storage

TPS reports

Document production

FIGURE 9.5

Data Processing Activities Common to Transaction Processing Systems

data editing

the process of checking data for validity and completeness

data correction

the process of reentering miskeyed or misscanned data that was found during data editing

data manipulation

the process of performing calculations and other data transformations related to business transactions

read the Universal Product Code (UPC) automatically. Reading the UPC bar codes is quicker and more accurate than having a cash register clerk enter codes manually. The scanner reads the bar code for each item and looks up its price in the item database. The point-of-sale transaction processing system uses the price data to determine the customer's bill. The number of units of this item purchased, the date, the time, and the price are also used to update the store's inventory database, as well as its database of detailed purchases. The inventory database is used to generate a management report notifying the store manager to reorder items whose sales have reduced the stock below the reorder quantity. The detailed purchases database can be used by the store (or sold to market research firms or manufacturers) for detailed analysis of sales (Figure 9.6).

Many grocery stores combine point-of-sale scanners and coupon printers. The systems are programmed so that each time a specific product—say, a box of cereal—crosses a checkout scanner, an appropriate coupon—perhaps a milk coupon—is printed. Other companies can also rent time on the system, which is then reprogrammed to print those companies' coupons if the customer buys a competitive brand. These TPSs help grocery stores to increase profits by improving their repeat sales and bringing in revenue from other businesses.

As another example, industrial data collection devices allow employees to scan their magnetized employee ID cards to enter data into the payroll TPS when they start or end a job. Not only do these devices provide important employee payroll information, they also help an organization determine how many labor hours are spent on each job or project so that staffing adjustments can be made or future projects can be accurately planned.

Data Editing

An important step in processing transaction data is to perform **data editing** for validity and completeness to detect any problems with the data. For example, quantity and cost data must be numeric and names must be alphabetic; otherwise, the data is not valid. Often the codes associated with an individual transaction are edited against a database containing valid codes. If any code entered (or scanned) is not present in the database, the transaction is rejected.

Data Correction

It is not enough to reject invalid data. The system should provide error messages that alert those responsible for the data edit function. These error messages must specify what problem is occurring so that corrections can be made. A **data correction** involves reentering miskeyed or misscanned data that was found during data editing. For example, a UPC that is scanned must be in a master table of valid UPCs. If the code is misread or does not exist in the table, the checkout clerk is given an instruction to rescan the item or key in the information manually.

Data Manipulation

Another major activity of a TPS is **data manipulation**, the process of performing calculations and other data transformations related to business transactions. Data manipulation can include classifying data, sorting data into categories, performing calculations, summarizing results, and storing data in

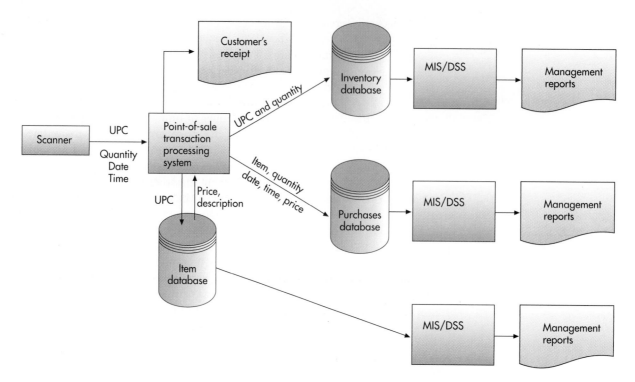

FIGURE 9.6

Point-of-Sale Transaction
Processing System

Scanning items at the checkout
stand results in updating a
store's inventory database and
its database of purchases.

data storage

the process of updating one
or more databases with new
transactions

document production

the process of generating output
records and reports

the organization's database for further processing. In a payroll TPS, for example, data manipulation includes multiplying an employee's hours worked by the hourly pay rate. Overtime calculations, federal and state tax withholdings, and deductions are also performed.

Data Storage

Data storage involves updating one or more databases with new transactions. Once the update process is complete, this data can be further processed and manipulated by other systems so that it is available for management decision making. Thus, although transaction databases can be considered a by-product of transaction processing, they have a pronounced effect on almost all other information systems and decision-making processes in an organization.

Document Production and Reports

TPSs produce important business documents. **Document production** involves generating output records and reports. These documents may be hard-copy paper reports or displayed on computer screens (sometimes referred to as *soft copy*). Paychecks, for example, are hard-copy documents produced by a payroll TPS, while an outstanding balance report for invoices might be a soft-copy report displayed by an accounts receivable TPS. Often, results from one TPS are passed downstream as input to other systems (as shown in Figure 9.6), where the results of updating the inventory database are used to create the stock exception report (a type of management report) of items whose inventory level is less than the reorder point.

In addition to major documents like checks and invoices, most transaction processing systems provide other useful management information and decision support, such as reports that help managers and employees perform various activities. These reports can be printed or displayed on a computer screen. A report showing current inventory is one example; another might be a document listing items ordered from a supplier to help a receiving clerk check the order for completeness when it arrives. A TPS can also produce reports required by local, state, and federal agencies, such as statements of tax withholding and quarterly income statements.

Throughout this chapter we will look at some ways companies have employed TPSs to help them meet organizational goals.

Control and Management Issues

Transaction processing systems are the backbone of any organization's information systems. They capture facts about the fundamental business operations of the organization—facts without which orders cannot be shipped, customers cannot be invoiced, and employees and suppliers cannot be paid. In addition, the data captured by the transaction processing systems flow downstream to the other systems of the organization. Like any structure, an organization's information systems are only as good as the foundation on which they are built. Indeed, most organizations would grind to a screeching halt if their transaction processing systems failed.

Business Resumption Planning

business resumption planning

the process of anticipating and providing for disasters

Business resumption planning is the process of anticipating and minimizing the effects of disasters. Disasters can be natural emergencies such as a flood, a fire, or an earthquake or interruptions in business processes such as labor unrest or erasure of an important file. Business resumption planning focuses primarily on two issues: maintaining the integrity of corporate information and keeping the information system running until normal operations can be resumed.

One of the first steps of business resumption planning is to identify potential threats or problems, such as natural disasters, employee misuse of personal computers, and poor internal control procedures. Business resumption planning also involves disaster preparedness. IS managers should occasionally hold an unannounced "test disaster"—similar to a fire drill—to ensure that the disaster plan is effective.

Disaster Recovery

disaster recovery

the implementation of the business resumption plan

Disaster recovery is the implementation of the business resumption plan. Although companies have known about the importance of disaster planning and recovery for decades, many do not adequately prepare. The primary tools used in disaster planning and recovery are backups for hardware, software and databases, telecommunications, and personnel.

A common backup for hardware is a similar or compatible computer system owned by another company or a specialized backup system provided by an organization from which a written hardware backup agreement is obtained. Some companies back up their hardware with duplicate systems and equipment of their own. For example, a company might use two identical mainframes—one for systems development and the other for running business applications. The systems development processor (used to create new programs and to modify existing ones) can be used to perform applications if the original applications processor fails. Two firms in different industries having compatible systems can also serve as hardware backups for one another.

Keeping a duplicate system that is operational or having immediate access to one through a specialized vendor is an example of a hot site. A hot site is often a compatible mainframe system that is operational and ready to use. If the primary mainframe has problems, the hot site can be used immediately as a backup.

Another approach is to use a cold site, also called a shell, which is a computer environment that includes rooms, electrical service, telecommunications links, data storage devices, and the like. If there is a problem with the primary mainframe, the primary computer hardware is brought into the cold site and the complete system is made operational. For both hot and cold sites, telecommunications media and devices are used to provide for fast and efficient transfer of processing jobs to the disaster facility.

Companies like Iron Mountain provide a secure, off-site environment for records storage. In the event of a disaster, vital data can be recovered.

(Source © 2000 Photodisc)

Software and databases can be backed up by making duplicate copies of all programs and data. At least two backup copies should be made. One backup copy can be kept in the information systems department in case of accidental destruction of the software. Another backup copy should be kept off-site in a safe, secure, fireproof, and temperature- and humidity-controlled environment in case of loss of the data processing facility. Service companies provide this type of backup environment.

Backup is also essential for the data and programs on users' desktop computers. The advent of more distributed systems, like client/server systems, means that many users now have important, and perhaps critical, data and applications on their desktop computers. Utility packages inexpensively provide backup features for desktop computers by copying data onto magnetic tape, CDs, and other storage devices.

Some business recovery plans call for the backup of vital telecommunications. Complex plans might call for recovering whole networks. In other plans, the most critical nodes on the network are backed up by duplicate components. Using such fault-tolerant networks, which will not break down when one node or part of the network malfunctions, can be a more cost-effective approach to telecommunications backup.

There should also be a backup for information systems personnel. This can be accomplished in a number of ways. One of the best approaches is to provide cross-training for IS and other personnel so that each individual can perform an alternate job if required. For example, a company might train employees in accounting, finance, or other IS departments to operate the system if a disaster strikes. The company could also make an agreement with another information systems department or an outsourcing company to supply IS personnel if necessary.

Transaction Processing System Audit

transaction processing system audit

an examination of the TPS in an attempt to answer whether the system meets the business need for which it was implemented, what procedures and controls have been established, and whether these procedures and controls are being used properly

A **transaction processing system audit** attempts to answer three basic questions:

• Does the system meet the business need for which it was implemented?
• What procedures and controls have been established?
• Are these procedures and controls being used properly?

In addition to these three basic auditing questions, other areas are typically investigated during an audit. These areas include the distribution of output documents and reports, the training and education associated with existing and new systems, and the time necessary to perform various tasks and to resolve problems and bottlenecks in the system. General areas of improvement are also investigated and reported during the audit.

Two types of audits exist. An internal audit is conducted by employees of the organization; an external audit is performed by accounting firms or companies and individuals not associated with the organization. In either case, a number of steps are performed. The auditor inspects all programs, documentation, control techniques, the disaster plan, insurance protection, fire protection, and other systems management concerns such as efficiency and effectiveness of the disk or tape library. This is accomplished by interviewing IS personnel and performing a number of tests on the computer system. External audits are important for stockholders and others outside the company, in addition to managers and employees inside the company. A number of Internet startup companies, for example, have overreported their income, which has resulted in high stock evaluations in some cases. An external audit by a reputable auditing company can help uncover these reporting problems.

In establishing the integrity of the computer programs and software, an audit trail must be established. The **audit trail** allows the auditor to trace any output from the computer system back to the source documents. With many of the real-time and time-sharing systems available today, it is extremely difficult to follow an audit trail. In many cases, no record of inputs to the system exist; thus, the audit trail is destroyed. In such cases, the auditor must investigate the actual processing in addition to the inputs and outputs of the various programs.

audit trail

documentation that allows the auditor to trace any output from the computer system back to the source documents

TRADITIONAL TRANSACTION PROCESSING APPLICATIONS

In this section we present an overview of several common transaction processing systems that support the order processing, purchasing, and accounting business processes (Table 9.2).

Order Processing Systems

order processing systems

systems that process order entry, sales configuration, shipment planning, shipment execution, inventory control, invoicing, customer interaction, and routing and scheduling

Order processing systems include order entry, sales configuration, shipment planning, shipment execution, inventory control, invoicing, customer interaction, and routing and scheduling. The business processes supported by these systems are so critical to the operation of the enterprise that the order processing systems are sometimes referred to as the "lifeblood of the organization." Figure 9.7 is a system-level flowchart that shows the various systems and the information that flows between them. A rectangle represents a system, a line represents the flow of information from one system to another, and a circle represents any entity outside the system—in this case, the customer.

Order Entry

order entry system

process that captures the basic data needed to process a customer order

The **order entry system** captures the basic data needed to process a customer order. Orders may come through the mail or via a telephone ordering system, be gathered by a staff of sales representatives, arrive via EDI transactions directly from a customer's computer over a wide area network, or be entered directly over the Internet by the customer using a data entry form on the firm's Web site. Figure 9.8 (p. 343) is a data flow diagram of a typical order entry system. The data flow diagram is more detailed than the system-level flowchart. It shows the various business processes that are supported by a system and the flow of data between processes. A rectangle with rounded corners represents a business process.

In most cases, the order originates with the customer. With more sophisticated order processing systems, however, orders can be initiated by the company. Safeway, for example, has a sophisticated software system that predicts when a family will need groceries, based on previous buying patterns and how

TABLE 9.2

The Systems That Support Order Processing, Purchasing, and Accounting Functions

Order Processing	Purchasing	Accounting
• Order entry	• Inventory control (raw materials, packing materials, spare parts, and supplies)	• Budget
• Sales configuration		• Accounts receivable
• Shipment planning	• Purchase order processing	• Payroll
• Inventory control (finished product)	• Receiving	• Asset management
• Invoicing and billing	• Accounts payable	• General ledger
• Customer interaction		
• Routing and scheduling		

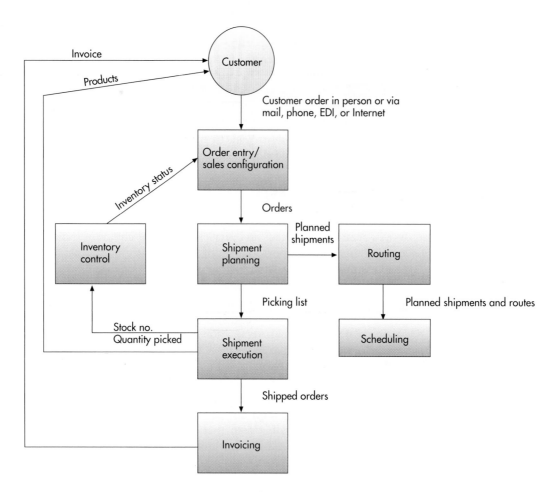

FIGURE 9.7

Order Processing Systems

long it has been since the family has been to the grocery store.[7] The software for Safeway, developed by IBM, has a database of ten million British customers that shop at Safeway. It predicts when a customer might need one of the more than 22,000 grocery items stored in Safeway's database. It then alerts the customer. Similar software is being developed for other industries.

With an on-line order processing system, such as one used by direct retailers, the inventory status of each inventory item (also called stock keeping unit, or SKU) on the order is checked to determine whether sufficient finished product is available. If an order item cannot be filled, a substitute item may be suggested or a back order is created—the order will be filled later, when inventory is replenished. Order processing systems can also suggest related items for order takers to mention to promote add-on sales. Order takers also review customer payment history data from the accounts receivable system to determine whether credit can be extended.

Once an order is entered and accepted, it becomes an open order. Typically, a daily sales journal (which includes customer information, products ordered, quantity discounts, and prices) is generated.

Electronic data interchange (EDI) can be an important part of the order entry TPS. With EDI, a customer or client organization can place orders directly from its purchasing TPS into the order processing TPS of another organization. Or, the order processing TPS of the supplier companies and the purchasing TPS of the customers could be linked indirectly through a third-party clearinghouse. In any event, this computer-to-computer link allows efficient and effective processing of sales orders and enables an organization to lock in customers and lock out competitors through enhanced customer service. With EDI, orders can be placed any time of the day or night, and immediate notification of order receipt

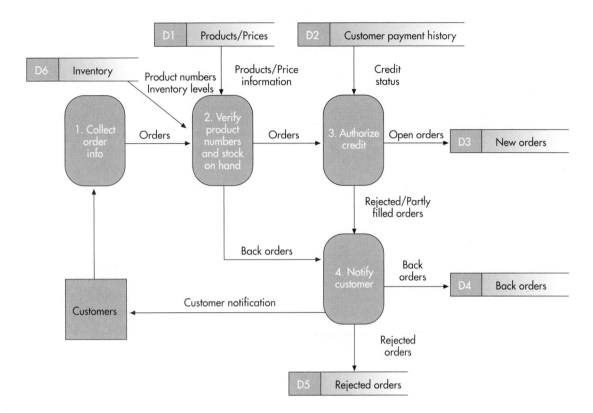

FIGURE 9.8

Data Flow Diagram of an Order Entry System

Orders are received by mail, phone, EDI, or the Internet from customers or sales reps and entered into the order processing system. This application affects accounting, inventory, warehousing, finance, and invoicing applications. Note that in an integrated order processing system, order entry personnel have access to back order, inventory, and customer information from separate data files or directly through the processing mechanism.
(Source: George W. Reynolds, *Information Systems for Managers*, 3rd ed., St. Paul, MN: West Publishing Co., 1995, p. 198. Reprinted with permission from Course Technology.)

sales configuration system

process that ensures that the products and services ordered are sufficient to accomplish the customer's objectives and will work well together

and processing can be made. Today, more and more companies are using electronic data interchange to make paperless business transactions a reality.

As discussed in Chapter 8, order processing is being done through e-commerce and Internet systems to a greater extent today. IBM, for example, sold more than $15 billion of computer systems and equipment over the Internet in 1999.[8] Seven-Eleven Japan Corporation is investing $375 million into an Internet ordering system.[9] After placing orders on the Internet, Seven-Eleven customers go to a local store to pick up and pay for their items. Even so, many companies have not yet launched an on-line order processing system.[10] One study reported that only 40 percent of large companies are able to receive orders over the Internet or on-line. Another study reported that more than 25 percent of on-line order transactions failed.[11] The Boston Consulting Group study defined failed transactions as ones in which customers couldn't find what they wanted, couldn't complete the transactions, or were not satisfied with the transactions. Common problems were Internet pages that took too long to load and Internet sites that were too confusing to use.

Sales Configuration

Another important aspect of order processing is sales configuration. The **sales configuration system** ensures that the products and services ordered are sufficient to accomplish the customer's objectives and will work well together. For example, using a sales configuration program, a sales representative knows that a computer printer needs a certain cable and a LAN card so that it can be connected to the LAN. Without a sales configuration program, a sales representative might sell a customer the wrong cable or forget the LAN card.

Sales configuration programs also suggest optional equipment. For example, if a customer orders a palmtop computer, the sales configuration program will suggest an AC adapter, backup software and cables, and a modem to allow the palmtop computer the ability to connect to the Internet. If a company is buying a 747 aircraft from Boeing, a sales configuration program can help the sales

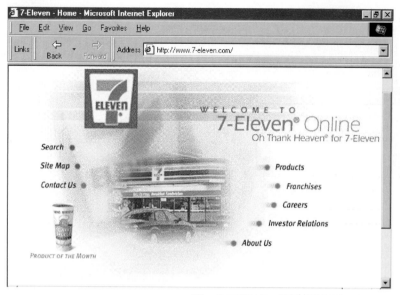

7-Eleven - Home - Microsoft Internet Explorer

File   Edit   View   Go   Favorites   Help

Links   Back   Forward   Address http://www.7-eleven.com/

WELCOME TO
7-Eleven® Online
Oh Thank Heaven® for 7-Eleven

Search
Site Map
Contact Us

Products
Franchises
Careers
Investor Relations
About Us

PRODUCT OF THE MONTH

Seven-Eleven Japan Corporation is investing $375 million into an Internet ordering system. After placing orders on the internet, Seven-Eleven customers go to a local store to pick up and pay for their items.

shipment planning system

system that determines which open orders will be filled and from which location they will be shipped

Many companies use a sophisticated warehouse management system to speed order processing time and improve inventory accuracy. (Source: Stone/Roger Tully)

representative work with the company to determine the number of seats that are needed, the most appropriate navigation systems to install, the type of landing gear that should be used, and hundreds of other available options that can be specified for the 747.

Sales configuration software can also solve customer problems and answer customer questions. For example, a sales configuration program can determine whether a factory robot made by one manufacturer can be controlled by a computer system developed by another manufacturer. Sales configuration programs can eliminate mistakes, reduce costs, and increase revenues. These advantages have led companies such as Hewlett-Packard, Boeing, and Silicon Graphics to implement these systems.

Shipment Planning

New orders received and any other orders not yet shipped (open orders) are passed from the order entry system to the shipment planning system. The **shipment planning system** determines which open orders will be filled and from which location they will be shipped. This is a trivial task for a small company with lots of inventory, only one shipping location, and a few customers concentrated in a small geographic area. But it is an extremely complicated task for a large global corporation with limited inventory (not all orders for all items can be filled), dozens of shipping locations (plants, warehouses, contract manufacturers, etc.), and tens of thousands of customers. The trick is to minimize shipping and warehousing costs while still meeting customer delivery dates.

The output of the shipment planning system is a plan that shows where each order is to be filled and a precise schedule for shipping with a specific carrier on a specific date and time. The system also prepares a picking list that is used by warehouse operators to select the ordered goods from the warehouse. These outputs may be in paper form or they may be computer records that are transmitted electronically. The picking list document, an example of which is shown in Figure 9.9, lists the customer name, number, order number, and all items that have been ordered. A description of all items, along with the number to be shipped, is also included.

Upgrading the order entry process and linking it with production scheduling has provided Anchor Glass with a competitive advantage. The company makes glass bottles for Coke, Pepsi, Smuckers, and other manufacturers that distribute their products in bottles. Before it implemented a new order processing system, the company was unable to provide a delivery date to customers who phoned in orders. But with the new system, order takers can check actual production schedules at the company's 15 factories, accept orders, and promise delivery—all during the initial phone call. Once the order is accepted, the system reserves production capacity for the order and the production schedule is updated, taking into account any changes in raw or packing material availability, plant capacity, and equipment downtime.

Shipment Execution

The **shipment execution system** coordinates the outflow of all products and goods from the organization, with the objective of

FIGURE 9.9

A Picking List

This document guides warehouse employees in locating items to fill an order. Note the second and third columns of the slip, which instruct the warehouse workers where to locate the items. Also note that in this instance, the third item ordered was back-ordered three cases. Once items are picked from inventory, this data is entered into the data transaction processing system, and a packing slip and shipping notice are generated.

shipment execution system

system that coordinates the outflow of all products from the organization, with the objective of delivering quality products on time to customers

inventory control system

system that updates the computerized inventory records to reflect the exact quantity on hand of each stock-keeping unit

FIGURE 9.10

An Inventory Status Report

This output from the inventory application summarizes all inventory items shipped over a specified time period.

INDUSTRIAL FASTENING PRODUCTS, INC.
3464 Seventh Street
Calgary, USA   (182) 997-4567

| CUSTOMER NO. 012345678 | ORDER NO. C-654321 | DATE 10/14/01 | PAGE 1 |

SOLD TO: DURA FURNITURE COMPANY, PO BOX 491, 478 ELM STREET, CINCINNATI, OHIO

SHIP TO:

CUSTOMER NO. 447918   SHIP VIA   WILSON FREIGHT -ATTENTION- JIM JONSON

LOC	LINK	ITEM NUMBER	DESCRIPTION	ORDERED	SHIPPED	B/O	COMMENT
8	105	10 L1L416028	FASTENING TOOL MODEL L	3 EACH	3		$
		20 S8276	STAPLE 3/4 INCH	15 CASE	15		$
		30 S8289	STAPLE 1 INCH	15 CASE	12	3	$
		40	SHIPPING CHARGE				

*** END OF REPORT ***

delivering quality products on time to customers. The shipping department is usually given responsibility for physically packaging and delivering all products to customers and suppliers.[12] This delivery can include mail services, trucking operations, and rail service. The system receives the picking list from the shipment planning system.

Sometimes orders cannot be filled exactly as specified. One reason is "out-of-stocks," meaning the warehouse does not have sufficient quantity of an item to fill a customer's order. Shortages can be caused if a production run did not produce the expected quantity of an item because of manufacturing problems. The company policy may be not to ship any of the item, ship as many units of the item as are available and create a back-order request for the remainder, or to substitute another item. Thus, as items are picked and loaded for shipment, warehouse operators must enter data about the exact items and quantity of each that are loaded for each order. When the shipment execution system processing cycle is complete, it passes the "shipped orders" business transactions "downstream" to the invoicing system. These transactions specify exactly what items were shipped, the quantity of each, and to whom the order was shipped. This data is used to generate a customer invoice. The shipment execution system also produces packing documents, which are enclosed with the items being shipped, to tell customers what items are in the shipment, what is back-ordered, and the exact status of all items in the order. Soft-copy data—such as that provided by advanced shipment notices and shipment tracking systems—is also made available to other business functions.

Inventory Control

For each item picked during the shipment execution process, a transaction providing the stock number and quantity picked is passed to the **inventory control system**. In this way, the computerized inventory records are updated to reflect the exact quantity on hand of each stock-keeping unit. Thus, when order takers check the inventory level of a product, they receive current information.

Once products have been picked out of inventory, other documents and reports are initiated by the inventory control application. For example, the inventory status report (Figure 9.10) summarizes all inventory items shipped over a specified time period. It can include stock numbers, descriptions, number of units on hand,

DATE 5 30 01   JANUS TOOL COMPANY   PAGE 2
INVENTORY STATUS/STOCK TAKE REPORT

PRD CLS	STOCK NUMBER — DESCRIPTION WH LOCN UNITS ON HAND	UNITS ON ORDER	UNITS RESERVED	UNIT AVERAGE COST	VALUE AVERAGE COST	PHYSICAL QUANTITY
10	1001 1/4" ELECTRIC DRILL .MO 1 0019 517.0000	0.0000	40.0000	29.50000	15,251.50	
40	1001 1/4" ELECTRIC DRILL .MO 1 0018 80.0000	30.0000	0.0000	29.00000	2,320.00	
10	1001 1/4" ELECTRIC DRILL .MO 1 0017 150.0000	25.0000	10.0000	28.75000	4,312.50	
20	1001 1/4" ELECTRIC DRILL .MO 1 0007 410.0000	25.0000	50.0000	10.25000	4,202.50	
20	1001 1/4" ELECTRIC DRILL .MO 1 0008 330.0000	0.0000	0.0000	27.45000	9,058.50	
25	1001 1/4" ELECTRIC DRILL .MO 1 0009 14,256.0000	0.0000	1,440.0000	0.45000	6,415.20	
32	1001 1/4" ELECTRIC DRILL .MO 1 0003 59.0000	40.0000	5.0000	41.50000	2,445.50	
100	1001 1/4" ELECTRIC DRILL .MO 1 0006 2,448.0000	0.0000	0.0000	4.10000	10,036.80	
25	1001 1/4" ELECTRIC DRILL .MO 1 0006 0.0000	0.0000	0.0000	1.75000	0.00	
30	1001 1/4" ELECTRIC DRILL .MO 1 0016 314.0000	100.0000	50.0000	5.85000	1,836.90	
15	1001 1/4" ELECTRIC DRILL .MO 1 0015 192.0000	240.0000	0.0000	2.05000	393.60-	
20	1001 1/4" ELECTRIC DRILL .MO 1 0012 183.0000	50.0000	35.0000	4.75000	869.25	
XX	1001 1/4" ELECTRIC DRILL .MO 1 0011 105.0000	50.0000	0.0000	2.65000	278.25	
50	1001 1/4" ELECTRIC DRILL .MO 1 0004 109.0000	10.0000	0.0000	16.50000	1,798.50	
100	1001 1/4" ELECTRIC DRILL .MO 1 0001 960.0000	240.0000	0.0000	4.50000	4,320.00	
GRAND TOTAL					62,754.80	

number of units ordered, back-ordered units, average costs, and related information. It is used to determine when to order more inventory and how much of each item to order, and it helps minimize "stockouts" and back orders. Data from this report is used as input to other information systems to help production and operations managers analyze the production process.

For almost all companies, inventory must be tightly controlled. One objective is to minimize the amount of cash tied up in inventory by placing just the right amount of inventory on the factory or warehouse floor.

To gain a competitive advantage, many manufacturing organizations are moving to real-time inventory control systems based on bar coding the finished product, scanners and radio display terminals mounted on forklifts, and wireless LAN communications to track each time an item is moved in the warehouse. One significant advantage is that the inventory data is more accurate and current for people performing order entry, production planning, and shipment planning. In addition, warehouse operations can be streamlined by providing directions to the forklift drivers.

In addition to being useful for physical goods such as automobiles and home appliances, inventory control is essential for industries in the service sector. Such organizations as hotels, airlines, rental car agencies, and universities, which primarily provide services, can use inventory applications to help them monitor use of rooms, airline seats, car rentals, and classroom capacity. Airlines face an especially difficult inventory problem. Empty airline seats (inventory) have absolutely no value after a plane takes off. Yet overbooking can result in too many seats being sold and customer complaints. Sophisticated reservation systems allow airlines to quickly update and add seating assignments.

FIGURE 9.11

An output of the invoicing system, a customer invoice reflects the value of the current invoice, as well as which products the customer purchased.

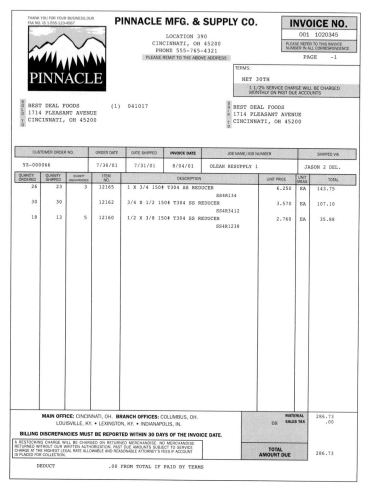

Invoicing

Customer invoices are generated based on records received from the shipment execution transaction processing system. This application encourages follow-up on existing sales activities, increases profitability, and improves customer service. Most invoicing programs automatically compute discounts, applicable taxes, and other miscellaneous charges (Figure 9.11). Because most computerized operations contain elaborate databases on customers and inventory, many invoicing applications require only information on the items ordered and the client identification number; the invoicing application does the rest. It looks up the full name and address of the customer, determines whether the customer has an adequate credit rating, automatically computes discounts, adds taxes and other charges, and prepares invoices and envelopes for mailing.

Invoicing in a service organization can be even more complicated than invoicing in manufacturing and retail firms. The trick is to match all services rendered with a specific customer and to include all appropriate rates and charges in calculating the bill. This is especially difficult if the data needed for billing has not been accurately and completely captured in a transaction processing system.

Customer Interaction

Winning new customers and keeping existing ones happy are key to financial success. It is often said that when customers have a pleasant experience with a firm, they may tell one or two people; however, when they have an unpleasant experience, they may tell 10 to 20 people! To keep current customers happy, some companies use a **customer interaction system** (Figure 9.12) to monitor and track each customer interaction with the company. The goal of such systems is to build customer loyalty. These software systems capture data whenever a customer contacts the company. Often the initial contact is a request for a proposal or a request for product information from a potential customer. Valuable data about the potential customer can be gathered at this time. Obviously, additional data is captured at the time of each sale. After the sale, a customer may contact the firm with a request for customer service or may have an idea or request other information. At other times the customer may contact the firm with a complaint or a product improvement idea.

The customer interaction system captures valuable data from each interaction and passes the data to others in the organization who can use it. The analysis of customer complaints provides ideas for market research and product development as well as useful measures for quality control. Sales and marketing employees gain a deeper understanding of what the customer wants through analysis of customer questions about the product and its use. The customer interaction data represents a gold mine of data that can be used to keep customers satisfied, generate new leads for future sales, and lead to new products or product improvements. For example, consumer goods company Procter & Gamble uses toll-free numbers on all its products for consumers to call for

customer interaction system

system that monitors and tracks each customer interaction with the company

FIGURE 9.12

Customer Interaction System

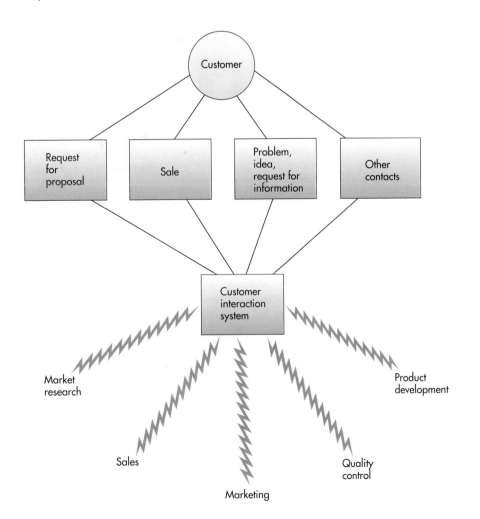

information on how to use the product, register a complaint, or find out where a new product is available. The next time the consumer calls, the service rep knows that the customer is a repeat caller and can look up information about his or her previous interaction(s) with the company. Consumer comments about the product are provided to product development and manufacturing. Comments regarding commercials and promotions are forwarded to the advertising department.

Routing and Scheduling

Many computer manufacturers and software firms have developed specialized transaction processing systems for companies in the distribution industry. Some distribution applications are for wholesale operations; others are for retail or specialized applications. Trucking firms, beverage distributors, electrical distributors, and oil and natural gas distribution companies are only a few examples.

routing system

system that determines the best way to get products from one location to another

Like airlines, distribution companies must also determine the best use of their resources. For example, a motor freight company might have 100 deliveries to make during the next week, including loads from Miami to Boston and Seattle to Salt Lake City. A **routing system** involves determining the best way to get products from one location to another. Waste Management Systems, for example, now has terminals on board its garbage collection trucks to facilitate routing. Drivers can be notified of changes in the routes after they are out on the streets. For an oil and natural gas company, routing means finding the fastest and cheapest pipelines to carry the product from the source to a distant final destination.

scheduling system

system that determines the best time to deliver goods and services

The **scheduling system** determines the best time to deliver goods and services. For example, trucks can be scheduled to deliver automobile transmission systems from California to Michigan during the second week of September, when oil and gas prices are low. Other objectives are to carry a profitable load on the return trip and to minimize total distance traveled, which can result in lower fuel, driver, and truck maintenance costs. For these reasons, many distribution companies have designed TPSs to help determine which routes will allow for efficient service, while making cost-effective use of drivers and trucks. For firms such as these, scheduling and routing programs are connected to the organization's order and inventory transaction processing system.

Purchasing Systems

purchasing transaction processing systems

systems that include inventory control, purchase order processing, receiving, and accounts payable

The **purchasing transaction processing systems** include inventory control, purchase order processing, receiving, and accounts payable (Figure 9.13).

Inventory Control

A manufacturing firm has several kinds of inventory, such as raw materials, packing materials, finished goods, and maintenance parts. We have already discussed the use of an inventory control system for finished product inventory. In addition, the firm needs to ensure that sufficient raw material, packing material, and maintenance parts are available. The same or a similar transaction processing system can be used to manage the inventory of these items.

Purchase Order Processing

purchase order processing system

system that helps purchasing departments complete their transactions quickly and efficiently

An organization's purchasing department typically has a number of employees who are responsible for all purchasing activities for the organization. Whenever materials or high-cost items are purchased, the purchasing department is involved. The **purchase order processing system** helps purchasing departments complete their transactions quickly and efficiently. Every organization has its own policies, practices, and procedures for purchasing supplies and equipment. IBM, for example, has a policy to use the Internet to streamline its

FIGURE 9.13

Purchasing Transaction
Processing System

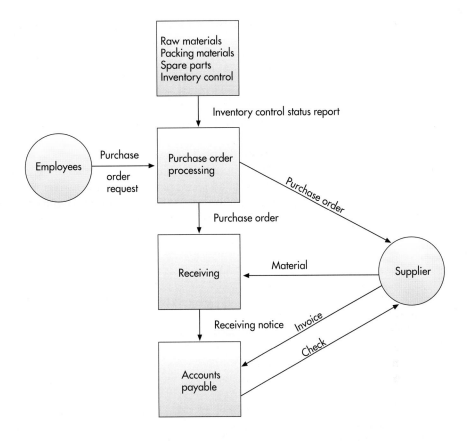

own operations.[13] In 1999, the company purchased more than $13 billion of supplies and equipment over the Internet, saving more than $270 million. As seen in the "E-Commerce" box, companies are increasingly purchasing needed supplies through the Internet or an Internet exchange.

The purchasing department can facilitate the buying process by keeping data on suppliers' goods and services.[14] The increased use of telecommunications has given many purchasing departments easier access to this information. For instance, technologies like the Internet and public networks allow purchasing managers to compare products and prices listed in Internet catalogs and large-scale consumer databases. Once the supplier is selected, the suppliers' computer systems might be directly linked to the systems of buyers. Orders can be sent via EDI, reducing purchasing costs and time spent and helping companies maintain low, yet adequate, inventory levels.

Companies are also using Internet exchanges to help them purchase materials and supplies at discounted prices. As discussed previously, an Internet exchange is formed by several companies in an industry and can be open to all companies in that industry. Giant consumer products companies Procter & Gamble, Kraft, Nabisco, and Pepsi Bottling Group are developing a multimillion-dollar exchange open to all companies in the consumer products industry.[15] One industry can have more than one exchange. Volkswagen, for example, is in the process of developing an exchange to help it purchase factory parts, tools, office equipment, and other products.[16] (Volkswagen has no plans to join with an auto exchange involving General Motors, Ford, and DaimlerChrysler.) Companies can use software, such as Ironside by Ironside Technology, to help them take advantage of purchasing materials and suppliers using multiple Internet exchanges.[17]

Instead of searching for the lowest prices from a list of suppliers, many companies form strategic partnerships with one or two major suppliers for important parts and materials. The partners are chosen based on prices and their

E-COMMERCE

Industrywide Processing and Exchanges

When most people think of doing business on the Internet, they visualize buying final products and services. Although order processing is important, it is not the only transaction processing application being placed on the Internet. Increasingly, companies are purchasing parts and services from each other on the Internet. This important business-to-business application is exploding.

There are many options in placing an important transaction processing application, like purchasing, on the Internet. At first, companies developed their own e-commerce applications on the Internet to perform purchasing. Today, however, more and more companies are investigating Internet partnerships to reap the benefits of the Internet, while sharing the costs and effort required. A massive partnership uniting General Motors, Ford Motor Company, and DaimlerChrysler is an example.

The three big auto companies have launched a massive effort to develop an Internet site to purchase auto parts on-line. According to Brian Kelley, vice president for Ford's e-commerce activities, "This will be the world's largest Internet company and trade exchange." With the huge company comes huge expectations. The new company, with a market capitalization that could range from $30 billion to $40 billion, is expected to have transaction and advertising revenues approaching $3 billion annually. Most auto executives believe that the vast majority of the $500 billion purchasing expenditures that these auto companies make annually will be conducted on the Internet.

In addition to the auto exchange, other exchanges are emerging. PaperExchange is used by International Paper and others for purchasing paper supplies. MyAircraft is

an aerospace purchasing exchange for parts and supplies. ChemConnect is an Internet exchange used by Dow Chemical, Eastman Chemical, and Rohm & Haas; and XSAg is an exchange for the agriculture industry. Although exchanges seem to be a current trend, some people believe that large purchasing sites are not the best approach for all industries. And some exchanges are limited to the strategic partners of the site, locking out other companies in the same industry. No matter what some critics say, large industrywide exchanges are here to stay.

Discussion Questions

1. What is an Internet exchange?
2. What are the advantages and disadvantages of an Internet exchange?

Critical Thinking Questions

3. Assume that you are the chief information officer of a European auto company that is not part of an existing auto exchange like the one discussed in this box. Would you recommend forming a partnership with other auto companies not currently in the exchange or developing your own Internet site just for your company?
4. In your opinion, what is the future of industry exchanges on the Internet over the next five years?

Sources: Adapted from Guy Matthews et al., "One Site May Not Meet All Industries' Needs," *The Wall Street Journal*, February 28, 2000, p. A16; Robert Simison et al., "Big Three Car Makers Plan Net Exchange," *The Wall Street Journal*, February 28, 2000, p. A3; and Tom Stundza, "Buyers Ready for E-Procurement," *Purchasing Magazine*, December 16, 1999, p. S23.

---

ability to deliver quality products on time consistently. For example, the automotive companies request that their suppliers have plants or offices close to their operations in Michigan. The ability to conduct electronic commerce following EDI standards is also a key factor.

Receiving

Like centralized purchasing, many organizations have a centralized receiving department responsible for taking possession of all incoming items, inspecting them, and routing them to the people or departments that ordered them. In addition, the receiving department notifies the purchasing department when items have been received. This notification may be done using a paper form called a receiving report or electronically through a business transaction created by entering data into the receiving transaction processing system.

An important function of many receiving departments is quality control by inspection. Inspection procedures and practices are set up to monitor the quality of incoming items. Any items that fail inspection are sent back to the supplier, or adjustments are made to compensate for faulty or defective products.

receiving system

system that creates a record of expected receipts

Many suppliers now send their customers advance shipment notices. This business transaction is input to the customer's **receiving system** to create a record of expected receipts. In addition, items are shipped with a bar code identification on

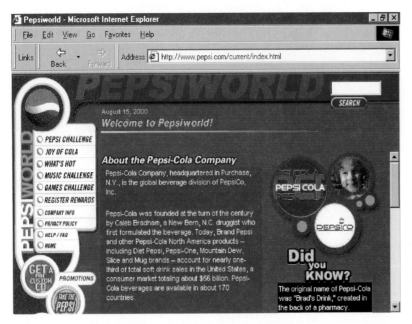

Giant consumer products companies Procter & Gamble, Kraft, Nabisco, and Pepsi Bottling Group are developing a multimillion-dollar exchange open to all companies in the consumer products industry.

accounts payable system

system that increases an organization's control over purchasing, improves cash flow, increases profitability, and provides more effective management of current liabilities

the container. At the receiving dock, the worker scans the bar code on each container and a transaction is sent to the receiving system, where the bar code identification number is matched against the file of expected receipt records. This improves the accuracy of the receiving process, eliminates the need to perform manual data entry, and reduces the manual effort required.

Accounts Payable

The **accounts payable system** attempts to increase an organization's control over purchasing, improve cash flow, increase profitability, and provide more effective management of current liabilities. Checks to suppliers for materials and services are the major outputs. Most accounts payable applications strive to manage cash flow and minimize manual data entry. Input from the purchase order processing system provides an electronic record to the accounts payable application that updates the accounts payable database to create a liability record showing that the firm has made a commitment to purchase a specific good or service. Once the accounts payable department receives a bill from a supplier, the bill is verified and checked for accuracy. Upon receiving notice that the goods and services have been delivered in a satisfactory manner, the data is entered into the accounts payable application. A typical check from an accounts payable application is shown in Figure 9.14. In addition to containing standard information found on any check, most accounts payable checks include the items ordered, invoice date, invoice numbers, amount of each item, any discounts, and the total amount of the check. This allows the company to consolidate several invoices and bills into a single payment. In addition to checks, companies can also pay their suppliers electronically using EDI, the Internet, or other electronic payment systems.

A common report produced by the accounts payable application is the purchases journal. As shown in Figure 9.15, this report summarizes all the organization's bill-paying activities for a particular period. Financial managers use this report to analyze bills that have been paid by the organization. This information

FIGURE 9.14

A Check Generated by an Accounts Payable Application

The check stub details items ordered, invoice dates, invoice numbers, cost of each item, discounts, and the total amount of the check.

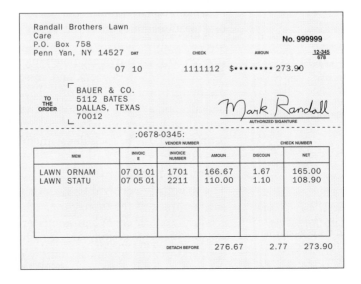

FIGURE 9.15

An Accounts Payable Purchases Journal

Generated by the accounts payable application, this report summarizes an organization's bill-paying activities for a particular period.

```
          DATE   07 10 01                     RANDALL BROTHERS LAWN CARE                                    PAGE

                                           ACCOUNTS PAYABLE PURCHASES JOURNAL

 TRANS    PYMT  REV  TRANS    INVOICE   INVOICE   INVOICE      BNK REFERENCE                INVOICE    DISCOUNT VENDOR BALANCE
CTRL NO    NO   CD   STATUS   NUMBER    DATE      DUE DATE     CD                           AMOUNT     ALLOWED  BEG.   */END **
                                          G/L NUMBER  G/L DESCRIPTION       G/L AMT


***       1002  ALFRED PAINTS, INC.
           3         INVOICE  91105     07 03 01    08 03 01         PAINT                  1,000.00    20.00          1,000.00 *
          17         CR-ME    CM6725    07 10 01                     DAMAGED MERCHANDISE     500.00      1.00
                     GL-DLS             1004200     INVENT.-TYPE 2                 500.00
                                                                                                                         500.00 **

                                                                                                                            .00 *
***       1002  ALFRED PAINTS, INC.
          15         INVOICE  7216      07 10 01   08 10 01  1 LAWN MACH & FURN         1,700.00    54.00
                     GL-DIS   CM6725    1004100     INVENT.-TYPE 1               1,350.00
                     GL-DLS             1004200     INVENT.-TYPE 2                 350.00
                                                                                                                       1,700.00 **


***       1002  ALFRED PAINTS, INC.
          16         INV-MP   2455      07 09 01    07 20 01     1  FASTENERS                  300.00     6.00
                     GL-DIS             5013000     BLDG, MAINT.                   300.00
          16         MLT-PY   2445      07 09 01    07 29 01     1  FASTENERS                  150.00     3.00
          16         MLT-PY   2445      07 09 01    07 29 01     1  FASTENERS                  150.00     3.00
                                                                                                                         300.00 **

***       1002  ALFRED PAINTS, INC.                                                                                         .00 *
          14         INVOICE  6211      07 10 01    08 10 01    1 LAWN MACH & FURN       1,050.00    54.00
                     GL-DIS             5018200     UTILITIES                   1,050.00
                                                                                                                       1,050.00 **
```

is also used to help analyze current and future cash flow needs. Data is summarized for each supplier or parts manufacturer. Invoice number, description, amounts, discounts, and total checking activity are included. In addition, many purchasing reports include the total amount of checks generated by the accounts payable application on a daily, weekly, or monthly basis.

The accounts payable application ties into other information systems, including cash flow analysis, which helps an organization ensure that sufficient funds are available for the accounts payable application and can show the best sources of funds for payments that must be made via the accounts payable application.

Accounting Systems

accounting systems

systems that include budget, accounts receivable, payroll, asset management, and general ledger

The primary **accounting systems** include the budget, accounts receivable, payroll, asset management, and general ledger (Figure 9.16).

Budget

budget transaction processing system

system that automates many of the tasks required to amass budget data, distribute it to users, and consolidate the prepared budgets

In an organization, a budget can be considered a financial plan that identifies items and dollar amounts that the organization estimates it will spend. In some organizations, budgeting can be an expensive and time-consuming process of manually distributing and consolidating information. The **budget transaction processing system** automates many of the tasks required to amass budget data, distribute it to users, and consolidate the prepared budgets. Automating the budget process allows financial analysts more time to manage it to meet organizational goals by setting enterprisewide budgeting targets, ensuring a consistent budget model and assumptions across the organization, and monitoring the status of each department's spending.

Accounts Receivable

accounts receivable system

system that manages the cash flow of the company by keeping track of the money owed the company on charges for goods sold and services performed

The **accounts receivable system** manages the cash flow of the company by keeping track of the money owed the company on charges for goods sold and services performed. When goods are shipped to a customer, the customer's accounts payable system receives a business transaction from the invoicing system, and the customer's account is updated in the accounts receivable system

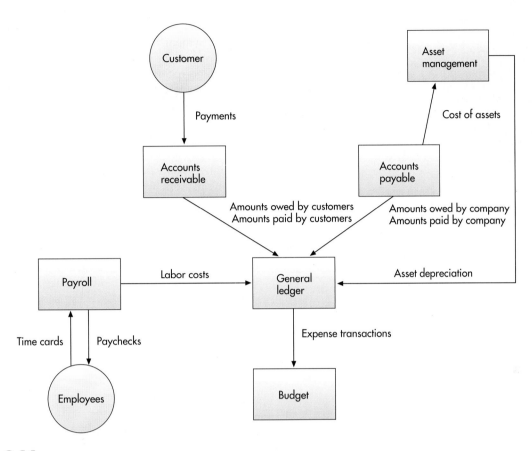

FIGURE 9.16

Financial Systems

of the supplier. A statement reflecting the balance due is sent to active customers. Upon receipt of payment, the amount due from that customer is reduced by the amount of payment.

The major output of the accounts receivable application is monthly bills or statements sent to customers.[18] As you can see in Figure 9.17, a bill sent to a customer can include the date items are purchased, descriptions, reference numbers, and amounts. In addition, bills can include amounts for various periods, totals, and allowances for discounts. The accounts receivable application should monitor sales activity, improve cash flow by reducing the time between a customer's receipt of items ordered and payment of bills for those items, and ensure that customers continue to contribute to profitability. Most systems can handle payment in a variety of ways, including standard bank checks, credit cards, money-wiring services, and electronic funds transfer via EDI. Increasingly, companies are using the Internet for their accounts receivables application. Using these Internet systems, customers can pay their bills while connected to the Internet on their home PC, at a retail store, or using a hand-held computer with a wireless connection to the Internet.[19]

The accounts receivable system is vital to managing the cash flow of the firm. One major way to increase cash flow is by identifying overdue accounts. Reports are generated that "age" accounts to identify customers whose accounts are overdue by more than 30, 60, or 90 days. Special action may be initiated to collect funds or reduce the customer's credit limit with the firm, depending on the amount owed and degree of lateness.

An important function of the accounts receivable application is to identify bad credit risks. Because a sizable amount of an organization's assets can be tied up in accounts receivable, one objective of an accounts receivable application is to minimize losses due to bad debts through early identification of potential bad-debt

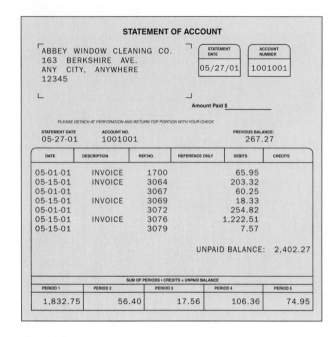

STATEMENT OF ACCOUNT

ABBEY WINDOW CLEANING CO.
163 BERKSHIRE AVE.
ANY CITY, ANYWHERE
12345

STATEMENT DATE	ACCOUNT NUMBER
05/27/01	1001001

Amount Paid $ _____

PLEASE DETACH AT PERFORATION AND RETURN TOP PORTION WITH YOUR CHECK

STATEMENT DATE	ACCOUNT NO.		PREVIOUS BALANCE:
05-27-01	1001001		267.27

DATE	DESCRIPTION	REF.NO.	REFERENCE ONLY	DEBITS	CREDITS
05-01-01	INVOICE	1700		65.95	
05-15-01	INVOICE	3064		203.32	
05-01-01		3067		60.25	
05-15-01	INVOICE	3069		18.33	
05-01-01		3072		254.82	
05-15-01	INVOICE	3076		1,222.51	
05-15-01		3079		7.57	
			UNPAID BALANCE:	2,402.27	

SUM OF PERIODS • CREDITS = UNPAID BALANCE				
PERIOD 1	PERIOD 2	PERIOD 3	PERIOD 4	PERIOD 5
1,832.75	56.40	17.56	106.36	74.95

FIGURE 9.17

An Accounts Receivable Statement

Generated by the accounts receivable application, a bill is sent to a customer (usually monthly) and details items purchased, dates of purchase, and amounts due.

customers. Thus, many companies routinely check a customer's payment history before accepting a new order. With advances in telecommunications, companies can search huge national databases for the names of firms and individuals who have been reported as delinquent on payments or as bad credit risks. When using external data like this in a TPS application, however, companies must be extremely cautious regarding the accuracy of the data.

The accounts receivable aging report, shown in Figure 9.18, is a valuable aspect of an accounts receivable application. In most cases, this report sorts all outstanding debts or bills by date. For unpaid bills that have remained outstanding for a predetermined amount of time, "reminder notices" can also be automatically generated. The accounts receivable aging report gives managers an immediate look into large bills that are long overdue so that they can be followed up to prevent more orders being sent to delinquent customers. This type of report can be produced on a customer-by-customer basis or in a summary format.

Payroll

The two primary outputs of the payroll system are the payroll check and stub, which are distributed to the employees, and the payroll register, which is a summary report of all payroll transactions. In addition, the payroll system prepares W-2 statements at the end of the year for tax purposes. In a manufacturing firm, hours worked and labor costs may be captured by job so that this information can be passed on to the manufacturing costs system.

Responsibility for running this TPS application can be outsourced to a service company. Rather than write their own payroll application, many firms rely on a purchased software application for payroll processing. Some of these packages are tailored for a specific industry; others can accommodate a wide range of uses. In

FIGURE 9.18

An Accounts Receivable Aging Report

The output from an accounts receivable application tells managers what bills are overdue, either customer by customer or in a summary format.

```
6588-DIAGE    TOOL DISTRIBUTORS INC.    0001 ACCOUNTS RECEIVABLE AGING ANALYSIS    SEP 25 2001    PAGE   1
                                             AS OF SEP 25 2001

TYPE OPEN-ITEM  ITM-DATE  REFERENCE INVOICE/PAYMENT CURRENT   31-60 DAYS  61-90 DAYS 91-120 DAYS OVER 120 DAYS
        NUMBER              NUMBER          AMOUNT

  CLASS 05    CUST 94367   NAME TOOLS OF AMERICA        3225 N PARKWAY    TEL 513-2864731  CONT B. BROWN

  INV  95361   03/17/01                 23,058.37                                               23,058.37
  DM   96853   03/20/01                  5,589.42                                                1,589.42
  INV 105395   07/04/01                  2,923.45               2,923.45
  DM  116594   07/06/01                    198.32                 198.32
  INV 123984   07/15/01                 23,087.28
  DSC 123984   08/17/01                  1,204.36
  CA  968351   08/19/01                 21,882.92
  INV 147296   07/23/01                 19,709.57
  CA   83495   07/31/01                 18,725.21
  CM  995473   08/19/01                    984.59
  INV 149384   08/29/01                 23,831.37    21,831.59
  INV 158439   09/10/01                 30,086.68    30,086.68
  INV 161236   09/23/01                 25,520.37    25,520.37

  TOTL RECEIVABLE FOR CUST 94367       107,208.20    79,438.64   3,121.77     0.00      0.00     24,647.79
     CA                                   3,121.77                3,121.77
     CA                                  22,640.01    22,640.01
     CM                                     279.84       279.84

  UNAPPLIED CREDITS                      26,041.62    22,919.85   3,121.77     0.00      0.00         0.00

  NET RECEIVABLE FOR CUST 94367         81,166.58    56,518.79       0.00     0.00      0.00     24,647.79

  TOTAL FOR CLASS    05    42   PRINTED        TOTAL RECEIVABLE  UNAPPLIED CREDITS  NET RECEIVABLE

                                        CURRENT    161,506.12    67,832.57    93,673.55
                                        31-60 DAYS  31,494.14    18,984.68    12,509.46
                                        61-90 DAYS  11,569.32     4,267.91     7,301.59
                                        91-120 DAYS 27,764.18         0.60    27,746.18

                                          TOTAL    238,823.97    93,232.49   145,581.08
```

FIGURE 9.19

A typical paycheck stub details the employee's hours worked for the period, salary, vacation pay, federal and state taxes withheld, and other deductions.

TENDER CARE DAY CARE CENTER, INC.
CINCINNATI, OH 45200

4207

Vacation Taken This Check		0.000	Sick Taken This Check	0.000
Vacation Available		0.667	Sick Available	0.667
	Earnings		Deductions	
Description	Hours	Amount	Description	Amount
Regular Pay	36.320	199.76	FICA Withheld	15.28
Overtime	0.000	0.00	Fed. Tax W/H	15.00
SICK TIME	0.000	0.00	State Tax W/H	1.67
PERSONAL	0.000	0.00	Other W/H #1	4.19
VACATION	0.000	0.00	Other W/H #2	0.00
HOLIDAY PAY	0.000	0.00	Other W/H #3	0.00
Gross Pay#4	0.000	0.00	Other W/H #4	0.00
Total		199.76	Total	36.14
Net Pay		**163.62		
YEAR TO DATE				
Total Earnings		397.76	FICA	30.43
Federal W/H		30.00	State W/H	3.32

most cases, the number of hours worked by each employee is collected using a variety of data entry devices, including time clocks, time cards, and industrial data collection devices. Once collected, payroll data is used to prepare weekly, biweekly, or monthly employee paychecks (Figure 9.19). Payroll systems can handle overtime, vacation pay, variable and multirate salary structures, incentive programs, and commissions. Most payroll applications automatically generate both federal and state tax forms related to payroll and process deductions, including tax-deferred annuities, savings plans, and U.S. government savings bonds. Often payroll applications have EDI arrangements with employees' banks to make direct deposits into employees' accounts.

As you can see in Figure 9.19, the payroll program has produced a weekly paycheck, which includes the employee's hourly rate, total hours worked, regular pay, premium pay, federal and state tax withholdings, and other deductions. In addition to paychecks, most payroll programs produce a **payroll journal**, shown in Figure 9.20. A typical payroll journal contains employees' names, the areas where employees worked during the week, hours worked, the pay rate, a premium factor for overtime pay, earnings, the earnings type, various deductions, and net pay calculations. Financial managers at the operational level use the payroll journal to monitor and control pay to individual employees. As you can see in Figure 9.20, the payroll journal also includes totals for hours worked, earnings, deductions, and net pay.

payroll journal

a report that contains employees' names, the area where employees worked during the week, hours worked, the pay rate, a premium factor for overtime pay, earnings, earnings type, various deductions, and net pay calculations

Payroll TPS applications also provide input into various weekly, quarterly, and yearly reports. Most of these reports are used by financial managers to help control payroll costs and cash flows. Payroll applications also provide necessary audit trails through the use of payroll master files and documents used by internal accountants and external auditors to make sure that the application is functioning as intended.

FIGURE 9.20

A Payroll Journal

Generated by the payroll application, this report helps managers monitor total payroll costs for an organization and the impact of those costs on cash flow.

DATE 12 25 01 — RANDALL BROTHERS LAWN CARE — PAGE 003

PAYROLL JOURNAL
PR020

EMPLOYEE NAME / ERRORS WARNINGS	EMPLOYEE NUMBER	COST CENTER	HOURS	RATE	PREM FACTOR	EARNINGS	EARN TYPE	DEDUCTIONS	DED TYPE	NET PAY	PAYMENT LOCATION
DANIEL JACKSON											
	112	30	40.00	6.00		240.00	001			44.19	CIN
	112	30	10.00	6.00	1.5	90.00	030				
	112							5.00	220		
	112							15.81	801		
	112							44.19	802		
HOWARD SIMPSON				5.25							
	126	30	32.00			168.00	001			12.82	CIN
	126							5.00	220		
	126							2.00	223		
	126							10.16	801		
	126							12.82	802		

Like many other transaction processing applications, the payroll application interfaces with other applications. All payroll entries are entered to the general ledger systems. Furthermore, there can be a direct link between payroll activities and production/inventory control operations. Direct links are often used for manufacturing operations or job-shop systems because data collected on hours worked from the payroll application helps determine the total cost of completing various jobs. For example, if an employee who earns $15 per hour spends 20 hours completing a particular job, the labor costs for that job are $300. This type of information from the payroll application is useful in determining the cost to produce a product or render a service and thus in determining its profitability.

Asset Management

Capital assets represent major investments for the organization whose value appears on the balance sheet under fixed assets. These assets have a useful life of several years or more, over which their value is depreciated, resulting in a tax reduction. The **asset management transaction processing system** controls investments in capital equipment and manages depreciation for maximum tax benefits. Key features of this application include efficient handling of a wide range of depreciation methods, country-specific tax reporting and depreciation structures for the various countries in which the firm does business, and workflow-managed processes to easily add, transfer, and retire assets.

General Ledger

Every monetary transaction that occurs within an organization must be properly recorded. Payment of a supplier's invoice, receipt of payment from a customer, and payment to an employee are examples of monetary transactions. A computerized **general ledger** system is designed to allow automated financial reporting and data entry. The general ledger application produces a detailed list of all business transactions and activities. Reports, including profit and loss (P&L) statements, balance sheets, and general ledger statements, can be generated (Figure 9.21). Furthermore, historical data can be kept and used to generate trend analyses and reports for various accounts and groups of accounts used in the general ledger package. Various income and expense accounts can be generated for the current period, year to date, and month to date as required. The reports generated by the general ledger application are used by accounting and financial managers to monitor the profitability of the organization and to control cash flows.

Financial reports that summarize sales by customer and inventory items can also be produced. These reports are used by marketing and financial managers to determine which customers are contributing to sales and inventory items that are selling as expected.

A key to the proper recording and reporting of financial transactions is the corporation's chart of accounts (Table 9.3). This chart provides codes for each type of expense or revenue. By entering transactions consistent with the chart of accounts, financial data can be reported in a simple and consistent fashion across all organizations of the enterprise, even if it is a multinational corporation.

ENTERPRISE RESOURCE PLANNING

Flexibility and quick response are hallmarks of business competitiveness. Access to information at the earliest possible time can help businesses serve customers better, raise quality standards, and assess market conditions. Enterprise resource planning (ERP) is a key factor in instant access.[20] Although some think that ERP systems are only for extremely large companies, this is not the case. Medium-sized companies can also benefit from the ERP approach. A few leading vendors of ERP systems are listed in Table 9.4.

---

asset management transaction processing system

system that controls investments in capital equipment and manages depreciation for maximum tax benefits

general ledger system

system designed to automate financial reporting and data entry

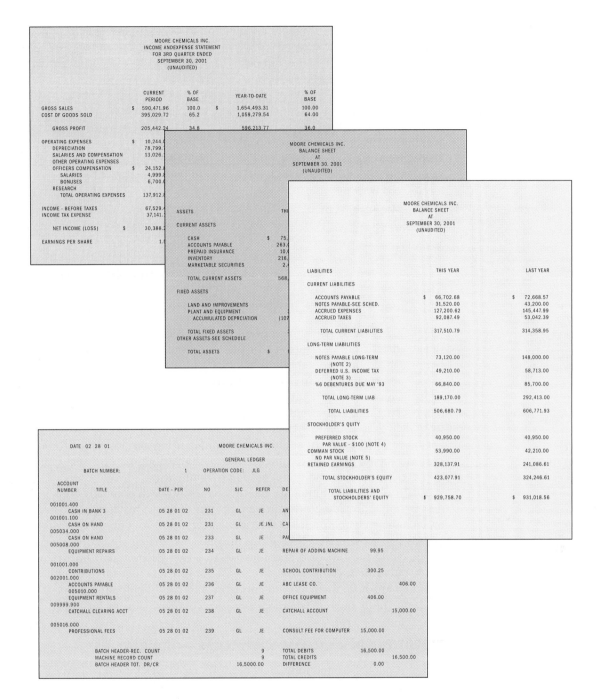

FIGURE 9.21

Outputs from a General Ledger System

An income statement (top) details sales, costs, and operating expenses to produce a statement of income for an organization. A balance sheet (center left and right) breaks down assets and liabilities so that managers can see at a glance whether income is covering expenses. A general ledger statement (bottom) track credits and debits.

Major Account Name	Type of Expense	Subaccount Code Used to Identify Transaction
Wages and Benefits	Management salaries and benefits	MSALSB
	Nonmanagement salaries and benefits	NMALSB
	Overtime	OVT
Travel and Training	Travel-related expenses	TRAVEL
	Tuition for training classes	TUITION
Professional Services	Fees paid to consultants, contractors, trainers, and other professionals	PROFSV
Maintenance Expense	Maintenance labor	MAINTL
	Maintenance parts	MAINTP
	Maintenance supplies	MAINTS

TABLE 9.3

Sample Partial Chart of Accounts

An Overview of Enterprise Resource Planning

The key to ERP is real-time monitoring of business functions, which permits timely analysis of key issues such as quality, availability, customer satisfaction, performance, and profitability. Financial and planning systems receive "triggered" information from manufacturing and distribution. When something happens on the manufacturing line that affects a business situation—for example, packing material inventory drops to a certain level, which affects the ability to deliver an order to a customer—a message is triggered for the appropriate person in purchasing. In addition to manufacturing and finance, ERP systems can also support human resources, sales, and distribution. This sort of integration breaks through traditional corporate boundaries.

ERP systems accommodate the different ways each company runs its business by either providing vastly more functions than one business could ever need or including customization tools that allow firms to fine-tune what should already be a close match. SAP R/3 is the undisputed king of the first approach.[21] R/3 is easily the broadest and most feature-rich ERP system on the market. Thus, rather than compete on size, most competitors focus on customizability. ERP systems have the ability to configure and reconfigure all aspects of the IS environment to support whatever way your company runs its business.

Advantages and Disadvantages of ERP

Increased global competition, new needs of executives for control over the total cost and product flow through their enterprises, and ever-more-numerous customer interactions are driving the demand for enterprisewide access to real-time information. ERP offers integrated software from a single vendor to help meet those needs. The primary benefits of implementing ERP include elimination of inefficient systems, easing adoption of improved work processes,

TABLE 9.4

Some ERP Software Vendors

Software Vendor	Name of Software
Oracle	Oracle Manufacturing
SAP America	SAP R/3
Baan	Triton
PeopleSoft	PeopleSoft
J. D. Edwards	World

improving access to data for operational decision making, and technology standardization. ERP vendors have also developed specialized systems for specific applications and market segments. SAP, for example, has developed a Customer Relationship Management (CRM) package for its ERP system.[22] Osram Sylvania, a lighting manufacturer, plans on using CRM to allow lighting buyers to place orders directly over the Internet. Of course, developing custom packages for every market need and segment would be a huge undertaking for ERP vendors.[23] As a result, major ERP vendors are increasingly seeking help from other software vendors to develop specialized programs to tie directly into their ERP systems. Even with the benefits of ERP, most companies have found it surprisingly difficult to justify implementation of an ERP system based strictly on cost savings.

Elimination of Costly, Inflexible Legacy Systems

Adoption of an ERP system enables an organization to eliminate dozens or even hundreds of separate systems and replace them with a single integrated set of applications for the entire enterprise. In many cases, these systems are decades old, the original developers are long gone, and the systems are poorly documented. As a result, the systems are extremely difficult to fix when they break, and adapting them to meet new business needs takes too long. They become an anchor around the organization that keeps it from moving ahead and remaining competitive. An ERP system helps match the capabilities of an organization's information systems to its business needs—even as these needs evolve.

Improvement of Work Processes

Competition requires companies to structure their business processes to be as effective and customer-oriented as possible. ERP vendors do considerable research to define the best business processes. They gather requirements of leading companies within the same industry and combine them with research findings from research institutions and consultants. The individual application modules included in the ERP system are then designed to support these **best practices**, the most efficient and effective ways to complete a business process. Thus, implementation of an ERP system ensures good work processes based on best practices. For example, for managing customer payments, the ERP system's finance module can be configured to reflect the most efficient practices of leading companies in an industry. This increased efficiency ensures that everyday business operations follow the optimal chain of activities, with all users supplied the information and tools they need to complete each step.

Increase in Access to Data for Operational Decision Making

ERP systems operate via an integrated database and use essentially one set of data to support all business functions. So, decisions on optimal sourcing or cost accounting, for instance, can be run across the enterprise from the start, rather than looking at separate operating units and then trying to coordinate that information manually or reconciling data with another application. The result is an organization that looks seamless, not only to the outside world but also to the decision makers who are deploying resources within the organization.

The data is integrated to provide excellent support for operational decision making and allows companies to provide greater customer service and support, strengthen customer and supplier relationships, and generate new business opportunities. For example, once a salesperson makes a new sale, the business data captured during the sale is distributed to related transactions for the financial, sales, distribution, and manufacturing business functions in other departments.

Upgrade of Technology Infrastructure

An ERP system provides an organization with the opportunity to upgrade and simplify the information technology it employs. In implementing ERP, a company must determine which hardware, operating systems, and databases it wants to use. Centralizing and formalizing these decisions enables the organization to eliminate the hodgepodge of multiple hardware platforms, operating systems, and databases it is currently using—most likely from a variety of vendors. Standardization on fewer technologies and vendors reduces ongoing maintenance and support costs as well as the training load for those who must support the infrastructure.

Expense and Time in Implementation

Getting the full benefits of ERP is not simple or automatic. Although ERP offers many strategic advantages by streamlining a company's transaction processing system, ERP is time-consuming, difficult, and expensive to implement. Some companies have spent years and tens of millions of dollars implementing ERP systems. And when there are problems with an ERP implementation, it can be expensive. A $350 million maker of home textile furnishings, for example, blamed a 24 percent drop in its sales in one quarter on problems with its ERP system.[24]

Difficulty Implementing Change

In some cases, a company has to make radical changes in how it operates to conform with the work processes (best practices) supported by the ERP. These changes can be so drastic to long-time employees that they retire or quit rather than go through the change. This exodus can leave a firm short of experienced workers.

Difficulty Integrating with Other Systems

Most companies have other systems that must be integrated with the ERP. These systems can include financial analysis programs, Internet operations, and other applications. Many companies have experienced difficulties making these other systems operate with their ERP.

Risks in Using One Vendor

The high cost to switch to another vendor's ERP system makes it extremely unlikely that a firm will do so. Thus, the initial ERP vendor knows it has a "captive audience," and there is less incentive to listen to and respond to customer issues. The high cost to switch also creates a high level of risk if the ERP vendor allows its product to become outdated or goes out of business. Picking an ERP system involves not just choosing the best software product but also choosing the right long-term business partner.

Even with the high cost, long installation times, and complexity, there is no indication that the enthusiasm for this powerful software is slowing.

Example of an ERP System

SAP R/3 has been called one of the most complex packages ever written for use in corporations. However, it is also the most widely used ERP solution in the world. The following sections provide a brief description of the fundamental design and architecture of the SAP R/3 ERP system.

SAP R/3 ERP software provides comprehensive solutions for companies of all sizes in all industry sectors.

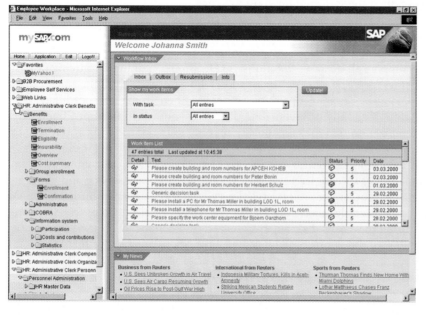

The SAP ERP system was developed from the perspective of a corporation as a whole rather than any specific business department. All data is entered only once in the system, and all SAP programs use the same database with little data redundancy. Each data item is clearly documented in a data dictionary. The software is flexible enough to be configured to meet the customer's business requirements. It is based on a three-level client/server architecture consisting of clients, application servers, and database servers (Figure 9.22). R/3 will run on a wide variety of hardware, from a small Windows NT server up to massively parallel systems.

Clients in the SAP System

The R/3 system typically supports hundreds or even thousands of clients. Clients are usually desktop computers with fast chips and at least 32 MB of RAM. Users of the clients request services from the application servers.

Application Servers in the SAP System

There are many application servers in a typical R/3 system. The servers are powerful midrange or even mainframe computers. The job of the server is to reply to all requests made of it, including requests for data, communication of messages, and update of master files. The request from the client travels along the network to the application server. A dispatcher program running on the application server manages the queues of user requests to determine which should be executed next. Application servers are grouped by the SAP R/3 system administrator into classes depending on the applications they run. One class might, for example, run the financial modules while another class might run the sales and distribution modules. The application server can contain

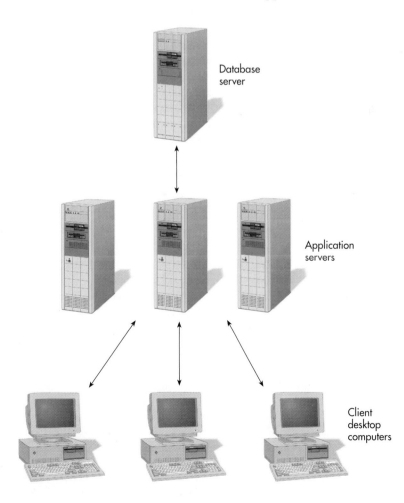

FIGURE 9.22

SAP Three-Tier Client/Server Architecture

either third-party or user-developed software, as long as it is written in ABAP/4, SAP's fourth generation programming language.

Business Application Programming Interfaces (BAPIs)

Business application programming interfaces, or BAPIs, are public interfaces. These interfaces were developed with SAP customers, software development organizations, and standards organizations to enable SAP customers to develop their own applications to interface with SAP. SAP then has the flexibility to change the underlying software, as long as the interface itself is not changed (Figure 9.23). Thus, new SAP software versions can be introduced without invalidating existing systems. An example of a BAPI is "customer order" to allow checking the status of the customer order.

Database Server in the SAP System

The database server in the R/3 system holds the data and is accessed and updated constantly. Depending on the hardware selected, the database may be distributed among multiple machines or reside on a single computer. The SAP update process is designed to accommodate hundreds or even thousands of users on a single database server and still provide satisfactory response times.

Objects in the SAP System

object

a collection of data and programs

Like many popular databases, SAP has adopted objects as one of its key implementation concepts. An SAP **object** is a collection of data and programs. "Purchase order" and "customer" are examples of SAP business objects used in business processes. Attributes contain the details of an object, like name, date of employment, and address of an employee.

Repository

All development objects of the ABAP/4 Development Workbench are stored in the ABAP/4 repository. These objects include ABAP/4 programs, screens, documentation, and other tools that a system developer would need. The SAP R/3 database makes constant use of the repository. The repository sits between the database and the application modules and provides the logical mappings of data relationships and conditions. The repository serves primarily as a tool to enter, manage, and evaluate information about a company's data. It is an active data dictionary, so new or changed data in the repository is immediately available to all system components. As a result, application programs and screens are always supplied with up-to-date information.

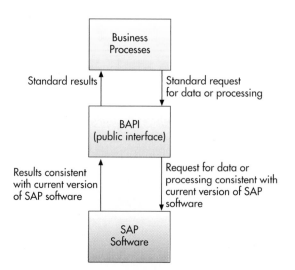

FIGURE 9.23

Business Application
Programming Interface (BAPI)

Tables

There are three major types of tables: system configuration tables, control tables, and application data tables. All are defined in the repository. System configuration tables are maintained primarily by SAP and define the structure of the system; clients do not change these tables. To customize the system, the client's ERP project team uses both control tables and application data tables. Control tables define functions that guide the user in his or her activities. For example, a control table might be set up to require a customer service representative to enter a line item to reference data about the product from the material master table before a purchase order is accepted. Application data tables are divided into two types—transactions and master data tables. Transaction tables are the largest, since they contain the daily operations data, such as orders, payments received, invoices, and shipments. Master data files describe sets of basic business entities such as customers, vendors, products, materials, and the like.

This section has focused on the key elements of the SAP R/3 ERP system; however, the systems from PeopleSoft, Oracle, and other vendors are based on similar basic design elements—the use of client/server architecture, standard business application interfaces, object-oriented programming, repository, and data tables.

● SUMMARY

PRINCIPLE ● An organization's TPS must support the routine, day-to-day activities that occur in the normal course of business and help a company add value to its products and services.

Transaction processing systems (TPSs) are at the heart of most information systems in businesses today. TPSs consist of all the components of a CBIS, including databases, telecommunications, people, procedures, software, and hardware devices to process transactions. All TPSs perform the following basic activities: data collection involves the capture of source data needed to complete a set of transactions; data edit checks for data validity and completeness; data correction involves providing feedback of a potential problem and enabling users to change the data; data manipulation is the performance of calculations, sorting, categorizing, summarizing, and storing for further processing; data storage involves placing transaction data into one or more databases; and document production involves outputting records and reports.

The methods of transaction processing systems include batch, on-line, and on-line with delayed processing. Batch processing involves the collection of transactions into batches, which are entered into the system at regular intervals as a group. On-line transaction processing (OLTP) allows transactions to be entered as they occur. Systems that use a compromise between batch and on-line processing are on-line with delayed-entry TPSs. Transactions

may be entered as they occur, but processing is not performed immediately.

Organizations expect TPSs to accomplish a number of specific objectives, including processing data generated by and about transactions, maintaining a high degree of accuracy, ensuring data and information integrity, compiling timely reports and documents, increasing labor efficiency, helping provide increased and enhanced service, and building and maintaining customer loyalty.

● ● ● ●

TPS applications are seen throughout an organization. The order processing systems include order entry, sales configuration, shipment planning, shipment execution, inventory control, invoicing, customer interaction, and routing and scheduling. Order entry captures the basic data needed to process a customer order. Once an order is entered and accepted, it becomes an open order. Sales configuration ensures that the products and services offered are sufficient. Shipment planning determines which open orders will be filled and from which location they will be shipped. The system prepares an order confirmation notice that is sent to the customer and a picking list used by warehouse operators to fill the order. The shipment execution system is used by the warehouse operators to enter data as to what was actually shipped to the customer. It passes shipped order transactions downstream to the invoicing system. For each item picked during the shipment execution process, a transaction providing stock number and quantity is passed to the finished product inventory control system. The invoicing system generates customer invoices based on the records received from the shipment execution system. The customer interaction system monitors and tracks each customer interaction. Routing and scheduling systems are used in distribution functions to determine the best use of a company's resources.

● ● ●

The purchasing information systems include inventory control, purchase order processing, accounts payable, and receiving. The inventory control system tracks the level of all packing materials and raw materials. It provides information to users on when to order additional materials. The purchase order processing system supports the policies, practices, and procedures of the purchasing department. The accounts payable system monitors and controls the outflow of funds to an organization's suppliers. The receiving system captures data about specific receipts of materials from suppliers so that approval for payment can be granted or refused.

● ● ●

The accounting systems include the budget, accounts receivable, payroll, asset management, and general ledger. The budget system automates many of the tasks required to amass budget data, distribute it to users, and consolidate the prepared budgets. The accounts receivable system manages the cash flow of the company by keeping track of the money owed the company. The payroll processing application processes employee paychecks and performs numerous calculations relating to time worked, deductions, commissions, and taxes. The outputs are used to help control payroll costs and cash flows and develop reports to the federal government. The asset management system controls investments in capital equipment and manages depreciation for maximum tax benefits. The general ledger system records every monetary transaction and enables production of automated financial reporting.

PRINCIPLE ● Implementation of an enterprise resource planning system enables a company to achieve numerous business benefits through the creation of a highly integrated set of systems.

Enterprise resource planning (ERP) software is a set of integrated programs that manage a company's vital business operations for an entire multisite, global organization. It must be able to support multiple legal entities, multiple languages, and multiple currencies. Although the scope of an ERP system may vary from vendor to vendor, most ERP systems provide integrated software to support manufacturing and finance. In addition to these core business processes, some ERP systems are capable of supporting additional business functions such as human resources, sales, and distribution.

Implementation of an ERP system can provide many advantages, including elimination of costly, inflexible legacy systems; providing improved work processes; providing access to data for operational decision making; and creating the opportunity to upgrade technology infrastructure. Some of the disadvantages associated with an ERP system are that they are time-consuming, difficult, and expensive to implement.

● **KEY TERMS**

accounting systems 352
accounts payable system 351
accounts receivable system 352
asset management transaction
 processing system 356
audit trail 341
batch processing system 331
best practices 359
budget transaction processing
 system 352
business resumption planning 339
customer interaction system 347
data collection 336

data correction 337
data editing 337
data manipulation 337
data storage 338
disaster recovery 339
document production 338
general ledger system 356
inventory control system 345
object 362
on-line transaction processing
 (OLTP) 331
order entry system 341
order processing systems 341

payroll journal 355
purchase order processing system 348
purchasing transaction processing
 system 348
receiving system 350
routing system 348
sales configuration system 343
scheduling system 348
shipment execution system 345
shipment planning system 344
transaction processing cycle 336
transaction processing system
 audit 340

● REVIEW QUESTIONS

1. Describe the basic activities common to all transaction processing systems.
2. List several characteristics of transaction processing systems.
3. What are the basic transaction processing activities?
4. What is an enterprise resource planning system?
5. List and briefly discuss the business objectives common to all TPSs.
6. What is the difference between batch processing, on-line processing, and on-line entry with delayed processing systems?
7. What is disaster recovery?
8. Describe an order entry system for taking orders over the Internet.

9. What systems are included in the order processing family of systems?
10. Identify the various stages consumers experience in the life cycle of a sale.
11. What systems are included in the purchasing family of systems?
12. Why is the general ledger application key to the generation of accounting information and reports?
13. Give an example of how transaction processing systems can be used to gain competitive advantage.
14. What is the purpose of a transaction processing system audit?

● DISCUSSION QUESTIONS

1. Assume that you are the owner of a small business. Describe the day-to-day transaction processing activities that you would encounter.
2. Your company is a medium-sized firm with sales of $500 million per year. It has been decided that the organization will implement an ERP system to support all operations at headquarters, three plants, and four distribution centers. What are some of the key questions that must be answered to further define the scope of this effort?
3. Imagine that you are the new IS manager for a Fortune 1000 company. Your internal information systems audit has revealed that your firm's systems are lacking disaster recovery plans and backup procedures. How would you justify to your manager spending a full year of IS staff time to implement these systems?
4. Is each of the various stages experienced by a consumer in the life cycle of a sale equally important for all kinds of products? Why or why not?

5. What is the advantage of implementing ERP as an integrated solution to link multiple business processes? What are some of the issues and potential problems?
6. You are the key user of the firm's purchase order processing system and have been asked to perform an information system audit of this system. Outline the steps you would take to complete the audit.
7. You are building your firm's first-ever customer interaction system. Discuss the features you would design into the system. How might you include suggestions from your customers into your design? Should the system be built using Internet technology?
8. You are in charge of a complete overhaul of your firm's order processing systems. How would you define the requirements for this collection of systems? What features would you want to include?
9. What steps would you use to plan for a potential disaster?

● PROBLEM-SOLVING EXERCISES

1. Assume that you own a small copy shop located near your school. Using a graphics program, draw a diagram that shows the basic ordering, purchasing, and accounting systems for your copy business.

2. The rental (order processing) application in your video store has three databases: video, title, and customer. The video database contains information about every tape available for

rental. The title database contains information about each specific tape title; any one tape title (for example, *Casablanca*) may have multiple videos associated with it because more than one copy may be available for rental. The customer database contains each customer's ID number and address.

Specific fields in each database are listed in the accompanying table. Key fields are indicated with an asterisk.

Title	Video	Customer
TapeNumber*	VideoNumber*	CustomerNumber*
Title	TapeNumber	CustomerName
Category	Status	Address
Rating		Phone Number
Rental Rate		Rental Amount Y-T-D
Rental Amount Y-T-D		

Build a simple transaction processing system to support the video store operation. Enter the complete database definitions into your database management software and create a data-entry screen to allow store personnel to efficiently enter customer rentals and returns. This screen must include basic information such as tape name and number, customer name and number, date out, date returned, and charge for rental per day. The data from the screen updates all appropriate data in each database.

a. Enter several of your favorite movies to create at least ten entries for the title database.

b. For each title entry, create one to five entries in the video database.

c. Make up at least ten entries for the customer database.

d. Enter the data necessary to handle the checkout of at least six specific videos by different customers. Can your simple transaction processing system handle a situation in which one customer checks out more than one video at a time? What happens if a customer wants to check out a video but there are no copies remaining?

e. Check to see whether the Y-T-D fields in the title and customer databases are updated correctly.

● TEAM ACTIVITIES

1. Assume that your team owns a T-shirt company. Assign each team member to a transaction processing activity. Describe the detailed activities that would be needed for your T-shirt company.

2. Your team should interview a business owner about the company's transaction processing system. Develop a report that describes this company's transaction processing system.

● WEB EXERCISES

1. A number of companies, including SAP, sell enterprise resource planning software to coordinate a company's transaction processing system. Search the Internet to get more information about ERP or one of the companies that makes and sells this powerful software. You may be asked to develop a report or send an e-mail message to your instructor about what you found.

2. Using the Internet, research several accounting firms that produce basic transaction processing reports. Describe each accounting package used. What are the advantages and disadvantages of each?

CASES

1 Starting a Procurement Business

As discussed in the "E-Commerce" box, companies are starting partnerships to form Internet exchanges. Although some exchanges are for general purpose, most are oriented toward the procurement function of transaction processing. In supporting these exchanges, businesses hope to save significant amounts of time and money in purchasing the parts, supplies, and services needed for these companies to manufacture their products—ranging from cars to agricultural products. In addition to companies forming these strategic alliances, others are developing procurement businesses and Internet exchanges for the general market. One example of this move is the alliance of Chase Manhattan Bank and Deloitte Consulting.

In early 2000, Chase Manhattan Bank and Deloitte Consulting decided to form a new company. The overall objective of the new firm is to offer Internet-based procurement services to larger, Fortune 1000 companies. Chase and Deloitte hope to make a profit from the new company and also save participating companies money. On average, Chase and Deloitte forecast that each company participating in the new Internet procurement system should save from $200 million to $350 million annually. The cost savings are expected to come from combining the purchasing needs of all participating companies to obtain better volume discounts.

The new Internet procurement firm, however, is still in the planning stage. The name of the new firm, the technology to be used, and the person who will head the new company have not been determined.

Discussion Questions

1. Will this new Internet procurement firm be attractive to the average Fortune 1000 company?
2. Do you think that other companies will also try to start Internet procurement companies?

Critical Thinking Questions

3. Assume that you are the manager of a Fortune 1000 manufacturing company. Discuss the advantages and disadvantages of this approach to procurement. Would you be willing to sign a long-term contract to be a part of this firm?
4. Assume that several Fortune 1000 manufacturing companies are considering forming a strategic partnership to form an Internet exchange. Compare and contrast this approach with the type of Internet procurement firm discussed in this case.

Sources: Adapted from Craig Stedman, "Chase, Deloitte Start Procurement Firm," *Computerworld*, February 14, 2000, p. 20; "Local Government Equals Big Business," *Computing*, March 2, 2000, p. 4; and Clinton Wilder et al., "Sabre to Open Internet Marketplace," *Information Week*, March 6, 2000.

2 ERP in Mergers and Acquisitions

As discussed in this chapter, enterprise resource planning can offer a company many benefits over traditional transaction processing approaches. By integrating a firm's TPS functions into a unified system, companies can process basic transactions faster, more efficiently, and at a lower cost, in most cases. In addition, because many ERP packages include best practices approaches to running a business, the ERP approach can help a company improve its overall approach to doing business. ERP implementations, however, can be costly and time-consuming.

Lyondell Chemical is a Houston-based company with 9,000 employees and annual revenues of $8 billion. From 1995 to 1997, Lyondell implemented SAP R/3. After the ERP system was operational in 1997, some believed that only maintenance of the system would be required in the future. But this forecast turned out not to be

the case. Like other companies, Lyondell is making acquisitions in its effort to grow. Because of the increased scope of the company, the result was two additional SAP implementations in 1999. According to Robert Tolbert, Lyondell's chief information officer, "If you're merging supply-chain organizations to be efficient, you need to be on the same system to share data." Lyondell decided to implement SAP R/3 in each of the merged companies. As a result, the costs and time required to implement an ERP system had to be incurred several times, and it is likely that Lyondell will continue to have ERP implementation costs as it continues to acquire other companies. Fortunately, Lyondell was able to learn how to implement the ERP systems more easily and efficiently. "After the first couple of times, you have some in-house experience. So the third time you don't need as many consultants," says Tolbert.

Lyondell is not alone. Other companies have also experienced multiple ERP implementations.

Celestia, for example, used one ERP program in Canada and another ERP program in its Asian operations. The company is an IBM spin-off based in Toronto with annual revenues of $3.8 billion. Celestia is now considering the possibility of standardizing on a single ERP system. As with Lyondell, this would require another ERP implementation in either Canada or Asia, with all the accompanying costs and time requirements.

Discussion Questions

1. What are the advantages of implementing ERP in a merger or acquisition?
2. What are some of the disadvantages?

Critical Thinking Questions

3. How would you minimize the cost of implementing multiple ERP systems over time?
4. Assume that you are the chief information officer of Celestia. What factors would you consider in deciding which ERP implementation to use in both Canada and Asia? How would you decide which system to use worldwide?

Sources: Adapted from Erik Sherman, "Early ERP Implementations Have Proved to Be Costly and Time-Consuming," *Computerworld*, February 14, 2000, p. 53; "Projects Hot Sheet," *Chemical Week*, January 19, 2000, p. 27; and "SAP Branches Out into Applications Hosting Market," *Network News*, March 1, 2000, p. 4.

 ### Setting Up Supply Networks

In the beginning of the U.S. automotive industry, there were many small start-up companies. The future was bright, and investors were lining up to get in on the ground floor. Many experts at the time thought that automobiles could change the world, and they were right. However, only three auto companies remain: Ford, GM, and DaimlerChrysler. Stiff competition and mergers cut the number of auto companies to just a few. Some early investors became wealthy; others lost everything.

Today we see another world-changing breakthrough with the Internet. Like the beginning of the automotive industry, the Internet industry is full of new, start-up companies. Stock prices have soared in the last few years, and many investors are hoping to make a small fortune. Yet, some seasoned investors are wondering whether the Internet will go the way of the automotive industry. How many Internet companies will remain in business in the next decade? Which ones will be successful? What can an Internet company do today to ensure a successful future? Some believe that getting in on the ground floor is the key.

In the year 2000, three petroleum companies have made separate deals with technology companies to develop online marketplaces, where oil companies can purchase supplies at discounted prices. Chevron, Shell, and Norway's Statoil are all hoping that their joint ventures will be successful. According to Bruce Richardson of AMR Research, "This is another land-grab, a space race, where everyone wants to be first in this market." Chevron is hoping that its Petrocosm Marketplace will be the dominant Internet site for procurement.

Chevron has teamed up with Ariba, a technology company. Statoil is teaming up with SAP, the ERP technology company, to develop another on-line oil supply marketplace. Meanwhile, Shell Oil is working with Commerce One to develop yet another marketplace for parts and supplies for the oil industry.

Discussion Questions

1. How important is being first in the market for an Internet company's long-term success?
2. What other industries could benefit from an Internet marketplace, like the ones discussed in this case?

Critical Thinking Questions

3. Comment on the following statement: Many oil company executives believe that it is critical to be part of one of the Internet marketplaces. Others fear that an exchange formed by one company will not be fair to other companies trying to use the service.
4. If you were an oil company executive for Amoco, how would you react to the three marketplaces discussed in this case? Would you try to form your own on-line marketplace, try to partner with one of the three companies discussed, or do nothing regarding an on-line marketplace for oil parts and supplies?

Sources: Adapted from Carol Sliwa, "Oil Firms Rush to Set Up Supply Nets," *Computerworld*, January 24, 2000, p. 4; "Chevron, Ariba to Join to Create Energy Marketplace," *Electronic Commerce World*, February 2000, p. 1092; and "Chevron to Share Rigs for Gulf Drilling," *Oil & Gas*, February 14, 2000, p. 28.

● NOTES

Sources for the opening vignette on page 328: Adapted from Alorie Gilbert et al., "A Question of Convenience," *Information Week*, February 21, 2000; Craig Stedman, "ERP, Production Data Can Be a Messy Mix," *Computerworld*, February 14, 2000, p. 60; and John Teresko, "Emerging Technologies," *Industry Week*, February 21, 2000, p. 20.

1. Jaikumar Vigayan, "Manufacturing Group Launches B-to-B Hub," *Computerworld*, March 20, 2000, p. 16.

2. Thomas Weber, "Net's Touch Question: Could You Show Me Some ID, Please," *The Wall Street Journal*, April 3, 2000, p. B1.

3. Linda Rosencrance, "Digital Nervous System Speeds Airline Data to Customers, Employees," *Computerworld*, June 5, 2000, p. 43.

4. Craig Stedman, "New Benchmarking Tools Planned," *Computerworld*, April 10, 2000, p. 42.

5. Arlene Weintraub, "A Ticket to Dot-Com Heaven?" *Business Week*, April 10, 2000, p. 87.

6. Mark Hall, "ASPs Answer Customer Service Message Deluge," *Computerworld*, March 13, 2000, p. 44.

7. Gary Anthes, "Order," *Computerworld*, March 20, 2000, p. 47.

8. Ian Sager, "Big Blue Gets Wired," *Business Week*, April 3, 2000, p. EB99.

9. Irene Kunni, "From Convenience Store to Online Behemoth?" *Business Week*, April 10, 2000, p. 64.

10. Julia King, "Companies Aren't Rushing to Conduct Business Online," *Computerworld*, March 27, 2000, p. 20.

11. Carol Sliwa, "More Than 25% of Online Transactions Fail," *Computerworld*, March 13, 2000, p. 10.

12. Steve Alexander, "E-Commerce Distribution," *Computerworld*, March 20, 2000, p. 58.

13. Ian Sager, "Big Blue Gets Wired," *Business Week*, April 3, 2000, p. EB99.

14. Kathleen Ohlson, "Lands' End Tailors to Corporate Clients," *Computerworld*, March 20, 2000, p. 20.

15. Julia King, "Consumer Products Giants Plan E-Market," *Computerworld*, March 20, 2000, p. 14.

16. Matthew Nelson, "VW Turns to a Marketplace to Streamline Purchasing," *Information Week Online*, April 4, 2000.

17. Julia King, "Supplier Network to Ease Access to Online Exchange," *Computerworld*, April 3, 2000, p. 20.

18. Jaikumar Vijayan, "Business-to-Business Billing No Easy Task," *Computerworld*, March 13, 2000, p. 20.

19. Bob Brewin, "MasterCard, Visa Vie for Wireless Victory," *Computerworld*, April 3, 2000, p. 16.

20. Alorie Gilbert, "ERP Vendors Look for Rebound after Slowdown," *Information Week*, February 14, 2000, p. 156.

21. Justin Fox, "Lumbering Toward B2B," *Fortune*, June 12, 2000, p. 257.

22. Craig Stedman, "Meshing CRM, Business Poses Challenges," *Computerworld*, March 27, 2000, p. 50.

23. Jaikumar Vijayan, "ERP Vendors Admit They Can't Do It All," *Computerworld*, April 3, 2000, p. 2.

24. Alorie Gilbert, "A Question of Convenience," *Information Week*, February 21, 2000, p. 34.

CHAPTER 10

Information and Decision Support Systems

*O*ur job was to give the Emergency Preparedness Division and the governor's office the information they needed to make decisions.

— Mark Hunter, assistant state maintenance engineer for the South Carolina Department of Transportation in describing the Hurricane Evacuation Decision Support System

Principles	Learning Objectives
The management information system (MIS) must provide the right information to the right person in the right fashion at the right time.	• *Outline and briefly describe the stages of a problem-solving process.* • *Define the term MIS and clearly distinguish the difference between a TPS and an MIS.* • *Discuss information systems in the functional areas of business organizations.*
Decision support systems (DSSs) are used when the problems are more unstructured.	• *List and discuss important characteristics of DSSs that give them the potential to be effective management support tools.* • *Identify and describe the basic components of a DSS.*
Specialized support systems, such as group decision support systems (GDSSs) and executive support systems (ESSs), use the overall approach of a DSS in situations such as group and executive decision making.	• *State the goal of a GDSS and identify the characteristics that distinguish it from a DSS.* • *Identify fundamental uses of an ESS and list the characteristics of such a system.*

Reuters Group

*Providing Information and Decision
Support on the Internet*

If information is power, then Reuters Group is one of the chief sources of that power. It is also one of the oldest information sources. Founded almost 150 years ago, the company originally was a news agency, feeding stories and information to newspapers. About twenty years ago, Reuters saw a need to provide financial information to banks, brokerage companies, and wealthy individuals. Companies and people needed information to support their investment and banking decisions, and Reuters decided to fill the need. Today, the company provides decision support for a half million users at about 60,000 companies around the world. Providing information and decision support on a wholesale basis to other companies and individuals generates revenues of $5 billion annually for the company. The average user pays $1,500 per month to get access to the vast financial information contained in Reuters's databases. In addition to financial information, Reuters still operates a news service business for newspapers, magazines, broadcasters, and Internet portals.

Unfortunately, the Internet has become the main competition for Reuters. Although the information on Internet sites may not be as complete and the decision support may not be as comprehensive as Reuters's databases, many Internet sites are free to users. Internet users have been happy to get free information, but Reuters stockholders have not. The stock price for Reuters has languished at about $50 per share, while other stock prices have generally soared. Until recently, Reuters did not have a coherent strategy for the future or a solution to its sagging stock price. The confusion ended, however, with Peter Job, the new chief executive of Reuters.

Peter Job decided that Reuters had to get on the Internet bandwagon. He decided to invest more than $800 million over four years to develop a comprehensive Internet site to sell and distribute information directly to individuals in addition to its traditional corporate customers. "Why does the individual have to get a different deal than the institution?" asked Job. The initiative has made potential individual customers and stockholders happy. The stock price of Reuters recently reached $137 a share after the announcement. Although most stockholders are happy with their new wealth, some are concerned. Will the move to the Internet mean that the company can't generate the hefty fees from its corporate clients?

As you read this chapter, consider the following:

- How valuable is the information and decision support provided by Reuters to its corporate clients? Do you think the information is worth the average $1,500 monthly cost to most users?

- What are the disadvantages of providing similar information and decision support to individuals over the Internet?

Information is the lifeblood of today's organizations. Thanks to information systems, managers and employees can obtain useful information in real time. As we saw in the preceding chapter, the TPS captures a wealth of data. When this data is filtered and manipulated, it can become powerful information and decision support for managers and employees. The ultimate goal of management information and decision support systems is to help managers and executives make better decisions and solve important problems. The results can include increased revenues, cost reductions, and the realization of corporate goals. We begin by investigating decision making and problem solving.

DECISION MAKING AND PROBLEM SOLVING

Every organization needs effective decision making to reach its objectives and goals. In most cases, strategic planning and the overall goals of the organization set the stage for value-added processes and the decision making required to make them work. Often, information systems assist with strategic planning and problem solving.

Decision Making as a Component of Problem Solving

In business, one of the highest compliments you can get is to be recognized by your colleagues and peers as a "real problem solver." Problem solving is a critical activity for any business organization. Once a problem has been identified, the problem-solving process begins with decision making. A well-known model developed by Herbert Simon divides the **decision-making phase** of the problem-solving process into three stages: intelligence, design, and choice. This model was later incorporated by George Huber into an expanded model of the entire problem-solving process (Figure 10.1).

The first stage in the decision-making process is the **intelligence stage**. During this stage, potential problems or opportunities are identified and defined. Information is gathered that relates to the cause and scope of the problem.

decision-making phase

the first part of problem solving, including three stages: intelligence, design, and choice

intelligence stage

the first stage of decision making, during which potential problems and opportunities are identified and defined

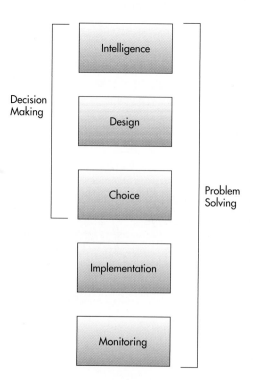

FIGURE 10.1

How Decision Making Relates to Problem Solving

The three stages of decision making—intelligence, design, and choice—are augmented by implementation and monitoring to result in problem solving.

During the intelligence stage, resource and environmental constraints are investigated. For example, exploring the possibilities of shipping tropical fruit from a farm in Hawaii to stores in Michigan would be done during the intelligence stage. The perishability of the fruit and the maximum price consumers in Michigan are willing to pay for the fruit are problem constraints. Aspects of the problem environment that must be considered in this case include federal and state regulations regarding the shipment of food products.

In the **design stage**, alternative solutions to the problem are developed. In addition, the feasibility of these alternatives is evaluated. In our tropical fruit example, the alternative methods of shipment, including the transportation times and costs associated with each, would be considered. During this stage the problem solver might determine that shipment by freighter to California and then by truck to Michigan is not feasible because the fruit would spoil.

The last stage of the decision-making phase, the **choice stage**, requires selecting a course of action. In our tropical fruit example, the Hawaiian farm might select the method of shipping by air to Michigan as its solution. The choice stage would then conclude with selection of the actual air carrier. As we will see later, various factors influence choice; the apparently easy act of choosing is not as simple as it might first appear.

Problem solving includes and goes beyond decision making. It also includes the **implementation stage**, when the solution is put into effect. For example, if the Hawaiian farmer's decision is to ship the tropical fruit to Michigan as air freight using a specific air freight company, implementation involves informing the farming staff of the new activity, getting the fruit to the airport, and actually shipping the product to Michigan.

The final stage of the problem-solving process is the **monitoring stage**. In this stage, decision makers evaluate the implementation to determine whether the anticipated results were achieved and to modify the process in light of new information. Monitoring can involve feedback and adjustment. For example, after the first shipment of fruit, the Hawaiian farmer might learn that the flight of the chosen air freight firm routinely makes a stopover in Phoenix, Arizona, where the plane sits exposed on the runway for a number of hours while loading additional cargo. If this unforeseen fluctuation in temperature and humidity adversely affects the fruit, the farmer might have to readjust his solution to include a new air freight firm that does not make such a stopover, or perhaps he would consider a change in fruit packaging.

Programmed versus Nonprogrammed Decisions

In the choice stage, various factors influence the decision maker's selection of a solution. One such factor is whether the decision can be programmed. **Programmed decisions** are made using a rule, procedure, or quantitative method. For example, to say that inventory should be ordered when inventory levels drop to 100 units is to adhere to a rule. Programmed decisions are easy to computerize using traditional information systems. It is simple, for example, to program a computer to order more inventory when inventory levels for a certain item reach 100 units or fewer. Most of the processes automated through transaction processing systems share this characteristic: the relationships between system elements are fixed by way of rules, procedures, or numerical relationships. Management information systems are

design stage

the second stage of decision making, during which alternative solutions to the problem are developed

choice stage

the third stage of decision making, which requires selecting a course of action

problem solving

a process that goes beyond decision making to include the implementation stage

implementation stage

the stage of problem solving during which a solution is put into effect

monitoring stage

final stage of the problem-solving process, during which decision makers evaluate the implementation

programmed decision

decisions made using a rule, procedure, or quantitative method

Ordering more inventory when inventory levels drop to specified levels is an example of a programmed decision.
(Source: Stone/Mitch Kezar)

nonprogrammed decision

decisions that deal with unusual or exceptional situations

also used to solve programmed decisions by providing reports on problems that are routine and in which the relationships are well defined (structured problems).

Nonprogrammed decisions, however, deal with unusual or exceptional situations. In many cases, these decisions are difficult to quantify. Determining the appropriate training program for a new employee, deciding whether to start a new type of product line, and weighing the benefits and disadvantages of installing a new pollution control system are examples. Each of these decisions contains many unique characteristics for which the application of rules or procedures is not so obvious. Today, decision support systems are used to solve a variety of nonprogrammed decisions, in which the problem is not routine and rules and relationships are not well defined (unstructured or ill-structured problems).

Optimization, Satisficing, and Heuristic Approaches

optimization model

a process to find the best solution, usually the one that will best help the organization meet its goals

satisficing model

a model that will find a good—but not necessarily the best—problem solution

heuristics

commonly accepted guidelines or procedures that usually find a good solution

In general, computerized decision support systems can either optimize or satisfice. An **optimization model** will find the best solution, usually the one that will best help the organization meet its goals. For example, an optimization model can find the appropriate number of products an organization should produce to meet a profit goal, given certain conditions and assumptions. Optimization models utilize problem constraints. A limit on the number of available work hours in a manufacturing facility is an example of a problem constraint. Some spreadsheet programs, such as Excel, have optimizing features (Figure 10.2).

A **satisficing model** is one that will find a good—but not necessarily the best—problem solution. Satisficing is usually used because modeling the problem properly to get an optimal decision would be too difficult, complex, or costly. Satisficing normally does not look at all possible solutions but only at those likely to give good results. Consider a decision to select a location for a new plant. To find the optimal (best) location, you would have to consider all cities in the United States or the world. A satisficing approach would be to consider only five or ten cities that might satisfy the company's requirements. Limiting the options may not result in the best decision, but it will likely result in a good decision, without spending the time and effort to investigate all cities. Satisficing is a good alternative modeling method because it is sometimes too expensive to analyze every alternative to get the best solution.

Heuristics, often referred to as "rules of thumb"—commonly accepted guidelines or procedures that usually find a good solution—are very often used in decision making. An example of a heuristic rule in e-commerce is "the first company to market will gain the greatest name recognition and ultimate profitability." A heuristic that baseball team managers use is to place batters most likely to get on base at the top of the lineup, followed by the power hitters who'll drive them in to score. An example of a heuristic used in business is to order four months' supply of inventory for a particular item when the inventory level drops to 20 units or fewer; even though this heuristic may not minimize total inventory costs, it may be a very good rule of thumb to avoid stockouts without too much excess inventory. One way to achieve better decision making and problem solving is through a management information system.

FIGURE 10.2

Some spreadsheet programs, such as Excel, have optimizing routines. This figure shows Solver, which can find an optimal solution given certain constraints.

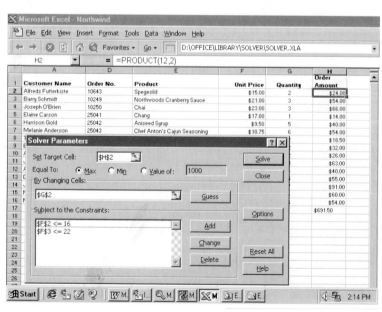

AN OVERVIEW OF MANAGEMENT INFORMATION SYSTEMS

Management information systems (MISs) can often give companies a competitive advantage by providing the right information to the right people in the right format and at the right time. In many cases, companies and individuals are willing to pay companies like Reuters for this type of information.

Management Information Systems in Perspective

The primary purpose of an MIS is to help an organization achieve its goals by providing managers with insight into the regular operations of the organization so that they can control, organize, and plan more effectively and efficiently. In short, an MIS provides managers with information and support for effective decision making and provides feedback on daily operations. In doing so, an MIS adds value to processes within an organization. A manufacturing MIS, for example, is a set of integrated systems that can help managers monitor a manufacturing process to maximize the value of raw materials as they are assembled into finished products. For the most part, this monitoring is accomplished through various summary reports produced by the MIS. These reports can be obtained by filtering and analyzing the detailed data contained in transaction processing databases and presenting the results to managers in a meaningful way. These reports support managers by providing them with data and information for decision making in a form they can readily use. Figure 10.3 shows the role of MISs within the flow of an organization's information. Note that business transactions can enter the organization through traditional methods or via the Internet or an extranet connecting customers and suppliers to the firm's transaction processing systems.

As Figure 10.3 shows, the summary reports from the MIS are just one of many sources of information available to managers. As previously discussed, the use of management information systems spans all levels of management. That is, they provide support to and are used by employees throughout the organization.

Each MIS is an integrated collection of subsystems, which are typically organized along functional lines within an organization. Thus, a financial MIS includes subsystems that address financial reporting, profit and loss analysis, cost analysis, and the use and management of funds. Many functional subsystems share certain hardware resources, data, and often even personnel. Some subsystems, however, are self-contained within one functional area and are suited to a specialized purpose. One of the IS manager's roles is to increase the overall efficiency of the MIS by improving the integration of subsystems. For instance, similar data may be collected and maintained by two distinct functional departments (e.g., customer lists maintained by both sales and accounting). Or perhaps hardware resources are only partially used by one functional area and might be shared by another. Like other corporate resources, the investment in the MIS should be maximized by decreasing waste and underutilization.

Even though increasing overall MIS efficiency is important, all managers (including IS managers) must appreciate that one important role of the MIS is to improve effectiveness by providing the right information to the right person in the right fashion at the right time. This aspect of an MIS is too often forgotten. In their zest to improve system efficiencies through integration, some IS managers overlook the basic problem-solving needs of functional managers.

Inputs to a Management Information System

Data that enters an MIS originates from both internal and external sources (Figure 10.3). The most significant internal source of data for an MIS is the organization's various TPSs. One of the major activities of the TPS is to capture

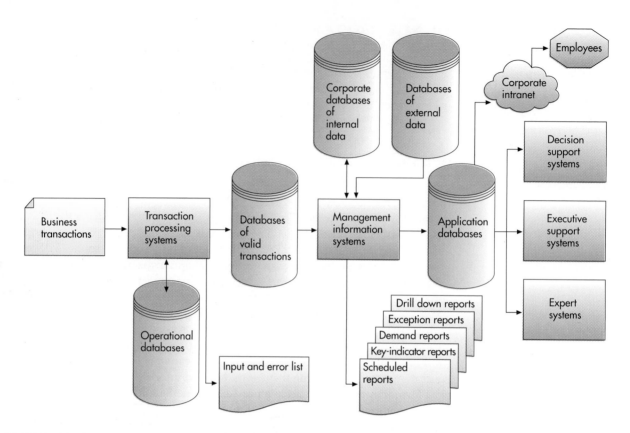

FIGURE 10.3

Sources of Managerial Information

The MIS is just one of many sources of managerial information. Decision support systems, executive support systems, and expert systems also assist in decision making.

and store the data resulting from ongoing business transactions. With every business transaction, various TPS applications make changes to and update the organization's databases. For example, the billing application helps keep the accounts receivable database up to date so that managers know who owes the company money. These updated databases are a primary internal source of data for the management information system. In companies that have implemented an ERP system, the collection of databases associated with this system are an important source of internal data for the MIS. Other internal data comes from specific functional areas throughout the firm. External sources of data can include customers, suppliers, competitors, and stockholders whose data is not already captured by the TPS, as well as other sources, such as the Internet. In addition, many companies have implemented extranets to link them to these entities and allow for the exchange of data and information.

The MIS uses the data obtained from these sources and processes it into information more usable to managers, primarily in the form of predetermined reports. For example, rather than simply obtaining a chronological list of sales activity over the past week, a national sales manager might obtain her organization's weekly sales data in a format that allows her to see sales activity by region, by local sales representative, by product, and even in comparison with last year's sales.

Outputs of a Management Information System

The output of most management information systems is a collection of reports that are distributed to managers. These reports include scheduled reports, key-indicator reports, demand reports, exception reports, and drill down reports (Figure 10.4).

scheduled reports

reports produced periodically, or on a schedule, such as daily, weekly, or monthly

Scheduled Reports

Scheduled reports are produced periodically, or on a schedule, such as daily, weekly, or monthly. For example, a production manager could use a weekly

FIGURE 10.4

Reports Generated by an MIS

The five types of reports are (a) scheduled, (b) key-indicator, (c) demand, (d) exception, and (e–h) drill down on p. 378. (Source: George W. Reynolds, *Information Systems for Managers*, 3rd ed., St. Paul, MN: West Publishing Co., 1995. Reprinted with permission from Course Technology.)

(a) Scheduled Report

Daily Sales Detail Report

Prepared: 08/10/XX

Order #	Customer ID	Salesperson ID	Planned Ship Date	Quantity	Item #	Amount
P12453	C89321	CAR	08/12/96	144	P1234	$3,214
P12453	C89321	CAR	08/12/96	288	P3214	$5,660
P12454	C03214	GWA	08/13/96	12	P4902	$1,224
P12455	C52313	SAK	08/12/96	24	P4012	$2,448
P12456	C34123	JMW	08/13/96	144	P3214	$ 720
....	....	....		....	....	....

(b) Key-Indicator Report

Daily Sales Key Indicator Report

	This Month	Last Month	Last Year
Total Orders Month to Date	$1,808	$1,694	$1,914
Forecasted Sales for the Month	$2,406	$2,224	$2,608

(c) Demand Report

Daily Sales by Salesperson Summary Report

Prepared: 08/10/XX

Salesperson ID	Amount
CAR	$42,345
GWA	$38,950
SAK	$22,100
JWN	$12,350
....	....
....	....

(d) Exception Report

Daily Sales Exception Report—Orders Over $10,000

Prepared: 08/10/XX

Order #	Customer ID	Salesperson ID	Planned Ship Date	Quantity	Item #	Amount
P12345	C89321	GWA	08/12/96	576	P1234	$12,856
P22153	C00453	CAR	08/12/96	288	P2314	$28,800
P23023	C32832	JMN	08/11/96	144	P2323	$14,400
....	....	....	....	....	....	....
....	....	....	....	....	....	....

summary report that lists total payroll costs to monitor and control labor and job costs. A manufacturing report produced once a day to monitor the production of a new product is another example of a scheduled report. Other scheduled reports can help managers control customer credit, the performance of sales representatives, inventory levels, and more.

A **key-indicator report** summarizes the previous day's critical activities and is typically available at the beginning of each workday. These reports can summarize inventory levels, production activity, sales volume, and the like. Key-indicator reports are used by managers and executives to take quick, corrective action on significant aspects of the business.

Demand Reports

Demand reports are developed to give certain information at a manager's request. In other words, these reports are produced on demand. For example, an executive may want to know the production of a particular item—a demand report can be generated to give the requested information. Other examples of

key-indicator report

summary of the previous day's critical activities; typically available at the beginning of each workday

demand reports

reports developed to give certain information at a manager's request

FIGURE 10.4 (continued)

Reports Generated by an MIS

(e) Manager sees actual earnings exceed forecast by 6.8 percent for 2nd quarter 1999. (f) Manager view sales and expenses for 2nd quarter 1999. (g) Manager views sales by division. (h) Manager views sales for the Health Care division.

(e) First-Level Drill Down Report

Earnings by Quarter (Millions)			
	Actual	Forecast	Variance
2nd Qtr. 1999	$12.6	$11.8	6.8%
1st Qtr. 1999	$10.8	$10.7	0.9%
4th Qtr. 1998	$14.3	$14.5	-1.4%
3rd Qtr. 1998	$12.8	$13.3	-3.8%

(f) Second-Level Drill Down Report

Sales and Expenses (Millions)			
Qtr: 2nd Qtr. 1999	Actual	Forecast	Variance
Gross Sales	$110.9	$108.3	2.4%
Expenses	$ 98.3	$ 96.5	1.9%
Profit	$ 12.6	$ 11.8	6.8%

(g) Third-Level Drill Down Report

Sales by Division (Millions)			
Qtr: 2nd Qtr. 1999	Actual	Forecast	Variance
Beauty Care	$ 34.5	$ 33.9	1.8%
Health Care	$ 30.0	$ 28.0	7.1%
Soap	$ 22.8	$ 23.0	-0.9%
Snacks	$ 12.1	$ 12.5	-3.2%
Electronics	$ 11.5	$ 10.9	5.5%
Total	$110.9	$108.3	2.4%

(h) Fourth-Level Drill Down Report

Sales by Product Category (Millions)			
Qtr: 2nd Qtr. 1999 Division: Health Care	Actual	Forecast	Variance
Toothpaste	$12.4	$10.5	18.1%
Mouthwash	$ 8.6	$ 8.8	-2.3%
Over-the-Counter Drugs	$ 5.8	$ 5.3	9.4%
Skin Care Products	$ 3.2	$ 3.4	-5.9%
Total	$30.0	$28.0	7.1%

demand reports include reports requested by executives to show the hours worked by a particular employee, total sales to date for a product, and so on.

Exception Reports

exception reports

reports automatically produced when a situation is unusual or requires management action

Exception reports are reports that are automatically produced when a situation is unusual or requires management action. For example, a manager might set a parameter that generates a report of all inventory items with fewer than the equivalent of five days of sales on hand. This unusual situation requires prompt action to avoid running out of stock on the item. The exception report generated by this parameter would contain only items with fewer than five days of sales in inventory. As with key-indicator reports, exception reports are most often used to monitor aspects important to an organization's success. In general, when an exception report is produced, a manager or executive takes action. Parameters, or *trigger points,* for an exception report should be set carefully. Trigger points that are set too low may result in an abundance of exception reports; trigger points that are too high could mean that problems requiring action are overlooked. For example, if a manager wants a report that contains all projects over budget by $100 or more, he may find that almost every company project exceeds its budget by at least this amount. The $100 trigger point is probably too low. A trigger point of $10,000 might be more appropriate.

drill down reports

reports providing increasingly detailed data about a situation

Drill Down Reports

Drill down reports provide increasingly detailed data about a situation. Through the use of drill down reports, analysts can see data at a high level first (similar to a bag of cookies), then at a more detailed level (say, an Oreo), and then at a very detailed level (an Oreo cookie's double-filling components).

Developing Effective Reports

Management information system reports can help managers develop better plans, make better decisions, and obtain greater control over the operations of the firm. It is important to recognize that various types of reports can overlap. For example, a manager can demand an exception report or set trigger points for items contained in a key-indicator report. Certain guidelines should be followed in designing and developing reports to yield the best results. Table 10.1 explains these guidelines.

Characteristics of a Management Information System

Scheduled, key-indicator, demand, exception, and drill down reports have all helped managers and executives make better, more timely decisions. When the guidelines for developing effective reports are followed, higher revenues and lower costs can be realized. In general, management information systems perform the following functions:

- *Provide reports with fixed and standard formats.* For example, scheduled reports for inventory control may contain the same types of information placed in the same locations on the reports. Different managers may use the same report for different purposes.
- *Produce hard-copy and soft-copy reports.* Some MIS reports are printed on paper and considered hard-copy reports. Most output soft copy, using visual displays on computer screens. Soft-copy output is typically formatted in a reportlike fashion. In other words, a manager might be able to call an MIS report up directly on the computer screen, but the report would still appear in the standard hard-copy format. Hard copy is still the most-often-used form of MIS report.
- *Use internal data stored in the computer system.* MIS reports use primarily internal sources of data that are contained in computerized databases. Some MISs

TABLE 10.1

Guidelines for Developing MIS Report.

Guidelines	Reason
Tailor each report to user needs.	The unique needs of the manager or executive should be considered, requiring user involvement and input.
Spend time and effort producing only reports that are useful.	Once instituted, many reports continue to be generated even though no one uses them anymore.
Pay attention to report content and layout.	Prominently display the information that is most desired. Do not clutter the report with unnecessary data. Use commonly accepted words and phrases. Managers can work more efficiently if they can easily find desired information.
Use management by exception reporting.	Some reports should be produced only when there is a problem to be solved or an action that should be taken.
Set parameters carefully.	Low parameters may result in too many reports; high parameters mean valuable information could be overlooked.
Produce all reports in a timely fashion.	Outdated reports are of little or no value.
Periodically review reports.	Review reports at least once a year to make sure all reports are still needed. Review report content and layout. Determine whether additional reports are needed.

use external sources of data about competitors, the marketplace, and so on. The Internet is a frequently used source for external data.

- *Allow end users to develop their own custom reports.* Although analysts and programmers may be involved in developing and implementing complex MIS reports that require data from many sources, end users are increasingly developing their own simple programs to query a database and produce basic reports. This, however, can result in several end users developing the same or similar reports. This can result in more total time expended and additional storage requirements, compared to having an analyst develop one report for all users.

- *Require user requests for reports developed by systems personnel.* When information systems personnel develop and implement MIS reports, a formal request to the information systems department may be required. If a manager, for example, wants a production report to be used by several people in his or her department, a formal request for the report is often required. End user–developed reports require much less formality.

FUNCTIONAL ASPECTS OF THE MIS

Most organizations are structured along functional lines or areas. This functional structure is usually apparent from an organization chart, which typically shows vice presidents under the president. Some of the traditional functional areas are accounting, finance, marketing, personnel, research and development (R&D), legal services, operations/production management, and information systems. The MIS can be divided along those functional lines to produce reports tailored to individual functions (Figure 10.5).

A Financial Management Information System

financial MIS

an information system that provides financial information to all financial managers within an organization

A **financial MIS** provides financial information not only for executives but also for a broader set of people who need to make better decisions on a daily basis. Finding opportunities and quickly identifying problems can mean the difference between a business's success and failure.[1] Deutsche Bank and Chase Manhattan, for example, started a joint trading and reporting system for currency.[2] Some banks use a reporting system that categorizes or grades customers. Excellent customers can get fees forgiven and red tape cut. First Union Bank has a Web-based computer system that can pull up customer records and rankings in about 15 seconds.[3]

Specifically, the financial MIS performs the following functions:

- Integrates financial and operational information from multiple sources, including the Internet, into a single MIS
- Provides easy access to data for both financial and nonfinancial users, often through use of the corporate intranet to access corporate Web pages of financial data and information
- Makes financial data available on a timely basis to shorten analysis turnaround time
- Enables analysis of financial data along multiple dimensions—such as time, geography, product, plant, and customer
- Analyzes historical and current financial activity
- Monitors and controls the use of funds over time

Figure 10.6 (p. 382) shows typical inputs, function-specific subsystems, and outputs of a financial MIS.

Depending on the organization and its needs, the financial MIS can include both internal and external systems that assist in acquiring, using, and controlling

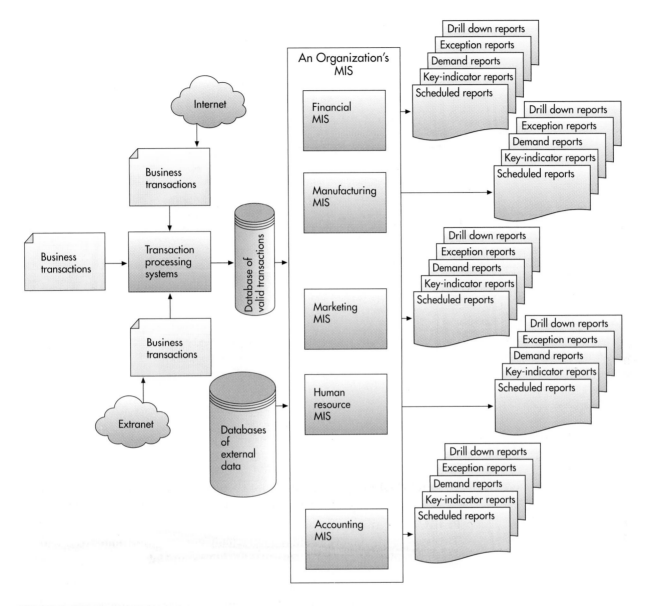

The MIS is an integrated collection of functional information systems, each supporting particular functional areas.

cash, funds, and other financial resources. These subsystems of the financial MIS have a unique role in adding value to a company's business processes. For example, a real estate development company might use a financial MIS subsystem to help with the use and management of funds. Suppose the firm takes $10,000 deposits on condominiums in a new development. Until construction begins, the company will be able to invest these surplus funds. By using reports produced by the financial MIS, finance staff can analyze investment alternatives. The company might invest in new equipment or purchase global stocks and bonds. The profits generated from the investment can be passed along to customers in different ways. The company can pay stockholders dividends, buy higher-quality materials, or sell the condominiums at a lower cost.

Other important financial subsystems include profit/loss and cost accounting and auditing. Each subsystem interacts with the TPS in a specialized way and has information outputs that assist financial managers in making better decisions. These outputs include profit/loss and cost systems reports, internal and external auditing reports, and uses and management of funds reports.

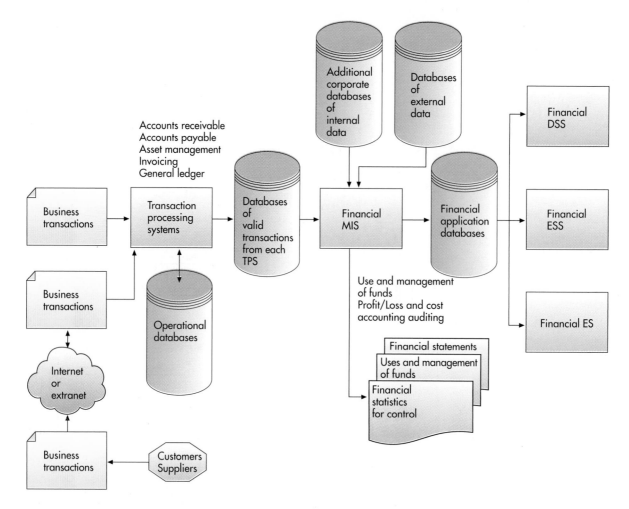

FIGURE 10.6

Overview of a Financial MIS

profit centers

departments within an organization that track total expenses and net profits

revenue centers

divisions within a company that track sales or revenues

cost centers

divisions within a company that do not directly generate revenue

auditing

process that involves analyzing the financial condition of an organization and determining whether financial statements and reports produced by the financial MIS are accurate

Profit/Loss and Cost Systems

Two specialized financial functional systems are profit/loss and cost systems, which organize revenue and cost data for the company. Revenue and expense data for various departments is captured by the TPS and becomes a primary internal source of financial information for the MIS.

Many departments within an organization are **profit centers**, which means they track total expenses and net profits. An investment division of a large insurance or credit card company is an example of a profit center. Other departments may be **revenue centers**, which are divisions within the company that primarily track sales or revenues, such as a marketing or sales department. Still other departments may be **cost centers**, which are divisions within a company that do not directly generate revenue, such as manufacturing or research and development. These units incur costs with little or no revenues. Data on profit, revenue, and cost centers is gathered (mostly through the TPS but sometimes through other channels as well), summarized, and reported by the profit/loss and cost subsystems of the financial MIS.

Auditing

Auditing involves analyzing the financial condition of an organization and determining whether financial statements and reports produced by the financial MIS are accurate. Because financial statements, such as income statements and balance sheets, are used by so many people and organizations (investors, bankers, insurance companies, federal and state government agencies, competitors, and customers), sound auditing procedures are important. Auditing can reveal potential fraud, such as credit card fraud.[4] It can also reveal false or misleading information.[5]

Departments such as manufacturing or research and development are cost centers since they do not directly generate revenue. (Source: © 2000 Photodisc)

internal auditing

auditing performed by individuals within the organization

external auditing

auditing performed by an outside group

Internal auditing is performed by individuals within the organization. For example, the finance department of a corporation may use a team of employees to perform an audit. Typically an internal audit is conducted to see how well the organization is doing in terms of meeting established company goals and objectives—no more than five weeks' inventory on hand, all travel reports completed within one week of returning from a trip, and so on. **External auditing** is performed by an outside group, like an accounting or consulting firm such as Arthur Andersen, Deloitte & Touche, or one of the other major, international accounting firms. The purpose of an external audit is to provide an unbiased picture of the financial condition of an organization. Auditing can also uncover fraud and other problems.

Uses and Management of Funds

Another important function of the financial MIS is funds usage and management. Companies that do not manage and use funds effectively often have lower profits or face bankruptcy. Outputs from the funds usage and management subsystem, when combined with other subsystems of the financial MIS, can locate serious cash flow problems and help the organization increase profits.

Internal uses of funds include use as additional inventory, new or updated plants and equipment, additional labor, the acquisition of other companies, new computer systems, marketing and advertising, raw materials, land, investments in new products, and research and development. External uses of funds are typically investment related. On occasion, a company might have excess cash from sales that is placed into an external investment. External uses of funds often include bank accounts, stocks, bonds, bills, notes, futures, options, and foreign currency.

A Manufacturing Management Information System

More than any other functional area, manufacturing has been revolutionized by advances in technology. As a result, many manufacturing operations have been dramatically improved over the last decade. Furthermore, with the emphasis on greater quality and productivity, having an efficient and effective manufacturing process is becoming even more critical. The use of computerized systems is emphasized at all levels of manufacturing—from the shop floor to the executive suite. The use of the Internet has also streamlined all aspects of manufacturing.[6] Figure 10.7 gives an overview of some of the manufacturing MIS inputs, subsystems, and outputs.

The subsystems and outputs of the manufacturing MIS monitor and control the flow of materials, products, and services through the organization. The objective of the manufacturing MIS is to produce products that meet customer needs—from the raw materials provided by suppliers to finished goods and services delivered to customers—at the lowest possible cost. The activities of the manufacturing MIS subsystems support value-added business processes. As raw materials are converted to finished goods, the manufacturing MIS monitors the process at almost every stage. Take a car manufacturer that converts raw steel, plastic, and other materials into a finished automobile. In this case, the manufacturer has added thousands of dollars of value to the raw materials used in assembling the car. If the manufacturing MIS also lets the manufacturer provide

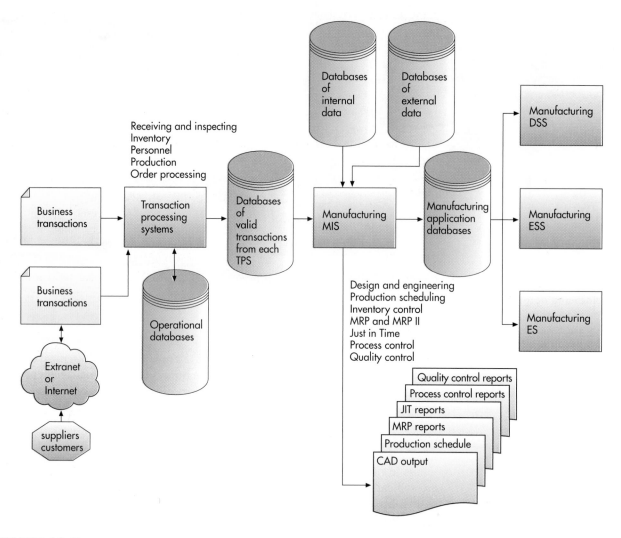

FIGURE 10.7

Overview of a Manufacturing MIS

customized paint colors on any of its models, it has further added value (although less tangible) by ensuring a direct customer fit. In doing so, the MIS helps provide the company the edge that can differentiate it from competitors. The success of an organization can depend on the manufacturing function. Some common information subsystems and outputs used in manufacturing are discussed next.

Design and Engineering

During the early stages of product development, engineering departments are involved in many aspects of design. The size and shape of parts, the way electrical components are attached to equipment, the placement of controls on a product, and the order in which parts are assembled into the finished product are decisions made with the help of design and engineering departments. In some cases, computer-assisted design (CAD) assists this process. Herman Miller, a furniture manufacturer, uses a 3-D CAD program to help salesmen select the right colors, styles, and configurations for corporate customers.[7] The design software is inexpensive and very effective in developing attractive and functional offices. CAD can be used to determine how an airplane wing or fuselage will respond to various conditions and stresses while in use. The data from design and engineering can also be used to identify problems with existing products and help develop new products. For example, Boeing uses a CAD system to develop a complete digital blueprint of an aircraft before it ever begins its manufacturing process. As mock-ups are built and tested, the digital blueprint is constantly revised to reflect the most current design. Using such technology helps Boeing reduce its manufacturing costs and the time required to design a new aircraft. Bottle manufacturers use CAD to develop three-dimensional models of their product so that it can be reviewed for aesthetics, stability, ease of handling, and other characteristics.

Master Production Scheduling

The overall objective of master production scheduling is to provide detailed plans for both short-term and long-range scheduling of manufacturing facilities (Figure 10.8). Master production scheduling software packages can include forecasting techniques that attempt to determine current and future demand for products and services. After current demand has been determined and future demand has been estimated, the master production scheduling package can determine the best way to engage the manufacturing facility and all its related equipment.

The result of the process is a detailed plan that reveals a schedule for every item that will be manufactured. Typically, the master production scheduling program determines total output for future periods by number of units or dollar equivalents. Most programs also perform **sensitivity analysis**, which allows a manager to determine how the production schedule would change with different assumptions concerning demand forecasts or cost figures.

The production schedule is critical to the entire manufacturing process. Information generated from this application is used with all aspects of production and manufacturing. Inventory control, labor force planning, product delivery, and maintenance programs depend on information generated from the master production schedule.

sensitivity analysis

process that allows a manager to determine how the production schedule would change with different assumptions concerning demand forecasts or cost figures

Inventory Control

An important key to the manufacturing process is inventory control. Great strides have been made in developing cost-effective inventory control programs and software packages that allow automatic reordering, forecasting, generation of shop documents and reports, determination of manufacturing costs, analysis of budgeted costs versus actual costs, and the development of master manufacturing schedules, resource requirements, and plans. A furniture company, for example, uses an approach called "simple, quick, and affordable (SQA)" to keep inventory levels and costs low.[8] Once an order is received, it is broken down into the inventory parts that are needed to successfully complete the order on time. An SQA Web site is used to make sure that the needed inventory is available to complete the order. Many inventory control programs contain mathematical inventory control equations. Most of these formulas attempt to determine how much and when to order. See the "E-Commerce" box for examples of how companies are using inventory control systems.

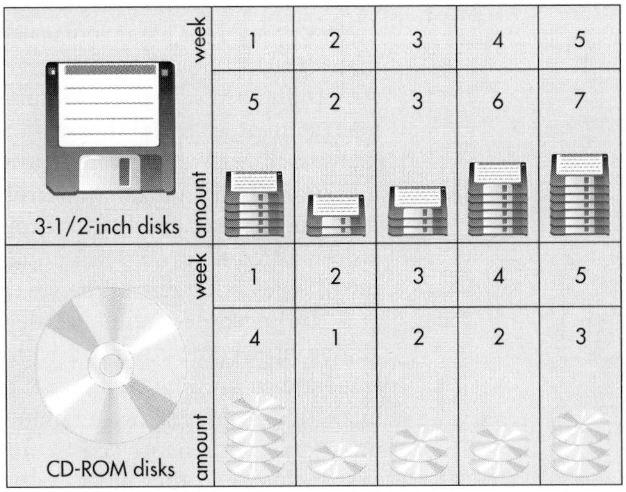

FIGURE 10.8

A master production schedule for computer disks and CDs indicates the quantity (in thousands) of each to be produced each week.

3-1/2-inch disks	week	1	2	3	4	5
	amount	5	2	3	6	7
CD-ROM disks	week	1	2	3	4	5
	amount	4	1	2	2	3

E-COMMERCE

Information and Inventory Control

Management information and decision support systems can be used to help managers control all aspects of business. From order processing to product delivery, they can reveal new revenue sources and slash costs. Controlling inventory is perhaps one of the most complex and difficult aspects of managing any company. Inventory control starts with purchasing raw materials and supplies needed to make products. Getting high-quality raw materials at low cost is challenging. Some companies use economic order quantity (EOQ) and reorder point (ROP) techniques to help them minimize inventory costs, while obtaining the needed materials on time. Manufacturing companies must then transform raw materials into finished products. Such techniques as material requirements planning (MRP) and just-in-time (JIT) inventory methods can be used. Finally, companies must efficiently ship products (finished inventory) to customers. Some companies solve these complex inventory problems internally, but others look to outside companies and consultants for help.

Manugistics offers software products to help companies control inventory. Using the Internet, Manugistics has helped companies such as General Electric and WalMart with all aspects of inventory control. Manugistics specializes in raw materials, production, and manufacturing. Manugistics also helps with shipping finished inventory to customers. In a joint effort with Burlington Northern Santa Fe (BNSF), Manugistics is giving customers the tools they need to help them decide between truck or rail in shipping products to their warehouses and retail stores. BNSF is also involved in

FreightWise, an online e-commerce site to help companies ship products to market.

Manugistics is also developing a system for an e-commerce site that will allow companies to bid on transportation alternatives. The system operates like an auction site. Shipping companies will have the ability to post their excess shipping capacity on the Internet. A trucking company, for example, will be able to advertise truck availability on-line. Railroad companies can advertise available space on rail cars on the site. Companies that need to ship products can then bid for this excess capacity.

Discussion Questions

1. Why is inventory control so important to companies?
2. How can an MIS be used in inventory control?

Critical Thinking Questions

3. Assume that you are the manager of a medium-sized manufacturing company. What features would you like to have in an on-line MIS for inventory control?
4. If you were a manager for Manugistics, would you try to integrate other systems, such as finance or marketing, into your inventory control system?

Sources: Adapted from Daniel Machalaba, "Manugistics Set to Expand E-Commerce Role," *The Wall Street Journal*, January 25, 2000, p. B4; "Supply Chain Vendor Feels the Pressure," *Computing*, September 23, 1999, p. 14; and Clinton Wilder et al., "Companies Invest in Business-to-Business Marketplaces," *InformationWeek*, December 20, 1999.

---

economic order quantity (EOQ)

the quantity of inventory that should be reordered to minimize total inventory costs

reorder point (ROP)

a critical inventory quantity level

One method of determining how much inventory to order is called the **economic order quantity (EOQ)**. This quantity is determined in such a way as to minimize the total inventory costs. The "When to order?" question is based on inventory usage over time. Typically, the question is answered in terms of a **reorder point (ROP)**, which is a critical inventory quantity level. A reorder point is an excellent example of a parameter for an exception report. When the inventory level for a particular item falls to the reorder point, or critical level, a report might be output so that an order is immediately placed for the EOQ of the product. As companies improve the speed and accuracy of their inventory management systems, they also aim to cut the lead time required to replace items. Bell Sports, a manufacturer of helmets for bicycling and other sports, uses a variety of inventory control techniques in producing helmets.[9] According to Robert Larioza, Bell's global planning manager, "We've set inventory rules—ordering frequency, ordering quantities, safety stock levels, service levels—and set all types of targets in the Inventory Planning module."

Some inventory items are dependent on one another. For example, most automobiles come equipped with four tires and a spare, an engine, a transmission, and so on. The engine can be further broken down into the fuel injector, intake manifold, exhaust manifold, engine blocks, and so forth. Each of these inventory components can be further broken down into a specific number of nuts, bolts, and other pieces of hardware. As soon as a company knows that it wants to produce one automobile, it can directly compute all the subparts and subassemblies needed to produce that automobile. Today, sophisticated inventory control programs help coordinate thousands of inventory items when the demand for one item is dependent on the demand for another. These inventory techniques are called **material requirements planning (MRP)**. The basic goal of MRP is to determine when finished products are needed, then to work backward in determining deadlines and resources needed to complete the final product on schedule. **Manufacturing resource planning (MRPII)** refers to an integrated, companywide system based on network scheduling that enables people to run their business with a high level of customer service and productivity, while lowering costs and inventories. MRPII is broader in scope than MRP; thus, the latter has been dubbed "little MRP." MRPII places a heavy emphasis on planning. This helps companies ensure that the right product is in the right place at the right time.

Just-in-Time Inventory and Manufacturing

High inventory levels on the factory floor mean higher costs, the possibility of damage, and an inefficient manufacturing process. Thus, one objective of a manufacturing MIS is to control inventory to the lowest levels without sacrificing the availability of finished products. One way to do this is to adopt the **just-in-time (JIT) inventory approach**. With this approach, inventory and materials are delivered just before they are used in a product. A JIT inventory system would arrange for a car windshield to be delivered to the assembly line only a few moments before it is secured to the automobile, rather than have it sitting around the manufacturing facility while the car's other components are being assembled.

Although JIT has many advantages, it also renders firms more vulnerable to process disruptions. If even one parts plant is shut down, an assembly plant might be left with no inventory to continue production. However, many organizations have had great success with JIT inventory control, and the overall concept has been expanded to include other aspects of the production process. Thus, materials (including raw materials and supplies) are delivered when they are needed rather than being stored for months or days in advance. The JIT manufacturing approach requires better coordination and cooperation between suppliers and manufacturing companies, substantially reducing inventory costs.

Process Control

Managers can use a number of technologies to control and streamline the manufacturing process. For example, the computer can be used to directly control manufacturing equipment, using systems called **computer-assisted manufacturing (CAM)**. CAM systems have the ability to control drilling machines, assembly lines, and more. Some of them operate quietly, are easy to program, and have self-diagnostic routines to test for difficulties with the computer system or the manufacturing equipment.

Computer-integrated manufacturing (CIM) involves the use of computers to link the components of the production process into an effective system. CIM's goal is to tie together all aspects of production, including order processing, product design, manufacturing, inspection and quality control, and shipping. CIM systems also increase efficiency by coordinating the actions of various production units. In some areas, CIM is used for even broader functions. For example, it can be used to integrate all organizational subsystems, not just the production systems. In automobile manufacture, design engineers can have their ideas evaluated by financial managers before new components are built to see if they are economically viable, saving not only time but also money.

A **flexible manufacturing system (FMS)** is an approach that allows manufacturing facilities to rapidly and efficiently change from making one product to

---

material requirements planning (MRP)

a set of inventory control techniques that help coordinate thousands of inventory items when the demand for one item is dependent on the demand for another

manufacturing resource planning (MRPII)

an integrated, companywide system based on network scheduling that enables people to run their business with a high level of customer service and productivity

just-in-time (JIT) inventory approach

a philosophy of inventory management whereby inventory and materials are delivered just before they are used in manufacturing a product

computer-assisted manufacturing (CAM)

a system that directly controls manufacturing equipment

computer-integrated manufacturing (CIM)

a system that uses computers to link the components of the production process into an effective system

flexible manufacturing system (FMS)

an approach that allows manufacturing facilities to rapidly and efficiently change from making one product to making another

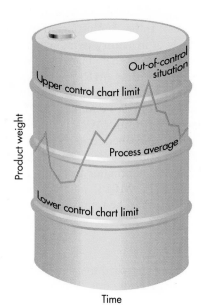

Product weight

Upper control chart limit

Out-of-control situation

Process average

Lower control chart limit

Time

FIGURE 10.9

Industrial Control Chart

This chart is used to monitor product quality in an industrial chemical application. Product variances exceeding certain tolerances cause those products to be rejected.

quality control

a process that ensures that the finished product meets the customers' needs

marketing MIS

information system that supports managerial activities in product development, distribution, pricing decisions, promotional effectiveness, and sales forecasting

making another. In the middle of a production run, for example, changes can be made to the production process to make a different product or change manufacturing materials. Often a computer is used to direct and implement the changes. By using an FMS, the time and cost to change manufacturing jobs can be substantially reduced, and companies can react quickly to market needs and competition.

FMS is normally implemented using computer systems, robotics, and other automated manufacturing equipment. New product specifications are fed into the computer system, and the computer then makes the necessary changes. Although few companies have a fully implemented FMS system, recent years have seen increasing use of the overall FMS approach.

Quality Control and Testing

With increased pressure from consumers and a general concern for productivity and high quality, today's manufacturing organizations are placing more emphasis on **quality control**, a process that ensures that the finished product meets the customers' needs. For a continuous process, control charts are used to measure weight, volume, temperature, or similar attributes (Figure 10.9). Then, upper and lower control chart limits are established. If these limits are exceeded, the manufacturing equipment is inspected for possible defects or potential problems.

When the manufacturing operation is not continuous, sampling plans can be developed that allow the producer or consumer to accept or reject one or more products. Acceptance sampling can be used for material as simple as nuts and bolts or as complex as airplanes. The development of the control chart limits and the specific acceptance sampling plans can be fairly complex. Therefore, a number of quality-control software programs have been developed to generate the appropriate acceptance-sampling plans and control chart limits.

Whether the manufacturing operation is continuous or discrete, the results from quality control are analyzed closely to identify opportunities for improvements. Teams using the total quality management (TQM) or continuous improvement process (see Chapter 2) often analyze this data to increase the quality of the product or eliminate problems in the manufacturing process. The result can be a cost reduction or increase in sales.

Information generated from quality-control programs can help workers locate problems in manufacturing equipment. Quality-control reports can also be used to design better products. With the increased emphasis on quality, workers should continue to rely on the reports and outputs from this important application.

A Marketing Management Information System

A **marketing MIS** supports managerial activities in product development, distribution, pricing decisions, promotional effectiveness, and sales forecasting. Marketing functions are increasingly being performed on the Internet.[10] Knight Ridder, a newspaper and publishing company, for example, places content and classified advertisements on the Internet.[11] The Adolph Coors Company, a Colorado brewery, advertises its products on the Internet.[12] In addition, a number of companies are developing Internet marketplaces to advertise and sell products.[13] Customer relationship management (CRM) programs, available from some ERP vendors, help a company manage all aspects of customer encounters.[14] CRM can help a company collect customer data, contact customers, educate customers on new products, and sell products to customers through an Internet site. Yet, not all marketing sites on the Internet are successful.[15] Rubbermaid, a company that makes plastic containers to store leftover food and other items, could not successfully sell products on its Internet site. Instead, the company stopped trying

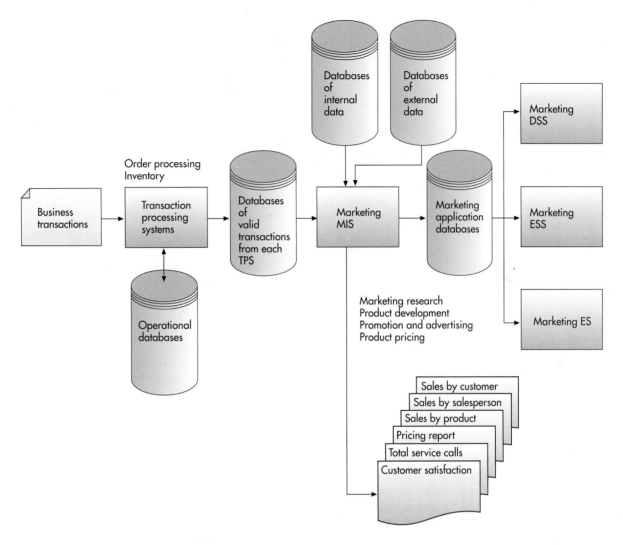

FIGURE 10.10

Overview of a Marketing MIS

to sell products and developed a site that helps customers build a wish-list of products that can be taken to a normal retail store. Figure 10.10 shows the inputs, subsystems, and outputs of a typical marketing MIS.

Subsystems for the marketing MIS include marketing research, product development, promotion and advertising, and product pricing. These subsystems and their outputs help marketing managers and executives increase sales, reduce marketing expenses, and develop plans for future products and services to meet the changing needs of customers.

Marketing Research

Surveys, questionnaires, pilot studies, and interviews are popular marketing research tools. The purpose of marketing research is to conduct a formal study of the market and customer preferences.[16] Marketing research can identify prospects (potential future customers) as well as the features that current customers really want in a good or service. Such attributes as style, color, size, appearance, and general fit can be investigated through marketing research. Pricing, distribution channels, guarantees and warranties, and customer service can also be determined. Once entered into the marketing MIS, data collected from marketing research projects is manipulated to generate reports on key indicators like customer satisfaction and total service calls. Reports generated by the marketing MIS help marketing managers be better informed to help the organization meet its performance goals.

Marketing research data yields valuable information for the development and marketing of new products.
(Source: Grandpix/Index Stock)

The Internet is changing the way many companies think about market research. Conventional methods of collecting data often cost millions of dollars. For a fraction of these costs, companies can put up Internet information servers and launch discussion lists on topics that their customers care about. These information sites must be well designed, or they won't be visited; a frequently visited site can provide feedback worth a fortune. FedEx, for example, has hired BetaSphere, an Internet marketing research firm, to gather and analyze customer feedback from FedEx's Web site.[17] Using this approach, FedEx often knows that a new product or feature will improve customer satisfaction before the new product or feature is launched. Companies that are viewed as credible—not just clever—will win enormous advantages. Presence and intelligent interaction—not just advertising—are the keys that will unlock commercial opportunities on-line. Some people, however, consider Internet marketing research to be a nuisance or even harmful. Some companies gather information on customers using cookies, which collect data on people's Internet surfing habits, and sell it to others.[18] To protect customer privacy and keep the valuable marketing research data to themselves, some companies, including General Motors, Ford, and Procter & Gamble, are starting to block Internet ad servers from getting the data. According to one General Motors spokesperson, "We've never given away anything."

Product Development

Product development involves the conversion of raw materials into finished goods and services and focuses primarily on the physical attributes of the product. Many factors, including plant capacity, labor skills, engineering factors, and materials are important in product development decisions. In many cases, a computer program is used to analyze these various factors and to select the appropriate mix of labor, materials, plant and equipment, and engineering designs. Make-or-buy decisions can also be made with the assistance of computer programs.

Promotion and Advertising

One of the most important functions of any marketing effort is promotion and advertising. Product success is a direct function of the types of advertising and sales promotion done. The size of the promotion budget and the allocation of this budget to various promotional campaigns are important factors in deciding on the type of campaigns that will be launched. Television coverage, newspaper ads, promotional brochures and literature, and training programs for salespeople are all components of these campaigns. Because of the time and scheduling savings they offer, computer programs are used to set up the original budget and to monitor expenditures and the overall effectiveness of various promotional campaigns.

Product Pricing

Product pricing is another important and complex marketing function. Retail price, wholesale price, and price discounts must be determined.[19] A major factor in determining pricing policy is an analysis of the demand curve, which attempts to determine the relationship between price and sales. Most companies try to develop pricing policies that will maximize total sales revenues. This is usually a function of price elasticity. If the product is very price sensitive, a reduction in price can generate a substantial increase in sales, which can result in higher revenues. A product that is relatively insensitive to price can have its price substantially increased without a large reduction in demand. Computer programs can help

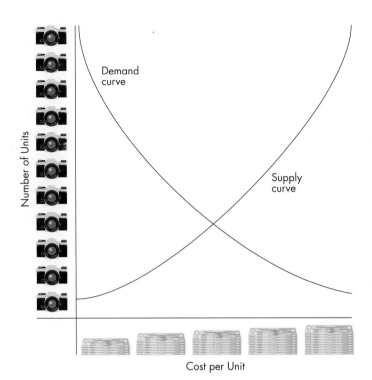

FIGURE 10.11

Typical Supply and Demand Curve for Pricing Analysis

determine price elasticity and various pricing policies, such as supply and demand curves for pricing analysis (Figure 10.11). Ford Motor Company, for example, used pricing analysis to help it earn more than $7 billion more than any automaker in history.[20] Ford's pricing analysis encouraged the company to cut the prices of high-margin vehicles, like the Ford Explorer and Crown Victoria, to dramatically increase sales and total profits. Typically, a marketing executive has the ability to make alterations in price on the computer system, which analyzes price changes and their impact on total revenues. The rapid feedback now obtainable through computer communications networks enables managers to determine the results of pricing decisions much more quickly than in the past. This ability facilitates more aggressive pricing strategies, which can be quickly adjusted to meet market needs.

Sales analysis is also important to identify products, sales personnel, and customers that contribute to profits and those that do not. Several reports can be generated to help marketing managers make good sales decisions (Figure 10.12). The sales-by-product report lists all major products and their sales for a period of time, such as a month. This report shows which products are doing well and which ones need improvement

(a) Sales by Product

Product	August	September	October	November	December	Total
Product 1	34	32	32	21	33	152
Product 2	156	162	177	163	122	780
Product 3	202	145	122	98	66	633
Product 4	345	365	352	341	288	1,691

FIGURE 10.12

Reports Generated to Help Marketing Managers Make Good Decisions

(a) This sales-by-product report lists all major products and their sales for the period from August to December. (b) This sales-by-salesperson report lists total sales for each salesperson for the same time period. (c) This sales-by-customer report lists sales for each customer for the period. Like all MIS reports, totals are provided automatically by the system to show managers at a glance the information they need to make good decisions.

(b) Sales by Salesperson

Salesperson	August	September	October	November	December	Total
Jones	24	42	42	11	43	162
Kline	166	155	156	122	133	732
Lane	166	155	104	99	106	630
Miller	245	225	305	291	301	1,367

(c) Sales by Customer

Customer	August	September	October	November	December	Total
Ang	234	334	432	411	301	1,712
Braswell	56	62	77	61	21	277
Celec	1,202	1,445	1,322	998	667	5,634
Jung	45	65	55	34	88	287

or should be discarded altogether. The sales-by-salesperson report lists total sales for each salesperson for each week or month. This report can also be subdivided by product to show which products are being sold by each salesperson. The sales-by-customer report is a tool to use to identify high- and low-volume customers.

A Human Resource Management Information System

human resource MIS

an information system that is concerned with activities related to employees and potential employees of an organization

A **human resource MIS**, also called the personnel MIS, is concerned with activities related to employees and potential employees of the organization. Because the personnel function relates to all other functional areas in the business, the human resource MIS plays a valuable role in ensuring organizational success. Some of the activities performed by this important MIS include workforce analysis and planning; hiring; training; job and task assignment; and many other personnel-related issues. Personnel issues can include offering new hires attractive stock option and incentive programs.[21] Interwoven, for example, is offering new engineers a two-year lease on a sporty BMW Z3 roadster as a signing bonus. An effective human resource MIS will allow a company to keep personnel costs at a minimum while serving the required business processes needed to achieve corporate goals. Figure 10.13 shows some of the inputs, subsystems, and outputs of the human resource MIS.

Human resource subsystems and outputs range from the determination of human resource needs and hiring through retirement and outplacement. Most medium-sized and large organizations have computer systems to assist with human resource planning, hiring, training and skills inventory, and wage and

FIGURE 10.13

Overview of a Human Resource MIS

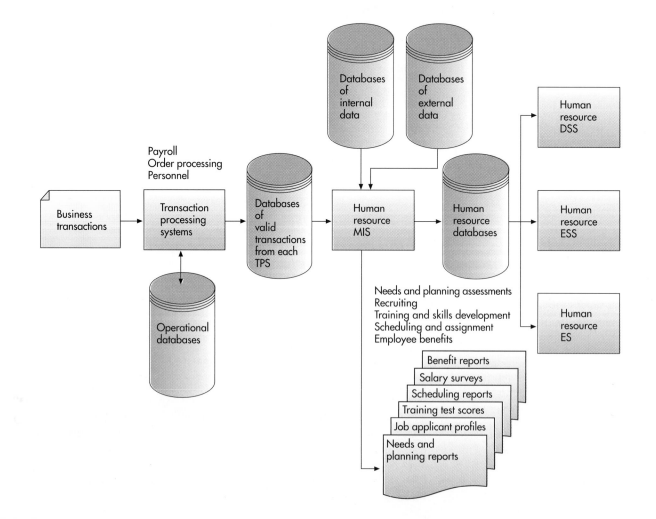

salary administration. Outputs of the human resource MIS include reports such as human resource planning reports, job application review profiles, skills inventory reports, and salary surveys.

Human Resource Planning

One of the first aspects of any human resource MIS is determining personnel needs. The overall purpose of this MIS subsystem is to put the right number and kinds of employees in the right jobs when they are needed. Effective human resource planning requires defining the future number of employees needed and anticipating the future supply of people for these jobs. For companies involved with large projects, such as military contractors and large builders, human resource plans can be generated directly from data on current and future projects.

Suppose a construction company obtains a contract from a group of investors to build a 250-unit apartment complex. Forecasting programs and project management software packages can be used to develop reports that describe what people are needed and when they are needed during the entire construction project. A typical output would be a human resource needs and planning report, which might specify that ten employees will be needed in August to pour concrete slabs and eight carpenters and four painters will be needed in October. Alternatively, a factory might use this report to list total number of employees needed, broken down according to skill level, such as highly technical, technical, semitechnical, and so on.

Personnel Selection and Recruiting

If the human resource plan reveals that additional personnel are required, the next logical step is recruiting and selection of personnel. This subsystem performs one of the most important and critical functions of any organization, especially in service organizations where employees can define the company's success. Companies seeking new employees often use computers to schedule recruiting efforts and trips and to test potential employees' skills. Some software companies, for example, use computerized testing to determine a person's programming skills and abilities. Management information systems can be used to help rank and select potential employees. For every applicant, the results of interviews, tests, and company visits can be analyzed by the system and printed. This report, called a job applicant review profile, can assist corporate recruiting teams in final selection. Some software programs can even analyze this data to help identify job applicants most likely to stay with the company and perform according to corporate standards.

Many companies now use the Internet to screen for job applicants. Applicants use a template to load their résumé onto the Internet. HR managers can then access these résumés and identify applicants they are interested in interviewing.

Training and Skills Inventory

Some jobs, such as programming, equipment repair, and tax preparation, require very specific training. Other jobs may require general training about the organizational culture, orientation, dress standards, and expectations of the organization. Today, many organizations conduct their own training, with the assistance of information systems and technology. Self-paced training can involve computerized tutorials, video programs, and CD-ROM books and materials.[22] Distance learning, whereby training and classes are conducted over the Internet, is also becoming a viable alternative to more traditional training and learning approaches. This text and supporting material, for example, can be used in a distance learning environment.

When training is complete, employees may be required to take computer-scored tests to reveal their mastery of skills and new material. The results of these tests are usually given to the employee's supervisor or boss in the form of training or skills inventory reports. In some cases, skills inventory reports are used for job placement. For instance, if a particular position in the company needs to be filled, managers might wish to hire internally before they recruit. The skills inventory report would help them evaluate current employees to determine their potential for the position. They can also be part of employee evaluations that can determine raises or bonuses. These types of tests, however, must be valid and reliable to avoid mistakes in job placement and bonuses.

Scheduling and Job Placement

Scheduling people and jobs can be relatively straightforward or extremely complex. For some small service companies, scheduling and job placements are based on which customers walk through the door. Determining the best schedule for flights and airline pilots, the placement of military recruits to jobs, and the truck drivers and equipment that should be used to transport materials across the country normally require sophisticated computer programs. In most cases, various schedules and job placement reports are generated. Employee schedules are developed for each employee, showing their job assignments over the next week or month. Job placements are often determined based on skills inventory reports, which show which employee might be best suited to a particular job.

Wage and Salary Administration

The last of the major human resource MIS subsystems involves determining wages, salaries, and benefits, including medical payments, savings plans, and retirement accounts. Wage data, such as industry averages for positions, can be taken from the corporate database and manipulated by the human resource MIS to provide wage information and reports to higher levels of management. These reports, called salary surveys, can be used to compare salaries with budget plans, the cost of salaries versus sales, and the wages required for any one department or office. The reports also help show backup of key positions in the company. Wage and salary administration also entails designing retirement programs for employees. Some companies use computerized retirement programs to help employees gain the most from their retirement accounts and options.

Other Management Information Systems

In addition to finance, manufacturing, marketing, and human resource MISs, some companies have other functional management information systems. For example, most successful companies have well-developed accounting functions and a supporting accounting MIS. Also, many companies make use of geographic information systems for presenting data in a useful form.

Accounting MISs

In some cases, accounting works closely with financial management. An **accounting MIS** performs a number of important activities, providing aggregate information on accounts payable, accounts receivable, payroll, and many other applications. The organization's TPS captures accounting data, which is also used by most other functional information systems.

Some smaller companies hire outside accounting firms to assist them with their accounting functions. These outside companies produce reports for the firm using raw accounting data. In addition, many excellent integrated accounting programs are available for personal computers in small companies. Depending on the needs of the small organization and its personnel's computer experience, using these computerized accounting systems can be a very cost-effective approach to managing information.

accounting MIS

information system that provides aggregate information on accounts payable, accounts receivable, payroll, and many other applications

Geographic Information Systems

geographic information system (GIS)

a computer system capable of assembling, storing, manipulating, and displaying geographic information (i.e., data identified according to its location)

Increasingly, managers want to see data presented in graphical form. A **geographic information system (GIS)** is a computer system capable of assembling, storing, manipulating, and displaying geographically referenced information—that is, data identified according to their locations. A GIS enables users to pair predrawn maps or map outlines with tabular data to describe aspects of a particular geographic region. For example, sales managers may want to plot total sales for each county in the states they serve. Using a GIS, they can specify that each county be drawn with a degree of shading that indicates the relative amount of sales—no shading or light shading represents no or little sales and deeper shading represents more sales. Because the GIS works with any data represented in tabular form, this capability is finding its way into spreadsheets. For example, Excel and Lotus include a mapping tool that lets you plot spreadsheet data as a demographic map. Such applications show up frequently in scientific investigations, resource management, and development planning. Retail, government, and utility organizations are frequent users of GISs. Retail chains, for example, need spatial analysis to determine where potential customers are located and where their competition is.

We saw earlier in this chapter that MISs provide useful summary reports to help solve structured and semistructured business problems. Decision support systems (DSSs) offer the potential to also assist in solving these problems.

AN OVERVIEW OF DECISION SUPPORT SYSTEMS

A DSS is an organized collection of people, procedures, software, databases, and devices used to support problem-specific decision making and problem solving. The focus of a DSS is on decision-making effectiveness when faced with unstructured or semistructured business problems. Decision support systems offer the potential to generate higher profits, lower costs, and better products and services. For example, healthcare organizations use DSSs to track and reduce costs. Visteon, a transportation equipment company, uses a decision support system at its Sterling plant to help the self-directed teams that oversee the design and development of front axles for Ford Expeditions and Explorers.[23] As a result, productivity has increased by 30 percent. The U.S. Forest Service uses a DSS optimization model to help it determine land usage.[24] As with a TPS and an MIS, a DSS should be designed, developed, and used to help the organization achieve its goals and objectives.

Decision support systems, although skewed somewhat toward the top levels of management, are used at all levels. To some extent, managers at all levels are faced with less structured, nonroutine problems, but the quantity and magnitude of these decisions increase as a manager rises higher in an organization. Many organizations face a bureaucracy of complex rules, procedures, and decisions. DSSs are used to bring more structure to these problems to aid the decision-making process. In addition, because of the inherent flexibility of decision support systems, managers at all levels can use DSSs to assist in some relatively routine, programmable decisions in lieu of more formalized management information systems.

Characteristics of a Decision Support System

Decision support systems have a number of characteristics that allow them to be effective management support tools. Of course, not all DSSs work the same—some are small in scope and offer only some of these characteristics. In general, a decision support system can perform the following functions:

Handle large amounts of data from different sources. For instance, advanced database management systems and data warehouses have allowed decision makers to search databases for information when using a DSS, even when some data sources reside in different databases stored in different computer systems or networks. Other sources of data may be accessed via the Internet or over a corporate intranet.

Provide report and presentation flexibility. Managers can get the information they want, presented in a format that suits their needs. Furthermore, output can be presented on computer screens or produced on printers, depending on the needs and desires of the problem solvers.

Offer both textual and graphical orientation. Today's decision support systems can produce text, tables, line drawings, pie charts, trend lines, and more. By using their preferred orientation, managers can use a DSS to get a better understanding of a true situation, if required, and to convey this understanding to others.

Support drill down analysis. A manager can get more levels of detail when needed by drilling down through data. For example, a manager can get more detailed information for a project if needed. He or she can view the overall project cost or drill down and see the cost for each project phase, activity, and task.

Perform complex, sophisticated analysis and comparisons using advanced software packages. Marketing research surveys, for example, can be analyzed in a variety of ways using analysis programs that are part of a DSS. Many of the analytical programs associated with a DSS are actually stand-alone programs. The DSS provides a means of bringing these together.

Support optimization, satisficing, and heuristic approaches. By supporting all types of decision-making approaches, a DSS gives the decision maker a great deal of flexibility in getting computer support for decision-making activities. For example, **what-if analysis**, the process of making hypothetical changes to problem data and observing the impact on the results, can be used to control inventory. Given the demand for products, such as automobiles, the computer can determine the necessary subparts and components, including engines, transmissions, windows, and so on. With what-if analysis, a manager can make changes to problem data (the number of automobiles needed for next month) and immediately see the impact on the requirements for subparts (engines, windows, and so on). Multicriteria decision making allows managers to take a number of important goals into account. A car manufacturing company, for example, may want to maximize profits, while keeping all its plants open for the next few months and avoiding a labor strike. Multicriteria decision-making approaches allow managers to seek several goals at the same time.

Simulation is the ability of the DSS to duplicate the features of a real system. In most cases, probability or uncertainty is involved. For example, the mean time between failure and the mean time to repair key components of a manufacturing line can be calculated to determine the impact on the number of products that can be produced each shift. Engineers can use this data to determine which components need to be reengineered to increase the mean time between failures and which components need to have an ample supply of spare parts to reduce the mean time to repair.

Goal-seeking analysis is the process of determining the problem data required for a given result. For example, a financial manager is considering an investment with a certain monthly net income. Furthermore, the manager might have a goal to earn a return of 9 percent on the investment. Goal seeking allows the manager to determine what monthly net income (problem data) is needed to have a return of 9 percent (problem result). Some spreadsheets can be used to perform goal-seeking analysis (see Figure 10.14).

what-if analysis

the process of making hypothetical changes to problem data and observing the impact on the results

simulation

the ability of the DSS to duplicate the features of a real system

goal-seeking analysis

the process of determining the problem data required for a given result

FIGURE 10.14

With a spreadsheet program, a manager can enter a goal, and the spreadsheet will determine the needed input to achieve the goal.
(Source: Courtesy of Lotus Development Corporation.)

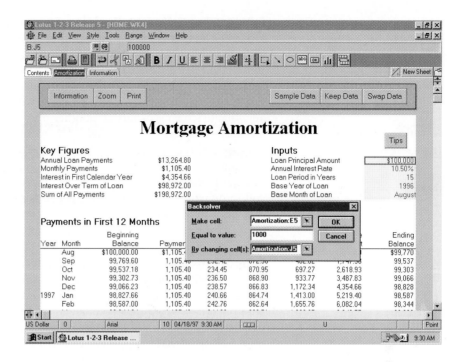

Capabilities of a Decision Support System

Developers of decision support systems strive to make them more flexible than management information systems and to give them the potential to assist decision makers in a variety of situations. See Table 10.2 for a few DSS applications. DSSs can assist with all or most problem-solving phases, decision frequencies, and different degrees of problem structure. DSS approaches can also help at all levels of the decision-making process. In this section we investigate these DSS capabilities. (An actual DSS may provide only a few of these capabilities, depending on the uses and scope of the DSS.)

Support for Problem-Solving Phases

The objective of most decision support systems is to assist decision makers with the phases of the problem-solving process. As previously discussed, these phases include intelligence, design, choice, implementation, and monitoring. A specific DSS might support only one or a few problem-solving phases.

TABLE 10.2

Selected DSS Applications

Company or Application	Description
Cinergy Corporation	The electric utility developed a DSS to reduce lead time and effort required to make decisions in purchasing coal.
RCA	The company developed a DSS called Industrial Relation Information System (IRIS) to help solve personnel problems and issues.
U.S. Army	It developed a DSS to help recruit, train, and educate enlisted forces. The DSS uses simulation that incorporates what-if features.
National Audubon Society	It developed a DSS called Energy Plan (EPLAN) to analyze the impact of U.S. energy policy on the environment.
Hewlett-Packard	The computer company developed a DSS called Quality Decision Management to help improve the quality of its products and services.
Virginia	The state of Virginia developed the Transportation Evacuation Decision Support System (TEDSS) to determine the best way to evacuate people in case of a nuclear disaster at its nuclear power plants.

Support for Different Decision Frequencies

Decisions can range on a continuum from one-of-a-kind to repetitive decisions. One-of-a-kind decisions are typically handled by an **ad hoc DSS**. An ad hoc DSS is concerned with situations or decisions that come up only a few times during the life of the organization; in small businesses, they may happen only once. For example, a company might be faced with a decision on whether to build a new manufacturing facility in another area of the country. Repetitive decisions are addressed by an institutional DSS. An **institutional DSS** handles situations or decisions that occur more than once, usually several times a year or more. An institutional DSS is used repeatedly and refined over the years. Examples of institutional DSSs include systems that support portfolio and investment decisions and production scheduling. These decisions may require decision support numerous times during the year. Between these two extremes are numerous decisions managers make several times, but not regularly or routinely.

Support for Different Problem Structures

As discussed previously, decisions can range from highly structured and programmed to unstructured and nonprogrammed. **Highly structured problems** are straightforward, requiring known facts and relationships. **Semistructured or unstructured problems**, on the other hand, are more complex. The relationships among the data are not always clear, the data may be in a variety of formats, and the data is often difficult to manipulate or obtain. In addition, the decision maker may not know the information requirements of the decision in advance.

Support for Various Decision-Making Levels

Decision support systems can offer help for managers at different levels within the organization. Operational-level managers can be assisted with daily and routine decision making. Tactical-level decision makers can be supported with analysis tools that assist in proper planning and control. At the strategic level, DSSs can help managers by providing analysis for long-term decisions requiring both internal and external information (Figure 10.15).

A Comparison of DSSs and MISs

A DSS differs from an MIS in numerous ways, including the type of problems solved; the support given to users; the decision emphasis and approach; and the type, speed, output, and development of the system used. Table 10.3 lists brief descriptions of these differences.

COMPONENTS OF A DECISION SUPPORT SYSTEM

At the core of a DSS are a database and a model base. In addition, a typical DSS contains a **dialogue manager**, which allows decision makers to easily access and manipulate the DSS and use common business terms and phrases. External database access allows the DSS to tap into vast stores of information contained in the corporate database, letting the DSS retrieve information on inventory,

---

ad hoc DSS

a DSS concerned with situations or decisions that come up only a few times during the life of the organization

institutional DSS

a DSS that handles situations or decisions that occur more than once, usually several times a year or more

highly structured problems

problems that are straightforward and require known facts and relationships

semistructured or unstructured problems

more complex problems in which the relationships among the data are not always clear, the data may be in a variety of formats, and the data is often difficult to manipulate or obtain

dialogue manager

user interface that allows decision makers to easily access and manipulate the DSS and use common business terms and phrases

FIGURE 10.15

Decision-Making Level

Strategic-level managers are involved with long-term decisions, which are often made infrequently. Operational-level managers are involved with decisions that are made more frequently.

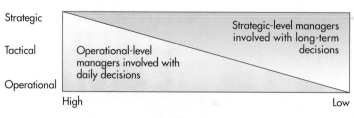

Factor	DSS	MIS
Problem Type	A DSS is good at handling unstructured problems that cannot be easily programmed.	An MIS is normally used only with more structured problems.
Users	A DSS supports individuals, small groups, and the entire organization. In the short run, users typically have more control over a DSS.	An MIS supports primarily the organization. In the short run, users have less control over an MIS.
Support	A DSS supports all aspects and phases of decision making; it does not replace the decision maker—people still make the decisions.	This is not true of all MIS systems—some make automatic decisions and replace the decision maker.
Emphasis	A DSS emphasizes actual decisions and decision-making styles.	An MIS usually emphasizes information only.
Approach	A DSS is a direct support system that provides interactive reports on computer screens.	An MIS is typically an indirect support system that uses regularly produced reports.
System	The computer equipment that provides decision support is usually on-line (directly connected to the computer system) and related to real time (providing immediate results). Computer terminals and display screens are examples—these devices can provide immediate information and answers to questions.	An MIS, using printed reports that may be delivered to managers once a week, may not provide immediate results.
Speed	Because a DSS is flexible and can be implemented by users, it usually takes less time to develop and is better able to respond to user requests.	An MIS's response time is usually longer.
Output	DSS reports are usually screen oriented, with the ability to generate reports on a printer.	An MIS, however, typically is oriented toward printed reports and documents.
Development	DSS users are usually more directly involved in its development. User involvement usually means better systems that provide superior support. For all systems, user involvement is the most important factor for the development of a successful system.	An MIS is frequently several years old and often was developed for people who are no longer performing the work supported by the MIS.

TABLE 10.3

Comparison of DSSs and MISs

sales, personnel, production, finance, accounting, and other areas. Finally, access to the Internet, networks, and other computer-based systems permits the DSS to tie into other powerful systems, including the TPS or function-specific subsystems. Internet software agents, for example, can be used in creating powerful decision support systems.[25] Figure 10.16 shows a conceptual model of a DSS. Since database and database management system (DBMS) concepts were thoroughly covered in Chapter 5 and networks and the Internet were covered in Chapters 6 and 7, we begin with a discussion of the model base.

model base

part of a DSS that provides decision makers access to a variety of models and assists them in decision making

model management software

software that coordinates the use of models in a DSS

financial model

model that provides cash flow, internal rate of return, and other investment analysis

The Model Base

The purpose of the **model base** in a DSS is to give decision makers access to a variety of models and to assist them in the decision-making process. The model base can include **model management software (MMS)** that coordinates the use of models in a DSS, including financial, statistical analysis, graphical, and project management models. Depending on the needs of the decision maker, one or more of these models can be used.

Financial Models

Financial models provide cash flow, internal rate of return, and other investment analysis. Spreadsheet programs such as Excel are often used for this purpose. In addition, more sophisticated financial planning and modeling programs can be

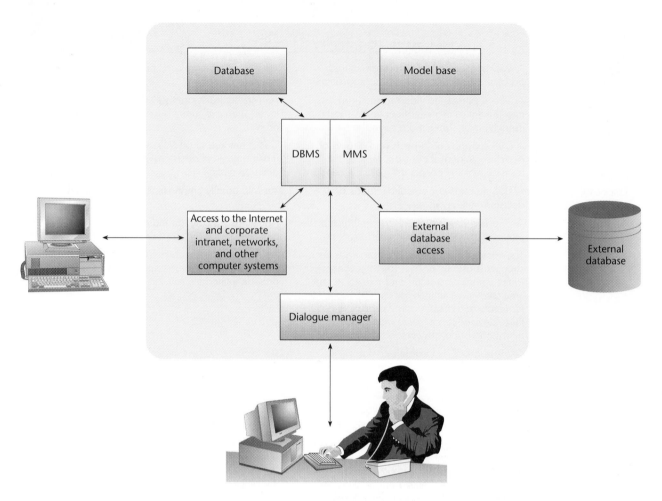

FIGURE 10.16

A Conceptual Model of a DSS

DSS components include a model base; database; external database access; access to the Internet and corporate intranet, networks, and other computer systems; and a dialogue manager.

statistical analysis model

model that can provide summary statistics, trend projections, hypothesis testing, and more

graphical modeling program

software package that assists decision makers in designing, developing, and using graphic displays of data and information

employed. Some organizations develop customized financial models to handle the unique situations and problems faced by the organization. However, as spreadsheet packages continue to increase in power, the need for sophisticated financial modeling packages may decrease. Individuals also develop financial models to help them plan for retirement or invest in the market. As seen in the "Ethical and Societal Issues" box, however, using financial models for such activities as on-line trading can lead to problems.

Statistical Analysis Models

Statistical analysis models can provide summary statistics, trend projections, hypothesis testing, and more. These programs are available on both personal and mainframe systems. Many software packages, including SPSS and SAS, provide outstanding statistical analysis for organizations of all sizes. These statistical programs can compute averages, standard deviations, correlation coefficients, and regression analysis; do hypotheses testing; and use many more techniques. Some statistical analysis programs also have the ability to produce graphic displays that reveal the relationship between variables or quantities. Hyperion Solutions Corporation and Cognos help statistical analysts determine which reports and graphs will generate the most meaningful decision support.[26]

Graphical Models

Graphical modeling programs are software packages that assist decision makers in designing, developing, and using graphic displays of data and information. Personal computer programs that can perform this type of analysis, such as PowerPoint, are available. In addition, sophisticated graphical analysis, including computer-assisted design (CAD), is available.

ETHICAL AND SOCIETAL ISSUES
The Dark Side of E-Trading

For decades, financial institutions and brokerage firms have developed sophisticated decision support systems for trading stocks and bonds. Some DSSs make use of the futures market and options to help companies hedge their stock bets and avoid risk. Financial DSSs can provide a wealth of investment information and discounted stock trades for corporations, which individual investors did not have access to previously. But this is changing rapidly. Additional investment information and discounted trades are now affordable and available to investors with few assets or financial resources, and discount brokers are aggressively advertising their services to the general public. But does this additional investment data and inexpensive trades result in better investment decisions?

You turn on your TV and see a young man with his face pushed against the glass of a copy machine. He pushes the copy button and makes a copy of his face. The secretary looks on in disgust. The boss walks by and calls the young man in for a talk. The boss looks concerned. Will the young man lose his job? Inside the office, the mood changes. The boss wants the young man to help him open an Internet trading account. The young man shows his boss how it is done and invites the boss to a party. The boss buys some stock. He might come to the party. With this and similar ads for discounted trading companies, it can appear that making money using discounted trading firms and free investment information is easy. Yet there are many potential problems, including fraud and large losses.

In Los Angeles, an employee of a company posted false information on an investment Web site and drove up the company's stock price by more than 30 percent.

Unsuspecting and inexperienced investors bought the stock on the way up and lost a substantial amount of money when the stock price plunged. The person posting the false information, like others involved with stock fraud, faces up to ten years in prison and up to $1 million in fines. Many stock frauds, however, go undetected, and investors go unprotected. In another case, an on-line trader lost about $750,000 through on-line trading. He was so upset that he took out a $500,000 life insurance policy on his wife and then pushed her off a balcony to collect the insurance money to help pay off his trading debts. When she didn't die, he climbed down and tried to suffocate her. The wife never knew of her husband's losses. After he landed in prison, he sued the on-line trading company.

Discussion Questions

1. How can a financial DSS be used to generate profits?
2. What are some of the potential problems of on-line trading for individual investors?

Critical Thinking Questions

3. What additional legislation, if any, would you suggest to help prevent stock fraud?
4. Is a trading company that aggressively advertises successful stock trading responsible to some extent for the trading activities of its clients?

Sources: Adapted from Rebbecca Buckman, "Heavy Losses," *The Wall Street Journal*, February 28, 2000, p. C1; David Rogers, "Online Stock Fraud," *The Wall Street Journal*, June 10, 1999, p. A6; and Michael Allen, "Bank Rule to Catch Money Launderers," *The Wall Street Journal*, March 23, 1999, p. A4.

---

project management model

model used to coordinate large projects and identify critical activities and tasks that could delay or jeopardize an entire project if they are not completed on time and cost-effectively

Project Management Models

Project management models are used to handle and coordinate large projects; they are also used to identify critical activities and tasks that could delay or jeopardize an entire project if they are not completed on time and cost-effectively. Some of these programs can also determine the best way to speed up a project by using additional resources, including cash, labor, and equipment. Project management models allow decision makers to keep tight control over projects of all sizes and types. These models can also help project managers prioritize user needs and requirements.[27]

The Dialogue Manager

The dialogue manager allows users to interact with the DSS to obtain information. It assists with all aspects of communications between the user and the hardware and software that constitute the DSS. In a practical sense, to most DSS users, the dialogue manager is the DSS. Upper-level decision makers are often less interested in where the information came from or how it was gathered than that the information is both understandable and accessible.

Computer-assisted design (CAD) programs are powerful tools in developing products.
(Source: Bob Krist/Corbis)

group decision support system (GDSS)

software application that consists of most elements in a DSS, plus software needed to provide effective support in group decision making

delphi approach

a decision-making approach in which group decision makers are geographically dispersed; this approach encourages diversity among group members and fosters creativity and original thinking in decision making

brainstorming

decision-making approach that often involves members offering ideas "off the top of their heads"

group consensus approach

decision-making approach that forces members in the group to reach a unanimous decision

nominal group technique

decision-making approach that encourages feedback from individual group members; the final decision is made by voting, similar to the way public officials are elected

THE GROUP DECISION SUPPORT SYSTEM

The DSS approach has resulted in better decision making for all levels of individual users. However, many DSS approaches and techniques are not suitable for a group decision-making environment. Although not all workers and managers are involved in committee meetings and group decision-making sessions, some tactical and strategic-level managers can spend more than half their decision-making time in a group setting. Such managers need effective approaches to assist with group decision making. A **group decision support system (GDSS)**, also called a *group support system* or a *computerized collaborative work system,* consists of most of the elements in a DSS, plus GDSS software needed to provide effective support in group decision-making settings (Figure 10.17).

Characteristics of a GDSS

A GDSS has a number of unique characteristics that go beyond the traditional DSS. Developers of these systems try to build on the advantages of individual support systems while realizing that new and additional approaches are needed in a group decision-making environment. For example, some GDSSs can allow the exchange of information and expertise among people without meetings or direct face-to-face interaction.[28] The following are some characteristics of a typical GDSS.

Special design. The GDSS approach acknowledges that special procedures, devices, and approaches are needed in group decision-making settings. These procedures must foster creative thinking, effective communications, and good group decision-making techniques.

Ease of use. Like an individual DSS, a GDSS must be easy to learn and use. Systems that are complex and hard to operate will seldom be used. Many groups have less tolerance than do individual decision makers for poorly developed systems.

Flexibility. Two or more decision makers working on the same problem may have different decision-making styles and preferences. Each manager makes decisions in a unique way, in part because of different experiences and cognitive styles. An effective GDSS not only has to support the different approaches that managers use to make decisions but also must find a means to integrate their different perspectives into a common view of the task at hand.

Decision-making support. A GDSS can support different decision-making approaches, including the **delphi approach**, in which group decision makers are geographically dispersed throughout the country or the world. This approach encourages diversity among group members and fosters creativity and original thinking in decision making. Another approach, called **brainstorming,** which often involves members offering ideas "off the top of their heads," fosters creativity and free thinking. The **group consensus approach** forces members in the group to reach a unanimous decision. With the **nominal group technique**, each decision maker can participate; this technique encourages feedback from individual group members, and the final decision is made by voting, similar to a system for electing public officials.

Anonymous input. Many GDSSs allow anonymous input, where the person giving the input is not known to other group members. For example, some organizations use a GDSS to help rank the performance of managers. Anonymous input allows the group decision makers to concentrate on the merits of the input without considering who gave it. In other words, input given by a top-level manager is given the same consideration as input from lower-level employees or other members of

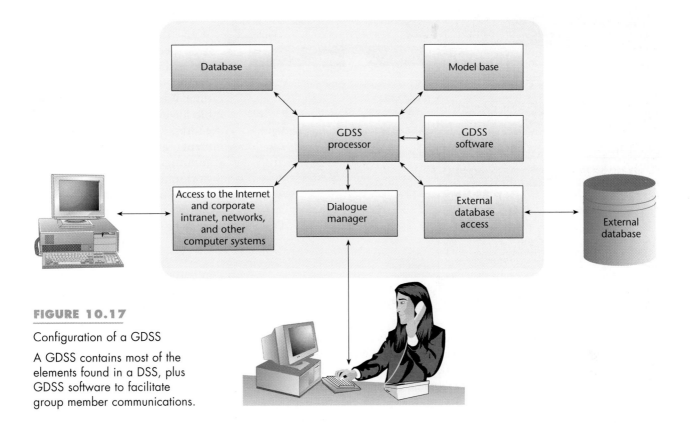

Configuration of a GDSS

A GDSS contains most of the elements found in a DSS, plus GDSS software to facilitate group member communications.

Software such as Microsoft Project can be used to keep track of project tasks and keep projects on schedule.
(Source: Courtesy of Microsoft Corporation)

the group. Some studies have shown that groups using anonymous input can make better decisions and have superior results compared with groups that do not use anonymous input. Anonymous input, however, can result in flaming, where an unknown team member posts insults or even obscenities on the GDSS system.[29]

Reduction of negative group behavior. One key characteristic of any GDSS is the ability to suppress or eliminate group behavior that is counterproductive or harmful to effective decision making. In some group settings, dominant individuals can take over the discussion, which can prevent other members of the group from presenting creative alternatives. In other cases, one or two group members can sidetrack or subvert the group into areas that are nonproductive and do not help solve the problem at hand. Other times, members of a group may assume they have made the right decision without examining alternatives—a phenomenon called *groupthink.* If group sessions are poorly planned and executed, the result can be a tremendous waste of time. Today, many GDSS designers are developing software and hardware systems to reduce these types of problems. Procedures for effectively planning and managing group meetings can be incorporated into the GDSS approach. A trained meeting facilitator is often employed to help lead the group decision-making process and to avoid groupthink.

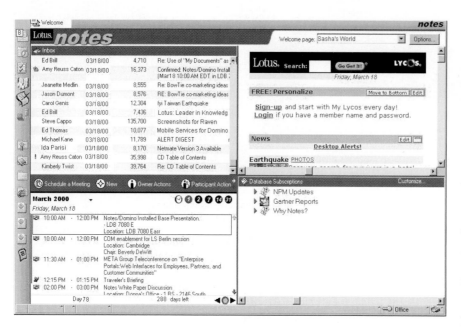

Workgroup software, such as Lotus Notes, allows people located around the world to work on the same project, documents, and files efficiently and at the same time.
(Source: Courtesy of Lotus Development Corporation)

Parallel communication. With traditional group meetings, people must take turns addressing various issues. One person normally talks at a time. With a GDSS, it is possible for every group member to address issues or make comments at the same time by entering them into a PC or workstation. These comments and issues are displayed on every group member's PC or workstation immediately. Parallel communication can speed meeting times and result in better decisions.

Automated record keeping. Most GDSSs have the ability to keep detailed records of a meeting automatically. Each comment that is entered into a group member's PC or workstation can be anonymously recorded. In some cases, literally hundreds of comments can be stored for future review and analysis. In addition, most GDSS packages have automatic voting and ranking features. After group members vote, the GDSS records each vote and makes the appropriate rankings.

GDSS Software

GDSS software, often called *groupware* or *workgroup software,* helps with joint work group scheduling, communication, and management. One popular package, Lotus Notes, can capture, store, manipulate, and distribute memos and communications that are developed during group projects. Microsoft's NetMeeting product supports application sharing in multiparty calls. It gains performance by sharing one task at a time. The user designated to be the "operator" must select one previously launched application, and each participant has to choose whether to "collaborate." Any collaborating participant can assume mouse control and work in the shared application while others watch. Exchange from Microsoft is another example of groupware. This software allows users to set up electronic bulletin boards, schedule group meetings, and use e-mail in a group setting. Other GDSS software packages include Collabra Share, OpenMind, and TeamWare. All these tools can aid in group decision making.[30]

GDSS Alternatives

Group decision support systems can take on a number of alternative network configurations, depending on the needs of the group, the decision to be supported, and the geographic location of group members. The frequency of GDSS use and the location of the decision makers are two important factors (Figure 10.18).

The Decision Room

decision room

a room that supports decision making, with the decision makers in the same building, combining face-to-face verbal interaction with technology to make the meeting more effective and efficient

The **decision room** is ideal for situations in which decision makers are located in the same building or geographic area and the decision makers are occasional users of the GDSS approach. In these cases, one or more decision rooms or facilities can be set up to accommodate the GDSS approach. Groups, such as marketing research teams, production management groups, financial control teams, and quality-control committees, can use the decision rooms when needed.

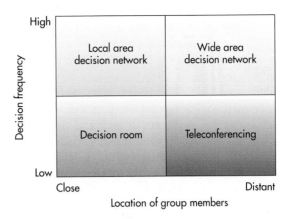

FIGURE 10.18

GDSS Alternatives

This figure demonstrates that the decision room may be the best alternative for group members who are located physically close together and who need to make infrequent decisions as a group. By the same token, group members who are situated at distant locations and who frequently make decisions together may require a wide area decision network to accomplish their goals.

The decision room alternative combines face-to-face verbal interaction with technology-aided formalization to make the meeting more effective and efficient. A typical decision room is shown in Figure 10.19.

The Local Area Decision Network
The local area decision network can be used when group members are located in the same building or geographic area and under conditions in which group decision making is frequent. In these cases, the technology and equipment of the GDSS approach is placed directly into the offices of the group members. Usually this is accomplished via a local area network (LAN).

The Teleconferencing Alternative
The teleconferencing alternative is used for situations in which the decision frequency is low and the location of group members is distant. These distant and occasional group meetings can tie together multiple GDSS decision-making rooms across the country or around the world. Using long-distance communications technology, these decision rooms are electronically connected in teleconferences and videoconferences. This alternative can offer a high degree of

FIGURE 10.19

The GDSS Decision Room

For group members located in the same physical location, the decision room is an optimal GDSS alternative. By use of networked computers and computer devices, such as project screens and printers, the meeting leader can pose questions to the group, instantly collect their feedback, and, with the help of the governing software loaded on the control station, process this feedback into meaningful information to aid in the decision-making process.

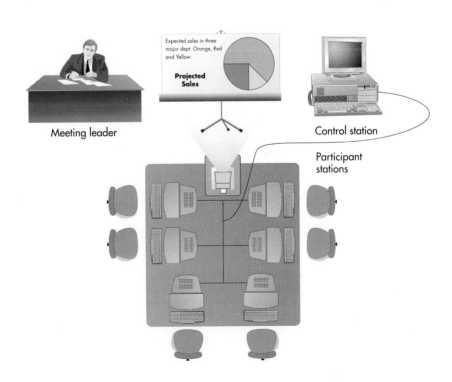

flexibility. The GDSS decision rooms can be used locally in a group setting or globally when decision makers are located throughout the world. GDSS decision rooms are often connected through the Internet.

The Wide Area Decision Network

The wide area decision network is used for situations in which the decision frequency is high and the location of group members is distant. In this case, the decision makers require frequent or constant use of the GDSS approach. This situation requires decision makers located throughout the country or the world to be linked electronically through a wide area network (WAN). The group facilitator and all group members are at geographically dispersed locations. In some cases, the model base and database are also geographically dispersed. This GDSS alternative allows people to work in **virtual workgroups**, where teams of people located around the world can work on common problems.

virtual workgroups

teams of people located around the world working on common problems

The Internet is increasingly being used to support wide area decision networks. As discussed in Chapters 7 and 8, a number of technologies—including video conferencing, instant messaging, chat rooms, and telecommuting—can be used to assist the GDSS process. In addition, many specialized wide area decision networks use the Internet for group decision making and problem solving.

THE EXECUTIVE SUPPORT SYSTEM

executive support system (ESS), or executive information system (EIS)

specialized DSS that includes all hardware, software, data, procedures, and people used to assist senior-level executives within the organization

Because top-level executives often require specialized support when making strategic decisions, many companies have developed systems to assist executive decision making. This type of system, called an **executive support system (ESS)**, is a specialized DSS that includes all hardware, software, data, procedures, and people used to assist senior-level executives within the organization. In some cases, an ESS, also called an **executive information system (EIS)**, supports the actions of members of the board of directors, who are responsible to stockholders. These top-level decision-making strata are shown in Figure 10.20.

An ESS can also be used by individuals further down in the organizational structure. Once targeted at the top-level executive decision makers, ESSs are now marketed to—and used by—employees at other levels in the organization. In the traditional view, ESSs give top executives a means of tracking critical success factors. Today, all levels of the organization share information from the same databases. However, for our discussion, we will assume that ESSs remain in the upper management levels, where they indicate important corporate issues, indicate new directions the company may take, and help executives monitor the company's progress.

Executive Support Systems in Perspective

An ESS is a special type of DSS, and, like a DSS, an ESS is designed to support higher-level decision making in the organization. The two systems are, however, different in important ways. DSSs provide a variety of modeling and analysis tools to enable users to thoroughly analyze problems—that is, they allow users to *answer* questions. ESSs present structured information about aspects of the organization that executives consider important—in other words, they allow executives to *ask* the right questions.[31]

The following are general characteristics of ESSs.

FIGURE 10.20

The Layers of Executive Decision Making

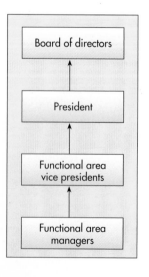

- *Tailored to individual executives.* ESSs are typically tailored to individual executives; DSSs are not tailored to particular users. As such, ESSs are truly representative of the overall objective of information systems: to deliver the right information to the right person at the right time in the right format. An ESS is an interactive, hands-on tool that allows an executive to focus, filter, and organize data and information.

- *Easy to use.* A top-level executive's most critical resource can be his or her time. Thus, an ESS must be easy to learn and use and not overly complex. Early ESSs were noted for their extreme ease of use, since they were targeted for decision makers who were often not technically oriented. The extensive use of color and graphics is common with an ESS.

- *Have drill down abilities.* An ESS allows executives to "drill down" into the company to determine how certain data was produced. Drill down allows an executive to get more detailed information if needed. For example, after seeing a report summarizing the progress and costs of a large project, an executive might want to drill down to see the status of a particular activity within the project.

- *Support the need for external data.* The data needed to make effective top-level decisions is often external—information from competitors, the federal government, trade associations and journals, consultants, and so on. This data is often "soft" as well. An effective ESS is able to extract data useful to the decision maker from a wide variety of sources, including the Internet and other electronic publishing sources such as Lexis/Nexis. Often they employ intelligent bots to search the Web for new information about competitors, suppliers, customers, and key business trends. Typically, decisions made at top levels are also complex and unstructured. Traditional computer programs are normally ineffective for these types of decisions. As a result, more advanced ESSs are required to help support all types of strategic decisions.

- *Can help with situations that have a high degree of uncertainty.* There is a high degree of uncertainty with most executive decisions. What will happen if a new plant is started? What will be the consequences of a potential merger? How will union negotiators take a new proposal? The answers to these questions are not known with certainty. Handling these unknown situations using modeling and other ESS procedures helps top-level managers measure the amount of risk in a decision.

- *Have a futures orientation.* Executive decisions are future oriented, meaning that decisions will have a broad impact for years or decades. The information sources to support future-oriented decision making are usually informal—from golf partners to members of social clubs or civic organizations.

- *Are linked with value-added business processes.* Like other information systems, executive support systems are linked with executive decision making about value-added business processes. For instance, executive support systems can be used by car rental companies to analyze trends. By detecting which firms generate enough business to be worth a certain discount, car-rental executives can ask questions to determine why these firms create high revenue and how this business can be continued or bettered. They might also use these factors to identify similar firms with business potential. Strategic planning of this type is just one of the functions that can be aided by an ESS.

Capabilities of an Executive Support System

The responsibility given to top-level executives and decision makers brings unique problems and pressures to their jobs. Following is a discussion of some of the characteristics of executive decision making that are supported through the ESS approach. As you will note, most of these are related to an organization's overall profitability and direction. An effective ESS should have the capability to support executive decisions with many of these capabilities, such as strategic planning and organizing, crisis management, and more.

Support for defining an overall vision. One of the key roles of senior executives is to provide a broad vision for the entire organization. This vision includes the organization's major product lines and services, the types of businesses it supports today and in the future, and its overriding goals.

strategic planning

determining long-term objectives by analyzing the strengths and weaknesses of the organization, predicting future trends, and projecting the development of new product lines

Support for strategic planning. ESSs also support strategic planning. **Strategic planning** involves determining long-term objectives by analyzing the strengths and weaknesses of the organization, predicting future trends, and projecting the development of new product lines. It also involves planning the acquisition of new equipment, analyzing merger possibilities, and making difficult decisions concerning downsizing and the sale of assets if required by unfavorable economic conditions.

Support for strategic organizing and staffing. Top-level executives are concerned with organization structure. For example, decisions concerning the creation of new departments or downsizing the labor force are made by top-level managers. Should the information systems department be placed under new leadership? Should the accounting and finance departments be joined under a new vice president of financial services? Should the marketing department be divided along the company's two major product lines? These and similar questions can profoundly affect the overall effectiveness of the organization and should be supported by an ESS.

Overall direction for staffing decisions and effective communication with labor unions are major decision areas for top-level executives. Middle- and lower-level managers make staffing decisions, but general decisions about the types and number of employees for various departments in the organization are determined at the top level of the organization. In addition, top-level managers are responsible for labor negotiation. Thus, ESSs can be employed to help analyze the impact of staffing decisions, potential pay raises, changes in employee benefits, and new work rules.

Support for strategic control. Another type of executive decision relates to strategic control, which involves monitoring and managing the overall operation of the organization. Goal seeking can be done for each major area to determine what performance these areas need to achieve to reach corporate expectations. Effective ESS approaches can help top-level managers make the most of their existing resources and control all aspects of the organization.

Support for crisis management. Even with careful strategic planning, a crisis can occur. Major disasters, including hurricanes, tornadoes, floods, earthquakes, fires, and sabotage, can totally shut down major parts of the organization. Handling these emergencies is another responsibility for top-level executives. In many cases, strategic emergency plans can be put into place with the help of an ESS. These contingency plans help organizations recover quickly if an emergency or crisis occurs.

Decision making is a vital part of managing businesses strategically. IS systems such as decision support, group decision support, and executive support systems help employees by tapping existing databases and providing them with current, accurate information. The increasing integration of all business information systems—from TPSs to MISs to DSSs—can help organizations monitor their competitive environment and make better-informed decisions.

SUMMARY

PRINCIPLE • The management information system (MIS) must provide the right information to the right person in the right fashion at the right time.

Problem solving begins with decision making. A well-known model developed by Herbert Simon divides the decision-making phase of the problem-solving process into three stages: intelligence, design, and choice. The three phases of decision making are augmented by implementation and monitoring to result in problem solving.

The first stage in the decision-making phase of the problem-solving process is the intelligence stage. During this stage, potential problems and opportunities are identified and defined. Information is gathered that relates to the cause and scope of the problem. Constraints on the possible solution and the problem environment are investigated. In the design stage, alternative solutions to the problem are developed. In addition, the feasibility and implications of these alternatives are evaluated. The last stage of the decision-making phase, the choice stage, requires selecting a course of action.

Problem solving includes and goes beyond decision making. It also includes the implementation stage, when the solution is put into effect. The final stage of the problem-solving process is the monitoring stage. In this stage, the decision makers evaluate the implementation of the solution to determine whether the anticipated results were achieved and to modify the process in light of new information learned during the implementation stage.

A management information system is an integrated collection of people, procedures, databases, and devices that provide managers and decision makers with information to help achieve organizational goals. An MIS can help an organization achieve its goals by providing managers with insight into the regular operations of the organization so that they can control, organize, and plan more effectively and efficiently. The primary difference between the reports generated by the TPS and those generated by the MIS is that MIS reports support managerial decision making at the higher levels of management.

Data that enters the MIS originates from both internal and external sources. The output of most management information systems is a collection of reports that are distributed to managers. These reports include scheduled reports, key indicator reports, demand reports, exception reports, and drill down reports. Scheduled reports are produced periodically, or on a schedule, such as daily, weekly, or monthly. A key-indicator report is a special type of scheduled report. Demand reports are developed to give certain information at a manager's request. Exception reports are automatically produced when a situation is unusual or requires management action. Drill down reports provide increasingly detailed data about situations.

Management information systems have a number of common characteristics, including producing scheduled, demand, exception, and drill down reports; producing reports with fixed and standard formats; producing hard-copy and soft-copy reports; using internal data stored in organizational computerized databases; and having reports developed and implemented by IS personnel or end users.

Most MISs are organized along the functional lines of an organization. Typical functional management information systems include accounting, manufacturing, marketing, and human resources. Each system is composed of inputs, processing subsystems, and outputs. The primary sources of input to functional MISs include the corporate strategic plan, data from the TPS, information from other functional areas, and external sources, including the Internet. The primary output of these functional MISs are summary reports that assist in managerial decision making.

A financial management information system provides financial information to all financial managers within an organization, including the chief financial officer (CFO). Subsystems are financial forecasting, profit/loss and cost systems, use and management of funds, and auditing.

A manufacturing MIS accepts inputs from the strategic plan, the TPS, and external sources. The TPSs involved support the business processes associated with the receiving and inspecting of raw material and supplies; inventory tracking of raw materials, work in process, and finished goods; labor and personnel management; management of assembly lines, equipment and machinery, inspection, and maintenance; and order processing. The subsystems involved are design and engineering, master production scheduling, inventory control, just-in-time inventory and manufacturing, process control, and quality control and testing.

A marketing MIS supports managerial activities in the areas of product development, distribution, pricing decisions, promotional effectiveness, and sales forecasting. Subsystems include product development and reporting, promotion and advertising, product pricing, and marketing research.

Human resource MISs are concerned with activities related to employees of the organization. Subsystems include human resource planning, personnel selection and recruiting, training and skills inventories, scheduling and job placement, and wage and salary administration.

An accounting MIS performs a number of important activities, providing aggregate information on accounts payable, accounts receivable, payroll, and many other applications. The organization's TPS captures accounting data, which is also used by most other functional information systems. Geographic information systems provide regional data in graphical form.

PRINCIPLE • Decision support systems (DSSs) are used when the problems are more unstructured.

A decision support system (DSS) is an organized collection of people, procedures, software, databases, and devices working to support managerial decision making. DSS characteristics include the ability to handle large amounts of data; obtain and process data from different sources; provide report and presentation flexibility; perform complex statistical analysis; offer textual and graphical orientations; support optimization, satisficing, and heuristic approaches; and perform what-if simulation, and goal-seeking analysis.

DSSs provide assistance through all phases of the decision-making process. The degree of problem structure and scope contributes to the complexity of the decision support system. Problems may be highly structured or unstructured, infrequent in occurrence or routine and repetitive. An ad hoc DSS addresses unique, infrequent decision situations; an institutional DSS handles routine decisions. A common database is often the link that ties together a company's TPS, MIS, and DSS.

• • •

The components of a DSS are the database, model base, dialogue manager, and a link to external databases, the Internet, the corporate intranet, extranets, networks, and other systems. The model base contains the models used by the decision maker, such as financial, statistical, graphical, and project management models. The dialogue manager provides a dialogue management facility to assist in communications between the system and the user. Access to other computer-based systems

permits the DSS to tie into other powerful systems, including the TPS or function-specific subsystems.

PRINCIPLE • Specialized support systems, such as group decision support systems (GDSSs) and executive support systems (ESSs), use the overall approach of a DSS in situations such as group and executive decision making.

A group decision support system, also called a computerized collaborative work system, consists of most of the elements in a DSS, plus GDSS software needed to provide effective support in group decision-making settings. GDSSs are typically easy to learn and use and can offer specific or general decision-making support. GDSS software, also called groupware, is specially designed to help generate lists of decision alternatives and perform data analysis. These packages let people work on joint documents and files over a network.

The frequency of GDSS use and the location of the decision makers will influence the GDSS alternative chosen. The decision room alternative supports users in a single location that meet infrequently. Local area networks can be used when group members are located in the same geographic area and users meet regularly. Teleconferencing is used when decision frequency is low and the location of group members is distant. A wide area network is used for situations where the decision frequency is high and the location of group members is distant.

• • •

Executive support systems (ESSs) are specialized decision support systems designed to meet the needs of senior management. They serve to indicate issues of importance to the organization, indicate new directions the company may take, and help executives monitor the company's progress. ESSs are typically easy to use, offer a wide range of computer resources, and handle a variety of internal and external data. In addition, the ESS performs sophisticated data analysis, offers a high degree of specialization, and provides flexibility and comprehensive communications abilities. An ESS also supports individual decision-making styles. Some of the major decision-making areas that can be supported through an ESS are providing an overall vision, strategic planning and organizing, staffing and labor relations, crisis management, and strategic control.

KEY TERMS

accounting MIS 394
ad hoc DSS 398
auditing 382
brainstorming 403
choice stage 373
computer-assisted manufacturing (CAM) 387
computer-integrated manufacturing (CIM) 387
cost centers 384
decision room 404
decision-making phase 372
delphi approach 403
demand reports 377
design stage 373
dialogue manager 398
drill down reports 379
economic order quantity (EOQ) 386
exception reports 378
executive information system (EIS) 406
executive support system (ESS) 406
external auditing 383
financial MIS 380
financial model 399

flexible manufacturing system (FMS) 387
geographic information system (GIS) 395
goal-seeking analysis 396
graphical modeling program 400
group consensus approach 403
group decision support system (GDSS) 402
heuristics 374
highly structured problems 398
human resource MIS 392
implementation stage 373
institutional DSS 398
intelligence stage 372
internal auditing 383
just-in-time (JIT) inventory approach 387
key-indicator report 377
manufacturing resource planning (MRPII) 387
marketing MIS 388
material requirements planning (MRP) 387
model base 399

model management software (MMS) 399
monitoring stage 373
nominal group technique 403
nonprogrammed decisions 374
optimization model 374
problem solving 373
profit centers 382
programmed decisions 373
project management model 401
quality control 388
reorder point (ROP) 386
revenue centers 382
satisficing model 374
scheduled reports 376
semistructured or unstructured problems 398
sensitivity analysis 385
simulation 396
statistical analysis model 400
strategic planning 408
virtual workgroup 406
what-if analysis 396

REVIEW QUESTIONS

1. Define the term *management information system (MIS)*.
2. What are the four basic kinds of reports produced by an MIS?
3. What guidelines should be followed in developing reports for management information systems?
4. Identify the functions performed by all MIS systems.
5. What are the functions performed by a financial MIS?
6. Describe the functions of a manufacturing MIS.
7. What is MRPII? How is it different from MRP?
8. What is a human resource MIS? What are its outputs?
9. List and describe some other types of MISs.
10. What are the five stages of problem solving?
11. What is a geographic information system?
12. Describe the difference between a structured and an unstructured problem and give an example of each.
13. Define *decision support system*. What are its characteristics?
14. What is the difference between what-if analysis and goal-seeking analysis?
15. What are the components of a decision support system?
16. Describe four models used in decision support systems.
17. What is meant by *groupthink?*
18. State the objective of a group decision support system (GDSS) and identify three characteristics that distinguish it from a DSS.
19. Identify three group decision-making approaches often supported by a GDSS.
20. What is an executive support system? Identify three fundamental uses for such a system.

● DISCUSSION QUESTIONS

1. What is the relationship between an organization's transaction processing systems and its management information systems? What is the primary role of management information systems?

2. How can management information systems be used to support the objectives of the business organization?

3. Describe a financial MIS for a Fortune 1000 manufacturer of food products. What are the primary inputs and outputs? What are the subsystems?

4. How can a strong financial MIS provide strategic benefits to a firm?

5. Why is there such keen interest in implementing an MRPII MIS system at many manufacturing firms?

6. What is the difference in roles played by an internal auditing group versus an external auditing group?

7. You have been hired to develop a management information system and a decision support system for a manufacturing company. Describe what information you would include in printed reports and what information you would provide using a screen-based decision support system.

8. Imagine that you are the CFO for a services organization. You are concerned with the integrity of the firm's financial data. What steps might you take to ascertain the extent of potential problems?

9. What functions do decision support systems support in business organizations? You have looked at the functions supported by TPSs and MISs in Chapter 9 and in this chapter. How does a DSS differ from these two types of systems?

10. How is decision making in a group environment different from individual decision making, and why are information systems that assist in the group environment different? What are the advantages and disadvantages of making decisions as a group?

11. You have been hired to develop group support software. Describe the features you would include in your new GDSS software.

12. The use of ESSs should not be limited to the executives of the company. Do you agree or disagree? Why?

13. Imagine that you are the vice president of manufacturing for a Fortune 1000 manufacturing company. Describe the features and capabilities of your ideal ESS.

● PROBLEM-SOLVING EXERCISES

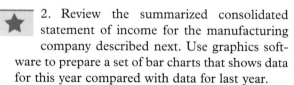

1. You have been asked to develop your company's employee training program. Make a list of the kind of information that must be available. Make a second list of the services that should be provided to the employee. Make a third list of the information you would provide managers, summarizing test results and performance information of employees taking the training course. Use your word processor or other software to describe the training program and all lists and reports.

2. Review the summarized consolidated statement of income for the manufacturing company described next. Use graphics software to prepare a set of bar charts that shows data for this year compared with data for last year.

a. Operating revenues increase by 3.5 percent while operating expenses increase 2.5 percent.

b. Other income and expenses decrease to $13,000.

c. Interest and other charges increase to $265,000.

Operating Results (in millions)

Operating Revenues	$2,924,177
Operating Expenses (including taxes)	2,483,687
Operating Income	440,490
Other Income and Expenses	13,497
Income before Interest and Other Charges	453,987
Interest and Other Charges	262,845
Net Income	191,142
Average Common Shares Outstanding	147,426
Earnings per share	$1.30

If you were a financial analyst tracking this company, what detail data might you need to perform a more complete analysis? Write a brief memo summarizing your data needs.

3. As the head buyer for a major supermarket chain, John is constantly being asked by manufacturers and distributors to stock their new products. Over 50 new items are introduced each week. Many times, these products are launched with national advertising campaigns

and special promotional allowances to both retailers, such as John's firm, and consumers. The store has only a limited amount of shelf and floor space to stock items. Thus, to add new products, the amount of shelf space allocated to existing products must be reduced, or items must be eliminated altogether.

Develop a spreadsheet that John can use to estimate the change in profits from adding or deleting an item from inventory. The spreadsheet will include input such as estimated weekly sales in units, shelf space allocated to stock item (measured in units), total cost per unit, sales price per unit, and promotional allowance earned per unit. The spreadsheet should calculate total profit by item and then sort the rows in descending order based on total profit. Because of the limited amount of shelf space, the spreadsheet should also calculate the accumulated shelf space based on the items stocked.

● TEAM ACTIVITIES

1. Divide the class into teams of three or four classmates and use the following role-playing scenario: Your consulting team is designing a management information system for a small manufacturing organization that produces ten different models of high-performance bicycles. These cycles are sold to distributors and bicycle shops throughout North America. One of the major problems this company faces is poor inventory control. It always seems that there are too many bikes in stock and yet the "hottest selling" model is out of stock. The owner has suggested that your team look at the feasibility of implementing a manufacturing MIS to help deal with this problem. Prepare a brief memo that describes the subsystems and outputs associated with the typical manufacturing MIS. Draw a systems-level flowchart, outline how these subsystems need to be integrated, and discuss where the inventory control system fits. What are some of the issues that will need to be addressed to develop a single integrated MIS to meet the needs of this organization? Are there some prerequisites that must be completed before the inventory control system can be built? If so, what are they? What additional benefits, besides better inventory control, can be expected?

2. Have your team work together in making a group decision, such as where to eat or the best classes to take next semester. After the meeting, have your team describe the most important features you would include in a software package to support the group's decision. What features described in this chapter would be the least important to your team in making its group decision?

3. Imagine that you and your team have decided to develop an ESS software product to support senior executives in the music recording industry. What are some of the key decisions these executives must make? Make a list of the capabilities that such a system must provide to be useful. Identify at least six sources of external information that will be useful to its users.

● WEB EXERCISES

1. Most companies typically have a number of functional MISs, such as finance. Find the sites of two finance companies, such as a bank or a brokerage company. Compare these sites. Which one do you prefer? How could these sites be improved? (*Hint:* If you are having trouble, try Yahoo. It should have a listing for "Business and Economy" on its home page. From there you can go to "companies" and then "finance." There will be several menu choices from there.) You may be asked to develop a report or send an e-mail message to your instructor about what you found.

2. FedEx and UPS both offer DSSs to help customers track their shipments. Research one of these companies using the Web and describe the Web pages and their features that can help customers track packages as they are shipped from one location to another. You may be asked to develop a report or send an e-mail message to your instructor about what you found.

3. Group decision support systems (GDSSs) permit collaboration and joint work. Groupware, an important software component, allows people to communicate and work on joint projects. Using an Internet search engine, find one or more companies that make groupware. Describe the features of these groupware products. You may be asked to develop a report or send an e-mail message to your instructor about what you found.

● CASES

 Collaborative Work Gives a Competitive Edge

In addition to superior people skills, good project management skills are the hallmark of a good manager. Controlling costs and delivering a project on time is no simple task. But some managers have to juggle hundreds of projects at the same time. Tom Liddell, vice president of operations at Medical Manager Midwest, for example, has responsibility for about 400 projects involving more than 150 people in over ten states. Other managers face similar project management workloads. How do these managers control all of these projects?

As discussed in this chapter, project management models can help a manager monitor and control projects. Software can help a manager determine the best and least expensive way to reduce the time it takes to complete a project. Often called *project crashing*, effectively reducing the amount of project completion time can involve optimization and decision support capabilities. Yet, good project management systems are typically not enough for many project managers. Good group support systems are also needed in many cases.

Because most projects involve meetings, collaboration is essential. As with Medical Manager Midwest, projects can involve over a hundred people who are dispersed over wide areas, requiring a wide area decision network or a teleconferencing approach to group support. Fortunately, software products combine some of the features of project management and group support packages. Medical

Manager Midwest, for example, uses Project Home Page and ActionPlan products to help manage projects using group support capabilities. Both products are available through Netmosphere. These programs link project teams and give them access to centralized information. These programs use the Internet to allow project members to work in different geographic locations, permitting real-time communication and discussion. According to Netmosphere CEO Kevin Nickles, "Most of the traditional views of project management are not collaborative." Project Home Page and ActionPlan attempt to combine the best of project management and group decision support systems.

Discussion Questions

1. Why is project management so important?
2. What features of a GDSS would be important to help manage projects?

Critical Thinking Questions

3. If you were managing a large project with people in the United States and France, what capabilities would you like to see in software to help you manage the project?
4. Search the Internet for project management software. Describe the features you found and the features that are lacking.

Sources: Adapted from Helen Johnson, "A Competitive Edge in Collaborative Work," *Computerworld*, January 31, 2000, p. 68; Andy Donoghue, "Lutus Is Banking on Raven," *Network News*, February 2, 2000, p. 12; and "Project Tracker," *Information Week*, February 7, 2000, p. 137.

 Market Research on the Internet

A marketing MIS is a key competitive tool for most companies. In addition to developing products, advertising products, and determining product prices, an effective marketing MIS requires good marketing research. Traditionally, companies have used surveys, questionnaires, pilot studies, and interviews to determine problems with existing products and services and to explore the potential of proposed products and services. Data collected from these marketing research approaches are typically entered into sophisticated statistical analysis programs. In the past, marketing research was done manually. For example, surveys and questionnaires were mailed out to customers or potential customers. After a few

weeks, the returned surveys and questionnaires were analyzed. Interviews were also conducted by companies that sent staff to shopping malls or other locations where they were likely to find people. Interviewing a few hundred people could take weeks and thousands of dollars. Like surveys and questionnaires, the data would then be entered into statistical analysis programs. Today, some companies are performing marketing research on the Internet.

Internet marketing research seems to be a winning strategy. It can be fast and inexpensive. Responses can take only a few days, and the raw data from the Internet site can be directly entered into statistical programs for analysis. With traditional surveys and questionnaires, the data often has to be manually entered into a computer through a keyboard. In addition to being fast, Internet marketing research can be inexpensive or even free. Questionnaires and surveys can be put on to Web pages. Mailing costs are eliminated for surveys and questionnaires, and a team of interviewers is not needed. With all these advantages, why would any company use more traditional ways to perform marketing research?

Polaroid Corporation decided to use the Internet to perform marketing research. The company asked over 1,500 people about their use of Polaroid cameras and image scanners. The results were amazing. About one-fourth of the people responding to the survey said they had scanned a Polaroid picture. That seemed to be wonderful news until the company uncovered a very disturbing fact when the results were forecast to the entire population. The director of marketing research for Polaroid, Bruce Godfrey, soon realized that "that is more than the number of people who have even taken a Polaroid picture last year!" Something was terribly wrong. People responding to the Internet survey either didn't understand the question or lied. In addition, for any survey to be accurate, it must be representative of the general population. Most people agree that those that surf the Web are not typical consumers. Most Internet users tend to be richer and more technically skilled than the population in general. Marketing research performed on the Internet could thus be seriously biased. To overcome these problems, some companies try to weight Internet results to make them more accurate. To others, however, weighting bad data only results in bad analysis.

Discussion Questions

1. What are the advantages of using the Internet to perform marketing research?
2. What are some of the disadvantages?

Critical Thinking Questions

3. If you were a marketing manager, under what circumstances would you use the Internet to perform marketing research?
4. Comment on the following statement: "The problems of performing marketing research on the Internet can be reduced by combining the results from the Internet with results from traditional marketing research methods."

Sources: Adapted from Erin White, "Market Research on the Internet Has Its Drawbacks," *The Wall Street Journal,* March 2, 2000, p. B4; "Outsourcing E-Commerce Website for Polaroid," *Washington Technology,* January 24, 2000, p. S4; and "Polaroid, Avery Form Marketing Alliance," *Photo Marketing Newsline,* January 19, 2000, p. 1.

 ## Investment Information from Financial Web Sites

The Internet has penetrated most aspects of business and marketing. As discussed in the opening vignette, the "Ethical and Societal Issues" box, and throughout the chapter, financial information is now available over the Internet. The Internet excels at providing and analyzing data. Although general news, sports, and entertainment normally contain pictures and depend on sometimes quirky, subjective personalities, financial information is based on large amounts of more objective raw data and the power to analyze it. These requirements for financial information are ideal for delivery on the Internet, and a number of companies specialize in providing this type of information.

Larry Kramer was tired of the newspaper business and believed the entire industry was in decline. In 1994, he decided to launch an Internet site specializing in providing financial information. The original Internet site was a part of Data Broadcasting, which provided real-time stock quotes. Data Broadcasting gave a 38 percent stake of the company to CBS in return for the CBS brand name and $30 million in promotional advertising on CBS's TV, radio, and billboard operations. Kramer's upstart Internet site became CBS MarketWatch. CBS MarketWatch then made a $21 million deal with America Online to get an even bigger audience. The strategy appears to have worked. Today, CBS MarketWatch offers hard news on stocks, bonds, and the market in general. CBS MarketWatch is the most popular investment information site on the Internet, with more visitors per month than other financial information Internet sites. In December 1999, for example, about 3.5 million people visited the site at least once. With more than 70 reporters and editors, CBS MarketWatch attempts to provide fast, reliable information. Although the market share for CBS MarketWatch has soared, its stock price has had a few bumps along the way. The initial public offering in 1999 started at $17 per share. The share price then climbed up to $120 per share before declining to about $40 per share.

In addition to CBS MarketWatch, a number of other Internet sites offer financial information. The second most visited site is The Motley Fool, which is known for its commentary and investment chat rooms. The Motley Fool had about two million visitors in December 1999. CNN FN, with about 1.3 million visitors, and TheStreet.com, with about 900,000 visitors, are next in terms of visitors. SmartMoney, CNBC, and The Wall Street Journal Online also have a significant number of visitors each month. Although some sites are free—notably, CBS MarketWatch and The Motley Fool—others charge for their news and financial information (e.g., The Wall Street Journal Online). Like many Internet companies, it is not unusual for financial information companies to lose money each year. The hope is to gain market share that one day will turn into profits.

The competition among these firms is fierce. One company charged that one of its competitors tried to drive its stock price down to buy it at a lower price. This charge, however, has not been substantiated.

Discussion Questions

1. Why is financial information ideally placed on the Internet compared with other information and news?
2. What information would you like to see on a financial information Internet site?

Critical Thinking Questions

3. Do you think some financial information Internet companies are trying to drive the price of their competitors' stocks down by reporting misleading or false information? How would you determine whether this was happening?
4. Use the Internet to research several of the companies described in this case. Which one would be best for your financial needs? Describe the features you like about this site and which features you would like to see in the future.

Sources: Adapted from Marc Gunther, "The Top Financial Website Is Nobody's Fool," *Fortune*, February 21, 2000, p. 54; Marcia Vickers, "Call Him The Streetfighter.com," *Business Week*, March 6, 2000, p. 8; and "Motley Falls at Starting Line on Survey," *Money Marketing*, November 11, 1999, p. 1.

● NOTES

Sources for the opening vignette on p. 371: Adapted from Richard Morais, "Instihot," *Forbes,* March 6, 2000, p. 68; "Setting Up Shop," *The Economist,* February 12, 2000; and Stanley Reed, "Reuters Jumps into the Net," *Business Week,* February 21, 2000, p. 55.

1. Maria Trombly, "New Bank's Net-Only Vision," *Computerworld,* April 3, 2000, p. 12.

2. Chip Cummins, "Deutsche Bank, Chase Manhattan Join Move Toward Online Currency Dealing," *The Wall Street Journal,* April 3, 2000, p. A43C.

3. Marcia Stepanek, "Webling," *Business Week,* April 3, 2000, p. EB26.

4. Dennis Berman, "Card Sharps," *Business Week,* April 3, 2000, p. EB68.

5. Dan Carney, "Fraud on the Net," *Business Week,* April 3, 2000, p. EB58.

6. Janet Ginsburg, "Selling Sofas Online Is No Snap," *Business Week,* April 3, 2000, p. 96.

7. David Rocks, "Reinventing Herman Miller," *Business Week,* April 3, 2000, p. EB89.

8. David Rocks, "Reinventing Herman Miller," *Business Week,* April 3, 2000, p. EB89.

9. Cade Metz, "Supply Chain from Excel to the Web," *PC Magazine,* April 18, 2000, p. 23.

10. Craig Stedman, "Online Exchange to Sell Oracle Software," *Computerworld,* March 13, 2000, p. 4.

11. Jay Greene, "No Magic in This Dot.Com Idea," *Business Week,* March 20, 2000, p. 46.

12. James Cope, "Coors Moves Branded Merchandise Online," *Computerworld,* March 13, 2000, p. 14.

13. Julia King, "Businesses Weigh Pros and Cons of Web Marketplaces," *Computerworld,* March 13, 2000, p. 28.

14. Robin Robinson, "Customer Relationship Management," *Computerworld,* February 28, 2000, p. 67.

15. Neil Weinberg, "Not.Coms," *Forbes,* April 17, 2000, p. 424.

16. Kathleen Melymuka, "Marketing Partnership," *Computerworld,* June 21, 1999, p. 70.

17. Linda Rosencrance, "BetaSphere Delivers FedEx Some Customer Feedback," *Computerworld,* April 3, 2000, p. 35.

18. Kathryn Kranhold et al., "Companies Bar Ad Networks from Selling Rich Data Collected as People Surf Sites," *The Wall Street Journal,* April 10, 2000, p. B1.

19. Craig Stedman, "Value-Based Pricing," *Computerworld,* March 13, 2000, p. 58.

20. Richard Dunman, "The Power of Smart Pricing," *Business Week,* April 10, 2000, p. 160.

21 Joan Hamilton, "The Panic Over Hiring," *Business Week,* April 3, 2000, p. EB130.

22. Monica Sambataro, "Just-In-Time Learning," *Computerworld,* April 3, 2000, p. 50.

23. Pfeil et al., "Viseton's Sterling Plant Uses Simulation-Based Decision Support," *Interfaces,* January 2000, p. 115.

24. Richard Curch et al., "Support System Development for Forest Ecosystem Management," *European Journal of Operations Research,* March 1, 2000, p. 247.

25. Traci Hess et al., "Using Autonomous Software Agents to Create the Next Generation of Decision Support Systems," *Decision Sciences,* Winter 2000, p. 1.

26. Peggy King, "Decision Support Grows Up and Out," *CIO,* November 15, 1999, p. 88.

27. Ed Yourdan, "The Value of Triage," *Computerworld,* March 20, 2000, p. 40.

28. Milam Aiken et al., "An Abductive Model of Group Support Systems," *Information & Management,* March 1, 2000, p. 87.

29. Milam Aiken et al., "Flaming Among First-Time Group Support System Users," *Information Management,* March 1, 2000, p. 95.

30. David Essex, "Teamware Offering That's Tailor-Made," *Computerworld,* April 10, 2000, p. 90.

31. Betty Vandenbosch, "An Empirical Analysis of the Association Between the Use of Executive Support Systems and Perceived Organizational Competitiveness," *Accounting, Organizations and Society,* January 1999, p. 7.

CHAPTER 11

Specialized Business Information Systems: Artificial Intelligence, Expert Systems, and Virtual Reality

*A*ccepting what a computer tells you, instead of thinking, is a big mistake.

— Herbert W. Lovelace, "But the Computer Said," *Information Week*, November 8, 1999.

Principles	Learning Objectives
Artificial intelligence systems form a broad and diverse set of systems that can replicate human decision making for certain types of well-defined problems.	• *Define the term* artificial intelligence *and state the objective of developing artificial intelligence systems.* • *List the characteristics of intelligent behavior and compare the performance of natural and artificial intelligence systems for each of these characteristics.* • *Identify the major components of the artificial intelligence field and provide one example of each type of system.*
Expert systems can enable a novice to perform at the level of an expert but must be developed and maintained very carefully.	• *List the characteristics and basic components of expert systems.* • *Identify at least three factors to consider in evaluating the development of an expert system.* • *Outline and briefly explain the steps for developing an expert system.* • *Identify the benefits associated with the use of expert systems.*
Virtual reality systems have the potential to reshape the interface between people and information technology by offering new ways to communicate information, visualize processes, and express ideas creatively.	• *Define the term* virtual reality *and provide three examples of virtual reality applications.*

Ask Jeeves

Search Engine Helps "Humanize" On-Line Experience

Ask Jeeves operates a consumer and corporate Web search service that allows users to seek information on the Internet using standard English questions. In addition, Microsoft, Dell, and other corporate customers employ Ask Jeeves to provide customer support and answer specific questions about their products and services. Using the Web site provides a customized, outsourced service for corporations that want an easier, more intuitive, and more intelligent way of interacting with customers.

Ask Jeeves at Ask.com is the thirteenth most visited Web property, according to Media Metrix, a Web industry watcher. Ask.com answers three million questions each day and receives more than 12 million unique visitors per month, presenting a tremendous opportunity for advertisers to tap into Ask Jeeves's broad and diverse user base.

Once the user asks a question of Jeeves, the process goes like this:

1. Ask Jeeves attempts to understand the precise nature of the question by using a question-processing engine. Ask Jeeves has been programmed to understand and respond to statements made in English, so it determines both the meaning of the words in the question (semantic processing) and the meaning of the grammar in the question (syntactic processing).

2. Ask Jeeves's answer-processing engine provides a question template response—a list of questions that users see after they ask Jeeves a question. When users click on a question that closely matches theirs, the answer-processing engine retrieves an answer template that contains links to locations that can answer the question.

The Ask Jeeves knowledge base contains links to more than seven million answers, which contain information about the most frequently asked questions on the Internet. Smart lists allow one question template to point to many answers (e.g., "What is the population of <city name>?").

To assure users that they'll get the best answers to their questions, Ask Jeeves also has partners who provide answers to supplement Jeeves's own answers. This additional resource guarantees that the user can find the best answer; it includes the top ten answers from other leading search engines.

As it does its searches, Ask Jeeves captures customer intelligence from the questions they ask, the language they use to ask those questions, and the items they select in the Ask Jeeves system. This information is used to better understand what customers need and want, what is successful on its system, and what Ask Jeeves can't provide. So Ask Jeeves is constantly improving its responses to people's questions.

A new extension of the technology is a decision support engine that enables users to solve complex problems such as buying a digital camera, selecting annuities, and choosing among healthcare programs. The decision support engine engages in a dialogue concerning preferences leading beyond simple questions and answers and document location.

As you read this chapter, consider the following:

- What does it take to design and implement an effective artificial intelligence system?

- What business needs can be addressed by an expert system?

- What are some practical applications of virtual reality?

At a Dartmouth College conference in 1956, John McCarthy proposed the use of the term *artificial intelligence (AI)* to describe computers with the ability to mimic or duplicate the functions of the human brain. Many AI pioneers attended this first conference; a few predicted that computers would be as "smart" as people by the 1960s. The prediction has not yet been realized, but the benefits of artificial intelligence in business and research can be seen today. Advances in AI have led to many practical applications of systems (like the one used at Ask Jeeves) that are capable of making complex decisions.

AN OVERVIEW OF ARTIFICIAL INTELLIGENCE

Science fiction novels and popular movies have featured scenarios of computer systems and intelligent machines taking over the world. Computer systems such as Hal in the classic movie *2001: A Space Odyssey* are futuristic glimpses of what might be. These accounts are fictional, but we see the real application of many computer systems that use the notion of AI. These systems help to make medical diagnoses, explore for natural resources, determine what is wrong with mechanical devices, and assist in designing and developing other computer systems. In this chapter we explore the exciting applications of artificial intelligence and virtual reality and look at what the future really might hold.

Artificial Intelligence in Perspective

artificial intelligence systems

people, procedures, hardware, software, data, and knowledge needed to develop computer systems and machines that demonstrate characteristics of intelligence

Artificial intelligence systems include the people, procedures, hardware, software, data, and knowledge needed to develop computer systems and machines that demonstrate characteristics of intelligence. Researchers, scientists, and experts on how humans think are often involved in developing these systems. The objective in developing contemporary AI systems is not to replace human decision making completely but to replicate it for certain types of well-defined problems. As with other information systems, the overall purpose of artificial intelligence applications in business is to help an organization achieve its goals.

Science fiction movies give us a glimpse of the future, but many practical applications of artificial intelligence exist today, among them medical diagnostics, mechanical diagnostics, and development of computer systems.
(Source: Photofest)

The Nature of Intelligence

From the early AI pioneering stage, the research emphasis has been on developing machines with **intelligent behavior**. Some of the specific characteristics of intelligent behavior include the ability to do the following:

Learn from experience and apply the knowledge acquired from experience. Being able to learn from past situations and events is a key component of intelligent behavior and is a natural ability of humans, who learn by trial and error. However, learning from experience is not natural for computer systems. This ability must be carefully programmed into the system. Today, researchers are developing systems that have this ability. For instance, computerized AI chess games can learn to improve their game while they play human competitors (Figure 11.1).

In addition to learning from experience, people apply what they have learned to new settings and circumstances. Often, individuals take what they learn and succeed with in one endeavor and apply it to another. For example, a company that developed a dish-washing product effective for cleaning greasy dishes developed a variation of the product for use in cleaning up messy highway spills. Although humans have the ability to apply what they have learned to new settings, this characteristic is not automatic with computer systems. Developing computer programs to allow computers to apply what they have learned can be difficult.

Handle complex situations. Humans are involved in complex situations. World leaders face difficult political decisions regarding conflict, global economic conditions, hunger, and poverty. In a business setting, top-level managers and executives are faced with a complex market, difficult and challenging competitors, intricate government regulations, and a demanding workforce. Even human experts make mistakes in dealing with these situations. Developing computer systems that can handle perplexing situations requires careful planning and elaborate computer programming.

Solve problems when important information is missing. The essence of decision

intelligent behavior

the ability to learn from experience and apply knowledge acquired from experience, handle complex situations, solve problems when important information is missing, determine what is important, react quickly and correctly to a new situation, understand visual images, process and manipulate symbols, be creative and imaginative, and use heuristics

FIGURE 11.1

Computers like Deep Blue attempt to learn from past chess moves. The powerful supercomputer's logic system was able to calculate the ramifications of up to 100 billion chess maneuvers within the allotted time for each move.
(Source: Photo courtesy of the Association for Computing Machinery.)

making is dealing with uncertainty. Quite often, decisions must be made even when we lack information or have inaccurate information, because obtaining complete information is too costly or impossible. You have probably seen movies in which computers have responded to human commands with statements like "Does not compute" and "Insufficient information." Today, AI systems can make important calculations, comparisons, and decisions even when missing information.

Determine what is important. Knowing what is truly important is the mark of a good decision maker. Every day we are bombarded with facts and must process large amounts of data, filtering out what is unnecessary. Determining which items are crucial can make the difference between good decisions and those that ultimately lead to problems or failures. Computers, on the other hand, do not have this natural ability. Developing programs and approaches to allow computer systems and machines to identify important information is not a simple task.

React quickly and correctly to a new situation. A small child can look over a ledge or a drop-off and know not to venture too close. The child reacts quickly and correctly to a new situation. Computers, on the other hand, do not have this ability without complex programming.

Understand visual images. Interpreting visual images can be extremely difficult, even for sophisticated computers. People and animals can look at objects interacting in our environment and understand exactly what is going on. For instance, we can see a man sitting at a table and know that he has legs and feet that we cannot see. Being able to understand and correctly interpret visual images is an extremely complex process for computer systems. Moving through a room of chairs, tables, and other objects can be trivial for people but extremely complex for machines, robots, and computers. Such machines require an extension of understanding visual images, called a **perceptive system**. Having a perceptive system allows a machine to approximate the way a human sees, hears, and feels objects.

perceptive system

a system that approximates the way a human sees, hears, and feels objects

Process and manipulate symbols. People see, manipulate, and process symbols every day. Visual images provide a constant stream of information to our brains. By contrast, computers have difficulty handling symbolic processing and reasoning. Although computers excel at numerical calculations, they aren't as good at dealing with symbols and three-dimensional objects. Recent developments in machine-vision hardware and software, however, allow some computers to process and manipulate symbols on a limited basis.

Be creative and imaginative. Throughout history, some people have turned difficult situations into advantages by being creative and imaginative. For instance, when shipped a lot of defective mints with holes in the middle, an enterprising entrepreneur decided to market these new mints as Lifesavers instead of returning them to the manufacturer. Ice cream cones were invented at the St. Louis World's Fair when an imaginative store owner decided to wrap ice cream with a waffle from his grill for portability. Developing new and exciting products and services from an existing (perhaps negative) situation is a human characteristic. Few computers have the ability to be truly imaginative or creative in this way, although software has been developed to enable a computer to write short stories.

Use heuristics. With some decisions, people use heuristics (rules of thumb arising from experience) or even guesses. In searching for a job, we may decide to rank companies we are considering according to profits per employee. Companies making more profits might pay their employees more. In a manufacturing setting, a corporate president may decide to look at only certain locations for a new plant. We make these types of decisions using general rules of thumb, without completely searching all alternatives and possibilities. Today, some computer systems also have this ability. They can, given the right programs, obtain good solutions that use approximations instead of trying to search for an optimal solution, which would be technically difficult or too time-consuming.

This list of traits only partially defines intelligence. Unlike virtually every other field of information systems research in which the objectives can be clearly defined, the term *intelligence* is a formidable stumbling block. One of the problems in artificial intelligence is arriving at a working definition of real intelligence with which to compare the performance of an artificial intelligence system.

The Difference between Natural and Artificial Intelligence

Since the term *artificial intelligence* was defined in the 1950s, experts have disagreed about the difference between natural and artificial intelligence. For instance, is there a difference between carbon life (human or animal life) and silicon life (a computer chip) in terms of behavior? Profound differences exist, but they are declining in number (Table 11.1). One of the driving forces behind AI research is an attempt to understand how humans actually reason and think. It is believed that the ability to create machines that can reason will be possible only once we truly understand our own processes for doing so.

The Major Branches of Artificial Intelligence

AI is a broad field that includes several specialty areas, such as expert systems, robotics, vision systems, natural language processing, learning systems, and neural networks (Figure 11.2). Many of these areas are related; advances in one can occur simultaneously with or result in advances in others.

Expert Systems

An *expert system* consists of hardware and software that stores knowledge and makes inferences, similar to a human expert. Because of their many business applications, expert systems are discussed in more detail in the next several sections of the chapter.

TABLE 11.1

A Comparison of Natural and Artificial Intelligence

Attributes	Natural Intelligence (Human)	Artificial Intelligence (Machine)
The ability to use sensors (eyes, ears, touch, smell)	High	Low
The ability to be creative and imaginative	High	Low
The ability to learn from experience	High	Low
The ability to be adaptive	High	Low
The ability to afford the cost of acquiring intelligence	High	Low
The ability to use a variety of information sources	High	High
The ability to acquire a large amount of external information	High	High
The ability to make complex calculations	Low	High
The ability to transfer information	Low	High
The ability to make a series of calculations rapidly and accurately	Low	High

FIGURE 11.2

A Conceptual Model of
Artificial Intelligence

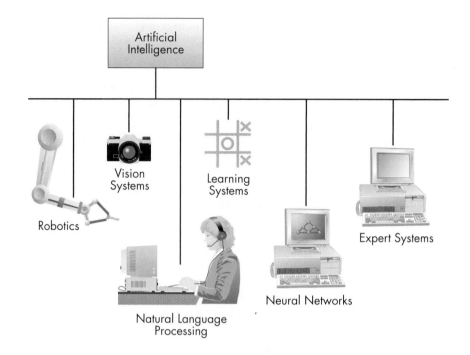

robotics

mechanical or computer devices
that perform tasks requiring a high
degree of precision or that are
tedious or hazardous for humans

FIGURE 11.3

Robots can be used in situations
that are hazardous, repetitive,
or difficult to do in other ways.
(Source: Courtesy of ABB Flexible
Automation.)

Robotics

Robotics involves developing mechanical or computer devices that can paint cars, make precision welds, and perform other tasks that require a high degree of precision or are tedious or hazardous for humans. Contemporary robotics combines both high-precision machine capabilities with sophisticated controlling software. The controlling software in robots is what is most important in terms of AI. The brain in an advanced industrial robot today works at about ten million instructions per second (MIPS)—no smarter than an insect. To achieve anything even approaching human intelligence, the robot brain must achieve 100 trillion operations per second.[1]

Many applications of robotics exist, and research into these unique devices continues (Figure 11.3). Manufacturers use robots to assemble and paint products. Welding robots have enabled firms to manufacture top-quality products and reduce labor costs while shortening delivery time to their customers.

In addition to their use in supporting manufacturing operations, robots have been applied in more unusual situations. Standardized Teleoperation System (STS) is a modular kit of robotic controls for converting a vehicle to unmanned operation. STS controls permit normal (or manual) vehicle operation and unmanned operation with the flick of a switch. The strap-on kit feature converts existing vehicles; vehicles can also be delivered with the STS installed. STS controls are used in many types and classes of vehicles, including bulldozers, tanks, and trucks used for bomb and landmine detection, neutralization, detonation, clearing, and route proofing. Some of the more unusual applications of robotics include a Ping-Pong ball server for the U.S. Olympic team, a toy soldier marching machine, and a bug sorter for museum display.

Although robots are essential components of today's automated manufacturing systems, future robots will find wider applications outside the factory in banks,

Dragon Systems' Naturally Speaking Preferred Edition 5 uses continuous voice recognition or natural speech, allowing the user to speak to the computer at a normal pace without pausing between words. The spoken words are transcribed immediately onto the computer screen.

(Source: Courtesy of Dragon Systems, Inc., a Lernout & Hauspie Company)

vision system

the hardware and software that permit computers to capture, store, and manipulate visual images and pictures

natural language processing

processing that allows the computer to understand and react to statements and commands made in a "natural" language, such as English

restaurants, homes, and hazardous working environments such as nuclear stations. A robot must not only execute tasks programmed by the user but must also be able to interact with its environment through its sensors and actuators, sense and avoid unforeseen obstacles, and perform its duties much the same way as humans.

Vision Systems

Another area of AI involves vision systems. **Vision systems** include hardware and software that permit computers to capture, store, and manipulate visual images and pictures. The U.S. Justice Department uses vision systems to perform fingerprint analysis, with almost the same level of precision as human experts. The speed with which the system can search through the huge database of fingerprints has brought quick resolution to many long-standing mysteries. Vision systems are also effective at identifying people based on facial features.

Vision systems can be used in conjunction with robots to give these machines "sight." Robots such as those used in factory automation typically perform mechanical tasks with little or no visual stimuli. Robotic vision extends the capability of these systems, allowing the robot to make decisions based on visual input. Generally, robots with vision systems can recognize black and white and some gray-level shades but do not have good color or three-dimensional vision. Other systems concentrate on only a few key features in an image, ignoring the rest. It may take years before a robot or other computer system can "see" in full color and draw conclusions from what it sees, the way humans do.

Natural Language Processing

As discussed in Chapter 4, **natural language processing** allows a computer to understand and react to statements and commands made in a "natural" language, such as English. There are three levels of voice recognition: command (recognizes dozens to hundreds of words), discrete (recognizes dictated speech with pauses between words), and continuous (recognizes natural speech). For example, a natural language processing system can be used to retrieve important information without typing in commands or searching for keywords. With natural language processing, it is possible to speak into a microphone connected to a computer and have the computer convert the electrical impulses generated from the voice into text files or program commands. With some simple natural language processors, you say a word into a microphone and type the same word on the keyboard. The computer then matches the sound with the typed word. With more advanced natural language processors, recording and typing words is not necessary.

Brokerage services are a perfect fit for voice-recognition technology to replace the existing "press 1 to buy or sell a stock" touch-pad telephone menu system. People buying and selling stock use a vocabulary too varied for easy access through menus and touch pads but small enough for software to process in real time. Several brokerages—including Charles Schwab, Fidelity Investments, DLJdirect, and TD Waterhouse Group—offer voice-recognition services. One of the big advantages is that the number of calls routed to the customer service department drops considerably once the new voice features are added. That is desirable to brokerages because market volatility means that the call centers are often either overstaffed or understaffed.

Within the next few years, voice recognition is expected to spread to all automated phone-answering systems. The main holdup is price and ease of installation—and the major vendors are working on both of these issues.[2]

learning system

a combination of software and hardware that allows the computer to change how it functions or reacts to situations based on feedback it receives

Learning Systems

Another part of AI deals with **learning systems**, a combination of software and hardware that allows the computer to change how it functions or reacts to situations based on feedback it receives. For example, some computerized games have learning abilities. If the computer does not win a particular game, it remembers not to make the same moves under the same conditions. Learning systems software requires feedback on the results of its actions or decisions. At a minimum, the feedback needs to indicate whether the results are desirable (winning a game) or undesirable (losing a game). The feedback is then used to alter what the system will do in the future.

neural network

a computer system that can simulate the functioning of a human brain

Neural Networks

An increasingly important aspect of AI involves neural networks. A **neural network** is a computer system that can simulate the functioning of a human brain. The systems use massively parallel processors in an architecture that is based on the human brain's own meshlike structure. In addition, neural network software can be used to simulate a neural network using standard computers. Neural networks can process many pieces of data at once and learn to recognize patterns. The systems then program themselves to solve related problems on their own. Some of the specific features of neural networks include the following:

- The ability to retrieve information even if some of the neural nodes fail
- Fast modification of stored data as a result of new information
- The ability to discover relationships and trends in large databases
- The ability to solve complex problems for which all the information is not present

Neural networks excel at pattern recognition. For example, neural network computers can be used to read bank check bar codes despite smears or poor-quality printing. Some hospitals use neural networks to determine a patient's likelihood of contracting cancer or other diseases. Neural nets work particularly well when it comes to analyzing detailed trends. Large amusement parks and banks use neural networks to figure out staffing needs based on customer traffic—a task that requires precise analysis, down to the half-hour. Increasingly, businesses are firing up neural nets to help them navigate ever-thicker forests of data and make sense of myriad customer traits and buying habits. Computer Associates has developed Neugents, neural intelligence agents, which "learn" patterns and behaviors and predict what will happen next. For example, Neugents can be used to track the habits of insurance customers and predict which ones will not renew, say, an automobile policy. They can then suggest to an insurance agent what changes might be made in the policy to get the consumer to renew it. The technology also can be employed to track individual users and their on-line preferences so that users at e-commerce sites don't have to input the same information each time they log on—their purchasing history and other data will be factored in each time they access a Web site.[3] Standard & Poor's (S&P) began using a neural network system called Decider from United Kingdom–based Neural Technologies to launch a credit-rating system on the Internet. The system, called CreditModel, allows S&P customers (banks and asset managers) to identify the firm for which they want a rating and then to launch the Decider engine, which develops a credit "score" on that company. It analyzes a list of variables, such as the company's debt load, revenue, and credit ratings that S&P collected in its files over the years—all at a fraction of the cost of a full rating.[4]

Read the "Ethical and Societal Issues" special interest box to learn how neural networks might help disabled people.

ETHICAL AND SOCIETAL ISSUES
Neural Networks Provide Hope

It has been estimated that computers that can exhibit humanlike intelligence—including musical and artistic aptitude, creativity, the ability to move physically through the world, and emotional responsiveness—require processing power on the order of 20 million billion calculations per second. Even more challenging than developing the computer hardware with such capacity is developing software to mimic a human. Perhaps it will be accomplished through essentially reverse-engineering the human brain, but we won't be able to crack the human brain's complexity anytime soon. It will be at least 2030 before we can design such computers. Much more promising is the use of neural network technology, which is being applied to assist disabled people.

The disabling effects of spinal injury or degenerative disease on voluntary movement can be permanent because damaged nerve cells and their "wiring" cannot regenerate. However, the motor areas of the brain that control body movements are usually left intact. Research scientists are exploring whether the activity of these motor areas can be used to operate robotic limbs.

John Chapin and his colleagues at MCP Hahnemann Medical College in Philadelphia trained a rat to press a lever for a food reward that was delivered by a robotic device. A 16-electrode array implanted in the rat's brain recorded the activity of 30 neurons that controlled the muscles involved in the lever press. The team then used these recordings to "train" a neural network to recognize the brain activity patterns that occur during a lever press. Control of the robotic device was switched from the lever to the neural network driven by the rat's thought patterns. Eventually the rat learned that pressing the lever with its paw was unnecessary and stopped its paw movements. Its brain activity continued to drive the robotic arm, bypassing nerves and muscles altogether.

The next stage of research involves the use of monkeys rather than rats and larger electrode arrays to record up to 130 movement-related neurons simultaneously. Such simultaneous recording is critical, because a neuron's activity is not specific to a particular muscle contraction,

so it cannot give complete directions for movements by itself. If that experiment can succeed, the team could encode directions for a more complex robotic device.

Although it now seems possible that direct brain control of robotic actions could assist those disabled by spinal damage, transferring the technology from animal experiments to human use will be extremely difficult. Use of a neural network to control a robotic limb would require paralyzed patients to learn, through trial and error, how to shape brain activity appropriate for driving it. Another complication is that although electrodes can be anchored to the skull, they cannot be "hardwired" to the neurons themselves. Without permanent connections, the electrode tips and neurons could move slightly relative to one another and short-circuit the whole system. Perhaps the biggest obstacle in creating neural signal-based actuators is the design of multielectrode arrays that are both stable and safe for humans over the long term.

Discussion Questions

1. How probable do you think it is that we will soon see a humanlike robot such as those depicted in the *Terminator* movies?
2. What are the biggest barriers to development of successful direct brain control of robotic actions?

Critical Thinking Questions

3. Would mastery of direct brain control of robots lead to development of artificial limbs stronger and faster than natural limbs? Why or why not?
4. Would there be a need for an "emergency short-circuit" of direct brain control robots in the event that their owners lost their temper and had brief flashes of terrible thoughts?

Sources: Adapted from Ray Kurzweil, "Will My PC Be Smarter Than I Am?" *Time*, June 19, 2000, pp. 82–84; and Mimi Zucker, "Mind over Matter," *Scientific American*, November 1999, accessed at http://www.sciam.com/1999/1199issue/1199techbus2.html.

AN OVERVIEW OF EXPERT SYSTEMS

As discussed, an expert system behaves similarly to a human expert in a particular field. Computerized expert systems have been developed to diagnose problems, predict future events, and solve energy problems. They have also been used to design new products and systems, determine the best use of lumber, and increase the quality of healthcare. Like human experts, computerized expert systems use heuristics, or rules of thumb, to arrive at conclusions or make suggestions. Expert systems have also been used to

FIGURE 11.4

Credit card companies often use expert systems to determine credit limits for credit cards. (Source: David Young Wolff/Tony Stone Images.)

determine credit limits for credit cards (Figure 11.4). The research conducted in AI during the past two decades is resulting in expert systems that explore new business possibilities, increase overall profitability, reduce costs, and provide superior service to customers and clients.

Characteristics of an Expert System

Expert systems have a number of characteristics and capabilities, including the following:

Can explain their reasoning or suggested decisions. A valuable characteristic of an expert system is the capability to explain how and why a decision or solution was reached. For example, the expert system can explain the reasoning behind the conclusion to approve a particular loan application. The ability to explain its reasoning processes can be the most valuable feature of a computerized expert system. The user of the expert system thus gains access to the reasoning behind the conclusion.

Can display "intelligent" behavior. Considering a collection of data, an expert system can propose new ideas or approaches to problem solving. A few of the applications of expert systems are an imaginative medical diagnosis based on a patient's condition, a suggestion to explore for natural gas at a particular location, and job counseling for workers.

Can draw conclusions from complex relationships. Expert systems can evaluate complex relationships to reach conclusions and solve problems. For example, one proposed expert system will work with a flexible manufacturing system to determine the best use of tools. Another expert system can suggest ways to improve quality control procedures.

Can provide portable knowledge. One unique capability of expert systems is that they can be used to capture human expertise that might otherwise be lost. A classic example of this is the expert system called DELTA (Diesel Electric Locomotive Troubleshooting Aid), which was developed to preserve the expertise of the retiring David Smith, the only engineer competent to handle many highly technical repairs of such machines.

Can deal with uncertainty. One of an expert system's most important features is its ability to deal with knowledge that is incomplete or not completely accurate. The system deals with this problem through the use of probability, statistics, and heuristics.

Even though these characteristics of expert systems are impressive, other characteristics limit their current usefulness. Many of these limiting characteristics are related to concerns of cost, control, and complexity. Some of these characteristics are as follows:

Not widely used or tested. Even though successes occur, expert systems are not used in a large number of organizations. In other words, they have not been widely tested in corporate settings.

Difficult to use. Some expert systems are difficult to control and use. In some cases, the assistance of computer personnel or individuals trained in the use of expert systems is required to help the user get the most from these systems. Today's challenge is to make expert systems easier to use by decision makers who have limited computer programming experience.

Limited to relatively narrow problems. Whereas some expert systems can perform complex data analysis, others are limited to simple problems. Furthermore, many problems solved by expert systems are not that beneficial in business settings. An expert system designed to provide advice on how to repair a machine, for example, is unable to assist in decisions about when or whether to repair it. In general, the narrower the scope of the problem, the easier it is to implement an expert system to solve it.

Cannot readily deal with "mixed" knowledge. Expert systems cannot easily handle a knowledge base that has a mixed representation. Knowledge can be represented through defined rules, through comparison to similar cases, and in various other ways. An expert system in one application might not be able to deal with knowledge that combined both rules and cases.

Possibility of error. Although some expert systems have limited abilities to learn from experience, the primary source of knowledge is a human expert. If this knowledge is incorrect or incomplete, it will affect the system negatively. Other development errors involve programming. Because expert systems are more complex than other information systems, the potential for such errors is greater.

Cannot refine own knowledge base. Expert systems are not capable of acquiring knowledge directly. A programmer must provide instructions to the system that determine how the system is to learn from experience. Also, some expert systems cannot refine their own knowledge bases—such as eliminating redundant or contradictory rules.

Difficult to maintain. Related to the preceding point is that expert systems can be difficult to update. Some are not responsive or adaptive to changing conditions. Adding new knowledge and changing complex relationships may require sophisticated programming skills. In some cases, a spreadsheet used in conjunction with an expert system shell can be used to modify the system. In others, upgrading an expert system can be too difficult for the typical manager or executive. Future expert systems are likely to be easier to maintain and update.

May have high development costs. Expert systems can be expensive to develop when using traditional programming languages and approaches. Development costs can be greatly reduced through the use of software for expert system development. **Expert system shells**, a collection of software packages and tools used to develop expert systems, can be implemented on most popular PC platforms to reduce development time and costs.

Raise legal and ethical concerns. People who make decisions and take action are legally and ethically responsible for their behavior. For example, a person can be taken to court and punished for a crime. When expert systems are used to make decisions or help in the decision-making process, who is legally and ethically responsible? The human experts used to develop the

expert system shell

a collection of software packages and tools used to develop expert systems knowledge base, the expert system developer, the user, or someone else? For example, if a doctor uses an expert system to make a diagnosis and the diagnosis is wrong, who is responsible? These legal and ethical issues have not been completely resolved.

Capabilities of Expert Systems

Compared with other types of information systems, expert systems offer a number of powerful capabilities and benefits. For example, one expert system, called XCON, is often used in designing computer system configurations because it consistently does a better job than human beings.

Expert systems can be used to solve problems in every field and discipline and can assist in all stages of the problem-solving process. Past successes have shown that expert systems are good at strategic goal setting, planning, design, decision making, quality control and monitoring, and diagnosis (Figure 11.5). Read the "E-Commerce" box to learn how bots are being used to support corporate buyers.

Strategic Goal Setting

Developing strategic goals for an organization is one of the most important functions of top-level decision makers. Strategic goals provide a framework for all other activities throughout the organization. Expert systems can suggest strategic goals and explore the impact of adopting them. Such goals can include identifying opportunities in the marketplace, analyzing the strengths of the existing organization, determining the power and position of competitors, and understanding the existing labor force. For example, say a California wine maker is currently perceived as a low-cost/low-quality producer. An expert system can help the company's top-level management determine costs and benefits involved in producing higher-quality wines and changing its image in the marketplace.

Planning

Expert systems have been employed to assist in the planning process. The ability to reach overall corporate objectives, the impact of plans on organizational resources, and the ways specific plans will help an organization compete in the marketplace can be investigated via expert systems. A manufacturing company, for example, might be exploring the possibility of building a new plant. An expert system can assist with this planning process by suggesting factors that should be considered in making the final decision, based on facts supplied by management.

Design

Designing new products and services requires experience, judgment, and an understanding of the marketplace. Some expert systems have been developed to assist in designing a variety of products, such as computer chips and systems. These types of expert systems use general design principles, an understanding of manufacturing procedures, and a collection of design rules.

Decision Making

Wouldn't it be nice to have an expert help us make our day-to-day decisions? Expert systems have provided this type of support for many individuals and organizations. Acting as advisors or counselors, these systems can suggest possible alternatives, ways of looking at problems, and logical approaches to the decision-making process. In addition, expert systems can improve the learning process for those who are not as experienced in decision making.

FIGURE 11.5

Solutions Offered by Expert Systems

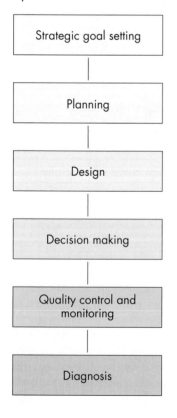

Strategic goal setting

Planning

Design

Decision making

Quality control and monitoring

Diagnosis

As previously discussed, bots (short for "knowledge robots") are programs that help users accomplish certain tasks. Bots can be written in a variety of programming languages and feature triggers that allow them to execute without human intervention. Most bots also can adapt to learn users' tendencies and preferences and so become personalized based on what they learn about users.

There are plenty of bots in use today. The beady-eyed paper clip seen in Windows 98 applications is an example of a bot. The most basic type of bot simply does the same routine task every day, such as backing up a hard drive for a server every night. More powerful bots can roam the Internet on your behalf and automatically perform intelligent searches, answer questions, tell you when an event occurs, provide individualized news delivery, and comparison shop. Indeed, experts see procurement as the killer application of business-to-business bots.

For example, the comparison shopping robot MySimon (http://www.mysimon.com/index.jhtml) doesn't sell anything; rather, it compares millions of products and prices at thousands of on-line stores that offer a wide range of products—appliances, credit cards, chocolates, electronics, and toys. Type in "Paul Garmirian cigars," and it lists 25 merchants. The results can be sorted by merchant, manufacturer, model, or price. You can also find out a product's price history. There are buyer guides for in-depth information and product suggestions based on your preferences. Now extend the MySimon model to corporate purchasing and procurement, and you can see why bots are a hot issue for businesses. Such databots use neural networks and exhibit characteristics usually associated with intelligent life forms, such as learning and creativity.

Much work remains before bots realize their potential. If bots are to collaborate and form a vast network of super corporate buyer agents, standards must be developed and adopted to address communication and security issues. The Defense Advanced Research Projects Agency is working on developing the Knowledge Query and Manipulation Language (KQML), which is both a message format and a message-handling protocol to support knowledge sharing among agents. KQML can be used as a language for an application program to interact with a bot or for two or more bots to share knowledge for cooperative problem solving. KQML provides a basic architecture for knowledge sharing through a special class of agent called *communication facilitators,* which coordinate the interactions of other bots.

In addition to complicated technical problems, there are legal and ethical issues as well. For example, in May 2000, U.S. District Court Judge Ronald Whyte barred Burlington, Massachusetts–based Bidder's Edge from using an automated system to search eBay on the grounds that it could slow the auction giant's site. The decision could also have broader implications for the openness of the Internet, because it relies on laws against trespass, not copyright infringement.

Discussion Questions

1. How might a communication facilitator bot assist a procurement bot in carrying out its role?
2. Why is it important for a procurement bot to exhibit the ability to learn?

Critical Thinking Questions

3. Why is it important for bots to be able to communicate with one another? What would be the impact of a "misinformation" robot that, when unleashed, provided incorrect information to other bots?
4. If the use of procurement bots proves highly successful, how might the role of the human purchasing agent change? Can you identify a set of products or services for which bots would not be effective?

Sources: Adapted from Steve Ulfelder, "Undercover Agents," *Computerworld,* June 5, 2000, p. 85; George Lawton, "Putting Agents to Work," *Knowledge Management,* November 1999, pp. 68–73; "What Is KQML," KQML Lab Web site, http://www.csee.umbc.edu/kqml/whats-kqml.html, accessed June 16, 2000; and Linda Rosencrance and Melissa Solomon, "U.S. Judge Blocks Web Bot from eBay Site," *Computerworld,* May 26, 2000, http://www.computerworld.com/.

Quality Control and Monitoring

Measuring the quality of products and services, determining whether an existing computer system is operating as intended, analyzing the efficiency of a manufacturing plant, and determining the overall effectiveness of a hospital or nursing home are some of the abilities of monitoring systems. Computerized expert systems can assist in monitoring various systems and proposing solutions to system problems. Expert systems can also be used to monitor product quality. When machines are malfunctioning, the expert system can assist in determining possible causes.

FIGURE 11.6

In a chemical plant, an expert system can be used to monitor machinery and predict potential problems.
(Source: Image © Copyright PhotoDisc, 1998.)

Diagnosis

Monitoring and diagnosis go hand in hand. Monitoring determines the current state of a system; diagnosis looks at the causes and proposes solutions. In medicine, expert systems have been employed to diagnose difficult patient conditions. An expert system can analyze test results and patient symptoms. Some systems put probability estimates on potential diseases, given the data and analysis performed. An expert system can provide the doctor with the probable cause of the medical problem and propose treatments or interventions. In a business setting, an expert system can diagnose potential problems of, for example, a chemical distillation facility that is not operating as expected or desired (Figure 11.6).

When to Use Expert Systems

Sophisticated expert systems can be difficult, expensive, and time-consuming to develop. This is especially true for large expert systems implemented on mainframes. Thus, it is important to make sure that the potential benefits are worth the effort and that various expert system characteristics are balanced in terms of cost, control, and complexity.

Following is a list of factors that normally make expert systems worth the expenditure of time and money:

- Provide a high potential payoff or significantly reduced downside risk
- Can capture and preserve irreplaceable human expertise
- Can develop a system more consistent than human experts
- Can provide expertise needed at a number of locations at the same time or in a hostile environment that is dangerous to human health
- Can provide expertise that is expensive or rare
- Can develop a solution faster than human experts can
- Can provide expertise needed for training and development to share the wisdom and experience of human experts with a large number of people

Components of Expert Systems

An expert system consists of a collection of integrated and related components, including a knowledge base, an inference engine, an explanation facility, a knowledge base acquisition facility, and a user interface. A diagram of a typical expert system is shown in Figure 11.7. In this figure, the user interacts with the user interface, which interacts with the inference engine. The inference engine interacts with the other expert system components. These components must work together in providing expertise.

The Knowledge Base

The *knowledge base* stores all relevant information, data, rules, cases, and relationships used by the expert system. A knowledge base must be developed for each unique application. For example, a medical expert system will contain facts about diseases and symptoms. The knowledge base can include generic knowledge from general theories that have been established over time and specific knowledge that comes from more recent experiences and rules of thumb. Knowledge bases, however, go far beyond simple facts, also

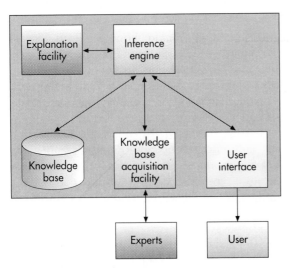

FIGURE 11.7

Components of an Expert System

if-then statements

rules that suggest certain conclusions

fuzzy logic

a special research area in computer science that allows shades of gray and does not require conditions to be black/white, yes/no, or true/false

storing relationships, rules or frames, and cases. For example, certain telecommunications network problems may be related or linked; one problem may cause another. In other cases, rules suggest certain conclusions, based on a set of given facts. In many instances, these rules are stored as **if-then statements**, such as "If a certain set of network conditions exists, then a certain network problem diagnosis is appropriate." Cases can also be used. This technique involves finding instances, or cases, that are similar to the current problem and modifying the solutions to these cases to account for any differences between the previously solved cases stored in the computer and the current situation or problem.

Purpose of a knowledge base. The overall purpose of the knowledge base is to hold the relevant facts and information for the specific expert system. A knowledge base is similar to the sum of a human expert's knowledge and experience gained through years of work in a specific area or discipline. The goal of the system is to capture as much experience and knowledge as possible.

Consider an expert system that locates hardware problems for a large mainframe computer system. A human expert will know a large number of facts about the system. The human expert will also look for specific problems that have occurred frequently in the past. Human experts use a number of rules to help locate problems. If the malfunctioning computer displays certain types of behaviors, then the human inspects certain parts of the hardware for potential problems. A knowledge base developed to identify hardware problems also contains important facts on the hardware, information on frequent problems, and relationships between computer performance and what may be wrong.

Assembling human experts. One challenge in developing a knowledge base is to assemble the knowledge of multiple human experts. Typically, the objective in building a knowledge base is to integrate the knowledge of individuals with similar expertise (e.g., many doctors may contribute to a medical diagnosis knowledge base). A knowledge base that contains information from numerous experts can be extremely powerful and accurate in terms of its predictions and suggestions. Unfortunately, human experts can disagree on important relationships and interpretations of data, presenting a dilemma for designers and developers of knowledge bases and expert systems in general. Some human experts are more expert than others; their knowledge, experience, and information are better developed and more accurately represent reality. When human experts disagree on important points, it can be difficult for expert systems developers to determine which rules and relationships to place in the knowledge base.

The use of fuzzy logic. Another challenge for expert system designers and developers is capturing knowledge and relationships that are not precise or exact. Computers typically work with numerical certainty; certain input values will always result in the same output. In the real world, as you know from experience, this is not always the case. To handle this dilemma, a specialty research area in computer science, called **fuzzy logic**, has been developed. Research into fuzzy logic has been going on for decades, but application to expert systems is just beginning to show results in a variety of areas.

Instead of the usual black and white, yes/no, or true/false conditions of typical computer decisions, fuzzy logic allows shades of gray, or what are known as "fuzzy sets." The criteria on whether a subject or situation fits into a set are given in percentages or probabilities. For example, a weather forecaster might state that "if it is very hot with high humidity, the likelihood of rain is 75 percent." The imprecise terms of "very hot" and "high humidity" are what fuzzy logic must determine to formulate the chance of rain. Fuzzy logic rules help computers evaluate the imperfect or imprecise conditions they encounter and make "educated guesses" based on the likelihood or probability of correctness of the decision. This ability to estimate whether a condition fits a situation more closely resembles the judgment a person makes when evaluating situations.

Fuzzy logic is used in embedded computer technology—for example, in auto-focus cameras, medical equipment that monitors patients' vital signs and makes automatic corrections, and temperature sensors attached to furnace controls.

Fuzzy logic was first applied in Japan. Seiji Yasunobu used it in an automatic control system for the city of Sendai's subway system. Even in a country famed for the precision of its underground railways, Sendai's is impressive. Each train stops to within 7 cm (3 in.) of the right spot on the platform. In addition, the trains travel more smoothly and use about 10 percent less energy than their human-controlled equivalents. The person in the driver's compartment is there for little more than reassurance. DaimlerChrysler used fuzzy logic to optimize the design process of Mercedes Benz truck components, such as gear boxes, axles, and steering. For this optimization, it was necessary to measure the "maturity" of the design process with a single parameter. Fuzzy logic was used to assess this single parameter from the numerous sources of information that describe various aspects of the design process. In a pilot study, a U.S. hospital used fuzzy logic to estimate the length of hospital stay of patients accepted to the hospital. The system uses the information that is provided by the doctor who admits the patient to the hospital. The fuzzy logic system considers the diagnosis, the patient's general condition, the likelihood of complications, the patient's previous medical history (if available), and other information. In yet another application of fuzzy logic, data analysis, the system must decide whether two similar entries in a database really represent the same person. This decision is harder than it appears at first glance because entries can vary in many ways—minor differences in spelling, order and grouping of words, and typos. Most humans, based on their experience comparing addresses, can come up with a pretty good guess whether two addresses in a database belong to the same person, but a mathematical model for the similarity of two addresses is hard to define.

rule

a conditional statement that links given conditions to actions or outcomes

The use of rules. A **rule** is a conditional statement that links given conditions to actions or outcomes. As we saw earlier, a rule is constructed using if-then constructs. If certain conditions exist, then specific actions are taken or certain conclusions are reached. In an expert system for a weather forecasting operation, for example, the rules could state that if certain temperature patterns exist with a given barometric pressure and certain previous weather patterns over the last 24 hours, then a specific forecast will be made, including temperatures, cloud coverage, and the wind-chill factor. Rules are often combined with probabilities, such as if the weather has a particular pattern of trends, then there is a 65 percent probability that it will rain tomorrow. Likewise, rules relating data and conclusions can be developed for any knowledge base. Most expert systems prevent users from entering contradictory rules. Figure 11.8 shows the use of expert system rules in helping to determine whether a person should receive a mortgage loan from a bank. In general, as the number of rules an expert system knows increases, the precision of the expert system increases.

The use of cases. As mentioned previously, an expert system can use cases in developing a solution to a current problem or situation. This process involves (1) finding cases stored in the knowledge base that are similar to the problem or

FIGURE 11.8

Rules for a Credit Application

Mortgage Application for Loans from $100,000 to $200,000

> If there are no previous credit problems and
>
> If monthly net income is greater than 4 times monthly loan payment and
>
> If down payment is 15% of the total value of the property and
>
> If net assets of borrower are greater than $25,000 and
>
> If employment is greater than three years at the same company

Then accept loan application

Else check other credit rules

---

situation at hand and (2) modifying the solutions to the cases to fit or accommodate the current problem or situation. Cases stored in the knowledge base can be identified and selected by comparing the parameters of the new problem with the cases stored in the computer system. For example, a company may be using an expert system to determine the best location of a new service facility in the state of New Mexico. Labor and transportation costs may be the most important factors. The expert system may identify two cases involving the location of a service facility where labor and transportation costs were also important—one in the state of Colorado and the other in the state of Nevada. The expert system will modify the solution to these two cases to determine the best location for a new facility in New Mexico. The result might be to locate the new service facility in the city of Santa Fe.

The Inference Engine

inference engine

part of the expert system that seeks information and relationships from the knowledge base and provides answers, predictions, and suggestions the way a human expert would

The overall purpose of an **inference engine** is to seek information and relationships from the knowledge base and to provide answers, predictions, and suggestions the way a human expert would. In other words, the inference engine is the component that delivers the expert advice.

The process of retrieving relevant information and relationships from the knowledge base is not simple. As you have seen, the knowledge base is a collection of facts, interpretations, and rules. The inference engine must find the right facts, interpretations, and rules and assemble them correctly. In other words, the inference engine must make logical sense out of the information contained in the knowledge base, the way the human mind does when sorting out a complex situation. The inference engine has a number of ways of accomplishing its tasks, including backward and forward chaining.

backward chaining

the process of starting with
conclusions and working
backward to the supporting facts

Backward chaining. **Backward chaining** is the process of starting with conclusions and working backward to the supporting facts. If the facts do not support the conclusion, another conclusion is selected and tested. This process is continued until the correct conclusion is identified.

Consider an expert system that forecasts product sales for next month. With backward chaining, we start with a conclusion, such as "Sales next month will be 25,000 units." Given this conclusion, the expert system searches for rules in the knowledge base that support the conclusion, such as "IF sales last month were 21,000 units and sales for competing products were 12,000 units, THEN sales next month should be 25,000 units or greater." The expert system verifies the rule by checking sales last month for the company and its competitors. If the facts are not true—in this case, if last month's sales were not 21,000 units or 12,000 units for competitors—the expert system would start with another conclusion and proceed until rules, facts, and conclusions matched.

forward chaining

the process of starting with the
facts and working forward to the
conclusions

Forward chaining. **Forward chaining** starts with the facts and works forward to the conclusions. Consider the expert system that forecasts future sales for a product. With forward chaining, we start with a fact, such as "The demand for the product last month was 20,000 units." With the forward-chaining approach, the expert system searches for rules that contain a reference to product demand. For example, "IF product demand is over 15,000 units, THEN check the demand for competing products." As a result of this process, the expert system might use information on the demand for competitive products. Next, after searching additional rules, the expert system might use information on personal income or inflation on a national basis. This process continues until the expert system can reach a conclusion using the data supplied by the user and the rules that apply in the knowledge base.

A comparison of backward and forward chaining. Forward chaining can reach conclusions and yield more information with fewer queries to the user than backward chaining, but this approach requires more processing and a greater degree of sophistication. Forward chaining is often used by more expensive expert systems. It is also possible to use mixed chaining, which is a combination of backward and forward chaining.

The Explanation Facility

explanation facility

component of an expert system
that allows a user or decision
maker to understand how the
expert system arrived at certain
conclusions or results

An important part of an expert system is the **explanation facility**, which allows a user or decision maker to understand how the expert system arrived at certain conclusions or results. A medical expert system, for example, may have reached the conclusion that a patient has a defective heart valve given certain symptoms and the results of tests on the patient. The explanation facility allows a doctor to find out the logic or rationale of the diagnosis made by the expert system. The expert system, using the explanation facility, can indicate all the facts and rules that were used in reaching the conclusion. This facility allows doctors to determine whether the expert system is processing the data and information in a correct and logical fashion.

The Knowledge Acquisition Facility

A difficult task in developing an expert system is the process of creating and updating the knowledge base. In the past, when more traditional programming languages were used, developing a knowledge base was tedious and time-consuming. Each fact, relationship, and rule had to be programmed into the knowledge base. In most cases, an experienced programmer was required to create and update the knowledge base.

**knowledge acquisition
facility**

part of the expert system that
provides convenient and efficient
means of capturing and storing all
components of the knowledge base

Today, specialized software allows users and decision makers to create and modify their own knowledge bases through the knowledge acquisition facility (Figure 11.9). The overall purpose of the **knowledge acquisition facility** is to

FIGURE 11.9

The knowledge acquisition facility acts as an interface between experts and the knowledge base.

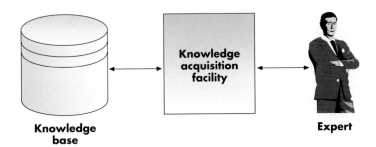

Knowledge base

Knowledge acquisition facility

Expert

provide a convenient and efficient means for capturing and storing all components of the knowledge base. Knowledge acquisition software can present users and decision makers with easy-to-use menus. After filling in the appropriate attributes, the knowledge acquisition facility correctly stores information and relationships in the knowledge base, making the knowledge base easier and less expensive to set up and maintain. Knowledge acquisition can be a manual process or a mixture of manual and automated procedures. Regardless of how the knowledge is acquired, it is important to validate and update the knowledge base frequently to make sure it is still accurate.

The User Interface

Specialized user interface software is employed for designing, creating, updating, and using expert systems. The main purpose of the user interface is to make the development and use of an expert system easier for users and decision makers. At one time, skilled computer personnel created and operated most expert systems; today, the user interface permits decision makers to develop and use their own expert systems. Because expert systems place more emphasis on directing user activities than do other types of systems, text-oriented user interfaces (using menus, forms, and scripts) may be more common in expert systems than the graphical interfaces often used with DSSs.

Expert Systems Development

Like other computer systems, expert systems require a systematic development approach for best results (Figure 11.10). This approach includes determining the requirements for the expert system, identifying one or more experts in the area or discipline under investigation, constructing the components of the expert system, implementing the results, and maintaining and reviewing the complete system.

The Development Process

Specifying the requirements for an expert system begins with identifying the system's objectives and its potential use. Identifying experts can be difficult. In some cases, a company will have human experts on hand; in other cases, experts outside the organization will be required. Developing the expert system components requires special skills. Implementing the expert system involves placing it into action and making sure it operates as intended. Like other computer systems, expert systems should be periodically reviewed and maintained to make sure they are delivering the best support to decision makers and users.

Many companies are only now beginning to use and develop expert systems. Expert system development is a team effort, but experienced personnel and users may be in high demand within an organization. Because development can take months to years, the cost of bringing in consultants for development can be high. It is critical, therefore, to find and assemble the right people to assist with development.

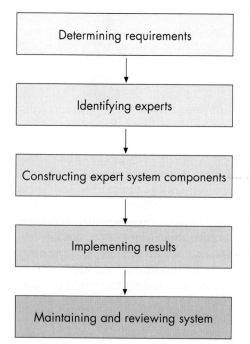

FIGURE 11.10

Steps in the Expert System
Development Process

domain expert

the individual or group whose
expertise or knowledge is captured
for use in the expert system

knowledge engineer

an individual who has training or
experience in the design,
development, implementation, and
maintenance of an expert system

knowledge user

the individual or group who uses
and benefits from the expert system

Participants in Developing and Using Expert Systems

Typically, several people are involved in developing and using an expert system (Figure 11.11).

The domain expert. Because of the time and effort involved in the task, an expert system is developed to address only a specific area of knowledge. This area of knowledge is called the *domain*. A careful evaluation of the domain of the expert system is needed to determine its stability and longevity, which should be weighed against its cost to implement. Many domains, such as the design of microcomputer chips, change quickly in their content and structure. Rapid changes in the knowledge or rules used to make decisions will quickly invalidate the system. On the other hand, the expert system should be built to be flexible so that new rules and knowledge can be added to the system—in effect, permitting the system to learn. The **domain expert** is the individual or group who has the expertise or knowledge one is trying to capture in the expert system. In most cases, the domain expert is a group of human experts. The domain expert (individual or group) usually has the ability to do the following:

- Recognize the real problem
- Develop a general framework for problem solving
- Formulate theories about the situation
- Develop and use general rules to solve a problem
- Know when to break the rules or general principles
- Solve problems quickly and efficiently
- Learn from past experience
- Know what is and is not important in solving a problem
- Explain the situation and solutions of problems to others

The knowledge engineer and knowledge users. A **knowledge engineer** is an individual who has training or experience in the design, development, implementation, and maintenance of an expert system, including training or experience with expert system shells. The **knowledge user** is the individual or group who uses and benefits from the expert system. Knowledge users do not need any previous training in computers or expert systems.

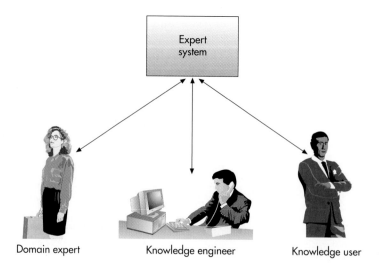

FIGURE 11.11

Participants in Expert Systems
Development and Use

Expert Systems Development Tools and Techniques

Theoretically, expert systems can be developed from any programming language. Since the introduction of computer systems, programming languages have become easier to use, more powerful, and increasingly able to handle specialized requirements. In the early days of expert systems development, traditional high-level languages, including Pascal, FORTRAN, and COBOL, were used (Figure 11.12). LISP was one of the first special languages developed and used for artificial intelligence applications. PROLOG, a more recent language, was also developed for AI applications. Since the 1990s, however, other expert system products (such as shells) are available that remove the burden of programming, allowing nonprogrammers to develop and benefit from the use of expert systems.

Expert system shells and products. As discussed, an expert system shell is a collection of software packages and tools used to design, develop, implement, and maintain expert systems. Expert system shells exist for both personal computers and mainframe systems. Some shells are inexpensive, costing less than $500. In addition, off-the-shelf expert system shells are available that are complete and ready to run. The user enters the appropriate data or parameters, and the expert system provides output to the problem or situation. For example, CLIPS is an expert system shell that supports the construction of rule- or object-based expert systems. It is used by all NASA sites and branches of the military, numerous federal bureaus, government contractors, universities, and many companies. CLIPS supports three different programming models: rule based, object oriented, and procedural. Rule-based programming allows knowledge to be represented as heuristics, or rules of thumb, which specify a set of actions to be performed for a given situation. Object-oriented programming allows complex systems to be modeled as modular components (which can be reused to model other systems or to create new components). The procedural programming capabilities provided by CLIPS are similar to capabilities found in languages such as C, Pascal, Ada, and LISP. CLIPS can be embedded within procedural code, called a subroutine, and integrated with languages such as C, FORTRAN, and Ada.[5] Other expert system shells in use today include Financial Advisor, 1st-Class Fusion, Knowledgepro, Leonardo, Personal Consultant, and MindWizard. These are summarized in Table 11.2.

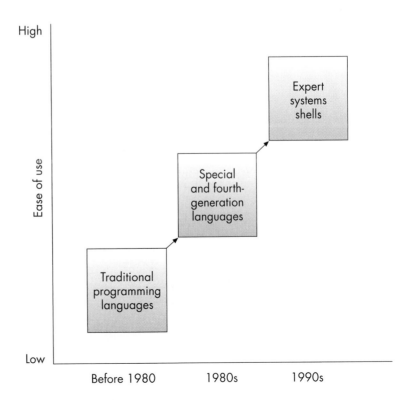

FIGURE 11.12

Software for expert systems development has evolved greatly since 1980, from traditional programming languages to expert system shells.

Financial Advisor can analyze financial investments in new equipment, facilities, and the like. The expert system requests the appropriate data and performs a complete financial analysis.

1st-Class Fusion offers a direct, easy-to-use link to the knowledge base. It also offers a visual rule tree, which graphically shows how rules are related. Because of its features, it has been used to develop models for large, sophisticated expert systems.

Knowledgepro, by Knowledge Garden, is a high-level language that combines functions of expert systems and hypertext. It allows the construction of classic if-then rules and can read dBase and Lotus 1-2-3 files.

Leonardo, an expert system shell that uses an object-oriented language, was used to develop an expert system called COMSTRAT, which can be used to help marketing managers analyze the position of their companies and products relative to their competition. COMSTRAT uses data on the company's operation and competitors' operations to create a marketing knowledge base that is used to give advice to marketing managers.

Personal Consultant (PC) Easy, an expert system shell developed by Texas Instruments, was used to develop an automated guidance expert system to route vehicles in warehouses and manufacturing plants. The expert system developed from PC Easy asks several questions and then determines the best use of automated guidance vehicles in warehouses and manufacturing facilities.

MindWizard, an inexpensive PC-based software package, enables development of compact expert systems ranging from simple models that incorporate their business decision rules to highly sophisticated models.

TABLE 11.2

Popular Expert System Shells

In addition to expert system shells, other expert system development tools make the development of expert systems easier and faster. These products help capture if-then rules for the rule base, assist in using tools such as spreadsheets and programming languages, interface with traditional database packages, generate the inference engine, and perform other functions.

Once developed, an expert system can be run by people with little or no computer experience. The expert system asks the user a series of questions. Subsequent questions are often based on answers to previous questions. After the user answers the system-generated questions, the expert system generates conclusions. Some expert systems with word processing capabilities can generate letters to users requesting additional information, if needed. The expert system can also access needed data and information from files and databases, instead of asking the user for this information.

Advantages of Expert System Shells and Products

Expert system shells and products are used to a greater extent than ever before. As discussed next, these newer programming approaches offer many advantages over traditional programming tools and techniques for developing expert systems.

Easy to develop and modify. As new facts and rules become available and as existing facts and rules need modification, the knowledge base must be updated. Systems developed using PROLOG and LISP are more difficult to modify than expert systems developed with shells. Shells have an editing facility that makes modification relatively easy and cost effective.

The use of satisficing. The traditional approach to problem solving attempts to find the optimal, or best, solution; advanced and symbolic languages can deal with more complex problems and return very good—but not necessarily optimal—decisions. As discussed earlier, this is called the satisficing approach. Good decisions that satisfy the requirements of the decision maker are found, whereas the optimal, or the best, solution would be too difficult or time-consuming to obtain.

The use of heuristics. As mentioned previously, expert systems must be able to handle imprecise relationships. Heuristics can help handle these situations and will often return a good solution that satisfies the decision maker. Heuristic rules are often easier to implement in an expert system shell than to code directly.

Development by knowledge engineers and users. With expert system shells, knowledge engineers and knowledge users can complete the development process. When developing expert systems with traditional programming languages, organizations often need to use systems analysis and computer programming, approaches that are expensive and usually require more time. In addition, it can be difficult to communicate the exact needs of the expert system to traditional programmers and computer personnel. Using expert system shells can result in a system that is less expensive to develop, requires less time to implement, and more accurately captures the needs of its intended decision makers and users. As expert system shells become easier to use, the role of the knowledge engineer will increasingly shift from system developer to consultant.

Expert Systems Development Alternatives

Expert systems can be developed from scratch by using an expert system shell or by purchasing an existing expert system package. The approach selected depends on the system benefits compared with the cost, control, and complexity of each alternative. A graph of the general cost and time of development is shown in Figure 11.13. It is usually faster and less expensive to develop an expert system using an existing package or an expert system shell. Note that there will be an additional cost of developing an existing package or acquiring an expert system shell if the organization does not already have this type of software.

In-house development: develop from scratch. Developing an expert system from scratch is usually more costly than the other alternatives, but an organization has more control over the features and components of the system.

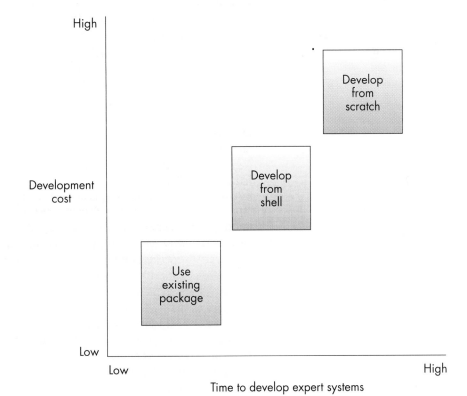

FIGURE 11.13

Some Expert Systems Development Alternatives and Their Relative Cost and Time Values

Such customization also has a downside; it can result in a more complex system, with higher maintenance and updating costs.

In-house development: develop from a shell. As you have seen, an expert system shell consists of one or more software products that assist in the development of an expert system. In some instances, the same shell can be used to develop many expert systems. Developing an expert system from a shell can be less complex and easier to maintain than developing one from scratch. However, the resulting expert system may need to be modified to tailor it to specific applications. In addition, the capabilities and features of the expert system can be more difficult to control.

Off-the-shelf purchase: use existing packages. Using an existing expert system package is the least expensive and fastest approach, in most cases. An existing expert system package is one that has been developed by a software or consulting company for a specific field or area, such as the design of a new computer chip or a weather forecasting and prediction system. The advantages of using an existing package can go beyond development time and cost. These systems can also be easy to maintain and update over time. A disadvantage of using an off-the-shelf package is that it may not be able to satisfy the unique needs or requirements of an organization.

Applications of Expert Systems and Artificial Intelligence

Expert systems and artificial intelligence are being used in a variety of ways. Some of the applications of these systems are summarized next.

Credit granting. Many banks employ expert systems to review an individual's credit application and credit history data from credit bureaus to make a decision on whether to grant a loan or approve a transaction.

Games. Some expert systems are used for entertainment. For example, Proverb is an expert system designed to solve standard American crosswords given the grid and clues.

Information management and retrieval. The explosive growth of information available to decision makers has created a demand for devices to help manage the information. Expert systems can aid this process through the use of bots. Businesses might use a bot to retrieve information from large distributed databases or a vast network like the Internet. Expert system agents help managers find the right data and information while filtering out irrelevant facts that might impede timely decision making.

Legal profession. SHYSTER is a case-based legal expert system, developed by James Popple. It provides advice in areas of case law that have been defined by a legal expert using a unique specification language.

AI and expert systems embedded in products. The antilock braking system on modern automobiles is an example of a rudimentary expert system. A processor senses when the tires are beginning to skid and releases the brakes for a fraction of a second to prevent the skid. AI researchers are also finding ways to use neural networks and robotics in everyday devices, such as toasters, alarm clocks, and televisions.

Plant layout. FLEXPERT is an expert system that uses fuzzy logic to perform plant layout. The software helps companies determine the best placement for equipment and manufacturing facilities.

Hospitals and medical facilities. Some hospitals use expert systems to determine a patient's likelihood of contracting cancer or other diseases. MYCIN is an expert system started at Stanford University to analyze blood infections. A medical

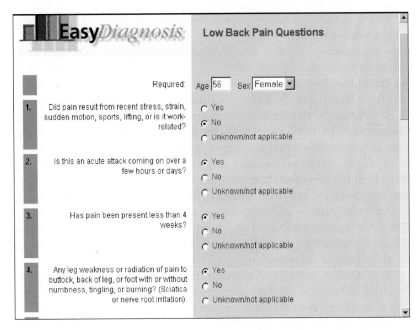

MatheMEDics' Easy Diagnosis is an online medical expert system. Patients answer a series of questions about their symptoms or signs, and the program presents the most likely conditions or diagnoses, in descending order of their probabilities.
(Source: Courtesy of MatheMEDics, Inc.)

expert system used by the Harvard Community Health Plan allows members of the HMO to get medical diagnoses via home personal computers. For minor problems, the system gives uncomplicated treatments; for more serious conditions, the system schedules appointments. The system is highly accurate, diagnosing 97 percent of the patients correctly (compared with the doctors' 78 percent accuracy rating). To help doctors in the diagnosis of thoracic pain, MatheMEDics® has developed THORASK™, a straightforward, easy-to-use program, requiring only the input of carefully obtained clinical information. The program helps the less experienced to distinguish the three principal categories of chest pain from each other. It does what a true medical expert system should do without the need for complicated user input. You answer basic questions about the patient's history and directed physical findings, and the program immediately displays a list of diagnoses. The diagnoses are presented in decreasing order of likelihood, together with their estimated probabilities. The program also provides concise descriptions of relevant clinical conditions and their presentations as well as brief suggestions for diagnostic approaches. For purposes of record-keeping, documentation, and data analysis, there are options for saving and printing cases.

Help desks and assistance. Expert systems are used by customer service help desks to provide timely and accurate assistance. Kaiser Permanente, a large HMO, uses an expert system and voice response to automate its help desk function. The automated help desk frees up staff to handle more complex needs, while still providing more timely assistance for routine calls.

Employee performance evaluation. An expert system by Austin-Hayne, called Employee Appraiser, provides managers with expert advice for use in employee performance reviews and career development discussions.

Loan analysis. KPMG Peat Marwick uses an expert system called Loan Probe to review loan loss reserves to determine whether sufficient funds have been set aside to cover the risk that some loans will become uncollectible.

Virus detection. IBM is using neural network technology to help create more advanced software for eradicating computer viruses, a major problem in American businesses. IBM's neural network software deals with "boot sector" viruses, the most prevalent type, using a form of artificial intelligence that mimics the human brain and generalizes by looking at examples. It

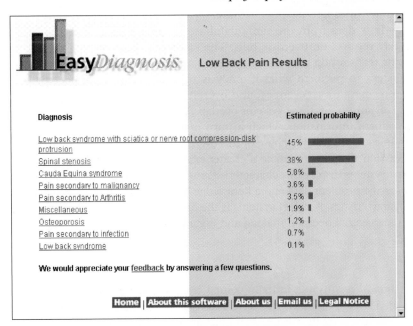

requires a vast number of training samples, which in the case of antivirus software are three-byte virus fragments.

Repair and maintenance. ACE is an expert system used by AT&T to analyze the maintenance of telephone networks. Nynex (New York and New England Telephone Exchange) has expert systems to help its workers locate and solve customer-related phone problems. IET-Intelligent Electronics uses an expert system to diagnose maintenance problems related to aerospace equipment. In the airline industry, prognosis can help reduce the high costs of unscheduled major component removals (such as engines) or failures in flight (such as in-flight engine shut-downs). Assessment of equipment health also supports early and better identification and in planning any optimal maintenance. General Electric Aircraft Engine Group uses an expert system to enhance maintenance performance levels at all sites, improve diagnostic accuracy and reduce ambiguity of the fault, advise on real-time repair action, provide clues to failures, and access and display relevant maintenance information.

Shipping. CARGEX-Cargo Expert System is used by Lufthansa, a German airline, to help determine the best shipping routes.

Marketing. CoverStory is an expert system that extracts marketing information from a database and automatically writes marketing reports.

Warehouse optimization. United Distillers uses an expert system to determine the best combinations of stocks to produce its blends of Scottish whiskey. This information is then supplemented with information about location of the casks for each blend. The system then optimizes the selection of required casks, keeping to a minimum the number of "doors" (warehouse sections) from which the casks must be taken and the number of casks that need to be moved to clear the way. Other constraints must be satisfied, such as the current working capacity of each warehouse and the maintenance and restocking work that may be in progress.

Integrating Expert Systems

As with the other information systems, an expert system can be integrated with other systems in an organization through a common database. An expert system that identifies late-paying customers who should not receive additional credit may draw data from the same database as an invoicing MIS that produces weekly reports on overdue bills. The same database—a by-product of the invoicing transaction processing system—might also be used by a decision support system to perform what-if analysis to determine the impact of late payments on cash flows, revenues, and overall profit levels.

In many organizations, these systems overlap. A TPS might be expanded to provide management information, which in turn may provide some DSS functions, and so on. In each progressive phase of this overlap, the information system assists with the decision-making process to a greater extent. Of all these information systems, expert systems display this characteristic most obviously, proposing decisions based on specific problem data and a knowledge base. Understanding the capabilities and characteristics of expert systems is the first step in applying these systems to support managerial decision making and organizational goals.

VIRTUAL REALITY

virtual reality system

system that enables one or more users to move and react in a computer-simulated environment

The term *virtual reality* was initially coined by Jaron Lanier, founder of VPL Research, in 1989. Originally, the term referred to *immersive virtual reality*, in which the user becomes fully immersed in an artificial, three-dimensional world that is completely generated by a computer. A **virtual reality system** enables one or more users to move and react in a computer-simulated environment.

Virtual reality simulations require special interface devices that transmit the sights, sounds, and sensations of the simulated world to the user. These devices can also record and send the speech and movements of the participants to the simulation program. Thus, users can sense and manipulate virtual objects much as they would real objects. This natural style of interaction gives the participants the feeling that they are immersed in the simulated world.

Interface Devices

To see in a virtual world, often the user will wear a head-mounted display (HMD) with screens directed at each eye. The HMD also contains a position tracker to monitor the location of the user's head and the direction in which the user is looking. Using this information, a computer generates images of the virtual world—a slightly different view for each eye—to match the direction in which the user is looking and displays these images on the HMD. With current technology, virtual-world scenes must be kept relatively simple so that the computer can update the visual imagery quickly enough (at least ten times a second) to prevent the user's view from appearing jerky and from lagging behind the user's movements.

Alternative concepts—BOOM and CAVE—were developed for immersive viewing of virtual environments to overcome the often uncomfortable intrusiveness of a head-mounted display.

The BOOM (Binocular Omni-Orientation Monitor) from Fakespace Labs is a head-coupled stereoscopic display device (Figure 11.14). Screens and optical systems are housed in a box that is attached to a multilink arm. The user looks into the box through two holes, sees the virtual world, and can guide the box to any position within the virtual environment. Head tracking is accomplished via sensors in the links of the arm that holds the box.

The Electronic Visualization Laboratory at the University of Illinois at Chicago introduced a room constructed of large screens on which the graphics are projected onto the three walls and the floor (Figure 11.15). The CAVE, as this room is called, provides the illusion of immersion by projecting stereo images on the walls and floor of a room-sized cube. Several people wearing lightweight stereo glasses can enter and walk freely inside the CAVE. A head-tracking system continuously adjusts the stereo projection to the current position of the leading viewer.

Users hear sounds in the virtual world through earphones. The information reported by the position tracker is also used to update audio signals. When a sound source in virtual space is not directly in front of or behind the user, the computer transmits sounds to arrive at one ear a little earlier or later than at the other and to be a little louder or softer and slightly different in pitch.

The *haptic* interface, which relays the sense of touch and other physical sensations in the virtual world, is the least developed and perhaps the most challenging to create. Currently, with the use of a glove and position tracker, the computer locates the user's hand and measures finger movements. The user can reach into the virtual world and handle objects; however, it is difficult to generate the sensations that are felt when a person taps a hard surface, picks up an object, or runs a finger across a textured surface. Touch sensations also have to be synchronized with the sights and sounds users experience.

Immersive Virtual Reality

In immersive virtual reality, the virtual world is presented in full scale and relates properly to humans' size. It may represent any three-dimensional setting, real or abstract, such as a building, an archaeological excavation site, the human anatomy, a sculpture, or a crime scene reconstruction. Virtual worlds can be animated, interactive, and shared. Through immersion, the user can gain a deeper understanding of the virtual world's behavior and functionality.

Other Forms of Virtual Reality

Virtual reality can also refer to applications that are not fully immersive, such as mouse-controlled navigation through a three-dimensional environment on a graphics monitor, stereo viewing from the monitor via stereo glasses, stereo projection systems, and others. Some virtual reality applications allow views of real environments with superimposed virtual objects. Motion trackers monitor the movements of dancers or athletes for subsequent studies in immersive virtual reality. Telepresence systems (e.g., telemedicine, telerobotics) immerse a viewer in a real world that is captured by video cameras at a distant location and allow for the remote manipulation of real objects via robot arms and manipulators. Many believe that virtual reality will reshape the interface between people and information technology by offering new ways to communicate information, visualize processes, and express ideas creatively.

Useful Applications

There are literally hundreds of applications of virtual reality, with more and more being developed as the cost of hardware and software declines and people's imaginations are opened to the potential of virtual reality. Here is a summary of some of the more interesting applications.

Medicine

Surgeons in France performed the first successful closed-chest coronary bypass operation in 1999. Instead of cutting open the patient's chest and breaking his breastbone, as is usually done, surgeons used a virtual reality system that enabled them to operate through three tiny half-inch holes between the patient's ribs. They inserted thin tubes to tunnel to the operating area and protect the other body tissue. Then three arms were inserted into the tubes. One was for a 3-D camera; the other two held tiny artificial wrists to which a variety of tools—scalpels, scissors, needle—were attached. The virtual reality system mimicked the movements of the surgeon's shoulders, elbows, and wrists. The surgeon sits at a computer workstation several feet from the operating table and moves instruments that control the ones inside the patient. The instruments inside the patient mimic the motion of the surgeon's hands so accurately that it is possible to sew up a coronary artery as thin as a thread. The surgeon watches his progress on a screen that enlarges the artery in 3-D to the size of a garden hose.[6]

Doctors use a technique called *palpation* to feel with their fingertips for telltale signs of abdominal injury or disease. By feeling the shape of the organs under the skin and noting how the skin springs back from the fingers' touch, skilled physicians can gain information they can't get from other instruments. In yet another medical application of virtual reality, researchers are trying to translate that expertise into biomechanical measurements that can be stored in a computer database. Data from actual patients will be gathered from sensors mounted on the fingertips of virtual-reality gloves to create a database that will be made available to examining physicians. Upon examination of a new patient, a feedback system hooked up to the virtual-reality gloves would capture sensations felt during the examination and compare them with data in the database. This approach could allow surgeons to go right into the operating room without having to obtain a CAT scan, thus saving time—and with many injuries, that's absolutely critical.

Education

Virtual environments are used in education to bring exciting new resources into the classroom. Students can stroll among digital bookshelves, learn anatomy on a simulated cadaver, or participate in historical events, all in virtual reality.

Third-grade students at John Cotton Tayloe School in Washington, North Carolina, can take an exciting virtual trip down the Nile for an integrated-curriculum lesson on ancient Egypt. This interactive virtual reality computer lesson integrates social studies, geography, music, art, science, math, and language arts. The software used to design the lesson was 3-D Website Builder (created by the Virtus Corporation), which allowed the designers to create a desert landscape complete with an oasis, camels, and a pyramid. Students could view the scenes from all angles, including front and top views, and could even enter a pyramid and view the sarcophagus holding the mummy in the middle of a room containing different Egyptian items. On one wall was a hieroglyphic message that was part of the lesson. On another wall was artwork depicting life in ancient Egypt, and against another were pieces of Egyptian furniture and a harp.[7]

Virtual technology has also been applied to aircraft maintenance training. The trainer uses virtual reality to simulate the aircraft. Feedback gives the user a sense of touch, while computer graphics give the senses of sight and sound. The user sees, touches, and manipulates the various parts of the virtual aircraft during training. The Virtual Aircraft Maintenance System simulates real-world maintenance tasks that are routinely performed on the AV8B vertical takeoff and landing aircraft used by U.S. Marines.

Real Estate Marketing

Virtual reality has been used to increase real estate sales in several powerful ways. From Web publishing to laptop display to a potential buyer, virtual reality provides excellent exposure for properties and attracts potential clients. Clients can take a virtual walk through properties and eliminate wrong choices without wasting valuable time. Virtual walk-throughs can be mailed on diskettes or posted on the Web as a convenience for nonlocal clients. A CD-ROM containing all virtual reality homes can also be sent to clients and other agents. Realatrends Real Estate Service, offering homes for sale in Orange County, California (http://www.realatrends.com/virtual_tours.htm), is just one of many real estate firms offering this service.

FIGURE 11.16

Computer-generated image technology is used widely in sports simulations. Shown here is NFL GameDay from Sony. (Source: Courtesy of Sony Computer Entertainment)

Computer-Generated Images

Computer-generated image technology, or CGI, has been around since the 1970s. A number of movies released in 2000 used this technology to bring realism to the silver screen. A team of artists rendered the roiling seas and crashing waves of *The Perfect Storm* almost entirely on computers using weather reports, scientific formulas, and their imagination. There was also *Dinosaur*, with its realistic talking reptiles, *Titan A.E.'s* beautiful 3-D space-scapes, and the casts of computer-generated crowds and battles in *Gladiator* and *The Patriot*. CGI can also be used for sports simulation, as Figure 11.16 shows, to enhance the viewers' knowledge and enjoyment of a game.

● SUMMARY

PRINCIPLE • Artificial intelligence systems form a broad and diverse set of systems that can replicate human decision making for certain types of well-defined problems.

The term *artificial intelligence* is used to describe computers with the ability to mimic or duplicate the functions of the human brain. The objective of building AI systems is not to replace human decision making completely but to replicate it for certain types of well-defined problems.

●   ●   ●

Intelligent behavior encompasses several characteristics, including the abilities to learn from experience and apply this knowledge to new experiences; handle complex situations and solve problems for which pieces of information may be missing; determine relevant information in a given situation, think in a logical and rational manner and give a quick and correct response; and understand visual images and processing symbols. The computer is better than humans at transferring information, making a series of calculations rapidly and accurately, and making complex calculations, but a human is better than a computer at all other attributes of intelligence.

●   ●   ●

Artificial intelligence is a broad field that includes several key components, such as expert systems, robotics, vision systems, natural language processing, learning systems, and neural networks. An expert system consists of the hardware and software to produce systems that act or behave like a human expert in a field or area (e.g., credit analysis). Robotics involves developing mechanical or computer devices to perform tasks that require a high degree of precision or are tedious or hazardous for humans (e.g., stacking cartons on a pallet). Vision systems include hardware and software that permit computers to capture, store, and manipulate images and pictures (face-recognition software). Natural language processing allows the computer to understand and react to statements and commands made in a "natural" language, such as English. Learning systems use a combination of software and hardware that allows the computer to change how it functions or reacts to situations based on feedback it receives (e.g., a computerized chess game). A neural network is a computer system that can simulate the functioning of a human brain (e.g., disease diagnostics system).

PRINCIPLE • Expert systems can enable a novice to perform at the level of an expert but must be developed and maintained very carefully.

Expert systems can explain their reasoning or suggested decisions, display intelligent behavior, manipulate symbolic information and draw conclusions from complex relationships, and deal with uncertainty. They are not yet widely used; some are difficult to use, are limited to relatively narrow problems, cannot readily deal with mixed knowledge, present the possibility for error, cannot refine their own knowledge base, are difficult to maintain, and may have high development costs. Their use also raises legal and ethical concerns.

An expert system consists of a collection of integrated and related components, including a knowledge base, an inference engine, an explanation facility, a knowledge acquisition facility, and a user interface. The knowledge base contains all the relevant data, rules, and relationships used in the expert system. The rules are often composed of if-then statements, which are used for drawing conclusions. Fuzzy logic allows expert systems to incorporate facts and relationships into expert system knowledge bases that may be imprecise or unknown.

The inference engine performs the processing of the rules, data, and relationships stored in the knowledge base to provide answers, predictions, and suggestions the way a human expert would. Two common methods for processing include forward and backward chaining. Backward chaining starts with a conclusion, then searches for facts to support it; forward chaining starts with a fact, then searches for a conclusion to support it. Mixed chaining is a combination of backward and forward chaining.

The explanation facility of an expert system allows the user to understand what rules were used in arriving at a decision. The knowledge acquisition facility helps the user add or update knowledge in the knowledge base. The user interface makes it easier to develop and use the expert system.

●   ●   ●

The steps involved in the development of an expert system include determining requirements, identifying experts, constructing expert system components, implementing results, and maintaining and reviewing the system.

Expert systems can be implemented in several ways. Previously, traditional high-level languages, including Pascal, FORTRAN, and COBOL, were used. LISP and PROLOG are two languages specifically developed for creating expert systems from scratch. A faster and less expensive way to acquire an expert system is to purchase an expert system shell or existing package. The shell program is a collection of software packages and tools used to design, develop, implement, and maintain expert systems. Advantages of expert system shells include ease of development and modification, use of satisficing, use of heuristics, and development by knowledge engineers and end users. The approach selected depends on the benefits compared with cost, control, and complexity considerations.

The individuals involved in the development of an expert system include the domain expert, the knowledge engineer, and the knowledge users. The domain expert is the individual or group who has the expertise or knowledge being captured for the system. The knowledge engineer is the developer whose job is the extraction of the expertise from the domain expert. The knowledge user is the individual who benefits from the use of the developed system.

* * *

Following is a list of factors that normally make expert systems worth the expenditure of time and money: a high potential payoff or significantly reduced downside risk, the ability to capture and preserve irreplaceable human expertise, the ability to develop a system more consistent than human experts, expertise needed at a number of locations at the same time, and expertise needed in a hostile environment that is dangerous to human health. Once the expert system is developed, the expert system solution can be obtained faster than the solution from human experts. An ES also provides expertise needed for training and development to share the wisdom and experience of human experts with a large number of people.

The benefits of using an expert system go beyond the typical reasons for using a computerized processing solution. Expert systems display "intelligent" behavior, manipulate symbolic information and draw conclusions, and can deal with uncertainty. Expert systems can be used to solve problems in many fields and disciplines and can assist in all stages of the problem-solving process. Past successes have shown that expert systems are good at strategic goal setting, planning, design, decision making, quality control and monitoring, and diagnosis.

PRINCIPLE • Virtual reality systems have the potential to reshape the interface between people and information technology by offering new ways to communicate information, visualize processes, and express ideas creatively.

A virtual reality system enables one or more users to move and react in a computer-simulated environment. Virtual reality simulations require special interface devices that transmit the sights, sounds, and sensations of the simulated world to the user. These devices can also record and send the speech and movements of the participants to the simulation program. Thus, users are able to sense and manipulate virtual objects much as they would real objects. This natural style of interaction gives the participants the feeling that they are immersed in the simulated world.

Virtual reality can also refer to applications that are not fully immersive, such as mouse-controlled navigation through a three-dimensional environment on a graphics monitor, stereo viewing from the monitor via stereo glasses, stereo projection systems, and others. Some virtual reality applications allow views of real environments with superimposed virtual objects.

● KEY TERMS

● REVIEW QUESTIONS

1. Define the term *artificial intelligence*.
2. What is a vision system? Discuss two applications of such a system.
3. What is a perceptive system?
4. What is natural language processing? What are the three levels of voice recognition?
5. Describe three examples of the use of robotics.
6. What is a learning system? Give a practical example of such a system.
7. What is a neural network? Describe two applications of neural networks.
8. What is meant when it is said that neural networks learn to program themselves?
9. What are the capabilities of an expert system?
10. Under what conditions is the development of an expert system likely to be worth the effort?
11. Identify the basic components of an expert system and describe the role of each.
12. What are fuzzy sets and fuzzy logic?
13. How are rules used in expert systems?
14. Expert systems can be built based on rules or cases. What is the difference between the two?
15. Describe the roles of the domain expert, the knowledge engineer, and the knowledge user in expert systems.
16. What are the primary benefits derived from the use of expert systems?
17. Identify three approaches for developing an expert system.
18. Identify three special interface devices developed for use with virtual reality systems.
19. Identify and briefly describe three specific virtual reality applications.

● DISCUSSION QUESTIONS

1. What are the requirements for a computer to exhibit human-level intelligence? How long will it be before we have the technology to design such computers? Do you think we should push to try to accelerate such a development? Why or why not?
2. What are some of the tasks at which robots excel? Which human tasks are difficult for them to master? What fields of AI are required to develop a truly perceptive robot?
3. Accuracy slip occurs when there are significant changes in the real world, making an expert system less accurate. Identify at least three expert system applications where accuracy slip could lead to the loss of human life. What process can be put in place to safeguard against accuracy slip for these critical expert system applications?
4. You have been hired to capture the knowledge of a brilliant attorney who has an outstanding track record for selecting jury members favorable to her clients during the pretrial jury selection process. This knowledge will be used as the basis

for an expert system to enable other attorneys to have similar success. Is this system a good candidate for an expert system? Why or why not?
5. Briefly explain why human decision making often does not lead to optimal solutions to problems.
6. What is the purpose of a knowledge base? How is one developed?
7. Describe an application that requires the concurrent use of more than one of the subfields of artificial intelligence.
8. Imagine that you are developing the rules for an expert system to select the strongest candidates for a medical school. What rules or heuristics would you include?
9. What skills does it take to be a good knowledge engineer? Would knowledge of the domain help or hinder the knowledge engineer in capturing knowledge from the domain expert?
10. Which interface is the least developed and most challenging to create in a virtual reality system? Why do you think this is so?

● PROBLEM-SOLVING EXERCISES

1. Imagine that you are a knowledge engineer and are developing an expert system to help consumers choose a used car that best meets their needs and their budget. You are going to your first interview with the owner of a used car lot who is the designated expert for this system. Use your word processing software and develop a list of questions that you would ask to begin to capture this individual's knowledge.

2. Assume you live in an area where there is a wide variation in weather from day to day. Develop a simple expert system to provide advice on the type of clothes to wear based on the weather. The system needs to help you decide which clothes and accessories (umbrella, boots, etc.) to wear for sunny, snowy, rainy, hot, mild, or cold days. Key inputs to the system include last night's weather forecast, your observation of the morning temperature and clouds, yesterday's weather, and the activities you have planned for the day. Using your word processing program, create seven or more rules that could be used in such an expert system. Create five cases and use the rules you developed to determine the best course of action.

3. Using your word processing software, document the rule for a simple expert system that picks which pair of shoes or boots you should wear today based on what other clothes you are wearing, the weather, and the activities you have planned.

● TEAM ACTIVITIES

1. With two or three of your classmates, do research to identify three real examples of expert systems in use. Discuss the problems solved by each of these systems. Regardless of how the knowledge is acquired, it is important to validate and update the knowledge base frequently to make sure it is still accurate. Failure to do so will result in "knowledge creep," whereby the knowledge base no longer matches reality. Identify any issues that may arise because of "knowledge creep." Choose which of the three systems provides the most benefit and state why you selected that system.

2. Form a team to debate other teams from your class on the following topic: "Are expert systems superior to humans when it comes to making objective decisions?" Develop several points supporting either side of the debate.

3. With members of your team, develop an expert system that makes suggestions about what to do if your car does not start. Develop a simple "dialogue" between the expert system and you, the end user. The expert system should suggest different actions to take in an attempt to diagnose and correct whatever may be wrong.

● WEB EXERCISES

1. Jess is an expert system shell written entirely in Java by Ernest Friedman-Hill. It enables you to build Java applets that have the capacity to "reason" based on knowledge you supply in the form of declarative rules. Visit the Jess Web site at http://herzberg.ca.sandia.gov/jess and read the FAQ page. Run the applet demo and see whether you can beat the expert system. Write a brief paper to your instructor summarizing your experience.

2. Visit the Web site http://ai.about.com. Explore one of the A1-related topics of interest to you. Write a brief paper summarizing your findings.

3. Take the Fuzzy Shower Challenge, a challenge faced each morning by everyone—trying to survive controlling the shower temperature. Visit the Web site at http://ai.iit.nrc.ca/fuzzy/shower/title.html and experience the use of a fuzzy logic system. Write a brief paper to your instructor summarizing your experience.

● CASES

 Fuzzy Logic System Designed to Help Travelers

Yatra.net, a Web travel site launched in July 2000, claims it can provide better service to businesses than travel agents by using a proprietary system that employs fuzzy logic. Its target customers are the 35,000 small and medium-sized U.S. companies that spend $0.5 million to $7 million annually on air travel. Yatra's goal is to deliver significant savings through large buying volume and strategic sourcing capabilities.

At the heart of Yatra.net is its patent-pending Cognizer technology that enables travelers and travel suppliers to make complex purchasing decisions like never before. Cognizer was developed by a five-member team over a two-year period and uses fuzzy logic that takes into account corporate, traveler, and supplier preferences to create simple business rules and a dynamic travel policy. It uses information from multiple sources to analyze each purchasing event independently by weighing more than 50 corporate policies and traveler preferences across factors of convenience, service, and price. For instance, why should a salesperson's trip for an important sales call be handled in the same way as a staff accountant's travel to an internal meeting? Yatra believes it shouldn't. In one case, convenience and service should weigh heavily to ensure on-time arrival, while in the other, price should be the primary consideration. Cognizer ensures that every travel decision draws from all available knowledge to identify the best travel options.

If travelers fail to choose from the recommended itineraries, Cognizer instantly tracks the purchase and notifies the company's travel services manager of the decision.

Suppliers of travel services will also be able to create their own rules to target potential customers. For example, an airline can offer specials such as first-class upgrades to travelers who fly the San Diego to Hong Kong route a certain number of times a year and aren't part of its frequent-flier program.

Most industry analysts agree that the on-line travel market is already crowded. Yatra faces a major challenge—it has to convince customers that their current relationships with travel agencies aren't working out, and that may be hard to do. The biggest opportunity is in lower fees, but if agencies begin reducing fees for bookings made on-line, then Yatra loses its primary advantage.

Discussion Questions

1. Identify some potential business rules and traveler preferences that might be implemented to enable Cognizer to analyze each purchasing event independently.
2. What are some of the advantages and disadvantages of this system from the traveler's perspective?

Critical Thinking Questions

3. Access Yatra's Web site and do other research to learn how successful this new venture has been. Write a brief report summarizing your findings.
4. In many small companies, employees are used to booking their own travel arrangements. What sort of issues might be raised if such a company converts to Yatra?

Sources: Adapted from Cheryl Rosen, "Fuzzy Logic Provides Clear Solutions," *Information Week,* May 15, 2000, p. 77; and Yatra home page, http://www.yatra.net/Default.htm, accessed July 5, 2000.

 Expert Systems Help Develop Business Plans

Unless you have easy access to lots of money, to sell an internal project or launch a new business you usually must create a business plan. Emerging enterprises, like Internet start-ups, are doubly challenged—to generate financial backing, they must move quicker than established businesses, and they often do not have people experienced in developing such plans. Even managers in established companies must develop a business plan to gain senior

management approval to launch a new product or service. Software has come to the rescue.

A number of software applications are designed to provide a standard template for data entry and guide you step by step in creating a business plan. As you proceed, the software offers additional assistance by providing descriptions of what is expected in each section and identifying issues to consider before writing a section. Use of such a product guarantees that you will produce a business plan with the key elements that most potential investors and senior managers expect to see; however, it provides no real feedback on the viability of the plan or the reasonableness of the business assumptions. Recognizing these shortcomings, a few software companies have developed an expert system that not only offers professional advice on how to create a viable business plan but also assesses the plan's probability for success.

Developing a plan with this software starts by answering a long series of questions asked by the inference engine. Some answers require a simple yes or no response; many of the questions relate to competitors and require a response on a scale of 1 to 10 to rate your relative strengths and weaknesses and other factors involved in the situation. When you are done answering the questions, the explanation facility provides a thumbnail analysis of your situation and some basic conclusions. It also provides a score of your chance of success and, most important, the option to trace its conclusions back to the answers you gave. The results are almost never based on a single answer but on the way your answers interrelate. That is the power of expert system technology—it recognizes patterns in combinations of factors. You may next wish to review and revise some of your original assumptions and answers to key questions to evaluate their impact on the probability of success of the project.

A good business planning tool will not generate eccentric advice but will look at business problems in standard, proven ways to provide objective conclusions. Most users of such software think that answering the questions and examining the connections between answers and conclusions provides a

deeper insight into the key factors and assumptions of their plan. Many times they realize that they simply do not have enough information about the situation and need to get more data before presenting a recommendation to management. The analysis delivered is often provocative and adds considerable value to initial planning.

Discussion Questions

1. What are the key advantages and disadvantages of using software such as that described in this case to evaluate business plans?
2. How might you choose between competing software packages to perform business planning?

Critical Thinking Questions

3. Imagine that you are the chief financial officer of a major corporation, and top executives have recommended use of business planning software based on expert system technology to prepare all business proposals for more than $1 million. Would you support this recommendation? Why or why not?
4. How might the use of expert system business planning software speed up the approval of a business recommendation? Is it possible that it could slow the process down? Why or why not?

Sources: Adapted from Jeff Angus, "Expert System Acts as Outside Business Consultant," *Information Week*, November 8, 1999, accessed at http://www.informationweek.com/760/expert.htm; and Kathy Yakal, "Business Resources Software Plan Write for Business 5.0," *PC Magazine*, April 24, 2000, accessed at http://www.brs-inc.com/reviews.asp.

 ### AI Software Used to Monitor and Control Networks

Companies have been quick to adopt distributed processing as discussed in Chapter 6. So, many IT organizations manage and control a variety of applications, hardware, software, and data communications protocols for other companies. Computer Associates developed enterprise management software called Unicenter TNG to meet this need. Through a variety of integrated core and optional applications, Unicenter TNG can manage a technology infrastructure including networks, systems, desktops, applications, databases, point-of-sale devices, automated teller machines, manufacturing devices, environmental controls, hospital equipment, and power lines.

Neugents is Computer Associates' software that uses neural network technology to predict failure in

the network or any of the devices connected to it. The software is offered as an option to the Unicenter TNG enterprise management software. Neugents analyzes system data collected by Unicenter and uses that data to learn what system behavior is acceptable and when a deviation in activity indicates a problem. For example, Neugents looks at 1,200 elements on a network server to determine what conditions foreshadow a failure. When similar conditions appear, Neugents predicts the likelihood of a slowdown or outage and sends Unicenter an alert when preselected thresholds are reached. For example, if Neugents' analysis indicates a 75 percent chance of a crash within an hour and 75 percent is the alert threshold, Neugents signals the Unicenter console. IT administrators can set Unicenter to automatically respond to Neugents' alerts.

Computer Associates promotes its Neugents technology to differentiate its support for e-business from its competitors. Computer Associates believes that Neugents will become even more critical as businesses become increasingly inundated with data and as back offices try to integrate seamlessly with Web storefronts.

NCI Information Systems typically required two engineers to support a 2,000-node network. After implementing Unicenter TNG and the Neugents application, the same two engineers can now support a 10,000-node network. At Allstate Insurance, Neugents traced a slowdown in a server's packet data transmissions to a bad network interface card that was then replaced before it could cause a system failure. Myers Industries, a plastic and rubber manufacturer, uses Neugents to review real-time information from its process controllers and production equipment on its manufacturing systems, giving the company early warnings of potential problems before faults begin corrupting processes.

To date, few Computer Associates clients have begun using Neugents to predict when various components of their network might fail because it is complicated to master.

Discussion Questions

1. What are some of the advantages of using neural network technology to monitor networks?
2. Why are effective and efficient network management and control key to the success of a company's e-commerce operation?

Critical Thinking Questions

3. Why do you think companies are finding it difficult to use software to predict when network components will fail?
4. Do a Web search to find at least two other companies that offer capabilities similar to those of Neugents to manage and control networks. How are the products of these companies similar? How are they different?

Sources: Adapted from George V. Hulme, "CA Heads into the Next Dimension," *Information Week*, May 8, 2000, accessed at http://www.informationweek.com/785/vaca.htm; Catherine Shull, "Computer Associates Teaches Brigham Young University How to Enhance E-Business with the Power of Neugents," Computer Associates press release, March 13, 2000, accessed at http://www.ca.com/press/2000/03/byu_neugents.htm; and Amy K. Larsen, "CA Offers Free Neugent Trial," *Information Week*, March 15, 1999, accessed at http://www.informationweek.com/story/IWK19990315S0007.

● NOTES

Sources for the opening vignette on page 419: Adapted from Dave Marino-Nachison, "The Motley Fool Interview with Ask Jeeves President and CEO Rob Wrubel," December 22, 1999, at http://www.fool.com/foolaudio/transcripts/1999/stock talk991230_askj.htm; and "What Is Ask Jeeves," Ask Jeeves Web site at http://www.ask.com/docs/about/whatIsAskJeeves.html, accessed June 14, 2000.

1. Gray H. Anthes, "The Robots Are Coming," *Computerworld*, May 22, 2000, p. 82.
2. Maria Trombly, "Voice Recognition Eases Call-In Trading," *Computerworld*, May 22, 2000, accessed at http://www.computerworld.com.
3. Nancy Weil, "CA World to Focus on E-Commerce," *Computerworld*, April 10, 2000, accessed at http://www.computerworld.com/cwi/.
4. Thomas Hoffman, "Neural Nets Spot Credit Risks," *Computerworld*, July 26, 1999, accessed at http://www.computerworld.com/cwi/story/0,1199,NAV47_STO36440,00.html.
5. "What Is CLIPS," accessed at http://www.ghg.net/clips/CLIPS.htm on July 7, 2000.
6. Jane Ellen Stevens, "Virtual Reality Hearts and Minds," accessed at http://www.msnbc.com/news on July 9, 2000.
7. Lynne Cox, "Cruising Down the Nile," "VR in the Schools," Vol. 4, no. 3, accessed at http://www.ecu.edu/vr/vrits/4-3Cox.htm.

PART IV

Systems Development

CHAPTER 12

Systems Investigation and Analysis

Large-scale software is being delivered on time and under budget. More important, a development project now sets 90 days for the key first deliverable [part of a new system] and no more than six months for full implementation.

— Peter Keen, consultant

Principles	Learning Objectives
Effective systems development requires a team effort of stakeholders, users, managers, systems development specialists, and various support personnel, and it starts with careful planning.	• *Identify the key participants in the systems development process and discuss their roles.* • *Define the term* information systems planning *and list several reasons for initiating a systems project.*
Systems development often uses tools to select, implement, and monitor projects, including net present value (NPV), prototyping, and rapid application development.	• *Identify important system performance requirements of transaction processing applications that run on the Internet or a corporate intranet.* • *Discuss three trends that illustrate the impact of enterprise resource planning software packages on systems development.* • *Discuss the key features, advantages, and disadvantages of the traditional, prototyping, rapid application development, and end-user systems development life cycles.* • *Identify several factors that influence the success or failure of a systems development project.*
Systems development starts with investigation and analysis of existing systems.	• *State the purpose of systems investigation.* • *State the purpose of systems analysis and discuss some of the tools and techniques used in this phase of systems development.*

Y2K Aftermath

IS Investments on the Rise

The Y2K problem threatened to be one of the most devastating systems development nightmares of the last century, so companies around the world invested heavily in new hardware and software to head it off. Some estimate that U.S. companies spent $100 billion or more to solve the problem. The Y2K glitch resulted from the old practice of storing dates as two digits in the computer to save expensive storage space. The year 1968 would be stored as "68," for example. Using these old programs, the year 2000 would be stored as "00" and retrieved as "1900" instead of "2000." If they hadn't fixed the problem, companies would have been forced to shut down, airlines might have been grounded, and the economy might have been damaged. Before the new year rolled by, some predicted total social disruption around the world. In this frenzied atmosphere, companies' IS budgets soared in 1999. Merrill Lynch's IS budget increased from $200 million to more than $500 million. Nabisco's IS budget increased from roughly $22 million to more than $40 million to solve the Y2K problem alone. Many other IS projects were put on hold in 1999 to devote needed resources to fixing the Y2K bug.

As the new millennium approached and passed, however, the dire predictions didn't come true. There were few glitches. Slot machines in Delaware were inoperable for a few hours. A mutual fund company had delays for a short time, and Harrah's reservation system put incorrect dates on a few reports. A few other companies also had minor problems, but there were no major computer outages. According to Tony Del Duca, vice president of enterprise supply chain systems at Nabisco, "It was a big yawn." After all of the Y2K expenses, one big surprise was the unleashing of an unprecedented number of systems development projects in 2000.

Instead of scaling back on systems development efforts and expenditures, many companies launched aggressive and impressive systems development projects. Sears, Roebuck, for example, launched a massive freight development project. The completion time for this and other IS projects for Sears has been mandated at an aggressive six months. "I think all of us wanted to do a lot of things last year, but we were reluctant to take the focus away from Y2K," commented Jerry Miller, chief information officer for Sears. Other companies also rushed into new systems development efforts. In mid-January 2000, Allstate lifted the ban on new systems development projects. Now the company is launching a completely new business model to include a number of important new systems, including development of new call centers and Internet initiatives. The Yankee Energy System also lifted its systems development ban in mid-January to complete a new merger deal with Northeast Utilities. "We'll spend the rest of 2000 focusing on integrating systems," said Scott Waleske, director of IT for Yankee. Railroad giant Union Pacific is also launching an aggressive, multimillion-dollar systems development project to complete a customer relationship management software project.

As you read this chapter, consider the following:

- What impact did Y2K have on systems development?
- Do you think that systems development projects will decrease, stay the same, or increase in the future?

When an organization needs to accomplish a new task or change a work process, how does it do it? It develops a new system or modifies an existing one. Systems development is the activity of creating or modifying existing business systems. It refers to all aspects of the process—from identifying problems to be solved or opportunities to be exploited to the implementation and refinement of the chosen solution. As seen in the opening vignette, many companies are launching systems development efforts. When it comes to developing Internet applications, multiple companies are often involved. Equiva Trading, a joint venture of Shell Oil, Texaco, and Saudi Arabian Oil, has invested more than $6 million in a systems development effort to build an oil trading site.[1]

Systems development can mean the success or failure of the entire organization. Successful systems development has resulted in huge increases in revenues and profits. Companies that don't innovate with new systems development initiatives or fail to successfully complete a systems development effort can lose millions. Understanding and being able to apply a systems development life cycle, tools, and techniques (discussed in this chapter and the next) will help ensure the success of the development projects in which you participate. It can also help your career and the financial success of the company.

AN OVERVIEW OF SYSTEMS DEVELOPMENT

Understanding systems development is important to all professionals, not just those in the field of information systems. In today's businesses, managers and employees in all functional areas work together and use business information systems. As a result, users of all types are helping with development and, in many cases, leading the way. This chapter and the next will provide you with a deeper appreciation of the systems development process and help you avoid costly failures.

Participants in Systems Development

Effective systems development requires a team effort. The team usually consists of stakeholders, users, managers, systems development specialists, and various support personnel.[2] This team, called the development team, is responsible for determining the objectives of the information system and delivering a system that meets these objectives to the organization. In the context of systems development, **stakeholders** are individuals who, either themselves or through the area of the organization they represent, ultimately benefit from the systems development project. **Users** are individuals who will interact with the system regularly. They can be employees, managers, customers, or suppliers. For large-scale systems development projects, where the investment in and value of a system can be quite high, it is common to have senior-level managers, including the company president and functional vice presidents (of finance, marketing, and so on), as part of the development team.

Depending on the nature of the systems project, the development team might also include systems analysts and programmers, among others. A **systems analyst** is a professional who specializes in analyzing and designing business systems. Systems analysts play various roles while interacting with the stakeholders and users, management, vendors and suppliers, external companies, software programmers, and other IS support personnel (Figure 12.1). Like an architect developing blueprints for a new building, a systems analyst develops detailed plans for the new or modified system. The **programmer** is responsible for modifying or developing programs to satisfy user requirements. Like a contractor constructing a new building or renovating an existing one, the programmer takes the plans from the systems analyst and builds or modifies the necessary software.

stakeholders

individuals who, either themselves or through the area of the organization they represent, ultimately benefit from the systems development project

users

individuals who will interact with the system regularly

systems analyst

professional who specializes in analyzing and designing business systems

programmer

specialist responsible for modifying or developing programs to satisfy user requirements

FIGURE 12.1

The systems analyst plays an important role in the development team and is often the only person who sees the system in its totality. The one-way arrows in this figure do not mean that there is no direct communication between other team members. Instead, they indicate the pivotal role of the systems analyst—an individual who is often called upon to be a facilitator, moderator, negotiator, and interpreter for development activities.

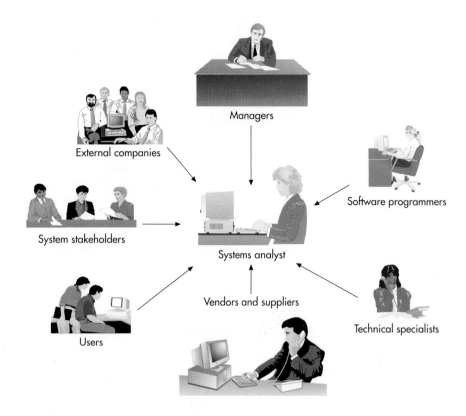

The other support personnel on the development team are mostly technical specialists, including database and telecommunications experts, hardware engineers, and supplier representatives. One or more of these roles may be outsourced to nonemployees or consultants. Depending on the magnitude of the systems development project and the number of IS systems development specialists on the team, the team may also include one or more IS managers. The composition of a development team may vary over time and from project to project. For small businesses, the development team may consist of a systems analyst and the business owner as the primary stakeholder. For larger organizations, formal IS staff can include hundreds of people involved in a variety of IS activities, including systems development. Every development team should have a team leader. As shown in Table 12.1, different types of team leaders should be considered for different projects and development teams.

Increasingly, companies are getting participants for their systems development efforts from other companies. Ford Motor, for example, is paying IBM about $300 million over five years to develop important software.[3] Wal-Mart

TABLE 12.1

Team Leaders for Different Systems Development Projects

Characteristics of Systems Development Project	Examples of Appropriate Team Leaders
Project involves new and advanced technology.	Individual from IS department
Impact will force critical changes in a functional area of the business.	Manager from that functional area
Project is extremely large and complex.	Specialist in project management
Project will have dramatic impact on personnel.	Individual from human resource department
Project will have a broad and important impact on the entire organization.	Senior management; leader should build a development team that includes personnel skilled in all affected areas
Project will involve two or more companies, such as an Internet exchange.	Senior executive or project manager with experience in starting and managing a new company; this team leader is often hired from outside

formed a partnership with Accel Partners, a venture capital firm, to develop an on-line retail Internet site.[4]

At some point in your career, you will likely be a participant in systems development. You could be involved in a systems development team—as a user, manager of a business area or project team, member of the IS department, maybe even as a CIO. With the increasing power and easy-use software suites and the Internet, you could also be involved with your own systems development project. You might develop a spreadsheet to analyze financial alternatives or a personal Web site.

Regardless of the specific nature of the project, systems development involves new or modified systems, which ultimately means change. Managing this change effectively requires development team members to communicate well. It is important that you learn communication skills, because you probably will participate in systems development during your career. You may even be the individual who initiates systems development.

Initiating Systems Development

Systems development efforts begin when an individual or group capable of initiating organizational change perceives a potential benefit from a new or modified system. Such individuals have a stake in the development of the system. Executives at Pets.com, for example, initiated a systems development project to help the company increase its market share on the Internet.[5] Not only did it cost the company to develop the system, but operating the Internet site was also expensive. Records reveal that the company may lose 57 cents on every dollar of sales. The company hopes to turn a profit when it increases market share and its presence on the Internet. Executives at A&P, a large grocer, initiated a systems development effort. They hope the project will save the company $300 million over four years.[6]

Equally important, however, is the ability of the individual or group to initiate organizational change. Many individuals cannot initiate change because they are constrained by hierarchical status, power, authority, and political standing within the organization. Managers are most often empowered by an organization to initiate change, and thus they most often initiate systems development projects. In addition, the degree of overall managerial support, especially senior-level managerial support, greatly influences the probable success or failure of a systems development effort.

Systems development initiatives arise from all levels of an organization and are both planned and unplanned. Solid planning and managerial involvement helps ensure that these initiatives support broader organizational goals. Systems development projects may be initiated for a number of reasons, as shown in Figure 12.2.

Mergers normally initiate many systems development projects.[7] In 1999, there were more than ten major corporate mergers, each with more than $15 billion in merged assets. These large mergers included telecommunications, drug, oil, and large banking companies. In addition, there were many smaller mergers. Because the information systems for these companies are normally different, a large systems development effort is typically required to unify systems. Even with similar information systems, the procedures, culture, training, and management of the information systems are typically different, requiring a realignment of the IS departments.

The external environment is another cause for new systems development. Changes can involve government or legal action. Chase Manhattan, for example, reached a settlement regarding the company's privacy policies with the

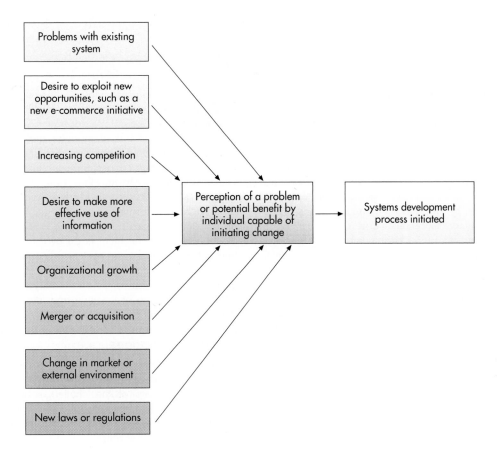

FIGURE 12.2

Typical Reasons to Initiate a
Systems Development Project

New York state attorney general's office.[8] In this case, Chase agreed not to share client information with other businesses unless it received clear permission. These types of agreements often require the IS department to initiate a systems development effort to change corporate information systems.

Information Systems Planning

Because an organization's strategic plan contains both organizational goals and a broad outline of steps required to reach them, the strategic plan affects the type of system an organization needs. For example, a strategic plan may identify as organizational goals a doubling of sales revenue within five years, a 20 percent reduction of administrative expenses over three years, acquisition of at least two competing companies within a year, or the capture of market leadership in a given product category. Organizational commitments to policies such as continuous improvement are also reflected in the strategic plan. Such goals and commitments set broad outlines of system performance. As discussed in the "E-Commerce" box, information systems planning can often involve two or more companies.

Often, a section of the strategic plan lists guidelines about how to meet organizational goals. Examples of these guidelines might be improving customer service for luxury car buyers, exploring the purchase of existing international distributors, and using a specific amount of money to buy back company stock. The strategic plan also provides direction to the functional areas within an organization, including marketing, production, finance, accounting, and human resources. For the information systems department, these directions are encompassed in the information systems plan.

E-COMMERCE

Developing an Internet Site for Suppliers and Dealers

Good information systems planning is a key for successful systems development. In most cases, IS planning starts with a strategic plan. The strategic perspective leads to general IS planning, which in turn can be converted into specific systems development initiatives. Today, strategic planning that can lead to systems development initiatives is often started by a partnership with two or more companies.

To stay competitive in the Internet age, Ford originally decided to form an alliance with Cisco. Cisco is one of the largest and most successful Internet companies; it provides the infrastructure for a large number of e-commerce operations. The strategic plan for the Ford-Cisco alliance was to build a comprehensive Internet site for suppliers. The plan called for Cisco to aid in wiring more than 40,000 suppliers and dealers to the Internet site. In return, Cisco would obtain an equity stake in the new auto exchange, along with Oracle, a database company. According to Sue Bostrom, vice president of the Internet business solutions group at Cisco, "We hear from customers across all industries. They want to move as quickly as possible to get connected." The specific systems development initiatives will be huge. According to a spokesperson for Cisco, the company has never done systems development on this scale.

More recently, the Ford-Cisco alliance has been joined by other companies and given an official name, Covisint. In addition to Ford, the Covisint exchange now includes General Motors, DaimlerChrysler, Nissan, and Renault. It is hoped that Covisint will save the auto industry billions of dollars and allow designers to save months in preproduction planning. The production cost savings per car could be over $2,000 on average if Covisint achieves its goals. Eventually, Covisint could allow a customer to custom order a car and to receive the car within weeks after the order is placed. "GM says that when somebody in a dealership orders a car with leather seats, the cow should wince," says Doug Grimm, a representative of an automotive supply company.

The Covisint systems development effort reveals a problem often faced by systems developers. Covisint can develop a fast, efficient Internet site, but will the thousands of suppliers also modify their systems? Although slow modems will be able to access the Covisint site, the real benefits to the auto companies and their suppliers will be achieved only if suppliers also upgrade their systems, including their connection to the Internet. If the suppliers fail to upgrade, the new system will be underutilized. According to Mark Duhaime, director of program release for the auto exchange, "There are a lot of suppliers that are using the site in a 'kick the tires' model. This announcement is an evolution from fingers and a browser to having system-to-system connectivity." The question remains, however, whether suppliers will get on the Internet bandwagon and modify their systems to take full advantage of the new and impressive systems development effort by Covisint.

Discussion Questions

1. What is the relationship between strategic planning and systems development initiatives?
2. What are some of the challenges facing systems developers when several companies form a partnership trying to connect thousands of suppliers to a comprehensive Internet site?

Critical Thinking Questions

3. Assume that you are in charge of this massive systems development effort. How would you ensure that suppliers upgraded their systems to take full advantage of the new auto exchange?
4. What types of company-supplier issues would have to be resolved in the IS planning stage to ensure a successful systems development effort?

Sources: Adapted from Fara Warner, "Ford, Cisco Team Up," *The Wall Street Journal*, February 10, 2000, p. A5; "Cisco Joins Oracle in Ford Exchange," *Computing*, February 17, 2000, p. 2; "Where E-Commerce Wins Hands Down," *The Economist*, February 26, 2000; and Steve Konicke, "Covisint's Rough Road," *Information Week*, August 2, 2000.

---

information systems planning

the translation of strategic and organizational goals into systems development initiatives

The term **information systems planning** refers to the translation of strategic and organizational goals into systems development initiatives (Figure 12.3). The Marriott hotel chain, for example, invites its chief information officer to board meetings and other top-level management meetings. Proper IS planning ensures that specific systems development objectives support organizational goals.[9] Not translating the strategic plan into a specific IS plan can be disastrous.[10]

One of the primary benefits of IS planning is a long-range view of information technology use in the organization.[11] Specific systems development initiatives may spring from the IS plan, but the plan must also provide a broad framework for future success. The IS plan should provide guidance on how the information systems infrastructure of the organization should be developed

FIGURE 12.3

Information systems planning transforms organizational goals outlined in the strategic plan into specific system development activities.

creative analysis

the investigation of new approaches to existing problems

over time. Another benefit of IS planning is that it ensures better use of information systems resources, including funds, IS personnel, and time for scheduling specific projects. The steps of IS planning are shown in Figure 12.4.

Overall IS objectives are usually distilled from the relevant aspects of the organization's strategic plan. IS projects can be identified either directly from the objectives determined in the first step or by others, such as managers within the various functional areas. Setting priorities and selecting projects typically requires the involvement and approval of senior management. When objectives are set, planners consider the resources necessary to complete the projects, including employees (systems analysts, programmers, and others), equipment (computers, network servers, printers, and other devices), expert advice (from specialists and other consultants), and software, among others.

Developing a Competitive Advantage

As part of translating the corporate strategic plan into the information systems plan, many companies seek systems development projects that will provide a competitive advantage. Thinking competitively usually requires creative and critical analysis. For example, a company may want to achieve a competitive advantage by improving the customer-supplier relationship.[12] Linking customers and suppliers can result in superior products and services. Pillsbury is developing software to help it analyze large volumes of data to give the food giant a competitive advantage.[13] The software will download and analyze customer feedback data from the Internet.

Creative analysis involves the investigation of new approaches to existing problems. By looking at problems in new or different ways and by introducing

FIGURE 12.4

The Steps of IS Planning

Some projects are identified through overall IS objectives, whereas additional projects, called unplanned projects, are identified from other sources. All identified projects are then evaluated in terms of their organizational priority.

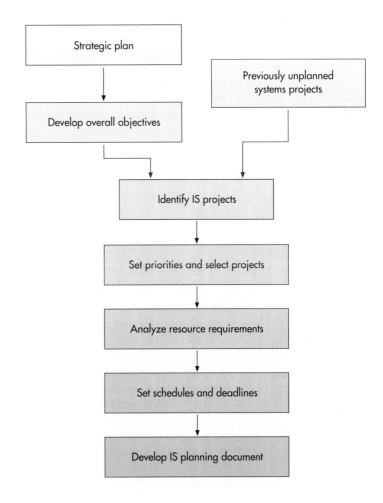

innovative methods to solve them, many firms have gained a competitive advantage. Typically, these new solutions are inspired by people and events not directly related to the problem.

critical analysis

the unbiased and careful questioning of whether system elements are related in the most effective and efficient ways

Critical analysis requires unbiased and careful questioning of whether system elements are related in the most effective and efficient ways. It involves considering the establishment of new or different relationships among system elements and perhaps introducing new elements into the system. Critical analysis in systems development involves the following actions:

- *Going beyond automating manual systems.* Many organizations use systems development simply to automate existing manual systems, which may result in relatively faster, more efficient systems. However, if the underlying manual system is flawed, automating it might just magnify its impact. In addition, automating existing manual systems might miss many opportunities by only considering doing things in the same old way. For example, rather than automate current customer service systems, many companies are implementing customer service systems based on use of the Internet and Web technology so that customers can provide a high degree of self-service. Critical analysis in systems development involves asking why things are done a certain way and considering alternative approaches.
- *Questioning statements and assumptions.* Questioning users about their needs and clarifying their initial responses can result in better systems and more accurate predictions. Too often, stakeholders and users specify certain system requirements because they assume their needs can be met only that way. Often, an alternative approach would be better. For example, a stakeholder may be concerned because there is always too much of some items in stock and not enough of other items. So, the stakeholder might request a totally new and improved inventory control system. An alternative approach is to identify the root cause for poor inventory management. This latter approach might determine that sales forecasting is inaccurate and needs improvement or that production is not capable of meeting the set production schedule. All too often, solutions are selected before a complete understanding of the nature of the problem itself is obtained.
- *Identifying and resolving objectives and orientations that conflict.* Different departments in an organization can have different objectives and orientations. The buying department may want to minimize the cost of spare parts by always buying from the lowest-cost supplier, but engineering might want to buy more expensive, higher-quality spare parts to reduce the frequency of replacement. These differences must be identified and resolved before a new purchasing system is developed or an existing one modified.

Establishing Objectives for Systems Development

The impact a particular system has on an organization's ability to meet its goals determines the true value of that system to the organization. Although all systems should support business goals, some systems are more pivotal in continued operations and goal attainment than others. These systems are called **mission-critical systems**. An order processing TPS, for example, is usually considered mission-critical. Without it, few organizations could continue daily activities, and they clearly would not meet set goals.

mission-critical systems

systems that play a pivotal role in an organization's continued operations and goal attainment

critical success factors (CSFs)

factors essential to the success of certain functional areas of an organization

The goals defined for an organization will in turn define the objectives set for a system. A manufacturing plant, for example, might determine that minimizing the total cost of owning and operating its equipment is a critical success factor in meeting a production volume and profit goals. **Critical success factors (CSFs)** are factors that are essential to the success of certain functional areas of an organization. This CSF would be converted into specific objectives for a proposed plant equipment maintenance system. One specific objective might be to alert maintenance planners when a piece of equipment is due for routine preventive maintenance (e.g., cleaning and lubrication). Another objective might be to alert the maintenance planners when the necessary cleaning materials, lubrication oils, or spare parts inventory levels are below specified limits. These objectives could be accomplished either through automatic stock replenishment via electronic data interchange or through the use of exception reports.

Regardless of the particular systems development effort, the development process should define a system with specific performance and cost objectives. The success or failure of the systems development effort will be measured against these objectives.

Performance Objectives

The extent to which a system performs as desired can be measured through its performance objectives. System performance is usually determined by such factors as the following:[14]

- *The quality or usefulness of the output.* Is the system generating the right information for a value-added business process or by a goal-oriented decision maker?
- *The quality or usefulness of the format of the output.* Is the output generated in a form that is usable and easily understood? For example, objectives often concern the legibility of screen displays, the appearance of documents, and the adherence to certain naming conventions.
- *The speed at which output is generated.* Is the system generating output in time to meet organizational goals and operational objectives? Objectives such as customer response time, the time to determine product availability, and throughput time are examples.

In some cases, the achievement of performance objectives can be easily measured (e.g., by tracking the time it takes to determine product availability). The achievement of performance objectives is sometimes more difficult to ascertain in the short term. For example, it may be difficult to determine how many customers are lost because of slow responses to customer inquiries regarding product availability. These outcomes, however, are often closely associated with corporate goals and are vital to the long-term success of the organization. Their attainment is usually dictated by senior management.

Cost Objectives

The benefits of achieving performance goals should be balanced with all costs associated with the system, including the following:

- *Development costs.* All costs required to get the system up and running should be included.
- *Costs related to the uniqueness of the system application.* A system's uniqueness has a profound effect on its cost. An expensive but reusable system may be preferable to a less costly system with limited use.
- *Fixed investments in hardware and related equipment.* Developers should consider costs of such items as computers, network-related equipment, and environmentally controlled data centers in which to operate the equipment.
- *Ongoing operating costs of the system.* Operating costs include costs for personnel, software, supplies, and such resources as the electricity required to run the system.

The costs of the current system or problem-solving method are subtracted from the costs associated with a given level of performance, control, and

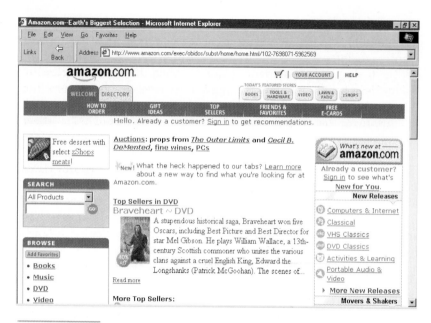

Amazon.com is a company that uses the Internet as its only sales channel.

complexity in the new or modified system. The resulting figure is an incremental cost differential, for which an objective is set.

Balancing performance and cost objectives within the overall framework of organizational goals can be challenging. Systems development objectives are important, however, in that they allow an organization to allocate resources effectively and efficiently and measure the success of a systems development effort.

Systems Development and E-Commerce

As seen in the many examples and cases in this and other chapters, companies are increasingly converting at least some portion of their business to run over the Internet, intranets, or extranets. Applications that are being moved to the Internet include those that support selling products to customers, placing orders with suppliers, and letting customers access information about production, inventory, orders, or accounts receivable. In addition, a number of companies such as Amazon.com sell their products and services only on the Internet. Internet technology provides a platform that enables companies to extend their information systems beyond their boundaries to their customers, suppliers, and partners, allowing them to conduct business much faster, to interact with more people, and to try to keep one step ahead of the competition. Some companies have been willing to realize sustained losses from their e-commerce initiatives to gain the boundary-expanding possibilities of the Internet.

Building a static Web site to display simple text and graphics is fairly straightforward. However, implementing a dynamic core business application that runs over the Web is much more complicated. Such applications must meet special business needs. They must be able to scale up to support highly variable transactions from potentially thousands of users. Ideally, they can scale up instantly when needed. They must be reliable and fault tolerant, providing continuous availability while processing all transactions accurately. They must also integrate with existing infrastructure, including customer and order databases, existing applications, and enterprise resource planning systems. Development and maintenance must be quick and easy, as business needs may require changing applications on the fly.

There are many tools available for building and running Web applications. The best tools provide components to support applications on an enterprise scale while speeding development. Several vendors—including NetDynamics, SilverStream, WebLogic, Novera Software, Netscape Communications, and IBM—provide what is known as an applications server to provide remote access to databases via a corporate intranet. Thomson Financial Services used an application server to build two applications—one tracks job candidates and the second monitors consultants' work hours.

Trends in Systems Development and Enterprise Resource Planning

Enterprise resource planning software has reached beyond business processes and is increasingly affecting systems development. Not only are planners considering different types of systems that include ERP software, but the ERP software

IBM provides consulting and implementation services for ERP packages. IBM also operates and manages ERP applications and complete ERP environments for clients.

that is already in place is driving planners to explore and develop different types of systems. Other ERP users are moving from just using the software to run the business to trying to use it to make business decisions.

An important trend in systems development and the use of ERP systems is that companies wish to stay with their primary ERP vendor (SAP, Oracle, PeopleSoft, etc.) instead of looking elsewhere for answers to their data warehousing and production planning needs or developing in-house solutions. Thus, they look to their original ERP vendor to provide these solutions.

A second trend is that many software vendors are building software that integrates with the ERP vendor's package. For example, Aspect Development, a leading supplier of electronic catalogs for high-tech components, has now entered the market for on-line catalogs of maintenance, repair, and operations (MRO) supplies. Its product Morocco is an MRO supplies catalog for users of SAP's ERP software. Morocco works in conjunction with the purchasing module of SAP and directs buyers to preferred suppliers specified by the purchasing department. Again, there is less in-house development and more dependence on ERP vendors and their strategic partners to provide enhancements and add-ons to the original ERP package.

A third interesting trend is the increase in the number of companies that, once they have successfully implemented their own company's ERP project, are branching out to provide consulting to other companies.

SYSTEMS DEVELOPMENT LIFE CYCLES

The systems development process is also called a systems development life cycle (SDLC) because the activities associated with it are ongoing. As each system is being built, the project has timelines and deadlines, until at last the system is installed and accepted. The life of the system continues as it is maintained and reviewed. If the system needs significant improvement beyond the scope of maintenance, if it needs to be replaced because of a new generation of technology, or if the IS needs of the organization change significantly, a new project will be initiated and the cycle will start over.

To help IS personnel acquire the skills needed to perform the activities required of the systems development life cycle, a number of companies and organizations offer training.[15] These training programs are offered all over the world in a variety of subject areas, including general systems development, determining system requirements, structured programming, network analysis and design, developing Internet content, auditing, project management, operating systems, financial considerations, and more.

A key fact of systems development is that the later in the SDLC an error is detected, the more expensive it is to correct (Figure 12.5). (This analysis is documented in a classic work by Barry Boehm.)[16] One reason for the mounting costs is that if an error is found in a latter phase of the SDLC, the previous phases must be reworked to some extent. Another reason is that the errors found late in the SDLC have an impact on more people. For example, an error found after a system is installed may require retraining users once a

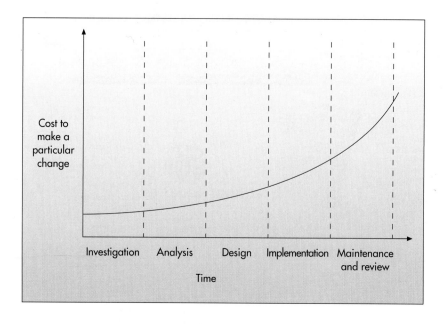

"workaround" to the problem has been found. Thus, experienced system developers prefer an approach that will catch errors early in the project life cycle.

Four common systems development life cycles exist: traditional, prototyping, rapid application development (RAD), and end-user development. In some companies, these approaches are formalized and documented so that system developers have a well-defined process to follow; in other companies, less formalized approaches are used. Keep Figure 12.5 in mind as you are introduced to alternative SDLCs in the next section.

The Traditional Systems Development Life Cycle

Traditional systems development efforts can range from a small project, such as purchasing an inexpensive computer program, to a major undertaking. The steps of traditional systems development may vary from one company to the next, but most approaches have five common phases: investigation, analysis, design, implementation, and maintenance and review (Figure 12.6).

In the *systems investigation* phase, potential problems and opportunities are identified and considered in light of the goals of the business. Systems investigation attempts to answer the question "What is the problem, and is it worth solving?" The primary result of this phase is a defined information system project for which business problems or opportunity statements have been created, to which some organizational resources have been committed and for which systems analysis is recommended. *Systems analysis* attempts to answer the question "What must the information system do to solve the problem?" This phase involves the study of existing systems and work processes to identify strengths, weaknesses, and opportunities for improvement. The major outcome of systems analysis is a list of requirements and priorities. *Systems design* seeks to answer the question "How will the information system do what it must do to obtain the problem solution?" The primary result of this phase is a technical design that either describes the new system or describes how existing systems will be modified. The system design details system outputs, inputs, and user interfaces; specifies hardware, software, database, telecommunications, personnel, and procedure components; and shows how these components are related. *Systems implementation* involves creating or acquiring the various system components

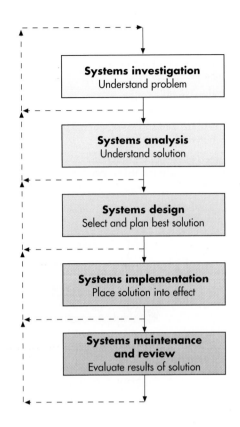

detailed in the systems design, assembling them, and placing the new or modified system into operation. An important task during this phase is to train the users. Systems implementation results in an installed, operational information system that meets the business needs for which it was developed. The purpose of *systems maintenance and review* is to ensure that the system operates and to modify the system so that it continues to meet changing business needs. As shown in Figure 12.6, a system under development moves from one phase of the traditional SDLC to the next.

The traditional SDLC allows for a large degree of management control. At the end of each phase, a formal review is performed and a decision is made whether to continue with the project, terminate the project, or perhaps repeat some of the tasks of the current phase. Use of the traditional SDLC also creates much documentation, such as entity relationship diagrams. This documentation, if kept current, can be useful when it is time to modify the system. The traditional SDLC also ensures that every system requirement can be related to a business need. In addition, resulting products can be reviewed to verify that they satisfy the system requirements and conform to organizational standards.

A major problem with the traditional SDLC is that the user does not use the solution until the system is nearly complete. Quite often, users get a system that does not meet their real needs because its development was based on the development team's understanding of the needs. The traditional approach is also inflexible. Changes in user requirements cannot be accommodated during development. In spite of its limitations, however, the traditional SDLC is still used for large, complex systems that affect entire businesses, such as TPS and MIS systems. It is also frequently employed on government projects because of the strengths mentioned previously. Table 12.2 lists advantages and disadvantages of the traditional SDLC.

Advantages	Disadvantages
Formal review at the end of each phase allows maximum management control.	Users get a system that meets the needs as understood by the developers; this may not be what was really needed.
This approach creates considerable system documentation.	Documentation is expensive and time-consuming to create. It is also difficult to keep current.
Formal documentation ensures that system requirements can be traced back to stated business needs.	Often, user needs go unstated or are misunderstood.
It produces many intermediate products that can be reviewed to see whether they meet the users' needs and conform to standards.	Users cannot easily review intermediate products and evaluate whether a particular product (e.g., data flow diagram) meets their business requirements.

TABLE 12.2

Advantages and
Disadvantages of Traditional
SDLC

prototyping

an iterative approach to the systems
development process

Prototyping

Prototyping takes an iterative approach to the systems development process. During each iteration, requirements and alternative solutions to the problem are identified and analyzed, new solutions are designed, and a portion of the system is implemented. Users are then encouraged to try the prototype and provide feedback (Figure 12.7). Prototyping begins with the creation of a preliminary model of a major subsystem or a scaled-down version of the entire system. For example, a prototype might be developed to show sample report formats and input screens.[17] Once developed and refined, the prototypical reports and input screens are used as models for the actual system, which may

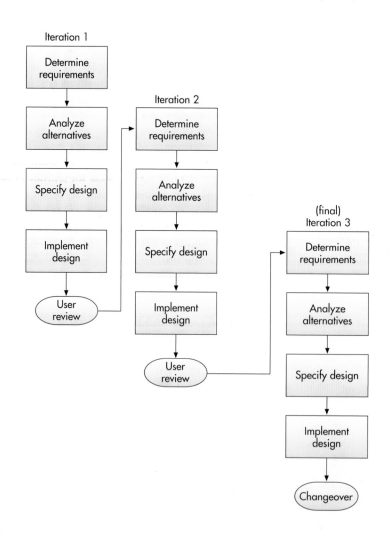

FIGURE 12.7

Prototype Is an Interactive
Approach to Systems
Development

Prototyping is a popular technique in systems development. Each generation of prototype is a refinement of the previous generation based on user feedback.

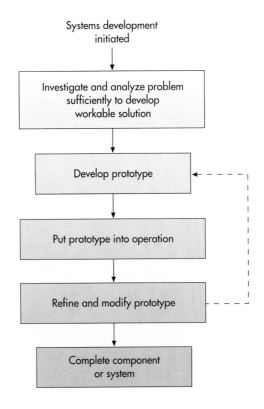

be developed using an end-user programming language such as SAS, Focus, or Visual Basic. The first preliminary model is refined to form the second- and third-generation models, and so on until the complete system is developed (Figure 12.8).

Types of Prototypes

Prototypes can be classified as operational or nonoperational. An **operational prototype** is a prototype that works—accesses real data files, edits input data, makes necessary computations and comparisons, and produces real output. Fully developed financial reports are examples. The operational prototype may access real files but perhaps does no editing of input. A **nonoperational prototype** is a mockup, or model. It typically includes output and input specifications and formats. The outputs include printed reports to managers and the screen layout of reports displayed on personal computers or terminals. The inputs reveal how data is captured, what commands users must enter, and how the system accesses other data files. The primary advantage of a nonoperational prototype is that it can be developed much faster than an operational prototype. Nonoperational prototypes can be discarded, and a fully operational system can be built based on what was learned from the prototypes. The advantages and disadvantages of prototyping are summarized in Table 12.3.

Rapid Application Development and Joint Application Development

Rapid application development (RAD) employs tools, techniques, and methodologies designed to speed application development.[18] RAD reduces paper-based documentation, automates program source code generation, and facilitates user participation in design and development activities. With RAD, entire systems are developed in less than six months. The ultimate goal is to accelerate the process so that applications can go into production much sooner than when using other approaches.

---

operational prototype

a functioning prototype that accesses real data files, edits input data, makes necessary computations and comparisons, and produces real output

nonoperational prototype

a mockup, or model, that includes output and input specifications and formats

rapid application development (RAD)

a systems development approach that employs tools, techniques, and methodologies designed to speed application development

Advantages	Disadvantages
Users can try the system and provide constructive feedback during development.	Each iteration builds on the previous iteration and further refines the solution. This makes it difficult to reject the initial solution as inappropriate and start over. Thus, the final solution will be only incrementally better than the first.
An operational prototype can be produced in weeks.	Formal end-of-phase reviews do not occur. Thus, it is very difficult to contain the scope of the prototype, and the project never seems to end.
Users become more positive about implementing the system as they see a solution emerging that will meet their needs.	System documentation is often absent or incomplete, since the primary focus is on development of the prototype.
Prototyping enables early detection of errors and omissions.	System backup and recovery, performance, and security issues can be overlooked in the haste to develop a prototype.

TABLE 12.3

Advantages and Disadvantages of Prototyping

joint application development (JAD)

process for data collection and requirements analysis

RAD makes extensive use of the **joint application development (JAD)** process for data collection and requirements analysis. Originally developed by IBM Canada in the 1970s, JAD involves group meetings in which users, stakeholders, and IS professionals work together to analyze existing systems, propose possible solutions, and define the requirements of a new or modified system. JAD groups consist of both problem holders and solution providers. A group normally requires one or more top-level executives who initiate the JAD process, a group leader for the meetings, potential users, and one or more individuals who act as secretaries and clerks to record what is accomplished and to provide general support for the sessions. Many companies have found that groups can develop better requirements than individuals working independently and have assessed JAD as a very successful development technique.

The use of software tools to support RAD has flourished. For example, Inprise Corporation is developing a Linux rapid application development tool, code named Kylix.[19] The new RAD tool will support the C and C++ programming languages. PowerBuilder, a RAD tool from Sybase, is popular in the federal government sector. In addition, such database vendors as Computer Associates International, Informix Software, and Oracle market fourth-generation languages and other products targeting the RAD market. Tools from such traditional vendors as Texas Instruments and Sterling Software are also being deployed as RAD products.

Throughout the RAD project, users and developers work together as one team. This teamwork promotes healthy risk taking and team-based decision making, resulting in better systems with shorter delivery dates. If the entire system is too large to be completed in less than six months, it is broken into subsystems and is delivered subsystem by subsystem. The first subsystem may be delivered in three to four months, with no delivery date more than six months after the last one. This subdividing leads to less waste, because even if there is a serious error in the system, only one subsystem has to be rebuilt. A team at the Louisiana Department of Natural Resources, for example, was able to use RAD in developing its Web site.[20] The use of RAD cut the

PowerBuilder, a RAD tool from Sybase, is used in both the public and private sectors. (Source: Courtesy of Sybase, Inc.)

Advantages	Disadvantages
For appropriate projects, this approach puts an application into production sooner than any other approach.	This intense SDLC can burn out systems developers and other project participants.
Documentation is produced as a by-product of completing project tasks.	This approach requires systems analysts and users to be skilled in RAD system development tools and RAD techniques.
RAD forces teamwork and lots of interaction between users and stakeholders.	RAD requires a larger percentage of stakeholders' and users' time than other approaches.

TABLE 12.4

Advantages and Disadvantages of RAD

development and deployment time in half. According to the director of the IS department, "Our vision was to come up with a system that would allow not only the citizens of Louisiana, but people from throughout the world, to access our data and not pay a fee." Although the department does not charge for the information on its Web site, it collects revenue from royalties. The Minerals Division collects more than $500 million on the state's oil and gas leases.

RAD should not be used on every software development project. In general, it is best suited for decision support and management information systems and less well suited for transaction processing applications. During a RAD project, the level of participation on the part of stakeholders and users is much higher than in other approaches. They become working members of the team and can be expected to spend more than 50 percent of their time producing project outcomes. This time commitment can be a problem if the users are also needed to perform their normal business role. For this reason, RAD team participants are often taken off their normal assignments and put full-time on the RAD project. Because of the full-time commitment and intense schedule deadlines, RAD is a high-pressure development approach that can easily result in employee burnout. Table 12.4 lists advantages and disadvantages of RAD.

end-user systems development

any systems development project in which the primary effort is undertaken by a combination of business managers and users

Many end users today are developing their own PC-based systems with technical assistance from IS personnel. (Source: © 2000 Photodisc)

The End-User Systems Development Life Cycle

Systems development initiatives arise from a wide variety of individuals and organizational areas, including users. The proliferation of general-purpose information technology and the flexibility of many packaged software programs have allowed non-IS employees to independently develop information systems that meet their needs. Such employees have believed that, by bypassing the formal requisitioning of resources from the IS department, they can develop systems more quickly. In addition, these individuals often believe that they have better insight into their own needs and can develop systems better suited for their purposes.

End-user-developed systems range from the very small (e.g., a software routine to merge form letters) to those of significant organizational value (such as customer contact databases). Like all projects, some end-user-developed systems fail, and others are successful. Initially, IS professionals discounted the value of these projects. As the number and magnitude of these projects increased, however, IS professionals began to realize that for the good of the entire organization, their involvement with these projects needed to increase.

Today, the term **end-user systems development** describes any systems development project in which the primary effort is undertaken by a combination of business managers and users. Rather than ignoring these initiatives, astute IS professionals encourage them by offering guidance and support. Technical assistance, communication of standards, and the sharing of "best practices" throughout the organization are just some of the ways IS professionals work with motivated managers and employees undertaking their own systems development. In this way, end-user-developed systems can be structured as complementary to, rather than in conflict with, existing and emerging information systems. In addition, this open communication among IS professionals, managers of the affected business area, and users allows the IS professionals to identify specific initiatives so that additional organizational resources, beyond the prerogative of the initiating business manager or user, are provided for its development.

Many end users are already demonstrating their systems development capability by designing and implementing their own PC-based systems. Sophisticated end users with a general knowledge of technology and the requisite analytical and communication skills may be ideal candidates for systems analyst positions.

FACTORS AFFECTING SYSTEMS DEVELOPMENT SUCCESS

Successful systems development means delivering a system that meets user and organizational needs—on time and within budget. Years of experience in completing systems development projects have resulted in the identification of factors that contribute to the success or failure of systems development. These factors include the degree of change involved with the project, the quality of the project planning, the use of project management tools, the use of formal quality assurance processes, and the use of CASE tools.

Degree of Change

A major factor that affects the quality of systems development is the degree of change associated with the project. The scope can vary from implementing minor enhancements to an existing system to major reengineering. The project team needs to recognize where they are on this spectrum of change.

Continuous Improvement versus Reengineering

As discussed in Chapter 2, continuous-improvement projects do not require significant business process or information system changes nor retraining of individuals; thus, they have a high degree of success. Typically, because continuous improvements involve minor improvements, they also have relatively modest benefits. On the other hand, reengineering involves fundamental changes in how the organization conducts business and completes tasks. The factors associated with successful reengineering are similar to those of any development effort, including top management support, clearly defined corporate goals and systems development objectives, and careful management of change. Major reengineering projects tend to have a high degree of risk but also a high potential for major business benefits. (See Figure 12.9.)

Managing Change

The ability to manage change is critical to the success of systems development. The systems created during systems development inevitably cause change. For example, the work environment and habits of users are invariably affected by the development of a new information system. Unfortunately, not everyone adapts easily. Managing change requires the ability to recognize existing or

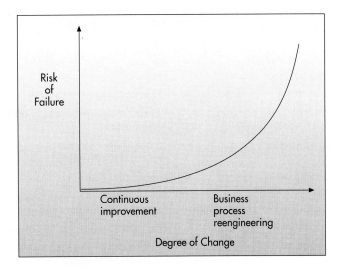

potential problems (particularly the concerns of users) and to deal with them before they become a serious threat to the success of the new or modified system. Although many problems can result from initiating new or modified systems, here are several of the most common:

- Fear that the employee will lose his or her job, power, or influence within the organization
- Belief that the proposed system will create more work than it eliminates
- Reluctance to work with "computer people"
- Anxiety that the proposed system will negatively alter the structure of the organization
- Belief that other problems are more pressing than those solved by the proposed system or that the system is being developed by people unfamiliar with "the way things need to get done"
- Unwillingness to learn new procedures or approaches

Preventing or dealing with these types of problems requires a coordinated effort involving stakeholders, users, managers, and information systems personnel. One positive step is simply to talk with all people concerned and learn what their biggest concerns are. Management can then deal with those concerns and try to eliminate them. Once the immediate concerns are addressed, these people can become part of the project team.

Quality of Project Planning

Another key success factor is the quality of project planning.[21] The bigger the project, the more likely that poor planning will lead to significant problems. A federal jury, for example, found the maker of navigational software partly responsible for an airplane crash near Cali, Colombia.[22] Poor systems development planning can be deadly.

Many companies find that large systems projects fall behind schedule, go over budget, and do not meet expectations. Although proper planning cannot guarantee that these types of problems will be avoided, it can minimize the likelihood of their occurrence.[23] Certain factors contribute to the failure of systems development projects. These factors and countermeasures to eliminate or alleviate the problem are summarized in Table 12.5.

Factor	Countermeasure
Solving the wrong problem	Establish a clear connection between the project and organizational goals.
Poor problem definition and analysis	Follow a standard systems development approach.
Poor communication	Communicate, communicate, communicate.
Project is too ambitious	Narrow the project focus to address only the most important business opportunities.
Lack of top management support	Identify the senior manager who has most to gain from the success of the project, and recruit this individual to champion the project.
Lack of management and user involvement	Identify and recruit key stakeholders to be active participants in the project.
Inadequate or improper system design	Follow a standard systems development approach.
Poor testing and implementation	Plan sufficient time for this activity.
Users are unable to use the system effectively	Develop a rigorous user training program and budget sufficient time in the schedule to execute it.
Lack of concern for maintenance	Include an estimate of people effort and costs for maintenance in the original project justification.

TABLE 12.5

Project Planning Issues Frequently Contributing to Project Failure

Good systems development is not automatic. Companies have lost millions of dollars and wasted years of effort because of faulty systems development. Because many development projects are large and expensive, a substantial amount of time and money is often on the line. Unfortunately, corporate America's track record for delivering these projects is not very good. An important part of following these methodologies is to consider the objectives of a proposed information system in terms of broader organizational goals.

Use of Project Management Tools

Project management involves planning, scheduling, directing, and controlling human, financial, and technological resources for a fixed-term task that will result in the achievement of specific goals and objectives. A **project schedule** is a detailed description of what is to be done. Each project activity, the use of personnel and other resources, and expected completion dates are described. A **project milestone** is a critical date for the completion of a major part of the project. The completion of program design, coding, testing, and release are examples of milestones for a programming project. The **project deadline** is the date the entire project is to be completed and operational—when the organization can expect to begin to reap the benefits of the project. BMC Software offers a 20 percent refund if it doesn't meet a client's project deadline.[24] In addition, any additional work done after the project deadline is performed free of charge.

Although the steps of systems development seem straightforward, larger projects can become complex, requiring literally hundreds or thousands of separate activities. For these types of systems development efforts, formal project management methods and tools become essential.

In systems development, each activity has an earliest start time, earliest finish time, and slack time, which is the amount of time an activity can be delayed without delaying the entire project. The **critical path** consists of all activities that, if delayed, would delay the entire project. These activities have zero slack. Any problems with critical path activities will cause problems for the entire project. To ensure that critical path activities are completed in a timely fashion, formalized project management approaches have been developed.

A formalized approach called **Program Evaluation and Review Technique (PERT)** involves creating three time estimates for an activity: shortest possible

project schedule

detailed description of what is to be done

project milestone

a critical date for the completion of a major part of the project

project deadline

date the entire project is to be completed and operational

critical path

all activities that, if delayed, would delay the entire project

Program Evaluation and Review Technique (PERT)

a formalized approach for developing a project schedule

Gantt chart

a graphical tool used for planning, monitoring, and coordinating projects

time, most likely time, and longest possible time. A formula is then applied to come up with a single PERT time estimate. A **Gantt chart** is a graphical tool used for planning, monitoring, and coordinating projects; it is essentially a grid that lists activities and deadlines. Each time a task is completed, a darkened line is placed in the proper grid cell to indicate the completion of a task (Figure 12.10).

Both PERT and Gantt techniques can be automated using project management software. This type of software monitors all project activities and determines whether activities and the entire project are on time and within budget. Project management software also has workgroup capabilities to handle multiple projects and to allow a team of people to interact with the same software. Project management software helps managers determine the best way to reduce project completion time at the least cost.[25] Several project management software packages are identified in Table 12.6.

Use of Formal Quality Assurance Processes

The development of information systems requires a constant trade-off of schedule and cost versus quality. Historically, the development of application software

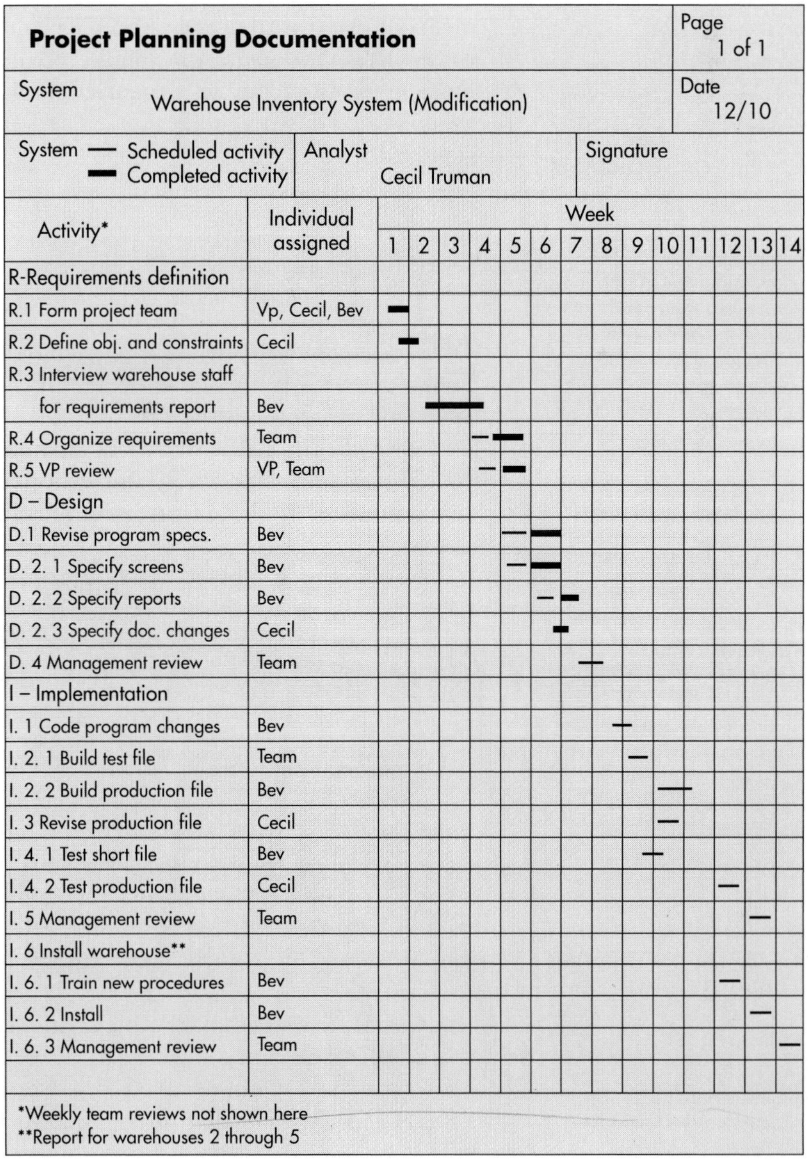

Project Planning Documentation		Page 1 of 1
System	Warehouse Inventory System (Modification)	Date 12/10
System —— Scheduled activity ━━ Completed activity	Analyst Cecil Truman	Signature

Activity*	Individual assigned	1	2	3	4	5	6	7	8	9	10	11	12	13	14
R-Requirements definition															
R.1 Form project team	Vp, Cecil, Bev	━													
R.2 Define obj. and constraints	Cecil		━												
R.3 Interview warehouse staff															
for requirements report	Bev				━	━									
R.4 Organize requirements	Team						━								
R.5 VP review	VP, Team						━								
D – Design															
D.1 Revise program specs.	Bev							━	━						
D. 2. 1 Specify screens	Bev							━							
D. 2. 2 Specify reports	Bev								━						
D. 2. 3 Specify doc. changes	Cecil								━						
D. 4 Management review	Team										━				
I – Implementation															
I. 1 Code program changes	Bev										━				
I. 2. 1 Build test file	Team											━			
I. 2. 2 Build production file	Bev											━			
I. 3 Revise production file	Cecil											━			
I. 4. 1 Test short file	Bev											━			
I. 4. 2 Test production file	Cecil													━	
I. 5 Management review	Team														━
I. 6 Install warehouse**															
I. 6. 1 Train new procedures	Bev														━
I. 6. 2 Install	Bev														━
I. 6. 3 Management review	Team														━

*Weekly team reviews not shown here
**Report for warehouses 2 through 5

FIGURE 12.10

Sample Gantt Chart

A Gantt chart shows progress through systems development activities by putting a bar through appropriate cells.

TABLE 12.6

Selected Project Management Software Packages

Software	Vendor
BeachBox '98	NetSQL Partners
Job Order	Management Software
OpenPlan	Welcom
Project	Microsoft
Project Scheduler	Scitor
Super Project	Computer Associates

computer-aided software engineering (CASE)

tools that automate many of the tasks required in a systems development effort and enforce adherence to the SDLC

upper-CASE tools

tools that focus on activities associated with the early stages of systems development

lower-CASE tools

tools that focus on the later implementation stage of systems development

integrated-CASE (I-CASE) tools

tools that provide links between upper- and lower-CASE packages, thus allowing lower-CASE packages to generate program code from upper-CASE package designs

ISO 9000 are international quality standards used by IS and other organizations to ensure quality of products and services.

has put an overemphasis on schedule and cost to the detriment of quality. Techniques, such as use of the ISO 9000 standards, have been developed to improve the quality of information systems. ISO 9000 are international quality standards originally developed in Europe in 1987. These standards help businesses define and document their own quality procedures for production and services. They can be used in any type of business and are accepted around the world as proof that a business can provide quality. Indeed, adherence to ISO 9000 is a requirement in many international markets.[26]

Many IS organizations have incorporated ISO 9000, total quality management, and statistical process control principles into the way they produce software. Often, to ensure the quality of the systems development process and finished product, an IS organization will form its own quality assurance groups to work with project teams and encourage them to follow established standards.

Use of Computer-Aided Software Engineering (CASE) Tools

Computer-aided software engineering (CASE) tools automate many of the tasks required in a systems development effort and enforce adherence to the SDLC, thus instilling a high degree of rigor and standardization to the entire systems development process.

CASE packages that focus on activities associated with the early stages of systems development are known as **upper-CASE tools**. These packages provide automated tools to assist with systems investigation, analysis, and design activities. Other CASE packages, called **lower-CASE tools**, focus on the later implementation stage of systems development and are capable of automatically generating structured program code. Some CASE tools provide links between upper- and lower-CASE packages, thus allowing lower-CASE packages to generate program code from upper-CASE package designs. These tools are called **integrated-CASE (I-CASE) tools**. Selected I-CASE tools are listed in Table 12.7. Total-CASE tools have also been developed for object-oriented programs and applications.

As with any team, coordinating the efforts of members of a systems development team can be a problem.

TABLE 12.7

Selected I-CASE Tools

Product	Vendor
ADW	KnowledgeWare
Bachman	Bachman Information Systems
Composer	TI Information Engineering
CorVision	Cortex Corporation
Developer's Studio	Microsoft
Envision	Future Tech Systems
Foundation	Andersen Consulting
HPS	Seer Technologies
Implementor	Implementors
Intelligent OOA	Kennedy Carter
Intersolv	Intersolv
KnowledgeWare	KnowledgeWare
Maestro II	Softlab
Methods Factory	VSF Ltd
Oracle CASE	Oracle
PACBASE	CGI
Paradigm Plus	ProtoSoft
Predict CASE	Software AG
Software through Pictures	Interactive Development Environments
System Architect	Popkin Software Systems
Systems Engineer	LBMS
Teamwork	ObjectTeam Cadre Technologies
ToolBuilder	IPSYS Software
TopWindows/TopCASE	TopSystems International
Westmount I-CASE	Westmount Technology

So, many CASE tools allow more than one person to work on the same system at the same time via a multiuser interface, which coordinates and integrates the work performed by all members of the same design team. With this facility, a person working on one aspect of systems development can automatically share his or her results with someone working on another aspect of the same system. Advantages and disadvantages of CASE tools are listed in Table 12.8.

Object-Oriented Systems Development

The success of a systems development effort can depend on the specific programming tools and approaches used. As mentioned in Chapter 4, object-oriented (OO) programming languages allow the interaction of programming objects—that is, an object consists of both data and the actions that can be performed on the data. So, an object could be data about an employee and all the operations (such as payroll, benefits, and tax calculations) that might be performed on the data.

Developing programs and applications using OO programming languages involves constructing modules and parts that can be reused in other programming projects. Chapter 4 discussed a number of programming languages that use the object-oriented approach, including Smalltalk, C++, and Java. These languages allow systems developers to take full advantage of the modular

TABLE 12.8

Advantages and
Disadvantages of CASE Tools

Advantages	Disadvantages
Produce systems with a longer effective operational life	Produce initial systems that are more expensive to build and maintain
Produce systems that more closely meet user needs and requirements	Require more extensive and accurate definition of user needs and requirements
Produce systems with excellent documentation	May be difficult to customize
Produce systems that need less systems support	Require training of maintenance staff
Produce more flexible systems	May be difficult to use with existing systems

object-oriented systems development (OOSD)

a systems development approach that combines the logic of the systems development life cycle with the power of object-oriented modeling and programming

approach, making program development faster and more efficient, resulting in lower costs. Modules can be developed internally or obtained from an external source. Once a company has the programming modules, programmers and systems analysts can modify them and integrate them with other modules to form new programs.

Object-oriented systems development (OOSD) combines the logic of the systems development life cycle with the power of object-oriented modeling and programming. OOSD follows a defined systems development life cycle, much like the SDLC. The life cycle phases can be, and usually are, completed with many iterations. Object-oriented systems development typically involves the following:

1. *Identifying potential problems and opportunities within the organization that would be appropriate for the OO approach.* This process is similar to traditional systems investigation. Ideally, these problems or opportunities should lend themselves to the development of programs that can be built by modifying existing programming modules.
2. *Defining what kind of system users require.* This analysis means defining all the objects that are part of the user's work environment (object-oriented analysis). The OO team must study the business and build a model of the objects that are part of the business (such as a customer, an order, or a payment). Many of the CASE tools discussed in the previous section can be used, starting with this step of OOSD.
3. *Designing the system.* This process defines all the objects in the system and the ways they interact (object-oriented design). Design involves developing logical and physical models of the new system by adding details to the object model started in analysis.
4. *Programming or modifying modules.* This implementation step takes the object model begun during analysis and completed during design and turns it into a set of interacting objects in a system. Object-oriented programming languages are designed to allow the programmer to create classes of objects in the computer system that correspond to the objects in the actual business process. Objects such as customer, order, and payment are redefined as computer system objects—a customer screen, an order-entry menu, or a dollar sign icon. Programmers then write new modules or modify existing ones to produce the desired programs.
5. *Evaluation by users.* The initial implementation is evaluated by users and improved. Additional scenarios and objects are added, and the cycle repeats. Finally, a complete, tested, and approved system is available for use.
6. *Periodic review and modification.* The completed and operational system is reviewed at regular intervals and modified as necessary.

Although it is still necessary to go through systems investigation and analysis to determine business objects and system requirements, systems design and implementation are often easier and cost less because predeveloped, pretested programming modules can be used. Also, maintenance can be simplified if programming objects can be reused to upgrade existing software and systems. Finally, because system objects are self-contained units, they can be changed or replaced with less disruption to the rest of the system. These benefits are the main reasons for the interest in object-oriented systems development.

SYSTEMS INVESTIGATION

As discussed earlier in the chapter, systems investigation is the first phase in the traditional SDLC of a new or modified business information system. The purpose is to identify potential problems and opportunities and consider them in light of the goals of the company. In general, systems investigation attempts to uncover answers to the following questions:

- What primary problems might a new or enhanced system solve?
- What opportunities might a new or enhanced system provide?
- What new hardware, software, databases, telecommunications, personnel, or procedures will improve an existing system or are required in a new system?
- What are the potential costs (variable and fixed)?
- What are the associated risks?

Initiating Systems Investigation

systems request form

document filled out by someone who wants the IS department to initiate systems investigation

Because systems development requests can require considerable time and effort to implement, many organizations have adopted a formal procedure for initiating systems development, beginning with systems investigation. The **systems request form** is a document that is filled out by someone who wants the IS department to initiate systems investigation. This form typically includes the following information:

- Problems in or opportunities for the system
- Objectives of systems investigation
- Overview of the proposed system
- Expected costs and benefits of the proposed system

The information in the systems request form helps to rationalize and prioritize the activities of the IS department. Based on the overall IS plan, the organization's needs and goals, and the estimated value and priority of the proposed projects, managers make decisions regarding the initiation of each systems investigation for such projects.

Participants in Systems Investigation

Once a decision has been made to initiate systems investigation, the first step is to determine what members of the development team should participate in the investigation phase of the project. Members of the development team change from phase to phase (Figure 12.11).

Ideally, functional managers are heavily involved during the investigation phase. Other members could include users or stakeholders outside management, such as an employee who helped initiate systems development. The technical and financial expertise of others participating in investigation would help the team determine whether the problem is worth solving. The members of the

The Systems Investigation Team

The team is made up of upper- and middle-level managers, IS personnel, users, and stakeholders.

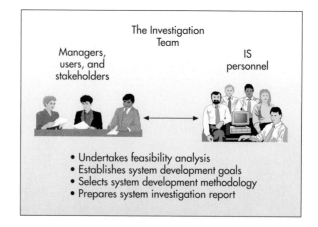

development team who participate in investigation are then responsible for gathering and analyzing data, preparing a report justifying systems development, and presenting the results to top-level managers.

Feasibility Analysis

feasibility analysis

assessment of the technical, operational, schedule, economic, and legal feasibility of a project

technical feasibility

assessment of whether the hardware, software, and other system components can be acquired or developed to solve the problem

operational feasibility

measure of whether the project can be put into action or operation

schedule feasibility

determination of whether the project can be completed in a reasonable amount of time

economic feasibility

determination of whether the project makes financial sense and whether predicted benefits offset the cost and time needed to obtain them

legal feasibility

determination of whether there are laws or regulations that may prevent or limit a systems development project

net present value

the net amount by which project savings exceed project expenses, after allowing for cost of capital and passage of time

A key step of the systems investigation phase is **feasibility analysis**, which assesses technical, operational, schedule, economic, and legal feasibility. **Technical feasibility** is concerned with whether the hardware, software, and other system components can be acquired or developed to solve the problem. **Operational feasibility** is a measure of whether the project can be put into action or operation. It can include logistical and motivational (acceptance of change) considerations. Motivational considerations are very important because new systems affect people and data flows and may have unintended consequences. As a result, power and politics may come into play, and some people may resist the new system. As seen in the "Ethical and Societal Issues" box, operational feasibility can also include legal and ethical considerations. **Schedule feasibility** determines whether the project can be completed in a reasonable amount of time—a process that involves balancing the time and resource requirements of the project with other projects. **Economic feasibility** determines whether the project makes financial sense and whether predicted benefits offset the cost and time needed to obtain them. Economic feasibility can involve cash flow analysis such as that done in net present value calculations. **Legal feasibility** determines whether there are laws or regulations that may prevent or limit a systems development project. For example, an Internet site that allowed users to share music without paying musicians or music producers was sued. Legal feasibility involves an analysis of existing and future laws to determine the likelihood of legal action against the systems development project and the possible consequences.

Net present value is the preferred approach for ranking competing projects and for determining economic feasibility.[27] The net present value represents the net amount by which project savings exceed project expenses, after allowing for the cost of capital and the passage of time. The cost of capital is the average cost of funds used to finance the operations of the business. It represents the minimum desired rate of return on an investment; thus, it is also called the hurdle rate. Net present value takes into account that a dollar returned at a later date is not worth as much as one received today, since the dollar in hand can be invested to earn profits or interest in the interim. Spreadsheet programs, such as Lotus and Excel, have built-in functions to compute the net present value

ETHICAL AND SOCIETAL ISSUES
Developing Legal and Ethical Drug Internet Sites

Few drugs have received the attention that Viagra, produced by Pfizer, has. The drug, which claims to be useful for solving erectile dysfunction in men, has been a big winner for Pfizer. Although the success of Viagra has increased revenues and profits for Pfizer, the drug has been a real problem for systems developers trying to establish Internet sites to provide information or market the successful drug. Regulations in Canada, for example, forbid companies from promoting prescription drugs directly to consumers. Yet Canadian pharmaceutical companies feel compelled to develop Web sites for many new drugs. Without an Internet site, drug companies could be deluged with calls seeking information. To help solve the problem, Pfizer Canada decided to develop an Internet site called the Canadian Erectile Dysfunction Resource Centre. The site discusses erectile dysfunction but does not mention any specific drugs or treatments.

In the United States, Internet drug sites have other problems. In one case, a 15-year-old boy filled out an order for a weight-loss drug that carried significant side effects, including bloating, cramps, and diarrhea. In a few days, the prescription arrived. Unfortunately for the drug company, the drug purchase was a sting operation set up by the Michigan attorney general's office. It is unknown how many underage customers are getting drugs illegally on the Internet, but the number is expected to be large. Some people try to obtain drugs directly on the Internet illegally without seeing a doctor or getting medical advice to avoid medical costs or to obtain drugs for recreational uses.

In addition to the traditional problems of implementing an Internet site, U.S. drug Internet sites must also have built-in screening processes and procedures to make sure that drugs are not sent to people who should not be getting them. To compound the problem, some illegal drug sites promise to send drugs through the mail without prescriptions or any screening. It has been estimated that more than 400 on-line drug sites tap into the more than $100 billion annually spent on retail prescriptions in the United States. With huge profits to be made on-line, systems developers for Internet drug sites must find a way to solve these problems.

Discussion Questions

1. What special problems do Canadian drug companies face in establishing an Internet site?
2. What are some of the problems facing U.S. drug companies wanting to establish an Internet site?

Critical Thinking Questions

3. Assume that you are a systems developer for a drug company in Canada who has been given the responsibility of establishing an Internet site. How would you determine whether the site was operationally feasible, legally and ethically?
4. The participants in a typical systems development effort were described in this chapter. What people should be involved in developing an Internet site for a drug company? How should these people interact to develop the best site legally possible?

Sources: Adapted from John Carey, "A Crackdown on E-Drugs," *Business Week*, February 7, 2000, p. 104; Wendy Cuthbert, "Regulations Hamper Web Efficacy," *Strategy*, January 17, 2000, p. 27; and "Prescription for Trouble," *PC World*, February 2000.

---

and internal rate of return. The net present value is the sum of the expected cash flow from each year of the project and can be expressed as follows:

$$\text{Net present value} = \sum_{t=1}^{n} (\text{CF}_t)/(1+k)^t$$

where CF_t is the expected cash flow in a period t and k is the project's cost of capital.

Since income tax payments represent a disbursement of cash, all cash flows affecting taxable income are credited to the project on an after-tax basis (i.e., computed by multiplying the tax complement by the pretax cash flow), with the notable exception of depreciation. Since depreciation expense does not involve actual cash disbursement, it is excluded from cash flows. However, depreciation is a deductible expense for federal tax purposes. Therefore, the amount of tax relief from depreciation is included in cash flows. It is computed by multiplying the federal tax rate by the depreciation expense. Table 12.9 illustrates these concepts for a sample project in a firm with a tax rate of 36 percent and a cost of capital of 20 percent.

Cash Flow (in thousands of dollars)	Year 1	Year 2	Year 3	Year 4
1. Cash inflow (gross savings)	25	105	125	200
2. Cash outflow (expenses)	–135	–25	–30	–35
3. Pretax cash flow (line 1 + 2)	–110	80	95	165
4. After-tax cash flow [line 3 × (1 – tax rate)]	–70	51	61	106
5. Depreciation	60	50	40	30
6. Tax relief from depreciation (line 5 × tax rate)	22	18	14	11
7. Net after-tax cash flow (line 4 + line 6)	–48	69	75	117
8. Discounted cash flow [line 7 ÷ (1 + cost of capital)^{Year}]	–40	48	43	56
9. Net present value (sum of amounts in row 8)	107			

TABLE 12.9

Sample Net Present Value Calculation

The key to making accurate estimates of the cash flows associated with a project is to involve the business managers responsible for the business functions served. Another key resource is your organization's financial manager, who should be very familiar with net present value analysis. If a systems development project is determined to be feasible, systems investigation will formally begin.

The Systems Investigation Report

systems investigation report

summary of the results of the systems investigation and the process of feasibility analysis; recommends a course of action

steering committee

an advisory group consisting of senior management and users from the IS department and other functional areas

The primary outcome of systems investigation is a **systems investigation report**. This report summarizes the results of systems investigation and the process of feasibility analysis and recommends a course of action: continue on into systems analysis, modify the project in some manner, or drop it. A typical table of contents for the systems investigation report is shown in Figure 12.12.

The systems investigation report is reviewed by senior management, often organized as an advisory committee, or **steering committee**, consisting of senior management and users from the IS department and other functional areas. These individuals help IS personnel with their decisions about the use of information systems in the business and give authorization to pursue further systems development activities. After review, the steering committee might agree with the recommendation of the systems development team or suggest a change in project focus to concentrate more directly on meeting a specific company objective. Another alternative is that everyone may decide that the project is not feasible for one reason or another and cancel the project.

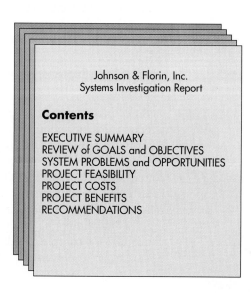

Johnson & Florin, Inc.
Systems Investigation Report

Contents

EXECUTIVE SUMMARY
REVIEW of GOALS and OBJECTIVES
SYSTEM PROBLEMS and OPPORTUNITIES
PROJECT FEASIBILITY
PROJECT COSTS
PROJECT BENEFITS
RECOMMENDATIONS

FIGURE 12.12

A Typical Table of Contents for a Systems Investigation Report

SYSTEMS ANALYSIS

After a project has been approved for further study, the next step is to answer the question "What must the information system do to solve the problem?" The process needs to go beyond mere computerization of existing systems. The entire system, and the business process with which it is associated, should be evaluated. Often, a firm can make great gains if it restructures both business activities and the related information system simultaneously. The overall emphasis of analysis is on gathering data on the existing system, determining the requirements for the new system, considering alternatives within these constraints, and investigating the feasibility of the solutions. The primary outcome of systems analysis is a prioritized list of systems requirements.

General Considerations

Systems analysis starts by clarifying the overall goals of the organization and determining how the existing or proposed information system helps meet them. A manufacturing company, for example, might want to reduce the number of equipment breakdowns. This goal can be translated into one or more informational needs. One need might be to create and maintain an accurate list of each piece of equipment and a schedule for preventive maintenance. Another need might be a list of equipment failures and their causes.

Analysis of a small company's information system can be fairly straightforward. On the other hand, evaluating an existing information system for a large company can be a long, tedious process. As a result, large organizations evaluating a major information system normally follow a formalized analysis procedure, involving these steps:

1. Assembling the participants for systems analysis
2. Collecting appropriate data and requirements
3. Analyzing the data and requirements
4. Preparing a report on the existing system, new system requirements, and project priorities

Participants in Systems Analysis

The first step in formal analysis is to assemble a team to study the existing system. This group includes members of the original development team—from users and stakeholders to IS personnel and management. Most organizations usually allow key members of the development team not only to analyze the condition of the existing system but also to perform other aspects of systems development, such as design and implementation.

Once the participants in systems analysis are assembled, this group develops a list of specific objectives and activities. A schedule for meeting the objectives and completing the specific activities is also developed, along with deadlines for each stage and a statement of the resources required at each stage, such as clerical personnel, supplies, and so forth. Major milestones are normally established to help the team monitor progress and determine whether problems or delays occur in performing systems analysis.

Data Collection

The purpose of data collection is to seek additional information about the problems or needs identified in the systems investigation report. During this process, the strengths and weaknesses of the existing system are emphasized.

FIGURE 12.13

Internal and External Sources of
Data for Systems Analysis

Internal Sources	External Sources
Users, stakeholders, and managers	Customers
Organization charts	Suppliers
Forms and documents	Stockholders
Procedure manuals and policies	Government agencies
Financial reports	Competitors
IS manuals	Outside groups
Other measures of business process	Journals, etc.
	Consultants

Identifying Sources of Data

Data collection begins by identifying and locating the various sources of data, including both internal and external sources (Figure 12.13).

Performing Data Collection

Once data sources have been identified, data collection begins. Figure 12.14 shows the steps involved. Data collection may require a number of tools and techniques, such as interviews, direct observation, and questionnaires.

In a **structured interview**, the questions are written in advance. In an **unstructured interview**, the questions are not written in advance; the interviewer relies on experience in asking the best questions to uncover the inherent problems of the existing system. An advantage of the unstructured interview is that it allows the interviewer to ask follow-up or clarifying questions immediately.

With **direct observation**, one or more members of the analysis team directly observe the existing system in action. One of the best ways to understand how the existing system functions is to work with the users to discover how data flows in certain business tasks. Determining the data flow entails direct observation of users' work procedures, their reports, current screens (if automated already), and so on. From this observation, members of the analysis team determine which forms and procedures are adequate and which are inadequate and need improvement. Direct observation requires a certain amount of skill. The observer must be able to see what is really happening and not be influenced by his or her own attitudes or feelings. This approach can reveal important problems and opportunities that would be difficult to obtain

structured interview

an interview for which the questions are written in advance

unstructured interview

an interview for which the questions are not written in advance

direct observation

watching the existing system in action by one or more members of the analysis team

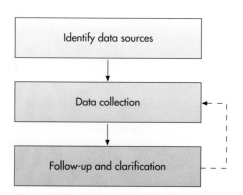

FIGURE 12.14

The Steps in Data Collection

Direct observation is a method of data collection. One or more members of the analysis team directly observe the existing system in action.
(Source: © 2000 Photodisc)

questionnaire

a tool for gathering data when the data sources are spread over a wide geographic area

statistical sampling

selection of a random sample of data and application of the characteristics of the sample to the whole group

data analysis

manipulation of the collected data so that it is usable for the development team members who are participating in systems analysis

using other data collection methods. An example would be observing the work procedures, reports, and computer screens associated with an accounts payable system being considered for replacement.

When many data sources are spread over a wide geographic area, **questionnaires** may be the best approach. Like interviews, questionnaires can be either structured or unstructured. In most cases, a pilot study is conducted to fine-tune the questionnaire. A follow-up questionnaire can also capture the opinions of those who do not respond to the original questionnaire.

A number of other data collection techniques can be employed. In some cases, telephone calls are an excellent method. In other cases, activities may be simulated to see how the existing system reacts. Thus, fake sales orders, stockouts, customer complaints, and data flow bottlenecks may be created to see how the existing system responds to these situations. **Statistical sampling**, which involves taking a random sample of data, is another technique. For example, suppose we want to collect data that describes 10,000 sales orders received over the last few years. Because it is too time-consuming to analyze each of the 10,000 sales orders, a random sample of 100 to 200 sales orders from the entire batch can be collected. The characteristics of this sample are then assumed to apply to the 10,000 orders.

Data Analysis

The data collected in its raw form is usually not adequate to determine the effectiveness and efficiency of the existing system or the requirements for the new system. The next step is to manipulate the collected data so that it is usable for the development team participating in systems analysis. This manipulation is called **data analysis**. Data and activity modeling, using data flow diagrams and entity-relationship diagrams, are useful during data analysis to show data flows and the relationships among various objects, associations, and activities. Other common tools and techniques for data analysis include application flowcharts, grid charts, and CASE tools.

Data Modeling

Data modeling, first introduced in Chapter 5, is a commonly accepted approach to modeling organizational objects and associations that employ both text and graphics. The exact way data modeling is employed, however, is governed by the specific systems development methodology.

Data modeling is most often accomplished through the use of entity-relationship (ER) diagrams. Recall from Chapter 5 that an entity is a generalized representation of an object type—such as a class of people (employee), events (sales), things (desks), or places (Philadelphia)—and that entities possess certain attributes. Objects can be related to other objects in numerous ways. An entity-relationship diagram, such as the one shown in Figure 12.15a, describes a number of objects and the ways they are associated. An ER diagram is not capable in and of itself of fully describing a business problem or solution, because it lacks descriptions of the related activities. It is, however, a good place to start, since it describes object types and attributes about which data may need to be collected for processing.

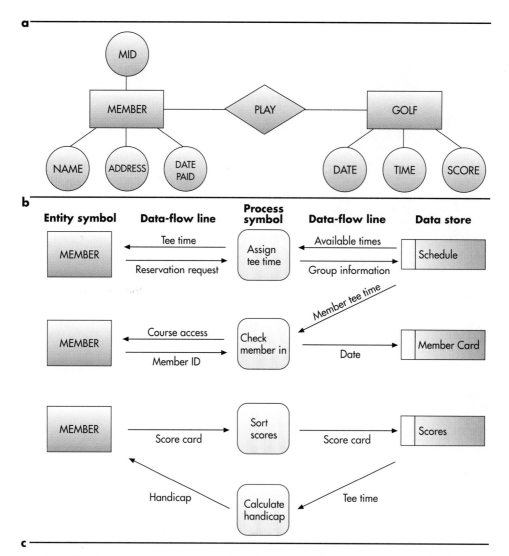

To play golf at the course, you must first pay a fee to become a member of the golf club. Members are issued member cards and are assigned member ID numbers. To reserve a tee time (a time to play golf), a member calls the club house at the golf course and arranges an available time slot with the reception clerk. The reception clerk reserves the tee time by writing the member's name and number of players in the group on the course schedule. When a member arrives at the course, he or she checks in at the reception desk where the reception clerk checks the course schedule and notes the date on the member's card. After a round of golf has been completed, the members leave their score card with the reception clerk. Member scores are tracked and member handicaps are updated on a monthly basis.

FIGURE 12.15

Data and Activity Modeling

(a) An entity-relationship diagram. (b) A data-flow diagram. (c) A semantic description of the business process.
Source: G. Lawrence Sanders, *Data Modeling* (Danvers, MA: Boyd & Fraser Publishing, 1995). Reprinted with permission from Course Technology.

data-flow diagram (DFD)

a model of objects, associations, and activities that describes how data can flow between and around various objects

Activity Modeling

To fully describe a business problem or solution, it is necessary to describe the related objects, associations, and activities. Activities in this sense are events or items that are necessary to fulfill the business relationship or that can be associated with the business relationship in a meaningful way.

Activity modeling is often accomplished through the use of data-flow diagrams. A **data-flow diagram (DFD)** models objects, associations, and activities by describing how data can flow between and around various objects. DFDs work on the premise that for every activity there is some communication, transference, or flow that can be described as a data element. DFDs describe what activities are occurring to fulfill a business relationship or accomplish a business task, not how these activities are to be performed. That is, DFDs show the logical sequence of associations and activities, not the physical processes. A system modeled with a DFD could operate manually or could be computer based; if computer based, the system could operate with a variety of technologies.

DFDs are easy to develop and easily understood by nontechnical people. Data-flow diagrams use four primary symbols, as illustrated in Figure 12.15b.

data-flow lines

arrows that show the direction of data element movement

process symbol

representation of a function that is performed

entity symbol

representation of either a source or destination of a data element

data store

representation of a storage location for data

- *Data flow.* The **data-flow line** includes arrows that show the direction of data element movement.
- *Process symbol.* The **process symbol** reveals a function that is performed. Computing gross pay, entering a sales order, delivering merchandise, and printing a report are examples of functions that can be represented with a process symbol.
- *Entity symbol.* The **entity symbol** shows either the source or destination of the data element. An entity can be, for example, a customer who initiates a sales order, an employee who receives a paycheck, or a manager who gets a financial report.
- *Data store.* A **data store** reveals a storage location for data. A data store is any computerized or manual data storage location, including magnetic tape, disks, a filing cabinet, or a desk.

Comparing entity-relationship diagrams with data-flow diagrams provides insight into the concept of top-down design. Figure 12.15a and b show an entity-relationship diagram and a data-flow diagram for the same business relationship—namely, a member of a golf course playing golf. Figure 12.15c provides a brief description of the business relationship for clarification.

Application Flowcharts

application flowcharts

diagrams that show relationships among applications or systems

Application flowcharts show the relationships among applications or systems. Assume that a small business has collected data about its order processing, inventory control, invoicing, and marketing analysis applications. Management is thinking of modifying the inventory control application. The raw facts collected, however, do not help in determining how the applications are related to each other and the databases required for each. These relationships are established through data analysis with an application flowchart (Figure 12.16). Using this tool for data analysis makes clear the relationships among the order processing, inventory control, invoicing, and marketing analysis applications.

In the simplified application flowchart in Figure 12.16, you can see that the sales ordering application provides important data to the inventory control and marketing analysis applications. The inventory control application provides data to the invoicing application. Any changes made to any one of these applications must take into account the other applications, which may provide data to or receive data from the others.

FIGURE 12.16

An Application Flowchart

The flowchart shows the relationships among various applications.

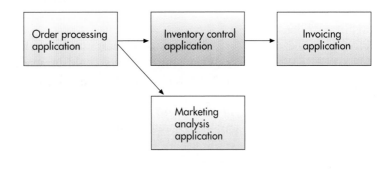

FIGURE 12.17

A Grid Chart

The chart shows the relationships among applications and databases.

Databases → Applications ↓	Customer database	Inventory database	Supplier database	Accounts receivable database
Order processing application	X	X		
Inventory control application		X	X	
Marketing analysis application	X	X		
Invoicing application	X			X

grid chart

table that shows relationships among the various aspects of a systems development effort

CASE repository

a database of system descriptions, parameters, and objectives

requirements analysis

determination of user, stakeholder, and organizational needs

Grid Charts

A **grid chart** is a table that shows relationships among various aspects of a systems development effort. For example, a grid chart can be used to reveal the databases used by the various applications (Figure 12.17).

The simplified grid chart shown in Figure 12.17 shows that the customer database is used by the order processing, invoicing, and marketing analysis applications. The inventory database is used by the order processing, inventory control, and marketing analysis applications. The supplier database is used by the inventory control application, and the accounts receivable database is used by the invoicing application. This grid chart shows which applications use common databases and reveals that, for example, any changes to the inventory control application must investigate the inventory and supplier databases.

CASE Tools

As discussed earlier, many systems development projects use upper-CASE tools to complete analysis tasks. Most computer-aided software engineering tools have generalized graphics programs that can generate a variety of diagrams and figures. Entity-relationship diagrams, data-flow diagrams, application flowcharts, and other diagrams can be developed using CASE graphics programs to help describe the existing system. During the analysis phase, a **CASE repository**—a database of system descriptions, parameters, and objectives—will begin to be developed.

Requirements Analysis

The overall purpose of **requirements analysis** is to determine user, stakeholder, and organizational needs. For an accounts payable application, the stakeholders could include suppliers and members of the purchasing department. Questions that should be asked during requirements analysis include the following:

- Are these stakeholders satisfied with the current accounts payable application?
- What improvements could be made to satisfy suppliers and help the purchasing department?

One of the most difficult procedures in systems analysis is eliciting user or systems requirements. In some cases, communications problems can interfere with the determination of these requirements. For example, an accounts payable manager may want a better procedure for tracking the amount owed by customers. Specifically, the manager would like to have a weekly report that shows all customers who owe more than $1,000 and are more than 90 days past due on their account. A financial manager might need a report that summarizes total amount owed by customers to look at the need to loosen or tighten credit limits. A sales manager might want to review the amount owed by a key customer relative to sales to that same customer. The purpose of requirements analysis is to capture these requests in detail. Numerous tools and techniques can be used to capture systems requirements. Often, various techniques are used in the context of a JAD session.

Asking Directly

One of the most basic techniques used in requirements analysis is asking directly. **Asking directly** is an approach that asks users, stakeholders, and other managers about what they want and expect from the new or modified system. This approach works best for stable systems in which stakeholders and users clearly understand the system's functions. Unfortunately, many individuals do not know exactly or are unable to adequately articulate what they want or need. The role of the systems analyst during the analysis phase is to critically and creatively evaluate needs and define them clearly so that the systems can best meet them.

Critical Success Factors

Another approach uses critical success factors (CSFs). Managers and decision makers are asked to list only those factors that are critical to the success of their area of the organization. A CSF for a production manager might be adequate raw materials from suppliers; a CSF for a sales representative could be a list of customers currently buying a certain type of product. Starting from these CSFs, the system inputs, outputs, performance, and other specific requirements can be determined.

The IS Plan

As we have seen, the IS plan translates strategic and organizational goals into systems development initiatives. The IS planning process often generates strategic planning documents that can be used to define system requirements. Working from these documents ensures that requirements analysis will address the goals set by top-level managers and decision makers (Figure 12.18). There are unique benefits to applying the IS plan to define systems requirements. Because the IS plan takes a long-range approach to using information technology within the organization, the requirements for a system analyzed in terms of the IS plan are more likely to be compatible with future systems development initiatives.

Screen and Report Layout

Developing formats for printed reports and screens to capture data and display information is one of the common tasks associated with developing systems. Screens and reports relating to systems output are specified first to verify that the desired solution is being delivered. Manual or computerized screen and report layout facilities are used to capture both output and input requirements.

Screen layout is a technique that allows a designer to quickly and efficiently design the features, layout, and format of a display screen. In general, users who interact with the screen frequently can be presented with more data and less descriptive information; infrequent users should have more descriptive information presented to explain the data that they are viewing (Figure 12.19).

asking directly

an approach to gather data that asks users, stakeholders, and other managers about what they want and expect from the new or modified system

screen layout

a technique that allows a designer to quickly and efficiently design the features, layout, and format of a display screen

FIGURE 12.18

Converting Organizational Goals into Systems Requirements

FIGURE 12.19

Screen Layouts

(a) A screen layout chart for frequent users who require little descriptive information.

(b) A screen layout chart for infrequent users who require more descriptive information.

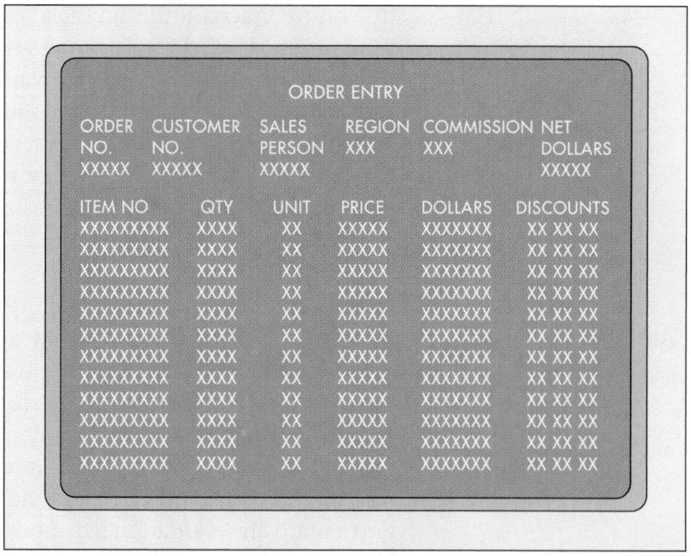

a

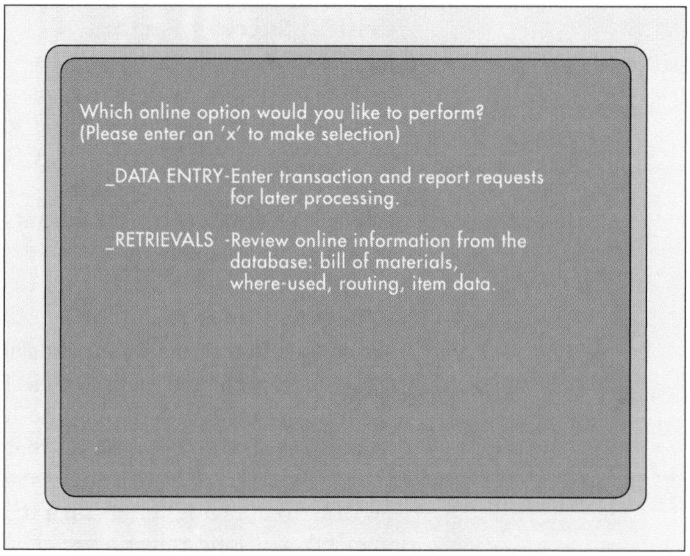

b

report layout

technique that allows designers to diagram and format printed reports

Report layout allows designers to diagram and format printed reports. These reports can be produced on either standard line printers (using 132 columns or positions) or personal computer printers (with 80 columns or positions). Reports can contain data, graphs, or both. Graphic presentations allow managers and executives to quickly view trends and take appropriate action, if necessary.

Screen layout diagrams can document the screen users desire for the new or modified application. Report layout charts reveal the format and content of various reports that the application will prepare. Other diagrams and charts can be developed to reveal the relationship between the application and outputs from the application.

Requirements Analysis Tools

A number of tools can be used to document requirements analysis. Again, CASE tools are often employed. As requirements are developed and agreed upon, entity-relationship diagrams, data-flow diagrams, screen and report layout forms, and other types of documentation will be stored in the CASE repository. These requirements might also be used later as a reference during the rest of systems development or for a different systems development project.

The Systems Analysis Report

Systems analysis concludes with a formal systems analysis report. It should cover the following elements:

1. The strengths and weaknesses of the existing system from a stakeholder's perspective
2. The user/stakeholder requirements for the new system (also called the *functional requirements*)
3. The organizational requirements for the new system
4. A description of what the new information system should do to solve the problem

Suppose analysis reveals that a marketing manager thinks a weakness of the existing system is its inability to provide accurate reports on product availability. These requirements and a preliminary list of the corporate objectives for the new system will be in the systems analysis report. Particular attention is placed on areas of the existing system that could be improved to meet user requirements. The table of contents for a typical report is shown in Figure 12.20.

The systems analysis report gives managers a good understanding of the problems and strengths of the existing system. If the existing system is operating better than expected or the necessary changes are too expensive relative to the benefits of a new or modified system, the systems development process can be stopped at this stage. If the report shows that changes to another part of the system might be the best solution, the development process might start over, beginning again with systems investigation. Or, if the systems analysis report shows that it will be beneficial to develop one or more new systems or to make changes to existing ones, systems design begins.

As the examples and discussion of this chapter have demonstrated, systems development can profoundly affect the current and future business activities of an organization. Certainly, information systems professionals are critical members of the development team, but people from all functional areas of an organization are involved, whether through providing input for data collection, testing of prototypes, or participating in any of the other steps. The goal of development activities is to create and maintain business systems that help an organization meet its strategic goals and remain competitive.

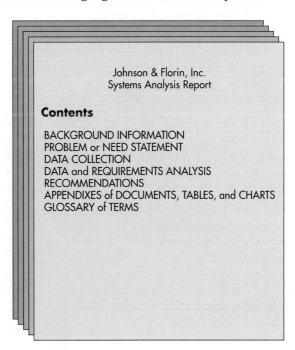

Johnson & Florin, Inc.
Systems Analysis Report

Contents

BACKGROUND INFORMATION
PROBLEM or NEED STATEMENT
DATA COLLECTION
DATA and REQUIREMENTS ANALYSIS
RECOMMENDATIONS
APPENDIXES of DOCUMENTS, TABLES, and CHARTS
GLOSSARY of TERMS

FIGURE 12.20

A Typical Table of Contents for a Report on an Existing System

● SUMMARY

PRINCIPLE ● Effective systems development requires a team effort of stakeholders, users, managers, systems development specialists, and various support personnel, and it starts with careful planning.

The systems development team consists of stakeholders, users, managers, systems development specialists, and various support personnel. The development team is responsible for determining the objectives of the information system and delivering to the organization a system that meets its objectives.

Stakeholders are individuals who, either themselves or through the area of the organization they represent, ultimately benefit from the systems development project. Users are individuals who will interact with the system regularly. They can be employees, managers, customers, or suppliers. Managers on development teams are typically representative of stakeholders or may be stakeholders themselves. In addition, managers are most capable of initiating and maintaining change. For large-scale systems development projects, where the investment in and value of a system can be quite high, it is common to have senior-level managers be part of the development team.

A systems analyst is a professional who specializes in analyzing and designing business systems. The programmer is responsible for modifying or developing programs to satisfy user requirements. Other support personnel on the development team include technical specialists, either IS department employees or outside consultants. Depending on the magnitude of the systems development project and the number of IS systems development specialists on the team, the team may also include one or more IS managers.

Information systems planning refers to the translation of strategic and organizational goals into systems development initiatives. Benefits of IS planning include a long-range view of information technology use and better use of information systems resources. Planning requires developing overall IS objectives; identifying IS projects; setting priorities and selecting projects; analyzing resource requirements; setting schedules, milestones, and deadlines; and developing the IS planning document.

Systems development projects may be initiated for a number of reasons, including the need to solve problems with an existing system, to exploit opportunities to gain competitive advantage, to increase competition, to make use of effective information, to create organizational growth, to settle a merger or corporate acquisition, and to address a change in the market or external environment.

PRINCIPLE ● Systems development often uses tools to select, implement, and monitor projects, including net present value (NPV), prototyping, and rapid application development.

Applications must be designed to meet special business needs. Transaction processing applications for the Internet must be able to scale up to support highly variable transactions from potentially thousands of users. Ideally, they can scale up dynamically when needed. They must be reliable and fault-tolerant, providing availability 24 hours a day, seven days a week, with extremely high transaction integrity. They must be able to integrate with existing infrastructure, including customer and order databases, legacy applications, and enterprise resource planning systems. Development and maintenance must be quick and easy, as business needs may require you to change applications on the fly.

The increasing use of enterprise resource planning software has affected systems development. The first trend is that companies wish to stay with their primary ERP vendor instead of looking elsewhere for answers to their data warehousing and production planning needs or developing in-house solutions. Thus, they look to their original ERP vendor to provide these solutions.

A second trend is that there is less in-house development and more dependence on ERP vendors and their strategic partners to provide enhancements and add-ons to the original ERP package.

A third interesting trend is the increase in the number of companies that, once they have successfully implemented their own company's ERP project, are branching out to provide consulting to other companies.

The five phases of the traditional SDLC are investigation, analysis, design, implementation, and maintenance and review. Systems investigation involves identifying potential problems and opportunities and considering them in light of organizational goals. Systems analysis seeks a general understanding of the solution required to solve the problem; the existing system is studied in detail and weaknesses are identified. Systems design involves creating new or modified system requirements. Systems implementation encompasses programming, testing, training, conversion, and operation of the system. Systems maintenance and review entails monitoring the system and performing enhancements or repairs.

Advantages of the traditional SDLC include the following: it provides for maximum management control, creates considerable system documentation, ensures that system requirements can be traced back to stated business needs, and produces many intermediate products for review. Its disadvantages include the following: users may get a system that meets the needs as understood by the developers; the documentation is expensive and difficult to maintain; users' needs go unstated or may not be met; and users cannot easily review the many intermediate products produced.

Prototyping is an iterative approach that involves defining the problem, building the initial version, having users utilize and evaluate the initial version, providing feedback, and incorporating suggestions into the second version. Prototypes can be fully operational or nonoperational, depending on how critical the system under development is and how much time and money the organization has to spend on prototyping.

Advantages of the prototyping approach include the following: users get an opportunity to try the system before it is completed; useful prototypes can be produced in weeks; users become positive about the evolving system; and errors and omissions can be detected early. Disadvantages include the following: the approach makes it difficult to start over if the initial solution misses the mark widely; it is difficult to contain the scope of the project; system documentation is often absent; and key operational considerations are often overlooked.

Rapid application development (RAD) uses tools and techniques designed to speed application development. Its use reduces paper-based documentation, automates program source code generation, and facilitates user participation in development activities. RAD makes extensive use of the joint application development (JAD) process to gather data and perform requirements analysis. JAD involves group meetings in which users, stakeholders, and IS professionals work together to analyze existing systems, propose possible solutions, and define the requirements for a new or modified system.

RAD has the following advantages: it puts an application into production quickly; documentation is produced as a by-product; and it forces good teamwork among users, stakeholders, and developers. Its disadvantages are the following: it is an intense process that can burn out the participants; it requires participants to be skilled in advanced tools and techniques; and a large percentage of time is required from stakeholders and users.

The end-user SDLC is used to support projects where the primary effort is undertaken by a combination of business managers and users.

The degree of change introduced by the project, the quality of project planning, the use of project management tools, the use of formal quality assurance processes, the use of CASE tools, and object-oriented systems development methods are all factors that affect the success of a project. The greater the amount of change, the greater the degree of risk and also frequently the amount of reward. The quality of project planning involves such factors as support from top management, strong user involvement, use of a proven methodology, clear project goals and objectives, concentration on key problems and straightforward designs, staying on schedule and within budget, good user training, and solid review and maintenance programs. The use of automated project management tools enables detailed development, tracking, and control of the project schedule. Effective use of a quality assurance process enables the project manager to deliver a quality system and to make intelligent trade-offs among cost, schedule, and quality. CASE tools automate many of the systems development tasks, thus reducing the time and effort required to complete them while ensuring good documentation. Object-oriented development focuses on developing modules that contain both data and the actions performed on the data. Because of this modular approach and use of iteration, OOSD often streamlines development efforts for an organization down the road.

PRINCIPLE • Systems development starts with investigation and analysis of existing systems.

In most organizations, a systems request form initiates the investigation process. The systems investigation is designed to assess the feasibility of implementing solutions for business problems. An investigation team follows up on the request and performs a feasibility analysis that addresses technical, economic, operational, schedule, and legal feasibility.

If the project under investigation is feasible, major goals are set for the system's development, including performance, cost, managerial goals, and procedural goals. Many companies choose a popular methodology so that new IS employees, outside specialists, and vendors will be familiar with the systems development tasks set forth in the approach. A systems development methodology must be selected. As a final step in the investigation process, a systems investigation report should be prepared to document relevant findings.

Systems analysis is the examination of existing systems, which begins once approval for further study is received from management. Additional study of a selected system allows those involved to further understand the systems' weaknesses and potential improvement areas. An analysis team is assembled to collect and analyze data on the existing system.

Data collection methods include observation, interviews, questionnaires, and statistical sampling. Data analysis manipulates the collected data to provide information. The analysis includes grid charts, application flowcharts, and CASE tools. The overall purpose of requirements analysis is to determine user and organizational needs.

Data modeling is used to model organizational objects and associations using text and graphical diagrams. It is most often accomplished through the use of entity-relationship (ER) diagrams. Activity modeling is often accomplished through the use of data-flow diagrams (DFDs), which model objects, associations, and activities by describing how data can flow between and around various objects. DFDs use symbols for data flows, processing, entities, and data stores.

● KEY TERMS

● REVIEW QUESTIONS

1. What is an information system stakeholder?
2. What is the goal of information systems planning? What steps are involved in IS planning?
3. What are the typical reasons to initiate systems development?
4. What actions can be used during critical analysis?
5. Identify each of the four systems development life cycles and summarize their strengths and weaknesses.
6. How is a Gantt chart developed? How is it used?
7. What are the steps of the systems development life cycle?
8. Why is it important to identify and remove errors early in the systems development life cycle?
9. Identify four reasons a systems development project may be initiated.
10. List factors that have a strong influence on project success.
11. What is the purpose of systems investigation?
12. Define the different types of feasibility.
13. What is the net present value of a project?
14. What is the purpose of systems analysis?
15. How does the JAD technique support the RAD systems development life cycle?
16. What are the advantages of OOSD?

● DISCUSSION QUESTIONS

1. Why is it important for business managers to have a basic understanding of the systems development process?
2. Briefly describe the role of a system user in the systems investigation and systems analysis stages of a project.
3. Assume that you are investigating a new sales marketing program for a clothing company. Use critical analysis to help you determine the major requirements for the new system.
4. During the systems investigation phase, how important is it to think creatively? What are some approaches to increase creativity?
5. For what types of systems development projects might prototyping be especially useful? What are the characteristics of a system developed with a prototyping technique?
6. Imagine that your firm has never developed an information systems plan. What sort of issues between the business functions and IS organization might exist?

7. Assume that you are responsible for a new payroll program. What steps would you take to ensure a high-quality payroll system?
8. What are some of the technical challenges for such applications?
9. How important are communications skills to IS personnel? Consider this statement: "IS personnel need a combination of skills—one-third technical skills, one-third business skills, and one-third communications skills." Do you think this is true? How would this affect the training of IS personnel?
10. Imagine that you are a highly paid consultant who has been retained to evaluate an organization's systems development processes. With whom would you meet? How would you make your assessment?
11. You are a senior manager of a functional area in which a mission-critical system is being developed. How can you safeguard this project from mushrooming out of control?

● PROBLEM-SOLVING EXERCISES

1. Develop a spreadsheet program to determine net present value using the form outlined below.

Cash Flow (in thousands of dollars)

	Year 1	Year 2	Year 3	Year 4
1. Cash inflow (gross savings)				
2. Cash outflow (expenses)				
3. Pretax cash flow (line 1 + 2)				
4. After-tax cash flow [line 3 × (1 - tax rate)]				
5. Depreciation				
6. Tax relief from depreciation (line 5 × tax rate)				
7. Net after-tax cash flow (line 4 + line 6)				
8. Discounted cash flow [line 7 ÷ (1 + cost of capital)^{Year}]				
9. Net present value (sum of amounts in row 8)				

Use the spreadsheet to select among two projects with the following cash flows. Project one has a gross savings of $100,000 per year and expenses of $25,000 per year, plus $15,000 per year in depreciation. Project two has a gross savings of $75,000 the first year with $125,000 per year thereafter. Expenses are $35,000 per year plus $18,000 per year in depreciation. Assume a capital cost rate of 15 percent and tax rate of 35 percent.

2. You are developing a new information system for the Fitness Center, a company that has five fitness centers in your metropolitan area, with about 650 members and 30 employees in each location. This system will be used by both members and fitness consultants to keep track of participation in various fitness activities, such as free weights, volleyball, swimming, stair climbers, and aerobic classes. One of the performance objectives of the system is that it help members plan a fitness program to meet their particular needs. The primary purpose of this system, as envisioned by the director of marketing, is to assist the Fitness Center in obtaining a competitive advantage over other fitness clubs.

Use word processing software to prepare a brief memo to the required participants in the development team for this systems development project. Be sure to specify what roles these individuals will play and what types of information you hope to obtain from them. Assume that the relational database model will be the basis for building this system. Use a database management system to define the various tables that will make up the database.

 3. Using a Gantt chart, determine what courses you need to take and how long it will take you to graduate. Use your presentation graphics program to develop your Gantt chart.

● TEAM ACTIVITIES

1. System development is more of an art and less of a science, with a wide variety of approaches in how companies perform this activity. You and the members of your team are to interview members of an information systems organization's development group. Prepare a list of questions to determine whether they follow the approach outlined in this chapter. During the course of your interview, when you find discrepancies between the approach they follow and the process suggested in the text, find out why there is a difference. Also learn what tools and techniques are most frequently employed. Prepare a short report on your findings.

2. Your team has been hired to determine the requirements and layout of the Web pages for a new company that sells fishing equipment over the Internet. Using the approaches discussed in this chapter, develop a rough sketch of at least five Web pages that you would recommend. Make sure to show the important features and the hyperlinks for each page.

● WEB EXERCISES

1. A number of companies were discussed in this chapter. Locate the Web site of one of these companies. What are the goods and services that this company produces? After visiting the company Web site, describe how systems development could be used to improve the goods or services it produces. You may be asked to develop a report or send an e-mail message to your instructor about what you found.

2. Locate a company on the Internet that sells products, such as books or clothes. Write a report describing the strengths and weaknesses of the Web pages you encountered. In your opinion, what are the most important steps of the systems development process that could be used to improve the Internet site?

● CASES

① IT Projects at Coca-Cola

There are many reasons companies initiate a systems development project. In some cases, a merger can cause a major systems development effort. In other cases, companies start systems development to take advantage of a new marketing opportunity. For some companies, a downturn in revenues and profits can cause a big shakeup, which often involves new systems development initiatives.

A 21-year-old student in Atlanta used to drink Diet Coke all day long. She has changed, though, saying, "As I've gotten older, I've realized that drinking five Diet Cokes a day isn't good for you." And she is not alone. More people are switching from colas and caffeine to water and noncarbonated beverages. What may be a good health move for some has meant a drop in sales for cola companies, including Coca-Cola. After decades of steady growth, Coca-Cola recently experienced flat sales. Net operating profits fell a staggering 20 percent, and Coca-Cola's stock price took a dip. But things are changing at the traditional cola company under the new leadership of Douglas Daft. According to

Steve Jones, chief marketing officer at Coca-Cola, "For us to achieve the growth rate that people are expecting, we have to become more diversified. We have to move beyond Coke and the carbs." In addition to moving into new products, Coca-Cola is moving into new IS projects through an aggressive systems development effort.

In addition to slashing 6,000 jobs, Coca-Cola is also aggressively launching new systems development initiatives. One involves an automated inventory system. The inventory systems development project involves the use of handheld computers on trucks for route drivers and sophisticated software to transfer data to computers to more accurately track inventory. The company is also developing systems to strengthen its relationship with local franchises. Coca-Cola sells concentrate to local bottlers, who add sugar or artificial sweeteners and water. The local bottlers then package and distribute the soft drink. According to chief operating officer Daft, "We've spent years building the brands, infrastructure, and technology needed to be successful at the local level."

Discussion Questions

1. Why did Coca-Cola initiate new systems development projects?
2. What was Coca-Cola trying to accomplish with its new inventory system?

Critical Thinking Questions

3. If you were a manager of a cola company, how big would a downturn in sales and profits have to be for you to cut costs and initiate systems development projects?
4. Would you suggest that Coca-Cola heavily invest in an Internet site? If so, what type of Internet site would you recommend? Would it be oriented toward customers or local bottlers and franchises?

Sources: Adapted from Stacy Collett, "IT Projects Part of Coca-Cola Realignment," *Computerworld*, January 31, 2000, p. 4; Dean Foust et al., "Coke Is No Longer It," *Business Week*, February 28, 2000, p. 148; and "Debunking Coke," *The Economist*, February 12, 2000.

2 Hackers Cause Reevaluation of Systems Development on the Web

As discussed in this chapter and explored in Case 1, the initiation of a systems development project can be caused by a number of different factors. In early 2000, a large number of Internet companies were very concerned when their sites were attacked. In many cases, the sites were shut down for hours or days. Such sites as Buy.com, eBay, and Yahoo! were just a few of the sites that were shut down by hackers. Most of the attacks were a denial of service, where a large number of Web site hits caused systems to jam and fail. According to Gregory Hawkins, chief executive of Buy.com, "We were hit with a coordinated denial of service attack that appears to be very similar to what happened to Yahoo." Indeed, many of the attacks were very similar or identical, suggesting that the attackers may have been the same individual or group.

The attacks sent the beleaguered companies to scramble for answers. Should a new systems development effort be started to help prevent future attacks? Were new systems needed? Yahoo! programmers and engineers quickly started to search for the source of the attack. The company aggressively investigated its own security measures, including firewalls. In addition to affecting Yahoo! itself, server computers run for the company by GlobalCenter and the telecommunications systems run by Global Crossing were affected. As a result of the attacks, Global Crossing initiated new procedures and installed new equipment, called rate limiters, to prevent a surge in traffic through networks and servers.

The multiple attacks had unexpected benefits. Internet companies started to cooperate with each other to find ways to prevent such attacks in the future. In addition, the federal government got involved. Louis Freeh, director of the FBI, stated, "Even though we have markedly improved our capabilities to fight cyberintrusions, the problem is growing even faster and thus we are falling further behind." Freeh urged Congress to consider using federal racketeering laws against hacker groups. President Clinton also held a summit on Internet security. Clinton, who called the shutdowns "very disturbing," ordered U.S. agencies to evaluate their computer systems and determine their vulnerability to online attacks. One of the most interesting results was a surge in new systems development efforts by Internet security firms. These companies realized that there is big money to be made in developing systems to thwart Internet attacks.

Discussion Questions

1. Do you think Internet attacks will increase or decline in the years to come?

2. What were some of the unexpected benefits resulting from the attacks?

Critical Thinking Questions

3. Assume that you are the CIO of a company that is now experiencing a shutdown as a result of an attack. What is the first thing you would do? What would you do that day? What would you do in the next week, the next month, and the next year?

4. Assume that you work for the FBI to deal with Internet attacks. What specific penalties, including jail time and fines, would you recommend for the types of attacks discussed in this case?

Sources: Adapted from Khan Tran, "Hackers Attack Major Internet Sites," *The Wall Street Journal*, February 9, 2000, p. A3; Brendan Koener, "Who Can Stop Cybervandals?" *U.S. News & World Report*, February 28, 2000, p. 54; and John Carey, "Look Out Hackers," *Business Week*, March 6, 2000, p. 32D.

 3 ## Mergers Impact Systems Development

The executives of Glaxo Wellcome and SmithKline Beecham are busy making plans for a huge $76 billion merger; the new company would be the largest pharmaceutical company in the world. The new merged company will have 7.3 percent of the total pharmaceutical market and over 100,000 employees.

While the executives of these two giant companies are planning the merger, the systems development staffs of both companies are scrambling to make plans to merge the IS departments and operations. Unfortunately, the companies are using different enterprise resource planning packages.

Glaxo Wellcome is currently using the SAP R/3 enterprise resource planning package. SmithKline Beecham, however, is using an ERP package from J.D. Edwards. Getting the systems to work together is as important as getting the two companies together. "The pharmaceutical business is an incredibly competitive market, and everybody is facing the same huge urgency to improve speed to market and globalize operations. You can't do that with fragmented systems," says Steve Shaha, an analyst for the Gartner Group. Shaha believes that the new company, to be called Glaxo SmithKline if the merger goes through, will need to plot a new IS strategy and have it operational in two to three years.

The two ERP systems now being used, however, will complicate any effort to integrate the information systems of the two companies. "Technically, this will not be a slam dunk," says Shaha. To make matters even more difficult, both companies have to make sure that any new or modified systems meet federal food and drug regulations. According to research, it is not uncommon for these governmental compliance requirements to take up to 40% of an IS department's total budget. A fragmented system could cost even more in compliance costs. "That's a scary number," says Roddy Martin, an industry analyst.

Discussion Questions

1. Why is the merger of Glaxo and SmithKline going to have a significant impact on the IS department of the merged company?
2. Would you recommend using a system from one of the existing companies or would you recommend that the merged companies explore new alternatives for integrating their technological infrastructure?

Critical Thinking Questions

3. Comment on the following: It appears that the decision to merge was made before any detailed considerations about integrating the two different information systems.
4. What impact do you think the merger will have on staffing for the IS area of the merged company? If you were a middle-level project manager in the IS department for one of the companies, would you be worried about your job security or job duties in the new company?

Sources: Adapted from Craig Stedman, "Market Pressures Will Make IT a Priority in Drug Merger," *Computerworld*, January 24, 2000, p. 2; Carol Sliwa, "Drug Giants' Merger to Bring Systems Integration Hurdles," *Computerworld*, January 3, 2000, p. 12; and Amy Barrett, "Addicted to Mergers?" *Business Week*, December 6, 1999, p. 84.

● NOTES

Sources from the opening vignette on page 457: Adapted from Linda Rosencrance, "IT Spending Lets Loose," *Computerworld*, January 17, 2000, p. 1; "A Last Check at Six Users," *Computerworld*, January 10, 2000, p. 14; and Lee Gomen, "Passage of Y2K May Unlock Tech Spending," *The Wall Street Journal*, January 3, 2000, p. A3.

1. Matthew Gallagher, "Oil Trading Headed for Cyberspace," *The Wall Street Journal*, March 6, 2000, p. 13B.

2. Julekha Dash, "Skills Shortage," *Computerworld*, March 27, 2000, p. 10.

3. Fara Warner, "Ford Reaches Software Pact with Big Blue," *The Wall Street Journal*, January 13, 1999, p. A8.

4. Dara Swisher, "A Matchmaker for 'Bricks and Clicks,'" *The Wall Street Journal*, March 13, 2000, p. B1.

5. Arlene Weintraub, "For Online Pet Stores, It's Dog-Eat-Dog," *Business Week*, March 6, 2000, p. 78.

6. Sami Lais, "A&P's $250M IT Plan Shunned by Wall Street," *Computerworld*, March 20, 2000, p. 4.

7. Julia King, "Big Merger Year Leaves Out IT," *Computerworld*, December 20, 1999, p. 20.

8. Jathon Sapsford, "Chase, New York Reach Accord on Privacy," *The Wall Street Journal*, January 20, 2000, p. A4.

9. Deborah Radcliff, "Aligning Marriott," *Computerworld*, April 10, 2000, p. 58.

10. Mike Simons, "Anatomy of an IT Disaster," *Computer Weekly*, February 3, 2000, p. 1.

11. Varun Grover et al., "The Role of Organizational and Information Technology Antecedents in Reengineering Initiation Behavior," *Decision Sciences*, Summer 1999, p. 749.

12. Gro Bjerknes et al., "Improving the Customer-Supplier Relation in IT Development," *Proceedings of the Hawaii International Conference on System Sciences*, January 4–7, 2000.

13. Rodger Crockett, "A Digital Doughboy," *Business Week*, April 3, 2000, p. EB79.

14. Raghunathan et al., "Dimensionality of the Strategic Grid Framework," *Information Systems Research*, December 1999, p. 343.

15. "Listing," *Computing*, November 11, 1999, p. 66.

16. Barry W. Boehm, *Software Engineering Economics* (Englewood Cliffs, NJ: Prentice-Hall, 1981).

17. Laurie Toupin, "Computer Productivity Tools," *Design News*, January 3, 2000, p. 100.

18. "Techniques: MFC for Non-Believers," *EXE*, February 1, 2000, pp. 24–33.

19. Amanda Wells, "Inprise Has Linus Project Under Way," *NZ Infotech Weekly*, October 18, 1999, p. 29.

20. Philip Gill, "RAD Tools Propel Web Applications from Development to Deployment," *Information Week*, August 23, 1999.

21. Harter et al., "Effects of Process Maturity on Quality, Cycle Time, and Effort in Software Product Development," *Management Science*, April 2000, p. 451.

22. Scott McCartney, "Makers of Flight Computer Software Found Partly Guilty for Colombia Crash," *The Wall Street Journal*, June 14, 2000, p. A8.

23. Howard Millman, "On Track and in Touch," *Computerworld*, June 26, 2000, p. 88.

24. Sami Lais, "BMC Backs Project Deadline Guarantee with 20% Rebate," *Computerworld*, April 10, 2000, p. 10.

25. "Project Tracker," *Information Week*, February 7, 2000, p. 137.

26. ISO 9000 Group Web page, http://www.commerce-associates.com/iso, accessed June 19, 1998.

27. "North Slope Alaska GTL Options Analyzed," *Oil and Gas Journal*, January 31, 2000, p. 74.

CHAPTER 13

Systems Design, Implementation, Maintenance, and Review

Application service providers, or ASPs, are clearly the future—and smart businesspeople have been using them for years.

— Bonny Georgia, author and consultant

Principles	Learning Objectives
Designing new systems or modifying existing ones should always be aimed at helping an organization achieve its goals.	• *State the purpose of systems design and discuss the differences between logical and physical systems design.* • *Outline key steps taken during the design phase.* • *Define the term RFP and discuss how this document is used to drive the acquisition of hardware and software.* • *Describe the techniques used to make systems selection evaluations.*
The primary emphasis of systems implementation is to make sure that the right information is delivered to the right person in the right format at the right time.	• *List the advantages and disadvantages of purchasing versus developing software.* • *Discuss the software development process and some of the tools used in this process.* • *State the purpose of systems implementation and discuss the various activities associated with this phase of systems development.*
Maintenance and review add to the useful life of a system but can consume large amounts of resources. These activities can benefit from the same rigorous methods and project management techniques applied to systems development.	• *State the importance of systems and software maintenance and discuss the activities involved.* • *Describe the systems review process.*

AT&T

Renting Software to Business Clients

An important part of systems implementation is the acquisition of software. In the past, most companies either developed software internally using their own analysts and programmers or purchased software from an outside company. Leasing was always an option, but most companies either purchased or developed systems.

Today, renting software is becoming increasingly attractive. One reason is the high cost of developing software internally. The salary of programmers and analysts continues to rise, and it can be difficult to find skilled programmers and analysts. If the company decides to purchase software, it runs the risk of investing a large amount of money into a product that may become obsolete—sometimes in the not-too-distant future. Some software packages cost hundreds of thousands of dollars or even more. Another big reason for the recent trend toward software rental is the growing number of companies that offer software services. Called *application service providers (ASPs)*, these companies are hoping to save other companies the time, effort, and expense of developing or purchasing software.

AT&T is known primarily for its long-distance telephone services. But in the year 2000, the company decided to get into the business of developing and renting software to others. The company plans to invest about $250 million in the new business. AT&T's strategy is to rent software over the Internet. The new system, called Ecosystem for ASPs, will use AT&T's long-distance network, the largest in the world. AT&T is hoping to provide fast Internet access with off-site data storage capabilities for companies that rent software on-line through AT&T. The storage capabilities being offered by AT&T will allow companies to save by not having to invest in additional hard disks for new software. Referring to the new system, Kathleen Earley, president of AT&T's Data and Internet Services, stated, "This announcement is profound. This is the first time you will see all of the players at the table." AT&T plans to work with IBM, EMC Corporation, and Cisco Systems in developing the software rental business.

AT&T will have plenty of competition, though. How big is the rental market? A spokesman for Sun Microsystems predicts that it could be as high as $23 billion by the year 2003. That is one reason so many companies, including SAP, PeopleSoft, Hewlett-Packard, and Microsoft, are getting into the ASP business. For software companies, the ASP approach can mean shorter software release cycles, easier upgrades, and larger potential markets. According to Sanjav Sharma of IBM, "Intense competition in the software application market is driving vendors towards the ASP model to bring down the cost to customers."

As you read this chapter, consider the following:

- How can a company acquire software?

- Why is renting software becoming more attractive?

The way an information system is designed, implemented, and maintained profoundly affects the daily functioning of an organization. Information systems must be continually updated or replaced as the needs of a business—including hardware, software, and the infrastructure components—change. As seen in the opening vignette, information system components, such as software, can be rented. Whatever the method used to ensure currency, it is important that the new or updated systems be flexible to meet evolving needs. Otherwise, companies are forced to repeat a costly cycle of building and replacing information systems. This chapter presents the basics of systems design, implementation, maintenance, and review. Both users and IS personnel need to be aware of these stages so that they can participate in good systems development in their organizations.

SYSTEMS DESIGN

The purpose of systems design is to answer the question "How will the information system solve a problem?" The primary result of the systems design phase is a technical design that details system outputs, inputs, and user interfaces; specifies hardware, software, databases, telecommunications, personnel, and procedures; and shows how these components are related. The new system should overcome shortcomings of the existing system and help the organization achieve its goals. Of course, the system must also meet certain guidelines, including user and stakeholder requirements and the objectives defined during previous development phases.

Design can range in scope from individual to multicorporate projects (see Table 13.1). Nearly all companies are continually involved in designing systems for individuals, workgroups, and the enterprise. Increasingly, companies undertake multicorporate design, where two or more companies form a partnership or an alliance to design a new system. See the "E-Commerce" box for an example of two companies that are cooperating on a retail Web site. Design is increasingly focusing on Internet applications.

Systems design is typically accomplished using the tools and techniques discussed in the previous chapter. Depending on the specific application, these methods can be used to support and document all aspects of systems design. Two key aspects of systems design are logical and physical design.

Logical and Physical Design

logical design

description of the functional requirements of a system

physical design

specification of the characteristics of the system components necessary to put the logical design into action

Information systems must be designed along two dimensions: logical and physical. The **logical design** refers to what the system will do. The **physical design** refers to how the tasks are accomplished, including how the components work together and what each component does.

TABLE 13.1

The Scope of Design

Design can range in scope from multicorporate to individual projects.

Multicorporate design projects	Increasingly
Enterprise design projects	Narrow
Workgroup design projects	Scope
Individual design projects	

E-COMMERCE
Retail Web Supply Exchange

Traditionally, systems design has been done within a single company. In most cases, the objective was to realize a new opportunity or to solve an existing problem. Although this type of systems development project is still the norm, companies are starting to form strategic alliances with partners and undertake joint systems development. This partnering is done to leverage the experience and resources of two or more companies. The recent announcement of a massive systems development effort by Sears and Carrefour SA is an example. Sears is the second largest U.S. retailer, and Carrefour SA is the world's second largest retailer. Together, the two companies have $80 billion of supply-chain purchases.

The joint systems development effort is to result in a new e-commerce company, called GlobalNetXchange. Like other Internet exchanges, GlobalNetXchange will provide an on-line trading site for retail stores. Although the biggest U.S. retailer, Wal-Mart, is not participating in the exchange at this time, there is still excitement about the new systems development effort. According to Thomas Tashjian, a retail analyst with Banc of America Securities, "This is one of the most dramatic changes in consumer-products distribution of the decade. It has very wide implications for cost savings." The exchange would make money by charging fees to companies that use the exchange.

The trend toward joint systems development raises interesting questions and concerns. When two or more companies engage in a joint systems development effort, which company will be in charge? How will important project decisions be made? Will the participants in the systems development effort be taken from the companies themselves or found externally?

Discussion Questions

1. Why are more companies starting to get involved in joint systems development efforts with other companies?
2. What are the potential problems of a joint systems development effort?

Critical Thinking Questions

3. Assume that you are the owner and manager of a marine supplies company. Do you think a joint exchange, like the one described in this case, would be useful to your company and the marine supplies industry?
4. How would you organize a joint systems development effort, where two or more companies form a partnership to start a new Internet site? How would the systems development team be structured? What issues do you think would surface, and how would you solve them?

Sources: Adapted from Calmetta Coleman, "Sears, Carrefour Plan Web Supply Exchange," *The Wall Street Journal,* February 29, 2000, p. A4; Alorie Gilbert, "Sears and French Retailer Team for Online Exchange," *InformationWeek,* March 6, 2000; and Simon Goodley, "Industry Giants Unite in Global Trade Venture," *Computing,* March 2, 2000, p. 1.

Logical Design

Logical design describes the functional requirements of a system. That is, it conceptualizes what the system will do to solve the problems identified through earlier analysis. Without this step, the technical aspects of the system (such as which hardware devices should be acquired) often obscure the solution. Logical design involves planning the purpose of each system element, independent of hardware and software considerations. The logical design specifications that are determined and documented include the following.

Output design. Output design describes all outputs from the system and includes the types, format, content, and frequency of outputs. For example, a requirement that all company invoices reference the customers' original invoice number is a logical design specification. Screen and layout tools can be used during output design to capture the output requirements for the system.

Input design. Once output design has been completed, input design can begin. Input design specifies the types, format, content, and frequency of input data. For example, the requirement that the system capture the customer's phone number from his or her incoming call and use that to automatically look up the customer's account information is a logical design specification. A variety of diagrams and screen and report layouts can be used to reveal the type, format, and content of input data.

Processing design. The types of calculations, comparisons, and general data manipulations required of the system are determined during processing design. For example, a payroll program requires gross and net pay computations, state and federal tax withholding, and various deductions and savings plans.

File and database design. Most information systems require files and database systems. The capabilities of these systems are specified during the logical design phase. For example, the ability to obtain instant updates of customer records is a logical design specification. In many cases, a database administrator is involved with this aspect of logical design. Data-flow and entity-relationship diagrams are typically used during file and database design.

Telecommunications design. During logical design, the network and telecommunications systems need to be specified. For example, a hotel might specify a client/server system with a certain number of workstations that are linked to a server. From these requirements, a hybrid topology might be chosen. Graphics programs and CASE tools can be used to facilitate logical network design.

Procedures design. All information systems require procedures to run applications and handle problems if they occur. These important policies are captured during procedures design. Once designed, procedures can be described by using text and word processing programs. For example, the steps to add a new customer account may involve a series of both manual and computerized tasks. Written procedures would be developed to provide an efficient process for all to follow.

Controls and security design. Another important part of logical design is to determine the required frequency and characteristics of backup systems. In general, everything should have a backup, including all hardware, software, data, personnel, supplies, and facilities. Planning how to avoid or recover from a computer-related disaster should also be considered in this stage of the logical design phase.

Personnel and job design. Some systems require additional employees; others may need modification of the tasks associated with one or more existing IS positions. The job titles and descriptions of these positions are specified during personnel and job design. Organization charts are useful during personnel design to diagram various positions and job titles. Word processing programs are also used to describe job duties and responsibilities.

Physical Design

Physical design specifies the characteristics of the system components necessary to put the logical design into action. In this phase, the characteristics of each of the following components must be specified.

Hardware design. All computer equipment, including input, processing, and output devices, must be specified by performance characteristics. For example, if the logical design specified that the database must hold large amounts of historical data, then the system storage devices must have large capacity.

Software design. All software must be specified by capabilities. For example, if the ability for dozens of users to update the database concurrently is specified in the logical design, then the physical design must specify a database management system that allows this to occur. In some cases, software can be purchased; in other situations it will be developed internally. Logical design specifications for program outputs, data inputs, and processing requirements are also considered during the physical design of the software. For example, the ability to access data stored on certain disk files that the program will use is specified.

Database design. The type, structure, and function of the databases must be specified. The relationships between data elements established in the logical design must be mirrored in the physical design as well. These relationships include such things as access paths and file structure organization. Fortunately, many excellent database management systems exist to assist with this activity.

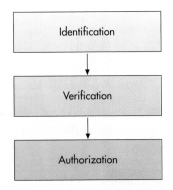

Telecommunications design. The necessary characteristics of the communications software, media, and devices must be specified. For example, if the logical design specifies that all members of a department must be able to share data and run common software, then the local area network configuration and the communications software that are specified in the physical design must possess this capability.

Personnel design. This step involves specifying the background and experience of individuals most likely to meet the job descriptions specified in the logical design.

Procedures and control design. How each application is to run, as well as what is to be done to minimize the potential for crime and fraud, must be specified. These specifications include auditing, backup, and output distribution methods.

Special System Design Considerations

A number of special system characteristics should be considered during both logical and physical design. These characteristics include sign-on procedures, interactive processing, interactive dialogue, error prevention and detection, and emergency alternate procedures.

Procedures for Signing On

System control methods are established during systems design. With almost any system, control problems may exist, such as criminal hackers breaking into the system or an employee mistakenly accessing confidential data. Sign-on procedures are the first line of defense against these problems. A **sign-on procedure** consists of identification numbers, passwords, and other safeguards needed for an individual to gain access to computer resources. A **system sign-on** allows the user to gain access to the computer system; an **application sign-on** permits the user to start and use a particular application, such as payroll or inventory control. These sign-on procedures help ensure that only authorized individuals can access a particular system or application.

The sign-on, also called a "logon," can identify, verify, and authorize access and usage (Figure 13.1). Identification means that the computer identifies the user as valid. If you must enter an identification number and password when logging on to a mainframe, you have gone through the identification process. For systems and applications that are more sensitive or secure, verification is used. Verification involves entering an additional code before access is given. Finally, authorization allows the user to gain access to restricted parts of a system or application. Consider a credit-checking application for a major credit card company. To grant credit, a clerk may be given only basic credit information on the screen. A credit manager, however, will have an authorization code to get additional credit information about a client.

Interactive Processing

Today, most computer systems allow interactive processing. With this type of system, people directly interact with the processing component of the system through terminals or networked PCs. The system and the user respond to each other in a real-time mode, which means within a matter of seconds. Interactive real-time processing requires special design features for ease of use, such as menu-driven systems, help commands, table lookup facilities, and restart procedures. With a **menu-driven system** (Figure 13.2), users simply pick what they want to do from a list of alternatives. Most people can easily operate these types of systems. They select their choice or respond to questions (or prompts) from the system, and the system does the rest.

FIGURE 13.2

A menu-driven system allows you to choose what you want from a list of alternatives.

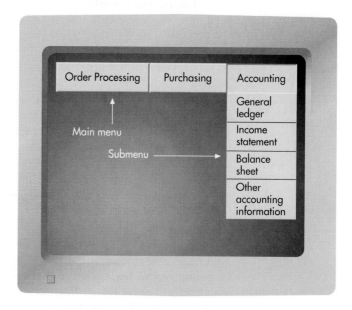

help facility

a program that provides assistance when a user is having difficulty understanding what is happening or what type of response is expected

lookup tables

tables containing data that are developed and used by computer programs to simplify and shorten data entry

restart procedures

simplified process to access an application from where it left off

Many designers incorporate a **help facility** into the system or applications program. When a user is having difficulty understanding what is happening or what type of response is expected, he or she can activate the help facility. The help screen relates directly to the problem the user is having with the software. The program responds with information on the program status, what possible commands or selections the user can give, and what is expected in terms of data entry.

Incorporating tables within an application is another very useful design technique. **Lookup tables** can be developed and used by computer programs to simplify and shorten data entry. For example, if you are entering a sales order for a company, you simply type in its abbreviation, such as ABCO. The program will then go to the customer table, normally stored on a disk, and look up all the information pertaining to the company abbreviated ABCO that is required to complete the sales order. This information is then displayed on a terminal screen for confirmation. The use of these tables can prevent wasting a tremendous amount of time entering the same data over and over again into the system.

If a problem occurs in the middle of an application—such as a temporary interruption of power or a printer running out of paper—the application currently being run is typically shut down. As a result, easy-to-use **restart procedures** are developed and incorporated into the design phase. With a restart procedure, it is very simple for an individual to restart an application where it left off.

Designing Good Interactive Dialogue

Dialogue refers to the series of messages and prompts communicated between the system and the user. From a user's point of view, good interactive dialogue from the computer system is essential, making data entry faster, easier, and more accurate. Poor dialogue from the computer can confuse the user and result in the wrong information being entered. If the computer system prompts the user to enter ACCOUNT, does it mean account number, type of account, or something else? Prototyping and an iterative approach are essential to good dialogue design, as users must have a great deal of input to ensure the system can be used properly. The following list covers some elements that create good interactive dialogue. These elements should be considered during systems design.

Clarity. The computer system should ask for information using language easily understood by users. Whenever possible, the users themselves should be involved in selecting the words and phrases used for dialogue with the computer system.

Response time. Ideally, responses from the computer system should approximate a normal response time from a human being carrying on the same sort of dialogue.

Consistency. The system should use the same commands, phrases, words, and function keys for all applications. After a user learns one application, all others will then be easier to use.

Format. The system should use an attractive format and layout for all screens. The use of color, highlighting, and the position of information on the screen should be considered carefully and consistently.

Jargon. All dialogue should be written in easy-to-understand terms. Avoid jargon known only to IS specialists.

Respect. All dialogue should be developed professionally and with respect. Dialogue should not talk down to or insult the user. Avoid statements such as "You have made a fatal error."

Preventing, Detecting, and Correcting Errors

The best and least expensive time to deal with potential errors is early in the design phase. During installation or after the system is operating, it is much more expensive and time-consuming to handle errors and related problems. Good systems design attempts to prevent errors before they occur, which involves recognizing what can happen and developing steps and procedures that can prevent, detect, and correct errors. This process includes developing a good backup system that can recover from an error. Table 13.2 lists the major causes of system errors.

Emergency Alternate Procedures and Disaster Recovery

As part of ongoing information systems assessment, organizations need to identify key information systems that control cash flow (e.g., invoicing, accounts receivable, payroll) and support other key business operations (e.g., inventory control, shipping customer products). Organizations typically develop a set of emergency procedures and recovery plans for these systems. The Royal Bank of Scotland, for example, uses data storage software to help it recover from disasters.[1] The software mirrors important data using fiber-optic connections between two sites.

In the event that a key information system becomes unusable, the end users need a set of emergency alternate procedures to follow to meet the needs of the business. End users should work with IS personnel to develop these procedures. They may be manual procedures to work around the unavailable automated work process. For example, if the order entry transaction processing system is unavailable, users could resort to special preprinted forms to capture basic order data for later entry when the system becomes available. In some cases, the emergency alternate procedures involve accessing a remote computer for computing resources.

TABLE 13.2

Major Causes of System Errors

Human	Natural	Technical
IS personnel	Wind	Hardware
Authorized users	Fire	Application software
Nonauthorized users	Earthquakes	Systems software
	Extreme temperatures	Database
	Floods	Communications
	Hurricanes	Electricity

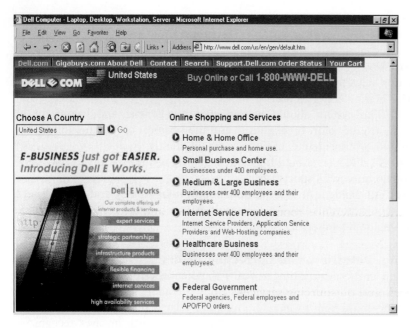

Companies such as Dell that rely on computer systems for their business may lose revenue if the system fails and they do not have a recovery plan.

Disaster planning is the process of anticipating and providing for disasters. A disaster can be an act of nature (a flood, fire, or earthquake) or a human act, error, labor unrest, or erasure of an important file. Disaster planning focuses primarily on two issues: maintaining the integrity of corporate information and keeping the information system running until normal operations can be resumed.

One of the first steps of disaster planning is to identify potential threats or problems, such as natural disasters, employee errors, and poor internal control procedures. Disaster planning also involves disaster preparedness. IS managers should occasionally hold an unannounced "test disaster"—similar to a fire drill—to ensure that the disaster plan is effective.

Disaster recovery is defined as the implementation of the disaster plan. Although companies have known about the importance of disaster planning and recovery for decades, many do not adequately prepare. The primary tools used in disaster planning and recovery are hardware, software and database, telecommunications, and personnel backup.

Hardware Backup

It is common for a company to form an arrangement with its hardware vendor or a disaster recovery company to provide access to a compatible computer hardware system in the event of a loss of hardware. Compaq, for example, has partnered with EverGreen Data Continuity to provide disaster recovery plans to help Compaq customers.[2] EverGreen Data provides Compaq customers a service, called Technology Protection, that helps them identify disaster risk and to formulate a comprehensive business-recovery plan.

A duplicate, operational hardware system (or immediate access to one through a specialized vendor) is an example of a **hot site**. A hot site is a compatible computer system that is operational and ready to use. If the primary computer has problems, the hot site can be used immediately as a backup. Another approach is to use a **cold site**, also called a *shell*, which is a computer environment that includes rooms, electrical service, telecommunications links, data storage devices, and the like. If a problem occurs with the primary computer, backup computer hardware is brought into the cold site, and the complete system is made operational. For both hot and cold sites, telecommunications media and devices are used to provide fast and efficient transfer of processing jobs to the disaster facility.

A number of firms offer disaster recovery services.[3] Arcus Data Security, for example, has 65 facilities worldwide to assist companies in recovering from emergencies. It provides off-site storage and security and has successfully responded to over 125 real disasters. Perm-A-Store is a firm that offers tape backup for critical data. BIA Professionals helps a company identify its critical risk areas to prepare for a potential disaster. SunGard offers Internet recovery services by providing duplicate locations for backing up critical data and Internet applications. The company provides hot sites and cold sites that are ready for extended recovery, strategically located around the country,

hot site

a duplicate, operational hardware system or immediate access to one through a specialized vendor

cold site

a computer environment that includes rooms, electrical service, telecommunications links, data storage devices, and the like; also called a *shell*

and available for quick response to multiple regional disasters. Their consulting arm analyzes a disaster's potential impact on a business and identifies the resources needed for recovery. Their expertise is available on disk, in the form of a Windows-based planning software, which can be used for disaster recovery planning.[4] Business Recovery Management offers a Disaster Recovery Center in Pittsburgh that provides the facilities, computer systems, and other equipment for recovery from unplanned business interruptions. To provide an effective, fully equipped workplace in time of disaster or special need, the center uses technology to ensure that a business stays connected both locally and globally, including fully equipped computer rooms for local and remote processing; wiring and PBX facilities to support voice, data, facsimile, and video telecommunications requirements; complete offices and services with conference rooms, private offices, and workstation areas to accommodate more than 250 people; and operational and technical support to assist in recovery.[5] Guardian Computer Support is an international computer support company providing a variety of computer services, including on-site hardware and software maintenance, contract staffing, disaster recovery services, and outsourcing services.[6]

Software and Database Backup

Software and databases can be backed up by making duplicate copies of all programs, files, and data. At least two backup copies should be made. One copy can be kept in the information systems department in case of accidental destruction of the software; the other should be kept off-site in a safe, secure, fireproof, and temperature- and humidity-controlled environment. A number of service companies provide this type of backup environment. Oracle, for example, offers its customers disaster recovery products for its database systems.[7] A key objective is to help companies keep their e-commerce applications running. Oracle Parallel Fail Safe, one of Oracle's disaster recovery products, can switch from a failed Web site to a parallel Web site and then bring the failed Web site back on-line within 30 seconds of its failure. A number of companies, such as on-line gold vendor Chipshot.com, use Oracle's fail-safe programs to keep their e-commerce sites running. According to a Chipshot.com official, "As our site grows, downtime becomes more expensive."

Backup is also essential for the data and programs on users' desktop computers. The advent of more distributed systems, like client/server systems, means that many users now have important, and perhaps mission-critical, data and applications on their desktop computers. Utility packages inexpensively provide backup features for desktop computers by copying data onto magnetic disk or tape. But software and database backup can be very difficult if an organization has a large amount of data. For some companies, making a backup of the entire database could take hours. A tight budget may also prohibit backing up significant quantities of data. As a result, some companies use **selective backup**, which involves creating backup copies of only certain files. For example, only critical files might be copied every night for backup purposes.

Another backup approach is to make a backup copy of all files changed during the last few days or the last week, a technique called **incremental backup**. This approach to backup uses an **image log**, which is a separate file that contains only changes to applications. Whenever an application is run, an image log is created that contains all changes made to all files. If a problem with a database occurs, an old database with the last full backup of the data, along with the image log, can be used to re-create the current database. Many disaster plans also include recovery of on-line hardware, software, and databases, which entails having another computer perform real-time backup of data at all times.

selective backup

creation of backup copies of only certain files

incremental backup

backup copy of all files changed during the last few days or the last week

image log

a separate file that contains only changes to applications

Telecommunications Backup

Some disaster recovery plans call for the backup of vital telephone communications. Complex plans might call for recovering whole networks. In other plans, the most critical nodes on the network are backed up by duplicate components. Using such fault-tolerant networks, which will not break down when one node or part of the network malfunctions, can be a more cost-effective approach to telecommunications backup.

Personnel Backup

Information systems personnel must also have backup. This can be accomplished in a number of ways. One of the best approaches is to provide cross-training for IS and other personnel so that each individual can perform an alternate job if required. For example, a company might train employees in accounting, finance, or other IS departments to operate the system if a disaster strikes. The company could also make an agreement with another information systems department or an outsourcing company to supply IS personnel if necessary.

Security, Fraud, and the Invasion of Privacy

Security, fraud, and the invasion of privacy can present disastrous problems. For example, because of an inadequate security and control system, a futures and options trader for a British bank lost about $1 billion. A simple system might have prevented a problem that caused the 200-year-old bank to collapse. In addition, from time to time, IRS employees have been caught looking at the returns of celebrities and others. Preventing and detecting these problems is an important part of systems design. Prevention includes the following:

• Determining potential problems
• Ranking the importance of these problems
• Planning the best place and approach to prevent problems
• Deciding the best way to handle problems if they occur

Every effort should be made to prevent problems, but companies must establish procedures to handle problems if they occur.

Systems Controls

systems controls

rules and procedures to maintain data security

closed shop

IS department in which only authorized operators can run the computers

open shop

IS department in which other people, such as programmers and systems analysts, are also authorized to run the computers

deterrence controls

rules and procedures to prevent problems before they occur

Most IS departments establish tight **systems controls** to maintain data security. Systems controls can help prevent computer misuse, crime, and fraud by employees and others. Most IS departments have a set of general operating rules that help protect the system. Some information systems departments are **closed shops**, in which only authorized operators can run the computers. Other IS departments are **open shops**, in which other people, such as programmers and systems analysts, are also authorized to run the computers. Other rules specify the conduct of the IS department.

These rules are examples of **deterrence controls**, which involve preventing problems before they occur. Making a computer more secure and less vulnerable to a break-in is another example. Good control techniques should help an organization contain and recover from problems. The objective of containment control is to minimize the impact of a problem while it is occurring, and recovery control involves responding to a problem that has already occurred.

Many types of system controls may be developed, documented, implemented, and reviewed. These controls touch all aspects of the organization, including the following.

Input controls. Input controls maintain input integrity and security. Some input controls involve the people who use the system; others relate to the data. The overall purpose is to reduce errors while protecting the computer system against improper or fraudulent input. Input controls range from using standardized input forms to eliminating data entry errors and using tight password and identification controls. For example, based on their logon identification, users are provided with access to a subset of the system and its capabilities. Some users can only view data; other users can update and view data. In addition, input controls can involve more sophisticated hardware and software that can use voice, fingerprints, and related techniques to identify and permit access to sensitive computer systems.

Processing controls. Processing controls deal with all aspects of processing and storage. In many cases, hardware and software are duplicated to provide procedures that ensure processing is as error-free as possible. In addition, storage controls prevent users from gaining access to or accidentally destroying data. The use of passwords and identification numbers, backup copies of data, and storage rooms that have tight security systems are examples of storage controls.

Output controls. Output controls are developed to ensure that output is handled correctly. In many cases, output generated from the computer system is recorded in a file that indicates the reports and documents generated, the time they were generated, and their final destinations.

Database controls. Database controls deal with ensuring an efficient and effective database system. These controls include the use of subschemas, identification numbers, and passwords, without which a user is denied access to certain data and information. Many of these controls are provided by database management systems.

Telecommunications controls. Telecommunications controls are designed to provide accurate and reliable data and information transfer among systems. Some telecommunications controls include hardware and software and other devices developed to ensure correct communications while eliminating the potential for fraud and crime. Examples are encryption devices and expert systems that can be used to protect a network from unauthorized access.

Personnel controls. Various personnel controls can be developed and implemented to make sure only authorized personnel have access to certain systems to help prevent computer-related mistakes and crime. Personnel controls can involve the use of identification numbers and passwords that allow only certain people access to certain data and information. ID badges and other security devices (such as "smart cards") can prevent unauthorized people from entering strategic areas in the information systems facility.

Once controls are developed, they should be documented in various standards manuals that indicate how the controls are to be implemented. They should then be implemented and frequently reviewed. It is common practice to measure the extent to which control techniques are used and to take action if the controls have not been implemented.

The Importance of Vendor Support

Whether an individual is purchasing a personal computer or an experienced company is acquiring an expensive mainframe computer, the system could be obtained from one or more vendors. In some cases, the vendor

Many companies use ID badges to prevent unauthorized access to sensitive areas in the information systems facility. (Source: Courtesy of Sensomatic Electronics Corporation)

simply provides hardware or software. In other cases, the vendor provides additional services. General Motors Corporation's Saturn Company, for example, hired a computer vendor to help the car company design, develop, and operate a customer-service and inventory-control Internet site.[8] The vendor will be paid about $300 million for 15 months of systems design and development efforts and five years of after-installation service. Some of the factors to consider in selecting a vendor are the following:

- The vendor's reliability and financial stability
- The type of service offered after the sale
- The goods and services the vendor offers and keeps in stock
- The vendor's willingness to demonstrate its products
- The vendor's ability to repair hardware
- The vendor's ability to modify its software
- The availability of vendor-offered training of IS personnel and system users
- Evaluations of the vendor by independent organizations

Generating Systems Design Alternatives

When additional hardware and software are not required, alternative designs are often generated without input from vendors. If the new system is complex, the original development team may want to involve other personnel in generating alternative designs. If new hardware and software are to be acquired from an outside vendor, a formal request for proposal (RFP) should be made.

The **request for proposal (RFP)** is one of the most important documents generated during systems development. It often results in a formal bid that is used to determine who gets a contract for new or modified systems. The RFP specifies in detail the required resources such as hardware and software. The Justice Ministry of New Zealand, for example, developed an RFP to get proposals on its disaster recovery facility for its large law enforcement system.[9] The system never had disaster recovery provisions previously. Only two companies were asked to submit proposals because, according to the CIO, "these companies have been assessed as the suppliers most capable of addressing the complex requirements for a disaster recovery service." The disaster facility will be based outside Auckland, New Zealand.

In some cases, separate RFPs are developed for different needs. For example, a company might develop separate RFPs for hardware, software, and database systems. The RFP also communicates these needs to one or more vendors, and it provides a way to evaluate whether the vendor has delivered what was expected. In some cases, the RFP is made part of the vendor contract. The table of contents for a typical RFP is shown in Figure 13.3.

request for proposal (RFP)

a document that specifies in detail required resources such as hardware and software

FIGURE 13.3

A Typical Table of Contents for a Request for Proposal

Johnson & Florin, Inc.
Systems Investigation Report

Contents

COVER PAGE (with company name and contact person)
BRIEF DESCRIPTION of the COMPANY
OVERVIEW of the EXISTING COMPUTER SYSTEM
SUMMARY of COMPUTER-RELATED NEEDS and/or PROBLEMS
OBJECTIVES of the PROJECT
DESCRIPTION of WHAT IS NEEDED
HARDWARE REQUIREMENTS
PERSONNEL REQUIREMENTS
COMMUNICATIONS REQUIREMENTS
PROCEDURES to BE DEVELOPED
TRAINING REQUIREMENTS
MAINTENANCE REQUIREMENTS
EVALUATION PROCEDURES (how vendors will be judged)
PROPOSAL FORMAT (how vendors should respond)
IMPORTANT DATES (when tasks are to be completed)
SUMMARY

Financial Options

When it comes to acquiring computer systems, three choices are available: purchase, lease, or rent. Cost objectives and constraints set for the system play a significant role in the alternative chosen, as do the advantages and disadvantages of each. Table 13.3 summarizes the advantages and disadvantages of these financial options.

TABLE 13.3

Advantages and
Disadvantages of
Acquisition Options

Renting (Short-Term Option)	
Advantages	**Disadvantages**
No risk of obsolescence	No ownership of equipment
No long-term financial investment	High monthly costs
No initial investment of funds	Restrictive rental agreements
Maintenance usually included	

Leasing (Longer-Term Option)	
Advantages	**Disadvantages**
No risk of obsolescence	High cost of canceling lease
No long-term financial investment	Longer time commitment than renting
No initial investment of funds	No ownership of equipment
Less expensive than renting	

Purchasing	
Advantages	**Disadvantages**
Total control over equipment	High initial investment
Can sell equipment at any time	Additional cost of maintenance
Can depreciate equipment	Possibility of obsolescence
Low cost if owned for a number of years	Other expenses, including taxes and insurance

Determining which option is best for a particular company in a given situation can be difficult. Financial considerations, tax laws, the organization's policies, its sales and transaction growth, marketplace dynamics, and the organization's financial resources are all important factors. In some cases, lease or rental fees can amount to more than the original purchase price after a few years. As a result, most companies prefer to purchase their equipment.

On the other hand, constant advances in technology can make purchasing risky. A company would not want to purchase a new multimillion-dollar computer only to have newer and more powerful computers available a few months later at a lower price. Some companies employ several people to determine the best option based on all the factors. This staff can also help negotiate purchase, lease, or rental contracts.

Evaluating and Selecting a System Design

The final step in systems design is to evaluate the various alternatives and select the one that will offer the best solution for organizational goals. Performance, cost, control, and complexity objectives must be considered and balanced during design evaluation. Depending on their weight, any of these objectives may result in the selection of one design over another. For example, financial concerns might make a company choose rental over equipment purchase. Specific performance objectives—say, that the new system must perform on-line data processing—may result in a complex network design for which control procedures must be established. Evaluating and selecting the best design involves a balance of system objectives that will best support organizational goals. Normally, evaluation and selection involves both a preliminary and a final evaluation before a design is selected.

The Preliminary Evaluation

preliminary evaluation

an initial assessment whose purpose is to dismiss unwanted proposals

A **preliminary evaluation** begins after all proposals have been submitted. The purpose of this evaluation is to dismiss unwanted proposals. Several vendors can usually be eliminated by investigating their proposals and comparing them with the original criteria. Those that compare favorably are asked to make a formal presentation to the analysis team. The vendors should also be asked to supply a list of companies that use their equipment for a similar purpose. The organization then contacts these references and asks them to evaluate their hardware, their software, and the vendor.

The Final Evaluation

final evaluation

a detailed investigation of the proposals offered by the vendors remaining after the preliminary evaluation

The **final evaluation** begins with a detailed investigation of the proposals offered by the remaining vendors. The vendors should be asked to make a final presentation and to fully demonstrate a system. The demonstration should be as close to actual operating conditions as possible. Such applications as payroll, inventory control, and billing should be conducted using a large amount of test data.

After the final presentations and demonstrations have been given, the organization makes the final evaluation and selection. Cost comparisons, hardware performance, delivery dates, price, flexibility, backup facilities, availability of software training, and maintenance factors are considered. Although it is good to compare computer speeds, storage capacities, and other similar characteristics, it is also necessary to carefully analyze whether the characteristics of the proposed systems meet the company's objectives. In most cases, the RFP captures these objectives and goals. Figure 13.4 illustrates the evaluation process.

Evaluation Techniques

Although the exact procedure used to make the final evaluation and selection varies from one organization to the next, four approaches are commonly considered: group consensus, cost/benefit analysis, benchmark tests, and point evaluation.

Group Consensus

group consensus

decision making by an appointed group given the responsibility of making the final evaluation and selection

In **group consensus**, a decision-making group is appointed and given the responsibility of making the final evaluation and selection. Usually, this group includes the members of the development team who participated in either systems analysis or systems design. This approach might be used to evaluate which of several screen layouts or report formats is best.

FIGURE 13.4

The Stages in Preliminary and Final Evaluations

Note that the number of possible alternatives decreases as the firm gets closer to making a final decision.

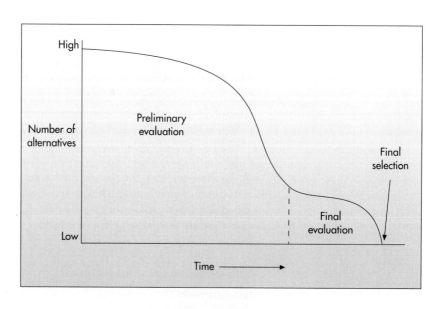

cost/benefit analysis

an approach that lists the costs and benefits of each proposed system

benchmark test

an examination that compares computer systems operating under the same conditions

point evaluation system

a process in which each evaluation factor is assigned a weight, in percentage points, based on importance; the system with the greatest total score is selected

Cost/Benefit Analysis

Cost/benefit analysis is an approach that lists the costs and benefits of each proposed system. Once expressed in monetary terms, all the costs are compared with all the benefits.[10] Table 13.4 lists some of the typical costs and benefits associated with the evaluation and selection procedure. This approach is used to evaluate options whose costs can be quantified, such as which hardware or software vendor to select.

Benchmark Tests

A **benchmark test** is an examination that compares computer systems operating under the same conditions. Although most computer companies publish their own benchmark tests, the best approach is for an organization to develop its own tests, then use them to compare the equipment it is considering. Several independent companies also rate computer systems. *Computerworld, Datamation, PC Week,* and *DataPro,* for example, not only summarize various systems but also evaluate and compare computer systems and manufacturers according to a number of criteria. This approach might be used to compare the end user system response time on two similar systems.

Point Evaluation

One of the disadvantages of cost/benefit analysis is the difficulty of determining the monetary values for all the benefits. An approach that does not employ monetary values is a **point evaluation system**. Each evaluation factor is

TABLE 13.4

Cost/Benefit Analysis Table

Costs	Benefits
Development costs	**Reduced costs**
Personnel	Fewer personnel
Computer resources	Reduced manufacturing costs
	Reduced inventory costs
	More efficient use of equipment
	Faster response time
	Reduced down- or crash time
	Less spoilage
Fixed costs	**Increased revenues**
Computer equipment	New products and services
Software	New customers
One-time license fees for software and maintenance	More business from existing customers
	Higher price as a result of better products and services
Operating costs	**Intangible benefits**
Equipment lease and/or rental fees	Better public image for the organization
Computer personnel (including salaries, benefits, etc.)	Higher employee morale
	Better service for new and existing customers
Electric and other utilities	The ability to recruit better employees
Computer paper, tape, and disks	Position as a leader in the industry
Other computer supplies	System easier for programmers and users
Maintenance costs	
Insurance	

FIGURE 13.5

An Illustration of the Point Evaluation System

In this example, software has been given the most weight (40 percent), compared with hardware (35 percent) and vendor support (25 percent). When system A is evaluated, the total of the three factors amounts to 82.5 percent. System B's rating, on the other hand, totals 86.75 percent, which is closer to 100 percent. Therefore, the firm chooses System B.

		System A			System B		
Factor's importance		Evaluation		Weighted evaluation	Evaluation		Weighted evaluation
Hardware	35%	95	35%	33.25	75	35%	26.25
Software	40%	70	40%	28.00	95	40%	38.00
Vendor support	25%	85	25%	21.25	90	25%	22.50
Totals	100%			82.5			86.75

assigned a weight, in percentage points, based on importance. Then each proposed information system is evaluated in terms of this factor and given a score that might range from 0 to 100, where 0 means the alternative does not address the feature at all and 100 means the alternative addresses that feature perfectly. The scores are totaled, and the system with the greatest total score is selected. When using point evaluation, literally hundreds of factors can be listed and evaluated. Figure 13.5 shows a simplified version of this process. This approach is used when there are many factors on which options are to be evaluated such as which software best matches a particular business's needs.

Because many elements must be considered before making a final selection, point evaluation can include a large number of factors. Performance concerns might include speed, storage capacity, and processing capabilities. Costs might include the deposit required on contract signing, payment schedules, lease and rental arrangements, maintenance costs, and availability of leasing companies. Complexity factors could include compatibility and ease of use, while control might include considerations such as training and maintenance offered by vendors, as well as system reliability and backup. When all these factors are added to the point evaluation system, a very large grid can result. The rows of the grid list the various factors important to the client company, and the columns of the grid represent the various vendors that responded to the request for proposal. Even if weights are not used, this type of chart can be very helpful. Some companies just use check marks to indicate which vendors have satisfied certain factors.

Freezing Design Specifications

Near the end of the design stage, an organization prohibits further changes in the design of the system. The design specifications are then said to be frozen. Freezing systems design specifications means that the user agrees in writing that the design is acceptable (Figure 13.6). Most system consulting companies insist on this step to avoid cost overruns and missed user expectations.

A problem that often arises during the implementation of any major project is that of "scope creep." As users more clearly understand how the system will work and what their needs are, they begin to request changes to the original design. Each change may be relatively minor, so the project team is strongly tempted to expand the scope of the project and incorporate the requested changes. However, the aggregate impact of many small changes can be significant; implementation of all minor changes can delay the project and/or increase the cost significantly. A common example is the request for a new report that was not included in the original design. If not managed carefully, the implementation of the first new report will lead to the request for "just one more report" and then another and another.

Prior to implementation, experienced project managers place formal controls on the project scope. A key component of the process is to assess the cost and schedule impact of each requested change, no matter how small, and to

FIGURE 13.6

Freezing Design Specifications

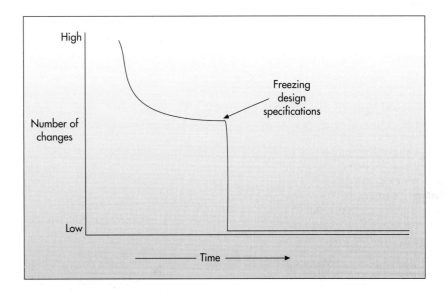

decide whether to include the change. Often the users and the project team decide to hold all changes until the original effort is completed and then prioritize the entire set of requested changes.

The Contract

One of the most important steps in systems implementation is to develop a good contract if new computer facilities are being acquired. Most computer vendors provide standard contracts; however, such contracts are designed to protect the vendor, not necessarily the organization buying the computer equipment.

More and more organizations are using outside consultants and legal firms to help them develop their own contracts. Such contracts stipulate exactly what they expect from the system vendor and what interaction will occur between the vendor and the organization. All equipment specifications, software, training, installation, maintenance, and so on are clearly stated. Furthermore, the contract stipulates deadlines for the various stages or milestones of installation and implementation, as well as actions the vendor will take in case of delays or problems. Some organizations include penalty clauses in the contract, in case the vendor is unable to meet its obligation by the specified date. Typically, the request for proposal becomes part of the contract. This saves a considerable amount of time in developing the contract, because the RFP specifies in detail what is expected from the vendors.

The Design Report

System specifications are the final results of systems design. They include a technical description that details system outputs, inputs, and user interfaces, as well as all hardware, software, databases, telecommunications, personnel, and procedure components and the way these components are related. The specifications are contained in a **design report**, which is the primary result of systems design. The design report reflects the decisions made for system design and prepares the way for systems implementation. The contents of the design report are summarized in Figure 13.7.

When developing a system, it is important to understand and thoroughly complete the systems development activities covered in this chapter. These phases provide the blueprints and groundwork for the rest of systems development. The

design report

the primary result of systems design, reflecting the decisions made for system design and preparing the way for systems implementation

FIGURE 13.7

A Typical Table of Contents for a Systems Design Report

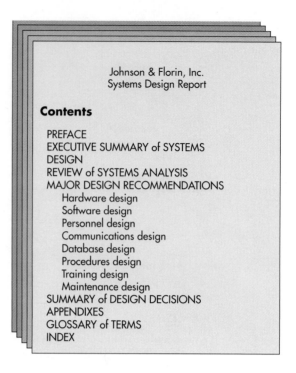

Johnson & Florin, Inc.
Systems Design Report

Contents

PREFACE
EXECUTIVE SUMMARY of SYSTEMS DESIGN
REVIEW of SYSTEMS ANALYSIS
MAJOR DESIGN RECOMMENDATIONS
    Hardware design
    Software design
    Personnel design
    Communications design
    Database design
    Procedures design
    Training design
    Maintenance design
SUMMARY of DESIGN DECISIONS
APPENDIXES
GLOSSARY of TERMS
INDEX

activities of the next phases will be easier, faster, and more accurate and will result in a more efficient, effective system if the design is complete and well thought out.

SYSTEMS IMPLEMENTATION

After the information system has been designed, a number of tasks must be completed before the system is installed and ready to operate. This process, called systems implementation, includes hardware acquisition, software acquisition or development, user preparation, hiring and training of personnel, site and data preparation, installation, testing, start-up, and user acceptance. The typical sequence of these activities is shown in Figure 13.8.

Choices and trade-offs are made at each step of systems implementation, which involve analyzing the benefits in terms of performance, cost, control, and complexity. Unfortunately, many organizations do not take full advantage of these steps or carefully analyze the trade-offs and hence never realize the full potential of new or modified systems. The hassles and carelessness must be avoided if organizations are to maximize the return on their information systems investment.

Acquiring Hardware from an Information Systems Vendor

To obtain the components for an information system, organizations can purchase, lease, or rent computer hardware and other resources. ABB, for example, signed a $250 million lease deal with IBM for about 75,000 desktop PCs, 25,000 laptop computers, and 9,000 PC servers.[11] The PC lease is expected to save ABB $35 million over the next three years. Some vendors, such as Hewlett-Packard, offer "pay-as-you-go" options.[12] The approach, which is being used for Internet applications, eliminates the need for a company to pay up front for a system. It also allows a company to change or upgrade its information system to meet changing business needs.

FIGURE 13.8

Typical Steps in Systems
Implementation

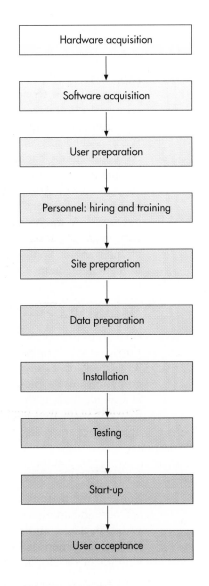

Hardware acquisition

↓

Software acquisition

↓

User preparation

↓

Personnel: hiring and training

↓

Site preparation

↓

Data preparation

↓

Installation

↓

Testing

↓

Start-up

↓

User acceptance

Computer dealers, such as
CompUSA, manufacture
build-to-order computer systems
and sell computers and
supplies from other vendors.
(Source: Courtesy of CompUSA, Inc.)

During implementation, organizations must identify and select one or more information systems vendors. An information systems vendor is a company that offers hardware, software, telecommunications systems, databases, information systems personnel, and/or other computer-related resources. Types of information systems vendors include general computer manufacturers (e.g., IBM and Hewlett-Packard), small computer manufacturers (e.g., Dell and Gateway), peripheral equipment manufacturers (e.g., Epson, Canon, and Piiceon), computer dealers and distributors (e.g., Canadian Communication Products, Dell, and Gateway), and leasing companies (e.g., National Computer Leasing, ECONOCOM-US, and Paramount Computer Rentals).

In addition, companies are increasingly turning to service providers to implement some or all of a systems development effort.[13] St. Joseph Health System, for example, used a service provider to help it develop and operate a computer system for 15 hospitals in California and Texas.[14] AT&T signed a $1 billion, seven-year contract to have a service provider develop and maintain about 150 of AT&T's applications, including billing, credit management, and customer care.[15] As discussed in Chapter 4, an application service provider (ASP) can help a company implement software and other systems.[16] The ASP can provide both end-user support and the computers on which to run the software.[17] ASPs often focus on high-end applications, such as database systems and enterprise resource planning packages.[18] As mentioned in Chapter 7, an Internet service provider assists a company in gaining access to the Internet. ISPs can also assist a company in setting up an Internet site. Some service providers specialize in specific systems or areas, such as marketing, finance, or manufacturing.

Acquiring Software: Make or Buy?

As with hardware, software can be acquired several ways. As previously mentioned, it can be purchased from external developers or developed in-house. This decision is often called the **make-or-buy decision**. In some cases, companies use a blend of external and internal software development. That is, off-the-shelf or proprietary software programs are modified or customized by in-house personnel. The advantages and disadvantages of these approaches were discussed in Chapter 4.

Externally Developed Software

Newmarket International, which invented hospitality sales and event management software, created a Web version of its software, called NetDelphi, through which users can store, access, share, and report on essential customer data, all via the Web. (Source: Courtesy of Newmarket International, Inc.)

Some of the reasons a company might purchase or lease externally developed software include lower costs, less risk regarding the features and performance of the package, and ease of installation. The cost of the software package is known, and there is little doubt that it will meet the company's needs. The amount of development effort is also less when software is purchased, compared with in-house development.

For example, a company may decide to purchase a general ledger software package developed by a major international consulting firm such as Ernst and Young that is used widely throughout their industry. If the company were to decide to build the software, it would take many months (even years), and there is a high degree of risk that when the system is implemented it may not meet the business needs as well as the software package.

Should a company choose off-the-shelf or contract software in its new systems, it must take the following steps:

Review needs and requirements. It is important to analyze the program's ability to satisfy user and organizational needs.

Acquire software. Many of the approaches discussed in previous sections, including the development of requests for proposals, performing financial analysis, and negotiating the software contract, should be undertaken.

Modify or customize software. Externally developed software seldom does everything the organization requires. Thus, it is likely that externally developed software will have to be modified to satisfy user and organizational needs. Some software vendors will assist with the modification, but others may not allow their software to be modified at all.

Acquire software interfaces. Usually, proprietary software requires a **software interface**, which consists of programs or program modifications that allow proprietary software to work with other software used in the organization. For example, if an organization purchases a proprietary inventory software package, software interfaces must allow the new software to work with other programs, such as sales ordering and billing programs.

Test and accept the software. Externally developed software should be completely tested by users in the environment in which it is to run before it is accepted.

Maintain the software and make necessary modifications. With many software applications, changes will likely have to be made over time. This aspect should be considered in advance because, as mentioned before, some software vendors do not allow their software to be modified.

software interface

programs or program modifications that allow proprietary software to work with other software used in the organization

In-House-Developed Software

Another option is to make or develop software internally. This requires the company's IS personnel to be responsible for all aspects of software development. Some advantages inherent with in-house-developed software include meeting user and organizational requirements and having more features and increased flexibility in terms of customization and changes. Software programs developed within a company also have greater potential for providing a competitive advantage because they are not easily duplicated by competitors in the short term. In addition, it can be possible to reuse software from other development efforts to reduce the time it takes to deliver in-house software. BankAmerica, for example, reuses previously developed software to deliver new software in 90 days or less.[19]

In some cases, a company that develops its own software internally decides to sell it to other companies.[20] General Electric, for example, developed a better system for order processing. Now it is looking to sell the software to other companies with similar problems. If successful, this new business for General Electric will help pay for the software development process. It may even turn into a profitable business.

Chief Programmer Teams

For software programming projects, the emphasis is on results—the finished package of computer programs. To get a smooth and efficient set of programs operating, the programming team must strive for the same overall objective. The **chief programmer team** is a group of skilled IS professionals with the task of designing and implementing a set of programs. This team has total responsibility for building the best software possible. Although the makeup of the chief programmer team varies with the size and complexity of the computer programs to be developed, a number of functions are common for all teams. A typical team has a chief programmer, a backup programmer, one or more other programmers, a librarian, and one or more clerks or secretaries (Figure 13.9).

chief programmer team

a group of skilled IS professionals with the task of designing and implementing a set of programs

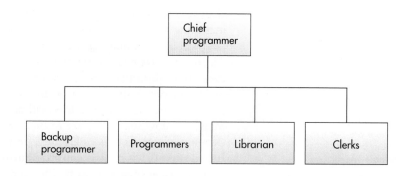

programming life cycle

a series of steps and planned
activities developed to maximize
the likelihood of developing good
software

The Programming Life Cycle

Developing in-house software requires a substantial amount of detailed planning. A series of steps and planned activities can maximize the likelihood of developing good software. These phases make up the **programming life cycle**, as illustrated in Figure 13.10 and described next.

Investigation, analysis, and design activities have already been completed. So the programmer has a detailed set of documents that describe what the system should do and how it should operate. An experienced programmer will begin with a thorough review of these documents before any code is written.

Language selection involves determining the best programming language for the application. Important characteristics to be considered are (1) the difficulty of the problem, (2) the type of processing to be used (batch or on-line), (3) the ease with which the program can be changed later, and (4) the type of problem, such as business or scientific. Often, a trade-off will need to be made between the ease of use of a language and the efficiency with which programs execute. Older, hard-to-write machine and assembly language programs are more efficient than easier-to-use, high-level language programs.

Program coding is the process of writing instructions in the language selected to solve the problem. Like a contractor building a house, the computer programmer follows the plans and documents developed in the previous steps. This careful attention to detail ensures that the software actually accomplishes the desired result.

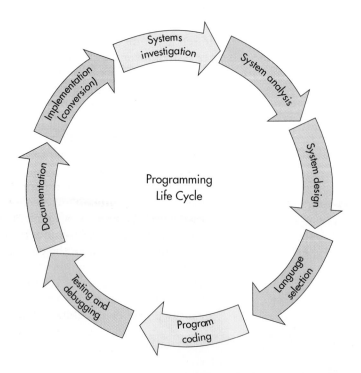

FIGURE 13.10

Steps in the Programming
Life Cycle

Testing and debugging are vital steps in developing computer programs. In general, testing is the process of making sure the program performs as intended; debugging is the process of locating and eliminating errors.

technical documentation

written details used by computer operators to execute the program and by analysts and programmers in case there are problems with the program or the program needs modification

user documentation

written description developed for individuals who use a program, showing users, in easy-to-understand terms, how the program can and should be used

Documentation is the next step, and it can include technical and user documentation. **Technical documentation** is used by computer operators to execute the program and by analysts and programmers in case there are problems with the program or the program needs modification. In technical documentation, each line of computer code is written out and explained with an English statement describing the function of the statement. Each variable is also described. **User documentation** is developed for the individuals who use the program. This type of documentation shows users, in easy-to-understand terms, how the program can and should be used. Incorporating a description of the benefits of the new application into user documentation may help stakeholders understand the reasons for the program and speed user acceptance. The software vendor often provides such documentation, or it may be obtained from a technical publishing firm. For example, Microsoft provides technical documentation for its spreadsheet package Excel, but there are literally hundreds of books and manuals for this software from other sources.

Implementation, or *conversion,* is the last step in developing new computer software. It involves installing the software and making it operational. Several approaches are discussed later in the chapter when we discuss installation.

The same basic steps can be followed for programming in both fourth-generation languages (4GLs) and traditional high-level programming languages. 4GLs, however, may be easier and faster to use. They are appropriate for iterative and prototyping development techniques, because prototypes can be developed quickly. The ease of coding with fourth-generation languages also allows more emphasis on creating programs to meet user and organizational needs.

object-oriented software development

approach to program development that uses a collection of existing modules of code, or objects, across a number of applications without being rewritten

Companies can also use **object-oriented software development** to develop programs. With this approach, a collection of existing modules of code, or objects, can be used across a number of applications. In most cases, minimal coding changes are required to mesh with the predeveloped objects or modules of code. Even though object-oriented software development does not require the use of object-oriented languages, most developers use them for the structure and ease they provide.

Tools and Techniques for Software Development

If software will be developed in-house, the chief programmer team can use a number of tools, techniques, and approaches. Options include structured design, structured programming, CASE tools, cross-platform development, integrated development environments, and structured walkthroughs.

Structured Design

Structured design is one well-known approach to designing and developing application software. The overall objective of structured design is to develop better software by reusing procedures and approaches that solve a variety of problems. Structured design breaks a large, difficult problem into smaller problems, each simple enough to manage and solve independently. These modules can then be reused in new and different programs. This building block, or modular, software usually costs less to develop and maintain and is easier to modify and update over time. For example, a module to generate a certain type of report can be developed independently, then plugged into numerous programs that require this type of report. One way to implement structured design is structured programming.

Structured Programming

For many programming projects, maintaining and updating the programs can take more time and effort than the original development process. Because structured programming makes program maintenance easier and faster, it continues to be an important programming technique.

A critical goal of structured programming is to untangle and reduce the complexity of computer programs. For example, consider the simple GOTO statement. This statement instructs the CPU to move, or "go to," another part of the program to read a different program statement. If a human analyst or programmer were attempting to follow the logic of such a program, too many uses of the GOTO statement would create confusion because of all the movements in the program from one location to another, often distant, location.

The basic idea behind structured programming is to improve the logical program flow by breaking the program into groups of statements. These groups are formed according to their function. One group may read data, while another group does a certain processing task. When using structured programming, statements in a group must conform to a standard structure. To begin with, there can be only one entering point into the block of statements and only one exit point from it. Therefore, you cannot branch to or from the middle of the structured group of statements. This restriction eliminates many programming errors and makes debugging much easier. Furthermore, there should be no groups of statements that cannot be reached or executed. Using the structured programming approach, each group can be tested separately.

As shown in Figure 13.11, only three types of structures are allowed when using structured programming. In the **sequence structure**, there must be definite starting and ending points. After starting the sequence, programming statements are executed one after another until all the statements in the sequence have been executed. Then the program either ends or continues on to another sequence. The **decision structure** allows the computer to branch, depending on certain conditions. Normally, there are only two possible branches. The final structure is the **loop structure**. Actually, there are two commonly used structures for loops. One is the do-until structure, and the other is the do-while structure. Both accomplish the same thing. In the do-until

sequence structure

a programming structure in which, after starting the sequence, programming statements are executed one after another until all the statements in the sequence have been executed; then the program either ends or continues on to another sequence

decision structure

a programming structure that allows the computer to branch, depending on certain conditions

loop structure

a programming structure with two commonly used structures for loops: do-until and do-while; in the do-until structure, the loop is done until a certain condition is met; for the do-while structure, the loop is done while a certain condition exists

FIGURE 13.11

The Three Structures Used in Structured Programming

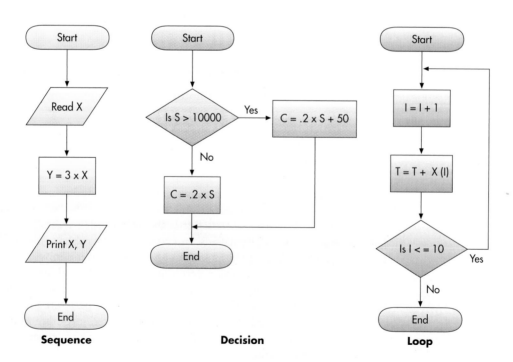

Sequence   Decision   Loop

TABLE 13.5

Characteristics of Structured Programming

Program code is broken into modules.

Each module has one and only one function. Such modules are said to have tight internal cohesion.

There is one and only one logic path into each module and one logic exit from each module.

The modules are loosely coupled.

GOTO statements are not allowed.

structure, the loop is done until a certain condition is met. For the do-while structure, the loop is done while a certain condition exists. Structured program code development can be the key to developing good program code. Some of the characteristics of structured programming are shown in Table 13.5.

Structured Programming: The Top-Down Approach

In general, a good approach to writing a large program is to start with the main module and work down to the other modules. This is called the **top-down approach** to programming and is used in structured program design and structured programming. Although the concept of top-down programming is simple, its use is beneficial in untangling or avoiding coding and debugging problems. The process begins by writing the main module. Then the modules at the next level are written. This procedure continues until all the modules have been written. Figure 13.12 can be used to visualize the top-down approach. In addition to program coding, the top-down approach should be used in testing and debugging. Thus, after the first, or main, module is written, it is tested and debugged. But the main module sends the computer to modules at the second level, which have not been written yet. Thus, simple modules at the second level are written to send the computer back to the main module so it can be fully and completely tested. If errors are found in the main module, they are corrected immediately.

top-down approach

a good general approach to writing a large program, starting with the main module and working down to the other modules

CASE Tools

CASE tools are often used during software development to automate some of the techniques. For example, source code can be automatically generated using CASE tools. Of the types of CASE tools previously discussed, lower-CASE tools are most likely to be used for software programming. Lower-CASE tools can provide a graphical programming environment and include compilers,

FIGURE 13.12

The Top-Down Approach to Writing, Testing, and Debugging a Modular Program

Level 1 (The main module)

a. Write the main module.
b. Write any necessary dummy modules at the second level.
c. Test the main module.
d. Debug the main module.

Level 2 (This procedure is done for each module one at a time.)

a. Write the module.
b. Write any necessary dummy modules for next lower level.
c. Test the module (this will automatically test all the modules that are above this one in the structure chart).
d. Debug the module.

Level N (This procedure is repeated for all levels.)

syntax checkers, and software modules that generate the actual program code. CASE tools may also have interfaces to the code generators of other vendors' CASE tools, a situation that allows a programmer to mix and match the program code generated. Using CASE can help increase programmer accuracy and productivity, particularly in terms of time spent on maintenance.

Cross-Platform Development

In the past, most applications were developed and implemented using mainframe computers. Today, many applications are developed on personal computers by users of the system. In response to the growth of end-user development, software vendors now offer more tools and techniques to PC users. One software development technique, called **cross-platform development**, allows programmers to develop programs that can run on computer systems having different hardware and operating systems, or platforms. With cross-platform development, the same program might be able to run on both a personal computer and a mainframe, or on two different types of PCs. Cross-platform development can be done on all sizes of computers, including PCs. One benefit of cross-platform development is that programs can run on both small and large systems, and users can perform software development activities on their PCs.

Integrated Development Environments

Software vendors also offer integrated development environments to assist with programming on personal computers. **Integrated development environments (IDEs)** combine the tools needed for programming with a programming language into one integrated package. IDE allows programmers to use simple screens, customized pull-down menus, and graphical user interfaces. Some even use different color text to allow a programmer to quickly locate sections, verbs, or errors in program code. In general, IDE can make programming software more intuitive as well as personal computer programmers more productive. Combining these tools with the language itself makes it easier for programmers to develop sophisticated programs on personal computers.

Turbo Pascal includes an IDE with a code editor, a program debugger, and a compiler. Visual C++ is another example of a programming language with an IDE. The MULTI Software Development Environment provides a framework of interacting tools to support program development by small or large workgroups. MULTI includes all of the tools needed to support major programming projects, including a program builder, version control, program editor, debugger, execution profiler, run-time error checker, and source-code control. JIG is a Java integrated development environment that helps programmers write, manage, and debug applets and applications written in Java. JIG is written entirely in Java.

Structured Walkthroughs

Regardless of the tools or techniques used, companies should review software throughout the development process. Companies often use a structured walkthrough technique, which is typically performed by chief programmer teams. As shown in Figure 13.13, a **structured walkthrough** is a planned and preannounced review of the progress of a program module, a structure chart, or a human procedure. The walkthrough helps team members review and evaluate the progress of components of a structured project. The structured walkthrough approach is also useful for programming projects that do not use the structured design approach.

cross-platform development

development technique that allows programmers to develop programs that can run on computer systems having different hardware and operating systems, or platforms

integrated development environments (IDEs)

software that combines the tools needed for programming with a programming language into one integrated package

structured walkthrough

a planned and preannounced review of the progress of a program module, a structure chart, or a human procedure

FIGURE 13.13

A structured walkthrough is a planned, preannounced review of the progress of a particular project objective.

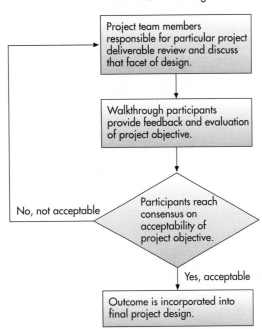

Walkthrough Planning and Preparation

Project team disseminates design documentation to relevant parties. → Walkthrough participants review design documentation in advance of actual walkthrough.

The Walkthrough

Project team members responsible for particular project deliverable review and discuss that facet of design.

Walkthrough participants provide feedback and evaluation of project objective.

Participants reach consensus on acceptability of project objective.

No, not acceptable

Yes, acceptable

Outcome is incorporated into final project design.

Acquiring Database and Telecommunications Systems

Acquiring or upgrading database systems can be one of the most important steps of a systems development effort. Because databases are a blend of hardware and software, many of the approaches discussed earlier for acquiring hardware and software also apply to database systems. For example, an upgraded inventory control system may require database capabilities, including more hard disk storage or a new DBMS. If so, additional storage hardware will have to be acquired from an information systems vendor. New or upgraded software might also be purchased or developed in-house.

Telecommunications is one of the fastest growing applications for today's businesses and individuals. Like database systems, telecommunications systems require a blend of hardware and software. For personal computer systems, the primary piece of hardware is a modem. For client/server and mainframe systems, the hardware can include multiplexers, concentrators, communications processors, and a variety of network equipment. Communications software will also have to be acquired from a software company or developed in-house. Again, the earlier discussion on acquiring hardware and software also applies to the acquisition of telecommunications hardware and software.

Providing users with proper training can help ensure that the information system is used correctly, efficiently, and effectively.
(Source: Eyewire)

User Preparation

user preparation

the process of readying managers, decision makers, employees, other users, and stakeholders for the new systems

User preparation is the process of readying managers, decision makers, employees, other users, and stakeholders for the new systems. With the growing trend to employee empowerment, system developers need to provide users with the proper training to make sure they use the information system correctly, efficiently, and effectively. User preparation can include active participation, marketing, training, documentation, and support. Top-management support in ensuring that sufficient time and resources are allocated to user preparation is absolutely essential to a successful system startup.

Informing and preparing users for new or modified systems can be done in a variety of ways. User preparation actually begins with user participation in system analysis. Some organizations also actively market new systems to users via brochures, newsletters, and seminars to promote them the way they would a new product or service.

Stakeholders, who benefit from the system but do not directly use or interact with it, need to be made aware of the results of the systems analysis and design effort. For example, suppose a toy manufacturer integrates a new inventory control system with the order processing and production planning systems so that it can quickly respond to changes in product demand. If a successful product promotion dramatically increases customer demand, the new application would alert managers to schedule larger or additional production runs to meet customer demand. Informed stakeholders—in this instance, customers—will likely shop at that chain, knowing that they can rely on finding stuffed toys in stock. Some companies even advertise their use of information systems to add product and service value.

Without question, training users is an essential part of user preparation, whether they are trained by internal personnel or by external training firms. In some cases, companies that provide software also train users at no charge or at a reasonable price. The cost of training can be negotiated during the selection of new software. Other companies conduct user training throughout the systems development process. Concerns and apprehensions about the new system must be eliminated through these training programs. Employees should be acquainted with the system's capabilities and limitations by the time they are ready to use it.

Continuing support provides assistance to users after a new or modified application has been installed. The overall purpose of support is to make sure users understand and benefit from the new or modified system. This support can be additional hardware, software, and service. Continuing support is provided by most vendors for a fee. Seminars, training programs, and consulting personnel are also popular. The preparation and distribution of user documentation is another important aspect of continuing support.

After a new or modified system has been installed, an organization may have to hire new IS personnel.
(Source: Eyewire)

IS Personnel: Hiring and Training

Depending on the size of the new system, an organization may have to hire and, in some cases, train new IS personnel. An information systems manager, systems analysts, computer programmers, data entry operators, and similar personnel may be needed for the new system.

As with users, the eventual success of any system depends on how it is used by the personnel within the organization. Training programs should be conducted for the IS personnel who will be using the computer system. These programs are similar to those for the users, although they may be more detailed in the technical aspects of the systems. Effective training will help IS personnel use the new system to perform their jobs and support other users in the organization.

Site Preparation

site preparation

preparation of the location of the new system

The location of the new system needs to be prepared in a process called **site preparation**. For a small system, site preparation can be as simple as rearranging the furniture in an office to make room for a computer. With a larger system, this process is not so easy because it may require special wiring and air-conditioning. One or two rooms may have to be completely renovated, and additional furniture may have to be purchased. A special floor may have to be built, under which the cables connecting the various computer components are placed, and a new security system may be needed to protect the equipment. For larger systems, additional power circuits may also be required.

Data Preparation

data preparation (data conversion)

conversion of manual files into computer files

If the organization is computerizing its work processes, all manual files must be converted to computer files in a process called **data preparation**, or **data conversion**. All permanent data must be placed on a permanent storage device, such as magnetic tape or disk. Usually the organization hires temporary, part-time data-entry operators or a service company to convert the manual data. Once the data has been converted, the temporary workers are no longer needed. A computerized database system or other software will then be used to maintain and update the computer files.

Installation

installation

the process of physically placing the computer equipment on the site and making it operational

Installation is the process of physically placing the computer equipment on the site and making it operational. Although normally the manufacturer is responsible for installing computer equipment, someone from the organization (usually the IS manager) should oversee the process, making sure that all equipment specified in the contract is installed at the proper location. After the system is installed, the manufacturer performs several tests to ensure that the equipment is operating as it should.

Testing

unit testing

testing of individual programs

system testing

testing the entire system of programs

volume testing

testing the application with a large amount of data

integration testing

testing all related systems together

acceptance testing

conducting any tests required by the user

Good testing procedures are essential to make sure that the new or modified information system operates as intended. Inadequate testing can result in mistakes and problems. A popular tax preparation company, for example, implemented a Web-based tax preparation system, but people could see one another's tax returns.[21] The president of the tax preparation company called it "our worst-case scenario." In another case, the London Stock Exchange experienced a crash that lasted almost eight hours.[22] Some believe that the London Stock Exchange can salvage its reputation, but others disagree. The head of a trading firm said that this is "the nail in the coffin for the London Stock Exchange in the way it exists now. The ramifications are going to be huge." Better testing might have prevented these types of problems.

Several forms of testing should be used, including testing each of the individual programs (**unit testing**), testing the entire system of programs (**system testing**), testing the application with a large amount of data (**volume testing**), and testing all related systems together (**integration testing**), as well as conducting any tests required by the user (**acceptance testing**). The sequence in which these testing activities normally occur is shown in Figure 13.14.

FIGURE 13.14

Types of Testing

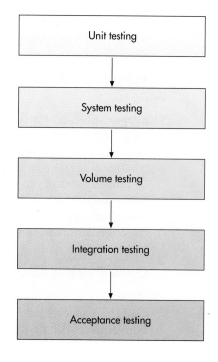

Unit testing is accomplished by developing test data that will force the computer to execute every statement in the program. In addition, each program is tested with abnormal data to determine how it will handle problems. System testing requires the testing of all the programs together. It is not uncommon for the output from one program to become the input for another. So system testing ensures that the output from one program can be used as input for another program within the system. Volume testing ensures that the entire system can handle a large amount of data under normal operating conditions. Integration testing ensures that the new programs can interact with other major applications. It also ensures that data flows efficiently and without error to other applications. For example, a new inventory control application may require data input from an older order processing application. Integration testing would be done to ensure smooth data flow between the new and existing applications. Integration testing is typically done after unit and system testing. Finally, acceptance testing makes sure that the new or modified system is operating as intended. Run times, the amount of memory required, disk access methods, and more can be tested during this phase. Acceptance testing ensures that all performance objectives defined for the system or application are satisfied. Involving users in acceptance testing may help them understand and effectively interact with the new system. Acceptance testing is the final check of the system before start-up.

Start-Up

Start-up begins with the final tested information system. When start-up is finished, the system is fully operational. Various start-up approaches are available (Figure 13.15). **Direct conversion** (also called *plunge* or *direct cutover*) involves stopping the old system and starting the new system on a given date. Direct conversion is usually the least desirable approach because of the potential for problems and errors when the old system is shut off and the new system is turned on at the same instant. The **phase-in approach** is a popular technique preferred by many organizations. In this approach, sometimes called a *piecemeal approach,* components of the new system are slowly phased in while components of the old one are slowly

start-up

the process of making the final tested information system fully operational

direct conversion

stopping the old system and starting the new system on a given date (also called *plunge* or *direct cutover*)

phase-in approach

slowly replacing components of the old system with those of the new one; this process is repeated for each application until the new system is running every application and performing as expected (also called *piecemeal approach*)

FIGURE 13.15

Start-Up Approaches

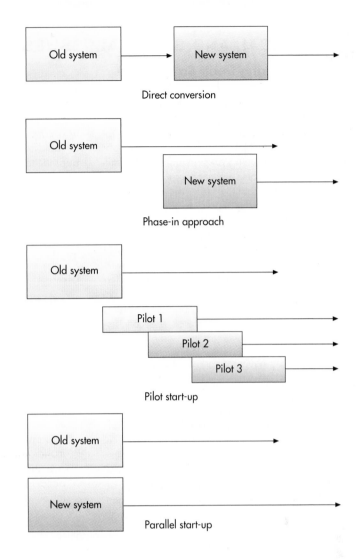

phased out. When everyone is confident that the new system is performing as expected, the old system is completely phased out. This gradual replacement is repeated for each application until the new system is running every application. **Pilot start-up** involves running the new system for one group of users rather than all users. For example, a manufacturing company with a number of retail outlets throughout the country could use the pilot start-up approach and install a new inventory control system at one of the retail outlets. When this pilot retail outlet runs without problems, the new inventory control system can be implemented at other retail outlets. **Parallel start-up** involves running both the old and new systems for a period of time. The output of the new system is compared closely with the output of the old system, and any differences are reconciled. When users are comfortable that the new system is working correctly, the old system is eliminated.

User Acceptance

Most mainframe computer manufacturers use a formal **user acceptance document**—a formal agreement signed by the user that states that a phase of the installation or the complete system is approved. This is a legal document that usually removes or reduces the information systems vendor from liability for problems that occur after the user acceptance document has been signed. Because this document is so important, many companies get legal assistance

pilot start-up

running the new system for one group of users rather than all users

parallel start-up

running both the old and new systems for a period of time and comparing the output of the new system closely with the output of the old system; any differences are reconciled; when users are comfortable that the new system is working correctly, the old system is eliminated

user acceptance document

formal agreement signed by the user that states that a phase of the installation or the complete system is approved

ETHICAL AND SOCIETAL ISSUES
Monitoring Debit Card Usage

William Scheurer, the president of a large company, was frustrated. His 15-year-old daughter was always running out of money, which required him to make frequent trips to an ATM machine. Not only was it frustrating—it was also time-consuming. Being president of a $400 million credit card service company, Scheurer knew there had to be something he could do for himself and the millions of parents who faced similar problems. Could an innovative systems development project give a solution?

His answer was to implement a special debit card, called PocketCard, just for teenagers. A special account would be set up. Then the parents could deposit or withdraw funds as needed and warranted. The most interesting part of the system was its connection to the Internet and online computer systems. With this on-line capability, parents could monitor their teenagers' spending, on a second-by-second basis as card charges worked their way through a Visa network. If a parent saw his or her kid purchasing alcohol at a liquor store or if the parent saw a charge at a forbidden nightclub, all the remaining money could be transferred out of the account, and the teenager could be disciplined when he or she arrived back home. The new teenager accounts could generate revenues for the card issuer by charging to activate the account, requiring a maintenance fee, and charging for each transaction. The systems development effort could cost from $5 million to $15 million. In addition to teenagers, the new debit card could be used with baby-sitters, secretaries, and even employees. With the new card, expenses could be monitored through the Internet or by making a simple phone call.

Although the system sounds like a blessing for parents, it poses ethical and societal concerns. If used with secretaries or employees, would the new card be an invasion of their privacy? Would the use of such a card in a corporate environment cause more problems than it would solve? These questions have yet to be resolved, but the novel approach is generating attention. Other companies, including SpendCash and Cobalt Card, are also getting into the debit-card business. SpendCash, for example, will be sold in $10, $20, $50, and $100 increments at retail and convenience stores. The company will post a 2.25 percent charge for each transaction.

Discussion Questions

1. What are the advantages and disadvantages of this new debit card?
2. Do you think this systems development effort will be acceptable to all involved?

Critical Thinking Questions

3. Assume that you are the systems development manager for the debit card project. How would you modify the teenager debit card system to make the card appropriate for use with employees, secretaries, and managers of a company?
4. Are there any privacy protections that you could build into the system during the systems implementation stage?

Sources: Adapted from Chana Schoenberger, "Big Brother (and Sister)," *Forbes*, January 24, 2000, p. 142; "Only Time Will Tell," *ITS World*, January 2000, p. 1086; and "Online Music Discounter to Promote PocketCard," *Card Fax*, February 25, 2000, p. 2.

---

before they sign the acceptance document. Stakeholders may also be involved in acceptance to make sure that the benefits to them are indeed realized. As seen in the "Ethical and Societal Issues" box, some systems development efforts may not be desirable or acceptable to everyone involved.

SYSTEMS MAINTENANCE

systems maintenance

stage of systems development that involves checking, changing, and enhancing the system to make it more useful in achieving user and organizational goals

Systems maintenance involves checking, changing, and enhancing the system to make it more useful in achieving user and organizational goals. Software maintenance is a major concern for organizations. In some cases, an organization encounters major problems that involve recycling the entire systems development process. In other situations, minor modifications are sufficient.

Reasons for Maintenance

Once a program is written, it is likely to need ongoing maintenance. To some extent, a program is like a car that needs oil changes, tune-ups, and repairs at certain times. Experience shows that frequent, minor maintenance to a program, if properly done, can prevent major system failures later. Some of the reasons for program maintenance are the following:

- Changes in business processes
- New requests from stakeholders, users, and managers
- Bugs or errors in the program
- Technical and hardware problems
- Corporate mergers and acquisitions
- Government regulations
- Change in the operating system or hardware on which the application runs

When it comes to making necessary changes, most companies modify their existing programs instead of developing new ones. That is, as new systems needs are identified, often the burden of fulfilling the needs falls on the existing system. Old programs are repeatedly modified to meet ever-changing needs. Over time, these modifications tend to interfere with the system's overall structure, reducing its efficiency and making further modifications more burdensome.

Types of Maintenance

slipstream upgrade

a minor upgrade—typically a code adjustment or minor bug fix—that usually requires recompiling all the code, and, in so doing, can create entirely new bugs

patch

a minor change to correct a problem or make a small enhancement, usually an addition to an existing program

release

a significant program change that often requires changes in the documentation of the software

version

a major program change, typically encompassing many new features

request for maintenance form

a form authorizing modification of programs

Software companies and many other organizations use four generally accepted categories to signify the amount of change involved in maintenance. A **slipstream upgrade** is a minor upgrade—typically a code adjustment or minor bug fix—not worth announcing. It usually requires recompiling all the code and, in so doing, it can create entirely new bugs. This maintenance practice can explain why the same computers sometimes work differently with what is supposedly the same software. A **patch** is a minor change to correct a problem or make a small enhancement. It is usually an addition to an existing program. That is, the programming code representing the system enhancement fix is usually "patched into," or added to, the existing code. For example, Microsoft released patches to remedy a glitch in several versions of Internet Explorer, a problem that opened a way for hackers and unscrupulous Web site operators to read the contents of files on users' computers. A new **release** is a significant program change that often requires changes in the documentation of the software. Finally, a new **version** is a major program change, typically encompassing many new features.

The Request for Maintenance Form

Because of the amount of effort that can be spent on maintenance, many organizations require a **request for maintenance form** to authorize modification of programs. This form is usually signed by a business manager who documents the need for the change and identifies the priority of the change relative to other work that has been requested. The IS group reviews the form and identifies the programs to be changed, determines the programmer who will be assigned to the project, estimates the expected completion date, and develops a technical description of the change. A cost/benefit analysis may be required if the change requires substantial resources.

Performing Maintenance

Depending on organizational policies, the people who perform systems maintenance vary. In some cases, the team that designs and builds the system also performs maintenance. This ongoing responsibility gives the designers and

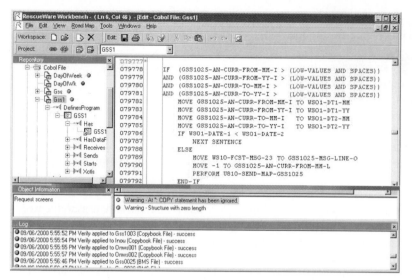

RescueWare provides companies with automated software that accelerates the process of transforming third-generation code, such as COBOL, to Internet or client/server platforms.
(Source: Courtesy of Relativity Technologies, Inc.)

maintenance team

a special IS team responsible for modifying, fixing, and updating existing software

programmers an incentive to build systems well from the outset: if there are problems, they will have to fix them. In other cases, organizations have a separate **maintenance team**. This team is responsible for modifying, fixing, and updating existing software. Because experience and skills are important in maintenance, some organizations utilize a specialized maintenance team or department. Java and the object-oriented programming languages hold the promise of reducing the program maintenance effort.

Regardless of who performs maintenance, the same tools and techniques used for earlier phases of systems development—CASE tools, flowcharts, structured programming, and so on—should be used. In addition, all maintenance should be fully documented. Unfortunately, documentation sometimes falls by the wayside. For example, if an order processing application crashes during peak hours, the company's main objective is to get the application running as soon as possible. Documenting the problem or any changes made to the application may be overlooked in the rush to restart the application. But lack of documentation can cause future problems if programmers reference outdated data-flow diagrams, layout charts, and so on when they perform maintenance for the system. Thus, it is essential that the tools used in maintenance allow easy documentation to accurately reflect the changes.

A number of vendors have developed tools to ease the software maintenance burden. Relativity Technologies recently unveiled RescueWare, a product that converts third-generation code such as COBOL to highly maintainable C++, Java, or Visual Basic object-oriented code. Using RescueWare, maintenance personnel download mainframe code to Windows NT or Windows 2000 workstations. They then use the product's graphical tools to analyze the original system's inner workings. RescueWare lets a programmer see the original system as a set of object views, which visually illustrate module functioning and program structures. IS personnel can choose one of three levels of transformation: revamping the user interface, converting the database access, and transforming procedure logic.

The Financial Implications of Maintenance

The cost of maintenance is staggering. For older programs, the total cost of maintenance can be up to five times greater than the total cost of development. In other words, a program that originally cost $25,000 to develop may cost $125,000 to maintain over its lifetime. The average programmer can spend over 50 percent of his or her time on maintaining existing programs instead of developing new ones. Furthermore, as programs get older, total maintenance expenditures in time and money increase, as illustrated in Figure 13.16. With the use of newer programming languages and approaches, including object-oriented programming, maintenance costs are expected to decline. Even so, many organizations have literally millions of dollars invested in applications written in older languages (such as COBOL), which are both expensive and time-consuming to maintain.

FIGURE 13.16

Maintenance Costs as a
Function of Age

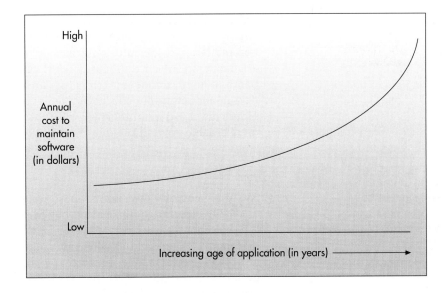

The financial implications of maintenance make it important to keep track of why systems are maintained, instead of simply keeping cost figures. This is another reason documentation of maintenance tasks is so crucial. A determining factor in the decision to replace a system is the point at which it is costing more to fix it than to enhance it.

The Relationship between Maintenance and Design

Programs are expensive to develop, but they are even more expensive to maintain. Programs that are well designed and documented to be efficient, structured, and flexible are less expensive to maintain in later years. Thus, there is a direct relationship between design and maintenance. More time spent on design up front can mean less time spent on maintenance later.

In most cases, it is worth the extra time and expense to design a good system. Consider a system that costs $250,000 to develop. Spending 10 percent more on design would cost an additional $25,000, bringing the total design cost to $275,000. Maintenance costs over the life of the program could be $1,000,000. If this additional design expense can reduce maintenance costs by 10 percent, the savings in maintenance costs would be $100,000. Over the life of the program, the net savings would be $75,000 ($100,000 − $25,000). This relationship between investment in design and long-term maintenance savings is graphically displayed in Figure 13.17.

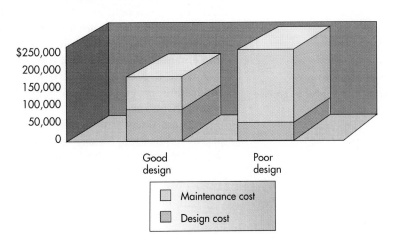

FIGURE 13.17

The Value of Investment in
Design

The need for good design goes beyond mere costs. There is a real risk in ignoring small system problems when they arise, as these small problems may become large in the future. As mentioned earlier, because maintenance programmers spend an estimated 50 percent or more of their time deciphering poorly written, undocumented program code, there is little time to spend on developing new, more effective systems. If put to good use, the tools and techniques discussed in this chapter will allow organizations to build longer-lasting, more reliable systems.

SYSTEMS REVIEW

systems review

the final step of systems development, involving the analysis of systems to make sure they are operating as intended

Systems review, the final step of systems development, is the process of analyzing systems to make sure they are operating as intended. This process often compares the performance and benefits of the system as it was designed with the actual performance and benefits of the system in operation. Cost, performance, control, and complexity factors investigated during design are revisited after the system has been operating. Problems and opportunities uncovered during systems review will trigger systems development and begin the process anew. For example, as the number of users of an interactive system increases, it is not unusual for system response time to increase. If the degradation in response time is too great, it may be necessary to redesign some of the system, modify databases, or increase the power of the computer hardware.

Systems review can be performed by internal employees, external consultants, or both. When the problems or opportunities are industrywide, people from several firms may get together. In some cases, this collaboration is done at an IS conference or in a private meeting involving several firms. For some situations, state or federal governments may be involved. For example, in early 2000, after a rash of Internet stoppages, President Clinton called a meeting with top Internet firms, the attorney general, the commerce secretary, and the national security advisor.[23] The purpose was to review the stoppages and develop a plan to help prevent this type of hacking from happening in the future.

Types of Review Procedures

event-driven review

review triggered by a problem or opportunity such as an error, a corporate merger, or a new market for products

There are two types of review procedures: event driven and time driven (Table 13.6). An **event-driven review** is triggered by a problem or opportunity such as an error, a corporate merger, or a new market for products. In some cases, companies wait until a large problem or opportunity occurs before a change is made, ignoring minor problems. In contrast, some companies use a continuous improvement approach to systems development. With this approach, an organization makes changes to a system even when small problems or opportunities occur. Although continuous improvement can keep the system current and responsive, doing the repeated design and implementation can be both time-consuming and expensive.

time-driven review

review performed after a specified amount of time

A **time-driven review** is performed after a specified amount of time. Many application programs are reviewed every six months to a year. With this approach, an existing system is monitored on a schedule. If problems or opportunities are uncovered, a new systems development cycle may be initiated. A payroll application, for example, may be reviewed once a year to make sure it is still operating as expected. If it is not, changes are made.

TABLE 13.6

Examples of Review Types

Event Driven	Time Driven
A problem with an existing system	Monthly review
A merger	Yearly review
A new accounting system	Review every few years
An executive decision that an upgraded Internet site is needed to stay competitive	Five-year review

Many companies use both approaches. A billing application, for example, might be reviewed once a year for errors, inefficiencies, and opportunities to reduce operating costs. This is a time-driven approach. In addition, the billing application might be redone if there is a corporate merger, if one or more new managers require different information or reports, or if federal laws on bill collecting and privacy change. This is an event-driven approach.

Factors to Consider during Systems Review

Systems review should investigate a number of important factors, such as the following:

Mission. Is the computer system helping the organization achieve its overall mission? Are stakeholder needs and desires satisfied or exceeded with the new or modified system?

Organizational goals. Does the computer system support the specific goals of the various areas and departments of the organization?

Hardware and software. Are hardware and software up to date and adequate to handle current and future processing needs?

Database. Is the current database up to date and accurate? Is database storage space adequate to handle current and future needs?

Telecommunications. Is the current telecommunications system fast enough, and does it allow managers and workers to send and receive timely messages? Does it allow for fast order processing and effective customer service?

Information systems personnel. Are there sufficient IS personnel to perform current and projected processing tasks?

Control. Are rules and procedures for system use and access acceptable? Are the existing control procedures adequate to protect against errors, invasion of privacy, fraud, and other potential problems?

Training. Are there adequate training programs and provisions for both users and IS personnel?

Costs. Are development and operating costs in line with what is expected? Is there an adequate information systems budget to support the organization?

Complexity. Is the system overly complex and difficult to operate and maintain?

Reliability. Is the system reliable? What is the mean time between failures (MTBF)?

Efficiency. Is the computer system efficient? Are system outputs generated by the right amount of inputs, including personnel, hardware, software, budget, and others?

Response time. How long does it take the system to respond to users during peak processing times?

Documentation. Is the documentation still valid? Are changes in documentation needed to reflect the current situation?

System Performance Measurement

system performance measurement

monitoring the system—the number of errors encountered, the amount of memory required, the amount of processing or CPU time needed, and other problems

system performance products

software that measures all components of the computer-based information system, including hardware, software, database, telecommunications, and network systems

Systems review often involves monitoring the system. This is called **system performance measurement**. The number of errors encountered, the amount of memory required, the amount of processing or CPU time needed, and other problems should be closely observed. If a particular system is not performing as expected, it should be modified, or a new system should be developed or acquired. In some cases, specialized software products have been developed that do nothing but monitor system performance. **System performance products** have been developed to measure all components of the computer-based information system, including hardware, software, database, telecommunications, and network systems. When properly used, system performance products can quickly and efficiently locate actual or potential problems.

Candle is a leading provider of mainframe performance monitoring and application availability management tools. Its products include Candle Command Center, advanced systems management tools for optimizing an organization's computing resources and maximizing business application availability, and OMEGAMON II performance monitors, for real-time and historical analysis of performance on a variety of systems.[24] Precise/Pulse is a product from Precise Software Solutions that provides around-the-clock performance monitoring for Oracle database applications. It detects and reports potential problems through systems management consoles. Precise/Pulse monitors the performance of critical database applications and issues alerts about inefficiencies before they turn into application performance problems.

Measuring a system is, in effect, the final task of systems development. The results of this process may bring the development team back to the beginning of the development life cycle, where the process begins again.

● SUMMARY

PRINCIPLE ● Designing new systems or modifying existing ones should always be aimed at helping an organization achieve its goals.

The purpose of systems design is to prepare the detailed design needs for a new system or modifications to the existing system. Logical systems design refers to the way the various components of an information system will work together. The logical design includes data specifications for output and input, processing, files and databases, telecommunications, procedures, personnel and job design, and controls and security design. Physical systems design refers to the specification of the actual physical components. The physical design must specify characteristics for hardware and software design, database and telecommunications, and personnel and procedures design.

● ● ● ●

A number of special design considerations should be taken into account during both logical and physical system design. A sign-on procedure consists of identification numbers, passwords, and other safeguards needed for individuals to gain access to computer resources. A system sign-on allows the user to gain access to the computer; an application sign-on permits the user to start and use a particular application.

If the system under development is interactive, the design approach must consider using menus, help facilities, table lookup facilities, and restart procedures. A good interactive dialogue will ask for information in a clear manner, respond rapidly, contain consistency between applications, and use an attractive format. Also, it will avoid use of computer jargon and treat the user with respect.

Error prevention, detection, and correction should be part of the system design process. Causes of errors include humans, natural phenomena, and technical problems. Designers should be alert to prevention of fraud and invasion of privacy.

At the end of the systems design step, the final specifications are frozen and no changes are allowed so that implementation can proceed. A final design report is developed.

. . .

If new hardware or software will be purchased from a vendor, a formal request for proposal (RFP) is needed. The RFP outlines the company's needs; in response, the vendor provides a written reply. In addition to responding to the company's stated needs, the vendor provides data on its operations. This data might include the vendor's reliability and stability, the type of postsale service offered, the vendor's ability to perform repairs and fix problems, vendor training, and the vendor's reputation.

RFPs from various vendors are reviewed and narrowed down to the few most likely candidates. In the final evaluation, a variety of techniques—including group consensus, cost/benefit analysis, point evaluation, and benchmark tests—can be used. After the vendor is chosen, contract negotiations can begin.

If externally developed software is to be used, a company should review software needs and requirements, acquire and customize software, and develop appropriate interfaces between the purchased software and existing software. Software should also be tested prior to its acceptance for use, and modifications and maintenance agreements should be reached with the vendor. Problems to avoid include paying for extras the organization will not use; underestimating the cost of the modifications, interfaces, implementation, and installation; and being one of the first to try out a new package.

. . .

There are four commonly used approaches to making a final system evaluation and selection: group consensus, cost/benefit analysis, benchmark tests, and point evaluation.

In group consensus, a decision-making group is appointed and given responsibility for making the final evaluation and selection. With cost/benefit analysis, all costs and benefits of the alternatives are expressed in monetary terms. Benchmarking involves comparing computer systems operating under the same condition. Point evaluation assigns weights to evaluation factors and each alternative is evaluated in terms of each factor and given a score from 0 to 100.

PRINCIPLE • The primary emphasis of systems implementation is to make sure that the right information is delivered to the right person in the right format at the right time.

Software can be purchased from external vendors or developed in-house—a decision termed the *make-or-buy decision*. A purchased software package usually has a lower cost, less risk regarding the features and performance, and easy installation. The amount of development effort is also less when software is purchased. Developing software can result in a system that more closely meets the business needs and has increased flexibility in terms of customization and changes. Developing software also has greater potential for providing a competitive advantage.

. . .

Software development is often performed by a chief programmer team—a group of IS professionals who design, develop, and implement a software program. Programming using traditional programming languages follows a life cycle that includes investigation, analysis, design, language selection, program coding, testing and debugging, documentation, and implementation (conversion). Documentation includes technical and user documentation. Implementation of the new computer program may be done in parallel, in small pilot tests, phased in slowly, or direct conversion.

There are many tools and techniques for software development. Structured design is a philosophy of designing and developing application software. Structured programming is not a new programming language; it is a way to standardize computer programming using existing languages. The top-down approach starts with programming a main module and works down to the other modules. Other tools, like cross-platform development and integrated development environments, make software development easier and more thorough. CASE tools are often used to automate some of these techniques.

Fourth-generation languages (4GLs) and object-oriented languages offer another alternative to in-house development. Development using these fast and easy-to-use languages requires several steps, much like the programming life cycle. The main difference is that, with object-oriented languages, programmers must identify and select objects and integrate them into an application, instead of using step-by-step coding.

. . .

The purpose of systems implementation is to install the system and make everything, including users, ready for its operation. Systems implementation includes hardware acquisition, software acquisition or development, user preparation, hiring and training of personnel, site and data preparation, installation, testing, start-up, and user acceptance.

Hardware acquisition requires purchasing, leasing, or renting computer resources from a vendor. Types of vendors include small and general computer manufacturers, peripheral equipment manufacturers, leasing companies, time-sharing companies, software companies, dealers, distributors, service companies, and others. Increasingly, companies are using service providers to acquire software, Internet access, and other IS resources.

Implementation must address personnel requirements. User preparation involves readying managers, employees, and other users for the new system. New IS personnel may need to be hired, and users must be well trained in the system's functions. Preparation of the physical site of the system must be done, and any existing data to be used in the new system will require conversion to the new format. Hardware installation is done during the implementation step, as is testing. Testing includes program (unit) testing, systems testing, volume testing, integration testing, and acceptance testing.

Start-up begins with the final tested information system. When start-up is finished, the system is fully operational. There are a number of different start-up approaches. Direct conversion (also called *plunge* or *direct cutover*) involves stopping the old system and starting the new system on a given date. With the phase-in approach, sometimes called a *piecemeal approach*, components of the new system are slowly phased in while components of the old one are slowly phased out. When everyone is confident that the new system is performing as expected, the old system is completely phased out. Pilot start-up involves running the new system for one group of users rather than all users. Parallel start-up involves running both the old and new systems for a period of time. The output of the new system is compared closely with the output of the old system, and any differences are reconciled. When users are comfortable that the new system is working correctly, the old system is eliminated.

PRINCIPLE • Maintenance and review add to the useful life of a system but can consume large amounts of resources. These activities can benefit from the same rigorous methods and project management techniques applied to systems development.

Systems maintenance involves checking, changing, and enhancing the system to make it more useful in obtaining user and organizational goals. Maintenance is critical for the continued smooth operation of the system. The costs of performing maintenance can well exceed the original cost of acquiring the system. Some major causes of maintenance are new requests from stakeholders and managers, enhancement requests from users, bugs or errors, technical or hardware problems, newly added equipment, changes in organization structure, and government regulations.

Maintenance can be as simple as a program patch to correct a small problem to the more complex upgrading of software with a new release from a vendor. Requests for maintenance should be documented with a request for maintenance form, a document that formally authorizes modification of programs. The development team or a specialized maintenance team may then make approved changes.

Systems review is the process of analyzing systems to make sure that they are operating as intended. It involves monitoring systems to be sure they are operating as designed. The two types of review procedures are event-driven review and time-driven review. An event-driven review is triggered by a problem or opportunity. A time-driven review is started after a specified amount of time.

Systems review involves measuring how well the system is supporting the mission and goals of the organization. System performance measurement monitors the system for number of errors, amount of memory and processing time required, and so on.

● KEY TERMS

● REVIEW QUESTIONS

1. What is the purpose of systems design?
2. What is procedures design?
3. What is interactive processing? What design factors should be taken into account for this type of processing?
4. What are some of the special design considerations that should be taken into account during both the logical and physical design?
5. What are some differences between emergency alternate procedures and a disaster recovery plan?
6. What are the different types of software and database backup? Describe the procedure you use to back up your homework files.
7. Identify specific controls that are used to maintain input integrity and security.
8. What is an RFP? What is typically included in one? How is it used?
9. What activities go on during the user preparation phase of system implementation?
10. What are the major steps of systems implementation?
11. What are some tools and techniques for software development?
12. Explain the three types of structures allowed in structured programming.
13. What are the financial options of acquiring hardware?
14. What are the steps involved in testing the information system?
15. What are some of the reasons for program maintenance? Explain the three types of maintenance.
16. Describe the point evaluation system for selection of the optimum system alternative.
17. How is systems performance measurement related to the systems review?

● DISCUSSION QUESTIONS

1. Describe the participants in the systems design stage. How do these participants compare with the participants of systems investigation?
2. Assume that you are the owner of a company that is about to start marketing and selling bicycles over the Internet. Describe your top three objectives in developing a new Web site for this systems development project.
3. Identify some of the advantages and disadvantages of purchasing versus leasing hardware.
4. Discuss the relationship between maintenance and system design.
5. Is it equally important for all systems to have a disaster recovery plan? Why or why not?
6. Four approaches were discussed to evaluate a number of systems alternatives. No one approach is always the best. How would you decide which approach to use in a particular instance?
7. Assume that you are starting an Internet site to sell office furniture. Describe how you would design the interactive processing system for this site. Draw a diagram showing the home Web page for the site. Describe the important features of this home page.

8. Identify the various forms of testing used. Why are there so many different types of tests?

9. Discuss the key responsibilities of end users and general management during software coding, data preparation, installation, and start-up of a new system.

10. What is the goal of conducting a systems review? What factors need to be considered during systems review?

11. What features and terms would you insist on in a software package contract?

12. What issues might you expect to arise if you initiate the use of a request for maintenance form where none had been required? How would you deal with these issues?

13. How would you go about evaluating a software vendor?

● PROBLEM-SOLVING EXERCISES

1. You have been hired to develop a new payroll program. Describe the tasks for developing this application. Develop a Gantt chart using a project management software package such as Microsoft Project or a spreadsheet program. Which of these tasks can be done concurrently and which must be done sequentially? Which of these tasks are likely to be relatively minor and take a lesser amount of effort? Which tasks are more involved and will take longer? How long will it take to complete the project?

2. A project team has estimated the costs associated with the development and maintenance of a new system. One approach requires a more complete design and will result in a slightly higher design and implementation cost but a lower maintenance cost over the life of the system. The second approach cuts the design effort, saving some dollars but with a likely increase in maintenance cost.

a. Enter the following data in the spreadsheet. Print the result.

The Benefits of Good Design

	Good Design	**Poor Design**
Design Costs	$14,000	$10,000
Implementation Cost	$42,000	$35,000
Annual Maintenance Cost	$32,000	$40,000

b. Create a stacked bar graph that shows the total cost, including design, implementation, and maintenance costs. Be sure that the chart has a title and that the costs are labeled on the chart.

c. Use your word processing software to write a paragraph that recommends which approach to take and why.

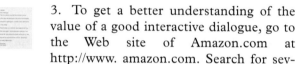

3. To get a better understanding of the value of a good interactive dialogue, go to the Web site of Amazon.com at http://www. amazon.com. Search for several books and periodicals based on different criteria. In searching for these materials, use the author as one criterion (e.g., R. M. Stair), then the subject (e.g., information systems) and the title (e.g., *Principles of Information Systems*). If you have access to another on-line library catalog, conduct the same searches using that system. Now evaluate the interactive dialogue of the systems, based on the elements discussed in the text. Write up your observations using your word processor. If you can, also send this evaluation to your instructor via e-mail.

● TEAM ACTIVITIES

1. Assume your project team has been working three months to complete the systems design of a new Web-based customer ordering system. There are two possible options that seem to meet all users' needs. The project team must make a final decision on which option to implement. The table that follows summarizes some of the key facts about each option.
   a. What process would you follow to make this important decision?
   b. Who needs to be involved?
   c. What additional questions need to be answered to make a good decision?
   d. Based on the data, which option would you recommend and why?
   e. How would you account for project risk in your decision making?

Factor	Option #1	Option #2
Annual gross savings	$1.5 million	$3.0 million
Total development cost	$1.5 million	$2.2 million
Annual operating cost	$0.5 million	$1.0 million
Time required to implement	9 months	15 months
Risk associated with project (expressed in probabilities)		
Benefits will be 50% less than expected	20%	35%
Cost will be 50% greater than expected	25%	30%
Organization will not/cannot make changes necessary for system to operate as expected	20%	25%
Does system meet all mandatory requirements?	Yes	Yes

2. Assume that your team is working for a medium-size consulting firm. You have been hired to develop an on-line tax preparation program to complete federal 1040 forms.
   a. Develop a project proposal for this effort including an estimate of the time required and stating potential project benefits.
   b. Develop a mock-up report showing what will be displayed to the senior consultants and partners if they want a project status report.
   c. Use a database management system to document the key forms to be included with the 1040, including descriptions and how the forms are to be linked to the 1040.

● WEB EXERCISES

1. EDS is just one of the many systems development outsourcing companies. Find the Web site of this company or another outsourcing company by entering the Internet address or using a search engine. If you used a search engine, did you discover any irrelevant information? Describe the company, its services, and any investor or employment information. You may be asked to develop a report or send an e-mail message to your instructor about what you found.

2. Accounting programs can be developed or purchased. Using the Internet and a search engine, such as Yahoo!, find one or more companies that sell accounting software, such as payroll, billing, order processing, or inventory control software. Describe the software package and the company that sells it. Describe the features of the software package. What are the advantages of purchasing this software package versus developing the software?

• CASES

Outsourcing: Benefits besides Huge Savings

Systems implementation involves acquiring the needed hardware, software, personnel, and equipment to start a new application or to enhance an existing one. Increasingly, companies are getting help during the implementation and maintenance stage. Today, companies often enter into outsourcing agreements with computer and consulting companies. The potential to save money is one advantage of outsourcing—companies can leave systems development to the consultant's experts and avoid the expense and headache of hiring and supervising a staff of programmers and analysts. Heilig-Meyers, a large $2 billion furniture retailer located in Richmond, Virginia, for example, expects to save $2 million per year by outsourcing a new payroll and human resources system to an outside company. These cost savings are not guaranteed, however.

Reid Hospital & Health Care Services in Richmond, Indiana, decided to outsource development of its materials management application. The 227-bed hospital was able to reduce its $5 million IS budget by only $150,000 per year. According to Craig Kinyon, chief financial officer for the hospital, "It's a small percentage." Although the cost savings were not substantial, there were other benefits to Reid. Outsourcing made upgrading to new systems faster and easier. The hospital could also concentrate on its core business by not having to hire and manage a staff of programmers and analysts. This was an important consideration for Reid, considering its location: "We're located out in a no-man's-land area, so recruiting is a challenging endeavor. It's hard to find people," according to Kinyon.

The benefits of outsourcing also depend on what is being outsourced. The more complicated and sophisticated the application, the greater the potential benefits. Outsourcing enterprise resource management programs, for example, can lead to many benefits. According to Karen Moser, an analyst at Aberdeen Group, it is possible to save "tons and tons of money" by outsourcing ERP systems. Global companies can also benefit from outsourcing. "It's hard to hire an IT staff to support fast rollouts in multiple countries," says Evret Jan Hoijtink, the chief information officer for a Netherlands-based pharmacy company.

Discussion Questions

1. In addition to cost savings, what are the advantages of outsourcing?
2. What types of applications can benefit the most from outsourcing?

Critical Thinking Questions

3. Assume that you are the chief information officer for a national retail store chain. Describe what key features you would like to see in an outsourcing contract to make sure your company can benefit the most from the outsourcing deal.
4. What are the potential problems with outsourcing? What could you put in an outsourcing contract to help reduce the impact of these potential problems?

Sources: Adapted from Crain Stedman, "Outsourcing Won't Always Pay Off with Huge Savings," *Computerworld*, February 7, 2000, p. 12; Karl Kunkel, "Mattress Makers Respond to Shifting Channels," *HNF*, November 8, 1999, p. 10; and "Heilig-Meyers," *HFN*, September 27, 1999, p. 28.

Developing a Wireless Net to Improve Customer Service

As a result of the deregulation of the electric power business in many states, power companies are striving to find ways to improve their business and customer service. Failure to take these actions could result in a substantial loss of business. In 1999, commercial electric customers in Illinois were free to choose their power company. In 2000, home users were free to choose their power company. According to Roger Koester, supervisor of energy delivery technology for Illinois Power Company, "A

few years ago, there weren't that many competitors. But since then, the competition has just exploded." Illinois Power, realizing the potential impact of increased competition from deregulation, decided to embark on a systems development project to build a wireless communications system to improve customer service.

Illinois Power is a subsidiary of Illinova Corporation, a $2.4 billion company. Illinois Power provides both gas and electric service to almost one million customers, primarily in central Illinois. Traditionally, Illinois Power received calls from customers requesting repairs or service over the phone at 26 field offices. This information was then placed on service and repair orders and given to dispatchers, who coordinated the work field crews. Although the system worked, it was slow and inefficient. After years of planning and millions of dollars invested in the systems development effort, Illinois Power was ready to roll out its new wireless system.

With the new wireless system, service and repair orders are captured at a centralized computer system. The work orders are then sent via wireless transmission directly to laptop computers of repair and service crews in the field. "The driver gets in his truck and instead of searching through paper, his day's work is loaded on the laptop," said Koester. In addition, drivers do not need to travel to a central office to get their work. They can wake up in the morning, get into their trucks, and download the day's service and repair work. Eliminating trips to a central office has saved the company thousands of dollars in gas, while making the crews more efficient. Each truck is also equipped with a global positioning system (GPS) to allow a central dispatcher to locate

the truck closest to an emergency repair. With the implementation of the new system, Illinois Power hopes to increase profits and market share.

Discussion Questions

1. Why did Illinois Power decide to invest in a new wireless system?
2. What are the benefits of the new system?

Critical Thinking Questions

3. From a customer's perspective, what could Illinois Power do to further increase customer service and satisfaction for home users?
4. Some believe that the power business has little room for increasing profits. What other systems development efforts might help Illinois Power increase its market share and profitability in its commercial power market?

Sources: Adapted from Matt Hamblen, "Wireless Net Helps Utility Improve Customer Service," *Computerworld*, January 24, 2000, p. 36; Dennis Berman, "Keeping the Home Fires Burning," *Business Week*, September 20, 1999, p. 6; and "EPA Bares Its Teeth," *Power Economics*, January 31, 2000, p. 17.

 ## Mergers Can Cause Systems Implementation Problems

Mergers are supposed to be happy marriages between two companies. The goal is usually greater efficiencies and higher profits. When there are problems with a merged information system, however, greater efficiencies and higher profits can be a dream that never comes true. In some cases, the resulting information system can even be a nightmare.

In a recent merger, Norfolk Southern Corporation acquired most of Consolidated Rail Corporation. The merged computer systems went on-line at the beginning of June, and that is when the systems implementation trouble started. The first problem was caused by human error. The wrong magnetic tape was loaded on a tape drive. As a result, detailed information about trainloads and schedules were completely wrong. But that was just the start of the problems. "As we started to experience some data problems, we ascribed them to the bad data dump. That was a wrong conclusion," said Henry Wolf, chief financial officer for Norfolk. The system soon started to generate wrong waybills, which contain information on where and when to move train cars. Some full train cars sat in place, and empty cars were moved as if they contained freight. At this point, problems started to compound. The wrong waybills caused

very congested train yards. The congestion eventually got so bad that some railcars blocked train crossings. Firefighters in one city literally had to crawl under train cars to get to fires. School children had to scramble under cars to get to school, and train crews sat idle in bewilderment. The crews that did work had to work longer hours and then be replaced with fresh crews as required by law. The financial consequences of this failed systems implementation effort were immediate and substantial.

Some companies, like United Parcel Service, started shipping products using trucks and alternate transportation modes. In total, Norfolk lost more than $40 million of business to trucking companies and other railroads. To make things worse, the company had to spend an additional $29 million on alternative transportation to satisfy service contracts with some of its bigger customers. Some chemical companies that had relied on rail service had to cut production in their plants because there was no way to ship their products to market. At the height of the chaos, there was a concern that chlorine for water treatment to a few East Coast cities wouldn't make it in time. "This created a lot of anxiety. Imagine if Philadelphia or New York didn't have clean water to drink," commented Tom Schick, an executive at the Chemical Manufacturers Association.

Discussion Questions

1. What started the problems with the implemented system?
2. What were some of the financial implications of the implementation problems discussed in this case?

Critical Thinking Questions

3. If you were a chemical company shipping your products via rail, how could you protect your company from rail problems like the ones discussed in this case?

4. Assume that you are the chief information officer for Norfolk. What procedures would you follow in all future systems implementation efforts to help avoid the types of problems discussed in this case? Do you think you would lose your job if another systems development project had problems?

Sources: Adapted from Kim Nash, "Merged Railroads Still Plagued by IT Snafus," *Computerworld*, January 17, 2000, p. 20; Doug Mohney, "Second Time Around," *Boardwatch Magazine*, January 2000, p. 108; and Louis Jacobson, "No Express Track for This Merger," *The National Journal*, February 26, 2000.

● NOTES

Sources for the opening vignette on p. 503: Adapted from Lisa Levenson, "AT&T Offers Service for Online Software Renters," *The Rocky Mountain News*, January 31, 2000, p. 21B; Steve Ulfelder, "Evaluate the ASP Phenomenon," *Computerworld*, January 2, 2000, p. S22; and Kavita Kaur, "Application Service Providers: Rent an App," *Computers Today*, February 29, 2000, p. 43.

1. John Leyden, "Bank Guards against Data Loss with Real Time Mirroring," *Network News*, January 26, 2000, p. 4.
2. Larry Greenemeir, "Deal Fills Out Compaq's Business Continuity Offerings," *Information Week*, February 14, 2000.
3. "2000 Buyer's Guide; Disaster Recovery," *Information Security*, December 1999, p. 115.
4. Products section of the SunGard Recovery Services Web site, http://www.sungard-recovery.com, accessed May 17, 2000.
5. Disaster Recovery Services at the Web site of Business Recovery Management, http://www.businessrecords.com, accessed May 17, 2000.
6. Disaster Recovery at the Guardian Computer Support Web site, http://www.guardiancomputer.com, accessed May 17, 2000.
7. Rick Whiting, "Oracle Offerings to Boost E-Biz Uptime," *Information Week*, January 24, 2000.
8. Julekha Dash, "Saturn to Move Customer Service to Web in $300M Deal," *Computerworld*, January 24, 2000, p. 6.
9. Amanda Wells, "Police System Tender Issued," *NZ Infotech Weekly*, January 17, 2000, p. 1.
10. "Measuring Costs and Benefits," *Aerospace America*, January 2000, p. 42.
11. Jaikumar Vijayan, "ABB, IBM Sign $250M Agreement," *Computerworld*, April 10, 2000, p. 6.
12. Jaikumar Vijayan, "HP Net Hosting Has Pay-As-You-Go Model," *Computerworld*, March 20, 2000, p. 8.
13. Mark Hall, "Service Providers Give Users More IT Options," *Computerworld*, February 7, 2000, p. 40.
14. Julekha Dash, "Hospital Signs $270M Deal," *Computerworld*, February 14, 2000, p. 4.
15. Julekha Dash, "AT&T Outsourcing Deal Will Cover 150 Apps," *Computerworld*, April 17, 2000, p. 24.
16. Kevin Fogarty, "Making the Right Bet on ASPs," *Computerworld*, April 3, 2000, p. 38.
17. Bonny Georgia, "Never Buy Software Again," *SmartBusinessMag.com*, May 2000, p. 168.
18. Julekha Dash, "ASPs Outsourcing Raises Risks for Clients," *Computerworld*, July 17, 2000, p. 39.
19. Thomas Hoffman, "Bank's Reuse Project Survives Mergers," *Computerworld*, January 25, 1999, p. 39.
20. Spikumar Rao, "General Electric, Software Vendor," *Forbes*, January 24, 2000, p. 144.
21. Darnell Little, "H&R Block Gets All Snarled Up in the Web," *Business Week*, April 17, 2000, p. 200.
22. "Computer Snag Halts London Market 8 Hours," *The Wall Street Journal*, April 6, 2000, p. A14.
23. "Clinton to Hold Internet Security Summit," *The Wall Street Journal*, February 11, 2000, p. A3.
24. Products and Services Section of the Candle Web site, http://www.candle.com, accessed May 17, 2000.

PART V

Information Systems in Business and Society

CHAPTER 14

Security, Privacy, and Ethical Issues in Information Systems and the Internet

*T*here are 13-year-old kids without degrees breaking into systems from their bedrooms. Security should be the subject of every Master and Ph.D. student's thesis.

— John V. Daley, telecom specialist at the Federal Aviation Administration and part-time community college professor

Principles	Learning Objectives
Policies and procedures must be established to avoid computer waste and mistakes.	• *Describe some examples of waste and mistakes in an IS environment, their causes, and possible solutions.* • *Identify policies and procedures useful in eliminating waste and mistakes.*
Computer crime is a serious and rapidly growing area of concern requiring management attention.	• *Explain the types and effects of computer crime* • *Identify specific measures to prevent computer crime.* • *Outline criteria for the ethical use of information systems.* • *Discuss the principles and limits of an individual's right to privacy.*
Jobs, equipment, and working conditions must be designed to avoid negative health effects.	• *List the important effects of computers on the work environment.* • *Identify specific actions that must be taken to ensure the health and safety of employees.*

Los Alamos National Laboratory

Compromise of Nuclear Weapon Secrets

The United States agreed in 1992 to end nuclear weapons testing after having developed such sophisticated computer models that weapons could be tested entirely on simulation software using data from decades of actual bomb testing. However, efforts to keep such classified data secure may have failed dangerously. In August 1999, U.S. Energy Secretary Bill Richardson confirmed that these classified nuclear weapon computer models and data were transferred by a Chinese lab employee at the Los Alamos National Laboratory to an unclassified computer system. Where they went from there is not known; however, the incident has been at the center of allegations that China pilfered U.S. nuclear secrets.

Although the software and data may not enable someone to make an exact copy of a U.S. weapon, they clearly would be enormously valuable to another country's nuclear weapons program. That's especially so for China, which also signed the test ban treaty and thus must rely on computer simulations. We may never know whether the secret files were given to China or what, if anything, China has done with them. But the possibilities are scary indeed.

This incident reinforces three basic computer security principles. First, computer security is often neglected compared with physical security because systems and attacks against them are often invisible—out of sight, out of mind. Second, the vast majority of computer crimes are committed by insiders unaffected by many of the standard protective measures, such as firewalls. A disgruntled employee is a bigger threat than a hacker because he or she already has access and knowledge of how a system works, whereas a hacker usually comes in blind. Third, the effort invested in protecting assets should be a product of two factors: the likelihood that attempts will be made to capture the assets and the impact on an organization if those attempts succeed. At Los Alamos, risk times cost equals infinity. Nevertheless, security efforts apparently fell short.

It is now clear that the Los Alamos monitoring systems were not as stringent as they needed to be. Moving files from a classified to an unclassified environment should have been an event that triggered an immediate audit for detection. The secretary of energy charged that his department wasn't getting the help it needed from Congress to develop such systems and that the fiscal year 2000 budget in Congress killed needed security funding. By Congress's denying $35 million in funds for cybersecurity upgrades, it was impossible to provide real-time cyberintrusion detection and protection for all DOE sites.

As you read this chapter, consider the following:

- What motivates an individual to commit a computer crime?

- What can be done to detect and avoid computer crime?

- Computer waste and mistakes
- Computer crime
- Privacy
- Health concerns
- Ethical issues
- Patent and copyright violations

Earlier chapters detailed the amazing benefits of computer-based information systems in business, including increased profits, superior goods and services, and higher quality of work life. Computers have become such valuable tools that today's businesspeople would have difficulty imagining work without them. Yet the information age has also brought some potential problems for workers, companies, and society in general (see Table 14.1).

To a large extent, this book has focused on the solutions—not the problems—presented by information systems. In this chapter we discuss detailed problems as a reminder of the social and ethical considerations underlying the use of computer-based information systems. No business organization, and hence no information system, operates in a vacuum. All IS professionals, managers, and users have a responsibility to see that the potential consequences of IS use are fully considered.

Managers and users at all levels play a major role in helping organizations achieve the positive benefits of IS. These individuals must also take the lead in helping to minimize or eliminate the negative consequences of poorly designed and improperly utilized information systems.

For managers and users to have such an influence, they must be properly educated. Many of the problems presented in this chapter, for example, should cause you to think back to some of the systems design and systems control issues we have already discussed. They should also help you look forward to how these issues and your choices might affect your future IS management considerations.

COMPUTER WASTE AND MISTAKES

Computer-related waste and mistakes are major causes of computer problems, contributing as they do to unnecessarily high costs and lost profits. Computer waste involves the inappropriate use of computer technology and resources. Computer-related mistakes refer to errors, failures, and other computer problems that make computer output incorrect or not useful, caused mostly by human error. In this section we explore the damage that can be done as a result of computer waste and mistakes.

Computer Waste

The U.S. government is the largest single user of information systems in the world. It should come as no surprise then that it is also perhaps the largest misuser. The government is not unique in this regard—the same type of waste and misuse found in the public sector also exists in the private sector. Some companies discard old software and even complete computer systems when they still have value. Others waste corporate resources to build and maintain complex systems never used to their fullest extent. A less dramatic, yet still relevant, example of waste is the amount of company time and money employees may waste playing computer games, sending unimportant e-mail, or accessing the Internet. Junk e-mail, also called spam, and junk faxes also cause waste. People receive hundreds of e-mail messages and faxes advertising products and services not wanted or requested. Not only does this waste time, but it also wastes paper and computer resources. When waste is identified, it typically points to one common cause: the improper management of information systems and resources.

Computer-Related Mistakes

Despite many people's distrust, computers themselves rarely make mistakes. Even the most sophisticated hardware cannot produce meaningful output if users do not follow proper procedures. Mistakes can be caused by unclear expectations and a lack of feedback. Or a programmer might develop a program that contains errors. In other cases, a data entry clerk might enter the wrong data. Unless errors are caught early and prevented, the speed of computers can intensify mistakes. As information technology becomes faster, more complex, and more powerful, organizations and individuals face greater risk of experiencing the results of computer-related mistakes. Take, for example, these cases from recent news:

- The fire brigade in Grampian, Scotland, said a combination of human and computer errors resulted in a budget shortfall of £688,000, or approximately $1.2 million. Significant human error and erroneous practices, as well as technological shortcomings, contributed to this problem. The financial information on which the senior officers and members of the joint fire board were reliant was fundamentally flawed in its construction and presentation.
- In Bernalillo County, New Mexico, a computer glitch turned portions of 50,000 absentee ballots into an illegible jumble. The ballots contained Spanish text for a constitutional amendment and state and county bond issues to be voted on in a November election. Words were misspelled, missing consonants or vowels, or they ran together. It looked like a language no human had ever seen.
- The wireless communications venture Globalstar Telecommunications lost 12 of its satellites when its rocket crashed in Kazakhstan. The Ukrainian-built Zenit-2 rocket crashed just minutes after liftoff from the Baikonur Cosmodrome. NPO Yuzhnoye, the company that manufactured the rocket, said that two computer glitches occurred in rapid succession. A company statement said that the computer "as a result sent an order to cut the engines." The satellites were valued at $185 million.
- A Utah retiree was ordered to pay back nearly $14,000 in excess retirement benefits he received. A retirement fund employee testified in the trial that a portion of the retiree's benefits were calculated by hand and another portion by computer, and the two figures were accidentally combined, resulting in double payments. The company overpaid in two ways—it paid a lump-sum distribution that was $14,000 too high, and it overpaid the retiree's monthly benefits.
- A recall of Toyota and Lexus cars, due to faulty on-board computers, may cost the company as much as $82.5 million. The California Air Resources Board ordered the recall of 330,000 1996–1998 Toyota and Lexus models. The ARB said that the computers fail to properly detect gasoline vapors. ARB estimates the cost to Toyota of fixing the computers will be approximately $250 each.
- Students in Chippewa Falls, Wisconsin, received free school lunches as a result of a computer glitch. New computer software that was supposed to run the lunch program did not communicate properly with the computer hardware. During tests, duplicate student identification numbers were issued. On the first day of the school year, students were sent home with notices saying that they were eligible for free lunches that week.

Preventing Computer-Related Waste and Mistakes

To remain profitable in a competitive environment, organizations must use all resources wisely. Preventing computer-related waste and mistakes like those just described should therefore be a goal. Today, most organizations use some type of CBIS. To employ IS resources efficiently and effectively, employees and managers alike should strive to minimize waste and mistakes. Preventing waste and mistakes involves (1) establishing, (2) implementing, (3) monitoring, and (4) reviewing effective policies and procedures.

Establishing Policies and Procedures

The first step to prevent computer-related waste is to establish policies and procedures regarding efficient acquisition, use, and disposal of systems and devices. Computers permeate organizations today, and it is critical for organizations to ensure that systems are used to their full potential. As a result, most companies have implemented stringent policies on the acquisition of computer systems and equipment, including requiring a formal justification statement before computer equipment is purchased, definition of standard computing platforms (operating system, type of computer chip, minimum amount of RAM, etc.), and the use of preferred vendors for all acquisitions.

Prevention of computer-related mistakes begins by identifying the most common types of errors, of which there are surprisingly few (see Table 14.2). To control and prevent potential problems caused by computer-related mistakes, companies have developed preventive policies and procedures that cover the following:

- Acquisition and use of computers, with a goal of avoiding waste and mistakes
- Training programs for individuals and workgroups
- Manuals and documents on how computer systems are to be maintained and used
- Approval of certain systems and applications before they are implemented and used to ensure compatibility and cost-effectiveness
- Requirement that documentation and descriptions of certain applications be filed or submitted to a central office, including all cell formulas for spreadsheets and a description of all data elements and relationships in a database system; such standardization can ease access and use for all personnel

Once companies have planned and developed policies and procedures, they must consider how best to implement them.

Implementing Policies and Procedures

Implementing policies and procedures to minimize waste and mistakes varies according to the business conducted. Most companies develop such policies and procedures with advice from the firm's internal auditing group or its external auditing firm. The policies often focus on the implementation of source data automation, the use of data editing to ensure data accuracy and completeness,

TABLE 14.2

Types of Computer-Related Mistakes

- Data entry or capture errors
- Errors in computer programs
- Errors in handling files, including formatting a disk by mistake, copying an old file over a newer one, and deleting a file by mistake
- Mishandling of computer output
- Inadequate planning for and control of equipment malfunctions
- Inadequate planning for and control of environmental difficulties (electrical problems, humidity problems, etc.)

TABLE 14.3

Useful Policies to Eliminate Waste and Mistakes

- Changes to critical tables should be tightly controlled, with all changes authorized by responsible owners and documented.
- A user manual should be available that covers operating procedures and that documents the management and control of the application.
- Each system report should indicate its general content in its title and specify the time period it covers.
- The system should have controls to prevent invalid and unreasonable data entry.
- Controls should exist to ensure that data input is valid, applicable, and posted in the right time period.
- Users should implement proper procedures to ensure correct input data.

and assigning clear responsibility for data accuracy within each information system. Table 14.3 lists some useful policies to minimize waste and mistakes.

Training is another key aspect of implementation. Many users are not properly trained in developing and implementing applications, and their mistakes can be very costly. Since more and more people use computers in their daily work, it is important that they understand how to use them. Training is often the key to acceptance and implementation of policies and procedures. Because of the importance of maintaining accurate data and of people understanding their responsibilities, companies converting to ERP systems invest weeks of training for key users of the system's various modules.

Monitoring Policies and Procedures

To ensure that users throughout the organization are following established procedures, the next step is to monitor routine practices and take corrective action if necessary. By understanding what is happening in day-to-day activities, organizations can make adjustments or develop new procedures. Many organizations implement internal audits to measure actual results against established goals for things such as percentage of end-user reports produced on time, percentage of data input errors rejected, number of input transactions entered per eight-hour shift, and so on.

Reviewing Policies and Procedures

The final step is to review existing policies and procedures and determine whether they are adequate. During review, people should ask the following questions:

- Do current policies cover existing practices adequately? Were any problems or opportunities uncovered during monitoring?
- Does the organization plan any new activities in the future? If so, does it need new policies or procedures on who will handle them and what must be done?
- Are contingencies and disasters covered?

This review and planning allows companies to take a proactive approach to problem solving, which can avert disasters. During such a review, companies are alerted to upcoming crises in information systems that could have a profound effect on many business activities. One such problem was the so-called year 2000 crisis. The original concerns that rallied the information systems community around the year 2000 problem during the early 1990s was that Y2K was a systemic problem with the potential to ripple across information systems and cause considerable problems. The issue was that older computer programs used only two digits to store the year in dates. For example, interest on a loan in the year 2003 could be computed incorrectly—starting in the year 1903—because the program would think that "03" was "1903" instead of "2003."

A U.S. Senate committee established in 1998 to monitor the year 2000 problem declared in February 2000 that the date bug was essentially dead and that the estimated $100 billion (others have estimated as much as $250 billion) spent in the United States to fix the problem was money well spent.[1] Although hundreds of Y2K problems were reported worldwide, they were relatively minor. Among the incidents cited were noncritical system failures at several nuclear power plants in Japan, Spain, the United States, and elsewhere. Failed credit card software rebilled accounts for a single charge. The Federal Reserve Bank of Chicago reported problems in transferring $700,000 in tax payments from customers of 60 financial institutions. The FDA reported 24 medical device failures. Heating systems went out in schools, and food stamp deliveries and Medicare payments were late. The Defense Department lost track of a spy satellite, prison terms were miscalculated in Italy, some e-mail systems shut down, and several ATM machines failed.[2] However, the problems reported were far fewer than even the optimists anticipated. Energy, air traffic control, water, heat, banking, medicine, transportation, government payment systems, and a host of other areas found and fixed glitches since the rollover. But no one died because of any Y2K problem, and global information system infrastructures held steady.[3]

Information systems professionals and users still need to be aware of the misuse of resources throughout an organization. Preventing errors and mistakes is one way to do so. Another is implementing in-house security measures and legal protections to detect and prevent a dangerous type of misuse: computer crime.

COMPUTER CRIME

Even good IS policies may not be able to predict or prevent computer crime. A computer's ability to process millions of pieces of data in less than a second can help a thief steal data worth thousands or millions of dollars. Compared with the physical dangers of robbing a bank or retail store with a gun, a computer criminal with the right equipment and know-how can steal large amounts of money from the privacy of a home. Computer crime often defies detection, the amount stolen or diverted can be substantial, and the crime is "clean" and nonviolent.

Here is a summary of recent computer crimes from 1999 to 2000. In March 1999, the Melissa virus caused an estimated $80 million in damage when it swept around the world, paralyzing e-mail systems. In December 1999, 300,000 credit card numbers were snatched from on-line music retailer CD Universe. In February 2000, hackers were able to crash the Web sites of several large e-commerce companies. In March, hackers-for-hire pleaded guilty to breaking into phone giants AT&T, GTE, and Sprint, among others, for calling card numbers that eventually made their way to organized crime gangs in Italy. According to the FBI, the phone companies were hit for an estimated $2 million.[4] In April, the U.S. energy secretary confirmed that classified nuclear weapon computer software at Los Alamos National Laboratory in New Mexico was transferred by a lab employee to an unclassified computer system.[5] In June, an embarrassing loss of computer disks containing additional classified nuclear information at the Los Alamos National Laboratory was uncovered. The computer disks reportedly contained information on how to disarm Russian and American nuclear devices.[6] In September, the FTC filed a case against individuals in Portugal and Australia who engaged in "pagejacking" and "mousetrapping" when they captured unauthorized copies of U.S.-based Web sites (including those of PaineWebber and *The Harvard Law Review*) and produced look-alike versions that were indexed by major search engines. The defendants diverted unsuspecting consumers to a sequence of porno sites that they couldn't exit.[7]

Although no one really knows how pervasive cybercrime is, most agree that it is growing rapidly. Almost all attacks go undetected—as many as 60 percent, according to security experts. What's more, of the attacks that are exposed, only an estimated 15 percent are reported to law enforcement agencies. Why? Companies don't want the bad press. When Russian organized crime used hackers to break into Citibank to steal $10 million—all but $400,000 was recovered—competitors used the news in marketing campaigns against the bank. Such publicity makes the job even tougher for law enforcement. Most companies that have been electronically attacked won't talk to the press. A big concern is loss of public trust and image—not to mention the fear of encouraging copycat hackers.[8]

The increase in computer security breaches is raising concern that the American workforce may not have enough troops to battle cybersnoops. Too few colleges offer security courses. Only about a half-dozen U.S. academic institutions have graduate programs in computer security—and that number hasn't changed much in ten years. Even more unsettling is that no quick solution is in sight. Computer assurance still isn't a recognized discipline at most colleges.[9]

Highlights of the annual Computer Crime and Security Survey are shown in Table 14.4. The survey is based on responses from 643 companies and government agencies. The Computer Security Institute, with the participation of the San Francisco Federal Bureau of Investigation (FBI) Computer Intrusion Squad, conducts this survey. The aim of the survey is to raise awareness of security, as well as to determine the scope of computer crime in the United States.

All told, the FBI estimates computer losses exceed $10 billion a year. Sadly, the biggest threat is from within. Law enforcement officials estimate that up to 60 percent of break-ins are from employees. For example, an entertainment company that was suspicious about an employee called in a digital detective from PricewaterhouseCoopers in Los Angeles. The employee, it turns out, was under financial pressure and had installed a program called Back Orifice on three of the company's servers. The program, which is widely available on the Internet, allowed him to take over those machines, gaining passwords and access to all the company's financial data. Fortunately, the employee was terminated before any damage could be done.[10]

Today, computer criminals are a new breed—bolder and more creative than ever. With the increased use of the Internet, computer crime is becoming global. It's not just on U.S. shores that law enforcement has to battle cybercriminals. Attacks from overseas, particularly eastern European countries, are on the rise. Indeed, the problem was so bad for America Online that it cut its connection to Russia in 1996. Nabbing bad guys overseas is a particularly thorny issue. Take Aye.Net, a small Internet service provider. In 1998, intruders broke into the ISP and knocked it off the Net for four days. Their director of systems engineering discovered the hackers and found messages in Russian. He reported it to the FBI, but no one has been able to track down the hackers.[11]

TABLE 14.4

Summary of Key Data from 2000 Computer Crime and Security Survey

Source: Data from "2000 CSI-FBI Survey Results," accessed at http://www.gocsi.com/prelea_000321.htm, August 6, 2000.

Incident	2000 Results
Companies reporting serious computer breaches	70%
Companies that acknowledge suffering financial losses from computer security breaches	74%
Companies that are able to quantify the financial losses from computer security breaches	42%
Companies with Web sites that had detected unauthorized access or misuse of Web site	19%
Companies reporting virus contamination	85%

Regardless of its nonviolent image, computer crime is different only because a computer is used. It is still a crime. Part of what makes computer crime so unique and hard to combat is its dual nature—the computer can be both the tool used to commit a crime and the object of that crime.

The Computer as a Tool to Commit Crime

A computer can be used as a tool to gain access to valuable information and as the means to steal thousands or millions of dollars. It is, perhaps, a question of motivation—many individuals who commit computer-related crime claim they do it for the challenge, not for the money. Credit card fraud—whereby a criminal illegally gains access to another's line of credit with stolen credit card numbers—is a major concern for today's banks and financial institutions. In general, criminals need two capabilities to commit most computer crimes. First, the criminal needs to know how to gain access to the computer system. Sometimes obtaining access requires knowledge of an identification number and a password. Second, the criminal must know how to manipulate the system to produce the desired result. Dallas FBI agent Mike Morris estimates that in at least a third of the cases he's investigated in his five years tracking computer crime, a critical computer password has been talked out of an individual, a practice called **social engineering**. Or, the attackers simply go through the garbage—**dumpster diving**—for important pieces of information that can help crack the computers or convince someone at the company to give them more access.[12] In addition, security experts estimate that there are 1,900 Web sites that offer the digital tools—for free—that will let people snoop, crash computers, hijack control of a machine, or retrieve a copy of every keystroke.[13]

Also, with today's sophisticated desktop publishing programs and high-quality printers, crimes involving counterfeit money, bank checks, traveler's checks, and stock and bond certificates are on the rise. As a result, the U.S. Treasury Department redesigned and printed new currency that is much more difficult to counterfeit.

social engineering
the practice of talking a critical computer password out of an individual

dumpster diving
searching through the garbage for important pieces of information that can help crack an organization's computers or be used to convince someone at the company to give them access to the computers

The Computer as the Object of Crime

A computer can also be the object of the crime, rather than the tool for committing it. Tens of millions of dollars of computer time and resources are stolen every year. Each time system access is illegally obtained, data or computer equipment is stolen or destroyed, or software is illegally copied, the computer becomes the object of crime. These crimes fall into several categories: illegal access and use, data alteration and destruction, information and equipment theft, software and Internet piracy, computer-related scams, and international computer crime.

Illegal Access and Use

Crimes involving illegal system access and use of computer services are a concern to both government and business. Federal, state, and local government computers are sometimes left unattended over weekends without proper security, and university computers are often used for commercial purposes under the pretense of research or other legitimate academic pursuits. A 28-year-old computer expert allegedly tied up thousands of US West computers in an attempt to solve a classic math problem. The individual reportedly obtained the passwords to hundreds of computers and diverted them to search for a new prime number, racking up ten years of computer processing time. The alleged hacking was discovered by a US West Intrusion Response Team after company officials noticed that computers were taking up to five minutes to retrieve telephone numbers, when normally they require only three to five seconds. At one

hacker

a person who enjoys computer technology and spends time learning and using computer systems

criminal hacker (cracker)

a computer-savvy person who attempts to gain unauthorized or illegal access to computer systems

script bunnies

wannabe crackers with little technical savvy who download programs—scripts—that automate the job of breaking into computers

insiders

employees, disgruntled or otherwise, working solo or in concert with outsiders to compromise corporate systems

virus

a program that attaches itself to other programs

worm

an independent program that replicates its own program files until it interrupts the operation of networks and computer systems

point, customer calls had to be rerouted to other states, and the delays threatened to close down the Phoenix Service Delivery Center.

Since the outset of information technology, computers have been plagued by criminal hackers. A **hacker** is a person who enjoys computer technology and spends time learning and using computer systems. A **criminal hacker**, also called a **cracker**, is a computer-savvy person who attempts to gain unauthorized or illegal access to computer systems. In many cases, criminal hackers are people who are looking for fun and excitement—the challenge of beating the system. In other cases, they are looking to steal passwords, files and programs, or even money. **Script bunnies** are wannabe crackers with little technical savvy/crackers who download programs—scripts—that automate the job of breaking into computers. **Insiders** are employees, disgruntled or otherwise, working solo or in concert with outsiders to compromise corporate systems.

Catching and convicting criminal hackers remains a difficult task. The method behind these crimes is often hard to determine. Even if the method behind the crime is known, tracking down the criminals can take a lot of time. It took years for the FBI to arrest one criminal hacker for the alleged "theft" of almost 20,000 credit card numbers that had been sent over the Internet. Table 14.5 provides some guidelines to follow in the event of a computer security incident.

Data Alteration and Destruction

Data and information are valuable corporate assets. The intentional use of illegal and destructive programs to alter or destroy data is as much a crime as destroying tangible goods. Most common of these types of programs are viruses and worms, which are software programs that, when loaded into a computer system, will destroy, interrupt, or cause errors in processing. There are more than 53,000 known computer viruses today, with more than 6,000 new viruses and worms being discovered each year. A **virus** is a program that attaches itself to other programs. A **worm** functions as an independent program, replicating its own program files until it interrupts the operation of networks and computer systems. In perhaps the most famous case involving a worm, an individual inserted a worm into the ARPANET that infected more than 6,000 computers on the network. In some cases, a virus or a worm can completely halt the operation of a computer system or network for days or longer until the problem is found and repaired. In other cases, a virus or a worm can destroy important

- Follow your site's policies and procedures for a computer security incident. (They are documented, aren't they?)
- Contact the incident response group responsible for your site as soon as possible.
- Inform others, following the appropriate chain of command.
- Further communications about the incident should be guarded to ensure intruders do not intercept information.
- Document all follow-up actions (phone calls made, files modified, system jobs that were stopped, etc.).
- Make backups of damaged or altered files.
- Designate one person to secure potential evidence.
- Make copies of possible intruder files (malicious code, log files, etc.) and store them off-line.
- Evidence, such as tape backups and printouts, should be secured in a locked cabinet, with access limited to one person.
- Get the National Computer Emergency Response Team involved if necessary.
- If you are unsure of what actions to take, seek additional help and guidance before removing files or halting system processes.

TABLE 14.5

How to Respond to a Security Incident

data and programs. If backups are inadequate, the data and programs may never be fully functional again.

Some viruses and worms attack personal computers, while others attack network and client/server systems. A personal computer can get a virus from an infected disk, an application, or e-mail attachments received from the Internet. A virus or worm that attacks a network or client/server system is usually more severe because it can affect hundreds or thousands of personal computers and other devices attached to the network. Workplace computer virus infections are increasing rapidly because of the increased spread of viruses in e-mail attachments. The number of infections per 1,000 PCs was 21.45 in 1997; in 1998 it had grown to 31.85, according to the International Computer Security Association. The primary ways to avoid viruses and worms are to install virus scanning software on all systems, update it routinely, and abstain from using disks or files from unknown or unreliable sources. You should also avoid opening files even from people you know unless you are expecting them. Many worms are sent as e-mails to people in the initial victim's address book so that it appears as a file received from someone you know.

The two most common kinds of viruses are application viruses and system viruses. **Application viruses** infect executable application files such as word processing programs. When the application is executed, the virus infects the computer system. Because these types of viruses normally attach themselves to application files, they can often be detected by checking the length or size of the file. If the file is larger than it should be, a virus may be attached. A **system virus** typically infects operating system programs or other system files. These types of viruses usually infect the system as soon as the computer is started.

Another type of program that can destroy a system is a **logic bomb**, an application or system virus designed to "explode" or execute at a specified time and date. Logic bombs are often disguised as a **Trojan horse**, a program that appears to be useful but actually masks the destructive program. Some of these programs execute randomly; others are designed to remain inert in software until a certain code is given. When it detects the cue, the bomb will explode months, or even years, after being "planted."

A **macro virus** is a virus that uses an application's own macro programming language to distribute itself. Unlike the viruses discussed so far, macro viruses do not infect programs; they infect documents. The document could be a letter created using a word processing application, a graphics file developed for a presentation, or a database file. WW6macro, for example, is a macro virus that attaches itself to Microsoft Word for Windows documents. Macro viruses that are hidden in a document file can be difficult to detect. As with other viruses, however, virus detection and correction programs can be used to find and remove macro viruses.

On April 1, 1999, a 31-year-old New Jersey programmer was arrested by federal and state officials and charged with creating and disseminating the Melissa virus, which began spreading across the Internet on March 26, 1999. Melissa is a typical macro virus that has an unusual payload. When a user opens an infected document, the virus attempts to e-mail a copy of this document to up to 50 other people, using Microsoft Outlook. The programmer was tracked down with the help of America Online and by traced phone calls. In a plea bargain with prosecutors, he pleaded guilty to one charge of computer theft. He acknowledged that the Melissa program caused more than $80 million in damage! This figure is related to the time spent by systems administrators to clear the virus off

application virus

a virus that infects executable application files such as word processing programs

system virus

a virus that typically infects operating system programs or other system files

logic bomb

an application or system virus designed to "explode" or execute at a specified time and date

Trojan horse

a program that appears to be useful but actually masks a destructive program

macro virus

a virus that infects documents by using an application's own macro programming language to distribute itself affected computers. Based on the agreement with defense attorneys, prosecutors recommended a sentence of ten years in prison, the maximum of what the law calls for in such crimes, and a fine of $150,000. This individual is considered to be one of the first people ever prosecuted for spreading a computer virus. "I did not expect or anticipate the amount of damage that took place . . . . I had no idea there would be such profound consequences to others," he said in court.[14]

Most macro viruses are written for Microsoft's Word for Windows and Excel for Windows. However, there are also macro viruses for Lotus AmiPro (APM/Greenstripe). If you count every single-bit difference as a virus variant, the total number is well above 2,000 and growing at the rate of a handful of new macro viruses every day.

Hoax, or false, viruses are another problem. Criminal hackers sometimes warn the public of a new and devastating virus that doesn't exist. Companies can spend hundreds of hours warning employees and taking preventive action against a nonexistent virus. Security specialists recommend that IS personnel establish a formal paranoia policy to thwart virus panic among gullible end users. Such policies should stress that before users forward an e-mail alert to colleagues and higher-ups, they should send it to the help desk or the security team. The corporate intranet can be used to explain the difference between real viruses and fakes, and it can provide links to Web sites to set the record straight. Table 14.6 lists some of the most informative sites about viruses. Table 14.7 lists the top 30 viruses according to McAfee, a provider of antivirus software. Note that several of those listed are hoaxes.

Information and Equipment Theft

Data and information represent assets or goods that can also be stolen. Individuals who illegally access systems often do so to steal data and information. To obtain illegal access, criminal hackers require identification numbers and passwords. Some criminals try different identification numbers and passwords until they find ones that work. Using password sniffers is another approach. A **password sniffer** is a small program hidden in a network or a computer system that records identification numbers and passwords. In a few days, a password sniffer can record hundreds or thousands of identification numbers and passwords. Using a password sniffer, a criminal hacker can gain access to computers and networks to steal data and information, invade privacy, plant viruses, and disrupt computer operations. LOpht, founded in 1992 by a group of hackers who provide computer and network security consulting, has developed a software product called AntiSniff that runs nonintrusive tests to determine whether a remote computer user is listening in on network communications.[15]

password sniffer
a small program hidden in a network or a computer system that records identification numbers and passwords

TABLE 14.6

Sources of Information about Viruses
Source: Data from Leslie Goff, "Resources," *Computerworld*, May 25, 1998.

Web Page	Address
CIAC Internet Hoaxes Page	http://hoaxbusters.ciac.org
Computer Virus Myths Homepage	http://www.vmyths.com
Trusecure Corp.	http://www.trusecure.com
The Truth About E-Mail Viruses	http://www.gerlitz.com/virushoax/
Dr. Solomon's Software	http://www.drsolomon.com
Symantec Corp.	http://www.symantec.com
Network Associates, Inc.	http://www.networkassociates.com

Virus Name	Date Discovered	Virus Type	Risk Assessment
WScript/Kak.worm.a	10/22/99	Virus	Medium
VBS/Loveletter.a	5/4/00	Virus	High
IRC/Stages.worm	5/26/00	Virus	High
Wobbler	10/13/98	Hoax	—
W97M/Resume.a@mm	5/26/00	Trojan	Medium
BackDoor-G	4/15/99	Trojan	Medium
KALI	3/8/00	Hoax	—
BackDoor-G2.svr.21	12/16/99	Trojan	Medium
California.IBM	6/29/00	Hoax	—
Bud Frogs Screen Saver Hoax	3/1/99	Hoax	—
W32/Pretty.worm.unp	2/15/00	Trojan	Medium
W32/Ska	1/27/99	Virus	Medium
W32/Pretty.Worm	5/26/99	Trojan	Medium
APStrojan.qa	1/18/00	Trojan	Medium
Win a Holiday Email Hoax	3/1/99	Hoax	—
W32/FunLove.4099	11/9/99	Virus	Medium
VBS/Netlog.worm.a	2/2/00	Trojan	Low
Simpsons	6/27/00	Trojan	Low
Your friend D@fit	3/19/00	Hoax	—
"Great Gas-Out" Hoax	3/31/00	Hoax	—
W95/CIH.1003	7/1/98	Virus	Medium
W97M/Ethan.a	1/21/99	Virus	Medium
Guts to Say Jesus	3/1/99	Hoax	—
VBS/COD.a	7/8/00	Trojan	Low
DOS/AntiOL	7/18/00	Trojan	Low
W97M/Inadd.d	7/21/00	Virus	Low
W97M/Berau	7/12/00	Virus	Low
VBS/Pica.g	7/3/00	Trojan	Low
CELCOM Screen Saver	5/25/99	Hoax	—
W32/Smash	4/21/00	Virus	Low

TABLE 14.7

Top 30 Viruses (Data Source: McAfee Web site)

In addition to theft of data and software, all types of computer systems and equipment have been stolen from offices. Computer theft is now second only to automobile theft, according to recent U.S. crime statistics. In the United Kingdom more than 30 percent of all reported thefts are computer related. Printers, desktop computers, and scanners are often targets. Portable computers such as laptops (and the data and information stored in them) are especially easy for thieves to take. In some cases, the data and information stored in these systems are more valuable than the equipment. Without adequate protection and security measures, equipment can easily be stolen.

Software and Internet Piracy

Each time you use a word processing program or access software on a network, you are taking advantage of someone else's intellectual property. Like books and movies—other intellectual properties—software is protected by copyright laws. Often, people who would never think of plagiarizing another author's written work have no qualms about using and copying software programs they

To fight computer crime, many companies use devices such as BookLock (shown here) which disables the disk drive and locks the computer to the desk. (Source: Kensington Notebook MicroSaver®)

software piracy

the act of illegally duplicating software

Internet piracy

illegally gaining access to and using the Internet

have not paid for. Such illegal duplicators are called pirates; the act of illegally duplicating software is called **software piracy**. Technically, software purchasers are granted the right only to use the software under certain conditions; they don't really own the software.

Internet piracy involves illegally gaining access to and using the Internet. Although not yet as prevalent as software piracy, the amount of Internet piracy is growing rapidly. Many companies on the Internet receive fees from customers for their information, services, and even products. Some investment firms, for example, offer market analysis and investment information for a monthly or annual fee. Other companies offer information on sports or provide research information on a variety of topics for a fee. For some services, the fees can be thousands of dollars annually. Typically, Internet companies give customers identification numbers or passwords. Like customers illegally copying software, some customers illegally share their identification numbers and passwords with others. In other cases, criminal hackers obtain these numbers illegally on the Internet or from other sources. When unauthorized people use these services, Internet firms lose valuable revenues. Internet piracy can also be directed against individuals. While users are surfing the Web, outsiders can download an applet to the browser's machine, use its processor to perform computations, and send the results back to a host. This technique is called *MIPs-sucking*.

Computer-Related Scams

People have lost hundreds of thousands of dollars on real estate, stock, and other business scams. Today, many of these same types of scams are being performed using computers. Using the Internet, scam artists offer get-rich-quick schemes involving real estate deals, bank fraud, telephone lotteries, penny stocks, and tax avoidance. In one Internet scam, a lottery investment of $129 was reported to guarantee a return that could range from $600 per week to $10,000 per week. The scam raised over $3 million in less than a year. In most cases, only the scam artists get rich. Here are some general tips to help you avoid becoming a victim:

- Don't agree to anything in a high-pressure meeting or seminar. Insist on having time to think it over and to discuss things with a spouse, partner, or even your lawyer. If a company won't give you the time you need to check it out and think things over, you don't want to do business with it. A good deal now will be a good deal tomorrow; the only reason for rushing you is if the company has something to hide.
- Don't judge a company based on appearances. Flashy Web sites can be created and put up on the Net in a matter of days. After a few weeks of taking money, a site can vanish without a trace in just a few minutes. You may find that the perfect money-making opportunity offered on a Web site was a money maker for the crook and a money loser for you.
- Avoid any plan that pays commissions simply for recruiting additional distributors. Your primary source of income should be your own product sales. If the earnings are not made primarily by sales of goods or services to consumers or sales by distributors under you, you may be dealing with an illegal pyramid.
- Beware of shills, people paid by the company to lie about how much they've earned and how easy the plan was to operate. Check with an independent source to make sure that you aren't having the wool pulled over your eyes.

- Beware of the company's claim that it can set you up in a profitable home-based business but that you must first pay up front to attend a seminar and buy expensive materials. Frequently, seminars are high-pressure sales pitches, and the material is so general that it is worthless.
- If you are interested in starting a home-based business, get a complete description of the work involved before you send any money. You may find that what you are asked to do after you pay is far different from what was stated in the ad. You should never have to pay for a job description or for needed materials.
- Get in writing the refund, buy-back, and cancellation policies of any company you deal with. Do not depend on oral promises.
- Do your homework. Check with your state attorney general and the National Fraud Information Center before getting involved, especially when the claims about the product or potential earnings seem too good to be true.

If you need advice about an Internet or on-line solicitation, or you want to report a possible scam, use the Online Reporting Form or Online Question & Suggestion Form features on the Web site for the National Fraud Information Center at http://fraud.org, or call the NFIC hotline at 1-800-876-7060.

International Computer Crime

Computer crime is also an international issue, and it becomes more complex when it crosses borders. Estimates of software piracy in the global marketplace indicate that more than one-quarter of software is pirated, adding up to more than $3 billion in lost revenue. In China alone, 96 percent of software is pirated, with lost revenue totaling over $1 billion. In Scotland, knife-wielding, masked criminals stole nearly $4 million worth of microchips. Another issue in international computer crime relates to illegally obtaining and selling restricted information in countries with less stringent laws. To avoid these problems, many countries require that computer equipment and software be registered with appropriate authorities before it can be brought into the country. With cash and funds being transferred electronically, some are concerned that international drug dealers and criminals are using information systems to launder illegally obtained funds. Computer terrorism is another aspect of international computer crime.

Preventing Computer-Related Crime

Because of increased computer use today, greater emphasis is placed on the prevention and detection of computer crime. Although more than 45 states have passed computer crime bills, some believe that they are not effective because companies do not always actively detect and pursue computer crime, security is inadequate, and convicted criminals are not severely punished. However, all over the United States, private users, companies, employees, and public officials are making individual and group efforts to curb computer crime, and recent efforts have met with some success.

Crime Prevention by State and Federal Agencies

State and federal agencies have begun aggressive attacks on computer criminals, including criminal hackers of all ages. In 1986, Congress enacted the Computer Fraud and Abuse Act, which mandates punishment based on the victim's dollar loss. The Department of Defense also supports the Computer Emergency Response Team (CERT), which responds to network security breaches and monitors systems for emerging threats. Law enforcement agencies are also increasing their efforts to stop criminal hackers, and many states are now passing new, comprehensive computer crime bills to help eliminate computer

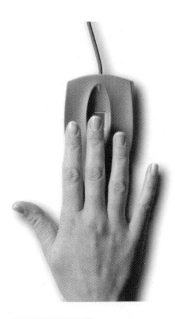

Bioscrypt Enterprise is a fingerprint authentication device that provides security in the PC environment by using fingerprint information instead of passwords.
(Source: Courtesy of Mytec Technologies Inc.)

biometrics

the measurement of a living trait, whether physical or behavioral

crimes. Recent court cases and police reports involving computer crime show that lawmakers are ready to introduce newer and tougher computer crime legislation. Several states have passed laws in an attempt to outlaw spamming, the practice of sending large amounts of unsolicited e-mail to overwhelm users' e-mail boxes or the e-mail servers on a network.

Crime Prevention by Corporations

Companies are also taking crime-fighting efforts seriously. Many businesses have designed procedures and specialized hardware and software to protect their corporate data and systems. Specialized hardware and software, such as encryption devices, can be used to encode data and information to help prevent unauthorized use. Using biometrics is another a way to protect important data and information systems. **Biometrics** involves the measurement of a living trait, whether physical or behavioral. Biometric techniques compare a person's unique characteristics against a stored set for the purpose of detecting differences between them. Biometric systems can scan fingerprints, faces, handprints, and retinal images to prevent unauthorized access to important data and computer resources. Most of the interest among corporate users is in fingerprint technology, followed by face recognition. Fingerprints hit the middle ground between price and usability. Iris and retina scans are more accurate, but they are more expensive, and they involve more equipment.

Many of the early adopters of biometric technologies have been healthcare organizations, because new federal legislation requires such organizations to protect the privacy of patients. Instead of typing in a password, a doctor or hospital employee puts his or her finger on a scanner connected to a personal computer and the image is matched against a set of authorized fingerprints. With some systems, instead of an image of a fingerprint, software uses an algorithm to create a personal identification number that equates to a person's fingerprint. This saves data storage and avoids having actual fingerprints stored on file.

Crime-fighting procedures usually require additional controls on the information system. Before designing and implementing controls, organizations must consider the types of computer-related crime that might occur, the consequences of these crimes, and the cost and complexity of needed controls. In most cases, organizations conclude that the trade-off between crime and the additional cost and complexity weighs in favor of better system controls. Having knowledge of some of the methods used to commit crime is also helpful in preventing, detecting, and developing systems resistant to computer crime (see Table 14.8). Some companies actually hire former criminals to

TABLE 14.8

Common Methods Used to Commit Computer Crimes

Even though the number of potential computer crimes appears to be limitless, the actual methods used to commit crime are limited.

Methods	Examples
Add, delete, or change inputs to the computer system.	Delete records of absences from class in a student's school records.
Modify or develop computer programs that commit the crime.	Change a bank's program for calculating interest to make it deposit rounded amounts in the criminal's account.
Alter or modify the data files used by the computer system.	Change a student's grade from C to A.
Operate the computer system in such a way as to commit computer crime.	Access a restricted government computer system.
Divert or misuse valid output from the computer system.	Steal discarded printouts of customer records from a company trash bin.
Steal computer resources, including hardware, software, and time on computer equipment.	Make illegal copies of a software program without paying for its use.

TABLE 14.9

How to Protect Your Corporate
Data From Hackers

- Install strong user authentication and encryption capabilities on your firewall.
- Install the latest security patches, which are often available at the vendor's Internet site.
- Disable guest accounts and null user accounts that let intruders access the network without a password.
- Do not provide overfriendly log-in procedures for remote users (e.g., an organization that used the word *welcome* on their initial log-on screen found they had difficulty prosecuting a hacker).
- Give an application (e-mail, file transfer protocol, and domain name server) its own dedicated server.
- Restrict physical access to the server and configure it so that breaking into one server won't compromise the whole network.
- Turn audit trails on.
- Consider installing caller ID.

thwart other criminals. Table 14.9 provides a set of useful guidelines to protect your computer from hackers.

Companies are also joining to fight crime. The Software Publishers Association (SPA), which was formed by a number of leading software companies, audits companies and checks for software licenses. Organizations that are found to be illegally using software can be fined or sued. Depending on the violation, the fine can be hundreds of thousands of dollars. While the SPA is an effective deterrent in the United States, efforts to curb software abuse in other countries is much more difficult.

Using Antivirus Programs

antivirus programs

programs or utilities that prevent
viruses and recover from them if
they infect a computer

Companies and individuals must protect their computer systems and networks from viruses. Most people use **antivirus programs** or utilities to prevent viruses and recover from them if they infect a computer. These programs range in cost from free (shareware) to a few hundred dollars. Antivirus programs are developed for different operating systems. Norton Antivirus for Windows 95/98 and NT workstations, Norton Antivirus for Macintosh, Quarterdeck Utility Pack for Windows 95/98, Dr. Solomon's Anti-Virus Toolkit for Windows 95/98, and Network Associates' Virex for Macintosh are just a few examples of antivirus programs. Proper use of antivirus software requires the following steps:

Antivirus software should be
used and updated often.

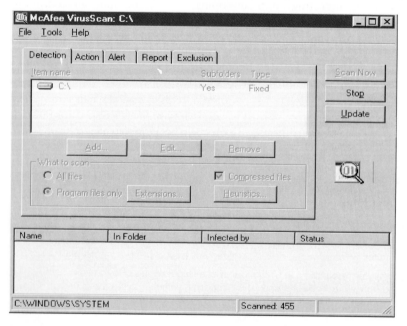

1. *Install a virus scanner and run it often.* Many of these programs automatically check for viruses each time you boot up your computer or insert a diskette, and some even monitor all transmissions and copying operations.

2. *Update the virus scanner often.* Old programs may fail to detect new viruses.

3. *Scan all diskettes before copying or running programs from them.* Hiding on diskettes, viruses often move between systems. If you carry document or program files on diskettes between computers at school or work and your home system, always scan them.

4. *Install software only from a sealed package produced by a known software company.* Even software publishers can unknowingly distribute viruses on their program diskettes or software downloads. Most scan their own systems, but viruses may still remain.

5. *Follow careful downloading practices.* If you download software from the Internet or a bulletin board, check your computer for viruses immediately after completing the transmission.

6. *If you detect a virus, take immediate action.* Early detection often allows you to remove a virus before it does any serious damage.

Despite careful precautions, viruses can still cause problems. They can elude virus-scanning software by lurking almost anywhere in a system. Future antivirus programs may incorporate "nature-based models" that check for unusual or unfamiliar computer code. The advantage of this type of virus program is the ability to detect new viruses that are not part of an antivirus database.

Internet Laws for Libel and Protection of Decency

The Telecommunications Act of 1996 included an act called the Communications Decency Act. One of the original provisions of this act is the ability of the government to jail or fine anyone up to $100,000 for sending indecent materials to a minor electronically. Many people and companies were very concerned about the free speech implications of this provision and many turned their Web pages to black in protest. Subsequently, the courts limited the Communications Decency Act. In an unrelated case, the German government forced CompuServe to suspend about 200 newsgroups on its service that Germany claimed violated German law. The law was Section 184 of the German Criminal Code, which deals with the distribution of pornographic materials to minors.

To help parents control what their children see on the Internet, some companies are developing software called filtering software to help screen Internet content. Many of these screening programs also prevent children from sending personal information over e-mail or through chat groups. This stops children from broadcasting their name, address, phone number, or other personal information over the Internet. The two approaches used are filtering, which blocks certain Web sites, and rating, which places a rating on Web sites. Examples include Border Manager, Choice Net, Click & Browse Junior, Cybersitter, Cyber Patrol, Net Nanny, Specs for Kids, Surf Guard, and SurfWatch.

With the increased popularity of networks and the Internet, libel and decency become important legal issues. A publisher, such as a newspaper, can be sued for libel, which involves publishing a written statement that is damaging to a person's reputation. Generally, a bookstore cannot be held liable for statements made in a newspaper or other publications it sells. On-line services, such as CompuServe and America Online, have control over who puts information on their service but may not have direct control over the content of what is published by others on their service. Can on-line services be sued for libel for content that someone else publishes on their service? Are on-line services more like a newspaper or a bookstore? This legal issue has not been completely resolved, but some court cases have been decided. The *Cubby, Inc. v. CompuServe* case ruled that CompuServe was more like a bookstore and not liable for content put on its service by others. In this case, the judge stated, "While CompuServe may decline to carry a given publication altogether, in reality, once it does decide to carry a given publication, it will have little or no editorial control over that publication's content."

Preventing Crime on the Internet

As mentioned in Chapter 7, Internet security can include firewalls and a number of methods to secure financial transactions. A firewall can include hardware

Net Nanny is a filtering software program that helps block unwanted Internet content from children and young adults.
(Source: Courtesy of Net Nanny Software International)

and software combinations that act as a barrier between an organization's information system and the outside world. A number of systems have been developed to safeguard financial transactions on the Internet.

To help prevent crime on the Internet, the following steps can be taken.

1. Develop effective Internet and security policies for all employees.
2. Use a stand-alone firewall (hardware and software) with network monitoring capabilities.
3. Monitor managers and employees to make sure they are using the Internet for business purposes only.
4. Use Internet security specialists to perform audits of all Internet and network activities.

Even with these precautions, computers and networks can never be completely protected against crime. One of the biggest threats is from employees. According to Rich Ayers, vice president of Network Information Management at Chase Manhattan Bank Corporation, "Just blocking everyone externally doesn't mean you are safe." Some believe that 60 percent or more of all computer attacks come from employees or managers inside the company. Although firewalls provide good perimeter control to prevent crime from the outside, procedures and protection measures are needed for personnel. Passwords, identification numbers, and tighter control of employees and managers also help prevent Internet-related crime. Read the "E-Commerce" box to gain an appreciation of the need to provide safeguards to protect Web sites.

PRIVACY

Another important social issue in information systems involves privacy. In 1890, U.S. Supreme Court Justice Louis Brandeis stated that the "right to be left alone" is one of the most "comprehensive of rights and the most valued by civilized man." Basically, the issue of privacy deals with this right to be left alone or to be withdrawn from public view. With information systems, privacy deals with the collection and use or misuse of data. Data is constantly being collected and stored on each of us. This data is often distributed over easily accessed networks and without our knowledge or consent. Concerns of privacy regarding this data must be addressed.

With today's computers, the right to privacy is an especially challenging problem. More data and information is produced and used today than ever before. A difficult question to be answered is, "Who owns this information and knowledge?" If a public or private organization spends time and resources in obtaining data on you, does the organization own the data and can it use the data in any way it desires? Government legislation answers these questions to some extent for federal agencies, but the questions remain unanswered for private organizations.

Privacy Issues

The issue of privacy is important because data on an individual can be collected, stored, and used without that person's knowledge or consent. When someone is born, takes certain high school exams, starts working, enrolls in a college course, applies for a driver's license, purchases a car, serves in the military, gets married, buys insurance, gets a library card, applies for a charge card or loan, buys a house, or merely purchases certain products, data is collected and stored somewhere in computer databases.

E-COMMERCE
Business Entrepreneurs Carve New Path

Scour was founded in December 1997 by five UCLA computer science students. Its core service is a search engine that specializes in finding multimedia on the Internet. Unlike conventional search engines that index text found on Web pages, Scour's unique search engine is used to find audio, video, still images, and animation, including music videos, movie trailers, and full-length movies. The search engine takes you directly to the individual multimedia files that you're seeking.

The sites linked from the Scour Web site are not under Scour's control, and Scour does not assume any responsibility or liability for any communications or materials available at such linked sites. Scour does offer a free application, the Scour Media Agent, to let you download files. If you want a copy of "Old Time Rock and Roll" by Bob Seger, Scour Exchange will connect you to a user with that track on his hard drive, and zap the song your way.

Scour wants to be viewed as a cooperative partner of the entertainment industry. It works with dozens of content owners to provide licensed premium content such as music downloads, music videos, radio stations, movie trailers, short films, animations, and much more. Scour states that it has worked with the leading copyright experts to design its service to conform to all applicable laws, including the Digital Millennium Copyright Act of 1998 (a summary of which may be found on the U.S. Copyright Office Web site at http://lcweb.loc.gov/copyright/legislation/dmca.pdf). This law was specifically designed to protect copyrights on-line. In accordance with the law, Scour provides a system for copyright owners to notify them of any sites, linked to by the Scour search engine, that have posted material without the proper permission of the copyright owner. Furthermore, Scour must respond expeditiously to all claims of copyright infringement. Scour's Web site contains procedures for submitting a claim of copyright infringement.

Most entertainment executives and artists in the movie and music business consider Scour illegal. Scour facilitates and makes copyright infringement possible. It provides software, an index, and connection to illegally copied files. As a result, the Motion Picture Association of America (MPAA), the Recording Industry Association of America (RIAA), and the National Music Publishers Association (NMPA), together with every major record label and movie studio, filed a lawsuit against Scour seeking to shut down Scour's search engine products and services. The law firm representing these associations is asking for damages of $150,000 for each illegally downloaded file. Scour has three million unique monthly visitors, so even assuming that 1 percent of them downloaded an illegally copied file, Scour would be forced to pay over $4.5 billion!

Scour thinks that it is fighting a battle that will affect the future of technology and the Internet. A bad result for Scour would have repercussions for almost every business on the Internet. This is why Scour has assembled a powerhouse legal team to defend itself and its right to develop new technologies for the Internet.

Discussion Questions

1. Read the Digital Millennium Copyright Act of 1998 and summarize the key points that apply to Scour.
2. Do you believe that Scour complies with the intent and spirit of this act? Why or why not?

Critical Thinking Questions

3. How might the results of this suit affect other Internet companies?
4. How might Scour modify its basic services to avoid future lawsuits?

Sources: Amy Kover, "Napster Who? Scour Raises Hollywood Ire, Investment," *Fortune*, p. 52, September 15, 2000; the Scour Technology Freedom Center at the Scour Web site, http://www.scour.com/, accessed September 19, 2000.

Privacy and the Federal Government

The federal government is perhaps the largest collector of data. Close to four billion records exist on individuals, collected by about 100 federal agencies, ranging from the Bureau of Alcohol, Tobacco, and Firearms to the Veterans Administration. Other data collectors include state and local governments and profit and nonprofit organizations of all types and sizes.

In recent years, a number of examples have caused concern for privacy of on-line data. Cookies were planted on the hard drives of visitors to the Web site of the White House drug enforcement office.[16] Echelon, a global surveillance network run by the National Security Agency and allied intelligence bureaus, listens to every electronic communication in the world—cell phone calls, satellite transmissions, e-mail messages.[17] In Maryland, a banker accessed medical records to find people diagnosed with cancer. Once he identified them, the bank called in their loans.[18]

As a result, the Federal Trade Commission is considering action, and about 70 laws are before Congress regarding various aspects of Internet privacy. The European Union has already passed a data-protection directive that requires firms transporting data across national boundaries to have certain privacy procedures in place. This directive affects virtually any company doing business in Europe, and it is driving much of the attention being given to privacy in the United States.

Most user companies and computer vendors are wary of having the federal government dictate Internet privacy standards. A group called the Online Privacy Alliance is developing a voluntary code of conduct. It is backed by companies such as AT&T, IBM, Dun & Bradstreet, Time-Warner, Walt Disney, the Lexis-Nexis division of Reed Elsevier, Microsoft, and Netscape. The alliance's guidelines will call on companies to notify users when they are collecting data at Web sites to gain consent for all uses of that data, to provide for the enforcement of privacy policies, and to have a clear process in place for receiving and addressing user complaints. The alliance's policy can be found at http://www.privacyalliance.org.

Privacy at Work

The right to privacy at work is also an important issue. Some experts believe that there will be a collision between workers who want their privacy and companies that demand to know more about their employees. Recently, companies that have been monitoring their employees have raised concerns. Workers may find that they are being closely monitored via computer technology. These computer-monitoring systems tie directly into computerized workstations; specialized computer programs can track every keystroke made by a user. This type of system can determine what workers are doing while at the keyboard. The system also knows when the worker is not using the keyboard or computer system. These systems can estimate what a person is doing and how many breaks he or she is taking. Needless to say, many workers consider this type of supervision very dehumanizing.

E-Mail Privacy Issues

E-mail also raises some interesting issues about work privacy. Federal law permits employers to monitor e-mail sent and received by employees. Furthermore, e-mail messages that have been erased from hard disks may be retrieved and used in lawsuits because the laws of discovery demand that companies produce all relevant business documents. On the other hand, the use of e-mail among public officials may violate "open meeting" laws. These laws, which apply to many local, state, and federal agencies, prevent public officials from meeting in private about matters that affect the state or local area.

There has been significant commentary from the legal community regarding the applicability of privileges—specifically, the attorney–client privilege—to communications made electronically, including electronic mail. Perhaps in response to this commentary, but more likely as a reflection of a technologically advanced society, the state of New York amended its Civil Practice Law and Rules and clarified the matter. A privileged communication does not lose its privileged character by virtue of its being communicated or transmitted electronically. Over a dozen more states have either put into effect a similar law or are debating this matter. Navigating these paradoxical privacy concerns will be a challenge for organizations in the information age. Read the "Ethical and Societal Issues" box to learn more about the monitoring of employee e-mail.

E-mail has changed how workers and managers communicate. With e-mail, people can communicate in the same building or around the world. E-mail, however, can be monitored and intercepted. As with other services—like cellular phones—the convenience of e-mail must be balanced with the potential of privacy invasion.
(Source: © PhotoDisc)

ETHICAL AND SOCIETAL ISSUES
Monitoring E-Mail—An Invasion of Privacy?

A two-month investigation by Dow Chemical, sparked by complaints from an employee, resulted in the firing of 50 workers for sending explicit pornographic images through the company's e-mail system. Another two hundred workers were disciplined by the chemical manufacturer for distributing, downloading, or saving pictures that were either pornographic or violent in nature. Pharmaceutical giant Merck fired or disciplined an undisclosed number of employees for inappropriate use of e-mail. The New York Times fired 23 workers because they had allegedly distributed offensive jokes on the company's e-mail system.

Such incidents soon become public, and some people protest the firings, saying they conflict with employees' rights to free speech. However, more and more companies are trying to gain greater control over employees' use of e-mail. Concerns about company e-mail abuse—and the decision about whether to monitor e-mail to avoid problems—are issues that businesses of all sizes wrestle with every day.

Companies' need to monitor e-mail reflects the realities of the modern workplace. Employers are concerned about how easily classified business information can be distributed via e-mail. They also worry about e-mail content, given what happened in the Microsoft antitrust trial, when e-mails thought to be private were dredged up in court and used against the software company. Employees sometimes write things in e-mails they would never say in public, including offensive remarks that can leave a company defenseless in a lawsuit. Indeed, courts have ruled that companies are liable for harassment charges if they institute monitoring of e-mail but don't uncover messages that led to the charges.

In most instances, employers have the legal right to monitor employee e-mail, as long as they notify workers that they might do so. State and federal courts have ruled in numerous lawsuits that employees can't always demand privacy in the workplace.

To monitor e-mail, companies install basic monitoring software to scan incoming and outgoing mail for words and key phrases that managers have compiled in lists. The software can also identify e-mail viruses. Messages that are flagged can be reviewed for possible policy violations. More sophisticated systems have advanced features, such as context-sensitive scanning that helps avoid false alarms and the ability to scan for viruses. But e-mail monitoring is still costly and labor-intensive.

The key to gaining employee support—or at least tolerance—of e-mail monitoring is to educate employees about why it's necessary and to provide carefully written e-mail usage policies. Written policies also help protect against potential legal problems. All workers need to be informed about the policies they're subject to on the job. Company managers should fully explain the details of their policies and the reasons for them. And employees need to understand that it's the company's e-mail—not theirs.

Discussion Questions

1. What are the key elements of an employee policy statement that explains the need for e-mail monitoring?
2. What steps must you take to successfully implement this policy?

Critical Thinking Questions

3. Obtain and read a copy of your company's or school's e-mail policy. Do you think the policy is clear and complete in addressing all the issues mentioned in the box? Who is responsible for enforcing this policy?
4. What steps would you take if you received "hate mail" or pornographic or violent pictures through your company's or school's e-mail system?

Sources: Adapted from Thomas York, "Invasion of Privacy? E-Mail Monitoring Is on the Rise," *Information Week*, February 21, 2000, pp. 142–146; Jennifer Disabatino, "Congress Weighs E-Mail, Net Monitoring Legislature," *Computerworld*, July 31, 2000, p. 30; Todd R. Weiss, "Dow Chemical Fires 50 Workers for Sending Pornographic E-mail," *Computerworld*, July 28, 2000, accessed at http://www.computerworld.com/cwi/story/0,1199, NAV47_STO47689,00.html; and Christopher Lindquist, "You've Got Dirty Mail," *Computerworld*, March 13, 2000, accessed at http://www.computerworld.com/cwi/story/0,1199,NAV47_STO42841,00.html.

Privacy of Hardware and Software Consumers

Recent actions by major hardware and software vendors have shown that there is a need for consumers to be on guard to protect their privacy.

Privacy groups protested in 1999 after Intel announced that it would release Pentium III chips with a processor serial number. The groups said the number could be used to track people's habits as they surfed the Internet. Intel said the number would be used to help information systems managers keep track of their computers, but it eventually agreed to work with personal computer makers to turn the feature off at the manufacturing level. Consumer groups worried that the presence of the number still left unanswered questions about privacy protection. So Intel announced in April 2000 that it would stop stamping serial numbers in its processors, starting with its Willamette chip due out in late 2000, but it will continue the number in its Pentium III chips.[19]

Microsoft acknowledged that Windows 98 and other Microsoft applications, such as Word and Excel, automatically record the author of electronic documents and the computer on which the documents were created. Documents created using Microsoft's popular Word and Excel programs in tandem with its Windows 98 operating system inserted into documents a hidden 32-digit number that was unique to the computer. This number, called a Globally Unique Identifier (GUID), is unique to each personal computer. It is based on the hardware components of the system, including the Ethernet card, if it's present. Since many offices and home businesses use Ethernet cards for local area networking and high-speed Internet access, the GUID can trace a document back to its originator, even if the author strives to maintain anonymity. The computer's GUID was sent to Microsoft when the owner ran Win98's registration wizard. So Microsoft knows your personal computer's ID, and it can trace documents you create.[20]

Responding to repeated customer requests and significant negative press, Microsoft released two software patches that remove the GUID from Windows 98 systems and from Office 97 documents and spreadsheets in mid-March 1999.[21]

Privacy and the Internet

Some people assume that there is no privacy on the Internet and that you use it at your own risk. Others believe that companies with Web sites should have strict privacy procedures and be accountable for privacy invasion. Regardless of your view, the potential for privacy invasion on the Internet is huge. People wanting to invade your privacy could be anyone from criminal hackers to marketing companies to corporate bosses. Personal and professional information on you can be seized on the Internet without your knowledge or consent. E-mail is a prime target, as discussed previously. Sending an e-mail message is like having an open conversation in a large room—people can listen to your messages. When you visit a Web site on the Internet, information about you and your computer can be captured. When this information is combined with other information, companies can know what you read, what products you buy, and what your interests are. According to an executive of an Internet software monitoring company, "It's a marketing person's dream."

Most people who buy products on the Web say it's very important for a site to have a policy explaining how personal information is used, and the policy statement must make people feel comfortable and be extremely clear about what information is collected and what will and will not be done with it. However, the Federal Trade Commission found privacy policies were posted at only about 14 percent of 1,400 Web sites it recently surveyed. A *Business Week* check of the top 100 Web sites found that 43 percent displayed privacy policies. Of the notices posted, some were difficult to find and inconsistent in explaining how data is tracked and used. The real issue that Internet users need to be concerned with is what content providers want with registration information. If a site requests that you provide your name and address, you have every right to know why and what will be done with it. If you buy something and provide a shipping address, will it be sold to other retailers? Will your e-mail address be sold on a list of active Internet shoppers? And if so, you should realize that it's no different than the lists compiled from the orders you place with catalog retailers. You have the right to be taken off any mailing list.

Fairness in Information Use

Selling information to other companies can be so lucrative that many companies will continue to store and sell the data they collect on customers, employees, and others. When is this information storage and use fair and reasonable to the individuals whose data is stored and sold? Do individuals have a right to know about data stored about them and to decide what data is stored and used? As shown in Table 14.10, these questions can be broken down into four issues that should be addressed: knowledge, control, notice, and consent.

Federal Privacy Laws and Regulations

In the past few decades, significant laws have been passed regarding an individual's right to privacy. Others relate to business privacy rights and the fair use of data and information.

The Privacy Act of 1974

The major piece of legislation on privacy is the Privacy Act of 1974 (PA74), enacted by Congress during Gerald Ford's presidency. PA74 applies only to certain federal agencies. The act, which is about 15 pages long, is straightforward and easy to understand.

The purpose of this act is to provide certain safeguards for individuals against an invasion of personal privacy by requiring federal agencies (except as otherwise provided by law) to do the following:

- Permit individuals to determine what records pertaining to them are collected, maintained, used, or disseminated by such agencies
- Permit individuals to prevent records pertaining to them from being used or made available for another purpose without their consent
- Permit individuals to gain access to information pertaining to them in federal agency records, to have a copy of all or any portion thereof, and to correct or amend such records
- Ensure that they collect, maintain, use, or disseminate any record of identifiable personal information in a manner that ensures that such action is for a necessary and lawful purpose, that the information is current and accurate for its intended use, and that adequate safeguards are provided to prevent misuse of such information

TABLE 14.10

The Right to Know and the Ability to Decide

Fairness Issues	Database Storage	Database Usage
The right to know	Knowledge	Notice
The ability to decide	Control	Consent

Knowledge. Should individuals have knowledge of what data is stored on them? In some cases, individuals are informed that information on them is stored in a corporate database. In others, individuals do not know that their personal information is stored in corporate databases.

Control. Should individuals have the ability to correct errors in corporate database systems? This is possible with most organizations, although it can be difficult in some cases.

Notice. Should an organization that uses personal data for a purpose other than the original purpose notify individuals in advance? Most companies don't do this.

Consent. If information on individuals is to be used for other purposes, should these individuals be asked to give their consent before data on them is used? Many companies do not give individuals the ability to decide if information on them will be sold or used for other purposes.

- Permit exemptions from this act only in cases where there is an important public need for such exemption, as determined by specific law-making authority
- Be subject to civil suit for any damages that occur as a result of willful or intentional action that violates any individual's rights under this act

PA74, which applies to all federal agencies except the CIA and law enforcement agencies, also established a Privacy Study Commission to study existing databases and to recommend rules and legislation for consideration by Congress. PA74 also requires training for all federal employees who interact with a "system of records" under the act. Most of the training is conducted by the Civil Service Commission and the Department of Defense. Another interesting aspect of PA74 concerns the use of social security numbers—federal, state, and local governments and agencies cannot discriminate against any individual for not disclosing or reporting his or her social security number.

Other Federal Privacy Laws

In addition to PA74, other pieces of federal legislation relate to privacy. A federal law that was passed in 1992 bans unsolicited fax advertisements. This law was upheld in a 1995 ruling by the Ninth U.S. Circuit Court of Appeals, which concluded that the law is a reasonable way to prevent the shifting of advertising costs to customers. Table 14.11 lists additional laws related to privacy.

State Privacy Laws and Regulations

TABLE 14.11

Federal Privacy Laws and their Provisions

Some states either have or are proposing their own privacy legislation. The use of social security numbers and medical records, the disclosure of unlisted

Law	Provisions
Fair Credit Reporting Act of 1970 (FCRA)	Regulates operations of credit-reporting bureaus, including how they collect, store, and use credit information
Tax Reform Act of 1976	Restricts collection and use of certain information by the Internal Revenue Service
Electronic Funds Transfer Act of 1979	Outlines the responsibilities of companies that use electronic funds transfer systems, including consumer rights and liability for bank debit cards
Right to Financial Privacy Act of 1978	Restricts government access to certain records held by financial institutions
Freedom of Information Act of 1970	Guarantees access for individuals to personal data collected about them and about government activities in federal agency files
Education Privacy Act	Restricts collection and use of data by federally funded educational institutions, including specifications for the type of data collected, access by parents and students to the data, and limitations on disclosure
Computer Matching and Privacy Act of 1988	Regulates cross-references between federal agencies' computer files (e.g., to verify eligibility for federal programs)
Video Privacy Act of 1988	Prevents retail stores from disclosing video rental records without a court order
Telephone Consumer Protection Act of 1991	Limits telemarketers' practices
Cable Act of 1992	Regulates companies and organizations that provide wireless communications services, including cellular phones
Computer Abuse Amendments Act of 1994	Prohibits transmissions of harmful computer programs and code, including viruses
Children's Online Privacy Protection Act of 1998	Sets standards for sites that collect information from children. The goal is to prohibit unfair or deceptive acts or practices in connection with the collection, use, or disclosure of personally identifiable information from and about children on the Internet

telephone numbers by telephone companies and credit reports by credit bureaus, the disclosure of bank and personal financial information, and the use of criminal files are some of the issues being considered by state legislators. Furthermore, many of these proposed legislative actions apply to both public and private organizations.

Corporate Privacy Policies

Even though privacy laws for private organizations are not very restrictive, most organizations are very sensitive to privacy issues and fairness. They realize that invasion of privacy problems can hurt their business, turn away customers, and dramatically reduce revenues and profits. Consider a major international credit card company. If the company sold confidential financial information on millions of customers to other companies, the results could be disastrous. In a matter of days, the firm's business and revenues could be reduced dramatically. Thus, most organizations maintain privacy policies, even though they are not required by law. Some companies even have a privacy bill of rights that specifies how the privacy of employees, clients, and customers will be protected. Corporate privacy policies should address a customer's knowledge, control, notice, and consent over the storage and use of information. They may also cover who has access to private data and when it may be used.

Individual Efforts to Protect Privacy

Although numerous state and federal laws deal with privacy, privacy laws do not completely protect individual privacy. In addition, not all companies have privacy policies. As a result, many people are taking steps to increase their own privacy protection. Some of the steps that individuals can take to protect personal privacy include the following:

- *Find out what is stored about you in existing databases.* Call the major credit bureaus to get a copy of your credit report for $8 (you can obtain one free if you have been denied credit in the last 60 days). The major companies are Equilan (800-392-1122), Trans Union (312-258-1717), and Equifax (800-685-1111). You can also submit a Freedom of Information Act request to a federal agency that you suspect may have information stored on you.
- *Be careful when you share information about yourself.* Don't share information unless it is absolutely necessary. Every time you give information about yourself through an 800, 888, or 900 call, your privacy is at risk. You can ask your doctor, bank, or financial institution not to share information about you with others without your written consent.
- *Be proactive to protect your privacy.* You can get an unlisted phone number and ask the phone company to block caller ID systems from reading your phone number. If you change your address, don't fill out a change-of-address form with the U.S. Postal Service; you can notify the people and companies that you want to have your new address. Be careful about sending personal e-mail messages over a corporate e-mail system. You can also avoid junk mail and telemarketing calls by contacting Direct Marketing Association at P.O. Box 3861, New York, NY 10163.

THE WORK ENVIRONMENT

The use of computer-based information systems has changed the makeup of the workforce. Jobs that require IS literacy have increased, and many less-skilled positions have been eliminated. Corporate programs, such as reengineering and continuous improvement, bring with them the concern that, as business processes are restructured and information systems are integrated within them, the people involved in these processes will be removed.

However, the growing field of computer technology and IS has opened up numerous avenues to professionals and nonprofessionals of all backgrounds. Enhanced telecommunications has been the impetus for new types of business and has created global markets in industries once limited to domestic markets. Even the simplest tasks have been aided by computers, making cash registers faster, smoothing order processing, and allowing people with disabilities to participate more actively in the workforce. As computers and other IS components drop in cost and become easier to use, more workers will benefit from the increased productivity and efficiency provided by computers.

Information systems, while increasing productivity and efficiency, can raise other concerns.

Health Concerns

Organizations can increase employee effectiveness by paying attention to the health concerns in today's work environment. For some people, working with computers can cause occupational stress. Workers' anxieties about job insecurity, loss of control, incompetence, and demotion are just a few of the fears they might experience. In some cases, the stress may become so severe that workers may sabotage computer systems and equipment. Monitoring of employee stress may alert companies to potential problems. Training and counseling can often help the employee and deter problems.

Computer use may affect physical health as well. Strains, sprains, tendonitis, and other problems account for more than 60 percent of all occupational illnesses and about a third of workers' compensation claims, according to the Joyce Institute in Seattle. The cost to U.S. corporations for these types of health problems is as high as $27 billion annually. Claims relating to **repetitive motion disorder**, which can be caused by working with computer keyboards and other equipment, have increased greatly. Also called **repetitive stress injury (RSI)**, the problems can include tendonitis, tennis elbow, the inability to hold objects, and sharp pain in the fingers. Also common is **carpal tunnel syndrome (CTS)**, which is the aggravation of the pathway for nerves that travel through the wrist (the carpal tunnel). CTS involves wrist pain, a feeling of tingling and numbness, and difficulty grasping and holding objects. It may be caused by a number of factors, such as stress, lack of exercise, and the repetitive motion of typing on a computer keyboard. Decisions on workers' compensation related to repetitive stress syndrome have been decided both for and against employees.

Other work-related health hazards involve emissions from improperly maintained and used equipment. Some studies show that poorly maintained laser printers may release ozone into the air; others dispute the claim. Numerous studies on the impact of emissions from display screens have also resulted in conflicting theories. Although some medical authorities believe that long-term exposure can cause cancer, studies are not conclusive at this time. In any case, many organizations are developing conservative and cautious policies.

Most computer manufacturers publish technical information on radiative emissions from their screens, and many companies pay close attention to this information. San Francisco was one of the first cities to propose a video display terminal (VDT) bill. The bill requires companies with 15 or more employees who spend at least four hours a day working with computer screens to give 15-minute breaks every two hours. In addition, adjustable chairs and workstations are required if requested by employees.

repetitive motion disorder (repetitive stress injury; RSI)

an injury that can be caused by working with computer keyboards and other equipment

carpal tunnel syndrome (CTS)

the aggravation of the pathway for nerves that travel through the wrist (the carpal tunnel)

A story initially distributed by *The Washington Post* in May 1999, which was published in the *Boston Globe* under the headline "Study suggests cell phones tied to cancer," cited "possible connections" between cell-phone use and cancer. It used a statistical study examining rates of a rare brain cancer called neurocytoma, as well as a laboratory study. The work was overseen by a Washington consulting firm, Wireless Technology Research LLC, under a six-year, $27 million contract that was funded by a blind trust established by cellular industry companies. The trust arrangement was aimed at enhancing the study's credibility. The scientists involved in the study were furious that preliminary data were leaked from their research in a way that, they say, falsely suggested that the phones may be linked to brain tumors. In fact, investigators say, the data show no clear link. For years, studies have overwhelmingly disputed a cancer risk from the low-level electromagnetic energy of cell phones. But with 70 million to 100 million Americans now using cell phones, any hint of danger attracts intense political and scientific interest.[22]

Avoiding Health and Environmental Problems

Many computer-related health problems are minor and caused by a poorly designed work environment. The computer screen may be hard to read, with problems of glare and poor contrast. Desks and chairs may also be uncomfortable. Keyboards and computer screens may be fixed in place or difficult to move. The hazardous activities associated with these unfavorable conditions are collectively referred to as *work stressors*. Although these problems may not be of major concern to casual users of computer systems, continued stressors such as repetitive motion, awkward posture, and eyestrain may cause more serious and long-term injuries. If nothing else, these problems can severely limit productivity and performance.

ergonomics

the study of designing and positioning computer equipment for employee health and safety

Prolonged use of keyboards and computer equipment may cause RSI and other problems.
(Source: © Steve Kahn 1993)

The study of designing and positioning computer equipment, called **ergonomics**, has suggested a number of approaches to reducing these health problems. The objective is to have "no pain" computing. The slope of the keyboard, the positioning and design of display screens, and the placement and design of computer tables and chairs have been carefully studied. Flexibility is a major component of ergonomics and an important feature of computer devices. People of differing sizes and preferences require different positioning of equipment for best results. Some people, for example, want to have the keyboard in their laps; others prefer to place the keyboard on a solid table. Because of these individual differences, computer designers are attempting to develop systems that provide a great deal of flexibility.

In addition to steps taken by companies, individuals can also reduce RSI and develop a better work environment. Here is what can be done.

• Maintain good posture and positioning. In addition to good equipment, good posture and work habits can eliminate or reduce the potential of RSI.
• Don't ignore pain or discomfort. Many workers ignore early signs of RSI, and as a result, the problem becomes much worse and more difficult to treat.

- Use stretching and strengthening exercises. Often, such exercises can prevent RSI.
- Find a good physician who is familiar with RSI and how to treat it.
- After treatment, start back slowly and pace yourself. Many people who are treated for RSI start back to work too soon and injure themselves again.

We have investigated how computers may be harmful to your health, but the computer can also be used to help prevent and treat general health problems. As discussed in Part III on business information systems, we have seen how computers can be used to assist doctors and other medical professionals by diagnosing medical problems and suggesting potential treatments. People can also use computers to get medical information. Special medical software for personal computers can help people get medical information and determine if they need to see a doctor. A wealth of information is also available on the Internet on a variety of medical topics. See Table 14.12 for a few examples.

Ethical Issues in Information Systems

As you've seen throughout the book in our "Ethical and Societal Issues" boxes, ethical issues deal with what is generally considered right or wrong. Some information systems professionals believe that their field offers many opportunities for unethical behavior. They also believe that unethical behavior can be reduced by top-level managers developing, discussing, and enforcing codes of ethics. Information systems professionals are usually more satisfied with their jobs when top management stresses ethical behavior.

According to one view of business ethics, the "old contract" of business, the only responsibility of business is to its stockholders and owners. According to another view, the "social contract" of business, businesses are responsible to society. At one point or another in their operations, businesses may have employed one or both philosophies.

Various organizations and associations promote ethically responsible use of information systems and have developed codes of ethics. These organizations include the following:

- The Association of Information Technology Professionals (AITP), formerly the Data Processing Management Association (DPMA)

TABLE 14.12

Medical Topics on the Internet

Internet Address	Description
http://www.neoforma.com	Enables users to conduct electronic commerce with healthcare vendors, automatically send out requests-for-proposals via broadcast e-mail, establish free e-mail accounts, post classified advertisements, obtain career information and job postings, and participate in topical discussion groups.
http://www.nlm.nih.gov/research/visible/visible_human.html	Anatomy and Medical Graphics Visible Human Project is a complete, anatomically detailed, three-dimensional representation of the male and female human body. The current phase of the project is collecting transverse CAT, MRI, and frozen section images of representative male and female cadavers at one-millimeter intervals.
http://www.WebMD.com	Provides access to reference material and on-line professional publications from Thomson Healthcare Information Group, Stamford, Connecticut.
http://www.cancer.org	Web site of the American Cancer Society.
http://www.mayo.edu	A tour of the Mayo Clinic.
http://www.looksmart.com	A comprehensive health center with information on a variety of topics including diet and nutrition, disease treatment and therapy, herbal remedies, baby care, and sex matters.
http://oncolink.upenn.edu	A University of Pennsylvania site that deals with cancer information.

- The Association for Computing Machinery (ACM)
- The Institute of Electrical and Electronics Engineers (IEEE)
- Computer Professionals for Social Responsibility (CPSR)

The AITP Code of Ethics

The AITP has developed a code of ethics, standards of conduct, and enforcement procedures that give broad responsibilities to AITP members (Figure 14.1). In general, the code of ethics is an obligation of every AITP member in the following areas:

- Obligation to management
- Obligation to fellow AITP members
- Obligation to society
- Obligation to college or university
- Obligation to the employer
- Obligation to country

For each area of obligation, standards of conduct describe the specific duties and responsibilities of AITP members. In addition, enforcement procedures stipulate that any complaint against an AITP member must be in writing, signed by the individual making the complaint, properly notarized, and submitted by certified or registered mail. Charges and complaints may be initiated by any AITP member in good standing.

The ACM Code of Professional Conduct

The ACM has developed a number of specific professional responsibilities. These responsibilities include the following:

- Strive to achieve the highest quality, effectiveness, and dignity in both the process and products of professional work
- Acquire and maintain professional competence
- Know and respect existing laws pertaining to professional work
- Accept and provide appropriate professional review
- Give comprehensive and thorough evaluations of computer systems and their impacts, including analysis of possible risks
- Honor contracts, agreements, and assigned responsibilities
- Improve public understanding of computing and its consequences
- Access computing and communication resources only when authorized to do so

FIGURE 14.1

ATP Code of Ethics (source: Courtesy of AIIP)

CODE OF ETHICS

I acknowledge:

That I have an obligation to management, therefore, I shall promote the understanding of information processing methods and procedures to management using every resource at my command.

That I have an obligation to my fellow members, therefore, I shall uphold the high ideals of AITP as outlined in its International Bylaws. Further, I shall cooperate with my fellow members and treat them with honesty and respect at all times.

That I have an obligation to society and will participate to the best of my ability in the dissemination of knowledge pertaining to the general development and understanding of information processing. Further, I shall not violate the privacy and confidentiality of information entrusted to my employer whose trust I hold, therefore, I shall endeavor to discharge this obligation to the best of my ability, to guard my employer's interests, and to advise him or her wisely and honestly.

That I have an obligation as a personal representative and as a member of this association. I shall actively discharge these obligations and I dedicate myself to that end.

The mishandling of the social issues discussed in this chapter—including waste and mistakes, crime, privacy, health, and ethics—can devastate an organization. The prevention of these problems and recovery from them are important aspects of managing information and information systems as critical corporate assets. Increasingly, organizations are recognizing that people are the most important component of a computer-based information system and that long-term competitive advantage can be found in a well-trained, motivated, and knowledgeable workforce.

● SUMMARY

PRINCIPLE • Policies and procedures must be established to avoid computer waste and mistakes.

Computer waste is the inappropriate use of computer technology and resources in both the public and private sectors. Computer mistakes relate to errors, failures, and other problems that result in output that is incorrect and without value. Waste and mistakes occur in government agencies as well as corporations. At the corporate level, computer waste and mistakes impose unnecessarily high costs for an information system and drag down profits. Waste often results from poor integration of IS components, leading to duplication of efforts and overcapacity. Inefficient procedures also waste IS resources, as do thoughtless disposal of useful resources and misuse of computer time for games and personal processing jobs. Inappropriate processing instructions, inaccurate data entry, mishandling of IS output, and poor systems design all cause computer mistakes.

Careful programming practices, thorough testing, flexible network interconnections, and rigorous backup procedures can help an information system prevent and recover from many kinds of mistakes. Companies should develop manuals and training programs to avoid waste and mistakes. Company policies should specify criteria for new resource purchases and user-developed processing tools to help guard against waste and mistakes.

PRINCIPLE • Computer crime is a serious and rapidly growing area of concern requiring management attention.

Some crimes use computers as tools (e.g., to manipulate records, counterfeit money and documents, commit fraud via telecommunications links, and make unauthorized electronic transfers of money). Other crimes target computer systems, including illegal access to computer systems by criminal hackers, alteration and destruction of data and programs by viruses (system, application, and document), and simple theft of computer resources. A virus is a program that attaches itself to other programs. A worm functions as an independent program, replicating its own program files until it destroys other systems and programs or interrupts the operation of computer systems and networks. Application viruses infect executable application files, and a system virus infects operating system programs. A macro virus uses an application's own macro programming language to distribute itself. Unlike other viruses, macro viruses

do not infect programs; they infect documents. A logic bomb is designed to "explode" or execute at a specified time and date. Because of increased computer use, greater emphasis is placed on the prevention and detection of computer crime. Software and Internet piracy may represent the most common computer crime. Computer scams have cost individuals and companies thousands of dollars. Computer crime is also an international issue.

Preventing computer crime is done by state and federal agencies, corporations, and individuals. Security measures, such as using passwords, identification numbers, and data encryption, help to guard against illegal access, especially when supported by effective control procedures. Virus scanning software identifies and removes damaging computer programs. Law enforcement agencies armed with new legal tools enacted by Congress now actively pursue computer criminals.

Ethics determine generally accepted and discouraged activities within a company and the larger society. Ethical computer users define acceptable practices more strictly than just refraining from committing crimes; they also consider the effects of their IS activities, including Internet usage, on other people and organizations. The Association for Computing Machinery and the Association of Information Technology Professionals have developed a code of ethics that provides useful guidance. To help prevent ethical problems, the Telecommunications Act of 1996 set a standard for decency issues on the Internet. Software products have been developed to help filter and rate Internet content. Many IS professionals join computer-related associations and agree to abide by detailed ethical codes.

Although most companies use data files for legitimate, justifiable purposes, opportunities for invasion of privacy abound. Privacy issues are a concern with government agencies, e-mail use, corporations, and the Internet. The Privacy Act of 1974, with the support of other federal laws, establishes straightforward and easily understandable requirements for data collection, use, and distribution by federal agencies; federal law also serves as a nationwide moral guideline for privacy rights and activities by private organizations. Some states supplement federal protections and limit activities within their jurisdictions by private organizations. A business should develop a clear and thorough policy about privacy rights for customers, including database access. That policy should also address the rights of employees, including electronic monitoring systems and e-mail. Fairness in information use for privacy rights emphasizes knowledge, control, notice, and consent for people profiled in databases. Individuals should have knowledge of the data that is stored about them and have the ability to correct errors in corporate database systems. If information on individuals is to be used for other purposes, these individuals should be asked to give their consent beforehand. Each individual has the right to know and the ability to decide.

PRINCIPLE ● Jobs, equipment, and working conditions must be designed to avoid negative health effects.

Computers have changed the makeup of the workforce and even eliminated some jobs, but they have also expanded and enriched employment opportunities in many ways. Computers and related devices affect employees' emotional and physical health, especially by causing repetitive stress injury (RSI). Some critics blame computer systems for emissions of ozone and electromagnetic radiation. Ergonomic design principles help to reduce harmful effects and increase the efficiency of an information system. RSI prevention includes keeping good posture, not ignoring pain or problems, performing stretching and strengthening exercises, and seeking proper treatment. In addition to these negative health consequences, information systems can be used to provide a wealth of information on health topics through the Internet and other sources.

● ● ●

The study of designing and positioning computer equipment, called ergonomics, has suggested a number of approaches to reducing these health problems. The slope of the keyboard, the positioning and design of display screens, and the placement and design of computer tables and chairs are essential for good health. It is essential to maintain good posture and positioning. In addition to good equipment, good posture and work habits can eliminate or reduce the potential of RSI. Don't ignore pain or discomfort. Use stretching and strengthening exercises.

● KEY TERMS

antivirus programs 566
application virus 560
biometrics 565
carpal tunnel syndrome (CTS) 576
criminal hacker (cracker) 559
dumpster diving 558
ergonomics 577
hacker 559

insider 559
Internet piracy 563
logic bomb 560
macro virus 560
password sniffer 561
repetitive motion disorder (repetitive stress injury; RSI) 576
script bunnies 559

social engineering 558
software piracy 563
system virus 560
Trojan horse 560
virus 559
worm 559

● REVIEW QUESTIONS

1. What was the impact of the Y2K problem?
2. What can organizations do to prevent computer-related waste and mistakes?
3. Identify four specific actions that can be taken to reduce crime on the Internet.
4. How might the computer be the object of crime?
5. What is the difference between a cracker and a script bunny? What are the major problems caused by criminal hackers?
6. What is the difference between a worm and a virus?
7. What are application viruses, system viruses, and macro viruses?
8. What is software piracy, and why is it so common?

9. What is ergonomics? How can it be applied to office workers?
10. What is Internet piracy, and how can companies avoid it?
11. What four issues should be addressed when considering the individual's right to privacy?
12. What are the provisions of the Privacy Act of 1974?
13. What specific actions can you take to avoid RSI?
14. What is the difference between CTS and RSI?
15. Under what conditions is the monitoring of e-mail not considered an invasion of privacy?
16. Describe the traditional views of ethics in business.
17. What is a code of ethics? Give an example.

● DISCUSSION QUESTIONS

1. How can a criminal use dumpster diving to enhance his opportunity of success at social engineering?
2. You are surprised when you receive a check from the IRS for a tax refund for $10,000 more than you are owed. How could this have happened? What would you do?
3. In some ways, the Y2K bug was like a logic bomb. Discuss.
4. Your marketing department has just opened a Web site and is requesting visitors to register at the site to enter a promotional contest where the chances of winning a prize are better than one in three. Visitors must provide the information necessary to contact them plus fill out a brief survey about the use of your company's products. What data privacy issues may arise?
5. What new laws do you think are necessary to improve the data privacy of Internet users? Is there a danger that some of these laws could be found unconstitutional? What portions of the Constitution would be most at risk of being violated?
6. Based on a number of recent workers' compensation cases, your employer has pledged to spend additional money for office furniture, equipment, and computers that are ergonomically designed so that employees avoid repetitive stress injuries.

Will this solve the problem? Why or why not?
7. How could you use the Internet to help improve your health?
8. In the chapter opening quote John Daley says: "There are 13-year-old kids without degrees breaking into systems from their bedrooms. Security should be the subject of every Master and Ph.D. student's thesis." While this may be taking the issue too far, what actions would you recommend?
9. Using information presented in this chapter on federal privacy legislation, identify which federal law regulates the following areas and situations: cross-checking IRS and social security files to verify the accuracy of information; use of data from youngsters gathered over the Internet; credit bureaus processing home loans; customer liability for debit cards; individuals' right to access data contained in federal agency files; the IRS obtaining personal information; the government obtaining financial records; and employers' access to university transcripts.
10. How can organizations ensure that their information systems are used ethically and morally? Who should audit the organization, departments, and employees to ensure that an ethical code of conduct is established and followed?

● PROBLEM-SOLVING EXERCISES

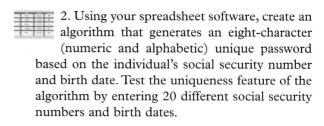

1. Access The CSI-FBI Survey Results for the past four years (start at http://www.gocsi.com/prelea_000321.htm) to get statistics to quantify the number of incidents and their dollar impact. Choose one of the parameters tracked by this survey and use the graphics routine in your spreadsheet software to graph the variation over time.

2. Using your spreadsheet software, create an algorithm that generates an eight-character (numeric and alphabetic) unique password based on the individual's social security number and birth date. Test the uniqueness feature of the algorithm by entering 20 different social security numbers and birth dates.

● TEAM ACTIVITIES

1. Your team has been hired as consultants to improve the security at the Los Alamos National Laboratory. You have been given a budget of $10 million and a time limit of 90 days. How would you identify opportunities for improvement? What actions is your team likely to take?
2. Your team is assigned the task of developing policies for the acquisition of computers and office equipment to provide the best ergonomic office

environment. Develop an initial draft of your policies. How would you select computer systems and office equipment to ensure good ergonomic features?
3. Have each member of your team access ten different Web sites and summarize their findings in terms of the existence of data privacy policy statements—did the site have such a policy, was it easy to find, was it complete and easy to understand?

● WEB EXERCISES

1. Visit the Web site of McAfee, Symantec, or another provider of computer security software. Develop a "top ten" list of the current viruses that are rated as having the highest risk assessment. Create a simple chart that identifies the symptoms or impact of each of these viruses.

2. Several computer-related organizations were discussed in this chapter, including AITP, ACM, IEEE, and CPSR. Locate the Web pages for any two of these associations. If you were an information systems professional, which organization would you join, given what you found at the Web sites? Write a brief paragraph explaining your answer.

3. Do Web research to identify any new federal acts that support ethics in the area of information technology. In your search, did you identify any such acts that have been overturned by the courts and ruled unconstitutional? Write a brief summary of your findings.

● CASES

 Predatory Hiring Practices

Executive employee defections to rival software vendors are commonplace. Senior managers are often enticed with interesting stock and salary incentives. When one company loses a number of key employees to another over a short period of time, however, lawsuits are likely.

SAP America brought two lawsuits against Siebel Systems over the defections of more than twenty-five SAP employees to the rival business-management software vendor. SAP alleged that Siebel had engaged in "predatory hiring practices" in a systematic effort to injure SAP's business and impede SAP's ability to compete with Siebel in the customer relationship management software marketplace. The parties eventually mutually agreed to settle the litigation, and the lawsuits were dismissed.

Microsoft and Borland International (now Inprise) clashed over the same issue. Borland sued Microsoft, alleging the software giant had systematically tried to raid its staff to gain a competitive advantage. This suit was also settled out of court.

Discussion Questions

1. How is the company that loses employees affected when a number of key executives leave for a rival firm?

2. Are key executives who are wooed away from their employer by a rival firm in violation of the AITP or ACM code of ethics? Why or why not?

Critical Thinking Questions

3. Every firm competes to employ the best people available. This need for skill and experience often means hiring people from rival firms. How would you distinguish between predatory hiring practices and "normal" hiring practices?

4. What can a firm do to protect itself from losing key executives to rival firms?

Sources: Adapted from Jack McCarthy, "SAP, Siebel Settle Recruitment Lawsuit," *Computerworld*, March 20, 2000, accessed at http://www.computerworld.com/cwi/story/ 0,1199,NAV47_STO41954,00.html; and Jack McCarthy, "SAP Files Suit against Siebel," *Computerworld*, November 15, 1999, accessed at http://www.computerworld.com/ cwi/story/0,1199,NAV47_STO37591,00.html.

② The Children's Online Privacy Protection Act of 1998

The Children's Online Privacy Protection Act of 1998 (COPPA) sets standards for sites that collect information from children. The goal of the legislation is to prohibit unfair or deceptive practices in the collection, use, or disclosure of personal information from and about children on the Internet. Under this rule, operators of Web sites directed to children must post prominent links on their Web sites to a notice of how they collect, use, and/or disclose personal information from the children; must notify parents that they wish to collect information from their children and obtain parental consent prior to collecting, using, and/or disclosing such information; cannot condition a child's participation in on-line activities on the provision of more personal information than is reasonably necessary to participate in the activity; must allow parents the opportunity to review and/or have their children's information deleted from the operator's database and prohibit further collection from the child; and must establish procedures to protect the confidentiality, security, and integrity of personal information they collect from children.

In June 2000, a federal appeals court ruled that COPPA is unconstitutional. The law, which would have made it a federal crime to use the World Wide Web to communicate "for commercial purposes" material considered "harmful to minors," included penalties of up to $150,000 for each day of violation and up to six months in prison. The decision upholds a lower court ruling from February 1999.

Passed by Congress in 1998, COPPA was quickly challenged in court by the American Booksellers Foundation for Free Expression (ABFFE), Powell's Books, and A Different Light Bookstores, which joined sixteen other companies and civil liberties groups in a lawsuit filed by the American Civil Liberties Union. The suit contended that the law would unlawfully restrict the ability of adults to obtain a wide range of constitutionally protected material on the Internet, including novels, poetry, art and photography books, and works on health and sex education.

In the COPPA ruling, the appeals court declared that the current definition of "harmful to minors" cannot be applied to cyberspace without censoring a wide variety of constitutionally protected materials. In the setting of a bricks-and-mortar store, when someone is accused of selling a work that is harmful to minors, a jury must decide whether the material is harmful by applying "contemporary community standards." However, the court noted that on the World Wide Web content might be viewed by people in many communities simultaneously. "Because of the peculiar geography-free nature of cyberspace," the court wrote in the 34-page decision, "a 'community standards' test would essentially require every Web communication to abide by the most restrictive community's standards," in essence forcing booksellers and others to eliminate any material that might be viewed as harmful in the most conservative community in the country. This, noted the court, "in and of itself, imposes an impermissible burden on constitutionally protected First Amendment speech."

In July 2000, the Federal Trade Commission (FTC) announced that it was sending e-mail messages to "scores of Web sites" that target children to alert companies that they could face legal action as early as September if they did not comply with COPPA. FTC staffers recently surfed the Internet to determine whether Web sites were in compliance with the act. They then sent e-mails to alleged offenders, according to an agency spokeswoman. Violators could face civil penalties of $11,000 per violation, and the commission said it has a number of private investigations under way.

Many sites have been slow to conform to COPPA. Organizations tried to become COPPA-compliant by the April 2000 deadline because they expected the FTC to crack down on them. But when the FTC failed to take action, those sites stopped further efforts to conform because they thought the law would not be enforced.

Discussion Questions

1. What is the contemporary community standards test? Can it be applied to material on the Internet? Why or why not?
2. What actions has the FTC taken to enforce COPPA? Do these actions seem reasonable to you? Why or why not?

Critical Thinking Questions

3. Do you agree with the June 2000 federal appeals court ruling that COPPA is unconstitutional? Why or why not? What can you find out about the most current status of COPPA?
4. Why have so many sites been slow to meet the COPPA requirements? Is their lack of action justifiable? Why or why not?

Sources: Adapted from Linda Rosencrance, "FTC Warns Sites to Comply with Children's Privacy Law," *Computerworld,* July 24, 2000, accessed at http://www.computerworld.com/cwi/story/0,1199,NAV47_STO47439,00.html; and Dan Cullen, "Appeals Court Upholds First Amendment in Cyberspace," Industry Newsroom, American Booksellers Association, June 30, 2000 accessed at http://www.bookweb.org/home/news/btw/3444.html.

3 Taxing Internet Sales

According to sales-tax law, a catalog company or other remote retail company is required to collect sales tax only from customers who reside in states where a company has a physical presence—referred to in legal terms as a *nexus*. Through recent court decisions, the concept of nexus has been interpreted to include a company's headquarters, distribution centers, retail stores, and other substantial operations. Businesses, consumers, and governments were willing to accept earlier tax laws regarding remote purchases, because mail order sales—and the resulting loss in tax revenue—were not great enough to prompt action. With the advent of e-commerce and Internet sales, the dollar value of remote purchases has increased dramatically.

Estimates vary widely on just how much revenue states lose from not taxing Internet sales—the estimated loss for 1998 is between $120 million and $4 billion. Yet e-commerce clearly offers a large and rapidly growing potential for increased tax revenue. Government agencies want to tap this source of income and eliminate the tax advantage remote sellers have over local businesses. They advocate assessing sales tax based on where the purchaser lives, rather than the seller. Making this change in the tax code would eliminate consumers' tax incentive to favor one type of retailer. But with roughly 6,500 taxing jurisdictions in the United States created by separate state, county, and city laws, such an approach would be a nightmare for retailers—they would have to calculate different taxes for each area, which would bury them in a mountain of tax forms. Adopting a uniform sales tax across all these jurisdictions is highly unlikely because local governments use taxes to address unique social, economic, and political issues for their constituencies.

The debate on Internet taxation began in earnest in October 1998, when Congress passed the Internet Tax Freedom Act. This act placed a three-year moratorium on domestic taxation authorities—federal, state, and local—from creating new taxes specially aimed at Internet transactions. It basically buys time for legislatures and the e-commerce industry to figure out the best approach for dealing with e-commerce sales taxes. An Advisory Commission on Electronic Commerce was created by Congress and charged with producing recommendations on e-commerce and tax policy. The commission completed its work in April 2000. In May 2000 the U.S. House of Representatives endorsed the conclusions of the commission by passing legislation to extend the moratorium for five years, prohibiting multiple or discriminatory taxes on e-commerce and eliminating taxes on Internet access fees. A three-year moratorium had been imposed in 1998 by the Internet Tax Freedom Act, which also grandfathered states taxing Internet access on or before October 1, 1998, by allowing them to continue doing so. The new legislation, if signed into law, would extend the moratorium to October 2006 and would eliminate the grandfather provision.

The United States is not alone in its struggle to define e-commerce taxation rules. Many other countries are also struggling to integrate this new form of retailing into their existing tax systems.

Discussion Questions

1. What is the current status of legislation with regard to taxing Internet purchases? Do research and write a short paragraph summarizing your findings.
2. Why do you think the Advisory Commission on Electronic Commerce decided to continue the moratorium on Internet taxes for five more years?

Critical Thinking Questions

3. Are you in favor of taxing Internet purchases, or do you oppose such taxes? Why?
4. What do you think will happen after the expiration of the five-year moratorium?

Sources: Adapted from Charles Waltner, "Internet Develops Its Own Tax Code," *Information Week,* December 6, 1999, pp. 110–116; and Advisory Commission on Electronic Commerce Press Room, "Advisory Commission on Electronic Commerce Concludes Business," May 4, 2000, accessed at http://www.ecommercecommission.org/releases/acec0525.htm August 2, 2000.

NOTES

Sources for the opening vignette on page 551: Adapted from Ann Harrison, "Audit Trails Might Have Fingered Los Alamos Insiders," *Computerworld*, May 24, 1999, http://www.computer-world.com/ cwi/story/0,1199,NAV47_STO35791,00.html; Gary H. Anthes, "Computer Security Bombs at Los Alamos," *Computerworld*, May 10, 1999, http://www.computerworld.com/ cwi/story/0,1199,NAV47_STO35589,00.html; and Patrick Thibodeau, "Energy Dept. Seeks More IT Security Funds," *Computerworld*, September 29, 1999, http:// www.computerworld.com/ cwi/story/0,1199,NAV47_STO29046,00.html.

1. Dewayne Lehman, "Senate: Y2K Fixes Worth Billions Spent," *Computerworld*, February 29, 2000, accessed at http://www.computerworld.com/cwi/story/0,1199,NAV47_S TO41578,00.html.

2. William Ulrich, "We Still Haven't Reached the Final Chapter of Y2K," *Computerworld*, January 24, 2000, accessed at http://www.computerworld.com/cwi/story/ 0,1199,NAV47_STO40850,00.html, August 4, 2000.

3. William Ulrich, "We Still Haven't Reached the Final Chapter of Y2K," *Computerworld*, January 24, 2000, accessed at http://www.computerworld.com/cwi/story/ 0,1199,NAV47_STO40850,00.html.

4. Ira Sage, "CyberCrime," *Business Week*, February 21, 2000, pp. 37–42.

5. Ann Harrison, "Audit Trails Might Have Fingered Los Alamos Insiders," *Computerworld*, May 24, 1999, accessed at http://www.computerworld.com/cwi/story/0,1199,NAV47_S TO35791,00.htm.

6. Jaikumar Vijayan, "Government Investigates Loss of Disks at Los Alamos," *Computerworld*, June 15, 2000, accessed at http://www.computerworld.com/cwi/story/0,1199,NAV47_S TO45843,00.html.

7. Ira Sage, "CyberCrime," *Business Week*, February 21, 2000, pp. 37–42.

8. Ira Sage, "CyberCrime," *Business Week*, February 21, 2000, pp. 37–42.

9. Nicole St. Pierre, "Who's Going to Train the Cyber Security Pros?" *Business Week*, February 16, 2000, p. 32.

10. Ira Sage, "CyberCrime," *Business Week*, February 21, 2000, pp. 37–42.

11. Ira Sage, "CyberCrime," *Business Week*, February 21, 2000, pp. 37–42.

12. Ira Sage, "CyberCrime," *Business Week*, February 21, 2000, pp. 37–42.

13. Ira Sage, "CyberCrime," *Business Week*, February 21, 2000, pp. 37–42.

14. Eric Luening, "Smith Pleads Guilty to Melissa Charges," CNET News.com, December 9, 1999, accessed at http://news.cnet.com/news/ 0-1005-200-1489249.html?tag=st.ne.1002.thed.1005-200-.

15. Ann Harrison, "Security Think Tank Releases Sniffer Tool", *Computerworld*, August 9, 1999, p. 28.

16. Robert Scheer, "Nowhere to Hide," *Yahoo! Internet Life*, October 2000, pp. 100–102.

17. Jeff Howe, "Global Eavesdroppers," *Yahoo! Internet Life*, October 2000, p. 103.

18. "Defend Your Data," ACLU Web site, http://www.aclu.org/ privacy/, accessed September 20, 2000.

19. Jack McCarthy, "Intel to Phase Out Processor Serial Numbers," *Computerworld*, April 28, 2000, accessed at http://www.computerworld.com/cwi/story/ 0,1199,NAV47_STO44013,00.html.

20. Dave Murphy, "Microsoft Documents Secretly Track Author," March 7, 1999, International Association of Information Technology Trainers Web site, accessed at http://itrain.org/itinfo/1999/it990307.html on August 9, 2000.

21. Dave Murphy, "Microsoft Patches Win98 GUID Privacy Bug," March 19, 1999, International Association of Information Technology Trainers Web site, accessed at http://itrain.org/itinfo/1999/it990319.html on August 9, 2000.

22. Peter J. Howe, "USA Researchers Decry Leak of Data on Cell Phones and Cancer," PRNewswire Web site, http://www.health.fgov.be/WHI3/krant/krantarch99/ kranttekstmei/990527m02prnewswire.htm, accessed September 20, 2000.

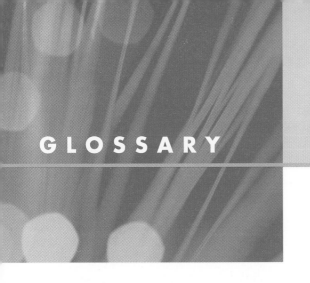

GLOSSARY

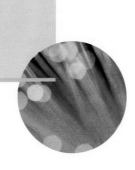

acceptance testing conducting any tests required by the user

accounting MIS information system that provides aggregate information on accounts payable, accounts receivable, payroll, and many other applications

accounting systems systems that include budget, accounts receivable, payroll, asset management, and general ledger

accounts payable system system that increases an organization's control over purchasing, improves cash flow, increases profitability, and provides more effective management of current liabilities

accounts receivable system system that manages the cash flow of the company by keeping track of the money owed the company on charges for goods sold and services performed

ad hoc DSS a DSS concerned with situations or decisions that come up only a few times during the life of the organization

analog signal a continuous, curving signal

antivirus programs programs or utilities that prevent viruses and recover from them if they infect a computer

applet small program embedded in Web pages

application flowcharts diagrams that show relationships among applications or systems

application program interface (API) interface that allows applications to make use of the operating system

application service provider a company that provides both end user support and the computers on which to run the software from the user's facilities

application sign-on procedures that permit the user to start and use a particular application

application software programs that help users solve particular computing problems

application virus a virus that infects executable application files such as word processing programs

arithmetic/logic unit (ALU) portion of the CPU that performs mathematical calculations and makes logical comparisons

ARPANET project started by the U.S. Department of Defense (DOD) in 1969 as both an experiment in reliable networking and a means to link DOD and military research contractors, including a large number of universities doing military-funded research

artificial intelligence (AI) a field in which the computer system takes on the characteristics of human intelligence

artificial intelligence systems people, procedures, hardware, software, data, and knowledge needed to develop computer systems and machines that demonstrate characteristics of intelligence

asking directly an approach to gather data that asks users, stakeholders, and other managers about what they want and expect from the new or modified system

assembly language second-generation language that replaced binary digits with symbols programmers could more easily understand

asset management transaction processing system system that controls investments in capital equipment and manages depreciation for maximum tax benefits

asynchronous communications communication in which the receiver gets the message minutes, hours, or days after it is sent

attribute a characteristic of an entity

audit trail documentation that allows the auditor to trace any output from the computer system back to the source documents

auditing process that involves analyzing the financial condition of an organization and determining whether financial statements and reports produced by the financial MIS are accurate

backbone one of the Internet's high-speed, long-distance communications links

backward chaining the process of starting with conclusions and working backward to the supporting facts

batch processing system method of computerized processing in which business transactions are accumulated over a period of time and prepared for processing as a single unit or batch

benchmark test an examination that compares computer systems operating under the same conditions

best practices the most efficient and effective ways to complete a business process

biometrics the measurement of a living trait, whether physical or behavioral

bit BInary digiT—0 or 1

bot a software tool that searches the Web for information, products, prices, and so on

brainstorming decision-making approach that often involves members offering ideas "off the top of their heads"

bridge connection between two or more networks at the media access control portion of the data link layer; the two networks must use the same communications protocol budget transaction processing system system that automates many of the tasks required to amass budget data, distribute it to users, and consolidate the prepared budgets

budget transaction processing system system that automates many of the tasks required to amass budget data, distribute it to users, and consolidate the prepared budgets

bus line the physical wiring that connects the computer system components

bus network a type of topology that contains computers and computer devices on a single line; each device is connected directly to the bus and can communicate directly with all other devices on the network; one of the most popular types of personal computer networks

business intelligence the process of getting enough of the right information in a timely manner and usable form and analyzing it so that it can have a positive impact on business strategy, tactics, or operations

business resumption planning the process of anticipating and providing for disasters

business-to-business (B2B) e-commerce a form of e-commerce in which the participants are organizations

business-to-consumer (B2C) e-commerce a form of e-commerce in which customers deal directly with the organization, avoiding any intermediaries

byte (B) eight bits together that represent a single character of data cache memory a type of high-speed memory that a processor can access more rapidly than main memory

cache memory a type of high-speed memory that a processor can access more rapidly than main memory

carpal tunnel syndrome (CTS) the aggravation of the pathway for nerves that travel through the wrist (the carpal tunnel)

CASE repository a database of system descriptions, parameters, and objectives

catalog management software software that automates the process of creating a real-time interactive catalog and delivering customized content to a user's screen

CD-rewritable (CD-RW) disk an optical disk that allows personal computer users to replace their diskettes with high-capacity CDs that can be written upon and edited over

CD-writable (CD-W) disk an optical disk that can be written upon but only once

central processing unit (CPU) the part of the computer that consists of three associated elements: the arithmetic/logic unit, the control unit, and the register areas

centralized processing processing alternative in which all processing occurs in a single location or facility

certificate authority (CA) a trusted third party that issues digital certificates

certification process for testing skills and knowledge that results in an endorsement by the certifying authority that an individual is capable of performing a particular job

change model representation of change theories that identifies the phases of change and the best way to implement them

character basic building block of information, consisting of uppercase letters, lowercase letters, numeric digits, or special symbols

chat room a facility that enables two or more people to engage in interactive "conversations" over the Internet

chief programmer team a group of skilled IS professionals with the task of designing and implementing a set of programs

choice stage the third stage of decision making, which requires selecting a course of action

clickstream data data gathered based on the Web sites you visit and what items you click on

client application the application that accepts objects from other applications

client/server an architecture in which multiple computer platforms are dedicated to special functions such as database management, printing, communications, and program execution

clock speed a series of electronic pulses produced at a predetermined rate that affect machine cycle time

closed shop IS department in which only authorized operators can run the computers

cold site a computer environment that includes rooms, electrical service, telecommunications links, data storage devices, and the like; also called a shell

collaborative computing software software that helps teams of people to work together toward a common goal

command-based user interface a user interface that requires that text commands be given to the computer to perform basic activities

common carriers long-distance telephone companies

communications software software that provides a number of important functions in a network, such as error checking and data security

compact disk read-only memory (CD-ROM) a common form of optical disk on which data, once it has been recorded, cannot be modified

competitive advantage a significant and (ideally) long-term benefit to a company over its competition

competitive intelligence a continuous process involving the legal and ethical collection of information, analysis, and controlled dissemination of information to decision makers

compiler a language translator that converts a complete program into a machine language to produce a program that the computer can process in its entirety

complementary metal oxide semi-conductor (CMOS) a semiconductor fabrication technology that uses special semiconductor material to achieve low power dissipation

complex instruction set computing (CISC) a computer chip design that places as many microcode instructions into the central processor as possible

computer literacy knowledge of computer systems and equipment and the ways they function; it stresses equipment and devices (hardware), programs and instructions (software), databases, and telecommunications

computer network the communications media, devices, and software needed to connect two or more computer systems and/or devices

computer programs sequences of instructions for the computer

computer system architecture the structure, or configuration, of the hardware components of a computer system

computer system platform the combination of a particular hardware configuration and systems software package

computer-aided software engineering (CASE) tools that automate many of the tasks required in a systems development effort and enforce adherence to the SDLC

computer-assisted manufacturing (CAM) a system that directly controls manufacturing equipment

computer-based information system (CBIS) a single set of hardware, software, databases, telecommunications, people, and procedures that are configured to collect, manipulate, store, and process data into information

computer-integrated manufacturing (CIM) a system that uses computers to link the components of the production process into an effective system

concurrency control a method of dealing with a situation in which two or more people need to access the same record in a database at the same time

content streaming a method for transferring multimedia files over the Internet so that the data stream of voice and pictures plays more or less continuously, without a break, or very few of them; enables users to browse large files in real time

continuous improvement constantly seeking ways to improve the business processes to add value to products and services

contract software software developed for a particular company

control unit part of the CPU that sequentially accesses program instructions, decodes them, and coordinates the flow of data in and out of the ALU, the registers, primary storage, and even secondary storage and various output devices

cookie a text file that an Internet company can place on the hard disk of a computer system

coprocessor part of the computer that speeds processing by executing specific types of instructions while the CPU works on another processing activity

cost centers divisions within a company that do not directly generate revenue

cost/benefit analysis an approach that lists the costs and benefits of each proposed system

counterintelligence the steps an organization takes to protect information sought by "hostile" intelligence gatherers

creative analysis the investigation of new approaches to existing problems

criminal hacker (cracker) a computer-savvy person who attempts to gain unauthorized or illegal access to computer systems

critical analysis the unbiased and careful questioning of whether system elements are related in the most effective and efficient ways

critical path all activities that, if delayed, would delay the entire project

critical success factors (CSFs) factors essential to the success of certain functional areas of an organization

cross-platform development development technique that allows programmers to develop programs that can run on computer systems having different hardware and operating systems, or platforms

cryptography the process of converting a message into a secret code and changing the encoded message back to regular text

culture a set of major understandings and assumptions shared by a group

customer interaction system system that monitors and tracks each customer interaction with the company

cybermall a single Web site that offers many products and services at one Internet location

data consists of raw facts, such as an employee's name and number of hours worked in a week, inventory part numbers, or sales orders

data analysis manipulation of the collected data so that it is usable for the development team members who are participating in systems analysis

data cleanup the process of looking for and fixing inconsistencies to ensure that data is accurate and complete

data collection the process of capturing and gathering all data necessary to complete transactions

data communications a specialized subset of telecommunications that refers to the electronic collection, processing, and distribution of data—typically between computer system hardware devices

data correction the process of reentering miskeyed or misscanned data that was found during data editing

data definition language (DDL) a collection of instructions and commands used to define and describe data and data relationships in a specific database

data dictionary a detailed description of all the data used in the database

data editing the process of checking data for validity and completeness

data entry process by which human-readable data is converted into a machine-readable form

data input process that involves transferring machine-readable data into the system

data integrity the degree to which the data in any one file is accurate

data item the specific value of an attribute

data manipulation the process of performing calculations and other data transformations related to business transactions

data manipulation language (DML) the commands that are used to manipulate the data in a database

data mart a subset of a data warehouse

data mining an information analysis tool that involves the automated discovery of patterns and relationships in a data warehouse

data model a diagram of data entities and their relationships

data preparation (data conversion) conversion of manual files into computer files

data redundancy duplication of data in separate files

data storage the process of updating one or more databases with new transactions

data store representation of a storage location for data

data warehouse a database that collects business information from many sources in the enterprise, covering all aspects of the company's processes, products, and customers

data-flow diagram (DFD) a model of objects, associations, and activities that describes how data can flow between and around various objects

data-flow lines arrows that show the direction of data element movement

database an organized collection of facts and information

database approach to data management an approach whereby a pool of related data is shared by multiple application programs

database management system (DBMS) a group of programs that manipulate the database and provide an interface between the database and the user of the database and other application programs

decentralized processing processing alternative in which processing devices are placed at various remote locations

decision room a room that supports decision making, with the decision makers in the same building, combining face-to-face verbal interaction with technology to make the meeting more effective and efficient

decision structure a programming structure that allows the computer to branch, depending on certain conditions

decision support system (DSS) an organized collection of people, procedures, software, databases, and devices used to support problem-specific decision making

decision-making phase the first part of problem solving, including three stages: intelligence, design, and choice

dedicated line a communications line that provides a constant connection between two points; no switching or dialing is needed, and the two devices are always connected

delphi approach a decision-making approach in which group decision makers are geographically dispersed; this approach encourages diversity among group members and fosters creativity and original thinking in decision making

demand reports reports developed to give certain information at a manager's request

design report the primary result of systems design, reflecting the decisions made for system design and preparing the way for systems implementation

design stage the second stage of decision making, during which alternative solutions to the problem are developed

deterrence controls rules and procedures to prevent problems before they occur

dialogue manager user interface that allows decision makers to easily access and manipulate the DSS and use common business terms and phrases

digital certificate an attachment to an e-mail message or data embedded in a Web page that verifies the identity of a sender or a Web site

digital computer camera input device used with a PC to record and store images and video in digital form

digital signal a signal represented by bits

digital signal processor (DSP) chips chips that improve the analog-to-digital-to-analog conversion process

digital signature encryption technique used to verify the identity of a message sender for the processing of on-line financial transactions

digital subscriber line (DSL) a communications line that uses existing phone wires going into today's homes and businesses to provide transmission speeds exceeding 500 Kbps at a cost of $20 or more per month

digital video disk (DVD) storage format used to store digital video or computer data

direct access retrieval method in which data can be retrieved without the need to read and discard other data

direct access storage device (DASD) device used for direct access of secondary storage data

direct conversion stopping the old system and starting the new system on a given date (also called plunge or direct cutover)

direct observation watching the existing system in action by one or more members of the analysis team economic feasibility determination of whether the project makes financial sense and whether predicted benefits offset the cost and time needed

disaster recovery the implementation of the business resumption plan

disk mirroring a process of storing data that provides an exact copy that protects users fully in the event of data loss

distance learning the use of telecommunications to extend the classroom

distributed database a database in which the data may be spread across several smaller databases connected via telecommunications devices

distributed processing processing alternative in which computers are placed at remote locations but connected to each other via telecommunications devices

document production the process of generating output records and reports

documentation text that describes the program functions to help the user operate the computer system

domain the allowable values for data attributes

domain expert the individual or group whose expertise or knowledge is captured for use in the expert system

downsizing reducing the number of employees to cut costs

drill down reports reports providing increasingly detailed data about a situation

dumpster diving searching through the garbage for important pieces of information that can help crack an organization's computers or be used to convince someone at the company to give them access to the computers

dynamic Web pages Web pages that contain variable information built in response to a specific Web visitor's request

e-commerce any business transaction executed electronically between parties such as companies (business-to-business), companies and consumers (business-to-consumer), business and the public sector, and consumers and the public sector

e-commerce software software that supports catalog management, product configuration, shopping cart facilities, e-commerce transaction processing, and Web traffic data analysis

e-commerce transaction processing software software that provides the basic connection between participants in the e-commerce economy, enabling communications between trading partners, regardless of their technical infrastructure

economic feasibility determination of whether the project makes financial sense and whether predicted benefits offset the cost and time needed to obtain them

economic order quantity (EOQ) the quantity of inventory that should be reordered to minimize total inventory costs

effectiveness a measure of the extent to which a system achieves its goals; it can be computed by dividing the goals actually achieved by the total of the stated goals

efficiency a measure of what is produced divided by what is consumed

electronic bill presentment a method of billing whereby the biller posts an image of your statement on the Internet and alerts you by e-mail that your bill has arrived

electronic cash an amount of money that is computerized, stored, and used as cash for e-commerce transactions

electronic data interchange (EDI) an intercompany, application-to-application communication of data in standard format, permitting the recipient to perform the functions of a standard business transaction

electronic document distribution process that involves transporting documents—such as sales reports, policy manuals, and advertising brochures—over communications lines and networks

electronic exchange an electronic forum where manufacturers, suppliers, and competitors buy and sell goods, trade market information, and run back-office operations

electronic retailing (e-tailing) the direct sale from business to consumer through electronic storefronts, typically designed around an electronic catalog and shopping cart model

electronic shopping cart a model commonly used by many e-commerce sites to track the items selected for purchase, allowing shoppers to view what is in their cart, add new items to it, and remove items from it

electronic software distribution process that involves installing software on a file server for users to share by signing onto the network and requesting that the software be downloaded onto their computers over a network

electronic wallet a computerized stored value that holds credit card information, electronic cash, owner identification, and address information

embedding procedure used when you want an object to become part of the client document

empowerment giving employees and their managers more responsibility and authority to make decisions, take certain actions, and have more control over their jobs

encapsulation the process of grouping items into an object

encryption the conversion of a message into a secret code

end-user systems development any systems development project in which the primary effort is undertaken by a combination of business managers and users

enterprise data modeling data modeling done at the level of the entire enterprise

enterprise resource planning (ERP) a set of integrated programs that manage a company's vital business operations for an entire multisite, global organization

enterprise sphere of influence sphere of influence that serves the needs of the firm in its interaction with its environment

entity generalized class of people, places, or things for which data is collected, stored, and maintained

entity symbol representation of either a source or destination of a data element

entity-relationship (ER) diagrams a data model that uses basic graphical symbols to show the organization of and relationships between data

ergonomics the study of designing and positioning computer equipment for employee health and safety

event-driven review review triggered by a problem or opportunity such as an error, a corporate merger, or a new market for products

exception reports reports automatically produced when a situation is unusual or requires management action

execution time (E-time) the time it takes to execute an instruction and store the results

executive support system (ESS), or **executive information system (EIS)** specialized DSS that includes all hardware, software, data, procedures, and people used to assist senior-level executives within the organization

expandable storage devices storage that uses removable disk cartridges to provide additional storage capacity

expert system a system that gives a computer the ability to make suggestions and act like an expert in a particular field

expert system shell a collection of software packages and tools used to develop expert systems

explanation facility component of an expert system that allows a user or decision maker to understand how the expert system arrived at certain conclusions or results

Extensible Markup Language (XML) markup language for Web documents containing structured information, including words, pictures, and other elements

external auditing auditing performed by an outside group

extranet a network based on Web technologies that allows selected outsiders, such as business partners and customers, to access authorized resources of the intranet of a company

feasibility analysis assessment of the technical, operational, schedule, economic, and legal feasibility of a project

feedback output that is used to make changes to input or processing activities

field typically a name, number, or combination of characters that describes an aspect of a business object or activity

file a collection of related records

file server an architecture in which the application and database reside on the one host computer, called the file server

file transfer protocol (FTP) a protocol that describes a file transfer process between a host and a remote computer and allows users to copy files from one computer to another

final evaluation a detailed investigation of the proposals offered by the vendors remaining after the preliminary evaluation

financial MIS an information system that provides financial information to all financial managers within an organization

financial model model that provides cash flow, internal rate of return, and other investment analysis

firewall a device that sits between your internal network and the outside Internet and limits access into and out of your network based on your organization's access policy

five-force model a widely accepted model that identifies five key factors that can lead to attainment of competitive advantage, including rivalry among existing competitors, the threat of new entrants, the threat of substitute products and services, the bargaining power of buyers, and the bargaining power of suppliers

flash memory a silicon computer chip that, unlike RAM, is nonvolatile and keeps its memory when the power is shut off

flat organizational structure organizational structure with a reduced number of management layers

flexible manufacturing system (FMS) an approach that allows manufacturing facilities to rapidly and efficiently change from making one product to making another

forecasting predicting future events to avoid problems

forward chaining the process of starting with the facts and working forward to the conclusions

front-end processor a special-purpose computer that manages communications to and from a computer system

fuzzy logic a special research area in computer science that allows shades of gray and does not require conditions to be black/white, yes/no, or true/false

Gantt chart a graphical tool used for planning, monitoring, and coordinating projects

gateway connection that operates at or above the OSI transport layer and links LANs or networks that employ different, higher-level protocols and allows networks with very different architectures and using dissimilar protocols to communicate

general ledger system system designed to automate financial reporting and data entry

general-purpose computers computers used for a wide variety of applications

geographic information system (GIS) a computer system capable of assembling, storing, manipulating, and displaying geographic information (i.e., data identified according to its location)

goal-seeking analysis the process of determining the problem data required for a given result

graphical modeling program software package that assists decision makers in designing, developing, and using graphic displays of data and information

graphical user interface (GUI) an interface that uses icons and menus displayed on screen to send commands to the computer system

grid chart table that shows relationships among the various aspects of a systems development effort

group consensus approach decision-making approach that forces members in the group to reach a unanimous decision

group decision support system (GDSS) software application that consists of most elements in a DSS, plus software needed to provide effective support in group decision making

groupware software that helps groups of people work together more efficiently and effectively

hacker a person who enjoys computer technology and spends time learning and using computer systems

hardware any machinery (most of which uses digital circuits) that assists in the input, processing, storage, and output activities of an information system

help facility a program that provides assistance when a user is having difficulty understanding what is happening or what type of response is expected

hertz one cycle or pulse per second

heuristics commonly accepted guidelines or procedures that usually find a good solution

hierarchical database model a data model in which data is organized in a top-down, or inverted tree, structure

hierarchical network a type of topology that uses a treelike structure with messages passed along the branches of the hierarchy until they reach their destination

hierarchy of data bits, characters, fields, records, files, and databases

highly structured problems problems that are straightforward and require known facts and relationships

home page a cover page for a Web site that has titles, graphics, and text

hot site a duplicate, operational hardware system or immediate access to one through a specialized vendor

HTML tags codes that let the Web browser know how to format text: as a heading, as a list, or as body text and whether images, sound, and other elements should be inserted

human resource MIS an information system that is concerned with activities related to employees and potential employees of an organization

hybrid network a network topology that is a combination of other network types

hypermedia tools that connect the data on Web pages, allowing users to access topics in whatever order they wish

hypertext markup language (HTML) the standard page description language for Web pages

iCOMP (Intel COmparative Microprocessor Performance) index Intel's speed benchmark

icon picture

if-then statements rules that suggest certain conclusions

image log a separate file that contains only changes to applications

implementation stage the stage of problem solving during which a solution is put into effect

in-house development development of application software using the company's resources

incremental backup backup copy of all files changed during the last few days or the last week

inference engine part of the expert system that seeks information and relationships from the knowledge base and provides answers, predictions, and suggestions the way a human expert would

information a collection of facts organized in such a way that they have additional value beyond the value of the facts themselves

information center a support function that provides users with assistance, training, application development, documentation, equipment selection and setup, standards, technical assistance, and troubleshooting

information service unit a miniature IS department

information system (IS) a set of interrelated components that collect, manipulate, and disseminate data and information and provide a feedback mechanism to meet an objective

information systems literacy knowledge of how data and information are used by individuals, groups, and organizations

information systems planning the translation of strategic and organizational goals into systems development initiatives

inheritance property used to describe objects in a group of objects taking on characteristics of other objects in the same group or class of objects

input the activity of gathering and capturing raw data

insiders employees, disgruntled or otherwise, working solo or in concert with outsiders to compromise corporate systems

installation the process of physically placing the computer equipment on the site and making it operational

instant messaging a method that allows two or more individuals to communicate on-line using the Internet

institutional DSS a DSS that handles situations or decisions that occur more than once, usually several times a year or more

instruction time (I-time) the time it takes to perform the fetch-instruction and decode-instruction steps of the instruction phase

integrated development environments (IDEs) software that combines the tools needed for programming with a programming language into one integrated package

integrated services digital network (ISDN) a technology that uses existing common-carrier lines to simultaneously transmit voice, video, and image data in digital form

integrated-CASE (I-CASE) tools tools that provide links between upper- and lower-CASE packages, thus allowing lower-CASE packages to generate program code from upper-CASE package designs joint application development (JAD) process for data collection

integration testing testing all related systems together

intellectual property music, books, inventions, paintings, and other special items protected by patents, copyrights, or trademarks

intelligence stage the first stage of decision making, during which potential problems and opportunities are identified and defined

intelligent behavior the ability to learn from experience and apply knowledge acquired from experience, handle complex situations, solve problems when important information is missing, determine what is important, react quickly and correctly to a new situation

internal auditing auditing performed by individuals within the organization

international network a network that links systems between countries

Internet the world's largest computer network, actually consisting of thousands of interconnected networks, all freely exchanging information

Internet piracy illegally gaining access to and using the Internet

Internet protocol (IP) communication standard that enables traffic to be routed from one network to another as needed

Internet service provider (ISP) any company that provides individuals and organizations with access to the Internet

interpreter a language translator that translates one program statement at a time into machine code

intranet an internal network based on Web technologies that allows people within an organization to exchange information and work on projects

inventory control system system that updates the computerized inventory records to reflect the exact quantity on hand of each stock-keeping unit

Java an object-oriented programming language based on C++ that allows small programs—applets—to be embedded within an HTML document

joining data manipulation that combines two or more tables

joint application development (JAD) process for data collection and requirements analysis

just-in-time (JIT) inventory approach a philosophy of inventory management whereby inventory and materials are delivered just before they are used in manufacturing a product

kernel the heart of the operating system, which controls the most critical processes

key a field or set of fields in a record that is used to identify the record

key-indicator report summary of the previous day's critical activities; typically available at the beginning of each workday

knowledge an awareness and understanding of a set of information and ways that information can be made

useful to support a specific task or reach a decision

knowledge acquisition facility part of the expert system that provides convenient and efficient means of capturing and storing all components of the knowledge base

knowledge base the collection of data, rules, procedures, and relationships that must be followed to achieve value or the proper outcome

knowledge engineer an individual who has training or experience in the design, development, implementation, and maintenance of an expert system

knowledge management the process of capturing a company's collective expertise wherever it resides—in computers, on paper, in people's heads—and distributing it wherever it can help produce the biggest payoff

knowledge user the individual or group who uses and benefits from the expert system

knowledge-based programming an approach to development of computer programs in which you do not tell a computer how to do a job but what you want it to do

language translator systems software that converts a programmer's source code into its equivalent in machine language

learning system a combination of software and hardware that allows the computer to change how it functions or reacts to situations based on feedback it receives

legal feasibility determination of whether there are laws or regulations that may prevent or limit a systems development project

linking data manipulation that combines two or more tables using common data attributes to form a new table with only the unique data attributes

linking procedure used when you want any changes made to the server object to automatically appear in all linked client objects

local area network (LAN) a network that connects computer systems and devices within the same geographic area

logic bomb an application or system virus designed to "explode" or execute at a specified time and date

logical design description of the functional requirements of a system

lookup tables tables containing data that are developed and used by computer programs to simplify and shorten data entry

loop structure a programming structure with two commonly used structures for loops: do-until and do-while; in the do-until structure, the loop is done until a certain condition is met; for the do-while structure, the loop is done while a certain condition exists

lower-CASE tools tools that focus on the later implementation stage of systems development

machine cycle the instruction phase followed by the execution phase

machine language the first generation of programming languages

macro virus a virus that infects documents by using an application's own macro programming language to distribute itself

magnetic disk common secondary storage medium, with bits represented by magnetized areas

magnetic tape common secondary storage medium, Mylar film coated with iron oxide with portions of the tape magnetized to represent bits

magneto-optical disk a hybrid between a magnetic disk and an optical disk

mainframe computer large, powerful computer often shared by hundreds of concurrent users connected to the machine via terminals

maintenance team a special IS team responsible for modifying, fixing, and updating existing software

make-or-buy decision the decision regarding whether to obtain the necessary software from internal or external sources

manufacturing resource planning (MRPII) an integrated, companywide system based on network scheduling that enables people to run their business with a high level of customer service and productivity

market segmentation the identification of specific markets and targeting them with advertising messages

marketing MIS information system that supports managerial activities in product development, distribution, pricing decisions, promotional effectiveness, and sales forecasting

management information system (MIS) an organized collection of people, procedures, software, databases, and decision makers

material requirements planning (MRP) a set of inventory control techniques that help coordinate thousands of inventory items when the demand for one item is dependent on the demand for another

megahertz (MHz) millions of cycles per second

menu-driven system system in which users simply pick what they want to do from a list of alternatives

meta tag a special HTML tag, not visible on the displayed Web page, that contains keywords representing your site's content, which search engines use to build indexes pointing to your Web site

meta-search engine a tool that submits keywords to several individual search engines and returns the results from all search engines queried

metadata the data that describes the contents of a database

microcode predefined, elementary circuits and logical operations that the processor performs when it executes an instruction

midrange computer formerly called minicomputer, a system about the size of a small three-drawer file cabinet that can accommodate several users at one time

MIPS millions of instructions per second

mission-critical systems systems that play a pivotal role in an organization's continued operations and goal attainment

model an abstraction or an approximation that is used to represent reality

model base part of a DSS that provides decision makers access to a variety of models and assists them in decision making

model management software software that coordinates the use of models in a DSS

modem a device that translates data from digital to analog and analog to digital

monitoring stage final stage of the problem-solving process, during which decision makers evaluate the implementation

Moore's Law a hypothesis that states that transistor densities on a single chip will double every 18 months

multidimensional organizational structure structure that may incorporate several structures at the same time

multifunction device a device that can combine a printer, fax machine, scanner, and copy machine into one device

multiplexer a device that allows several telecommunications signals to be transmitted over a single communications medium at the same time

multiprocessing simultaneous execution of two or more instructions at the same time

multitasking capability that allows a user to run more than one application at the same time

music device a device that can be used to download music from the Internet and play the music

natural language processing processing that allows the computer to understand and react to statements and commands made in a "natural" language, such as English

net present value the net amount by which project savings exceed project expenses, after allowing for cost of capital and passage of time

network computer a cheaper-to-buy and cheaper-to-run version of the personal computer that is used primarily for accessing networks and the Internet

network management software software that enables a manager on a networked desktop to monitor the use of individual computers and shared hardware (like printers), scan for viruses, and ensure compliance with software licenses

network model an expansion of the hierarchical database model with an owner-member relationship in which a member may have many owners

network operating system (NOS) systems software that controls the computer systems and devices on a network and allows them to communicate with each other

network topology logical model that describes how networks are structured or configured

networks connected computers and computer equipment in a building, around the country, or around the world to enable electronic communications

neural network a computer system that can simulate the functioning of a human brain

newsgroups on-line discussion groups that focus on specific topics

nominal group technique decision-making approach that encourages feedback from individual group members; the final decision is made by voting, similar to the way public officials are elected

nonoperational prototype a mockup, or model, that includes output and input specifications and formats

nonprogrammed decision decisions that deal with unusual or exceptional situations

object a collection of data and programs

object code machine language code

object linking and embedding (OLE) a software feature that allows you to copy text from one document to another or embed graphics from one program into another program or document

object-oriented languages languages that allow interaction of programming objects, including data elements and the actions that will be performed on them

object-oriented software development approach to program development that uses a collection of existing modules of code, or objects, across a number of applications without being rewritten

object-oriented systems development (OOSD) a systems development approach that combines the logic of the systems development life cycle with the power of object-oriented modeling and programming

object-relational database management system (ORDBMS) a DBMS capable of manipulating audio, video, and graphical data

off-the-shelf software existing software program

on-line analytical processing (OLAP) software that allows users to explore data from a number of different perspectives

on-line transaction processing (OLTP) computerized processing in which each transaction is processed immediately, without the delay of accumulating transactions into a batch

open database connectivity (ODBC) standards that ensure that software written to comply with these standards can be used with any ODBC-compliant database

open shop IS department in which other people, such as programmers and systems analysts, are also authorized to run the computers

open source software software that is freely available to anyone in a form that can be easily modified

Open Systems Interconnection (OSI) model a standard model for network architectures that divides data communications functions into seven distinct layers to promote the development of modular networks that simplify the development, operation, and maintenance of complex telecommunications networks

opensourcing extending software development beyond a single organization by finding others who share the same problem and involving them in a common development effort

operating system (OS) a set of computer programs that controls the computer hardware and acts as an interface with application programs

operational feasibility measure of whether the project can be put into action or operation

operational prototype a functioning prototype that accesses real data files, edits input data, makes necessary computations and comparisons, and produces real output

optical disk a rigid disk of plastic onto which data is recorded by special lasers that physically burn pits in the disk

optical processors computer chips that use light waves instead of electrical current to represent bits

optimization model a process to find the best solution, usually the one that will best help the organization meet its goals

order entry system process that captures the basic data needed to process a customer order

order processing systems systems that process order entry, sales configuration, shipment planning, shipment execution, inventory control, invoicing, customer interaction, and routing and scheduling

organization a formal collection of people and other resources established to accomplish a set of goals

organizational change the responses that are necessary for for-profit and nonprofit organizations to plan for, implement, and handle change

organizational culture the major understandings and assumptions for a business, a corporation, or an organization

organizational learning concept according to which organizations adapt to new conditions or alter their practices over time

organizational structure organizational subunits and the way they relate to the overall organization

output production of useful information, usually in the form of documents and reports

outsourcing contracting with outside professional services to meet specific business needs

paging process of swapping programs or parts of programs between memory and one or more disk devices

parallel processing a form of multiprocessing that speeds processing by linking several processors to operate at the same time, or in parallel

parallel start-up running both the old and new systems for a period of time and comparing the output of the new system closely with the output of the old system; any differences are reconciled; when users are comfortable that the new system is working correctly, the old system is eliminated

password sniffer a small program hidden in a network or a computer system that records identification numbers and passwords

patch a minor change to correct a problem or make a small enhancement, usually an addition to an existing program

payroll journal a report that contains employees' names, the area where employees worked during the week, hours worked, the pay rate, a premium factor for overtime pay, earnings, earnings type, various deductions, and net pay calculations

perceptive system a system that approximates the way a human sees, hears, and feels objects

personal computer (PC) relatively small, inexpensive computer system, sometimes called a microcomputer

personal productivity software software that enables users to improve their personal effectiveness, increasing the amount of work they can do and its quality

personal sphere of influence sphere of influence that serves the needs of an individual user

phase-in approach slowly replacing components of the old system with those of the new one; this process is repeated for each application until the new system is running every application and performing as expected (also called piecemeal approach)

physical design specification of the characteristics of the system components necessary to put the logical design into action

pilot start-up running the new system for one group of users rather than all users

pipelining a form of CPU operation in which there are multiple execution phases in a single machine cycle

pixel a dot of color on a photo image or a point of light on a display screen

planned data redundancy a way of organizing data in which the logical database design is altered so that certain data entities are combined, summary totals are carried in the data records rather than calculated from elemental data, and some data attributes are repeated in more than one data entity to improve database performance

plotter a type of hard-copy output device used for general design work

point evaluation system a process in which each evaluation factor is assigned a weight, in percentage points, based on importance, the system with the greatest total score is selected

point-of-sale (POS) device terminal used in retail operations to enter sales information into the computer system

point-to-point protocol (PPP) a communications protocol that transmits packets over telephone lines

polymorphism a process allowing the programmer to develop one routine or set of activities that will operate on multiple objects

preliminary evaluation an initial assessment whose purpose is to dismiss unwanted proposals

primary key a field or set of fields that uniquely identifies the record

primary storage (main memory; memory) part of the computer that holds program instructions and data

private branch exchange (PBX) a communications system that can manage both voice and data transfer within a building and to outside lines

problem solving a process that goes beyond decision making to include the implementation stage

procedures the strategies, policies, methods, and rules for using a CBIS process a set of logically related tasks performed to achieve a defined outcome

process a set of logically related tasks performed to achieve a defined outcome

process symbol representation of a function that is performed

processing converting or transforming data into useful outputs

product configuration software software used by buyers to build the product they need on-line

productivity a measure of the output achieved divided by the input required

profit centers departments within an organization that track total expenses and net profits

Program Evaluation and Review Technique (PERT) a formalized approach for developing a project schedule

program-data dependence concept according to which programs and data developed and organized for one application are incompatible with programs and data organized differently for another application

programmed decision decisions made using a rule, procedure, or quantitative method

programmer specialist responsible for modifying or developing programs to satisfy user requirements

programmer specialist responsible for modifying or developing programs to satisfy user requirements

programming languages coding schemes used to write both systems and application software

programming life cycle a series of steps and planned activities developed to maximize the likelihood of developing good software

project deadline date the entire project is to be completed and operational

project management model model used to coordinate large projects and identify critical activities and tasks that could delay or jeopardize an entire project if they are not completed on time and cost-effectively

project milestone a critical date for the completion of a major part of the project

project organizational structure structure centered on major products or services

project schedule detailed description of what is to be done

projecting data manipulation that eliminates columns in a table

proprietary software a one-of-a-kind program for a specific application

protocols rules that ensure communications among computers of different types and from different manufacturers

prototyping an iterative approach to the systems development process questionnaire a tool for gathering data when the data sources are spread over a wide geographic area

public network services systems that give personal computer users access to vast databases and other services, usually for an initial fee plus usage fees

purchase order processing system system that helps purchasing departments complete their transactions quickly and efficiently

purchasing transaction processing systems systems that include inventory control, purchase order processing, receiving, and accounts payable

push technology automatic transmission of information over the Internet rather than making users search for it with their browsers

quality the ability of a product (including services) to meet or exceed customer expectations

quality control a process that ensures that the finished product meets the customers' needs

query languages programming languages used to ask the computer questions in Englishlike sentences

questionnaire a tool for gathering data when the data sources are spread over a wide geographic area

random access memory (RAM) a form of memory in which instructions or data can be temporarily stored

rapid application development (RAD) a systems development approach that employs tools, techniques, and methodologies designed to speed application development

read-only memory (ROM) a nonvolatile form of memory

receiving system system that creates a record of expected receipts

record a collection of related data fields

reduced instruction set computing (RISC) a computer chip design based on reducing the number of microcode instructions built into a chip to an essential set of common microcode instructions

redundant array of independent/inexpensive disks (RAID) method of storing data that generates extra bits of data from existing data, allowing the system to create a "reconstruction map" so that if a hard drive fails, it can rebuild lost data

reengineering (process redesign) the radical redesign of business processes, organizational structures, information systems, and values of the organization to achieve a breakthrough in business results

register high-speed storage area in the CPU used to temporarily hold small units of program instructions and data immediately before, during, and after execution by the CPU

relational model a database model that describes data in which all data elements are placed in two-dimensional tables, called relations, that are the logical equivalent of files

release a significant program change that often requires changes in the documentation of the software

reorder point (ROP) a critical inventory quantity level

repetitive motion disorder (repetitive stress injury; RSI) an injury that can be caused by working with computer keyboards and other equipment

replicated database a database that holds a duplicate set of frequently used data

report layout technique that allows designers to diagram and format printed reports request for maintenance form a form authorizing modification of programs

request for maintenance form a form authorizing modification of programs

request for proposal (RFP) a document that specifies in detail required resources such as hardware and software

requirements analysis determination of user, stakeholder, and organizational needs

restart procedures simplified process to access an application from where it left off

return on investment (ROI) one measure of IS value that investigates the additional profits or benefits that are generated as a percentage of the investment in information systems technology

reusable code the instruction code within an object that can be reused in different programs for a variety of applications

revenue centers divisions within a company that track sales or revenues

ring network a type of topology that contains computers and computer devices placed in a ring, or circle; there is no central coordinating computer; messages are routed around the ring from one device or computer to another

robotics mechanical or computer devices that perform tasks requiring a high degree of precision or that are tedious or hazardous for humans

router connection that operates at the network level of the OSI model and features more sophisticated addressing software than bridges; whereas bridges simply pass along everything that comes to them, routers can determine preferred paths to a final destination

routing system system that determines the best way to get products from one location to another

rule a conditional statement that links given conditions to actions or outcomes

safe harbor principles a set of principles that address the e-commerce data privacy issues of notice, choice, and access

sales configuration system process that ensures that the products and services ordered are sufficient to accomplish the customer's objectives and will work well together

satisficing model a model that will find a good—but not necessarily the best—problem solution

scalability the ability of the computer to handle an increasing number of concurrent users smoothly

schedule feasibility determination of whether the project can be completed in a reasonable amount of time

scheduled reports reports produced periodically, or on a schedule, such as daily, weekly, or monthly

scheduling system system that determines the best time to deliver goods and services

schema a description of the entire database

screen layout a technique that allows a designer to quickly and efficiently design the features, layout, and format of a display screen

script bunnies wannabe hackers with little technical savvy who download programs—scripts—that automate the job of breaking into computers

search engine a Web search tool

secondary storage (permanent storage) devices that store larger amounts of data, instructions, and information more permanently than allowed with main memory

secure sockets layer (SSL) a communications protocol used to secure all sensitive data

selecting data manipulation that eliminates rows according to certain criteria

selective backup creation of backup copies of only certain files

semistructured or unstructured problems more complex problems in which the relationships among the data are not always clear, the data may be in a variety of formats, and the data is often difficult to manipulate or obtain

sensitivity analysis process that allows a manager to determine how the production schedule would change with different assumptions concerning demand forecasts or cost figures

sequence structure a programming structure in which, after starting the sequence, programming statements are executed one after another until all the statements in the sequence have been executed, then the program either ends or continues on to another sequence

sequential access retrieval method in which data must be accessed in the order in which it is stored

sequential access storage device (SASD) device used to sequentially access secondary storage data

serial line Internet protocol (SLIP) a communications protocol that transmits packets over telephone lines

server application the application that supplies objects you place into other applications

shipment execution system system that coordinates the outflow of all products from the organization, with the objective of delivering quality products on time to customers

shipment planning system system that determines which open orders will be filled and from which location they will be shipped

sign-on procedure identification numbers, passwords, and other safeguards needed for an individual to gain access to computer resources

simulation the ability of the DSS to duplicate the features of a real system

site preparation preparation of the location of the new system

slipstream upgrade a minor upgrade—typically a code adjustment or minor bug fix—that usually requires recompiling all the code, and, in so doing, it can create entirely new bugs

smart card a credit card–sized device with an embedded microchip to provide electronic memory and processing capability

social engineering the practice of talking a critical computer password out of an individual

software the computer programs that govern the operation of the computer

software bug a defect in a computer program that keeps it from performing in the manner intended

software interface programs or program modifications that allow proprietary software to work with other software used in the organization

software piracy the act of illegally duplicating software

software suite a collection of single-application software packages in a bundle

source code high-level program code written by the programmer

source data automation capturing and editing data whereby the data is originally created and in a form that can be directly input to a computer, thus ensuring accuracy and timeliness

spam e-mail sent to a wide range of people and Usenet groups indiscriminately

special-purpose computers computers used for limited applications by military and scientific research groups

sphere of influence the scope of problems and opportunities addressed by a particular organization

split-case distribution a distribution system that requires cases of goods to be opened on the receiving dock and the individual items from the cases are stored in the manufacturer's warehouse

stakeholders individuals who, either themselves or through the area of the organization they represent, ultimately benefit from the systems development project

star network a type of topology that has a central hub or computer system, and other computers or computer devices are located at the end of communications lines that originate from the central hub or computer

start-up the process of making the final tested information system fully operational

static Web pages Web pages that always contain the same information

statistical analysis model model that can provide summary statistics, trend projections, hypothesis testing, and more

statistical sampling selection of a random sample of data and application of the characteristics of the sample to the whole group

steering committee an advisory group consisting of senior management and users from the IS department and other functional areas

storage area network (SAN) technology that uses computer servers, distributed storage devices, and networks to tie the storage system together

storefront broker companies that act as middlemen between your Web site and on-line merchants that have the products and retail expertise

strategic alliance (strategic partnership) an agreement between two or more companies that involves the joint production and distribution of goods and services

strategic planning determining long-term objectives by analyzing the strengths and weaknesses of the organization, predicting future trends, and projecting the development of new product lines

structured interview an interview for which the questions are written in advance

structured query language (SQL) a standardized language often used to perform database queries and manipulations

structured walkthrough a planned and preannounced review of the progress of a program module, a structure chart, or a human procedure

subschema a file that contains a description of a subset of the database and identifies which users can view and modify the data items in the subset

supercomputers the most powerful computer systems, with the fastest processing speeds

superconductivity a property of certain metals that allows current to flow with minimal electrical resistance

supply chain management a key value chain composed of demand planning, supply planning, and demand fulfillment

switch a device that routes or switches data to its destination

switched line a communications line that uses switching equipment to allow one transmission device to be connected to other transmission devices

synchronous communications communication in which the receiver gets the message instantaneously

syntax a set of rules associated with a programming language

system a set of elements or components that interact to accomplish goals

system boundary defines the limits of a system and distinguishes it from everything else (the environment)

system parameter a value or quantity that cannot be controlled, such as the cost of a raw material

system performance measurement monitoring the system—the number of errors encountered, the amount of memory required, the amount of processing or CPU time needed, and other problems

system performance products software that measures all components of the computer-based information system, including hardware, software, database, telecommunications, and network systems

system performance standard a specific objective of the system

system sign-on procedures that allow the user to gain access to the computer system

system testing testing the entire system of programs

system variable a quantity or item that can be controlled by the decision maker

system virus a virus that typically infects operating system programs or other system files

systems analysis a stage of systems development during which the problems and opportunities of the existing system are defined

systems analyst professional who specializes in analyzing and designing business systems

systems controls rules and procedures to maintain data security

systems design a stage of systems development that determines how the new system will work to meet the business needs defined during systems analysis

systems development the activity of creating or modifying existing business systems

systems implementation a stage of systems development during which the various system components (hardware, software, databases, etc.) defined in the design step are created or acquired and then assembled and the new system is put into operation

systems investigation a stage of systems development that has as its goal to gain a clear understanding of the problem to be solved or opportunity to be addressed

systems investigation report summary of the results of the systems investigation and the process of feasibility analysis; recommends a course of action

systems maintenance and review stage of systems development that involves checking, changing, and enhancing the system to make it more useful in achieving user and organizational goals

systems request form document filled out by someone who wants the IS department to initiate systems investigation

systems review the final step of systems development, involving the analysis of systems to make sure they are operating as intended

systems software the set of programs designed to coordinate the activities and functions of the hardware and various programs throughout the computer system

T1 carrier a line or channel developed by AT&T and used in North America to increase the number of voice calls that can be handled through existing cables

team organizational structure structure centered on work teams or groups

technical documentation written details used by computer operators to execute the program and by analysts and programmers in case there are problems with the program or the program needs modification

technical feasibility assessment of whether the hardware, software, and other system components can be acquired or developed to solve the problem

technology acceptance model (TAM) an explanation of the factors that can lead to higher acceptance and usage of technology in an organization, including the perceived usefulness of the technology, the ease of its use, the quality of the information system, and the degree to which the organization supports the use of the information system

technology diffusion a measure of how widely technology is spread throughout an organization

technology infrastructure all the hardware, software, databases, telecommunications, people, and procedures that are configured to collect, manipulate, store, and process data into information

technology infusion the extent to which technology is deeply integrated into an area or department

technology-enabled relationship management the use of detailed information about a customer's behavior, preferences, needs, and buying patterns to set prices, negotiate terms, tailor promotions, add product features, and otherwise customize the entire relationship with that customer

telecommunications the electronic transmission of signals for communications; enables organizations to carry out their processes and tasks through effective computer networks

telecommunications medium anything that carries an electronic signal and interfaces between a sending device and a receiving device

telecommuting a work arrangement whereby employees work away from the office using personal computers and networks to communicate via e-mail with other workers and to pick up and deliver results

Telnet a terminal emulation protocol that enables users to log on to other computers on the Internet to gain access to public files

terminal-to-host an architecture in which the application and database reside on one host computer, and the user interacts with the application and data using a "dumb" terminal

time-driven review review performed after a specified amount of time

time-sharing capability that allows more than one person to use a computer system at the same time

top-down approach a good general approach to writing a large program, starting with the main module and working down to the other modules

total cost of ownership (TCO) measurement of the total cost of owning computer equipment, including desktop computers, networks, and large computers

total quality management (TQM) a collection of approaches, tools, and techniques that offers a commitment to quality throughout the organization

traditional approach to data management an approach whereby separate data files are created and stored for each application program

traditional organizational structure organizational structure in which major department heads report to a president or top-level manager

transaction any business-related exchange such as payments to employees, sales to customers, and payments to suppliers

transaction processing cycle the process of data collection, data editing, data correction, data manipulation, data storage, and document production

transaction processing system (TPS) an organized collection of people, procedures, software, databases, and devices used to record completed business transactions

transaction processing system audit an examination of the TPS in an attempt to answer whether the system meets the business need for which it was implemented, what procedures and

controls have been established, and whether these procedures and controls have been established, and whether these procedures and controls are being used properly

Transmission Control Protocol/Internet Protocol (TCP/IP) the primary communications protocol of the Internet, originally developed to link defense research agencies

transport control protocol (TCP) widely used transport layer protocol that is used in combination with IP by most Internet applications

Trojan horse a program that appears to be useful but actually masks a destructive program

tunneling the process by which VPNs transfer information by encapsulating traffic in IP packets over the Internet

uniform resource locator (URL) an assigned address on the Internet for each computer

unit testing testing of individual programs

unstructured interview an interview for which the questions are not written in advance

upper-CASE tools tools that focus on activities associated with the early stages of systems development

Usenet a system closely allied with the Internet that uses e-mail to provide a centralized news service; a protocol that describes how groups of messages can be stored on and sent between computers

user acceptance document formal agreement signed by the user that states that a phase of the installation or the complete system is approved

user documentation written description developed for individuals who use a program, showing users, in easy-to-understand terms, how the program can and should be used

user interface element of the operating system that allows individuals to access and command the computer system

user preparation the process of readying managers, decision makers,

employees, other users, and stakeholders for the new systems

users individuals who will interact with the system regularly

utility programs programs used to merge and sort sets of data, keep track of computer jobs being run, compress data files before they are stored or transmitted over a network, and perform other important tasks

value chain a series (chain) of activities that includes inbound logistics, warehouse and storage, production, finished product storage, outbound logistics, marketing and sales, and customer service

value-added carriers companies that have developed private telecommunications systems and offer their services for a fee

version a major program change, typically encompassing many new features

very long instruction word (VLIW) a computer chip design based on further reductions in the number of instructions in a chip made by lengthening each instruction

video compression a process that reduces the number of bits required to represent a single video frame by using mathematical formulas

videoconferencing a telecommunication system that combines video and phone call capabilities with data or document conferencing

virtual memory memory that allocates space on the hard disk to supplement the immediate, functional memory capacity of RAM

virtual private network (VPN) a secure connection between two points across the Internet

virtual reality immersive virtual reality, which means the user becomes fully immersed in an artificial, three-dimensional world that is completely generated by a computer

virtual reality system system that enables one or more users to move and react in a computer-simulated environment

virtual workgroups teams of people located around the world working on common problems

virus a program that attaches itself to other programs

vision system the hardware and software that permit computers to capture, store, and manipulate visual images and pictures

visual programming languages languages that use a mouse, icons, or symbols on the screen and pull-down menus to develop programs

voice mail technology that enables users to leave, receive, and store verbal messages for and from other people around the world

voice-over-IP (VOIP) technology that enables network managers to route phone calls and fax transmissions over the same network they use for data

voice-recognition device an input device that recognizes human speech

volume testing testing the application with a large amount of data

Web appliance a device that can connect to the Internet, typically through a phone line

Web auction an Internet site that matches people who want to sell products and services with people who want to purchase these products and services

Web browser software that creates a unique, hypermedia-based menu on your computer screen that provides a graphical interface to the Web

Web log file a file that contains information about visitors to a Web site

Web page construction software software that uses Web editors and extensions to produce both static and dynamic Web pages

Web site development tools tools used to develop a Web site, including HTML or visual Web page editor, software development kits, and Web page upload support

Web site hosting companies companies that provide the tools and services required to set up a Web page and conduct e-commerce within a matter of days and with little up-front cost

Web site traffic data analysis software software that processes and analyzes data from the Web log file to provide useful information to improve Web site performance

what-if analysis the process of making hypothetical changes to problem data and observing the impact on the results

wide area network (WAN) a network that ties together large geographic regions using microwave and satellite transmission or telephone lines

wordlength the number of bits the CPU can process at any one time

workflow system rule-based management software that directs, coordinates, and monitors execution of an interrelated set of tasks arranged to form a business process

workgroup two or more people who work together to achieve a common goal

workgroup sphere of influence sphere of influence that serves the needs of a workgroup

workstation computer that fits between high-end personal computers and low-end midrange computers in terms of cost and processing power

World Wide Web (WWW, or W3) a collection of tens of thousands of independently owned computers that work together as one in an Internet service

worm an independent program that replicates its own program files until it interrupts the operation of networks and computer systems

INDEX

A boldface page number indicates a key term and the location where its definition can be found.